Preface to the Second Edition

The title 'Principles of Digital Communications' was the outcome of authors long experience in teaching communication and related topics, that is why this book received wide appreciation from the readers, making us to go for this edition of the book. Unfortunately the authors are no longer available, so we got it thoroughly checked by experts in the field for any errors or omissions. The original formate and style of the book was largely retained. We hope that this revised edition will further surve engineering students and senior profesionals interested to become familiar with recent developments in digital techniques.

Publisher

Principles of

DIGITAL COMMUNICATION

Signal Representation, Detection, Estimation and Information Coding

Principles of
DIGITAL COMMUNICATION

Signal Representation, Detection, Estimation and Information Coding

(SECOND EDITION)

J Das
S K Mullick
P K Chatterjee
Department of Electrical Engineering
Indian Institute of Technology
Kanpur, Uttar Pradesh

Publishing Globally

NEW AGE INTERNATIONAL (P) LIMITED, PUBLISHERS
LONDON • NEW DELHI • NAIROBI
Bangalore • Chennai • Cochin • Guwahati • Hyderabad • Kolkata • Lucknow • Mumbai
Visit us at **www.newagepublishers.com**

Published by New Age International (P) Ltd., Publishers
First Edition: 1986
Second Edition: 2012
Reprint: 2016

GLOBAL OFFICES

- **New Delhi** **NEW AGE INTERNATIONAL (P) LIMITED, PUBLISHERS**
  7/30 A, Daryaganj, New Delhi-110002, (INDIA)
  Tel.: (011) 23253771, 23253472, **Telefax:** 23267437, 43551305
  Visit us at www.newagepublishers.com • e-mail: contactus@newagepublishers.com
- **London** **NEW AGE INTERNATIONAL (UK) LTD.**
  27 Old Gloucester Street, London, WC1N 3AX, UK
  Visit us at www.newacademicscience.co.uk • e-mail: info@newacademicscience.co.uk
- **Nairobi** **NEW AGE GOLDEN (EAST AFRICA) LTD.**
  Ground Floor, Westlands Arcade, Chiromo Road (Next to Naivas Supermarket), Westlands, Nairobi, KENYA
  Tel.: 00-254-713848772, 00-254-725700286
  Visit us at www.newagepublishers.com • e-mail: kenya@newagepublishers.com

BRANCHES

- **Bangalore** 37/10, 8th Cross (Near Hanuman Temple), Azad Nagar, Chamarajpet, Bangalore- 560 018
  Tel.: (080) 26756823, **Telefax:** 26756820, **E-mail: bangalore@newagepublishers.com**
- **Chennai** 26, Damodaran Street, T. Nagar, Chennai-600 017, **Tel.:** (044) 24353401, **Telefax:** 24351463
  E-mail: chennai@newagepublishers.com
- **Cochin** CC-39/1016, Carrier Station Road, Ernakulam South, Cochin-682 016, **Tel.:** (0484) 2377303, **Telefax:** 4051304
  E-mail: cochin@newagepublishers.com
- **Guwahati** Hemsen Complex, Mohd. Shah Road, Paltan Bazar, Near Starline Hotel, Guwahati-781 008,
  Tel.: (0361) 2513881, **Telefax:** 2543669, **E-mail: guwahati@newagepublishers.com**
- **Hyderabad** 105, 1st Floor, Madhiray Kaveri Tower, 3-2-19, Azam Jahi Road, Near Kumar Theater, Nimboliadda Kachiguda, Hyderabad-500 027, **Tel.:** (040) 24652456, **Telefax:** 24652457
  E-mail: hyderabad@newagepublishers.com
- **Kolkata** RDB Chambers (Formerly Lotus Cinema) 106A, 1st Floor, S N Banerjee Road, Kolkata-700 014
  Tel.: (033) 22273773, **Telefax:** 22275247, **E-mail: kolkata@newagepublishers.com**
- **Lucknow** 16-A, Jopling Road, Lucknow-226 001, **Tel.:** (0522) 2209578, 4045297, **Telefax:** 2204098
  E-mail: lucknow@newagepublishers.com
- **Mumbai** 142C, Victor House, Ground Floor, N.M. Joshi Marg, Lower Parel, Mumbai-400 013
  Tel.: (022) 24927869, **Telefax:** 24915415, **E-mail: mumbai@newagepublishers.com**
- **New Delhi** 22, Golden House, Daryaganj, New Delhi-110 002, **Tel.:** (011) 23262368, 23262370, **Telefax:** 43551305
  E-mail: sales@newagepublishers.com

ISBN: 978-81-224-3347-0
C-16-05-9393

Printed in India at Ajit Printing Press, Delhi.
Typeset at Innovative, Delhi.

NEW AGE INTERNATIONAL (P) LIMITED, PUBLISHERS
7/30 A, Daryaganj, New Delhi-110002
Visit us at **www.newagepublishers.com**
(CIN: U74899DL1966PTC004618)

Foreword

Man's development and the growth of civilizations have depended, in the main, on progress in a few activities—the discovery of fire, dimestication of animals, the division of labour; but, above all, in the evolution of means to receive, to communicate, and to record his knowledge, and especially in the development of phonetic writing. Man is essentially a communicating animal; communication is one of his oldest activities.

Communication is essentially a social affair. Man has evolved a host of different systems of communication which render his social life possible...Most prominent among all these systems of communication is, of course, human speech and language...—but human language is vastly more than a complicated system of clucking.

Colin Cherry
'On Human Communication'

Preface to the First Edition

For more than a decade, the national and international trends in communication have been to progressively digitalize all communication networks, except perhaps in the entertainment field. Recent developments uninitiated information technology, e.g., videotex, electronic mail, teleconferencing and interactive CATV, and the parallel developments in electronic switching systems, digital radio and fibre-optics communication, are eventually leading to a fairly integrated general-purpose 'C and C' (communication and computer) network, otherwise known as Integrated Service Digital Network (ISDN), as designated by CCITT. The net result of this atmost universal digital connectivity would be the rapid development of assimilated offices and 'wired household', where the future generation will spend most of their productive time. It is, therefore, essential that our electrical engineering students, specialising in communication, must have a strung base in digital communication including information theory and including e.g., Information Theory and Reliable Communication by R.G. Gallugher, Error-correcting codes by W.W. Peterson and E.J. Weldon, Principles of Digital Communication and Coding by A.J. Viterbi and R.K. Omura, and others. Unfortunately, there are no comprehensive text books, covering both the theory and the system aspects of digital communication volume on 'Introduction of Digital Communication Systems', in meant to fulfil this need.

The book is an outgrowth of the authors' teaching and research activities in the area of communication over the last two decades. This is addressed to the senior undergraduate and junior graduate students in an university. The book is also addressed to the senior professionals in the communication area, who are not very familiar with the recent developments in digital techniques; and it is very necessary that they update their knowledge in this modern area. It is hoped that the book will fulfil this need as well.

Originally, the intentions of the authors had been to write a comprehensive text book on 'Digital Communication—Theory and Systems'. But as the present volume progressed, it was felt that it will be too voluminous to include all the materials in one volume, and as such, it was decided to break up the material into two volumes, the present one discussing only the principles and the theory. The companion volume on 'Introduction to Digital Communication Systems' will discuss the current systems and concepts, including some of the complex systems using spreads-spectrum techniques and transmission through optical fibres. Hopefully, this will provide the students as well as the professionals the understanding and background required for the evaluation and design of futuristic communication systems.

The present volume starts with the mathematical and conceptual background for random signals and Gaussian-derived processes, and then discuss the digital representation of signals, optimum detection and estimation, and error-correcting codes in subsequent chapters. The concluding chapter is on information theory which includes a discussion on the evaluation of practical communication systems. The major emphasis has been on digital techniques, and as such, analog communication has not been discussed in the text, except marginally in the chapter on Detection and Estimation.

In contrast to the treatment of the topics in more advanced books, the present book places more emphasis on conceptual understanding of the subject matter than on mathematical rigour. The usual practice of 'theorem and proof' method of treating coding and information theory topics has been avoided as far as possible and attempts have been made to clarify the concepts and basic results with suitable examples from the engineering field. Specially while discussing optimum detection and estimation, an unified approach has been taken for both radar and communication signals.

Assuming that the students are already familiar with probability and random signal theory, the book can be covered selectively in a one-semester course. But it is advisable to take two semesters to cover the topics along with some systems design examples from the companion volume. With the background of Chapter I, the other chapters can be read somewhat independently; and that is an advantage to the professionals who can pick and choose the material as their interest grows. Once the basic concepts in detection and coding theory are understood, the reader can easily move to the companion volume on systems, and perhaps, go back and forth to fully appreciate the complexities involved in the design of modern communication systems.

Since the present volume is the outcome of the authors' experiences in teaching the topics using the existing books on Communication systems, Information theory, Decision theory and Coding, it is natural that the concepts and ideas for the present volume have grown out of the materials in these books; and as such, the present authors are indebted to a large extent to their predecessors. However, it is not practicable to acknowledge the debt individually, but the acknowledgments have been made in proper place in the text. The first author has spent a lifetime at the Indian Institute of Technology, Kharagpur, and as such, he is largely indebted to this innumerable students, some of them colleagues later on, for the intellectual enrichment he derived through their interaction in class rooms, interateries and research projects over all these years.

The authors are grateful to the authorities of the Indian Institute of Technology, Kanpur, for providing financial support to the project through Quality Improvement Program. The authors also wish to thank Mr. Joseph John for correcting the manuscript with care and patience, and Mr. C.M. Abraham for excellent typing of the manuscript and for general secretarial help.

January. 1986

J. Das
S. K. Mullick
P.K. Chatterjee

Contents

Introduction

> It is a very inconvenient habit of kittens (Alice had once made the remark) that, whatever you say to them, they always purr. 'If they would only purr for 'yes' and mew for 'no' or any rule of that sort,' she had said, 'so that one could keep up a conversation! But how can you talk with a person if they always say the same thing.'
>
> LEWIS CARROLL (1832–1898)
> *'Through the Looking Glass'*

The art and science of communication have made tremendous progress over last four decades. Simultaneous development of a digital transmission equipment in the form of PCM in late 40's and the formulation of communication theory by Shannon in 1948 made a significant impact on the trends of activities for information processing, coding and system development for communication. This resulted into a series of activities for both theoreticians and practical designers to evaluate futuristic communication systems, to evolve new codes for signal design and to optimize information transfer in the transmitter-channel-receiver path. All these researches, led by Wiener, Shannon, Kotel'nikov, Woodward and others [1–4], developed into important sub-disciplines, e.g., theory of detection and estimation including decision theory, coding theory including signal design and information theory and related topics. It is now generally accepted that the statistical communication theory or generalised information theory encompasses all the above subdisciplines and a student of communication systems has to be familiar with one and all to be able to perform as a successful designer/researcher. To appreciate the interplay of different theories in the final optimization of complete communication systems. Figure 0.1 shows a tree depicting the important concepts in statistical communication theory and their interrelations, leading to complex signal design and optimum receivers. The cross-reference chart shows the similarity between the problems in radar reception and communication of signals and brings in evidence the common aspects of the design of matched filters, orthogonal and other codes and the possible use of ambiguity functions for the optimum signal design for radars and random-access communication systems.

Traditionally the communication systems have been mostly analog, but during the last decade, there have been new developments in information technology and signal processing which have led to widespread use of digital techniques in transmission and reception. The trends in USA, Canada, Japan and in European countries clearly indicate the predominance of digital approach in new long-haul transmission facilities. In the forecast for 1990's, it has been shown that terrestrial microwave, satellite communication and fibre-optic systems will be predominantly digital networks, the ratio between digital to analog being approximately 10 : 1, if not more. Consider now the outline of a modern telecommunication network, shown in Fig. 0.1, where the local exchanges, trunk exchanges and trunk circuits are already digital through Electronic Switching Systems (ESS) and PCM. The message signal, e.g. speech, TV, telemetry signal etc., may be efficiently converted to the digital form by PCM/ADM and digital techniques for source encoding, bandwidth compression, error correction and data formating, may be efficiently used. The digital long-haul trunks are becoming economical through recent developments in digital radio, TDMA-satellites and fibre-optic cables. Further, the ESS and signal-processing are now computer-controlled; and as such, an Integrated Service Digital Network (ISDN) has become the national and global goal for all communication networks. The study of such digital networks then calls for the understanding of various topics in generalized information theory, as mentioned earlier.

In a generalized communication link, the information is generated in a source, transformed and coded into a suitable signal, transmitted through a channel and transferred to a sink through a decoder-receiver, as shown in Fig. 0.2. For efficient transmission of information, it is necessary to transform the information at many stages as dictated by the characteristics of the channel and the optimum receiver. Since physical signals, e.g., speech and TV, have considerable redundancy, the Encoder I seeks to remove this redundancy and then, the derived signal is protected against the channel noise by recoding this through the error-correcting/orthogonal codes in Encoder II. The longhaul trunks are either band-limited lowpass or bandpass channels and the coded signals (mostly lowpass) are now transformed through a suitable modulation scheme to match the channel. This matching of raw data (signals) to the channel specifications involves the various techniques of signal representation and approximation, data compression, source encoding, FEC (forward-error-correction) and wideband modulation techniques.

The transmission channels, e.g., radio, LOS and satellite links, cables, and optical fibres, are disturbed by Gaussian and impulsive noise, and their characteristics may have to be treated as incoherent and time-varying when considered over a longer time span. This calls for further sophistication in signal design and coding in the form of time-frequency diversity and in optimum receivers using correlators/matched filters (FM). The receiver now performs the complementary signal processing through the techniques of optimum detection/estimation and decoding. The overall objective is to maximize the information transfer in the channel and equivalently to minimize the error rate at the receiver output (sink) (or in some case to maximize the signal-to-noise ratio (SNR) at the sink). To this end, the optimum receivers are defined on the basis of various decision criteria, e.g., mean-square-error, maximum-likelihood, Bayes criteria etc. Thus the overall signal processing in an optimum communication system depends heavily on the techniques and theorems of information theory, coding theory, decision theory and theory of approximation. These can be efficiently utilized only through digital techniques and sometimes through the use of large computers.

The organization of the book follows to a large extent the signal flow in an optimum communication systems, as shown in Fig. 0.2. Since most of the signal processing for coding and optimum detection is best done digitally, Chapter 2 on 'Digital Representation of Signals' follows Chapter 1 on 'Mathematical Review'. Chapter 3 on 'Optimum Detection and Estimation' then introduces the concepts of optimum receivers and discusses the techniques that are available for detecting/estimating a signal in noise. Next

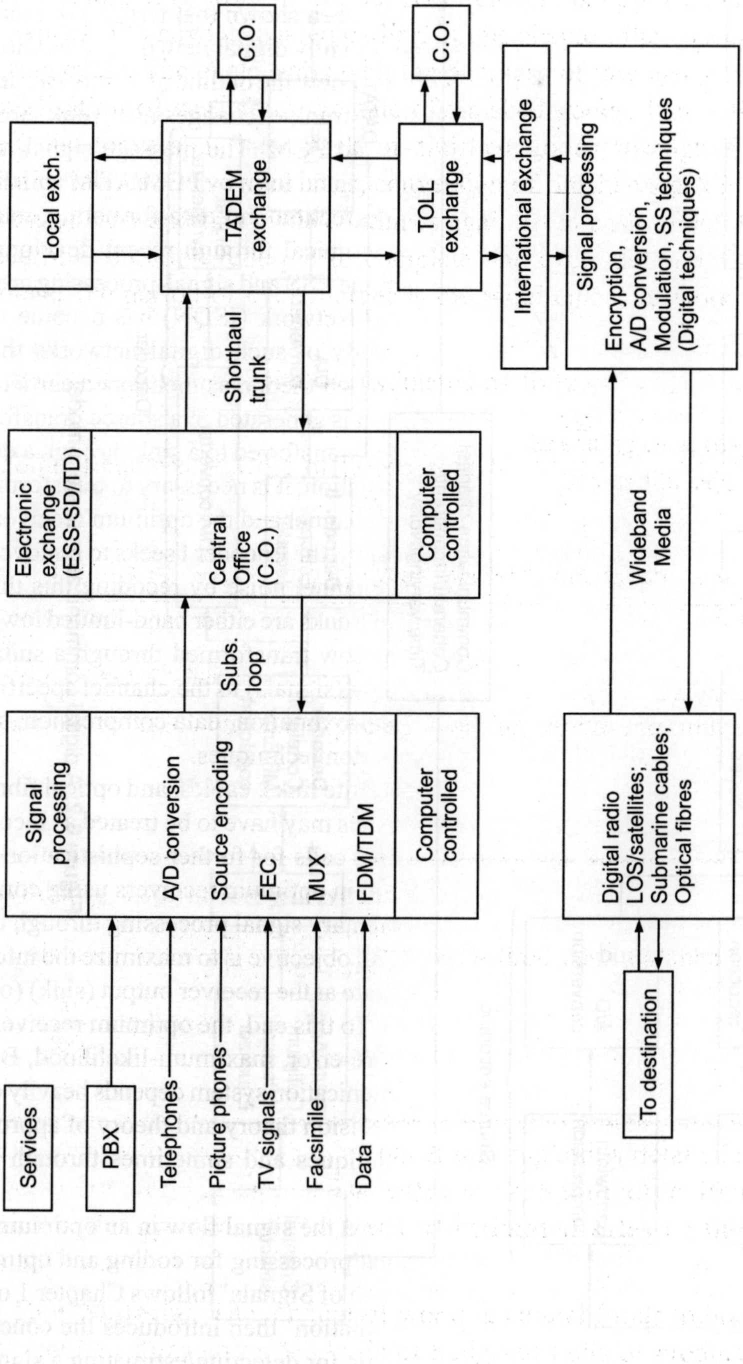

Figure 0.1: An outline of a modern communication network

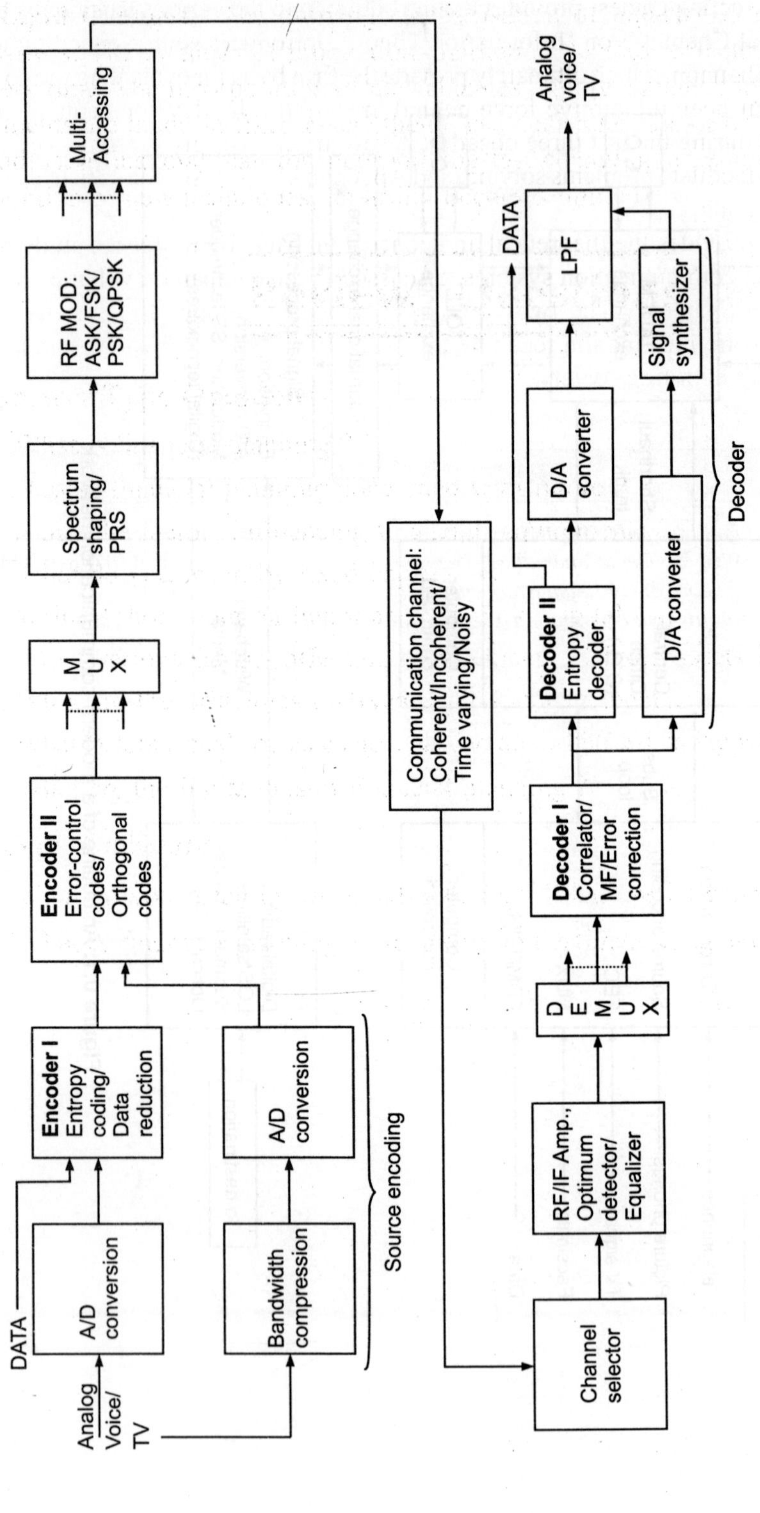

Figure 0.2: A digital communication system-model

Chapter 4 on 'Error-correcting Codes' provides the introduction to the various techniques for protection of signals in noise. Final Chapter 5 on 'Information Theory' introduces source encoding and 'Channel Capacity' theorems of Shannon, which ultimately provide the firm basis for evaluating the communication systems and have, so far been the motive force behind the spectacular developments in coding theory and optimum receivers during the last three decades. A few appendices giving some useful tables have been added at the end to facilitate problems solving, and Appendix A discusses the statistical characteristics of speech and image signals.

The present volume provides the theoretical and conceptual background for the study of the present day and futuristic digital communication systems, discussed in the companion volume on 'Introduction to Digital Transmission Systems'. The two volumes together will, hopefully, provide the necessary understanding of the exotic information technology and its twin ISDN, which are surely and steadily invading the present and the future society.

REFERENCES

1. N. Wiener, *Extrapolation, Interpolation, and Smoothing of Stationary Time Series*, John Wiley, NY, 1950.
2. C.E. Shannon and W. Weaver, *The Mathematical Theory of Communication*, Univ. of Ill. Press, Urbana, 1949.
3. V.A. Kotel'nikov, *The Theory of Optimum Noise Immunity*, McGraw-Hill, NY, 1959.
4. P.M. Woodword, *Probability and Information Theory with Application to Radar*, Pergamon Press, Oxford, 1953.

CHAPTER

1

Mathematical Review

1.0 INTRODUCTION

The description or representation of signals and system by abstract mathematical models is basic to communication theory. They provide us 'information' about the 'state' or behaviour of a physical system because this information is obtained as a pattern or variation of some sort, i.e. a function. The independent variable in this mathematical representation is usually time or space. Thus a speech signal as seen on oscilloscope would be represented as a function of time and a picture would be represented as a brightness function of two spatial variables. The independent variable in the function model of a signal may be either continuous or discrete. Accordingly the signals may be called *continuous-time* or *discrete-time* signals. Furthermore the signal amplitude may take on a continuous or a discrete set of values. *Digital signals* are those for which both time and amplitude are discrete. Continuous time, continuous amplitude signals are often referred to as *analog* signals. Discrete-time signals generally arise by sampling a continuous time signal. Alternatively they may also be the outcome of a discrete-time process. No matter how they originate, their processing by digital systems has lately become a very powerful option for many reasons, the most important being their flexible realization using digital computers and other digital hardware. They can be used to simulate analog systems to any desired degree of accuracy and they can realize signal transformations almost impossible to realize with analog hardware.

The present chapter starts with a brief review of the concepts of continuous and discrete-time systems and their representations both in frequency and time domains. Special attention is paid to narrowband signals and systems. This is followed by comprehensive review of random processes and their characterization. Here Gaussian and related processes are singled out for detailed discussion for their relevance to later chapters.

1.1 CONTINUOUS AND DISCRETE SIGNALS AND SYSTEMS [1–6]

For analytical purposes, signals are modelled either as discrete sequences with n (the set of natural integers) as the independent variable or as continuous functions with t (real time) as the variable. There is a set of signals which are frequently used in the theoretical discussion of systems as probing signals. We list them for future reference.

Continuous	Discrete
1. Unit impulse: $\delta(t)$ (Dirac delta)	Unit sample sequence $\delta(n) = \begin{cases} 0 & n \neq 0 \\ 1 & n = 0 \end{cases}$
2. Unit step: $u(t)$ $u(t) = \begin{cases} 1 & t > 0 \\ 0 & t < 0 \end{cases}$	Unit sequence: $u(n)$ $u(n) = \begin{cases} 1 & n \geq 0 \\ 0 & n < 0 \end{cases}$
3. Sinusoid $x(t) = A \cos (\omega_0 t + \phi)$	$x(n) = A \cos (\omega_0 n + \phi)$
4. Real exponential $x(t) = e^{\sigma t} = a^t$	$x(n) = a^n$
5. Complex exponential $x(t) = e^{st} = e^{(\sigma + j\omega_0)t}$	$x(n) = e^{(\sigma + j\omega_0)n}$
6. Periodic with period T $x(t) = x(t + T)$ for all t	Periodic with period N $x(n) = x(n + N)$ for all n

A theoretically useful representation of an arbitrary signal is obtained by expressing it as an integral or sum of scaled, delayed impulses or unit samples as

$$x(t) = \int_{-\infty}^{\infty} x(t)\, \delta(t - \tau) d\tau$$

$$x(n) = \sum_{k=-\infty}^{\infty} x(k) \delta(n - k) \tag{1.1}$$

An input-output description (model) of a system defines the system mathematically as a unique transformation or operator that maps input signals into output signals, as given by

$$y(t) = T\,[x(t)]$$

$$y(n) = T\,[x(n)] \tag{1.2}$$

and represented as in Fig. 1.1.

Figure 1.1: Input-output model of a system

Depending upon the characteristics of the transformation $T\,[\;\;]$, a system may be linear and time-invariant (LTI) or nonlinear. A linear system is defined as: If $x_1(\cdot)$ and $x_2(\cdot)$ are specific inputs and $y_1(\cdot)$, $y_2(\cdot)$ the corresponding outputs, then the system T is linear if and only if

$$T\,[ax_1(\cdot) + bx_2(\cdot)] = aT\,[x_1(\cdot)] + bT\,[x_2(\cdot)]$$

$$= ay_1(\cdot) + by_2(\cdot) \tag{1.3}$$

for arbitrary constants a and b. The representation of an arbitrary signal as in (1.1) together with the linearity postulate (1.3) suggests that a linear system can be completely characterized by its response to a unit impulse for a continuous system or its unit sample response for a dicrete system. Let $h(t, \tau)$ be the response of the linear system to the impulse $\delta(t - \tau)$, i.e., an impulse occurring at $t = \tau$. Then

$$y(t) = T\left[\int_{-\infty}^{\infty} x(\tau)\,\delta(t-\tau)d\tau\right] = \int_{-\infty}^{\infty} x(\tau)\,h(t,\tau)d\tau \tag{1.4}$$

For a discrete system, the corresponding input-output relation in terms of the system response $h(n, k)$ to a unit sample $\delta(n - k)$ occuring at $n = k$ is

$$y(n) = \sum_{k=-\infty}^{\infty} x(k)h(n,k) \tag{1.5}$$

The class of time or shift-invariant (LTI/LSI) systems is characterized by the property that if $y(t)$ is the response to $x(t)$, then $y(t - \tau)$ is the response to $x(t - \tau)$ where τ is any real number, positive or negative. This implies that if $h(t)$ is the response to $\delta(t)$, then response to $\delta(t - \tau)$ is simply $h(t - \tau)$. For a linear time-invariant (LTI) system, therefore, eq. (1.4) and eq. (1.5) may be written as:

$$y(t) = \int_{-\infty}^{\infty} x(\tau)\,h(t-\tau)d\tau$$

$$y(n) = \sum_{k=-\infty}^{\infty} x(k)h(n-k) \tag{1.6}$$

Thus any linear time (or shift) invariant system is completely characterized by its impulse (or unit sample) response. Equation (1.6) is called the *convolution integral/convolution sum*. The convolution operation is often symbolized as

$$y = x * h \tag{1.7}$$

It may easily be verified that convolution operation is commutative, associative and distributive. These properties are listed below:

$$\begin{aligned} &\bullet\quad x * y = y * x && \text{(commutative)} \\ &x * (y * z) = (x * y) * z && \text{(associative)} \\ &(ax + by) * z = a(x * z) + b(y * z) && \text{(distributive)} \end{aligned} \tag{1.8}$$

An LTI system is said to be *causal* or *realizable* if the output at $t = t_0$ is dependent only on values of the input for $t \le t_0$ i.e., if $x_1(t) = x_2(t)$ for $t \le t_0$, then the corresponding outputs also agree up to that instant: $y_1(t) = y_2(t)$, $t \le t_0$. It may be shown that an LTI system is causal if and only if its impulse response is zero for $t < 0$. [Many idealized systems like ideal low pass filter or ideal differentiator are not causal. In fact a large part of filter theory is concerned with methods of approximating non-realizable ideal systems by realizable ones. A physical system is always causal, if t denotes real time.]

A system is said is to be *stable* if every bounded input produces a bounded output. It may be shown that for linear time (shift)- invariant systems, this leads to the following necessary and sufficient condition on impulse (unit pulse) response for stability, given by:

$$\int_{-\infty}^{\infty} |h(t)|\, dt < \infty$$

$$\sum_{-\infty}^{\infty} |h(n)| < \infty \tag{1.9}$$

As a result of the above constraints on LTI systems, the eq. (1.6) reduces to:

$$y(t) = x(t) * h(t) = \int_{-\infty}^{t} x(\tau)h(t-\tau)\, d\tau$$

$$= \int_{0}^{\infty} h(\tau)x(t-\tau)\, d\tau \tag{1.10}$$

and

$$y(n) = x(n) * h(n) = \sum_{k=-\infty}^{n} x(k)\cdot h(n-k)$$

$$= \sum_{k=0}^{\infty} h(k)x(n-k) \tag{1.11}$$

A significant property of a signal $x(t)$ is its *auto-correletion function* defined as:

$$R_{xx}(\tau) = \lim_{T\to\infty} \frac{1}{2T}\int_{-T}^{T} x(t)x(t-\tau)dt$$

$$= \lim_{T\to\infty} \frac{1}{2T}\int_{-T}^{T} x(t)x(t+\tau)dt,$$

$$x(t) \neq 0 \text{ for } -T < t < T \tag{1.12}$$

If $x(t) = x_1(t) + x_2(t)$, then, it may be shown that:

$$R_{xx}(\tau) = R_{x_1x_1}(\tau) + R_{x_1x_2}(\tau) + R_{x_2x_1}(\tau) + R_{x_2x_2}(\tau) \tag{1.13}$$

where $R_{x_1x_2}(\tau)$ is defined as the *cross-correlation* between $x_1(t)$ and $x_2(t)$, given by

$$R_{x_1x_2}(\tau) = \lim_{T\to\infty} \frac{1}{2T}\int_{-T}^{T} x_1(t)x_2(t+\tau)dt \tag{1.13a}$$

Thus the cross-correlation between the input $x(t)$ and the output $y(t)$ of a system, characterised by the convolution integral of eq. (1.6), may be obtained as

$$R_{xy}(\tau) = \lim_{T\to\infty} \frac{1}{2T}\int_{-T}^{T} x(t-\tau)dt \int_{-\infty}^{\infty} h(\lambda)x(t-\lambda)d\lambda$$

$$= \int_{-\infty}^{\infty} h(\lambda)R_{xx}(t-\lambda)d\lambda \tag{1.14}$$

where $h(\lambda)$ is the impulse response of the system.

The autocorrelation of the system output $y(t)$ may now be obtained in terms of $h(\lambda)$ and $R_{xx}(\tau)$ as

$$R_{yy}(\tau) = \int_{-\infty}^{\infty} h(\lambda_1)\int_{-\infty}^{\infty} h(\lambda_2)R_{xx}(\tau+\lambda_1-\lambda_2)\, d\lambda_2\, d\lambda_1 \tag{1.15}$$

Some of the useful properties of $R(\tau)$ are as follows:

(a) $R_{xx}(\tau)$ is an even function and as such $R_{xx}(\tau) = R_{xx}(-\tau)$.
(b) $R_{xx}(0)$ = the average power of $x(t)$, and $R_{xx}(0) \geq |R_{xx}(\tau)|$.
(c) If $x(t)$ contains periodic components (or a d.c. value), then $R_{xx}(\tau)$ contains components of the same periods (or a d.c. component).
(d) If $x(t)$ contains only random components, then $R_{xx}(\tau) \to 0$ for $\tau \to \infty$.
(e) Although a given $x(t)$ has a unique $R_{xx}(\tau)$, a given $R_{xx}(\tau)$ may correspond to an infinite number of different time functions.
(f) The cross-correlation $R_{x_1x_2}(\tau)$ is not an even function, but $R_{x_1x_2}(\tau) = R_{x_2x_1}(-\tau)$.
(g) $R_{x_1x_2}(\tau)$ is not necessarily maximum at $\tau = 0$.

The above relations constitute the time-domain characterisation of LTI systems and are used extensively for system analysis. ($R_{xx}(\tau)$ and $R_{x_1x_2}(\tau)$ of random signals will be discussed later.)

Example 1.1: Consider a sinewave $x(t) = A \sin \omega_0 t$, the autocorrelation function $R_{xx}(\tau)$ of $x(t)$ is:

$$R_{xx}(\tau) = \lim_{T\to\infty} \frac{1}{2T}\int_{-T}^{T} A^2 \sin \omega_0 t \cdot \sin \omega_0 (t-\tau)dt$$

Since the integrand is periodic, the integral and limiting steps can be replaced by the integral over one period only, and thus,

$$R_{xx}(\tau) = \frac{\omega_0 A^2}{2\pi}\int_0^{2\pi/\omega_0} \sin \omega_0 t \cdot \sin \omega_0 (t-\tau)dt \cdot \omega_0 = 2\pi / T$$

$$= \frac{A^2}{2}\cos \omega_0 \tau$$

Now if $x(t)$ is a periodic time function given by the Fourier series expansion

$$x(t) = \sum_{n=-\infty}^{\infty} a_n e^{jn\omega_0 t},$$

then,

$$R_{xx}(\tau) = \sum_{n=-\infty}^{\infty} |a_n|^2 e^{jn\omega_0 \tau},$$

Thus $R_{xx}(\tau)$ has the same periodic nature as $x(t)$.

But for $x(t) = \begin{cases} A & 0 < t < T \\ 0 & \text{otherwise} \end{cases}$

$$R_{xx}(\tau) = \frac{A^2}{2}\left(1 - \left|\frac{\tau}{T}\right|\right),$$ which is a triangle with a base width of $2T$.

1.2 FREQUENCY DOMAIN (SPECTRAL) REPRESENTATION OF SIGNALS AND SYSTEMS

If $f(t)$ is periodic with period T, then the signal is represented in the frequency domain by the Fourier series as:

$$f(t) = \sum_{n=-\infty}^{\infty} a_n e^{jn\omega_0 t}, \quad \omega_0 = 2\pi / T$$

where

$$a_n = \frac{1}{T}\int_{-T/2}^{T/2} f(t)\cdot e^{-jn\omega_0 t}\, dt \tag{1.16}$$

If the frequency response function of LTI system is $H(\omega)$, then the output $y(t)$ is

$$y(t) = \sum_{n=-\infty}^{\infty} a_n H(n\omega_0) e^{jn\omega_0 t} \tag{1.17}$$

For non-periodic signals, $f(t)$ is represented by the Fourier integral defined as

$$F(\omega) = \mathcal{F}[f(t)] = \int_{-\infty}^{\infty} f(t) e^{-j\omega t}\, dt \tag{1.18*}$$

and

$$f(t) = \mathcal{F}^{-1}[F(\omega)] = \frac{1}{2\pi}\int_{-\infty}^{\infty} F(\omega) e^{j\omega t}\, d\omega$$

The above equations constitute the Fourier Transform pair which is often denoted symbolically as

$$f(t) \leftrightarrow F(\omega)$$

In general, $F(\omega)$ is complex and is expressed as

$$F(\omega) = A(\omega) e^{j\phi(\omega)} \tag{1.19}$$

where $A(\omega)$ is the relative amplitude of the frequency spectrum and $\phi(\omega)$ is the relative phase. For real signals, $f(t)$, $F(\omega)$ may be written as

$$F(\omega) = R(\omega) + jX(\omega) \tag{1.20}$$

where

$$R(\omega) = \int_{-\infty}^{\infty} f(t) \cos \omega t\, dt$$

$$X(\omega) = -\int_{-\infty}^{\infty} f(t) \sin \omega t\, dt$$

---

*The theory of Fourier integral pertains mostly to integral representation of functions which are square integrable over the infinite interval. The set of functions $f(t)$ with this property viz.

$$\int_{-\infty}^{\infty} |f(t)|^2\, dt < \infty$$

are said to belong to L_2. For such functions, $F(\omega)$ defined by eq. (1.18) exists.

Thus, $R(-\omega) = R(\omega)$, $X(-\omega) = -X(\omega)$, and therefore

$$F^*(\omega) = F(-\omega) \tag{1.21}$$

Further, if $f(t) \leftrightarrow R(\omega) + jX(\omega)$, then

$$f(-t) \leftrightarrow R(\omega) - jX(\omega)$$

Therefore it follows that

$$f_e(t) \leftrightarrow R(\omega)$$

$$f_o(t) \leftrightarrow jX(\omega) \tag{1.22}$$

where $f_e(t)$ and $f_o(t)$ are respectively the even and odd parts of $f(t)$, i.e.

$$f_e(t) = \frac{f(t) + f(-t)}{2}$$

$$f_o(t) = \frac{f(t) - f(-t)}{2}$$

For a real causal function, $f(t) = 0$, for $t < 0$ and hence

$$f(t) = 2f_e(t) = 2f_o(t) \quad \text{for } t > 0$$

It follows from the above discussion that a real causal function is uniquely determined in terms of $R(\omega)$ or $X(\omega)$. Thus,

$$f(t) = \frac{2}{\pi}\int_0^\infty R(\omega)\cos\omega t\, d\omega = -\frac{2}{\pi}\int_0^\infty X(\omega)\sin\omega t\, d\omega \tag{1.23}$$

Furthermore, as will be shown in Section 1.4, $R(\omega)$ and $X(\omega)$ constitute a *Hilbert transform pair*. Since the impulse response of physically realizable LTI system is causal, therefore the real and imaginary parts of the transfer function of the system form a Hilbert transform pair.

Example 1.2: Let $f(t)$ be a pulse $p_T(t)$ given by

$$p_T(t) = \begin{cases} 1 & -T/2 \le t \le T/2 \\ 0 & \text{otherwise} \end{cases}$$

Then

$$F(\omega) = \int_{-T/2}^{T/2} e^{-j\omega t}\, dt$$

$$= T\frac{\sin\omega T/2}{\omega T/2} = T\,\text{sinc}\left(\frac{\omega T}{2}\right),$$

where $\text{sinc}\, x \triangleq \dfrac{\sin x}{x}$.

Example 1.3: A Gaussian pulse is given by

$$f(t) = A\exp(-t^2/2a^2)$$

$$F(\omega) = A\int_{-\infty}^{\infty} \exp(-t^2/2a^2)\cdot e^{-j\omega t}\, dt = B\cdot\exp(-\omega^2/2b^2)$$

where $b = \frac{1}{a}$, $B = \sqrt{2\pi}aA$.

We see that the Fourier transform of a Gaussian pulse is again Gaussian, as shown in Fig. 1.2.

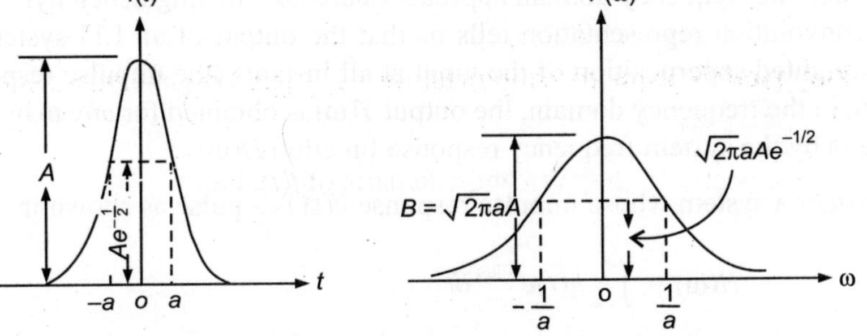

Figure 1.2: Fourier transform of a Gaussian pulse

A very useful and distinctive property of the Fourier transform pair is that it maps the convolution of two time functions into the product of their transforms, i.e., if

$$f_1(t) \leftrightarrow F_1(\omega)$$

and

$$f_2(t) \leftrightarrow F_2(\omega),$$

then

$$f_1(t) * f_2(t) \leftrightarrow F_1(\omega)\, F_2(\omega) \tag{1.24}$$

As a corollary,

$$\int_{-\infty}^{\infty} |f(t)|^2\, dt = \frac{1}{2\pi}\int_{-\infty}^{\infty} |F(\omega)|^2\, d\omega,$$

where $|F(\omega)|^2$ is called the *energy density spectrum* of the signal.

As a consequence of eq. (1.24), the input and the output of the LTI system of Fig. 1.3 is related by

$$y(t) = h(t) * x(t)$$

and

$$Y(\omega) = H(\omega)\, X(\omega) \tag{1.25}$$

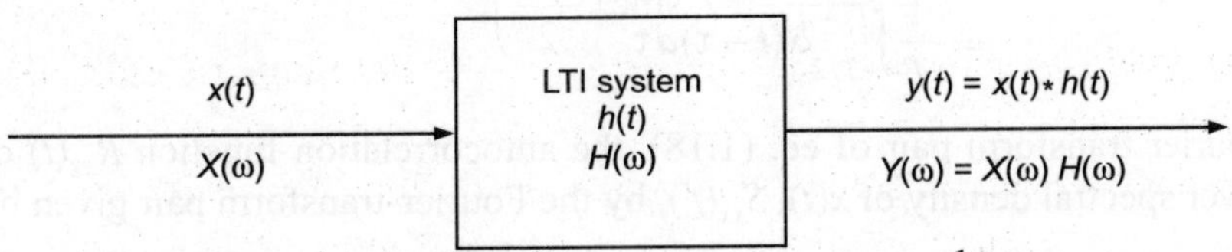

$$X(\omega) = \int_{-\infty}^{\infty} x(t)e^{-j\omega t}\, dt \qquad y(t) = \frac{1}{2\pi}\int_{-\infty}^{\infty} Y(\omega)e^{j\omega t}\, d\omega$$

Figure 1.3: Input-output characterization of LTI systems

where $H(\omega)$ is the system frequency response (equivalently the transfer function of the network) given by

$$H(\omega) = \int_{-\infty}^{\infty} h(t)e^{-j\omega t}\,dt = |H(\omega)|e^{j\phi(\omega)}. \tag{1.25a}$$

The time domain and the frequency domain approaches are now distinguished by:

Whereas the convolution representation tells us that the output of an LTI system at any instant is obtained as the weighted superposition of the input at all instants (the impulse response $h(t)$ being the weighting factor); in the frequency domain, the output $Y(\omega)$ is obtained for any ω by merely scaling the input at the same ω by the system frequency response function $H(\omega)$.

Example 4: Consider a system whose impulse response $h(t)$ is a pulse as shown in Fig. 1.4.

$$H(\omega) = \int_{-\infty}^{\infty} h(t)e^{-j\omega t}dt$$

$$= \frac{1}{T}\int_{-T/2}^{T/2} e^{-j\omega t}\,dt = \frac{\sin \omega T/2}{\omega T/2}$$

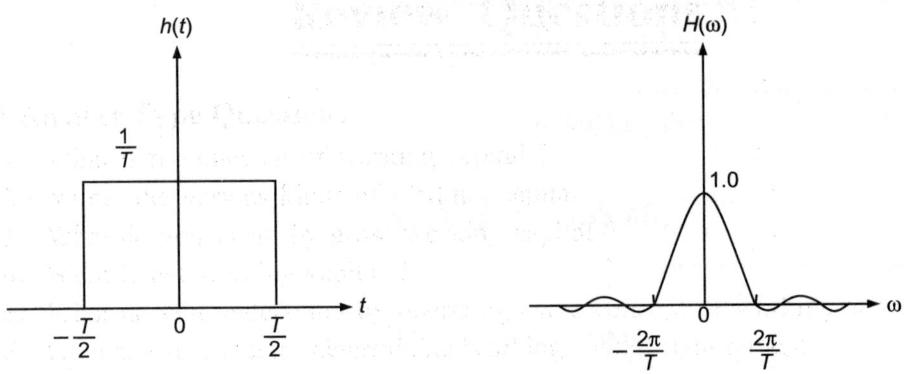

Figure 1.4: Impulse response $h(t)$ and the corresponding transfer function $H(\omega)$

In time domain, the output $y(t)$ for an input $x(t)$ to such a system is obtained as

$$y(t) = x(t) * h(t) = \frac{1}{T}\int_{t-T/2}^{t+T/2} x(\tau)d\tau$$

$$= \frac{1}{T}\int_{-T/2}^{T/2} x(t-\tau)d\tau$$

Similar to the Fourier transform pair of eq. (1.18), the autocorrelation function $R_{xx}(t)$ of eq. (1.12) is related to the power spectral density of $x(t)$, $S_{xx}(f)$, by the Fourier-transform pair given by

$$S_{xx}(f) = \int_{-\infty}^{\infty} R_{xx}(\tau)e^{-j\omega\tau}\,d\tau$$

$$R_{xx}(\tau) = \frac{1}{2\pi}\int_{-\infty}^{\infty} S_{xx}(f)e^{j\omega\tau}\,d\omega \tag{1.26}$$

For a periodic signal of eq. (1.16), $S_{xx}(f) = \sum_{n=-\infty}^{\infty} |a_n|^2 \delta(f - nf_0)$.

This relation will be used in later sections.

1.2.1 Discrete Systems [4, 5]

Similar to the representation of the continuous systems, given by eq. (1.25), the discrete systems are represented by

$$Y(z) = X(z)\,H(z) \tag{1.27}$$

where the z-transforms of the variables, $y(n)$, $x(n)$ and $h(n)$ of eq. (1.11) have been used. The z-transform is defined by

$$Z\{x(n)\} = X(z) = \sum_{n=-\infty}^{\infty} x(n) \cdot z^{-n} \tag{1.28}$$

and is seen as a delay operator such that for causal systems, the discrete values of $x(n)$ may be expressed in the z-domain as (with $x(n) = 0$ for $n < 0$),

$$X(z) = x_0 + x_1 z^{-1} + x_2 z^{-2} + \ldots x_n z^{-n},$$

where $x_n = x(n)$ and $|z| > 1$.

Assuming that the analytical restrictions on $|z|$ are fulfilled and the series (1.28) converges, the integral inversion formula gives, (somewhat similar to the formula for inversion of Laplace transforms)

$$x(n) = \frac{1}{2\pi j} \oint_c X(z) z^{n-1}\, dz$$

The system response to $x(n)$ is then

$$y(n) = \frac{1}{2\pi} \oint_{c_0} X(z) \cdot H(z) z^{n-1}\, dz \tag{1.29}$$

Example 1.5: Let $x(n) = u(n)$, and $h(n) = a^n u(n)$, then,

$$X(z) = \frac{z}{z-1}, \quad |z| > 1$$

and

$$H(z) = \frac{z}{z-a}, \quad |z| > a$$

$$Y(z) = \frac{z^2}{(z-1)(z-a)} = \frac{1}{(1-a)}\left[\frac{z}{(z-1)} - \frac{az}{(z-a)}\right],$$

$|z| > a$ for $a > 1$, $|z| > 1$ for $a < 1$

or $$y(n) = \frac{1}{1-a}[1 - a^{n+1}]u(n) \ldots \text{ for either case.}$$

As in the case of continuous systems, the most important property of the z-transform is that it maps the convolution of two sequences into the product of their transforms, i.e., if

$$y(n) = x(n) * h(n)$$

then

$$Y(z) = X(z) \cdot H(z) \tag{1.30}$$

The sample-domain and the z-transform approaches and their equivalence in the input-output characterization of discrete LSI system are depicted in Fig. 1.5.

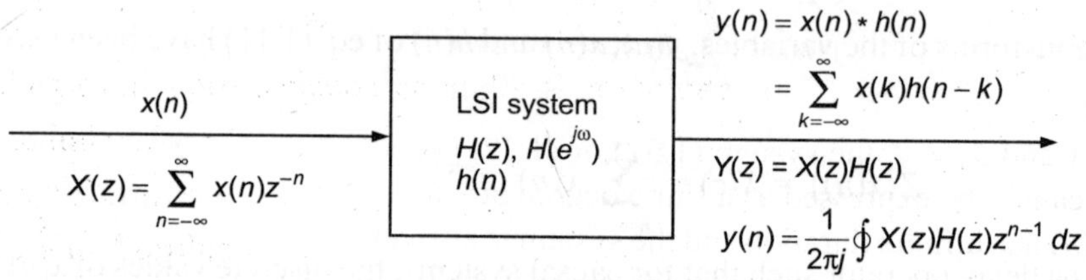

Figure 1.5: Input-output characterization of a discrete LSI system

Difference or Recursion Equations

An important sub-class of discrete LSI systems is the one for which the output and the input sequences satisfy an N-th order linear difference equation with constant coefficients, given by

$$\sum_{k=0}^{N} a_k y(n-k) = \sum_{r=0}^{M} b_r x(n-r)$$

Normalizing it by setting $a_0 = 1$, we may write this difference equation as a recursion equation,

$$y(n) = -\sum_{k=1}^{N} a_k y(n-k) + \sum_{r=0}^{M} b_r x(n-r) \tag{1.31}$$

With the assumption of causality [i.e. $x(n) = 0$ for $n < n_0$ implies $y(n) = 0$ for $n < n_0$] and if eq. (1.31) holds for all n, then we note that $y(n)$ is a sequence uniquely obtained for a given $x(n)$.

To obtain the functional form of $y(n)$, we take the Z-transform of both sides of eq. (1.31) to obtain

$$Y(z) + a_1 z^{-1}\, Y(z) \ldots a_N z^{-N}\, Y(z) = b_0 X(z) + \ldots b_M z^{-M} X(z)$$

or

$$Y(z) = H(z)X(z) \tag{1.32}$$

where

$$H(z) = \frac{b_0 + b_1 z^{-1} + \cdots b_M z^{-M}}{1 + a_1 z^{-1} + \cdots a_N z^{-N}} = \frac{N(z)}{D(z)} \tag{1.32a}$$

The solution $y(n)$ is obtained either by inverting eq. (1.32) or by obtaining $h[n]$ by inverting $H(z)$ and then convolving it with $x(n)$. The solution is valid only for causal systems.

The linear difference equations with constant coefficients are important in the context of digital signal processing because when written as a recursion relation such as eq. (1.31), they provide an explicit relationship between the input and output sequences and thus render the realization problem particularly simple. Such systems are easily realized by using delay elements, multipliers and summers. Conceptually

M and/or *N* can be finite or infinite, but any practical realization would necessarily mean finite *M* and *N*. When a_k's are all zero, the system is *non-recursive*. This non-recursive system is also called a *finite duration impulse response (FIR) filter, all zero or moving average (MA) filter,* and a *transversal filter*. Figure 1.6 shows the symbolic realizaton of such a non-recursive system with

$$H(z) = b_0 + b_1 z^{-1} + \cdots b_M z^{-M} = \sum_{k=0}^{M} b_k z^{-k} \tag{1.33}$$

and

$$y(n) = \sum_{k=0}^{M} b_k x(n-k)$$

When $N > 1$ and $a_N \neq 0$, the system (1.31) or (1.32a) is called an *N*th order recursive filter. In general, when $H(z)$ cannot be expressed as a finite degree polynomial in z^{-1}, then the corresponding unit sample response $h(n)$ has infinite duration and the system is also referred to as *infinite duration impulse response (IIR) filter*. A realization corresponding to $N(z) = 1$ for such a filter is shown in Fig. 1.7. This case is also called an *all-pole or auto-regressive (AR) realization*.

In general, the transfer function $H(z)$ corresponding to (1.32a) is a rational function in z or z^{-1}. The order of $H(z)$ is the largest of the exponents of z in any non-zero term of $H(z)$ i.e. max (M, N). One realization known as the direct form, for the case $M = N$, is shown in Fig. 1.8 which is suggested by writting $H(z)$ as

$$H(z) = \frac{1}{D(z)} \sum_{k=0}^{M} b_k z^{-k}$$

and combining Figs. 1.6 and 1.7.

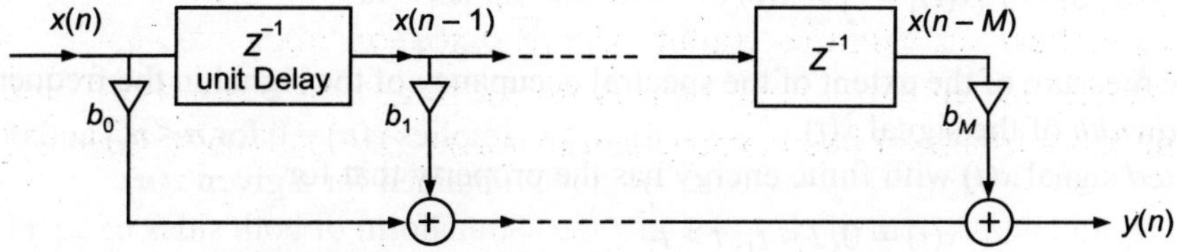

Figure 1.6: Realization of a non-recursive system

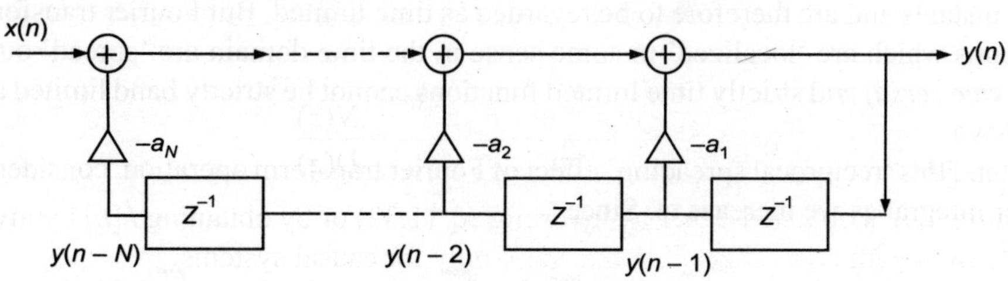

Figure 1.7: Realization of an all-pole recursive system, $y(n) = x(n) - \sum_{k=1}^{N} a_k y(n-k)$

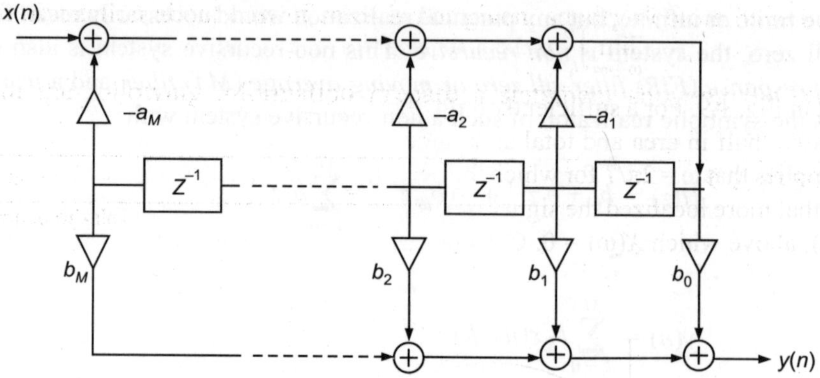

Figure 1.8: Direct form realization of a general discrete system of order M

$$H(z) = \frac{b_0 + b_1 z^{-1} + \cdots + b_M z^{-M}}{1 + a_1 z^{-1} + \cdots + a_M z^{-M}}$$

Direct form realization is canonic in the sense that it requires the minimum number of unit delay elements namely max (M, N). There are other canonic forms which are suggested by writing $H(z)$ as a product (cascade form), or sum (parallel form) of rational functions. Still another form is obtained by the continued fraction expansion of $H(z)$ (ladder structure).

1.3 BAND-LIMITED AND TIME-LIMITED SIGNALS [2, 7, 8]

A finite energy signal $x(t)$ is said to be *band-limited* if there is a constant Ω such that for $|\omega| > \Omega$, its Fourier transform $X(\omega)$ is zero, i.e.,

$$X(\omega) = \mathcal{F}[x(t)] = \int_{-\infty}^{\infty} x(t)\, e^{-j\omega t}\, dt = 0 \;\text{ for } |\omega| > \Omega \tag{1.34}$$

Ω, which is the measure of the extent of the spectral occupancy of the signal in the frequency domain, is called the *bandwidth* of the signal $x(t)$.

A *time limited* signal $x(t)$ with finite energy has the property that for

$$x(t) = 0, t < t_1, t > t_2 \tag{1.35}$$

Such band limited and time limited functions belonging to L_2 are of special importance in signal theory because they constitute very useful models of physical signals. Physical signals do start and stop at some time instants and are therefore to be regarded as time limited. But Fourier transform theory tells us that functions which are 'localized' in some sense in the time domain are 'spread' out in frequency domain (and *vice versa*) and strictly time limited functions cannot be strictly band limited and *vice versa*, as defined above.

To understand this 'reciprocal spreading' effect of Fourier transform operation, consider the behaviour of the Fourier integral as we increase ω. Since

$$X(\omega) = \int_{-\infty}^{\infty} x(t) e^{-j\omega t}\, dt = \int_{-\infty}^{\infty} x(t) \cos \omega t\, dt - j \int_{-\infty}^{\infty} x(t) \sin \omega t\, dt$$

and assuming $x(t)$ to be absolutely integrable over a finite interval $[t_1, t_2]$, Riemann's lemma tells us that

$$\lim_{\omega \to \infty} \int_{t_1}^{t_2} x(t) \cos \omega t \, dt = \lim_{\omega \to \infty} \int_{t_1}^{t_2} x(t) \sin \omega t \, dt = 0 \tag{1.36}$$

This is illustrated in Fig. 1.9. For a sufficiently high ω each positive half cycle of $\cos \omega t$ or $\sin \omega t$ cancels the adjacent negative half in area and total area under the curve is very close to zero. The sufficiently high frequency implies that $\omega = 2\pi/T$ for which the period T is sufficiently small compared to the duration of $x(t)$. It is clear that more localized the signal $x(t)$, the higher is the boundary frequency of its transform or spectrum $X(\omega)$, above which $X(\omega) = 0$. Consider now the following example of a rectangular and a Gaussian pulse.

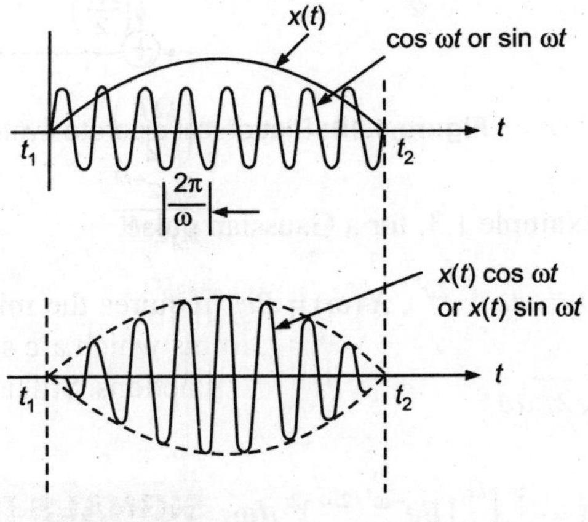

Figure 1.9: Illustration of eq. (1.36)

Example 1.6: Let us obtain the fraction $\eta(\Omega)$ of the total energy lying in the frequency band $(0, \Omega)$ for a rectangular pulse and a Gaussian pulse.

(a) Rectangular pulse

$$E(0, \Omega) = \frac{1}{\pi} \int_0^{\Omega} T^2 \text{sinc}^2\left(\frac{\omega T}{2}\right) d\omega$$

$$= \frac{T^2}{\pi} \frac{2}{T} \int_0^{\Omega T/2} \text{sinc}^2 x \, dx$$

Since the total energy of the pulse is T, therefore

$$\eta\left(\frac{\Omega T}{2}\right) = \frac{2}{\pi} \int_0^{\Omega T/2} \text{sinc}^2 x \, dx = \frac{2}{\pi}\left[\text{Si}\left(\frac{\Omega T}{2}\right) - \frac{\Omega T}{2} \text{sinc}^2\left(\frac{\Omega T}{2}\right)\right]$$

where $\text{Si}(x) \triangleq \int \frac{\sin x}{x} dx$ denotes the sine integral.

The function $\eta(\Omega T/2)$ is plotted in Fig. 1.10. It is seen that when $\Omega T = 2\pi$ (or $BT = 1$, where $\Omega = 2\pi B$), 90% energy of the pulse is contained in the frequency range from 0 to $B = 1/T$.

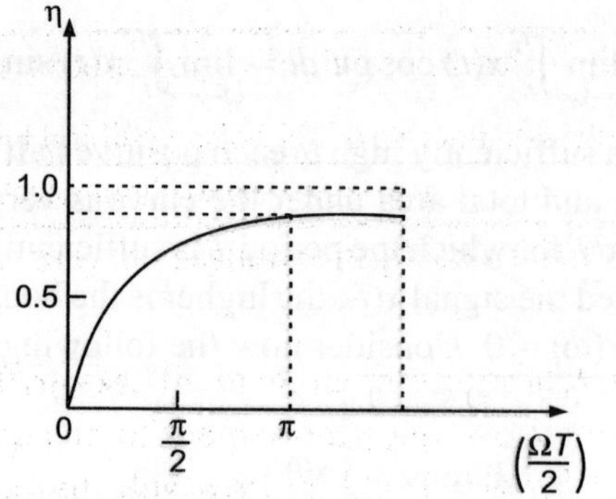

Figure 1.10: Plot of $\eta\left(\frac{\Omega T}{2}\right)$

(b) Gaussian pulse: From Example 1.3, for a Gaussian pulse,

$$x(t) = Ae^{-t^2/2a^2}, X(\omega) = Be^{-\omega^2/2b^2}$$

where $b = \frac{1}{a}$ and $B = \sqrt{2\pi}aA$.

$$E(0, \Omega_1) = \frac{1}{\pi}\int_0^{\Omega_1} [Be^{-\omega^2/2b^2}]^2\, d\omega$$

$$= 2a^2A^2\int_0^{\Omega_1} e^{-a^2\omega^2/2}\, d\omega = 2a^2A^2 \cdot \frac{\sqrt{2}}{a}\int_0^{a\Omega/\sqrt{2}} e^{-x^2}\, dx$$

$$= aA^2\sqrt{2\pi}\frac{2}{\sqrt{\pi}}\int_0^{a\Omega_1/\sqrt{2}} e^{-x^2}\, dx$$

$$= \sqrt{2\pi}aA^2\phi\left(\frac{a\Omega_1}{\sqrt{2}}\right)$$

where $\phi(z) = \frac{2}{\sqrt{\pi}}\int_0^z e^{-x^2}\, dx$

Since the total energy of the Gaussian pulse is $aA^2/\sqrt{\pi}$, therefore,

$$\eta(\Omega_1) = \sqrt{2}\phi\left(\frac{a\Omega_1}{\sqrt{2}}\right)$$

The function $\phi(z)$ is plotted in Fig. 1.11. It is seen that to obtain 90% of the total energy, $z = \frac{a\Omega_1}{\sqrt{2}} = 0.72$ or the product of the pulse duration $2a$ and $W_1 = \frac{\Omega_1}{2\pi}$ should be equal to

$$2aW_1 \simeq \frac{0.72 \times \sqrt{2}}{\pi} \simeq 0.3.$$

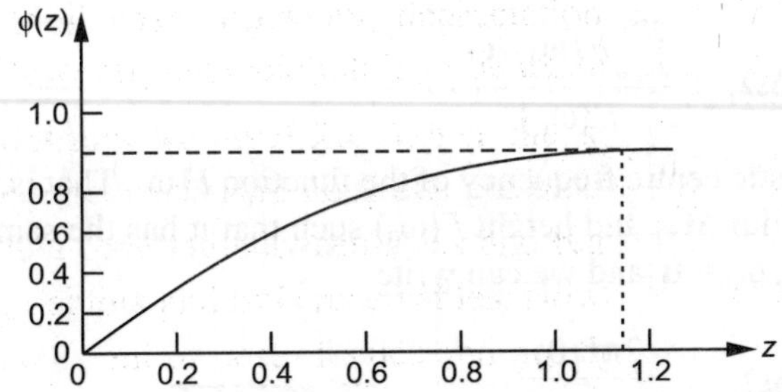

Figure 1.11: Plot of $\phi(z)$

The above results are very useful because they provide us a measure of the bandwidth W_1 needed in a channel for transmission of such a pulse for a given ratio of pulse energy utilization.

It is seen from the above example that for certain reasonable measures of the bandwidth and time duration T of a signal, the product $\Omega T \geq$ constant. This is known as the 'Uncertainty Relation' for band/time-limited signals. To show analytically that $x(t)$ cannot be both time-limited and band-limited, consider the following properties of $x(t)$ [2]. Let $x(t)$ be a band-limited function, band-limited to Ω, then,

(i) $x(t) = \dfrac{1}{2\pi}\int_{-\Omega}^{\Omega} x(\omega)e^{j\omega t}\, d\omega$

(ii) $\int_{-\Omega}^{\Omega} |X(\omega)|\, d\omega < \infty$

(iii) For any t, real or complex,

$$|x(t)| \leq \sqrt{\frac{\Omega E}{\pi}}\, e^{\Omega|t|} \tag{1.37}$$

Thus $x(t)$ is a function of so called 'exponential type' for which $|x(t)| < Ae^{\sigma|t|}$ for two constants A and σ.

(iv) $x(t)$ is an entire function of t i.e., it is analytic in the entire t-plane. Moreover, if a function $x(t)$ has finite energy, is an entire function of t and is of exponential type, as in (iii), then it is Ω-band-limited, i.e., $X(\omega)$ exists and $X(\omega) = 0$ for $|\omega| > \Omega$ (Paley-Wiener theorem).

From the theory of complex variables, the entire functions are completely known (through analytic continuation) if known only over any finite interval and they cannot vanish on any interval unless they vanish everywhere. Thus a band-limited function $x(t)$ cannot be time-limited, unless $x(t)$ is identically zero everywhere. The converse also follows from the time-frequency duality of the Fourier transform.

1.3.1 Uncertainty Relationship [9]

The 'reciprocal spreading' property of the Fourier transform and the result that a function $f(t)$ cannot be both time-limited and band-limited is often expressed more succinctly in terms of various uncertainty relationships depending on the choice of measure of the duration and bandwidth of the signal. Let us consider some of these measures and the corresponding uncertainty relationships for real signals.

(a) Equivalent rectangular bandwidth $\Delta\Omega_1$ and duration ΔT_1. They are defined as follows:

Let $f(t) \leftrightarrow F(\omega)$. Then

$$\Delta\Omega_1 = \frac{\int_{-\infty}^{\infty} F(\omega)\, d\omega}{F(\omega_0)} \tag{1.38}$$

where ω_0 is some characteristic centre frequency of the function $F(\omega)$. That is, we are replacing $F(\omega)$ by a rectangular function of width $\Delta\Omega_1$ and height $F(\omega_0)$ such that it has the same area as under the curve $F(\omega)$. For the low pass case, $\omega_0 = 0$, and we can write

$$\Delta\Omega_1 = \frac{2\pi f(0)}{\int_{-\infty}^{\infty} f(t)\, dt} \tag{1.39}$$

for the Fourier transform pair-relationship.

Similarly, we define

$$\Delta T_1 = \frac{\int_{-\infty}^{\infty} f(t)\, dt}{f(t_c)} \tag{1.40}$$

where t_0 is some characteristic time instant denoting the centre of the pulse. For $t_0 = 0$, it is seen from (1.39) and (1.40) that

$$(\Delta T_1)\,(\Delta\Omega_1) = 2\pi. \tag{1.41}$$

Example 1.7: Let $f(t) = Ae^{-at}$; $a > 0$. Then

$$F(\omega) = \frac{2aA}{a^2 + \omega^2}$$

$$\Delta T_1 = 2\int_0^{\infty} e^{-at}\, dt = \frac{2}{a}$$

and

$$\Delta\Omega_1 = 2a\int_0^{\infty} \frac{a}{a^2 + \omega^2}\, d\omega = a\pi$$

Therefore,

$$(\Delta T_1)(\Delta\Omega_1) = 2\pi.$$

(b) Second-Moment-Bandwidth $\Delta\Omega_2$ and Duration ΔT_2.

Let E be the energy of the signal $f(t)$ given by

$$E = \int_{-\infty}^{\infty} |f(t)|^2\, dt = \frac{1}{2\pi}\int_{-\infty}^{\infty} |F(\omega)|^2\, d\omega$$

and let

$$\bar{\omega} = \frac{\int_{-\infty}^{\infty} \omega |f(\omega)|^2\, d\omega}{\int_{-\infty}^{\infty} |F(\omega)|^2\, d\omega} \tag{1.42}$$

$$\bar{t} = \frac{\int_{-\infty}^{\infty} t|f(t)|^2\, dt}{\int_{-\infty}^{\infty} |f(t)|^2\, dt} \tag{1.43}$$

be the mean or the first moments of $|F(\omega)|^2$ and $|f(t)|^2$ respectively. Then the second moment or (variance) bandwidth $\Delta\Omega_2$ is given by

$$(\Delta\Omega_2)^2 = \frac{1}{2\pi E}\int_{-\infty}^{\infty} (\omega - \bar{\omega})^2 |F(\omega)|^2\, d\omega \tag{1.44}$$

and the second moment duration by

$$(\Delta T_2)^2 = \frac{1}{E}\int_{-\infty}^{\infty} (t - \bar{t})^2 |f(t)|^2\, dt \tag{1.45}$$

Without any loss of generality, we can let $E = 1$ and similarly let $t = 0$ and $\bar{\omega} = 0$ because any translation of the time t-axis and frequency ω-axis will leave $|F(\omega)|^2$ and $|f(t)|^2$ unchanged respectively, and thus not affect $\Delta\Omega_2$ or ΔT_2.

From Fourier transform properties, one can easily show that

$$\int_{-\infty}^{\infty} f'(t) f^*(t)\, dt = \frac{1}{2\pi}\int_{-\infty}^{\infty} \omega^2 |F(\omega)|^2\, d\omega$$

Therefore the product $\Delta\Omega_2 \Delta T_2$ may be written as

$$(\Delta\Omega_2)^2 (\Delta T_2)^2 = \int_{-\infty}^{\infty} tf(t) f*(t)\, dt \int_{-\infty}^{\infty} f'(t) f*(t)\, dt \tag{1.46}$$

Using one form of Schwartz inequality, one may write

$$(\Delta\Omega_2)^2\, (\Delta T_2)^2 \geq \frac{1}{4}\int_{-\infty}^{\infty} [tf^*(t) f'(t) + tf(t) f^{*\prime}(t)\, dt]^2$$

$$= \frac{1}{4}\left|\int_{-\infty}^{\infty} t \frac{d}{dt}[f(t) f^{*\prime}(t)]\, dt\right|^2$$

On integration by parts

$$(\Delta\Omega_2)^2 (\Delta T_2)^2 \geq \frac{1}{4}\left[t|f(t)|^2\Big|_{-\infty}^{\infty} - \int_{-\infty}^{\infty} |f(t)|^2\, dt\right] \tag{1.47}$$

Requiring that $|f(t)|$ vanish at infinity at least as fast as $1/\sqrt{t}$, we obtain

$$(\Delta\Omega_2)(\Delta T_2) \geq \frac{1}{2} \tag{1.48}$$

It may be shown easily that the unique pulse for which the lower bound of $\frac{1}{2}$ is obtained is a Gaussian pulse.

$$f(t) = \sqrt{\frac{k}{\pi}}\ \exp(-kt^2/2) \tag{1.49}$$

where k is a real constant.

Note: In the above discussion, for real signals, $\bar{\omega}$ given by (1.42) is always zero. Therefore, for band pass signals, $\Delta\Omega_2$ is not a meaningful measure where the bandwidth measure should be defined about $\omega = \omega_0$ rather than $\omega = 0$. For this case, Kay and Silverman have shown [10] that, with second moment bandwidth $\Delta\Omega_+$ defined as

$$E(\Delta\Omega_+)^2 = 2\int_{-\infty}^{\infty} (\omega - \omega_0)^2 |F(\omega)|^2\, d\omega \tag{1.50}$$

and with E, the signal energy normalized to unity as before, the uncertainty relationship may be written as

$$(\Delta\Omega_+)(\Delta T_2) \geq \frac{1}{2}\,[1 - 2|F(0)|^2\omega_0] \tag{1.51}$$

In this case, the equality cannot be achieved for any real signal. For odd pulses, $F(0) = 0$ and

$$(\Delta\Omega_+)(\Delta T_2) \geq \frac{1}{2} \tag{1.52}$$

For a more general uncertainty relationship for real signals and the optimum signal which attains the lower bound, the reader is referred to [11].

From the above discussions, we conclude that for any reasonable measures of duration and bandwidth of a finite energy signal, the time-bandwidth product is lower bounded by a number which is of the order of unity. This, so-called, uncertainty relationship is of great importance to communication theory with its following implications.

(a) Decreasing the duration of a pulse, say for the purpose of increasing the accuracy of the measurement of the instant of its occurrence, as in detection of radar signals; inevitably results in increase of its bandwidth and therefore requires that the bandwidth of the measuring instrument be correspondingly increased. Similarly, to locate it sharply in the frequency space, as in spectrum analysis, the duration for which the signal is available should be made larger.

(b) If one observes the signals received through a channel modelled as linear system having a bandwidth, ΔF, then the shortest duration signal which may be measured or whose presence or absence at the input may be discussed by observing the output is ΔT where $\Delta F \cdot \Delta T$, is of the order unity. Consider the practical case of a digital communication system using binary signalling in which a pulse of duration τ is used for mark (or 1) and the absence of a pulse for the same duration signifies space (or 0). Even if the channel used for transmission has no attenuation, because of its finite bandwidth, not all of the transmitted energy will be received and the pulse will also suffer some distortion. If the bandwidth is too small, the transmitted pulse sequence will be received as a meaningless waveform in which the identity of individual pulses will be lost. Let us assume that the bandwidth is at least f_{min} such that a considerable fraction say $\eta = 90\%$ of the signal energy is received and there is little distortion. From previous discussion and Figs. 1.10 and 1.11, we know that for a given η, (f_{min}, τ) is a constant depending on the pulse shape. If the number of ones and zeros transmitted per sec defines the signalling rate, then it is clear that it is proportional to f_{min}. Thus the uncertainty relationship leads to the conclusion that the maximum signalling rate possible is proportional to the bandwidth of the channel.

Similar conclusion holds for an analog signal such as a speech or video waveform. Such a signal lasting for a time T, say may be transmitted in time $T/2$ (thereby doubling the rate of information transmission) by first recording the waveform on a magnetic tape and then running the tape at twice the speed for transmission. This process is reversed at the receiving end. However, since the transmitted signal is compressed in time by a factor of two, it requires twice the bandwidth for transmission when using this procedure.

(c) The uncertainty relationship tells us that our mathematical model of a physical signal as continuous function $f(t)$ of time is unrealistic because it implies that we can assign an independent value to $f(t)$ at every instant of time. However, an instant of time can be communicated through a channel of bandwidth ΔF, (or observed with an instrument of bandwidth ΔF) to an accuracy of no better than $\Delta T = 1/\Delta F$. Conversely, any signal with a spectral width of ΔF, requires a time of at least ΔT for its transmission. Alternatively, one may say that the joint uncertainty associated with both an instant (or an epoch) and frequency of a signal can be no less than a certain constant of the order of unity. (Gabor calls such limits of uncertainty $\Delta F \cdot \Delta T$ logons [12] with reference, in particular to signals of Gaussian form. See equation (1.49) in this connection.)

This immediately suggests that a 'continuous' function $f(t)$ representing a signal waveform as received through a channel or observed with an instrument of bandwidth B Hz may be characterized discretely in terms of its samples spaced $1/2B$ seconds apart. This is the classical Sampling Theorem which will be taken up for detailed discussion in Section 1.5. Knowing these sample values and a knowledge of the bandwidth, $f(t)$ can be reconstructed using an appropriate interpolation formula. In the communications context, this leads to the conclusion that it is impossible to transmit more than $2B$ independent data per second through a channel bandwidth of B Hz.

(d) According to the uncertainty principle, the time bandwidth product of a finite energy signal is lower bounded by a number k of the order of unity. There is no upper bound on k. In fact, one can design signals whose time-bandwidth product is arbitrarily large. Such signals have been termed as spread-spectrum signals [13] as compared to narrow band signals, considered in examples before, for which the product is of order of unity. To see the consideration involved in the design of such signals, let us consider a signal $f(t)$ and its Fourier transform.

$$F(\omega) = A(\omega)e^{j\phi(\omega)}$$

since

$$tf(t) \leftrightarrow j\frac{dF(\omega)}{d\omega} = j\left[\frac{dA(\omega)}{d\omega} + jA(\omega)\frac{d\phi(\omega)}{d\omega}\right]e^{j\phi(\omega)}$$

$$\int_{-\infty}^{\infty} t^2 |f(t)|^2 \phi\, dt = \frac{1}{2\pi}\int_{-\infty}^{\infty}\left[\left(\frac{dA}{d\omega}\right)^2 + A^2\left(\frac{d\phi(\omega)}{d\omega}\right)^2\right]d\omega \qquad (1.53)$$

From (1.44) and (1.45), it is seen that whereas the second moment duration depends on both $\phi(\omega)$ and $A(\omega)$, the corresponding bandwidth depends only on $A(\omega)$. Necessarily therefore, if $f(t)$ is 'simple', then its associated phase spectrum $\phi(\omega)$ is smooth. Conversely keeping $A(\omega)$ constant and choosing $\phi(\omega)$ such that $d\phi(\omega)/d(\omega)$ is large we obtain a spread-spectrum signal whose bandwidth duration product can be made as large as we like.

Example 1.8: Consider the following example. Let

$$f(t) = e^{-\alpha t^2} e^{j\beta t^2}$$

$$T(\omega) = \sqrt{\frac{j\pi}{b + j\alpha}} \exp\left[-\frac{j\omega^2}{4\beta^2}\right] \exp\left[-\alpha_1\left(\frac{\omega}{2\beta}\right)^2\right]$$

where

$$x_1 = \frac{\alpha}{1+\alpha^2/\beta^2},\ \beta_1 = \beta\left(1+\frac{\alpha^2}{\beta^2}\right)$$

If $\beta \gg \alpha$, then

$$F(\omega) \approx \sqrt{\frac{j\pi}{\beta}} e^{j\omega^2/4\beta} e^{-\alpha\omega^2/4\beta^2}$$

Evaluating from (1.44) and (1.45), we obtain

$$(\Delta\Omega_2)^2 = \frac{\beta^2}{\alpha} \text{ and } (\Delta T_2)^2 = \frac{1}{4\alpha}$$

so that,

$$(\Delta\Omega_2)(\Delta T_2) = \frac{\beta}{2\alpha} \tag{1.54}$$

and therefore with $\beta \gg \alpha$, $f(t)$ of above example is a spread-spectrum signal. Such spread-spectrum signals are commonly employed in radar systems where high resolution in range as well as Doppler or velocity is desired [14].

1.4 NARROWBAND SIGNALS AND SYSTEMS [15, 16, 17]

Very often the signals and the systems for processing them, in a communication system satisfy the so-called 'narrowband' assumption. This means that for such signals and systems, the spectrum of the signal and the transfer function of the systems are significant over a relatively small band of frequencies around a 'carrier' or center frequency, relatively away from the zero frequency. Most of the narrowband signals of interest in communications arise either from modulation of some carrier signal by a lowpass information bearing signal or from passage of wideband signals through narrowband systems. For example, a radar system may operate at 1 GHz carrier frequency and have a bandwidth of few MHz. Voice signals requiring about 4 kHz bandwidth are transformed to MHz range for broadcast transmission and to hundreds of MHz range for microwave communication.

Narrowband signals usually have the appearance shown in Fig. 1.12. Apart from the rapid oscillations (at the carrier frequency) one discerns a slowly varying envelope of these oscillations (amplitude modulation). Furthermore, the zero crossings of the oscillations may not be uniformly spaced in time and are changing slowly (phase modulation). In modulated systems, such changes are purposely introduced and are uniquely related to the information bearing signal to be transmitted. They also arise inadvertently because of the noise affecting the signal in its passage through the communication channel or by nonideal transmission characteristics.

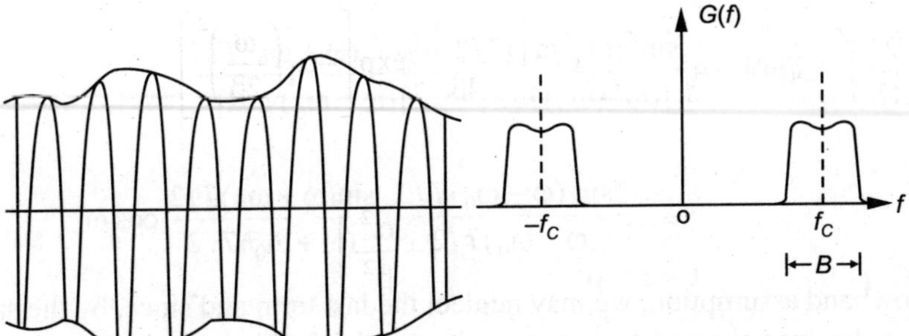

Figure 1.12: A narrowband signal and its spectrum

The most familiar representation for narrowband signals is

$$g(t) = r(t) \cos [2\pi f_c t + \phi(t)] \tag{1.55}$$

The narrowband assumption requires that the spectrum of the signal is confined to a band $B \ll f_c$.

Writing (1.55) in alternative way

$$g(t) = x(t) \cos 2\pi f_c t - y(t) \sin 2\pi f_c t \tag{1.56}$$

where

$$x(t) = r(t) \cos \phi(t) \quad \text{and} \quad y(t) = r(t) \sin \phi(t) \tag{1.57}$$

$x(t)$ and $y(t)$ are called the 'inphase' and 'quadrature' components respectively of the narrowband signal $f(t) \cdot r(t)$ the envelope or the amplitude modulation component and $\phi(t)$ the phase modulation component are slowly varying compared to cos $2\pi f_c t$ or sin $2\pi f_c t$ and hence $x(t)$ and $y(t)$ are lowpass in nature.

In the following sections we introduce representations of such signals and systems which allow us to study the signal processing operations with such signals in a conceptually and computationally simpler way.

Narrowband Signal Representation

To introduce the topic, it is instructive considering the following example.

Example 1.9: Assume a finite duration signal

$$\begin{aligned} g(t) &= A \cos \omega_0 t \qquad 0 \le t \le T \\ &= 0 \qquad \text{elsewhere} \end{aligned}$$

and furthermore $\dfrac{2\pi}{\omega_0} \ll T.$

Its Fourier transform $G(\omega)$ is

$$\begin{aligned} G(\omega) &= \int_0^T A \cos \omega_0 t e^{-j\omega t}\, dt \\ &= \frac{AT}{2} e^{-j(\omega+\omega_0)T/2} \frac{\sin (\omega+\omega_0)T/2}{(\omega+\omega_0)T/2} + \frac{AT}{2} e^{-j(\omega-\omega_0)T/2} \frac{\sin (\omega-\omega_0)T/2}{(\omega-\omega_0)T/2} \end{aligned} \tag{1.58}$$

It is seen that $G(\omega)$ is symmetrical about $\pm\omega_0$. Taking the magnitude square and normalising (1.55)

$$\left(\frac{2}{AT}\right)^2 |G(\omega)|^2 = \frac{\sin^2(\omega-\omega_0)T/2}{[(\omega-\omega_0)T/2]^2} + \frac{\sin^2(\omega+\omega_0)T/2}{[(\omega+\omega_0)T/2]^2}$$

$$= \frac{2\sin(\omega-\omega_0)T/2}{(\omega-\omega_0)T/2}\frac{\sin(\omega+\omega_0)T/2}{(\omega+\omega_0)T/2}\cos\omega_0 T \qquad (1.59)$$

Invoking the narrowband assumption, we may neglect the last term and since, by the same assumption, the first term dominates for values of ω near ω_0 and second for ω near $-\omega_0$, we may write

$$|G(\omega)|^2 \approx |G_+(\omega)|^2 + |G_-(\omega)|^2$$

where

$$|G_\pm(\omega)|^2 = \left(\frac{AT}{2}\right)^2 \frac{\sin^2(\omega\mp\omega_0)T/2}{[(\omega\mp\omega_0)]T/2]^2} \qquad (1.60)$$

This is depicted in Fig. 1.13. This signal like any real signal contains both positive and negative frequency components. (From property of the Fourier transform, the magnitude spectrum has even and phase spectrum; odd symmetry).

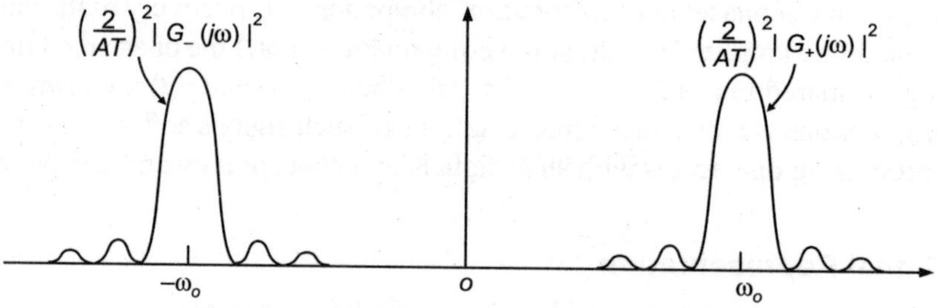

Figure 1.13: Spectrum of a finite duration sinusoid

The narrowband signal $g(t)$ may be written as

$$g(t) = \text{Re}[\tilde{g}(t)\exp(j\omega_c t)] \qquad (1.61)$$

where $$\tilde{g}(t) = x(t) + jy(t) = r(t)\exp[j\phi(t)] \qquad (1.62)$$

Since $x(t)$ and $y(t)$ are lowpass functions, $G(\omega)$ is also seen to be a lowpass function and hence $g(t)$ is slowly varying with time. $g(t)$ is called the *complex envelope* of the narrowband process $g(t)$. The magnitude of the complex envelope $r(t)$ is called the envelope or the *amplitude modulation* of the narrowband signal. The argument or phase of $g(t)$ is called the *phase modulation* component.

The envelope/phase components are

$$r(t) = |\tilde{g}(t)| = \sqrt{x^2(t) + y^2(t)}$$

$$\phi(t) = \tan^{-1}\frac{y(t)}{x(t)} \qquad (1.63)$$

Further,

$$\tilde{g}(t) = \frac{1}{2\pi}\int_{-\infty}^{\infty} \tilde{G}(\omega) \exp(j\omega t)\, d\omega$$

where
$$\tilde{G}(\omega) = \begin{cases} 2G(\omega+\omega_c) & -\omega_c < \omega \\ 0 & \omega < -\omega_c \end{cases} \tag{1.64}$$

Utilizing the concept of the complex envelope and its Fourier transform we can express any bandpass signal spectrum in terms of the equivalent lowpass spectrum of the complex envelope as follows:

$$G(\omega) = \frac{1}{2}\tilde{G}(\omega - \omega_c) + \frac{1}{2}\tilde{G}^*(-\omega - \omega_c) \tag{1.65}$$

This may be proved by noting that since $g(t) = \text{Re}\,[\tilde{g}(t) \exp(j\omega_c t)]$

$$\begin{aligned} G(\omega) &= \int_{-\infty}^{\infty} g(t) \exp(-j\omega t)\, dt = \int_{-\infty}^{0} \text{Re}\,[\tilde{g}(t) \exp(j\omega_c t)] \exp(-j\omega t)\, dt \\ &= \frac{1}{2}\left[\int_{-\infty}^{\infty} (\tilde{g}(t) \exp(j\omega_c t) + \tilde{g}^*(t) \exp(-j\omega_c t)) \exp(-j\omega t)\, dt\right] \\ &= \frac{1}{2}\int_{-\infty}^{\infty} \tilde{g}(t) \exp[-j(\omega - \omega_c)t]\, dt + \frac{1}{2}\int_{-\infty}^{\infty} \tilde{g}^*(t) \exp[+j(-\omega - \omega_c)t]\, dt \\ &= \frac{1}{2}\tilde{G}(\omega - \omega_c) + \frac{1}{2}\tilde{G}^*(-\omega - \omega_c) \end{aligned}$$

The signal energy E may now be calculated from above as

$$\begin{aligned} E &= \frac{1}{2}\int_{-\infty}^{\infty} |\tilde{g}(t)|^2\, dt + \frac{1}{2}\int_{-\infty}^{\infty} |\tilde{g}(t)|^2 \cos[2\omega_c t + 2\phi(t)]\, dt \\ &= \frac{1}{2}\int_{-\infty}^{\infty} |\tilde{g}(t)|^2\, dt \end{aligned} \tag{1.66}$$

In the above discussion, the choice of ω_c is arbitrary and therefore, in most of the problems ω_c may be chosen to simplify the results.

1.4.1 Narrowband System Representation

Just as for narrowband signals, the impulse response and the transfer function of a narrowband system can be usefully represented in terms of complex low frequency equivalents as shown below.

Let the impulse response and the corresponding transfer function of a narrowband LTI system be $h(t)$ and $H(\omega)$ respectively. Thus,

$$\begin{aligned} h(t) &= \int_{-\infty}^{\infty} H(\omega) \exp(j\omega t)\, df \\ &= \int_{0}^{\infty} H(\omega) \exp(j\omega t)\, df + \int_{0}^{\infty} H(-\lambda) \exp(-j\lambda t)\, d\lambda \end{aligned}$$

Since $h(t)$ is real $H(-\omega) = H^*(\omega)$, and therefore

$$h(t) = 2\,\mathrm{Re}\left[\int_0^{\infty} H(\omega)\exp(j\omega t)\,df\right]$$

$$-2\,\mathrm{Re}\left[\frac{1}{2}\int_0^{\infty} \tilde{H}(\omega-\omega_c)\exp(j\omega t)\,df\right] \tag{1.67}$$

where $\tilde{H}(\omega)$ is defined as in (1.64).

Since $\tilde{H}(\omega-\omega_c) = 0$ when $\omega < 0$

$$h(t) = \mathrm{Re}\left[\int_{-\infty}^{\infty} \tilde{H}(\omega-\omega_c)\exp(j\omega t)\,df\right]$$

Letting $\lambda = \omega - \omega_c$

$$h(t) = \mathrm{Re}\left[\exp(j\omega_c t)\frac{1}{2}\int_{-\infty}^{\infty} H(\lambda)\exp(j\lambda t)\,d\lambda\right] = \mathrm{Re}\,[\tilde{h}(t)\exp(j\omega_c t)]$$

where $\tilde{h}(t) \leftrightarrow \tilde{H}(\omega)$.

Since $\tilde{H}(\omega)$ is the equivalent lowpass system transfer function, $\tilde{h}(t)$ is the equivalent lowpass impulse response.

Thus to summarize

$$h(t) = \mathrm{Re}\,[\tilde{h}(t)\exp(j\omega_c t)]$$

$$H(\omega) = \frac{1}{2}\tilde{H}(\omega-\omega_c) + \frac{1}{2}\tilde{H}^*(-\omega-\omega_c) \tag{1.68}$$

Figure 1.14 depicts the bandpass lowpass equivalence in an explicit manner for ease of reference.

Example 1.10: Let us find out the response of the high Q bandpass filter tuned to f_c as shown in Fig. 1.15, to a finite duration sinusoidal current pulse $i(t)$ of frequency f_c. The input is thus a narrowband signal

$$x_{BP}(t) = i(t) = \begin{cases} \mathrm{Re}\,\{A_c \exp(j\omega_c t)\} & -T/2 \le t \le T/2 \\ 0 & \text{otherwise} \end{cases}$$

The transfer function of the filter is

$$H_{BP}(\omega) = \frac{1}{C}\frac{j\omega}{(j\omega)^2 + \dfrac{\omega_c}{Q}j\omega + \omega_c^2}$$

where $\omega_c = \dfrac{1}{\sqrt{LC}}$; $Q = \omega_c RC = R\sqrt{C/L}$.

Using the high Q assumption, $Q \gg 1$, we write the transfer function as

$$H_{BP}(\omega) = \frac{1}{2C}\left[\frac{1-j/2Q}{j(\omega+\omega_c)+\omega_c/2Q} + \frac{1+j/2Q}{j(\omega-\omega_c)+\omega_c/2Q}\right]$$

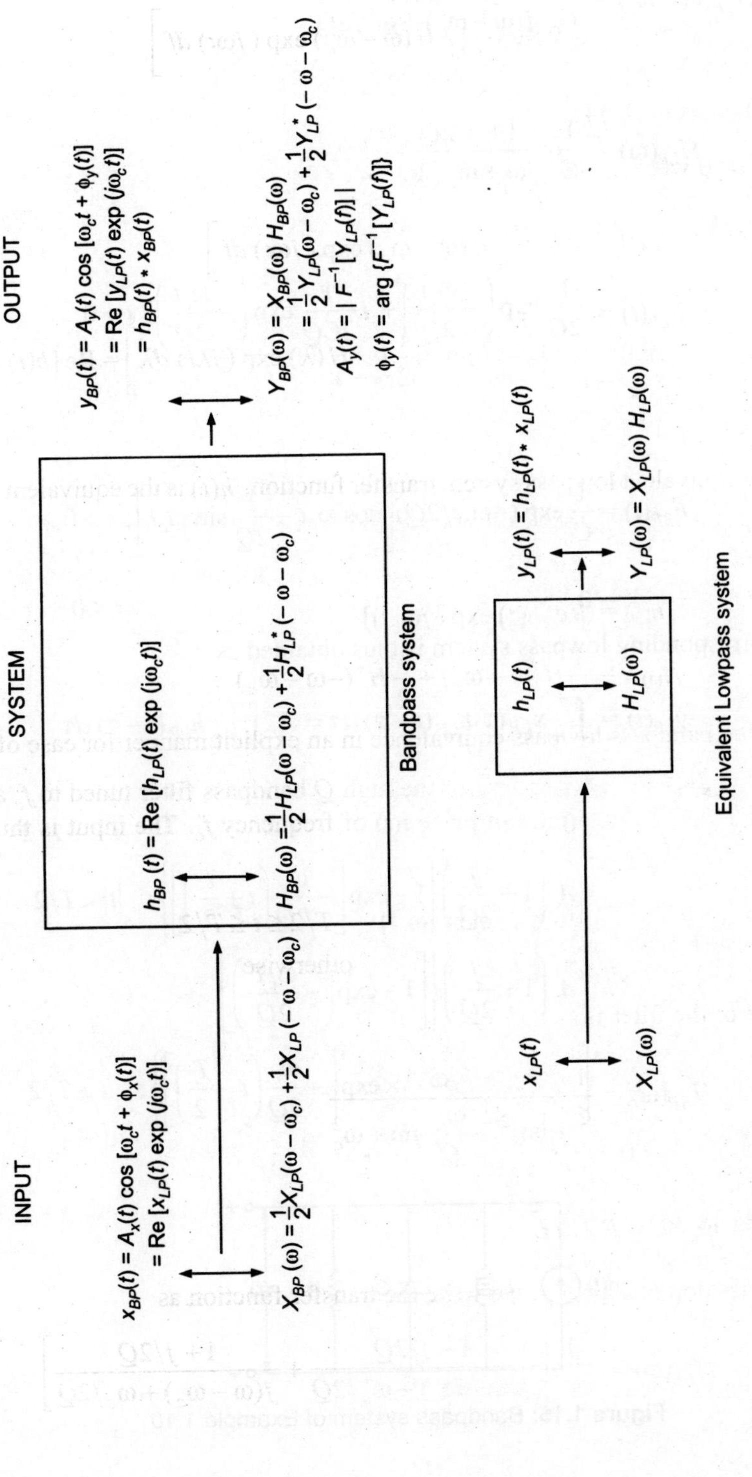

Figure 1.14: Bandpass-lowpass equivalence

Therefore,

$$H_{LP}(\omega - \omega_c) = \frac{1}{C} \cdot \frac{1 + j/2Q}{j(\omega - \omega_c) + \omega_c/2Q}$$

and

$$H_{LP}(\omega) = \frac{1}{C} \cdot \frac{1 + j/2Q}{j\omega + \omega_c/2Q}$$

Hence

$$h_{LP}(t) = \frac{1}{2C} \exp\left(-\frac{\omega_c t}{2Q}\right) + j\frac{1}{2CQ} \exp\left(-\frac{\omega_c t}{2Q}\right); \quad t > 0$$
$$= 0 \qquad t < 0$$

and

$$h_{BP}(t) = \frac{1}{C} \exp\left(-\omega_c t/2Q\right)\left[\cos \omega_c t - \frac{1}{2Q} \sin \omega_c t\right]; \quad t > 0$$
$$= 0 \qquad t < 0$$

The output of the corresponding lowpass system is thus obtained as

$$y_{LP}(t) = \int_{-\infty}^{\infty} x_{LP}(\tau) h_{LP}(t-\tau)\, d\tau = A_c \int_{-T/2}^{T/2} h_{LP}(t-\tau)\, d\tau$$

$$= \begin{cases} 0 & t < -T/2 \\ A_c\left(1 + \dfrac{j}{2Q}\right)\left\{1 - \exp\left[-\dfrac{\omega_c}{2Q}\left(t + \dfrac{T}{2}\right)\right]\right\}; & |t| < T/2 \\ A_c\left(1 + \dfrac{j}{2Q}\right)\left\{\left[1 - \exp\left(-\dfrac{\omega_c T}{2Q}\right)\right] \right. & \\ \qquad\qquad \left. \times \exp\left[-\dfrac{\omega_c}{2Q}\left(t - \dfrac{T}{2}\right)\right]\right\}; & t > T/2 \end{cases}$$

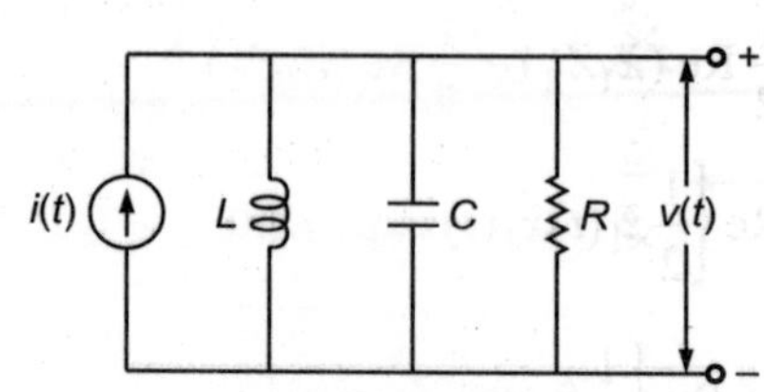

Figure 1.15: Bandpass system of Example 1.10

Finally, the output of the bandpass system is obtained as

$$y_{BP}(t) = \text{Re}\,[y_{LP}(t)\exp(j\omega_c t)]$$

$$= \begin{cases} 0 & t < -T/2 \\ A_c\left\{1-\exp\left[-\dfrac{\omega_c}{2Q}\left(t+\dfrac{T}{2}\right)\right]\right\} \\ \quad\times\left(\cos\omega_c t - \dfrac{1}{2Q}\sin\omega_c t\right); & |t| < T/2 \\ A_c\left\{\left[1-\exp\left(-\dfrac{\omega_c T}{2Q}\right)\right]\exp\left[-\dfrac{\omega_c}{2Q}\left(t-\dfrac{T}{2}\right)\right]\right\} \\ \quad\times\left(\cos\omega_c t - \dfrac{1}{2Q}\sin\omega_c t\right); & t > T/2 \end{cases}$$

1.4.2 Operation with Bandpass Signals [17]

(i) Sum of B.P. Signals

The sum of two bandpass signals $x_1(t)$ and $x_2(t)$ of the same carrier frequency f_c can be expressed as before in terms of the corresponding complex envelope as shown below.

$$\begin{aligned} x(t) &= x_1(t) + x_2(t) \\ &= \text{Re}\,\{\tilde{x}_1(t)\exp(j2\pi f_c t)\} + \text{Re}\,\{\tilde{x}_2(t)\exp(j2\pi f_c t)\} \\ &= \text{Re}\,\{\tilde{x}(t)\exp(j2\pi f_c t)\} \end{aligned} \tag{1.69}$$

where $\tilde{x}(t) = \tilde{x}_1(t) + \tilde{x}_2(t)$

and

$$\begin{aligned} |\tilde{x}(t)| &= |\tilde{x}_1(t) + \tilde{x}_2(t)| \\ &= [r_1^2(t) + r_2^2(t) + 2r_1(t)\,r_2(t)\cos(\theta_1(t) - \theta_2(t))]^{1/2} \end{aligned}$$

(ii) Product of B.P. Signals

$$x_1(t)x_2(t) = \text{Re}\,\{\tilde{x}_1(t)\exp(j2\pi f_1 t)\}\,\text{Re}\,\{\tilde{x}_2(t)\exp(j2\pi f_2 t)\}$$

Since $\text{Re}\,(Z_1)\,\text{Re}\,(Z_2) = \dfrac{1}{2}\text{Re}\,(Z_1 Z_2) + \dfrac{1}{2}\text{Re}\,(Z_1 Z_2^*)$,

$$\begin{aligned} x_1(t)x_2(t) &= \text{Re}\left\{\frac{1}{2}\tilde{x}_1(t)\tilde{x}_2(t)\exp[j2\pi(f_1+f_2)t]\right\} \\ &\quad + \text{Re}\left\{\frac{1}{2}\tilde{x}_1(t)\tilde{x}_2^*(t)\exp[j2\pi(f_1-f_2)t]\right\} \end{aligned} \tag{1.70}$$

Defining the first terms containing the sum frequency as x_{SF} and the second terms containing the difference frequency as x_{DF}, we may write

$$\tilde{x}_{SF}(t) = \frac{1}{2}\tilde{x}_1\tilde{x}_2 \leftrightarrow \tilde{X}_{SF}(t) = \frac{1}{2}\int_{-\infty}^{\infty} \tilde{X}_1(\lambda)\tilde{X}_2(f-\lambda)\, d\lambda$$

$$\tilde{x}_{DF}(t) = \frac{1}{2}\tilde{x}_1\tilde{x}_2^* \leftrightarrow \tilde{X}_{DF}(f) = \frac{1}{2}\int_{-\infty}^{\infty} \tilde{X}_1(\lambda)\tilde{X}_2^*(f+\lambda)\, d\lambda \qquad (1.71)$$

(iii) Correlation of B.P. Signals

Consider two bandpass signals $x(t)$ and $y(t)$ with same frequency f_c.

$$x(t) = \text{Re}\{\tilde{x}(t)\exp(j2\pi f_c t)\}$$

$$y(t) = \text{Re}\{\tilde{y}(t)\exp(j2\pi f_c t)\}$$

Then from eq. (1.13a),

$$R_{xy}(\tau) = \int_{-\infty}^{\infty} x(t)y(t+\tau)\, dt$$

Making use of the results (1.70), we can write

$$R_{xy}(\tau) = \text{Re}\left\{\exp(j2\pi f_c\tau)\int_{-\infty}^{\infty}\frac{1}{2}x(t)y(t+\tau)\exp(j4\pi f_c t)\right\} dt$$

$$+\,\text{Re}\left\{\exp(j2\pi f_c\tau)\int_{-\infty}^{\infty}\frac{1}{2}\tilde{x}(t)\tilde{y}^*(t+\tau)\, dt\right\}$$

$$= \text{Re}\left\{\exp(j2\pi f_c\tau)\int_{-\infty}^{\infty}\frac{1}{2}\tilde{x}(t)\tilde{y}^*(t+\tau)\, dt\right\}$$

The last result follows from the narrow-band assumption. Thus,

$$R_{xy}(\tau) = \text{Re}\{R_{\tilde{x}\tilde{y}}(\tau)\exp(j2\pi f_c\tau)\} \qquad (1.72)$$

where

$$R_{\tilde{x}\tilde{y}}(\tau) = \frac{1}{2}\int_{-\infty}^{\infty}\tilde{x}(t)\tilde{y}^*(t+\tau)\, dt \qquad (1.73)$$

Thus $R_{\tilde{x}\tilde{y}}(\tau)$ is lowpass in nature and complex, in general. Its Fourier Transform then defines the cross-energy density spectrum, i.e. (eq. (1.26))

$$R_{\tilde{x}\tilde{y}}(t) \leftrightarrow S_{\tilde{x}\tilde{y}}(f) = \frac{1}{2}\tilde{X}(f)\,\tilde{Y}^*(f) \qquad (1.74)$$

where $\tilde{x}(t) \leftrightarrow X(f)$ and $\tilde{y}(t) \leftrightarrow \tilde{Y}(f)$. Thus

$$R_{\tilde{x}\tilde{y}}(0) = \int_{-\infty}^{\infty} S_{\tilde{x}\tilde{y}}(f)\, df \tag{1.75}$$

and

$$\int_{-\infty}^{\infty} \tilde{x}(t)\tilde{y}^*(t)\, dt = \int_{-\infty}^{\infty} \tilde{X}(f)\tilde{Y}^*(f)\, df \tag{1.76}$$

is then the Parseval's relationship for complex envelopes of signals $x(t)$ and $y(t)$.

Setting $\tilde{x}(t) = \tilde{y}(t)$, we get from eq. (1.70),

$$\int_{-\infty}^{\infty} \frac{1}{2}|\tilde{x}(t)|^2\, dt = \int_{-\infty}^{\infty} \frac{1}{2}|\tilde{X}(f)|^2\, df = \text{energy of } \tilde{x}(t).$$

Since

$$R_{xx}(\tau) = \int_{-\infty}^{\infty} x(t)x(t+\tau)\, dt = \text{Re}\,\{R_{\tilde{x}\tilde{x}}(\tau)\exp(j2\pi f_c\tau)\} \tag{1.77}$$

where

$$R_{\tilde{x}\tilde{x}}(\tau) = \frac{1}{2}\int_{-\infty}^{\infty} \tilde{x}(t)\tilde{x}*(t+\tau)\, dt \tag{1.78}$$

is called the complex autocorrelation function of the complex envelope $\tilde{x}(t)$. Also,

$$R_{\tilde{x}\tilde{x}}(t) \leftrightarrow S_{\tilde{x}\tilde{x}}(f) = \frac{1}{2}|\tilde{X}(f)|^2 \tag{1.79}$$

This lowpass spectrum is not quite the same as an ordinary energy density spectrum $S_{\tilde{x}\tilde{x}}(f)$ is real, nonnegative but in general it is not an even function of frequency.

As before, we may express the energy density spectrum $S_{xx}(f)$ of the bandpass signal $x(t)$ in terms of the corresponding spectrum of its complex envelope. Thus, from eq. (1.65),

$$S_{xx}(f) = \frac{1}{2}S_{\tilde{x}\tilde{x}}(f-f_c) + \frac{1}{2}S^*_{\tilde{x}\tilde{x}}(-f-f_c) \tag{1.80}$$

These results may be summarized as below with reference to Fig. 1.16.

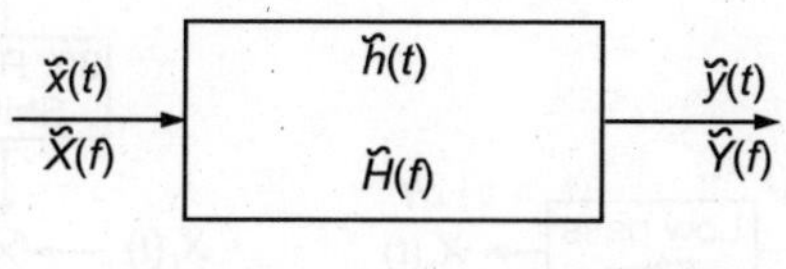

Figure 1.16: Bandpass filtering

$$S_{\tilde{x}\tilde{x}}(f) = \frac{1}{2}|\tilde{X}(f)|^2$$

$$S_{\tilde{y}\tilde{y}}(f) = \frac{1}{2}|\tilde{Y}(f)|^2$$

$$\tilde{y}(t) = \int_{-\infty}^{\infty} \tilde{x}(\tau)\tilde{h}(t-\tau)d\tau$$

$$\tilde{Y}(f) = \tilde{H}(f)\tilde{X}(f)$$

$$S_{\tilde{y}\tilde{y}}(f) = \frac{1}{2}\left|\tilde{H}(f)\tilde{X}(f)\right|^2 = \frac{1}{2}\left|\tilde{H}(f)\right|^2\left|\tilde{X}(f)\right|^2$$

$$= \left|\tilde{H}(f)\right|^2 S_{\tilde{x}\tilde{x}}(f) \tag{1.81}$$

$$S_{\tilde{y}\tilde{y}}(f) = \frac{1}{2}S_{\tilde{y}\tilde{y}}(f - f_c) + \frac{1}{2}S_{\tilde{y}\tilde{y}}(-f - f_c)$$

$$= \frac{1}{2}\left|\tilde{H}(f - f_c)\right|^2 S_{\tilde{x}\tilde{x}}(f - f_c) + \frac{1}{2}\left|\tilde{H}(-f - f_c)\right|^2 S_{\tilde{x}\tilde{x}}(-f - f_c) \tag{1.82}$$

1.4.3 Complex Envelope, Hilbert Transform and Analytic Signal [9, 18]

We have seen earlier that a bandpass signal $x_{BP}(t)$ and its complex envelope or lowpass equivalent $x_{LP}(t)$ are related by (1.61) and (1.62). Thus we have

$$x_{BP}(t) = r(t) \cos [\omega_c t + \phi(t)] = \text{Re } [X_{LP}(t) \exp (j\omega_c t)]$$

or

$$x_{LP}(t) = r(t) \exp [j\phi(t)]$$

How do we, in practice, obtain $x_{LP}(t)$ from $x_{BP}(t)$? One practical way of doing this is suggested by writing $x_{LP}(t)$ from eq. (1.62) as

$$x_{LP}(t) = x_r(t) + jx_i(t) = r \cos \phi + jr \sin \phi$$

The component x_r and x_i are easily obtained by using a mixer or multiplier-lowpass filter combination and in phase and quadrature carriers at ω_c as shown in Fig. 1.17(a). The bandpass signal $x_{BP}(t)$ can be reconstructed using the scheme suggested in Fig. 1.17(b). It is assumed here that the signal $x_{LP}(t)$ is restricted to frequencies $|\omega| < \omega_c$.

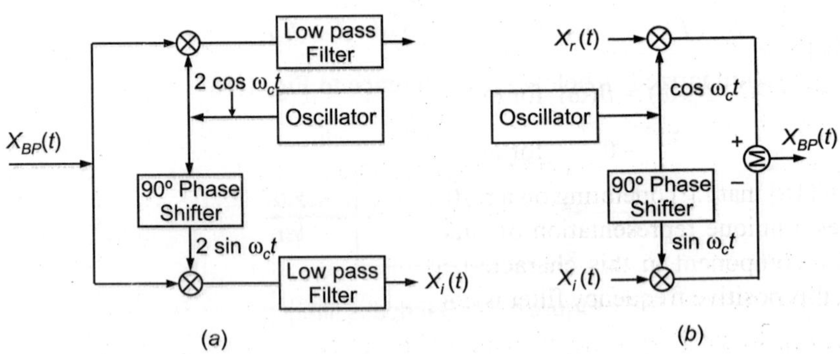

Figure 1.17: (***a***) Scheme for obtaining the in-phase and quadrature components of a bandpass signal; (***b***) Scheme for reconstructing the bandpass signal

Another approach is to look at the bandpass-lowpass transformation as a linear operator $L : x_{LP} = L[x_{BP}]$ and look for a characterization of this operator $L[\cdot]$. To do this we first define a LTI operator with the transfer function

$$B(\omega) = \begin{cases} 2 & \omega > 0 \\ 1 & \omega = 0 \\ 0 & \omega < 0 \end{cases} \tag{1.83}$$

Let $\hat{x}(t)$ be the output of this filter for input $x_{BP}(t)$. Then L is characterized by

$$x_{LP}(t) = L[x_{BP}(t)] = \hat{x} \exp(-j\omega_c t) \tag{1.84}$$

This characterization is depicted in Fig. 1.18(*a*). The operator L (which is seen to be time varing) may be defined in time

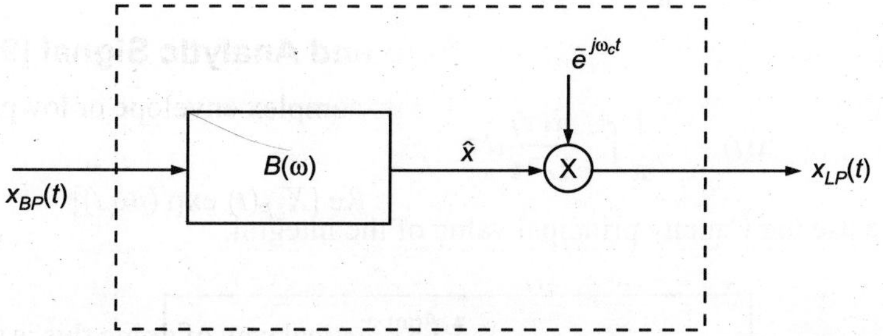

Figure 1.18 (*a*): Linear time varying operator *L* characterizing the bandpass-lowpass transformation

$$L[x_{BP}(t)] = \exp(-j\omega_c t) \lim_{n\to\infty}[b_n(t) * x_{BP}(t)] \tag{1.85}$$

where

$$b_n(t) = \frac{2\sin(n/2)t}{\pi} \exp\left(-j\frac{n}{2}t\right)$$

and corresponds to

$$B_n(\omega) = B(\omega) \text{ for } \omega \le n; \ (n \text{ any + integer})$$

$$= 0 \qquad \text{for } \omega > n \tag{1.86}$$

It may be shown [18] that $L[\cdot]$ operating on a restricted domain of bandpass signals, namely narrowband signals, provides a unique representation of such signals in terms of the associated complex lowpass signals. The key component in this characterization is the filter with transfer function $B(\omega)$. Since $B(-\omega) \neq B*(\omega)$, this positive frequency filter is not physically realizable as it is not a real parameter filter. Writing

$$B(\omega) = B_r(\omega) + jB_i(\omega)$$

where

$$B_r(-\omega) = B_r^*(\omega) \text{ and } B_i(-\omega) = B_i^*(\omega)$$

we thus obtain $B_r(\omega) = 1$ and

$$B_i(\omega) = \begin{cases} -j & \omega > 0 \\ 0 & \omega = 0 \\ j & \omega < 0 \end{cases} \tag{1.87}$$

The transfer function $B_i(\omega)$ of (1.87) represents the frequency domain definition of the Hilbert transform operator. It corresponds to a 90° phase shift network (also called 'quadrature filter'). Its magnitude and phase response is depicted in Fig. 1.18(b). The corresponding time domain description in terms of the impulse response is $b_i(t) = 1/\pi t$. Formally, we define the Hilbert transform $\hat{x}(t)$ of a signal $x(t)$ as

$$\hat{x}(t) = \frac{1}{\pi}\int_{-\infty}^{\infty} \frac{x(\tau)}{t-\tau}\,d\tau \tag{1.88}$$

and

$$x(t) = -\frac{1}{\pi}\int_{-\infty}^{\infty} \frac{\hat{x}(\tau)}{t-\tau}\,d\tau \tag{1.89}$$

In this definition, we use the Cauchy principal value of the integral.

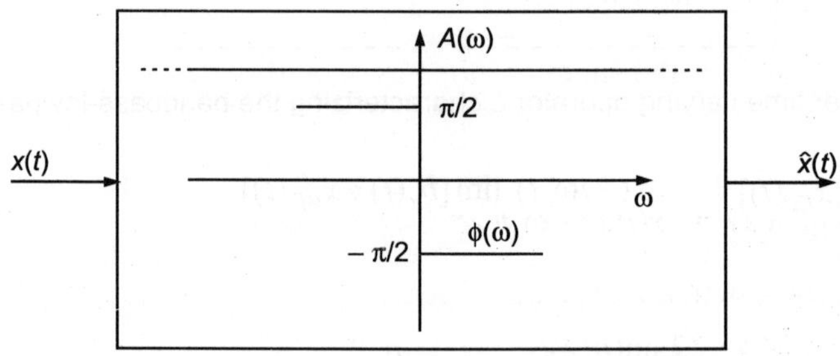

Figure 1.18 (*b*): Hilbert transformer or quadrature filter

Alternatively, we may write

$$\hat{x}(t) = x(t) * \frac{1}{\pi t} \tag{1.90}$$

or the corresponding F.T. $\hat{X}(f)$ of $\hat{x}(t)$ as

$$\hat{X}(f) = \begin{cases} -jX(f) & f \geq 0 \\ +jX(f) & f < 0 \end{cases} \tag{1.91}$$

Some of the important properties of Hilbert transforms are given below:

1. $\hat{x}(t) = x(t) * \frac{1}{\pi t}$; $\hat{X}(f) = -jX(f)\,\text{sgn}\, f$
2. For $x(t) = \cos(\omega t + \phi)$,

$$\hat{x}(t) = \sin(\omega t + \phi)$$

and for $x(t) = \sin(\omega t + \phi)$,

$$\hat{x}(t) = -\cos(\omega t + \phi).$$

3. Power in the signal is unaltered in taking its Hilbert transform, i.e.,

$$P_x = P_{\hat{x}} \text{ or } \lim_{T\to\infty} \frac{1}{T}\int_{-T/2}^{T/2} x^2(t)\,dt = \lim_{T\to\infty}\frac{1}{T}\int_{-T/2}^{T/2} \hat{x}^2(t)\,dt$$

4. $x(t)$ and $\hat{x}(t)$ are orthogonal, i.e.,

$$\lim_{T\to\infty}\frac{1}{T}\int_{-T/2}^{T/2} x(t)\hat{x}(t)\,dt = 0$$

5. $\hat{\hat{x}}(t) = -x(t)$
6. If $y(t) = v(t) * x(t)$, then

   $\hat{y}(t) = v(t) * \hat{x}(t) = x(t) * \hat{v}(t)$
7. If $X(\omega)$ is bandlimited to $|\omega| < W$, then

$$\widehat{[x(t)\cos\omega_c t]} = x(t)\sin\omega_c t \text{ as long as } \omega_c > W$$

   and

$$\widehat{[x(t)\sin\omega_c t]} = -x(t)\cos\omega_c t$$

8. The autocorrelation function $R_{xx}(\tau)$ and p.s.d $S_{xx}(\omega)$ of a signal $x(t)$ is unaltered for the Hilbert transformed signal $\hat{x}(t)$, i.e.

$$R_{xx}(\tau) = R_{\hat{x}\hat{x}}(\tau)$$

$$S_{xx}(\omega) = S_{\hat{x}\hat{x}}(\omega)$$

9. $R_{\hat{x}x}(\tau) = \hat{R}_{xx}(\tau)$

$$R_{x\hat{x}}(\tau) = -\hat{R}_{xx}(\tau)$$

The notion of Hilbert transform of a signal is instrumental in formalizing the description of the 'envelope' and 'phase' of a narrowband process. This is done by defining the so-called 'analytic signal' $z_x(t)$ (also called 'pre-envelope') corresponding to a real physical signal $x(t)$ as follow:

$$z_x(t) = x(t) + j\hat{x}(t) \tag{1.92}$$

This is visualized in signal space as shown in Fig. 1.19 and represents a formal generalized version of the concept of a rotating phasor. Using eq. (1.62), $z_x(t)$ may be written as

$$z_x(t) = r(t)\exp(j\theta(t)) = r(t)\exp(j(\omega_c t + \phi(t)))$$

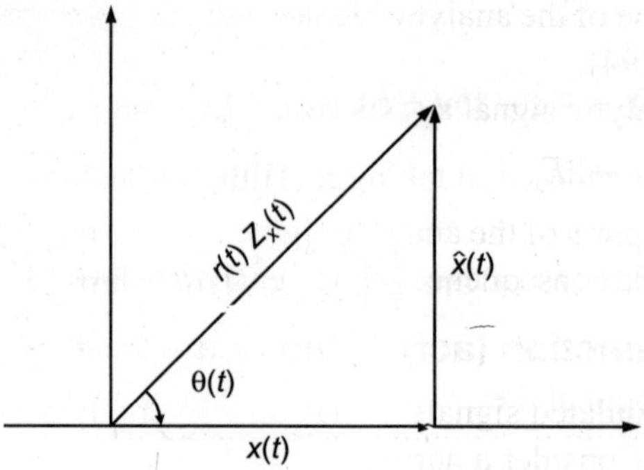

Figure 1.19: Analytic signal as a generalised rotating phasor

$$= \hat{x}(t) \exp (j\omega_c t) \tag{1.93}*$$

where $\hat{x}(t)$ is the complex envelope as given by (1.62). Clearly, then

$$r(t) = [x^2(t) + \hat{x}^2(t)]^{1/2} = |z_x(t)|$$

and

$$\theta(t) = \tan^{-1}\frac{\hat{x}(t)}{x(t)} = \angle z_x(t) \tag{1.94}$$

and the envelope or a.m. component and phase or p.m. component of the signal $x(t)$.

The concept of Hilbert transform and analytic signal [19] are quite useful and advantageous in the description and analysis of complex bandpass signals, such as modulated signals. The basic properties of analytic signals are summarised below:

(a) The Fourier transform of the analytic signal vanishes for negative frequencies, while that of its complex conjugate vanishes for positive frequencies, i.e. analytic signals have one sided spectrum.

$$Z_x(\omega) = \begin{cases} 2X(\omega) & \omega \geq 0 \\ 0 & \omega < 0 \end{cases} \tag{1.95}$$

---

*When the narrowband signal $x(t)$ is represented as

$$x(t) = r(t)\cos\theta(t) = r(t)\cos(\omega_c t + \phi(t))$$

there is an ambiguity in this representation because

$$r_1(t)\cos\theta_1(t) = r_2(t)\cos\theta_2(t)$$

can be true without $r_1 \exp(j\theta_1) = r_2 \exp(j\theta_2)$. For example $f(t) = A_0 \cos\omega_c t$ can also be expressed as $f(t) = r(t) \cos \omega t$ where $\omega = \omega_c + \Delta\omega$. Equating the two, we have

$$r(t) = \frac{A_0}{\cos\Delta\omega t - \sin\Delta\omega t \tan\omega_c t}$$

One can easily verify that $r(t)$ is not an 'envelope' in the usual sense, since it can intersect the curve $f(t)$ (instead of touching it as points where $x(t)$ is maximum). However, using the analytical signal $z_x(t)$, there is no such ambiguity.

(b) The magnitude and phase of the analytic signal describe envelope and phase of the corresponding real signal as in Eq. (1.94).

(c) The energy E_z of an analytic signal $z_x(t)$ is twice the energy E_x of thc original signal $f(t)$

$$E_z = 2E_x \tag{1.96}$$

(d) The real and imaginary parts of the analytic signal have identical correlation functions and power spectra. This is simply the consequence of the fact that the two are related via the Hilbert transform.

1.4.4 Autocorrelation Function (acf) of Modulated Signals

The calculation of a.c.f. of modulated signals is made considerably easier if the corresponding analytic signals representation is used. Consider a narrowband signal

$$x(t) = a(t)\cos\theta(t) = a(t)\cos(\omega_c t + \phi(t))$$

and its a.c.f.

$$R_{xx}(\tau) = \int_{-\infty}^{\infty} x(\tau)x(t-\tau)\,dt$$

(We assume that $x(t)$ is absolutely integrable.)

Let $x(t)$ be expressed as

$$x(t) = \frac{1}{2}[z_x(t) + z_x^*(t)] \tag{1.97}$$

This follows from (2.6.48) since $x(t)$ is real. Therefore,

$$R_{xx}(\tau) = \frac{1}{4}\int_{-\infty}^{\infty} z_x(t)z_x(t-\tau)\,dt + \int_{-\infty}^{\infty} z_x^*(t)z_x^*(t-\tau)\,dt$$

$$+\int_{-\infty}^{\infty} z_x^*(t)z_x(t-\tau)\,dt + \int_{-\infty}^{\infty} z_x(t)z_x^*(t-\tau)\,dt \tag{1.98}$$

Now

(i) $z_x(t)\,z_x(t-\tau) = a(t)a(t-\tau)\exp(j[\theta(t) - \theta(t-\tau)])$
$= a(t)\,a(t-\tau)\exp(j[\phi(t) + \phi(t-\tau) - \omega_c\tau]) \times \exp(j2\omega_c t)$

(ii) $z_x^*(t)z_x^*(t-\tau) = a(t)\,a(t-\tau)\exp(-j[\phi(t) + \phi(t-\tau) - \omega_c t]) \times \exp(-j2\omega_c)t$

(iii) $z_x^*(t)z_x(t-\tau) = a(t)\,a(t-\tau)\exp(-j[\phi(t) - \phi(t-\tau) + \omega_c\tau])$

(iv) $z_x(t)z_x^*(t-\tau) = a(t)\,a(t-\tau)\exp(j[\phi(t) - \phi(t-\tau) + \omega_c\tau])$

The first and the second integrands in (1.98) contain the rapidly oscillating product time $\exp(\pm j2\omega_c t)$ compared to slowly varying $a(t)$, $a(t-\tau)$, $\phi(t)$, $\phi(t-\tau)$ and hence their integrals may be neglected compared to contributions from third and fourth integrands. Thus,

$$R_{xx}(\tau) = \frac{1}{4}\left[\int_{-\infty}^{\infty} z_x^*(t)z(t-\tau)\,dt + \int_{-\infty}^{\infty} z_x(t)z_x^*(t-\tau)\,dt\right]$$

$$= \frac{1}{2}\mathrm{Re}\left[\int_{-\infty}^{\infty} z_x(t)z_x^*(t-\tau)\,dt\right]$$

$$= \frac{1}{2}\int_{-\infty}^{\infty} a(t)\,a(t-\tau)\cos[\omega_c\tau + \theta(t) - \theta(t-\tau)]\,dt. \tag{1.99}$$

Let us apply these results to (i) pure a.m., and (ii) combined a.m. and f.m.

For pure a.m. signal,

$$x(t) = a(t)\cos(\omega_c t + \theta_0)$$

and

$$R_{xx}(\tau) = \frac{1}{2}\cos\omega_c\tau \int_{-\infty}^{\infty} a(t)\, a(t-\tau)\, dt$$

$$= \frac{1}{2} R_{aa}(\tau)\cos\omega_c\tau \tag{1.100}$$

r an r.f. pulse with a rectangular envelope, the a.c.f. is shown in Fig. 1.20. $R_{xx}(\tau)$ does not depend on the initial phase θ_0 and its envelope coincides with the a.c.f. of a rectangular pulse.

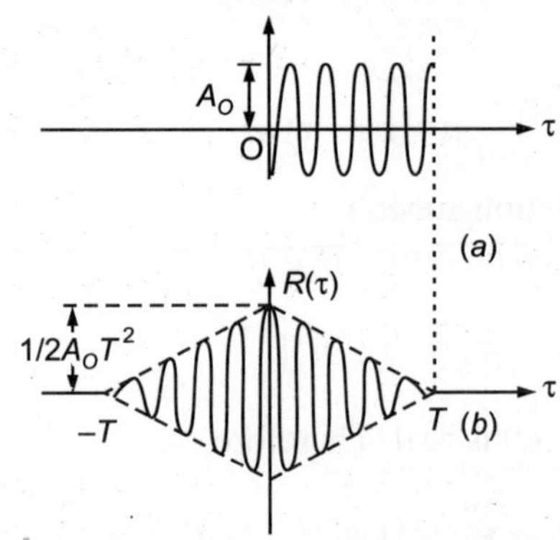

Figure 1.20: a.c.f. of an r.f. pulse

Example 1.11: For the combined AM-FM case, we consider the so-called 'chirp' signal as used in radar signalling. This is expressed as

$$x(t) = \begin{cases} A_0 \cos\left(\omega_c t + \dfrac{\beta t^2}{2}\right); & -T/2 \le t \le T/2 \\ 0 & \text{otherwise} \end{cases} \tag{1.101}$$

The corresponding analytic signal is

$$z_x(t) = A_0 \exp[j(\omega_c t + \beta t^2/2)]; \quad -T/2 \le t \le T/2$$

Using (1.78), we can write

$$R_{xx}(\tau) = \frac{A^2}{2}\text{Re}\left\{\int_{-T/2}^{T/2} \exp(j(\omega_c t + \beta t^2/2)) \exp(-j[\omega_c(t-\tau) + \frac{1}{2}\beta(t-\tau)^2])\, dt\right\}$$

This integral can be put in standard form and evaluated as

$$R_{xx}(\tau) = \begin{cases} A_0^2 \dfrac{\sin\left(\frac{1}{2}\beta T\tau - \frac{1}{2}\beta\tau^2\right)}{\beta\tau}\cos\omega_c\tau; & |\tau| \le T/2 \\ 0 & ; \ |\tau| > T/2 \end{cases} \tag{1.102}$$

Since $\beta T^2 = 2\omega_d T \,\Delta\, 2\pi m$, where m is a parameter and ω_d is the peak angular frequency deviation from ω_c, we can write (1.102) in a convenient form

$$R_{xx}(\tau) = \frac{1}{2} A_0^2 T \frac{\sin[\pi m(\tau/T)(1-\tau/T)]}{\pi m \tau/T} \cos \omega_c \tau \tag{1.103}$$

$$\frac{A_0^2 T}{2} = R_{xx}(0) = E_x$$

is the total energy of the r.f. pulse.

Therefore,

$$\frac{R_{xx}(\tau)}{R_{xx}(0)} = \frac{\sin[\pi m \tau/T(1-\tau/T)]}{\pi m \tau/T} \cos \omega_c \tau \tag{1.104}$$

The plot of this function is shown in Fig. 1.21, for the parameter $m = 100$, and assuming $\omega_c T$ is quite large. The envelope of the correlation function is seen to have a very sharp peak ($m \gg 1$) while the oscillation frequency is the same as the centre frequency ω_c of the r.f. pulse.

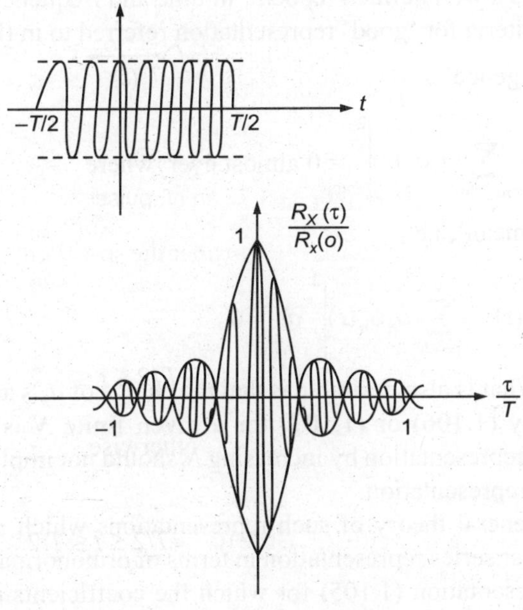

Figure 1.21: a.c.f. of a 'chirp' signal

1.5 DISCRETE REPRESENTATION OF CONTINUOUS SIGNALS—SAMPLING

In Section 2.3, we discussed the frequency domain or spectral representation of continuous signals and emphasized the tremendous usefulness of such a choice of representation, namely the Fourier and Laplace Transform, for the study of interaction of signals with LTI systems. Such a description maps one continuum to another continuum. In the special case of periodic signals, the Fourier series representation is a mapping from continuous to discrete and is therefore a 'simpler' description made possible by the periodicity property of the signal. The next logical question to ask is: Does there exist a sufficiently broad class of signals which realistically represent or approximate physical signals and for which discrete representation is possible. The answer is yes, the class of band limited signals. In Section 1.3 we have discussed such signals and their properties in brief. For further details, refer to [2]. Since our concern is mainly with linear representations, the signal $x(t)$ may be expressed in the form of a weighted sum of preselected waveforms or basis functions $\phi_n(t)$, i.e.

$$x(t) = \sum_n a_n \phi_n(t) \tag{1.105}$$

The $\phi_n(t)$'s are also sometimes called 'interpolation functions'. Some or all of the desirable properties of the series representation (1.105) may be stated as follows:

(a) a_n's and ϕ_n's are obtained in a relatively 'simple' way, i.e., the computation and/or generation of a_n's and ϕ_n's can be easily implemented.
(b) The number of terms required to get a 'good' representation of $x(t)$ should be reasonably small, i.e., the choice of $\phi_n(t)$'s should result in reasonably fast convergence.
(c) Since, inasmuch as the physical signals are essentially 'time limited' and 'band limited', each of the $\phi_n(t)$'s should have a well defined 'epoch' in time and frequency.
(d) Some of the useful criteria for 'good' representation referred to in (b) are

1. 'least squares convergence'

$$\lim_{N\to\infty} \left| x(t) - \sum_{n=-N}^{N} a_n \phi_n(t) \right|^2 = 0 \text{ almost everywhere} \tag{1.106}$$

2. 'convergence in the mean', i.e.,

$$\lim_{N\to\infty} \int_{-\infty}^{\infty} \left| x(t) - \sum_{n=-N}^{N} a_n \phi_n(t) \right|^2 dt = 0 \tag{1.107}$$

These are limiting criteria. What is also desirable is that the choice of a_n's and ϕ_n's be such that the error in representation implied by (1.106) or (1.107) for a given finite N is minimized. Moreover, the improvement of accuracy of representation by increasing N should not imply recalculation of all the a_n's obtained from lower degree representation.

Without going into the general theory of such representations which involve the study of Hilbert spaces and generalized Fourier series representation in terms of orthonormal basis functions [20], at this stage, let us consider a representation (1.105) for which the coefficients are obtained in the simplest possible manner, by periodically sampling the waveform. Such a representation known as the 'sampling theorem' manifests as a natural fall-out if we study the relationship between the spectrum of the analog signal and that of its periodically sampled sequence [5].

Let $x_a(t)$ be an analog signal with Fourier representation $X_a(\Omega)$, given by

$$x_a(t) = \frac{1}{2\pi} \int_{-\infty}^{\infty} X_a(\Omega) e^{j\Omega t} \, d\Omega$$

$$X_a(\Omega) = \int_{-\infty}^{\infty} x_a(t) e^{-j\Omega t} \, dt \tag{1.108}$$

Let $x[n]$ be the sequence $x[n] = x_a(nT_s)$, derived from $x_a(t)$ by periodic sampling with sampling period T_s. $F_s = 1/T_s$ is called the sampling frequency or sampling rate. From eq. (1.108),

$$x[n] = x_a(nT_s) = \frac{1}{2\pi}\int_{-\infty}^{\infty} X_a(\Omega)\exp(j\Omega nT_s)\,d\Omega \tag{1.109}$$

$x[n]$ also has a following representation:

$$x[n] = \frac{1}{2\pi}\int_{-\pi}^{\pi} X(e^{j\omega})e^{j\omega n}\,d\omega \tag{1.110}$$

To relate (1.108) and (1.110), let (1.109) be written as a sum of integrals over intervals of length $2\pi/T_s$, as in

$$x[n] = \frac{1}{2\pi}\sum_{r=-\infty}^{\infty}\int_{(2r-1)/\pi/T_s}^{(2r+1)\pi/T_s} X_a(\Omega)\exp(j\Omega nT_s)\,d\Omega$$

Changing the variable Ω to $\Omega + 2\pi r/T_s$,

$$x[n] = \frac{1}{2\pi}\sum_{r=-\infty}^{\infty}\int_{-\pi/T_s}^{\pi/T_s} X_a\left(\Omega + \frac{2\pi r}{T_s}\right)\exp(j\Omega nT_s)\exp(j2\pi rn)\,d\Omega \tag{1.111}$$

Since $\exp(j2\pi rn) = 1$, we get, on interchanging the order of summation and integration,

$$x[n] = \frac{1}{2\pi}\int_{-\pi/T_s}^{\pi/T_s}\left[\sum_{r=-\infty}^{\infty} X_a\left(\Omega + \frac{2\pi r}{T_s}\right)\right]\exp(j\Omega nT_s)\,d\Omega$$

Writing $\Omega = \dfrac{\omega}{T_s}$, we get

$$x[n] = \frac{1}{2\pi}\int_{-\pi}^{\pi}\left[\left(\frac{1}{T_s}\right)\sum_{r=-\infty}^{\infty} X_a\left(\frac{\omega}{T_s} + \frac{2\pi r}{T_s}\right)\right]e^{j\omega n}\,d\omega \tag{1.112}$$

From (1.110) and (1.112),

$$X(e^{j\omega}) = \frac{1}{T}\sum_{r=-\infty}^{\infty} X_a\left(\frac{\omega}{T_s} + \frac{2\pi r}{T_s}\right)$$

or

$$X(e^{j\Omega T_s}) = \frac{1}{T_s}\sum_{r=-\infty}^{\infty} X_a\left(\Omega + \frac{2\pi r}{T_s}\right) \tag{1.113}$$

(In the above discussion Ω is the analog frequency variable and ω is the frequency variable in the discrete case.)

Equations (1.113) establish the relationship between the Fourier transform of a continuous time function and the Fourier transform of its sampled version. This is illustrated in Fig. 1.22, for a band-limited signal, band-limited to $(-\Omega_0/2, \Omega_0/2)$. It is seen that if the sampling period is large so that $\Omega_0 2 > \pi/T_s$, the shifted versions of $X_a(\omega/T_s)$ overlap. For this case, the higher frequencies in $X_a(\Omega)$ get reflected into lower frequencies in $X(e^{j\omega})$. This phenomenon, where, in effect a high frequency component in $X_a(\Omega)$ possess as a lower frequency, is appropriately called *aliasing*. This can also be seen from Fig. 1.23, which shows a waveform reconstructed from grossly sampled high frequency sinusoidal signal (Stroboscopic effect).

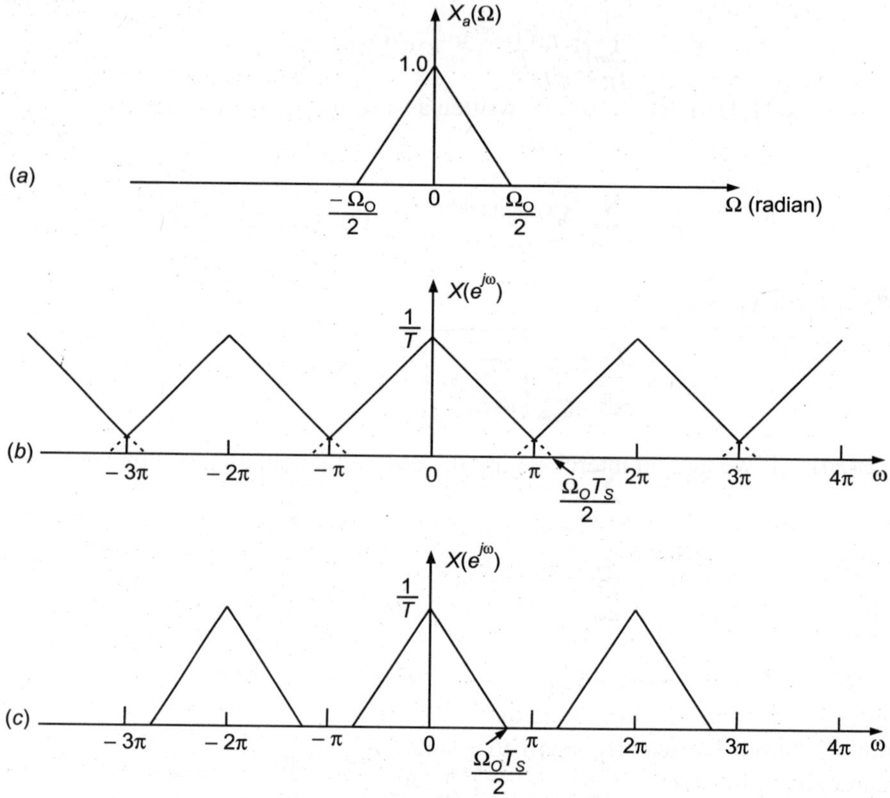

Figure 1.22: **(*a*)** Fourier transform of analog signal $X_a(t)$; **(*b*)** Fourier transform of sampled signal $T_z > 2\pi/\Omega_0$; **(*c*)** Fourier tansform of sampled signal: $T_s < 2\pi/\Omega_0$

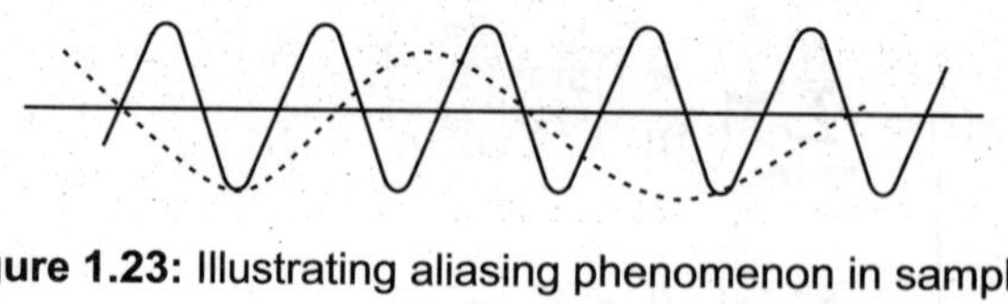

Figure 1.23: Illustrating aliasing phenomenon in sampling

It is also clear from Fig. 1.22 that if $\Omega_0/2 < \pi/T_s$, i.e., we sample at a rate at least twice the highest frequency of $X_a(\Omega)$, then $X(e^{j\omega})$ is identical to $X_a(\omega/T_s)$ in the interval $-\pi \le \omega \le \pi$. In this case it is clear that one can recover $x_a(t)$ from its samples by passing them through an idealized lowpass filter or by an appropriate interpolation formula. This sampling rate is called the *Nyquist rate*.

To derive the interpolation formula, assuming $\Omega_0/2 < \pi/T_s$, we see that

$$X(e^{j\Omega T_s}) = \frac{1}{T_s} X_a(\Omega); \; -\frac{\pi}{T_s} \le \Omega \le \frac{\pi}{T_s}$$

Since

$$x_a(t) = \frac{1}{2\pi}\int_{-\pi/T_s}^{\pi/T_s} X_a(\Omega)e^{j\Omega t}\, d\Omega$$

$$= \frac{1}{2\pi}\int_{-\pi/T_s}^{\pi/T_s} T_s \cdot X(e^{j\Omega T_s}) \cdot e^{j\Omega t}\, d\Omega \tag{1.114}$$

and

$$X(e^{j\Omega T_s}) = \sum_{n=-\infty}^{\infty} x_a(nT_s)e^{-j\Omega nT} \tag{1.115}$$

it follows that

$$x_a(t) = \frac{T_s}{2\pi}\int_{-\pi/T_s}^{\pi/T_s}\left[\sum_{n=-\infty}^{\infty} x_a(nT_s)e^{-j\Omega nT}\right]e^{j\Omega t}\, d\Omega$$

$$= \sum_{n=-\infty}^{\infty} x_a(nT_s)\left[\frac{T_s}{2\pi}\int_{-\pi/T_s}^{\pi/T_s} \exp\,[j\Omega(t-nT_s)]\, d\Omega\right]$$

$$= \sum_{n=-\infty}^{\infty} x_a(nT_s)\frac{\sin\,(\pi/T_s)(t-nT)}{(\pi/T_s)(t-nT)} \tag{1.116}$$

Equation (1.116) is the appropriate interpolation formula for recovering continuous time signal $x_a(t)$ from its samples. This representation called the 'sampling representation' is only valid for band limited functions with T_s chosen sufficiently small to avoid aliasing. Therefore, formally, we may state the sampling theorem as follows.

A band limited signal $x(t)$ of finite energy whose spectrum vanishes for frequencies greater than W Hz, i.e., $X(f) = 0, |f| > W$, is completely specified by or recovered from its sample values taken $1/2W$ seconds apart. The reconstruction formula is

$$x(t) = \sum_{n=-\infty}^{\infty} x\left(\frac{n}{2W}\right)\frac{\sin 2\pi W(t - n/2W)}{2\pi W(t - n/2W)}$$

$$= \sum_{n=-\infty}^{\infty} x\left(\frac{n}{2W}\right)\text{sinc}\; 2\pi W\left(t - \frac{n}{2W}\right) \tag{1.117}$$

Discussion

(a) Equation (1.116) is a special case of general signal representation of the form (1.105), where

$$a_n = x_a(nT_s)$$

and

$$\phi_n(t) = \frac{\sin(\pi/T_s)(t - nT_s)}{(\pi/T_s)(t - nT_s)}$$

As mentioned earlier, in general there are many classes of orthogonal or linearly independent functions $\phi_n(t)$ that may be used to express a continuous time function in the above form. In any representation of this type, the sequence of coefficients a_n can be regarded as discrete time signal that represents the continuous time signal $x_a(t)$. Not all such representations are equally useful. Sampling representation has the advantage that the coefficients are obtained simply by sampling the continuous time signal. Another advantage is that it preserves convolution in the sense that if $y_a(t)$ is the convolution integral of $x_a(t)$ and $h_a(t)$ then $y_a(nT_s)$ will be the discrete convolution or convolution sum of $x_a(nT_s)$ and $h_a(nT_s)$ provided there is no aliasing. The advantage of a representation that preserves convolution is that it permits the simulation or implementation of continuous time linear time-invariant system by a discrete time linear-shift invariant system.

(b) It is also seen that (1.117) can be interpreted as the output of a filter whose input is the periodic sample train $x(n/2W)$ and whose impulse response is sinc $2\pi Wt$. A filter with this impulse response is simply an ideal lowpass filter of bandwidth W with zero delay, as shown in Fig. 1.24(a). Figure 1.24(b) depicts the process of reconstruction of $x(t)$ from its samples. Equation (1.117) is also an interpolation formula, as the identity of the samples at the sampling instants is preserved, because the interpolation function has zeros at all other sampling instants except when $t = n/2W$.

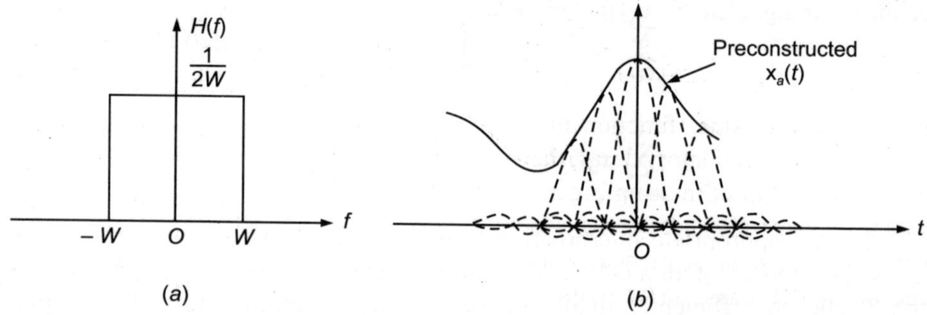

Figure 1.24: (*a*) Ideal lowpass filter of bandwidth *W*; **(*b*)** Illustrating reconstruction of $x_0(t)$ from its samples

Of course, an ideal lowpass filter cannot be used in practice, as it is physically unrealizable. To ease the problem of physical realization of an acceptable approximation to ideal lowpass filter, the sampling rate is generally made somewhat higher that the Nyquist rate. When we do this, it is clear from Fig. 1.22, that the frequency band from $\frac{\Omega_0 T_s}{2}$ to π is available to accommodate the transition from passband to stopband. When the sampling rate $\frac{1}{T_s}$ is less than Nyquist's the aliasing occurs as shown in Fig. 1.22(b) and no linear filtering operation can recover the original signal from its sampled version without distortion.

(c) For a finite duration, say $(0, T)$ segment of the signal, we may write

$$x(t) \simeq \sum_{n=0}^{2WT-1} x\left(\frac{n}{2W}\right) \text{sinc } 2\pi W\left(t - \frac{n}{2W}\right),\ 2WT \gg 1 \qquad (1.118)$$

Thus we may state that a signal which is essentially duration and bandwidth limited can be represented by a set of $2WT$ samples. Equation (1.118), of course, is an approximation, for we realize that a signal which is exactly band limited (for which the sampling theorem is valid) cannot be exactly time limited.

For a function which is exactly time limited, the approximation (1.118) does not provide the best fit for a given number of terms. The best approximation in the mean square sense is provided by an expansion in terms of 'prolate spheroidal wave functions' (for a detailed discussion see [2, 21]).

(d) For a finite sum sampling approximation $x_N(t)$ of a W band limited signal $x(t)$,

$$x_N(t) = \sum_{n=-N}^{N} x(nT_s)\frac{\sin 2\pi W(t-nT_s)}{2\pi W(t-nT_s)} \tag{1.119}$$

$$T_s = 1/2W; \qquad N \gg 1$$

the resulting mean square error e_N is given by

$$e_N = \int_{-\infty}^{\infty}|x(t)-x_N(t)|^2\,dt = T\sum_{|n|>N}|x(nT_s)|^2 \tag{1.120}$$

Since the error signal $x(t) - x_N(t)$ is also band limited with energy e_N, it may be shown [22] that

$$|x(t)-x_N(t)| \le \sqrt{2We_N} \tag{1.121}$$

(e) The choice of the 'sinc' function in the sampling expansion has some disadvantages. Firstly, because of its slow oscillatory decay, the infinite summation does not converge absolutely. Secondly, in the presence of jitter in sampling instants, there is a significant amount of 'crosstalk' or 'inter-symbol influence' (ISI) between the samples. To demonstrate this, let us consider the problem of transmitting a periodic sequence of numbers $a[n]$ at the rate v per second. For transmission over a band limited channel, an obvious choice of the basic waveform for transmission will be the 'sinc' function:

$$x(t) = \sum_n a[n]\frac{\sin \pi(vt-n)}{\pi(vt-n)} \tag{1.122}$$

$a[n]$'s can be recovered from $x(t)$ by sampling at the right instants. However, if the sampling time is in error by a fraction ε of the sample interval $1/v$, we get

$$x\left(\frac{k+\varepsilon}{v}\right) = a[k]\frac{\sin\pi}{\pi} + \frac{\sin\pi\varepsilon}{\pi\varepsilon}\left[\frac{a[k-1]}{1-\varepsilon} - \frac{a[k+1]}{1+\varepsilon} - \frac{a[k-2]}{2-\varepsilon} + \frac{a[k+2]}{2+\varepsilon} + \cdots\right] \tag{1.123}$$

Apart from a negligible error in $a[k]$ from first term, we see that the terms in the bracket, arising from other samples, contribute substantially to the error. [In fact, for arbitrary $a[n]$, these terms may not even be summable.] To show the plausibility of doing better with other choices of functions,

Lerner [23] suggests using $\phi(t) = \dfrac{\sin \pi t}{\sinh \pi t}$. This function is plotted in Fig. 1.25 and is seen to decay to zero exponentially with large t. With this choice, we generate

$$x(t) = \sum a[n] \frac{\sin \pi(vt - n)}{\sin h\, \pi(vt - n)} \tag{1.124}$$

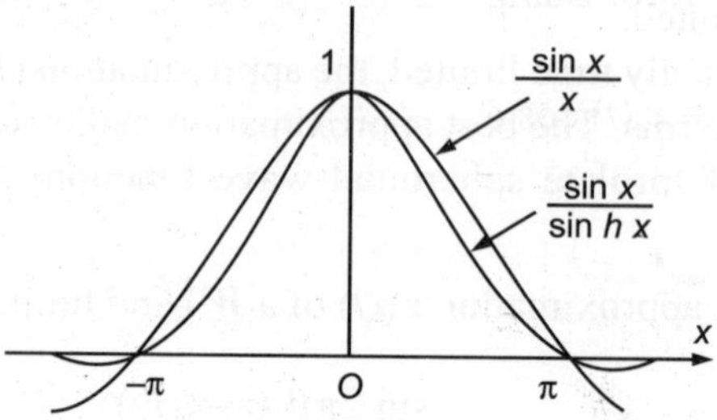

Figure 1.25: Two interpolation kernels for sampled signal

Again, as before a's may be recovered by periodic sampling. However, in this case, the presence of 'jitter' in sampling instants results in

$$x\left(\frac{k+\varepsilon}{v}\right) = b[k]\frac{\sin \pi\varepsilon}{\sinh \pi\varepsilon} + \sin \pi\varepsilon\left[\frac{a[k-1]}{\sinh \pi(1-\varepsilon)} - \frac{a[k+1]}{\sinh \pi(1+\varepsilon)} - \cdots\right]$$

$$= a[k]\frac{\sin \pi\varepsilon}{\sin h\, \pi\varepsilon} + \sin \pi\varepsilon\,[0.086a[k-1] - 0.0036a[k-2] + \ldots]e^{\pi\varepsilon}$$

$$-\sin \pi\varepsilon[0.086a[k+1] - 0.0036a[k+2] + \cdots]e^{-\pi} + \cdots;\ |\varepsilon| < 0.5 \tag{1.125}$$

and obviously the 'crosstalk' between samples is quite small. The contribution mainly is from the adjacent samples. Moreover, the spectrum of $\frac{\sin x}{\sinh x}$ is only slightly wider than that of $\frac{\sin x}{x}$ and goes exponentially to zero outside the passband.

(f) Because of the time-frequency symmetry in Fourier transform relationship, a sampling theorem in frequency domain for strictly time limited signals may be expressed as follows:

$$Y(f) = \sum_{n=-\infty}^{\infty} Y(n/2T)\frac{\sin 2\pi T(f - n/2T)}{2\pi T(f - n/2T)} \tag{1.126}$$

where $Y(f) \leftrightarrow y(t)$ and $y(t)$ is strictly time limited to $(-T, T)$.

1.5.1 Sampling of Bandpass Signals [8, 24, 25, 26]

As mentioned in Section 1.4, narrowband signals are frequently encountered in communication situation and therefore their discrete representation in terms of periodic samples is also very important. Two approaches to this problem are possible. One approach is based on the sampling of the equivalent lowpass envelope and phase functions or equivalently inphase and quadrature components of the narrowband signal as discussed in Section 1.4. The other involves direct sampling of bandpass signals. We shall take up the equivalent lowpass representation first.

Quadrature Sampling of Bandpass Signals

Consider a strictly bandlimited, bandpass signal $x(t)$ whose Fourier spectrum $X(f)$ vanishes for $W_1 \leq |f| \leq W_2$ as shown in Fig. 1.26. Using the inphase and quadrature lowpass components $x_c(t)$ and $x_s(t)$ as in (1.56), we write

$$x = x_c(t) \cos \omega_c t - x_s(t) \sin \omega_c t \tag{1.127}$$

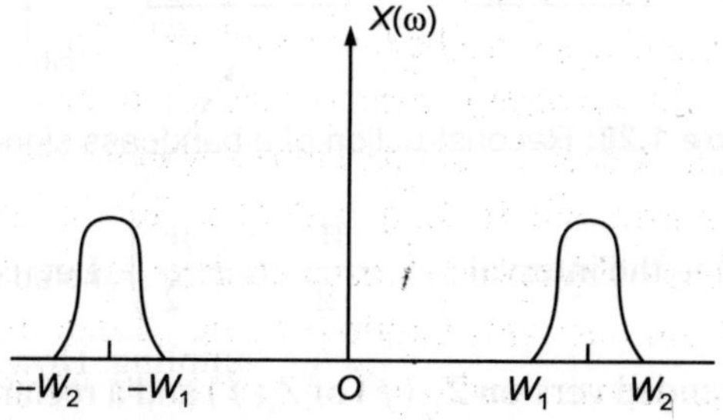

Figure 1.26: Bandpass signal spectrum

Here $f_c = \dfrac{\omega_c}{2\pi}$ is the reference frequency. If for convenience, we choose $f_c = W_1 + \dfrac{W_2 - W_1}{2}$ $= W_1 + \dfrac{W}{2}$ and $\dfrac{W}{f_c} \ll 1$, $x_c(t)$ and $x_c(t)$ are lowpass functions of bandwidth $W/2$ and the magnitude spectrum of $x_c(t)$ and $x_s(t)$ is identical to that of $x(t)$ except that it is centered at the origin. Therefore, $x_c(t)$ and $x_s(t)$ are completely determined by a sequence of samples spaced $1/W$ apart for each case. Thus a bandpass function of bandwidth W is completely specified by a set of $2W$ samples per sec. The process of extracting the inphase and quadrature components $x_c(t)$ and $x_s(t)$ and their subsequent sampling, is shown in Fig. 1.27. Figure 1.28 shows the reconstruction scheme of the bandpass signals from the sampled sequences. This process may also be understood by considering the corresponding analytic signal representation discussed in Section 2.6. The analytic signal $z_x(t)$ corresponding to $x(t)$ of (1.127) is given by

$$z_x(t) = x(t) + j\,\hat{x}(t) \tag{1.128}$$

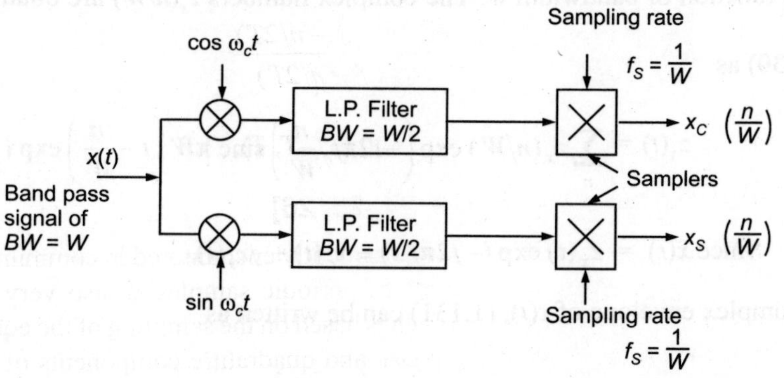

Figure 1.27: Inphase and quadrature sampling of a bandpass function, $W < 2f_c$

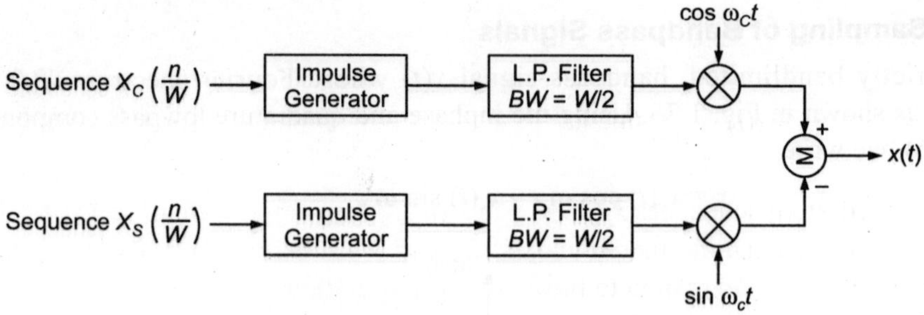

Figure 1.28: Reconstruction of a bandpass signal

$Z_x(f)$, the F.T. of $z_x(t)$, is confined to the interval $\left(f_c - \frac{W}{2}, f_c + \frac{W}{2}\right)$. Let us consider $Z_x(f)$ to be obtained as a product of a periodically extended version $Z_{xp}(f)$ of $Z_x(f)$ and a rectangular window $w\left(\frac{f - f_c}{W}\right)$ in frequency, centered at f_c, i.e.,

$$Z_x(f) = Z_{xp}(f) w\left(\frac{f - f_c}{W}\right)$$

or

$$z_x(t) = \mathcal{F}^{-1}\,[Z_{xp}(f)]^*\,\mathcal{F}^{-1}\left[w\left(\frac{f - f_c}{W}\right)\right] \tag{1.129}$$

Since $Z_{xp}(f)$ is periodic, its inverse F.T. is simply the Fourier series representation. Hence, we may write

$$z_x(t) = \sum_{n=-\infty}^{\infty} z_x(n/w)\,\text{sinc}\,\pi W\left(t - \frac{n}{W}\right)\exp\left[j2\pi f_c\left(t - \frac{n}{W}\right)\right] \tag{1.130}$$

and it is seen that $z_x(t)$ and therefore $x(t) = \text{Re}\,[z_x(t)]$ is completely determined by the complex valued samples of $z_x(n/W)$ taken at regular intervals of $1/W$. This, again, means that $2W$ real values/sec completely specify a bandpass function of bandwidth w. The complex numbers $z_x(n/W)$ are obtained indirectly as follows:

We express (1.130) as

$$z_x(t) = \sum_n z_x(n/W)\exp\left(-i2\pi f_c \frac{n}{W}\right)\text{sinc}\,\pi W\left(t - \frac{n}{W}\right)\exp(j2\pi f_c t) \tag{1.131}$$

$$\text{Since } \tilde{x}(t) = z_x(t)\exp(-j2\pi f_c t) = x_c(t) + jx_s(t)$$

where $\tilde{x}(t)$ is the complex envelope of $x(t)$, (1.131) can be written as

$$z_x(i) = \sum_n \tilde{x}(n/W)\,\text{sinc}\,\pi W\left(t - \frac{n}{W}\right)\exp(j2\pi f_c t) \tag{1.132}$$

On taking the real part of (1.132),

$$x(t) = \sum_n \text{sinc}\, \pi W\left(t - \frac{n}{W}\right)\left[x_c\left(\frac{n}{W}\right)\cos 2\pi f_c t - x_s\left(\frac{n}{W}\right)\sin 2\pi f_c t\right] \quad (1.133)$$

The inphase and quadrature sequences $x_c(n/W)$ and $x_s(n/W)$ respectively are obtained as shown in Fig. 1.27. Equivalently, we can sample the envelope $a(t)$ and phase modulation $\phi(t)$, both being lowpass functions of bandwidth $W/2$ and related to those of $x_c(t)$ and $x_s(t)$ by

$$a(n/W) = [x_c^2(n/W) + x_s^2(n/W)]^{1/2}$$

$$\phi(n/W) = \tan^{-1}\left[\frac{x_s(n/W)}{x_c(n/W)}\right]$$

The reconstruction formula in this case is [8]

$$x(t) = \sum_{n=-\infty}^{\infty} a(n/W)\frac{\sin \pi W[t-(n/W)]}{\pi W[t-(n/W)]}\cos[\omega_c(t-n/W)+\theta(n/W)] \quad (1.134)$$

where

$$\theta(n/W) = \omega_c\, n/W + \phi(n/W)$$

is the sample of total phase $\theta(t) = \omega_c t + \phi(t)$ of $x(t)$ at the sampling instant $t = n/W$.

1.5.2 Direct Sampling of Bandpass Signals

The above mentioned indirect method will not work without distortion if the bandpass signal has a bandwidth W larger than $2f_c$. Of course, one can get around this difficulty by first heterodyning the signal to a sufficiently high frequency band and then choosing the reference carrier frequency to satisfy the above mentioned constraint.

Another obvious choice is to directly sample the bandpass signal at a rate at least twice the highest frequency of the signal which is upper band-limiting frequency W_2 (see Fig. 1.26). This, of course, will be quite wasteful unless W_1 is close to zero. It will be shown that it is not necessary to sample this fast but the minimum unambiguous sampling frequency depends on W_1, W_2 and the bandwidth $W_2 - W_1$.

To show this, let us consider the direct sampling of the bandpass function whose spectrum is band limited as shown in Fig. 1.29(a). The spectrum of the sampled signal will then be periodic with the period F_s, the sampling frequency, as in (1.113). For convenience in visualisation, we consider the positive and negative frequency parts of the spectrum separately and show their periodic repetition in Fig. 1.29 (b, c, d, e).

The total spectrum of the sampled sequence $X_s(f)$ is then obtained as sum of $X_{s+}(f)$ and $X_{s-}(f)$. If the reconstruction of the bandpass signal from the sampled values is to be affected by passing the sample values through an ideal bandpass filter covering the frequency band $(-W_2, -W_1)$ and (W_1, W_2), it is clear that no overlap between $X_{s+}(f)$ and $X_{s-}(f)$ can be allowed to occur in this band. To achieve this objective, the sampling frequency must be chosen large enough to leave guard bands at least W wide between the bands of $X_{s+}(f)$ to accommodate the bands of $X_{s-}(f)$. This means that $f_s \geq 2W$ is a necessary condition.

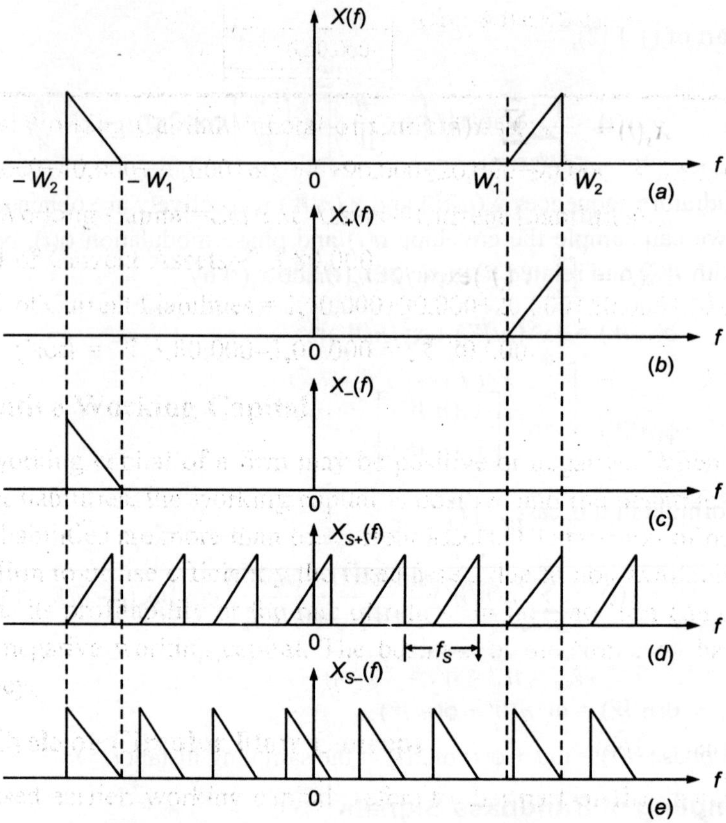

Figure 1.29: ***(a)*** **Bandpass signal spectrum of bandwidth** $W = W_2 - W_1$**;** ***(b)*** **Positive frequency part of** $X(f)$**;** ***(c)*** **Negative frequency part of** $X(f)$**;** ***(d)*** **Sampled signal spectrum corresponding to** $X_+(f)$**;** ***(e)*** **Sampled signal spectrum corresponding to** $X\_(f)$

Since $X_s(f)$ is a periodic, expanding it in Fourier series, we write

$$X_s(f) = \sum_{n=-\infty}^{\infty} c(n) \exp(-j2\pi nf/F_s)$$

where

$$c(n) = \frac{1}{f_s}\int_0^{f_s} X_s(f) \exp(j2\pi nf/F_s)\, df$$

$$= \frac{1}{f_s}\int_{-\infty}^{\infty} X(f) \exp(j2\pi nf/F_s)\, df$$

$$= \frac{1}{f_s} x(n/F_s)$$

Therefore,

$$X_s(f) = \frac{1}{f_s}\sum_{n=-\infty}^{\infty} x(n/F_s)\exp(-j2\pi nf/F_s) \tag{1.135}$$

From $x(t) \leftrightarrow X(f)$,

$$x(t) = \int_{-\infty}^{\infty} X(f)\exp(j2\pi ft)\,df$$

$$= \int_{-W_2'}^{-W_1'} X_s(f)\exp(j2\pi ft)\,df + \int_{W_1'}^{W_2'} X_s(f)\exp(j2\pi ft)\,df \tag{1.136a}$$

where

$$W_1 + W - \frac{F_s}{2} \le W_1' \le W_1$$

and

$$W_2 \le W_2' \le W_2 + \frac{f_s}{2} - W$$

Substituting (1.135) in (1.136a),

$$x(t) = \frac{1}{F_s}\sum_{n=-\infty}^{\infty} x(n/F_s)\left[\int_{-W_2}^{-W_1'} \exp\left(j2\pi\left(t-\frac{n}{F_s}\right)\right)df + \int_{W_1'}^{W_2'} \exp\left(j2\pi\left(t-\frac{n}{F_s}\right)\right)df\right] \tag{1.136}$$

The expression in square brackets is equivalently written as

$$2W'\frac{\sin(W'/F_s)(F_st-n)}{\pi(W'/F_s)(F_st-n)}\cos 2\pi\frac{f_c'}{F_s}(F_st-n)$$

where $W' = W_2' - W_1'$ and $f_c' = \dfrac{W_2' + W_1'}{2}$.

If we let $f_c = \dfrac{W_2 + W_1}{2} = \left(k + \dfrac{1}{4}\right)F_s$ where k is an integer then it can be seen that the limits on W' and f_c' will be

$$W \le W' \le f_s - W:\ (F_s \ge 2W)$$

and

$$f_c - (F_s - 2W)/2 \le f_c' \le f_c + 2(F_s - 2W) \tag{1.137}$$

Therefore, the sampling theorem in this case may be written as

$$x(t) = 2\frac{W'}{F_s}\sum_{n=-\infty}^{\infty} x\left(\frac{n}{F_s}\right)\frac{\sin\pi(W'/F_s)(F_st-n)}{\pi(W'/F_s)(F_st-n)}\cos 2\pi\frac{f_c'}{F_s}(F_st-n) \tag{1.137a}$$

If we let $W' = W = F_s/2$ and $f_c = f_c'$, then

$$x(t) = \sum_{n=-\infty}^{\infty} x(n/F_s)\frac{\sin\frac{1}{2}\pi(F_s t - n)}{\frac{1}{2}\pi(F_s t - n)}\cos\frac{1}{2}\pi(4f_c t - n) \tag{1.138}$$

The conditions (1.137) are sufficient conditions for sampling a bandpass signal without running into aliasing problems. They result in a choice of sampling frequency F_s less than $2W_2$. However, the question remains: how small can F_s be in this direct method of sampling of bandpass signals? Observing Fig. (1.29), we may obtain the following necessary conditions:

$$F_s \geq 2W$$

$$kF_s \geq 2W_2 \text{ where } k \text{ is an integer, } k = 1, 2, 3, \ldots$$

$$2W_1 = 2(W_2 - W) \geq (k-1)F_s$$

$$1 \leq k \leq \frac{W_2}{W}$$

These constraints are more succinctly expressed as

$$1 \leq k \leq \frac{W_2}{W} \leq \frac{kF_s}{2W}$$

and

$$(k-1)\frac{F_s}{2W} \leq \frac{(W_2 - W)}{W} = \frac{W_1}{W} \tag{1.139}$$

A graphical description of these constraints is illustrated in Fig. 1.30. It is seen that the sampling rate F_s has certain permissible values which lie in disjoint intervals, shown by clear areas in the figure. The shaded areas represent the forbidden zone. Figure 1.30 also shows the locus of the minimum sampling rate $F_{s\,\min}$ as solid lines $F_{s\,\min}$ is given by

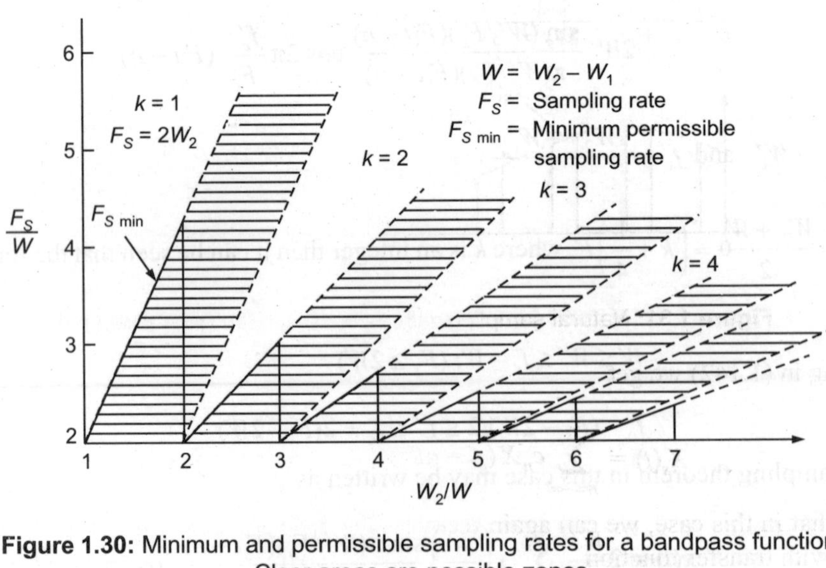

Figure 1.30: Minimum and permissible sampling rates for a bandpass function. Clear areas are possible zones

$$F_{s\,\min} = 2W\left(1+\frac{m}{M'}\right) \tag{1.140}$$

where $m = W_2/W$ and M' is the largest integer $\le W_2/W$. Equation (1.140) should not be interpreted to mean that any $F_s > F_{s\,\min}$ will be acceptable. The choice must still fall in the clear area. In any case $F_{s\,\min}$ will never exceed $4W$ and whenever W_2 is an integer multiple of W_1, $F_{s\,\min} = 2W$. Also, as long as $W \ll W_2$ (the narrowband case), $F_{s\,\min} \approx 2W$.

1.5.3 Practical Sampling

An ideal sampler provides the exact values of the signal $x(t)$ at the sampling instants nT_s to produce a sequence $x[nT_s]$. In practice, ideal sampling can be approximated closely. To see this, consider the sampling process as multiplication or modulation operation as depicted in Fig. 1.31. Ideal sampling results when $s(t)$ is a train of delta functions. In practice $s(t)$ would be a train of short duration pulses with a period of T_s seconds. Since the sampled signal is

$$x_s(t) = x(t)s(t) \tag{1.141}$$

its spectrum $X_s(f)$ will be given by

$$X_s(f) = X(f) * S(f) \tag{1.142}$$

Representing the periodic signal $s(t)$ by Fourier series

$$s(t) = \sum_{n=-\infty}^{\infty} c_n \exp\,(jn2\pi F_s t);\; F_s = \frac{1}{T_s} \tag{1.143}$$

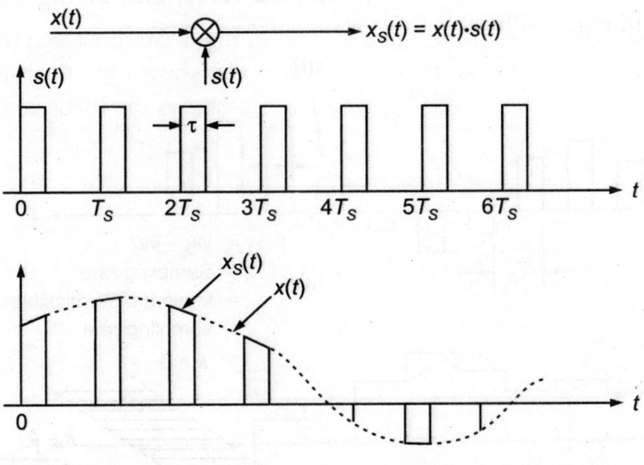

Figure 1.31: Natural sampling with a periodic rectangular pulse train

and substituting in (1.142) we get

$$X_s(f) = \sum_{n=-\infty}^{\infty} c_n X(f - nF_s) \tag{1.144}$$

Thus we see, that in this case, we can again recover $X(f)$ from $X_s(f)$, to within a scale factor of c_0 by a lowpass filter with transfer function

$$H(f) = 1 \quad \text{for } |f| < W$$
$$= 0 \quad \text{for } |f| > F_s - W$$

as long as the signal is band limited to W and $F_s > 2W$.

Sample and Hold

Quite often, the sampling is done with sampling devices of the sample and hold variety which produce a sampled signal of the form

$$x_s(t) = \sum x(nTs)p(t-nTs)$$
$$= p(t) * \sum_n x(nT_s)\delta(t-nT_s) \tag{1.145}$$

and shown in Fig. 1.32(a). The spectrum of the sampled signal in this case is

$$X_s(f) = p(f)\left[F_s \sum_n X(f-nF_s)\right]$$
$$= p(f)X_{s\delta}(f) \tag{1.146}$$

where $X_{s\delta}(f)$ symbolizes the spectrum of corresponding ideally sampled signal. Here $P(f) \leftrightarrow p(t)$, the basic sampling pulse. In sample and hold method, the value of the signal at the leading edge of the sampling pulse is held by the sampling pulse for the duration of the sample time unlike the multiplicative sampling discussed earlier which essentially views the signal through a time aperture provided by the sampling pulse. The hold time determines the pulse width and when this is equal to $1/F_s$, the sampling technique is referred to as 'boxcar' sampling. That is, the previous sample is held until the next sample is acquired, at which time, the output takes the new sampled value, thus giving a staircase shape to the sampled signal as shown in Fig. 1.32b.

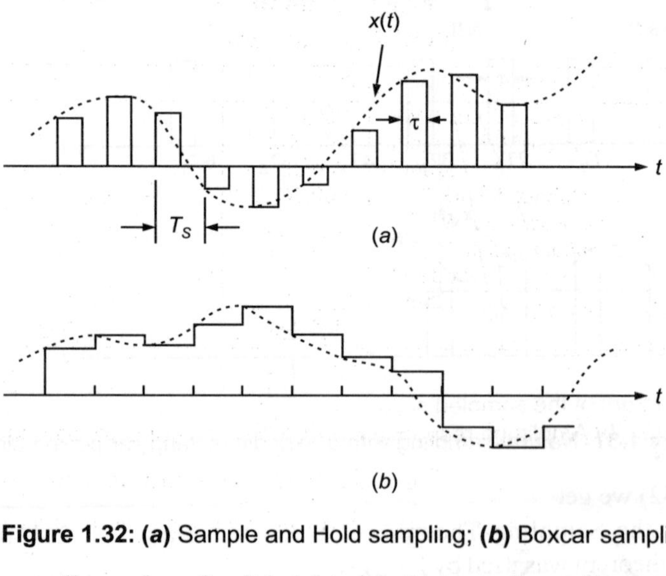

Figure 1.32: (a) Sample and Hold sampling; (b) Boxcar sampling

For a rectangular sampling pulse of unit height, duration τ and sampling at the leading edge of the pulse, (1.146) becomes

$$X_s(f) = \tau \operatorname{sinc}(\pi f\tau)e^{-j\pi f\tau} X_{s\delta}(f) \tag{1.147}$$

The delay factor $e^{-j\pi f\tau}$ is unimportant for reconstruction of $x(t)$. However, the factor, sinc $\pi f\tau$ affects it in two ways. If $X(f)$ is in a band near $f = 1/\tau$, the reconstructed signal will be small in amplitude. Secondly $P(f)$, in general or sinc $(\pi f\tau)$ in this acts as a filter introducing some distortion in the reconstructed signal. Both these effects are reduced by making τ small.

For 'boxcar' sampling, the spectrum of the sampled signal can be obtained from (1.147) by setting $\tau = T_s$. Thus,

$$X_s(f) = \text{sinc}(\pi f T_s)\exp(-j\pi f T_s)\sum_n X(f - nf_s) \tag{1.148}$$

The magnitude spectrum is sketched in Fig. 1.33 (b) for the two cases. It is clear that as long as the signal is band-limited and there is no aliasing, it can be recovered from the boxcar sequence without any significant distortion. However, the boxcar spectrum $X_s(f)$ at the band edge is approx. 3 dB below the value at $f = 0$. Thus the recovered signal $x(t)$ will have a 3 dB loss at $f/F_s \simeq \frac{1}{2}$. This is now corrected by a suitable amplitude equalizer. In practice, the zero-order hold is often followed by a simple lowpass filter to attenuate the undesired high frequency components and to provide a smooth output.

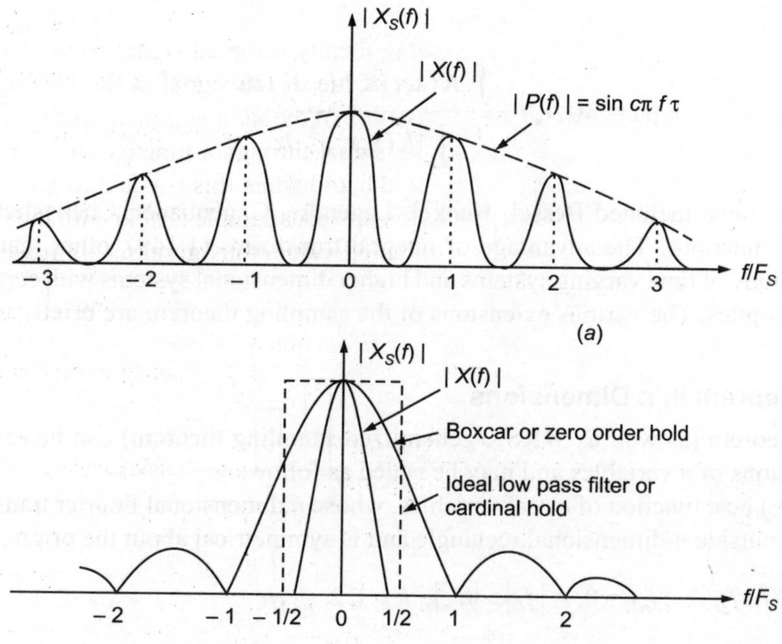

Figure 1.33: (*a*) Spectrum of the sampled signal, with rectangular sampling pulse of duration τ and period $1/f_S$, **(*b*)** Amplitude response of Boxcar sampling or zero-order hold

1.5.4 Discussion

Historically, the origin of the Sampling Theorem is traced to Whittakers [27]. In the context of communication theory, the theorem was used by Nyquist [28] and in its present form was enunciated and used extensively by Shannon [29]. In the Russian literature, the theorem was first discussed and used in communication theory by Kotelnikov [30]. Hence in a recent and an excellent tutorial review [31] of the whole subject, Jerri refers to it as WKS theorem to acknowledge the three pioneers.

The sampling theorem has been generalized and extended in various ways in the last twentyfive years and its fascination has not yet worn off. The earlier attempts at generalization included more general, finite limit integral transforms besides Fourier transform. This generalization is stated by Kramer [32] as follows: (Jerri refers to it as WKSK sampling theorem).

Let I be an interval and $L_2(I)$ the class of functions $\phi(t)$ for which $\int_I |\phi(x)|^2\, dx < \infty$. Supoose that for each real t,

$$f(t) = \int_I K(x, t) g(x)\, dx \tag{1.149}$$

where $g(x) \in L_2(I)$. Suppose that for each real t, $K(x, t) \in L_2(I)$ and that there exists a countable set $E = \{t_n\}$ such that $K(x, t_n)$ is a complete orthogonal set on I. Then

$$f(t) = \lim_{N\to\infty} \sum_n f(t_n) s_n(t) \tag{1.150}$$

where

$$s_n(t) = s(t, t_n) = \frac{\int_I K(x, t) K^*(x, t_n)\, dx}{\int_I |K(x, t_n)|^2\, dx} \tag{1.151}$$

The kernels $K(x, t)$ have included Bessel, Hankel, Legendre, Gegenbauer, Chebyshev, Laguerre and prolate spheroidal functions. The advantage of integral transforms (1.149), other than the Fourier, is manifested in the study of time varying systems and higher dimensional systems with certain symmetries, as for example, in optics. The various extensions of the sampling theorem are briefly mentioned in the following:

A. Sampling Theorem in *n* Dimensions

WKS sampling theorem (as well as WKSK generalized sampling theorem) can be easily extended to band limited functions of n variables and may be stated as follows:

Let $x(t_1, t_2, ..., t_n)$ be a function of n real variables, whose n dimensional Fourier transform exists and is identically zero outside n-dimensional rectangle and is symmetrical about the origin, i.e.,

$$X(f_1, f_2, \ldots, f_n) = 0, \quad |f_k| > |W_k|,\ k = 1, 2, \ldots, n \tag{1.152}$$

Then

$$x(t_1, t_2, \ldots t_n) = \sum_{m_1=-\infty}^{\infty} \cdots \sum_{m_n=-\infty}^{\infty} x\left(\frac{m_1}{2W_1}, \ldots, \frac{m_n}{2W_n}\right) \prod_{i=1}^{n} \operatorname{sinc} 2\pi W_i \left(t_i - \frac{m_i}{2W_i}\right) \tag{1.153}$$

For a detailed treatment of sampling theorem in n dimensions, see [33].

B. Sampling Theorem for Random Processes

WKS sampling theorem was used for representation of random processes with a continuous time parameter by Balakrishnan [34] and its extension to multi-dimensional random fields along with host of applications has been discussed in detail by Middleton and Peterson [35, 36]. For one dimensional case of random processes, we can state it as follows:

Let $x(t)$, $-\infty < t < \infty$, be a real or a complex valued stochastic process, stationary in 'wide sense', possessing a spectral density which vanishes outside the interval $[-W, W]$. Then $x(t)$ has a sampling representation

$$x(t) = \lim_{N\to\infty} \sum_{m=-N}^{N} x\left(\frac{n}{2W}\right) \text{sinc}\, 2\pi W\left(t - \frac{n}{2W}\right) \tag{1.154}$$

for every t. Here lim stands for 'limit in the mean square'. We shall consider this representation in some detail in Section 1.9 and present its proof.

C. Other Extensions

Other extensions of the sampling theorem concern sampling with values of its function and its derivatives [37, 38], interlaced sampling [39, 40], sampling in terms of samples of the outputs of m, LTI systems [2, 41], sampling with nonuniformly spaced sampling points [42], random sampling [43, 44] and sampling theorems for nonband-limited signals and causal signals [45, 46].

In conclusion, it may be said again that the sampling theorem still holds its magic spell on mathematicians and communication theorists. Anyone who would like to delve deeper into its mysteries will find Jerri's review [31] still the best source of references.

1.6 RANDOM PROCESSES

Communication theory largely deals with the theory of random processes. The reason is that the information bearing signals like voice, television, telegraph, data etc. which the communication system is expected to handle, are by their very nature, unpredictable. Moreover, very often, the channel itself introduces noise in the systems and in this sense, particularly from the view point of the receiver, all signals must be viewed as random. Thus the study and characterization of such signals is fundamental to the understanding of communication systems.

In what follows, it is assumed that the reader is familiar with basic probability theory, particularly random variables and their characterization. If not, he is referred to the excellent text by Parzen [47].

1.6.1 Notion of a Random Process

In the notation of probability theory, the appropriate mathematical model for dealing with random quantities, say, a noise voltage, involves the concept of a random variable x, defined as the mapping or assignment of a real number $x(\xi)$, to each point ξ of the sample space Ω in such a way that a probability distribution function F_x exists. To extend this notion of a random variable to a random process, such as a noise voltage waveform, we now assign to each point $\xi \in \Omega$ a real time function $x(\xi, t)$. As an example consider the situation depicted in Fig. 1.34, which may represent, say the signals observed on a multichannel (in this case 3-channel) radio link. A hypothetical case may be considered where a link is chosen randomly with some probability assignment $P[\xi_i]$, $i = 1, 2, 3 \ldots$ and for each choice, a time waveform is observed. Thus our sample space Ω has three points ξ_i, $i = 1, 2, 3$ and to each ξ_i a corresponding waveform or time function $x(\xi_i, t)$ is assigned. If we look at this set of waveforms at some time instant $t = t_1$, since each point $\xi_i \in \Omega$ has a real number $x(\xi_i, t_1)$ associated with it with probability $P[\xi_i]$, the collection $x(\xi_i, t_1)$, $i = 1, 2, 3$ forms a random variable. It is clear that each choice of t defines a different random variable. Though, in this example, we have considered a discrete finite sample space, in general, it could be infinite or continuous and the entire collection of time functions together with the probability assignment will define a random variable for each choice of observation instant.

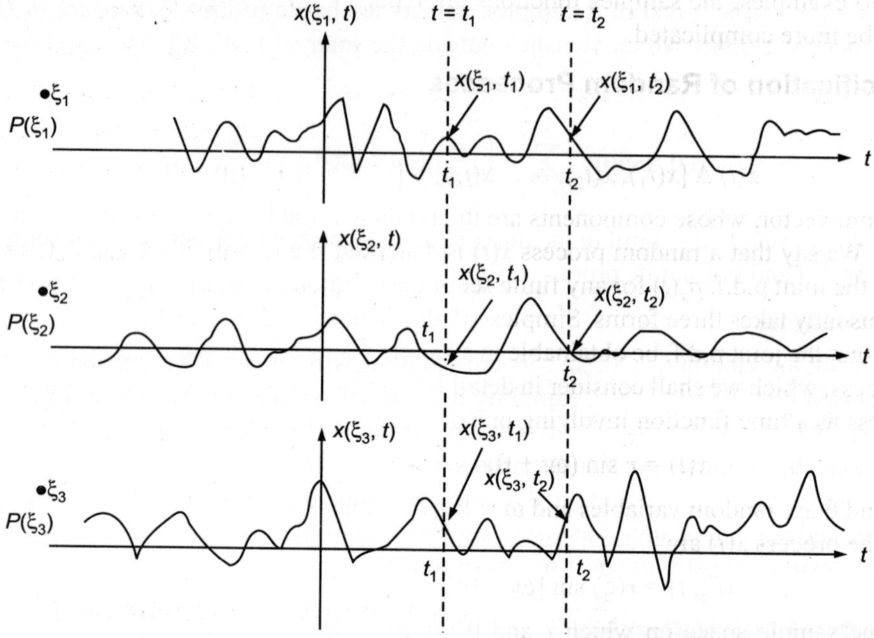

Figure 1.34: Illustrating the concept of a random process

This probability system made up of a sample space, a set of waveforms and a probability measure is called a random process and denoted as $x(\xi, t)$ or usually abbreviated as $x(t)$. $x(\xi_i, t)$ is called a *sample function* or a sample realization, $x(\xi, t_i)$ is a random variable and $x(\xi_i, t_i)$ is merely a number. In short, a random or stochastic process is a collection $\{x(t), t \in T\}$ of random variables. The set T is called the index set of the process. When $T = \{0, \pm 1, \pm 2, \ldots\}$ or $T = \{0, 1, 2, \ldots,\}$ the random process is referred to as *discrete parameter process* and when $\{T = t; -\infty < t < \infty\}$ or $\{T = t, t \geq 0,\}$ it is called a *continuous parameter process*.

Examples:

1. $x(t) = A \cos(\omega t + \theta)$
   where A and ω are constants but θ is a random variable with a given probability density function (p.d.f). Each possible value of θ determines one of the possible sample functions of the random process, e.g.
   $x(t) = A \cos(\omega t + \pi/2)$
   $x(t) = A \cos(\omega t - 3\pi/4)$ etc.
2. For a coin tossing experiment, we define $x(t)$, $t > 0$, such that
   $x(t) = e^{-t}$ if ξ = heads
   $x(t) = \log t$ if ξ = tails
3. $x(t) = \sum_{i=1}^{10} a_i \phi_i(t)$

   where a_i's are random variables, say Gaussian p.d.f. and $\phi_i(t)$ are a predetermined set of orthogonal functions.

In the first two examples, the samples functions are regular curves whereas the waveforms of the last example will be more complicated.

1.6.2 Specification of Random Processes

Let

$$x(t) \Delta [x(t_1), x(t_2), \ldots, x(t_n)] = [x_1, x_2, \ldots, x_n] \tag{1.155}$$

denote a random vector, whose components are the random variables $x(t_i)$ defined for n time instants t_i, $i = 1, \ldots, n$. We say that a random process $x(t)$ is specified if and only if we can determine from the specification, the joint p.d.f. $p_x(t)$ for any finite set of observation instants $(t_1, t_2, \ldots, t_n)$. In practice, this specification usually takes three forms. Simplest of all is when one can state the probability rule directly. This requires that the joint p.d.f. be obtainable in a simple way from the specification. An example is the Gaussian process, which we shall consider in detail later in Section 1.9. The second method specifies the random process as a time function involving one or more parameters serving as random variable e.g.,

$$x(t) = r \sin(\omega t + \theta)$$

where say r and θ are random variables and ω is fixed and joint p.d.f. $p_{r,\theta}$ is specified. Then the sample functions of the process $x(t)$ are

$$x(\xi, t) = r(\xi) \sin[\omega t + \theta(\xi)], \;\; \xi \in \Omega$$

where Ω is the sample space on which r and θ are defined. In theory, then, any p.d.f. $p_x(t)$ can be obtained from the knowledge of $p_{r,\theta}$. As we shall see later, a useful joint p.d.f., which arises in the study of narrowband noise processes is

$$p_{r,\theta}(\alpha, \beta) = \begin{cases} \alpha/2\pi e^{-\alpha^2/2} & 0 \le \alpha < \infty, \; 0 \le \beta < 2\pi \\ 0 & \text{otherwise} \end{cases}$$

Here, the sample functions of the process may be associated, conveniently with points ξ of a two-dimensional sample space Ω as shown in Fig. 1.35.

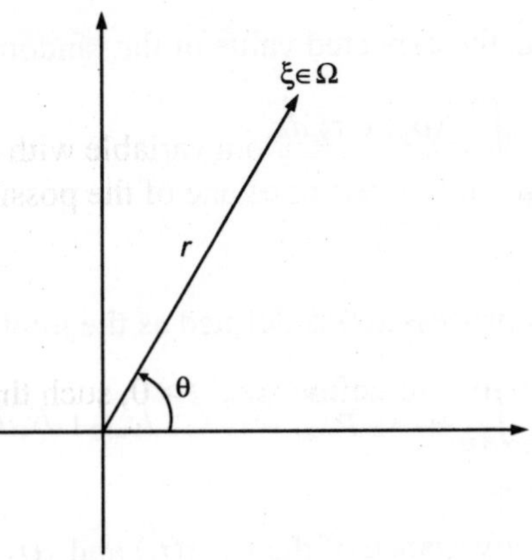

Figure 1.35: Sample space of the random process $x(t) = r \sin(\omega t + \theta)$

The third method of specification of random processes, generates the ensemble by transforming the sample functions of a known process by an explicitly stated rule or functional mapping. Simple examples are the amplitude scaling and time translation of a given process.

$$y(\xi, t) = ax(\xi, t)$$

$$y(\xi, t) = x(\xi, t + T)$$

The joint p.d.f. of the new processes will, in this case, be

$$p_y(t) = ap_x(t)$$

and

$$p_y(t) = p_x(t + T)$$

A most useful example of this method is linear time invariant filtering of the signals in a communication system as depicted in Fig. 1.36. The sample functions of the input and the output process, in this case, are related through the convolution integral

$$y(\xi, t) = \int_{-\infty}^{\infty} h(t - \tau)x(\xi, \tau)d\tau \qquad \xi \in \Omega \tag{1.156}$$

or more succinctly

$$y(t) = \int_{-\infty}^{\infty} h(t - \tau)x(\tau)d\tau \tag{1.157}$$

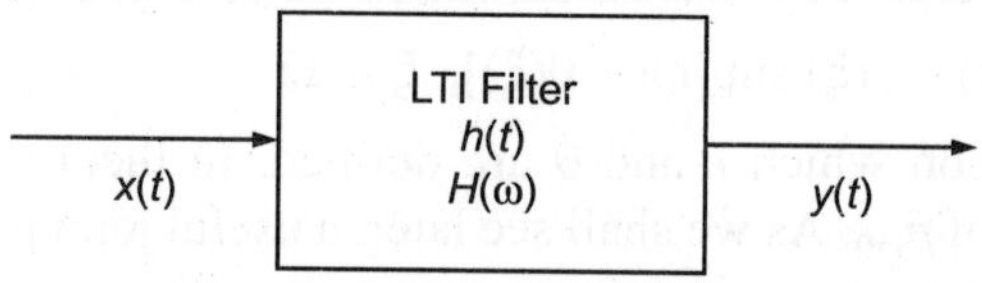

Figure 1.36: Linear time invariant filtering of a random process

Here, it will be difficult to obtain p.d.f. of the process $y(t)$, in general. However, as we shall see, if $x(t)$ is a Gaussian process then the p.d.f. of $y(t)$ is obtained in a simple way.

1.6.3 Statistics of a Random Process

Mean or Expectation

The *mean* $m_x(t)$ of a process $x(t)$ is the expected value of the random variable $x(t)$:

$$m_x(t) \triangleq E[x(t)] = \int_{-\infty}^{\infty} xp_x(x, t)\, dx \tag{1.158}$$

It is, in general, a function of time.

Autocorrelation

The *autocorrelation* $R_x(t_1, t_2)$ of a process $x(t)$ is defined as the joint moment of r.v. $x(t_1)$ and $x(t_2)$:

$$R_{xx}(t_1, t_2) \triangleq E[x(t_1)\, x(t_2)] = \int_{-\infty}^{\infty} x_1, x_2\ p_{x_1x_2}(x_1, x_2;\ t_1, t_2)\, dx_1dx_2 \tag{1.159}$$

and is a function of t_1 and t_2.
The *autocovariance* of $x(t)$ is the covariance of the r.v. $x(t_1)$ and $x(t_2)$:

$$K_x(t_1, t_2) = \begin{cases} \text{cov}\,[x(t_1), x(t_2)] \\ E[\{x(t_1) - m_x(t_1)\}\,\{x(t_2) - m_x(t_2)\}] \\ R_{xx}(t_1, t_2) - m_x(t_1)\, m_x(t_2) \end{cases} \tag{1.160}$$

The *variance* of $x(t)$ is variance of the r.v. $x(t)$;

$$\text{var}\,[x(t)] = \sigma_x^2(t) = K_{xx}(t,t) = R_{xx}(t,t) - m_x^2(t) \tag{1.161}$$

Example 1.12: Let $x(t)$ represents the position of a particle in motion with uniform velocity v and let

$$x(t) = x_0 + vt$$

where x_0 and v are r.v.'s representing initial position and velocity respectively. The sample function of the process thus consists of straight lines. Its mean value, autocorrelation, autocovariance and variance are given by

$$m_x(t) = E[x(t)] = E[x_0] + tE[v]$$

$$R_{xx}(t_1, t_2) = E[(x_0 + vt_1)(x_0 + vt_2)]$$

$$= E[(x_0^2) + E[x_0](t_1 + t_2) + E[v^2]\, t_1 \cdot t_2]$$

$$K_{xx}(t_1, t_2) = R_{xx}(t_1, t_2) - m_x(t_1)\, m_x(t_2)$$

$$= E[x_0^2] - E^2[x_0] + (t_1 + t_2)\,\{E[x_0 v] - E[v]\,E\,[x_0]\} + t_1 t_2\{E[v^2] - E^2[v]\}$$

$$\sigma_x^2 = E[x_0^2] - E^2[x_0] + 2t\{E[x_0 v] - E[v]\,E[x_0]\} + t^2\{E[v^2] - E^2[v]\}$$

$$= \text{var}\,[x_0] + 2t\,\text{cov}\,[x, v] + t^2\,\text{var}\,[v]$$

It is to be noted that in order to obtain the mean and the autocovariance of $x(t)$, one need not know the joint probability law of x_0 and v but only their means, variances and covariances.

1.6.4 Examples of Random Processes

Before we discuss further the properties of random processes, it would help to gain some familiarity with certain specific random processes which arise in a number of situations of interest to communication engineers.

1. Sinewave with Random Phase

Such a signal is written as

$$x(t) = A\cos(\omega_c t + \phi) \tag{1.162}$$

where A and ω_c are constants, whereas ϕ is a random variable, which we take to be uniformly distributed in the range $-\pi$ to π, i.e.

$$p_\phi(\phi) = \begin{cases} \dfrac{1}{2\pi} & -\pi \le \phi \le \pi \\ 0 & \text{elsewhere} \end{cases} \tag{1.163}$$

The autocorrelation function a.c.f. of $x(t)$ is

$$R_{xx}(t_1, t_2) = E[x(t_1)\, x(t_2)]$$

$$= E[A^2 \cos(\omega_c t_1 + \phi)\cos(\omega_c t_2 + \phi)]$$

$$= \frac{A^2}{2}\; E[\cos\{\omega_c(t_1 + t_2) + 2\phi\} + \cos\{\omega_c(t_1 - t_2)\}]$$

$$= \frac{A^2}{2}\left[\int_{-\pi}^{\pi} \frac{1}{2\pi} \cos\{\omega_c(t_1+t_2)+2\phi\}\, d\phi + \cos\omega_c(t_1-t_2)\right]$$

$$= \frac{A^2}{2}\cos\omega_c(t_1-t_2) \tag{1.164}$$

If we let $t_1 = t + \tau$, $t_2 = t$, then a.c.f. of the process (1.162) becomes

$$R_{xx}(\tau) = \frac{A^2}{2}\cos\omega_c\tau \tag{1.165}$$

2. Poisson and the Wiener Processes [48, 49, 50]

These two processes serve as very useful models or building block of models for many important phenomena manifesting in the physical world. In order to define them we need the notion of a random process with *independent increment*.

A continuous parameter random process $x(t)$, $0 \le t \le \infty$ is an independent increments random process if $x(0) = 0$ and for all choices of indices $t_0 < t_1 \ldots < t_n$, the n random variables

$$x(t_1) - x(t_0),\ x(t_2) - x(t_1),\ \ldots,\ x(t_n) - x(t_{n-1}) \tag{1.166}$$

are statistically independent (s.i.). Furthermore the process is said to be one with *stationary independent increment*, if in addition, $x(t_2+\tau) - x(t_1+\tau)$ has the same p.d.f. as $x(t_2) - x(t_1)$ for all choices of indices t_1 and t_2 and all $\tau > 0$. It may be easily proved that for a process with independent increments the joint characteristic function

$$\phi_{x(t)}(u) = \phi_{x(t_1)}(u_1 + u_2 + \ldots u_n)\prod_{k=2}^{n}\phi_{x(t_k)-x(t_{k-1})}(u_k + \ldots u_n) \tag{1.167}$$

where

$$u \triangleq [u_1, u_2, \ldots u_n] \tag{1.168}$$

Thus, knowing the probability laws of $x(t)$ and $x(t) - x(s)$ for $s, t \ge 0$, one can obtain the joint probability law for $x(t)$ and thus specify such a process.

3. Poisson Process

For an extensive list of the physical phenomena for which the Poisson process serves as a useful model, see [48, 49]. Some of these are shot noise in electron devices, radioactive decay, spatial distribution of plants, animals and galaxies, arrival of calls at a telephone exchange, arrival of customers for service, occurrence of accidents, errors, machine failures etc.

For modelling these phenomena as random processes, one considers the so called *counting function* $x(t)$, $t > 0$, which represents the number of events that have occurred during an interval $[0, t]$, be it a time or space interval. A typical function $x(\cdot)$ is sketched in Fig. 1.37. For any t, $x(t)$ is a random variable taking on any of the integer values 0, 1, 2, ... The family of random variables $\{x(t), t \ge 0\}$ constitutes an integer valued random process. An integer valued random process $\{x(t), t \ge 0\}$ is said to be a Poisson process, with mean rate or density λ if the following holds:

(i) $\{x(t), t \ge 0\}$ is a process with stationary independent increments.

(ii) For any times t_1 and t_2 such that $t_1 < t_2$, the number $x(t_2) - x(t_1)$ of counts or points in the interval t_1 to t_2 is Poisson distributed with mean $\lambda(t_2 - t_1)$, i.e.

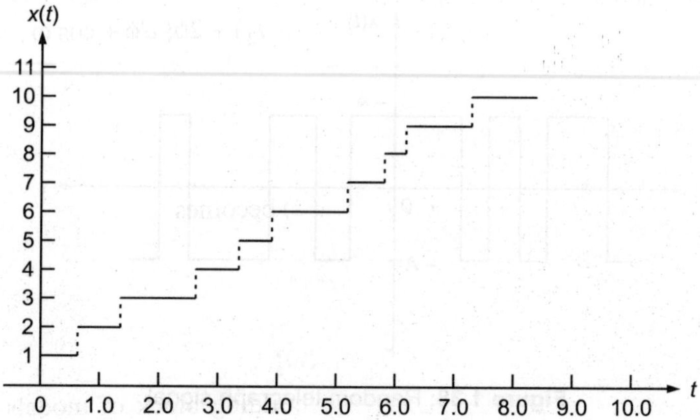

Figure 1.37: A sample function of a Poisson counting process

$$P[x(t_2) - x(t_1) = k] = e^{-\lambda(t_2 - t_1)} \frac{[\lambda(t_2 - t_1)]^k}{k!} \tag{1.169}$$

$$E[x(t_2) - x(t_1)] = \lambda(t_2 - t_1) \tag{1.170}$$

$$E[x(t_2) - x(t_1)]^2 = \lambda^2(t_2 - t_1) + \lambda\,(t_2 - t_1) \tag{1.171}$$

$$\text{var}\,[x(t_2) - x(t_1)] = \lambda(t_2 - t_1) \tag{1.172}$$

It may be easily seen that the means and a.c.f. of a Poisson process with parameter λ is given by

$$m_x(t) = \lambda t \tag{1.173}$$

and

$$R_{xx}(t_1, t_2) = \begin{cases} \lambda t_2 + \lambda^2 t_1 t_2 & t_1 \geq t_2 \\ \lambda t_1 + \lambda^2 t_1 t_2 & t_1 \leq t_2 \end{cases} \tag{1.174}$$

4. Random Telegraph Signal

The sample function of such a process is shown in Fig. 1.38. It is a two valued random process taking either of the values $+A$ or $-A$ with equal probability. The process changes its value from $+A$ to $-A$ or from $-A$ to $+A$ and the probability that n such changes occur in a time interval is given by Poisson distribution with parameter ν, i.e., ν is the average number of such changes per unit time.

$$P[n, \tau] = \frac{(\nu\tau)^n}{n!} e^{-\nu\tau} \tag{1.175}$$

Clearly the mean for such a process is zero. The a.c.f. is obtained as follows:

$$R_{xx}(t_1, t_2) = E[x(t_1)x(t_2)]$$

Both $x(t_1)$ and $x(t_2)$ are two valued discrete random variables, taking values $+A$ and $-A$. The product therefore will be A^2 or $-A^2$ depending on whether $x(t_1) = x(t_2)$ or $x(t_1) = -x(t_2)$. This in turn depends on whether the number of changes during the interval $t_1 - t_2 = \tau > 0$, is even or odd.

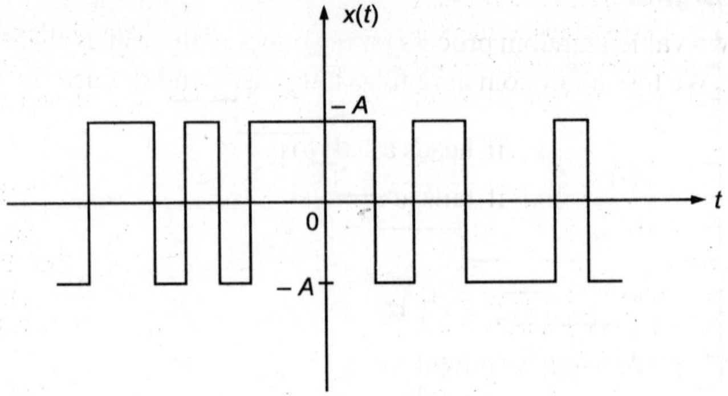

Figure 1.38: Random telegraph signal

The probabilities of these two possible outcomes are obtained as follows. From (1.175)

$$P[n = \text{even}] = P[x(t_2) = x(t_1)]$$

$$= e^{-\nu\tau} \sum_{n \text{ even}}^{\infty} \frac{(\nu\tau)^n}{n!} = e^{-\nu\tau} \cosh \nu\tau$$

and

$$P[n = \text{odd}] = P[x(t_2) = -x(t_1)]$$

$$= e^{-\nu\tau} \sum_{n \text{ odd}}^{\infty} \frac{(\nu\tau)^n}{n!} = e^{-\nu\tau} \sinh \nu\tau$$

Thus the autocorrelation function is given by

$$R_{xx}(t_1, t_2) = A^2 P[x(t_2) = x(t_1)] + (-A^2)\, P[x(t_2) = -x(t_1)]$$

$$= A^2 e^{-\nu\tau} [\cosh \nu\tau - \sinh \nu\tau]$$

$$= A^2 e^{-2\nu\tau} = \mathrm{A}^2 e^{-2\nu(t_1 - t_2)}$$

Reversing the role of t_1 and t_2, i.e., if $t_1 < t_2$, we conclude that

$$R_{xx}(t_1, t_2) = A^2 e^{-2\nu|\tau|} = A^2 e^{-2\nu|t_1 - t_2|} \tag{1.176}$$

This is plotted in Fig. 1.39.

$$R_{xx}(\tau) = A^2 e^{-2\nu|\tau|}$$

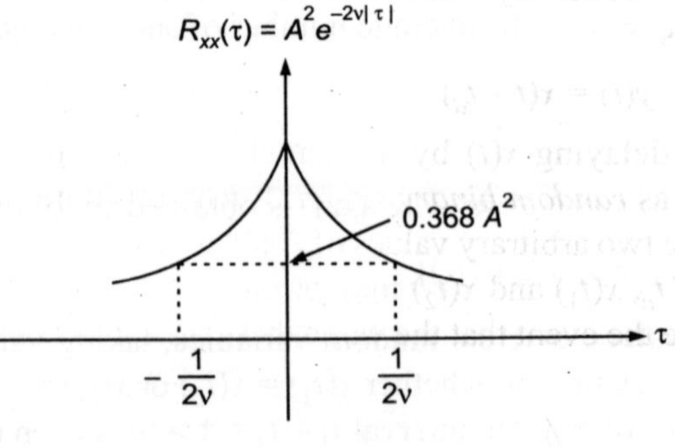

Figure 1.39: Autocorrelation function of a random telegraph signal

5. Random Binary Signal

Another example of a two valued random process taking either of the values +1 or –1 with equal probability is generated as follows. We toss a fair coin at regular intervals T and defined the process $x(t)$, by

$$x(t) = \begin{cases} +1 & \text{if heads at } n\text{th toss} \\ -1 & \text{if tails at } n\text{th toss} \end{cases}$$

where

$$(n-1)\,T < t < nT \tag{1.177}$$

A sample function of such a process is shown in Fig. 1.40.

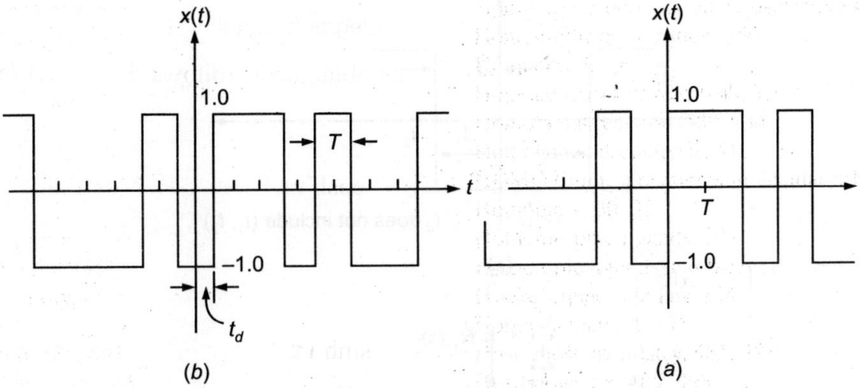

Figure 1.40: (*a*) Semirandom binary signal **(*b*)** Random binary signal

It is clear that

$$E[x(t)] = 0;\; E[x^2(t)] = 1$$

and

$$R_{xx}(t_1, t_2) = E[x(t_1)x(t_2)]$$

$$= \begin{cases} 1 & (n-1)T < t_1, t_2 < nT \\ 0 & \text{otherwise} \end{cases} \tag{1.178}$$

Let t_d be a random variable uniformly distributed in the interval $(0, T)$ and statistically independent of $x(t)$. Now, form the process

$$y(t) = x(t - t_d) \tag{1.179}$$

Thus $y(t)$ is obtained by delaying $x(t)$ by an amount t_d. The sample function of $y(t)$ is shown in Fig. 1.40(b) and is known as *random binary signal*. It may again be easily verified that $E[x(t)] = 0$. To find its a.c.f., let t_1 and t_2 be two arbitrary values of t and assume that $0 < t_1 < t_2 < T$ and $|t_1 - t_2| < T$. Now, depending on the value of t_d, $x(t_1)$ and $x(t_2)$ may or may not be in the same pulse interval as shown in Fig. 1.40(c). Let $E(t_1, t_2)$ be the event that the value that t_d assumes is such that t_1 and t_2 occur during the same pulse. In this case

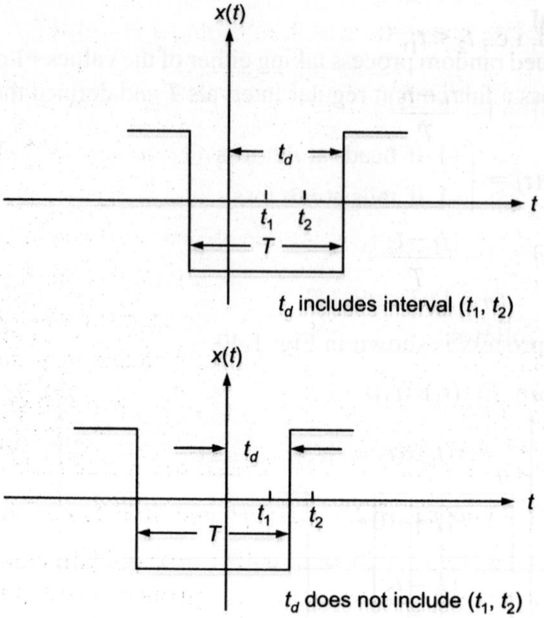

(c)

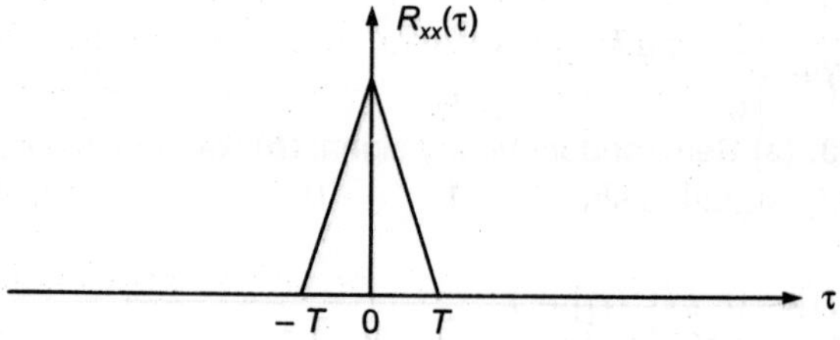

(d)

Figure 1.40: (*c*) Timing diagram to calculate a.c.f.; **(*d*)** Autocorrelation function of random binary signal

$$P[E(t_1, t_2)] = P[t_d < t_1 \text{ or } t_d > t_2]$$

$$= \frac{t_1}{T} + \frac{T - t_2}{T}$$

$$= 1 - \frac{(t_2 - t_1)}{T}$$

Similarly for the complement event $\overline{E}(t_1, t_2)$

$$P[\overline{E}(t_1, t_2)] = P[t_1 < t_d < t_2]$$

$$= \frac{t_2 - t_1}{T}$$

If role of t_1 and t_2 is changed, i.e., $t_2 < t_1$,

$$P[E(t_1, t_2)] = 1 - \frac{t_1 - t_2}{T}$$

and therefore, in general

$$P[E(t_1, t_2)] = 1 - \frac{|t_1 - t_2|}{T}$$

$R_{xx}(t_1, t_2)$ is now obtained as follows:

$$R_x(t_1, t_2) = E(x(t_1)x(t_2))$$

$$= E[x(t_1)x(t_2)/E]P[E] + E[x(t_1)x(t_2)/\overline{E}]P[\overline{E}]$$

$$= 1 \cdot P[E] + 0 \cdot P[\overline{E}]$$

$$= 1 - \frac{|t_1 - t_2|}{T}$$

Writing $\tau = |t_1 - t_2|$, we have

$$R_{xx}(\tau) = \begin{cases} 1 - |\tau|/T & |\tau| < T \\ 0 & |\tau| > T \end{cases} \tag{1.180}$$

The a.c.f. of the random binary signal is shown in Fig. 1.40d.

Wiener Processes

A stochastic process $\{x(t), t \geq 0\}$ is said to be a Wiener process if

(i) it is a process with stationary independent increments,
(ii) for every $t > 0$, it is normally distributed with $E[x(t)] = 0$,
(iii) $x(0) = 0$

Since $x(t)$ has independent increments and $x(0) = 0$, the probability law of $x(t)$ is uniquely characterized by the probability law of the $x(t) - x(s)$ for any $s < t$. Because of (ii) $x(t) - x(s)$ is also normally distributed and therefore determined by its mean and variance. Since mean is again zero, the characteristic function

$$\phi_{x(t)-x(s)}(u) = \exp\{-1/2u^2 \text{ var}[x(t) - x(s)]\} \tag{1.181}$$

It may be shown that

$$\text{var}[x(t) - x(s)] = \sigma^2(t - s) \tag{1.182}$$

where σ^2 is a positive constant, an empirical characteristic of the process and determined from observations of the process.

Though Wiener process originated as a model for Brownian motion and thermal noise in electric circuits, it constitutes along with the Poisson process, 'two fundamental species of stochastic processes in both theory and application'. It can be derived as a limiting form of another important discrete random process known as 'random walk'. The simplest random walk in one dimension is described by taking a step to the right or left, if at every tossing of a coin at regular interval, a head or a tail shows. If the position after t seconds after the start is denoted by $x(t)$, a sample function of the random process so constructed is shown in Fig. 1.41. It may be shown that

$$E[x(nT)] = 0, E[x^2(nT)] = ns^2 \tag{1.183}$$

where s is the step length.

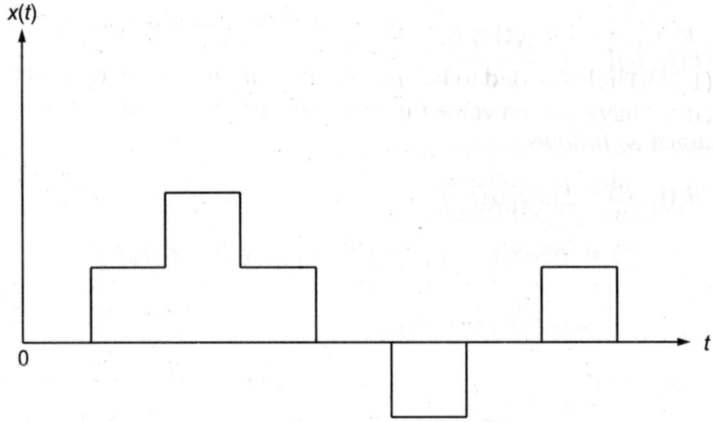

Figure 1.41: Sample function of a 'random walk'

For the discussion on Wiener process as a limiting form of random walk, see [50, 51]. In the context of digital communication the random walk and Wiener process play a fundamental role in the theory of synchronization [52].

1.6.5 Stationary Processes

In the study of random processes, it is useful to divide them into two broad categories: stationary and nonstationary or evolutionary. Intuitively *stationary process* is one whose statistical description is invariant with time, because the underlying physical mechanism generating the process is not changing with time. A process which is not stationary and therefore does not exhibit the time invariance of its statistical properties is said to be *nonstationary*.

Since all the statistical descriptions are derived from probability density or distribution functions a (strictly) stationary process is formally defined as the one for which

$$p_x(t + T) = p_x(t) \tag{1.184}$$

for every finite set of time instants t_i, $i = 1, 2, \ldots, k$ for all and for every constant T. (The notation is defined in (1.155)).

A strictly stationary process of order k is defined as one for which (1.184) holds for a given positive integer k. In particular, for $k = 1$, if

$$p_x(x_1, t) = p_x(x_1, t + T) \tag{1.185}$$

the first order probability density function is independent of t and hence

$$E[x(t)] = m_x(t) = m_x;\ \text{a constant} \tag{1.186}$$

and the process is said to be stationary in mean. Similarly for $k = 2$, not only (1.185) and hence (1.186) is true but also

$$p_x(x_1, x_2; t_1, t_2) = p_x(x_1, x_2; t_1 + T, t_2 + T)$$

This implies

$$p_x(x_1, x_2; t_1, t_2) = p_x(x_1, x_2; \tau) \tag{1.187}$$

where $\tau = t_1 - t_2$. Hence the autocovariance $K_{xx}(t_1, t_2)$ and the autocorrelation functions have the property that

$$\begin{aligned} K_{xx}(t_1, t_2) &= K_{xx}(\tau) = K_{xx}(-\tau) \\ R_{xx}(t_1, t_2) &= R_{xx}(\tau) = R_{xx}(-\tau) \end{aligned} \tag{1.188}$$

A process for which (1.188) holds is said to be *covariance stationary*. It is to be noted that a covariance stationary process need not have a mean value function independent of time. A process for which (1.187) and (1.188) hold, is called *stationary in the wide sense* or *weakly stationary*.

In general, it is quite difficult to prove that a given process is strictly stationary, for it involves verification of a large number of conditions. This difficulty is well illustrated in proving the stationarity of even very simple examples of stationary processes.

Examples:

1. Let

$$x(t) = g(t + \tau)$$

where $g(t)$ is any periodic function with period T and τ is a uniformly distributed random variable, over the interval $[0, T]$. Then $x(t)$ may be shown to be a stationary process. See [53].

2. A particular case of 1, and of special relevance in the communications context, is

$$x(t) = A \cos (\omega t + \theta)$$

where A and ω are constants and θ is uniformly distributed r.v. over the interval $[0, 2\pi]$.

3. The process

$$x(t) = a \cos \omega t + b \sin \omega t$$

is stationary in the wide sense iff a and b are uncorrelated r.v.'s with zero mean and same variance.

4. For zero mean uncorrelated random variables a_i and b_i with variance σ_i^2 the processes

$$x(t) = \sum_{i=1}^{N} a_i \exp (j\omega_i t)$$

and

$$x(t) = \sum_{i=1}^{N} (a_i \cos \omega_i t + b_i \sin \omega_i t)$$

are wide sense stationary.

5. The random binary signal described by (1.177) and (1.179) in section 1.6.4 is another example of a wide sense stationary process.

6. A simple example of a nonstationary random process is provided by

$$x(t) = \sin 2\pi f t$$

where f is a r.v. with p.d.f.

$$p_f(\alpha) = \begin{cases} \dfrac{1}{W} & 0 \le \alpha \le W \\ 0 & \end{cases}$$

To see that $x(t)$ is nonstationary, observe that every waveform in the ensemble is zero at $t = 0$, positive for $0 < t < \frac{1}{2}W$ and negative for $-\frac{1}{2}W < t < 0$. Therefore, the p.d.f. of $x(t_1)$ for $t_1 = \frac{1}{4}W$ is identically zero for negative t and p.d.f. of $x(t_2)$ for $t_2 = -\frac{1}{4}W$ is nonzero only for negative t. Hence $p_x(t_1) \neq p_x(t_2)$.

The stationary random processes are quite important for many reasons. Firstly many observed random processes are indeed stationary or may be modelled as such over appropriate observation intervals and some of their useful properties may be well described by first and second moments. And most important of them all, a spectral theory can be developed easily for such processes as discussed in Section 1.7. It is also true that the so-called ergodic theorem was obtained first for stationary processes as discussed in next section.

1.6.6 Ergodic Processes

Whatever we have discussed so far in the context of characterization or classification of random processes, has involved ensemble averages, for computation of which the process in its entirety is required. This clearly is not feasible in most of the practical situations. One, instead, has a single finite record or just one sample realization over a finite interval. Therefore, one is forced to ask the question: under what conditions is it possible to obtain estimates of ensemble averages, from this record, which may reliably be used to characterize the process. Before we answer this question, we must understand the notion of an integral of a random process. Given a random process $x(t)$, let

$$s = \int_a^b x(t)\, dt \tag{1.189}$$

Assuming that this integral exists in the usual Riemann sense for every sample function $x(t, \xi)$ of the process $x(t)$, it defines a number $s(\xi)$. Therefore s is a random variable. If the integral does not exist in this sense, we may define it in another way, say in the mean square sense as follows:

$$\lim_{n\to\infty} E\left[s - \sum_{t=1}^{n} x(t_i)\Delta t_i\right]^2 = 0 \quad \text{for } \Delta t_k \to 0 \tag{1.190}$$

In our discussion, we assume that all integrals exist in this sense. Writing the integral as limits of the sum and using the linearity property of the expectation operator, one can easily prove that

$$E[s] = \int_a^b E[x(t)]\, dt = \int_a^b m_x(t)\, dt \tag{1.191}$$

and

$$\begin{aligned} \text{var}\,[s] = \sigma_s^2 &= \int_a^b \int_a^b [R_{xx}(t_1, t_2) - m_x(t_1)m_x(t_2)]\, dt_1\, dt_2 \\ &= \int_a^b \int_a^b K_{xx}(t_1, t_2)\, dt_1\, dt_2 \end{aligned} \tag{1.192}$$

Let us consider $x(t)$ to be a stationary process and

$$s = \frac{1}{2T}\int_{-T}^{T} x(t)\,dt \tag{1.193}$$

Clearly, for the stationary process $x(t)$

$$E[s] = m_x$$

$$\text{var}\,[s] = \sigma_s^2 = \frac{1}{4T^2}\int_{-T}^{+T}\int_{-T}^{T} K_{xx}(t_1 - t_2)\,dt_1\,dt_2$$

The double integral above may be written as a single integral as

$$\text{var}\,[s] = \sigma_s^2 = \frac{1}{2T}\int_{-2T}^{2T}\left(1 - \frac{|\tau|}{2T}\right)K_{xx}(\tau)\,d\tau \tag{1.194}$$

When $x(t)$ is real, $K_{xx}(\tau)$ is even and we may write

$$\text{var}\,[s] = \frac{1}{2T}\int_{0}^{2T}\left(1 - \frac{\tau}{2T}\right)K_{xx}(\tau)\,d\tau$$

$$= \frac{1}{2T}\int_{0}^{2T}\left(1 - \frac{\tau}{2T}\right)[R_{xx}(\tau) - m_x^2]\,d\tau \tag{1.195}$$

Let us now define the time averages

$$\langle x(t)\rangle = \lim_{T\to\infty}\frac{1}{2T}\int_{-T}^{T} x(t)\,dt \tag{1.196}$$

and

$$\langle x(t)x(t+\tau)\rangle = \lim_{T\to\infty}\int_{-T}^{T} x(t)x(t+\tau)\,dt \tag{1.197}$$

Now let us repeat our question and ask, under what conditions are the time averages (1.196) and (1.197) are the same as ensemble averages for the stationary process $x(t)$ under consideration. The answer is provided by the following *ergodic theorem*

$$\langle x(t)\rangle = \lim_{T\to\infty}\frac{1}{2T}\int_{-T}^{T} x(t)\,dt = E[x(t)] = m_x \tag{1.198}$$

iff

$$\lim_{T\to\infty}\frac{1}{2T}\int_{0}^{2T}\left(1 - \frac{\tau}{2T}\right)[R_{xx}(\tau) - m_x^2]\,d\tau = 0 \tag{1.199}$$

or equivalently

$$\lim_{T\to\infty}\frac{1}{2T}\int_{-T}^{T} R_{xx}(\tau)\,d\tau = m_x^2 \tag{1.200}$$

A random process for which (1.198) and (1.199) or (1.200) holds, is called an *ergodic* (*in the mean*) *process*. For such a process, the limit of the sequence of sample means (1.196) converges with probability 1 to the ensemble mean because the var [s] of the sequence goes to zero.

Similarly one can talk of the ergodicity of the higher moments like autocorrelation and ergodicity of the distribution functions. For details see [50]. Thus in general

> 'For an ergodic process, the ensemble average of a function of the random process is equal, with probability 1, to the time average over all possible time translations of a particular sample function, except for a possible subset of a sample function of measure zero.'

Stated mathematically, for any ergodic process, every function $g\{x(t)\}$ satisfies

$$\begin{aligned} E[g\{x(t)\}] &= \lim_{T\to\infty} \frac{1}{2T}\int_{-T}^{T} g[x(t+\tau)]\, d\tau \\ &= \langle g\{x(t+\tau)\}\rangle \end{aligned} \tag{1.201}$$

Though, we have stated the ergodic theorem for ergodicity in the mean for a wide-sense stationary process, one must not confuse the notion of stationarity with ergodicity. Ergodic processes are always stationary but the converse is not true. However, it is difficult to come up with counter examples. One simple example is a random process whose sample functions are constant with time (in other words, a random variable with nonzero variance). For each sample function, the time average is a constant. Let $x(t) = a$ be the process where a is a r.v. with p.d.f.

$$\begin{aligned} P_a(\alpha) &= 1 \quad 0 < \alpha < 1 \\ &= 0 \quad \text{otherwise} \end{aligned}$$

For any sample function, say for one in which $\alpha = 0.3$

$$\langle x(t)\rangle = 0.3$$

However, for any t,

$$E[x(t)] = \int_0^1 \alpha p_a(\alpha)\, d\alpha = 1/2$$

Clearly the process is stationary but not ergodic. Thus, though ergodicity seems to imply a very restrictive class of random processes, and very difficult to prove, none the less, we shall often assume that the random processes encountered in our study of communication systems are ergodic, to simplify matters. After all, in the real world we are, but forced to work, very often with just one sample function of a process and must obtain the relevant statistics from one such realization.

1.6.7 Further Classification of Random Processes

1. A two-dimensional random process consists of a pair of processes $x(t)$ and $y(t)$ and is characterized by the joint p.d.f. of the random vectors $x(t)$ and $y(t)$ and is characterized by the joint p.d.f. of the random vectors $x(t) = [x(t_1), x(t_2), \ldots, x(t_n)]$ and $y(t) = [y(t_1'), y(t_2'), \ldots, y(t_m')]$ for any $t_1, t_2, \ldots, t_n$; $t_2', t_2', \ldots, t_m'$.
2. A complex random process $z(t) = x(t) + jy(t)$ is defined in terms of two dimensional real processes $x(t)$ and $y(t)$. For such processes, the definitions of second order moments are accordingly modified as follows:

$$\text{Autocorrelation}: R_{xx}(t_1, t_2) = E[x(t_1)x^*(t_2)] \tag{1.202}$$

$$\text{Autocovariance}: K_{xx}(t_1, t_2) \triangleq \text{conv}\,[x(t_1), x^*(t_2)]$$

$$= R_{xx}(t_1, t_2) - m_x(t_1)\, m_x^*(t_2)] \quad (1.203)$$

Variance : $$K_{xx}(t, t) = R_{xx}(t, t) - |m_x(t)|^2 \quad (1.204)$$

Cross-correlation : $$R_{xy}(t_1, t_2) = E[x(t_1)y^*(t_2)] \quad (1.205)$$

$$= E[\{x(t_1) - m_x(t_1)\}\{y^*(t_2) - m_y^*(t_2)\}]$$

$$= R_{xy}(t_1, t_2) - m_x(t_1)\, m_y^*(t_2) \quad (1.206)$$

3. Two processes $x(t)$ and $y(t)$ are *uncorrelated*, if for any t_1, t_2, we have

$$R_{xy}(t_1, t_2) = E[x(t_1)]E[y^*(t_2)] \quad (1.207)$$

They are *orthogonal* if, for all t_1, t_2

$$R_{xy}(t_1, t_2) = 0 \quad (1.208)$$

and statistically independent if $x(t) = [x(t_1), x(t_2), \ldots, x(t_n)]$ is statistically independent of $y(t)$ $= [y(t_1'), y(t_2')\ldots, y(t_m')]$ for any $t_1, t_2, \ldots, t_n$ and $t_1', t_2' \ldots, t_m'$.

4. Two random processes $x(t)$ and $y(t)$ are jointly wide-sense stationary if they are individually wide-sense stationary and if their cross-correlation function depends only upon the difference of its two arguments, i.e.

$$R_{xy}(t_1, t_2) = R_{xy}(t_1 - t_2)$$

$$R_{yx}(t_1, t_2) = R_{yx}(t_1 - t_2) \quad (1.209)$$

1.7 POWER DENSITY SPECTRUM

In Section 1.2 we have seen how the spectral description of signals and linear systems plays a central role in their study. Since, in the study of communication systems, we are very much concerned with the processing of random signal by linear systems, it is important to ask the question: can one define for a random process $x(t)$, $\infty < t < \infty$, a fourier transform

$$X(\omega) = \int_{-\infty}^{\infty} x(t)e^{-j\omega t}\, d\omega$$

This is clearly not possible, in general, as the integral does not exist for most of random processes whose sample functions are nonperiodic undamped functions and therefore do not belong to the class of functions for which Fourier theory is valid. Nevertheless a spectral or harmonic description is possible based on averages as was first shown by Wiener [54]. Briefly the arguments (without going into details of the proofs) go as follows.

For a finite sample $x(t)$; $0 \le t \le T$ of the process, define a sample correlation function

$$R_T(\tau) = \begin{cases} \dfrac{1}{T}\displaystyle\int_0^{T-\tau} x(t)x(t+\tau)\, d\tau; & 0 \le \tau \le T \\ 0 \quad ; & \tau > T \\ R_T(-\tau) \quad ; & \tau < 0 \end{cases} \quad (1.210)$$

and a *sample spectral density function* (also called 'periodogram')

$$X_T(\omega) = \frac{1}{T}\left|\int_0^T x(t)e^{-j\omega t}dt\right|^2; \quad -\infty < \omega < \infty \tag{1.211}$$

It may be shown that $R_T(\tau) \leftrightarrow X_T(\omega)$, i.e. they form a Fourier transform pair,

$$X_T(\omega) = \int_{-T}^{T} R_T(\tau)e^{-j\omega t}\, d\tau \tag{1.212}$$

$$R_T(\tau) = \frac{1}{2\pi}\int_{-\infty}^{\infty} X_T(\omega)e^{j\omega\tau}\, d\omega \tag{1.213}$$

Taking the expectation in (1.213), we get

$$E[R_T(\tau)] = \frac{1}{2\pi}\int_{-\infty}^{\infty} E[X_T(\omega)]e^{j\omega\tau}\, d\omega \tag{1.214}$$

Let us assume that there exists a continuous function $R_{xx}(\tau)$ such that

$$\lim_{T\to\infty} E[R_T(\tau)] = R_{xx}(\tau) \tag{1.215}$$

For a wide sense stationary process with zero mean, (1.215) holds with $R_{xx}(\tau)$ as the correlation function of the process, because

$$E[R_T(\tau)] = \frac{1}{T}\int_0^{T-\tau} R_{xx}(\tau)\, d\tau = \left(1-\frac{\tau}{T}\right)R_{xx}(\tau)$$

If (1.215) holds then one can show that there exists a bounded non-decreasing function $P_{xx}(\omega)$ and called the power spectral distribution function such that

$$R_{xx}(\tau) = \frac{1}{2\pi}\int_{-\infty}^{\infty} e^{j\omega\tau}dP_{xx}(\omega) \tag{1.216}$$

If $P_{xx}(\omega)$ is everywhere differentiable and the derivative is denoted by $S_{xx}(\omega)$, then

$$R_{xx}(\tau) = \frac{1}{2\pi}\int_{-\infty}^{\infty} S_{xx}(\omega)e^{j\omega\tau}\, d\omega \tag{1.217}$$

and $S_{xx}(\omega)$ is called the *power spectral density function* (p.s.d.). A sufficient condition for $P_{xx}(\omega)$ to be differentiable is that

$$\int_{-\infty}^{\infty} |R_x(\tau)|\, d\tau < \infty \tag{1.218}$$

Explicitly, $S_{xx}(\omega)$ is obtained as

$$S_{xx}(\omega) = \int_{-\infty}^{\infty} R_{xx}(\tau)e^{-j\omega\tau}\, d\tau \tag{1.219}$$

Equations (1.217) and (1.218) are known as *Wiener-Khintchine relations* and are basic to the spectral theory of random processes.

From eqs. (1.211, 1.214, 1.215 and 1.218), it is also clear that the power spectral density function $S_{xx}(\omega)$ is the limiting value for $T \to \infty$ of the mean of the sample power spectral density function or periodogram, i.e.

$$S_{xx}(\omega) = \lim_{T\to\infty} E[X_T(\omega)] \tag{1.220}$$

The Cross-Power Density Spectrum

Analogous to power density spectrum, the cross-power density spectrum of two random processes $x(t)$ and $y(t)$ is the Fourier transform of crosscorrelation function $R_{xy}(\tau)$:

$$S_{xy}(\omega) = \int_{-\infty}^{\infty} R_{xy}(\tau)e^{-j\omega\tau}\, d\tau \tag{1.221}$$

since

$$R_{xy}(\tau) = R^*_{yx}(-\tau),$$

we have

$$S_{xy}(\omega) = S^*_{yx}(\omega), \tag{1.222}$$

Some Properties of the Power Spectral Density Functions

The following properties of the power spectral density function and correlation function of a w.s.s. process follow easily from the definitions and Wiener-Khintchine relations

1. $R^*_{xx}(\tau) = R_{xx}(-\tau)$
2. $R_{xy}(\tau) = R^*_{yx}(-\tau)$
3. $R_{xx}(0) \geq |R_{xx}(t)|$
4. $R^2_{xy}(\tau) \leq R_{xx}(0)\, R_{yy}(0),\ 2|R_{xy}(\tau)| \leq R_{xx}(0) + R_{yy}(0)$
5. $S_{xx}(\omega) \geq 0$
6. For a real process, $S_{xx}(\omega)$ is even; $S_{xx}(\omega) = S_{xx}(-\omega)$
7. $S_{xy}(\omega) \geq S^*_{yx}(\omega)$
8. $S_x(\omega)$ can also be expressed in terms of its second order density $p(x_1, x_2; \tau)$. To do this let $G_x(\omega)$ be F.T. of $p(x_1, x_2; \tau)$ with respect to τ:

$$G(x_1, x_2; \omega) = \int_{-\infty}^{\infty} p(x_1, x_2; \tau)e^{-j\omega\tau}\, d\tau$$

since

$$R_{xx}(\tau) = \int_{-\infty}^{\infty}\int_{-\infty}^{\infty} x_1 x_2\, p(x_1, x_2; \tau)\, dx_1\, dx_2$$

and

$$S_{xx}(\omega) = \int_{-\infty}^{\infty} R_{xx}(\tau)e^{-j\omega\tau}\, d\tau$$

Therefore,

$$S_{xx}(\omega) = \int_{-\infty}^{\infty}\int_{-\infty}^{\infty} x_1, x_2 G(x_1, x_2; \omega)\, dx_1\, dx_2 \tag{1.223}$$

9. If $z(t) = x(t) + y(t)$ and $x(t)$ and $y(t)$ are zero mean jointly stationary processes, then

$$S_{zz}(\omega) = S_{xx}(\omega) + S_{yx}(\omega) + S_{xy}(\omega) + S_{yy}(\omega) \tag{1.224}$$

Examples:

1. Consider the a.c.f of a random telegraph signal evaluated in Section 3.3, eq. (1.176)

   $$R_{xx}(\tau) = e^{-2\nu|\tau|}$$

   Its p.s.d. is obtained as

   $$S_{xx}(\omega) = \frac{4\nu}{\omega^2 + 4\nu^2} \tag{1.225}$$

2. Similarly for the random binary signal $x(t)$ of Fig. 1.40(a),

   $$R_{xx}(\tau) = \begin{cases} 1 - \dfrac{|\tau|}{T} & |\tau| < T \\ 0 & |\tau| > T \end{cases} \tag{1.226}$$

   and

   $$S_{xx}(\omega) = \frac{2\sin^2 \omega T/2}{\omega^2 T}.$$

3. For a sine wave with random phase (eq. 1.162),

   $$R_{xx}(\tau) = \frac{A^2}{2} \cos \omega_c \tau$$

   $$S_{xx}(\omega) = \frac{A^2}{4}[\delta(\omega - \omega_c) + \delta(\omega + \omega_c)] \tag{1.227}$$

   In general, a w.s.s. process is said to be periodic or cyclo-stationary if its mean and a.c.f. is periodic. For such a process, $R_{xx}(\tau)$ can be expanded in Fourier series as

   $$R_{xx}(\tau) = \sum_{n=-\infty}^{\infty} a_n \exp(j\omega_0 n\tau); \quad \omega_0 = \frac{2\pi}{T}$$

   and

   $$S_x(\omega) = \sum_{n=-\infty}^{\infty} a_n \delta(\omega - n\omega_0) \tag{1.228}$$

1.8 TRANSFORMATIONS OF STOCHASTIC PROCESSES

In general, a transformation of a stochastic process may be represented as

$$y(t) = T[x(t)] \tag{1.229}$$

T is given the interpretation of an operator or a system with domain or input $x(t)$ and range or output $y(t)$. T is assumed to be deterministic. This means that if the sample functions $x(\xi_1, t)$ and $x(\xi_2, t)$ corresponding to two points ξ_1, ξ_2 are same, i.e,

$$x(\xi_1, t) = x(\xi_2, t)$$

then

$$y(\xi_1, t) = y(\xi_2, t)$$

T will be random, if there exist two distinct points ξ_1, ξ_2 such that $x(\xi_1, t) = x(\xi_2, t)$ but $y(\xi_1, t) \neq y(\xi_2, t)$. In the following, we shall deal only with deterministic systems.

1.8.1 Zero Memory Systems

For such a system, the output $y(t_1)$ at any instant $t = t_1$ depends only on the input at that instant $x(t_1)$ and not on its past or future values. For such a system, if we write

$$y(t) = g[x(t)] \tag{1.230}$$

then $g(x)$ is a function only of x, e.g. $g(x) = \sin x$. Such a system is time invariant, for $y(t+\varepsilon) = g[x(t+\varepsilon)]$. For zero memory systems, the p.d.f. $p_y(y, t)$ of the output $y(t)$ is obtained from p.d.f. of the input $x(t)$ and the knowledge of the transformation $g(x)$ using the methods of transformations of random variables.

In general, the joint density of $y = [y(t_1), y(t_2) \ldots y(t_n)]$ is obtained as

$$p_y(y_1, y_2, \ldots, y_n; t_1, t_2, \ldots, t_n) = \frac{p_x(x_1, x_2, \ldots, x_n; t_1, t_2, \ldots, t_n)}{|J|} \tag{1.231}$$

where J is the Jacobian of the transformation of the system of equations $y_i = g(x_i)$, $i = 1, \ldots, n$.

Since there is no explicit dependence on t_i in (1.231), if $x(t)$ is strict sense stationary or stationary of order n, then $y(t)$ is also stationary. It is to be noted that if $x(t)$ is only w.s.s. then $y(t)$ need not be stationary in any sense.

The first and second order statistics of the output process $y(t)$ are given by

$$E[y(t)] = \int_{-\infty}^{\infty} g(x) p_x(x; t)\, dx \tag{1.232}$$

$$E[y(t_1)y(t_2)] = \int_{-\infty}^{\infty}\int_{-\infty}^{\infty} g(x_1) g(x_2) p_x(x_1, x_2; t_1, t_2)\, dx_1\, dx_2 \tag{1.233}$$

Example 1.12:

Let $y(t) = x^2(t)$

Since the equation $y = x^2$ has no real solution for $y < 0$ and for $y > 0$, has two solutions;

$$x = \pm\sqrt{y} \text{ and } \frac{dy}{dx} = 2\sqrt{y}$$

The p.d.f. for the process $y(t)$ is obtained as

$$p_y(y; t) = \frac{p_x(\sqrt{y}; t) + p_x(-\sqrt{y}; t)}{2\sqrt{y}} \qquad y \geq 0$$

$$= 0 \qquad y < 0 \tag{1.234}$$

To obtain the joint p.d.f. $p_y(y_1, y_2; t_1, t_2)$, we note that for $y_1 > 0$ and $y_2 > 0$, the system

$y_1 = x_1^2, y_2 = x_2^2$ has four solutions

$$x_1 = \pm\sqrt{y_1}, x_2 = \pm\sqrt{y_2}$$

and

$$J = \begin{vmatrix} \pm 2\sqrt{y_1} & 0 \\ 0 & \pm 2\sqrt{y_2} \end{vmatrix} = \pm 4\sqrt{y_1}\sqrt{y_2}$$

$$|J| = 4\sqrt{y_1}\sqrt{y_2}$$

Hence,

$$p_y(y_1, y_2) = \frac{p_x(\sqrt{y_1}, \sqrt{y_2}) + p_x(-\sqrt{y_1}, \sqrt{y_2}) + p_x(\sqrt{y_1}, -\sqrt{y_2}) + p_x(-\sqrt{y_1}, -\sqrt{y_2})}{4\sqrt{y_1}\sqrt{y_2}}$$

$$\text{if } \; y_1, y_2 > 0$$

$$= 0 \quad \text{if } y_1 < 0 \text{ or } y_2 < 0 \tag{1.235}$$

Since power law transformations, as exemplified above are very often used in the detection of modulated signals in communication system, their study is of great importance in evaluating the performance of such systems. In a later section, we shall evaluate the first and second order statistics of such transformation for the Gaussian input processes. For details see [9].

Linear Time Invariant Systems

For an LTI system with impulse response $h(t)$ and transfer function $H(f)$, the input and output are related through the convolution integral

$$y(t) = \int_{-\infty}^{\infty} x(\tau)h(t-\tau)d\tau \tag{1.236}$$

For an input $x(t)$ which is a random process, the output $y(t)$ is also a random process. Let $x(t)$ be a stationary process. Then $y(t)$ will also be a stationary process. Its mean will be given by

$$m_y = E[y(t)] = \int_{-\infty}^{\infty} E[x(t-\tau)]h(\tau)\, d\tau$$

$$= m_x \int_{-\infty}^{\infty} h(\tau)\, d\tau = m_x H(o) \tag{1.237}$$

The a.c.f. of the output $R_{yy}(\tau)$ is obtained as follows by first determining the cross-correlation function $R_{yx}(\tau)$ as

$$E[y(t)x^*(t-\tau)] = \int_{-\infty}^{\infty} E[x(t-\beta)x^*(t-\tau)]h(\beta)\, d\beta$$

$$= \int_{-\infty}^{\infty} R_x(\tau-\beta)h(\beta)\, d\beta$$

or

$$R_{yx}(\tau) = R_{xx}(\tau)^* h(\tau) \tag{1.238}$$

Now consider

$$E[y(t+\tau)y^*(t)] = \int_{-\infty}^{\infty} E[y(t+\tau)x^*(t-\beta)h^*(\beta)\,d\beta]$$

or

$$R_{yy}(\tau) = \int_{-\infty}^{\infty} R_{yx}(\tau+\beta)h^*(\beta)\,d\beta$$

or

$$= R_{yx}(\tau) * h^*(-\tau) \tag{1.239}$$

Similarly one gets

$$R_{xy}(\tau) = R_{xx}(\tau) * h^*(-\tau)$$

$$R_{yy}(\tau) = R_{xy}(\tau) * h^*(\tau) \tag{1.240}$$

and therefore

$$R_{yy}(\tau) = R_{xx}(\tau) * h^*(-\tau) * h(\tau) \tag{1.241}$$

In the frequency domain, taking F.T. of (1.239) and (1.240) etc., we get

$$S_{xy}(\omega) = S_{xx}(\omega)\,H^*(\omega) \tag{1.242}$$

$$S_{yy}(\omega) = S_{xy}(\omega)H(\omega) \tag{1.243}$$

and

$$S_{yy}(\omega) = S_{xx}(\omega)|H(\omega)|^2 \tag{1.244}$$

(1.244) is a fundamental relation relating the power spectral densities of the input and output process of a LTI system with transfer function $H(\omega)$. The above mentioned relationships are summarized in Fig. 1.42.

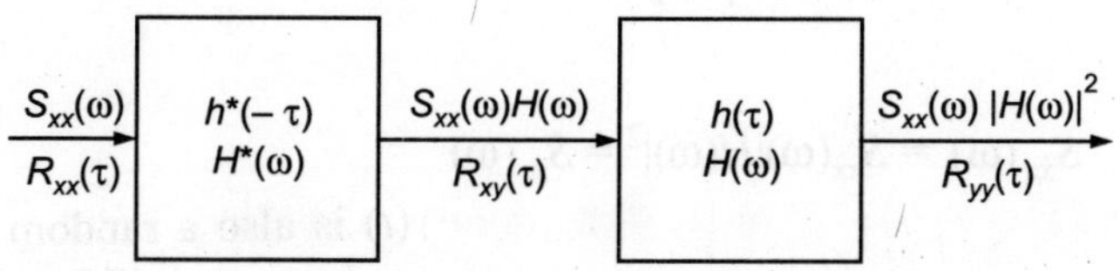

Figure 1.42: Power spectral density relations in LTI systems

Examples:

1. *Smoothing*: Consider the process $y(t)$ obtained from $x(t)$ by the operation

$$y(t) = \frac{1}{2\pi}\int_{t-T}^{t+T} x(\alpha)d\alpha$$

i.e. $x(t)$ is averaged over an interval $(-T, T)$ centered at t. The system realization of this operation is L.T. I. system with an impulse response

$$h(t) = \begin{cases} \dfrac{1}{2}T & |t| < T \\ 0 & |t| > T \end{cases}$$

The corresponding transfer function is

$$H(\omega) = \frac{1}{2\pi}\int_{-T}^{T} e^{j\omega t}\, dt = \frac{\sin \omega T}{\omega T}$$

Therefore, the p.s.d of the output $y(t)$ is

$$S_{yy}(\omega) = S_{xx}(\omega) \cdot \left(\frac{\sin \omega T}{\omega T}\right)^2$$

and the a.c.f. of the output is

$$R_{yy}(\tau) = \frac{1}{2T}\int_{-2T}^{2T}\left(1 - \frac{|\alpha|}{2T}\right) R_{xx}(\tau - \alpha)\, d\alpha.$$

2. *Hilbert Transform:* We have seen in Section 1.4 on bandpass systems, that a quadrature filter whose transfer function is

$$H(\omega) = \begin{cases} -j & \omega > 0 \\ j & \omega < 0 \end{cases}$$

and impulse response is

$$h(t) = 1/\pi t$$

implements the Hilbert transform operation

$$\hat{x}(t) = 1/\pi \int_{-\infty}^{\infty} \frac{x(\tau)}{t - \tau}\, d\tau$$

Therefore

$$S_{\hat{x}x}(\omega) = S_{xx}(\omega)|H(\omega)|^2 = S_{xx}(\omega)$$

Also

$$S_{\hat{x}x}(\omega) = S_{xx}(\omega)\, H(\omega) = \begin{cases} -jS_{xx}(\omega) & \omega > 0 \\ jS_{xx}(\omega) & \omega < 0 \end{cases}$$

Hence

$$R_{\hat{x}x}(\tau) = R_{xx}(\tau)$$

and

$$R_{\hat{x}x}(\tau) = R_{xx}(\tau) * h(t) = \hat{R}_{xx}(\tau)$$

Also, since $S_{xx}(\omega)$ is purely imaginary, $R_{\hat{x}x}(\tau)$ is odd. Therefore,

$$R_{\hat{x}x}(-\tau) = -R_{\hat{x}x}(\tau)$$

Similarly

$$R_{x\hat{x}}(\tau) = R_{\hat{x}x}(-\tau) = -R_{\hat{x}x}(\tau)$$

We note that

$$R_{x\hat{x}}(o) = 0$$

This means that, for a given t, $x(t)$ and $\hat{x}(t)$ are orthogonal. (This does not mean that the processes $x(t)$ and $\hat{x}(t)$ are orthogonal.)

3. *Analytic Signal:* The analytical signal corresponding to $x(t)$ is $z(t) = x(t) + j\hat{x}(t)$. It can be shown that

$$R_{zz}(\tau) = 2[R_{xx}(\tau) + j\hat{R}_{xx}(\tau)] \tag{1.245}$$

and thus, $1/2R_{zz}(\tau)$ is the analytic signal corresponding to $R_{xx}(\tau)$. Also

$$S_{zz}(\omega) = 2[S_{xx}(\omega) + jS_{xx}(\omega)] = \begin{cases} 4S_{xx}(\omega) & \omega > 0 \\ 0 & \omega < 0 \end{cases} \tag{1.246}$$

Since

$$x(t) = \text{Re}[z(t)]$$

Therefore

$$R_{xx}(\tau) = \frac{1}{2}\text{Re}[R_{zz}(\tau)] \tag{1.247}$$

4. *White Noise:* A *white noise* is a w.s.s. random process with constant power spectral density:

$$S_{xx}(\omega) = N_{02} = \frac{N_{01}}{2} \tag{1.248}$$

[N_{01} is, the 'one sided' power density spectrum]. Its a.c.f. is, therefore,

$$R_{xx}(\tau) = N_{02}\,\delta(\tau) \tag{1.249}$$

and its Hilbert transform is

$$\widehat{R}_{xx}(\tau) = N_{02}\cdot\frac{1}{\pi\tau} \tag{1.250}$$

From example 3, the p.s.d. of the analytic signal corresponding to white noise process $x(t)$ is

$$S_{zz}(\omega) = \begin{cases} 4N_{02} & \omega > 0 \\ 0 & \text{otherwise} \end{cases} \tag{1.251}$$

and the corresponding a.c.f. is

$$R_{zz}(\tau) = 2N_{02}\left[\delta(\tau) + \frac{j}{\pi\tau}\right] \tag{1.252}$$

5. *Noise Bandwidth:* Noise bandwidth is a useful concept in evaluating the signal to noise ratio (SNR) performance of linear systems. Consider such a system with a low pass transfer function

$H(\omega)$. Let the input to this system be white noise of p.s.d. $N_0/2$, where N_0 is a real positive constant. The total average power at the output of the system

$$P_y = \frac{1}{2\pi}\int_{-\infty}^{\infty} S_{yy}(\omega)\, d\omega = \frac{1}{2\pi}\int_{-\infty}^{\infty} \frac{N_0}{2} |H(\omega)|^2\, d\omega \tag{1.253}$$

$$= \frac{N_0}{2\pi}\int_{0}^{\infty} |H(\omega)|^2\, d\omega \tag{1.254}$$

for a system with real impulse response. If we now consider an ideal lowpass system which is equivalent to the given system in the sense that both produce the same output average power when excited by the same white noise source and have the same d.c. power value $|H(0)|^2$, then the equivalent ideal system is to be attributed a bandwidth W_N, so that

$$\frac{N_0}{2\pi}\int_{0}^{W_N} |H(0)|^2\, d\omega = \frac{N_0}{2\pi}\int_{0}^{\infty} |H(\omega)|^2\, d\omega$$

or

$$\frac{N_0 |H(0)|^2 W_N}{2\pi} = \frac{N_0}{2\pi}\int_{0}^{\infty} |H(\omega)|^2\, d\omega$$

and therefore

$$W_N = \frac{\int_{0}^{\infty} |H(\omega)|^2\, d\omega}{|H(0)|^2} \tag{1.255}$$

W_N is called the *noise bandwidth* of the system.

1.9 BANDLIMITED PROCESSES

Most of the concepts and results pertaining to deterministic signals and discussed in Section 1.1 to 1.4 are easily extended to random processes if the power spectral density characterization is substituted for the Fourier transform description in terms of magnitude and phase versus frequency functions. Similarly, in time domain, all time averages are replaced by ensemble averages in theoretical developments. When the observed process is assumed ergodic, the ensemble averages are easily obtained as time averages from one sufficiently long duration of the sample realization.

Bandlimited processes are processes whose p.s.d. is zero outside certain regions of the frequency axis. Such processes are easily modelled as the output of LTI systems or filters excited by white noise. Thus lowpass and bandpass processes are easily defined as one whose p.s.d $S(\omega)$ has the characteristic:

Low pass: $\quad S(f) = 0; \quad |f| > W \qquad (1.256)$

Bandpass: $\quad S(f) \neq 0; \quad f_c - W_1 \le |f| \le f_c - W_1 + 2W \qquad (1.257)$

This is illustrated in Fig. 1.43.

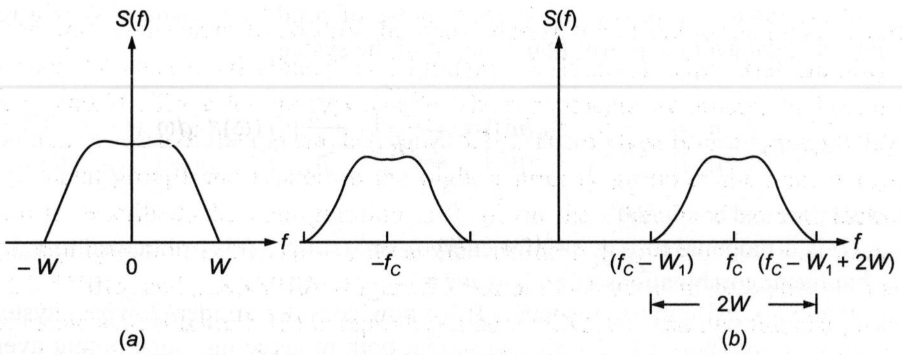

Figure 1.43: (a) p.s.d. of a lowpass process **(b)** p.s.d. of a bandpass process

1.9.1 Sampling Theorem for Bandlimited Processes

The sampling theorems, as stated earlier in Section 1.5 for lowpass and bandpass deterministic signals are also valid for such random processes as long as the convergence of the sampling expansion is understood in the mean square sense. The theorem states that if $x(t)$ is a lowpass band-limited process satisfying (1.256), then it has a sampling representation

$$x(t) \doteq \sum_{n=-\infty}^{\infty} x\left(\frac{n}{2W}\right)\frac{\sin 2\pi W(t-n/2W)}{2\pi W(t-n/2W)} \tag{1.258}$$

where $\doteq$ means that $E\left\{\left|x(t)-x_{\infty}(t)\right|^2\right\}=0$. Where

$$x_{\infty}(t) = \sum_{h=-\infty}^{\infty} x\left(\frac{n}{2W}\right)\frac{\sin 2\pi W(t-n/2W)}{2\pi W(t-n/2W)} \tag{1.259}$$

To prove the theorem, we first apply the deterministic sampling theorem to the autocorrelation function $R_{xx}(\tau)$ of the w.s.s. bandlimited process $x(t)$. Thus

$$R_{xx}(\tau) = \sum_{n=-\infty}^{\infty} R_{xx}\left(\frac{n}{2W}\right)\frac{\sin 2\pi W(\tau-n/2W)}{2\pi W(\tau-n/2W)} \tag{1.260}$$

Since $R_x(\tau-\alpha)$ is also a W-bandlimited process, hence

$$R_x(\tau-\alpha) = \sum_{n=-\infty}^{\infty} R_{xx}\left(\frac{n}{2W}-\alpha\right)\frac{\sin 2\pi W(\tau-n/2W)}{2\pi W(\tau-n/2W)} \tag{1.261}$$

Making a change of variables, we may also write

$$R_{xx}(\tau) = \sum_{n=-\infty}^{\infty} R_{xx}\left(\frac{n}{2W}-\alpha\right)\frac{\sin 2\pi W(\tau+\alpha-n/2W)}{2\pi W(\tau+\alpha-n/2W)} \tag{1.262}$$

To prove that $E\left\{|x(t)-x_\infty(t)|^2\right\}=0$, consider

$$E\left\{|x(t)-x_\infty(t)|x\left(\frac{m}{2W}\right)\right\} = R_x\left(t-\frac{m}{2W}\right)-\sum_{n=-\infty}^{\infty}R_{xx}\left(\frac{n-m}{2W}\right)\times\frac{\sin 2\pi W(t-n/2W)}{2\pi W(t-n/2W)}=0$$

using (1.261), with $\tau = t$ and $\alpha = m/2W$.

Thus $x(t) - x_\infty(t)$ is orihogonal to $x(m/2W)$ for every m or $x(t) - x_\infty(t)$ is orthogonal to $x_\infty(t)$, because $x_\infty(t)$ is nothing but linear combinations of r.v.'s $x(m/2W)$.

Hence,

$$E\left\{\left[x(t)-x_\infty(t)\right]x_\infty(t)\right\}=0$$

and since

$$E\left\{\left[x(t)-x_\infty(t)\right]x(t)\right\}$$

$$=R_{xx}(0)-\sum_{n=-\infty}^{\infty}R_{xx}\left(\frac{n}{2W}-t\right)\frac{\sin 2\pi W(t-n/2W)}{2\pi W(t-n/2W)}$$

$$=0$$

from (1.262) with $\tau = 0$, and $\alpha = t$, we have

$$E\left\{\left[x(t)-x_\infty(t)\right]^2\right\}=E\left\{\left[x(t)-x_\infty(t)\right]x(t)\right\}-E\left\{\left[x(t)-x_\infty(t)\right]x_\infty(t)\right\}$$

and the theorem is proved.*

Similarly the sampling theorems for bandpass, narrowband deterministic signals discussed in Sec. 1.5 can be extended to the random signal once we establish the representation of such signals in terms of the equivalent lowpass inphase and quadrature components or the complex envelope etc. This is what we intend to do in the next section.

1.9.2 Narrowband Random Process

A bandpass process, whose power spectral density bandwidth is much smaller compared to the carrier frequency ω_c about which it is centered, is called a *narrowband process*. (Fig. 1.43(b)). The sample function of such a process may be represented by any of the following these equivalent forms:

$$x(t) = a(t)\cos[\omega_c t + \phi(t)] \tag{1.263}$$

$$= x_c(t)\cos\omega_c t - x_s(t)\sin\omega_c t \tag{1.264}$$

$$= \text{Re}\,[x(t)\exp(j\omega_c t)] \tag{1.265}$$

where $a(t)$, $\phi(t)$, $x_c(t)$, $x_s(t)$ and $\tilde{x}(t)$ have the same interpretation as before, being the envelope, phase, inphase, quadrature components and the complex envelope respectively, of $x(t)$. As before they are all lowpass processes and one can easily establish the following properties about the representation (1.264).

If $x(t)$ is a bandlimited w.s.s. real random process with zero mean and a p.s.d. satisfying (1.257), then

1. $x_c(t)$ and $x_s(t)$ are also zero mean jointly w.s.s. processes
2. var $[x_c(t)]$ = var $[x_s(t)]$ = var $[x(t)]$ (1.266)

* It may be easily shown that sampling theorem also represents the solution to the linear least mean square error estimation problem of estimating $x(t)$ in terms of its periodic samples $x(n/2W)$.

3. $R_{x_cx_c}(\tau)=R_{x_sx_s}(\tau)=\frac{1}{\pi}\int_0^{\infty}S_{xx}(\omega)\cos[(\omega-\omega_c)\tau]d\omega$ (1.267)

4. $R_{x_cx_s}(\tau)=\frac{1}{\pi}\int_0^{\infty}S_{xx}(\omega)\sin[(\omega-\omega_c)\tau]d\omega$ (1.268)

5. $R_{x_sx_c}(\tau)=-R_{x_cx_s}(\tau)=R_{x_cx_s}(-\tau)$ (1.269)

6. $R_{x_cx_s}(0)=R_{x_sx_c}(0)=E[x_c(t)x_s(t)]=0$ (1.270)

7. $R_{xx}(\tau)=R_{x_cx_c}(\tau)\cos\omega_c\tau-R_{x_sx_c}(\tau)\sin\omega_c\tau$ (1.271)

8. $S_{x_cx_c}(\omega)=S_{x_sx_s}(\omega)=L[S_{xx}(\omega-\omega_c)+S_{xx}(\omega+\omega_c)]$ (1.272)

9. $S_{x_c}(\omega)=-S_{x_s}(\omega)=jL[S_{xx}(\omega-\omega_c)-S_{xx}(\omega+\omega_c)]$ (1.273)

Here ω_c is any convenient frequency within the band of $S_{xx}(\omega)$ and L is the bandpass to lowpass operator of (1.84).

Similarly for the representation (1.265), one can prove the following properties. For the complex envelope $\tilde{x}(t)=x_c(t)+jx_s(t)$:

10. $R_{\tilde{x}\tilde{x}}(t)=R^*_{\tilde{x}\tilde{x}}(-\tau)$ (1.274)

11. $R_{\tilde{x}\tilde{x}}(\tau)=2[R_{xx}(\tau)+jR_{x_sx_c}(\tau)]$ (1.275)

12. $R_{xx}(\tau)=\text{Re}[R_{\tilde{x}\tilde{x}}(\tau)\exp(j\omega_c t)]$ (1.276)

13. $S_{xx}(\omega)=\frac{1}{2}[S_{\tilde{x}\tilde{x}}(\omega-\omega_c)+S_{\tilde{x}\tilde{x}}(-\omega-\omega_c)]$ (1.277)

From property 5 we conclude that $R_{x_cx_s}(\tau)$ is an odd function of τ and therefore $R_{x_cx_s}(o)=0$ (property 6). This means that the inphase and quadrature components are uncorrelated for $\tau=0$. For these processes to be completely uncorrelated, we need $R_{x_cx_s}(\tau)=0$. If this were so then, $R_{\tilde{x}\tilde{x}}(\tau)$ would be real and p.s.d. $S_{xx}(\omega)$ would satisfy

$$S_{\tilde{x}\tilde{x}}(\omega)=S_{\tilde{x}\tilde{x}}(-\omega)$$

and the even symmetry of $S_{\tilde{x}\tilde{x}}(\omega)$ would mean $R_{x_cx_s}(\tau)=0$, i.e., complete uncorrelatedness or orthogonality of the inphase and quadrature components of the bandpass process $x(t)$.

Example 1.13:

Consider the ideal bandpass process whose p.s.d. $S_{xx}(\omega)$ is shown in Fig. 1.44(a). Figure 1.44(b) shows the shifted version $S_{xx}(\omega-\omega_c)$ and $S_{xx}(\omega+\omega_c)$. Figure 1.44(c) shows how $S_{xx}(\omega)$ or $S_{x_sx_c}(\omega)$ is obtained according to property 8 eqn. (1.267). Similarly the cross spectrum $S_{x_cx_s}(\omega)$ is sketched in Fig. 1.44(d) and (e) according to property 9 eqn. (1.273).

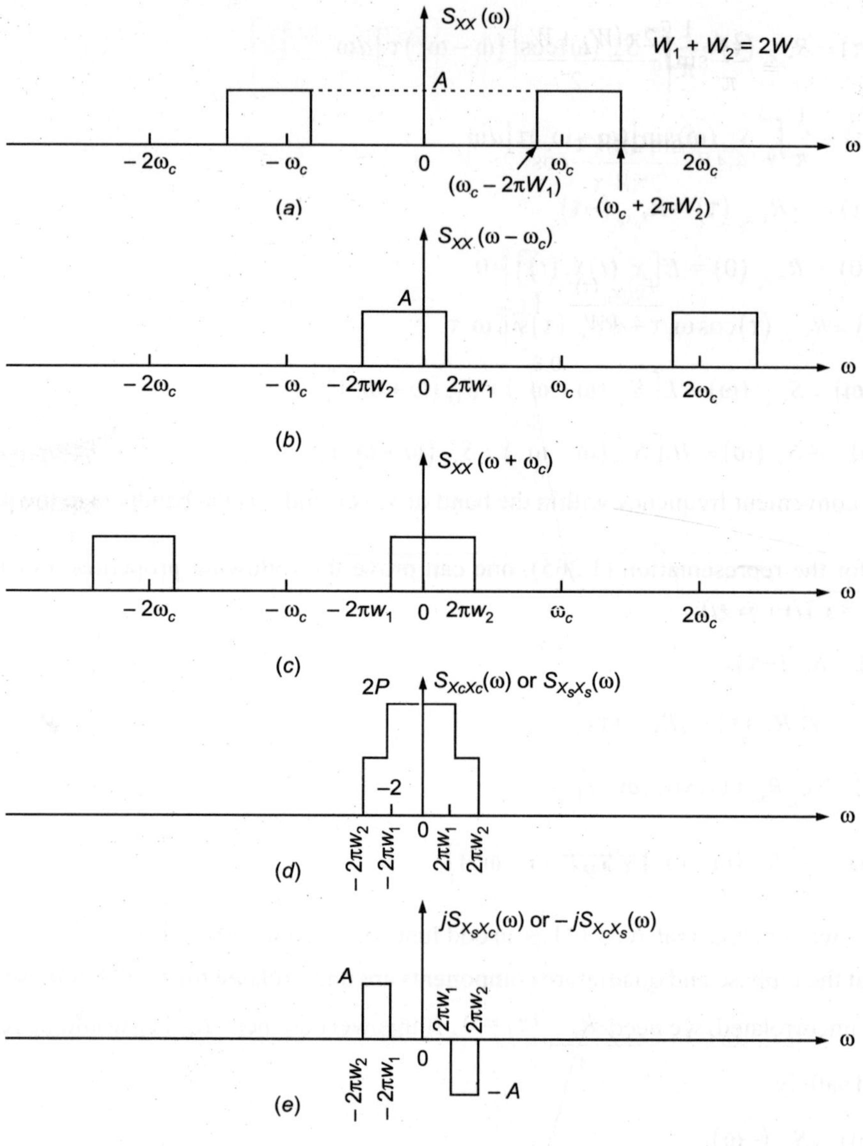

Figure 1.44: Various p.s.d. functions for an ideal bandpass process

Using (1.267), $R_{X_cX_c}(\tau)$ is obtained as

$$R_{X_cX_c}(\tau) = \frac{1}{\pi}\int_{\omega_c-2\pi W_1}^{\omega_c+2\pi W_2} A\cos\left[(\omega-\omega_c)\tau\right]d\omega = \frac{A}{\pi\tau}\int_{-2\pi W_1\tau}^{2\pi W_2\tau}\cos x\,dx$$

$$= \frac{A}{\pi}\left[\sin(2\pi W_2\tau) + \sin(2\pi W_1\tau)\right]$$

$$= \frac{2A}{\pi}\sin\left[\frac{2\pi(W_2+W_1)\tau}{2}\right]\cos\left[\frac{2\pi(W_2-W_1)\tau}{2}\right]$$

$$= 4AW\frac{\sin 2\pi W\tau}{2\pi W\tau}\cos\left[2\pi(W-W_1)\tau\right]$$

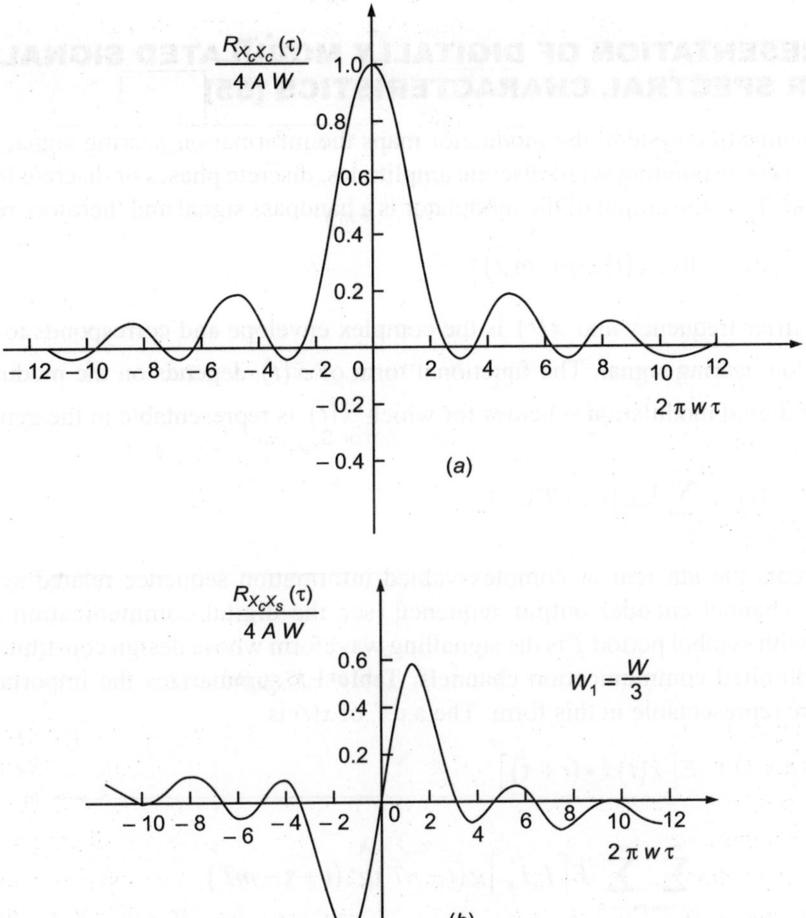

Figure 1.45: ***(a)*** Autocorrelation; ***(b)*** Cross correlation functions of the inphase and quadrature components of the bandpass process of Fig. 1.44

Similarly, using (1.268) we can obtain the cross correlation function

$$R_{x_c x_s}(\tau) = 4AW\frac{\sin 2\pi W\tau}{2\pi W\tau}\sin\left[2\pi(W-W_1)\tau\right]$$

These functions are plotted in Fig. 1.45(a) and (b) for the case $W_1 = W/3$. It is to be noted that for $W_1 = W_2 = W$, i.e., for a bandpass spectrum symmetric about $\pm\omega_c$, we get

$$R_{x_c x_c}(\tau) = R_{x_s x_s}(\tau) = 4AW \cdot \frac{\sin 2\pi W\tau}{2\pi W\tau}$$

and

$$R_{x_c x_c}(\tau) = R_{x_c x_c}(\tau) = 0 \text{ for all } \tau$$

i.e., $x_c(t)$ and $x_s(t)$ are orthogonal processes.

1.10 REPRESENTATION OF DIGITALLY MODULATED SIGNALS AND THEIR SPECTRAL CHARACTERISTICS [55]

In a digital communication system, the modulator maps the information bearing signal, a sequence of binary digits, into a corresponding set of discrete amplitudes, discrete phases or discrete frequency shifts of the carrier signal. Thus the output of the modulator is a bandpass signal and therefore representable as

$$x(t) = \text{Re}\left[\tilde{x}(t)\exp(j\omega_c t)\right] \tag{1.278}$$

where ω_c is the carrier frequency and $\tilde{x}(t)$ is the complex envelope and corresponds to the equivalent lowpass information bearing signal. The functional form of $\tilde{x}(t)$ depends on the modulation scheme. We shall consider digital modulation schemes for which $\tilde{x}(t)$ is representable in the general form

$$\tilde{x}(t) = \sum_{n=0}^{\infty} I_n g(t - nT), \tag{1.279}$$

where I_n's represents the nth real or complex-valued information sequence related symbols usually determined from channel encoder output sequence (see the digital communication system model Fig. 0.2) and $g(t)$ with symbol period T is the signalling waveform whose design constitutes an important problem for bandlimited communication channels. Table 1.1 summarizes the important modulation schemes which are representable in this form. The a.c.f. of $x(t)$ is

$$R(t; t+\tau) = E\left[\tilde{x}(t)\tilde{x}*(t+\tau)\right]$$

$$= \sum_{n=-\infty}^{\infty}\sum_{m=-\infty}^{\infty} E\left[I_n I_m^*\right] g(t-nT) g^*(t+\tau-mT) \tag{1.280}$$

We assume that the sequence I_n is a w.s.s. process with mean μ_i and a.c.f., given by

$$R_I(m) = E\left[I_n I_{n+m}^*\right] \tag{1.281}$$

Therefore

$$R_{\tilde{x}\tilde{x}}(t; t+\tau) = \sum_{n=-\infty}^{\infty}\sum_{m=-\infty}^{\infty} R_I(m-n) g(t-nT) g^*(t+\tau-mT)$$

$$= \sum_{m=-\infty}^{\infty} R_I(m) \sum_{n=-\infty}^{\infty} g(t-nT) g^*(t+\tau-nT-mT) \tag{1.282}$$

Table 1.1: Representation of digitally modulated signals

Here a_n is the M-ary sequence at the channel encoder output and $a_n \in \{0, 1, ..., M-1\}$ unless otherwise stated. $u_0(t)$ is the basic baseband signalling waveform of duration T. Ideally $u_0(t)$ is a rectangular pulse, but for bandwidth conservation, and ISI reduction, $u_0(t)$ is shaped by an appropriate lowpass filter (For details of modulation formats see Chapter 3)

Modulation	I_n	$g(t)$	Remarks
PAM/DSB-SC	a_n	$u_0(t)$	For binary case with $a_n = \pm 1$, this represents BPSK
PAM/SSB	a_a	$u_0(t) \pm j\hat{u}_0(t)$	$\hat{u}_0(t)$ is the Hilbert transform of $u_0(t)$
QAM	$a_n + jb_n$	$u_0(t)$	a_n and b_n correspond to two separate sequences obtained from channel encoder output
MPSK	$\exp(j\theta_n)$	$u_0(t)$	$\{a_n\}$ is mapped to $\{\theta_n\}$ where $\theta_n \in \{2\pi m/M\}$ $m = 0, 1, ..., M-1$
DPSK	$\exp(j\theta_n)$	$u_0(t)$	$I_n = a_n \cdot a_{n-1}$ is now mapped to $\theta_n = \pm \pi$
MFSK	$\exp\{ja_n\omega t\}$	$u_0(t)$	$a_n \in \{0, \pm 1, \pm 2, ..., \pm M-1)\}$
CPFSK	$\exp\{j\omega_d[d_nT + (t-nT)a_n]\}$	$u_0(t)$	$d_n = \sum_{k=0}^{n-1} a_n$

In the above expression, the inner summation is periodic in t with period T. Hence $R_{\tilde{x}\tilde{x}}(t, t+\tau)$ is also periodic in t with the same period T. Also, the mean of $\tilde{x}(t)$

$$E[\tilde{x}(t)] = \mu_i \sum_{n=-\infty}^{\infty} g(t-nT) \tag{1.283}$$

is periodic with period T. A process such as $\tilde{x}(t)$ whose mean and a.c.f. is periodic is called a *cyclostationary process*.

To compute the p.s.d. of such a process, we will have to average $R_{\tilde{x}\tilde{x}}(t, t+\tau)$, over a period to eliminate this dependence. thus

$$R_{\tilde{x}\tilde{x}}(\tau) = \frac{1}{T}\int_{-T/2}^{T/2} R_u(t, t+\tau)\,dt$$

$$= \sum_{m=-\infty}^{\infty} R_I(m) \sum_{n=-\infty}^{\infty} \frac{1}{T}\int_{-T/2}^{T/2} g(t-nT)\,g^*(t+\tau-nT-mT)\,dt$$

$$= \sum_{m=-\infty}^{\infty} R_I(m) \sum_{n=-\infty}^{\infty} \int_{-T/2-nT}^{T/2-nT} g(t)\,g^*(t+\tau-mT)\,dt$$

or,

$$R_{\tilde{x}\tilde{x}}(\tau) = \frac{1}{T}\sum_{m=-\infty}^{\infty} R_I(m) R_g(\tau-mT)$$

$$= \frac{1}{T}\cdot R_g(\tau) * \sum_{m=-\infty}^{\infty} R_I(m)\cdot\delta(t-mT) \tag{1.284}$$

where $R_g(\tau)$ is time a.c.f. of the deterministic signal $g(t)$, i.e.,

$$R_g(\tau) = \int_{-\infty}^{\infty} g(t)g^*(t+\tau)dt \tag{1.285}$$

Taking the Fourier transform of (1.284) yields

$$S_{\tilde{x}\tilde{x}}(\omega) = \frac{1}{T}S_I(\omega)\left|G(\omega)\right|^2 \tag{1.286}$$

where $S_I(\omega)$ is the p.s.d. of the sequence I_n, defined as

$$S_I(\omega) = \sum_{m=-\infty}^{\infty} R_I(m)e^{-j\omega mT} \tag{1.287}$$

Clearly $S_I(\omega)$ is periodic in ω with period $\frac{2\pi}{T}$, and $R_I(m)$ represents the Fourier coefficients in the Fourier series representation of $S_I(\omega)$ with

$$R_I(m) = \frac{T}{2\pi}\int_{-\pi/T}^{\pi/T} S_I(\omega)e^{j\omega mT}d\omega \tag{1.288}$$

Equation (1.286) dictates that the p.s.d. of complex envelope $\tilde{x}(t)$ corresponding to the bandpass signal $x(t)$ is the product of the p.s.d. of the information related sequence I_n and the energy density spectrum of the signalling pulse $g(t)$. Thus the overall spectral characteristic of $\tilde{x}(t)$ can be tailored by designing an appropriate $g(t)$ and the correlation characteristics of the sequence I_n.

To understand the dependence of p.s.d. of $\tilde{x}(t)$ on the correlation properties of the sequence I_n, let us consider the case of real-valued sequence with uncorrelated terms. For such a sequence,

$$R_I(m) = \begin{cases} \sigma_I^2 + \mu_I^2 & m = 0 \\ \mu_I^2 & m \neq 0 \end{cases} \tag{1.289}$$

where σ_I^2 is the variance of the w.s.s. sequence I_n. The corresponding power spectrum, therefore, is obtained as

$$S_I(\omega) = \sigma_I^2 \sum_{m=-\infty}^{\infty} e^{-j\omega mT}$$

$$= \sigma_I^2 + \frac{2\pi}{T}\mu_I^2 \sum_{m=-\infty}^{\infty} \delta\left(\omega - \frac{2\pi m}{T}\right) \tag{1.290}$$

Thus, for the uncorrelated I_n sequence, we obtain

$$S_z(\omega) = \frac{\sigma_I^2}{T}|G(\omega)|^2 + \frac{2\pi\mu_I^2}{T^2}\sum_{m=-\infty}^{\infty}\left|G\left(\frac{2\pi m}{T}\right)\right|^2 \delta\left(\omega - \frac{2\pi m}{T}\right) \tag{1.291}$$

It is to be seen from (1.291) that the spectrum of the digitally modulated signal with uncorrelated information symbols consists of a continuous part, depending on the signalling pulse $g(t)$ and a periodic discrete part, each spectral line at m/T Hz having power proportional to $|G(m/T)|^2$. It is also to be noted that the discrete components vanish for a zero mean sequence. This is a desirable situation for digital modulation schemes and is achieved for equally likely and symmetrically placed signal sets in the signal space.

Example 1.14: For the information bearing sequence satisfying (1.289), let the signalling pulse $g(t)$ be an ideal rectangular pulse of duration τ and amplitude A shown in Fig. 1.46(a). Then

$$|G(\omega)|^2 = (A\tau)^2\left\{\frac{\sin(\omega\tau/2)}{\omega\tau/2}\right\}^2 \tag{1.292}$$

and

$$S_{\tilde{x}\tilde{x}}(\omega) = \sigma_I^2/T\cdot(A\tau)^2 \sin c^2\frac{\omega\tau}{2} + 2\pi\mu_I^2 A^2(\tau/T)^2$$

$$\sum_{m=-\infty}^{\infty}\sin c^2\left(\pi m\frac{\tau}{T}\right)\delta\left(\omega - \frac{2\pi m}{T}\right) \tag{1.293}$$

Case 1: $\tau/T < 1$ (no overlapping). The spectrum for this case is illustrated in Fig. 1.46(b) and consists of a sum of a continuous and discrete part.

Case 2: $\tau/T = 1$.

In this case, except, for $m = 0$, all other terms in the summation are zero as the discrete components coincide in position with the spectral nulls of $|G(\omega)|^2$. Therefore,

$$S_{\tilde{x}\tilde{x}}(\omega) = A^2\sigma_I^2 T \sin c^2\frac{\omega T}{2} + 2\pi\mu_I^2 A^2\cdot\delta(\omega) \tag{1.294}$$

The spectrum is shown in Fig. 1.46(c).

Example 1.15: Let the sequence I_n be such that successive symbols are correlated with a.c.f. given by

$$R_I(m) = E\left[I_j\cdot I_{j+m}\right] = \begin{cases}\sigma_I^2 & m=0\\ \rho\sigma_I^2 & m=\pm 1\\ 0 & \text{otherwise}\end{cases} \tag{1.295}$$

and zero-mean.

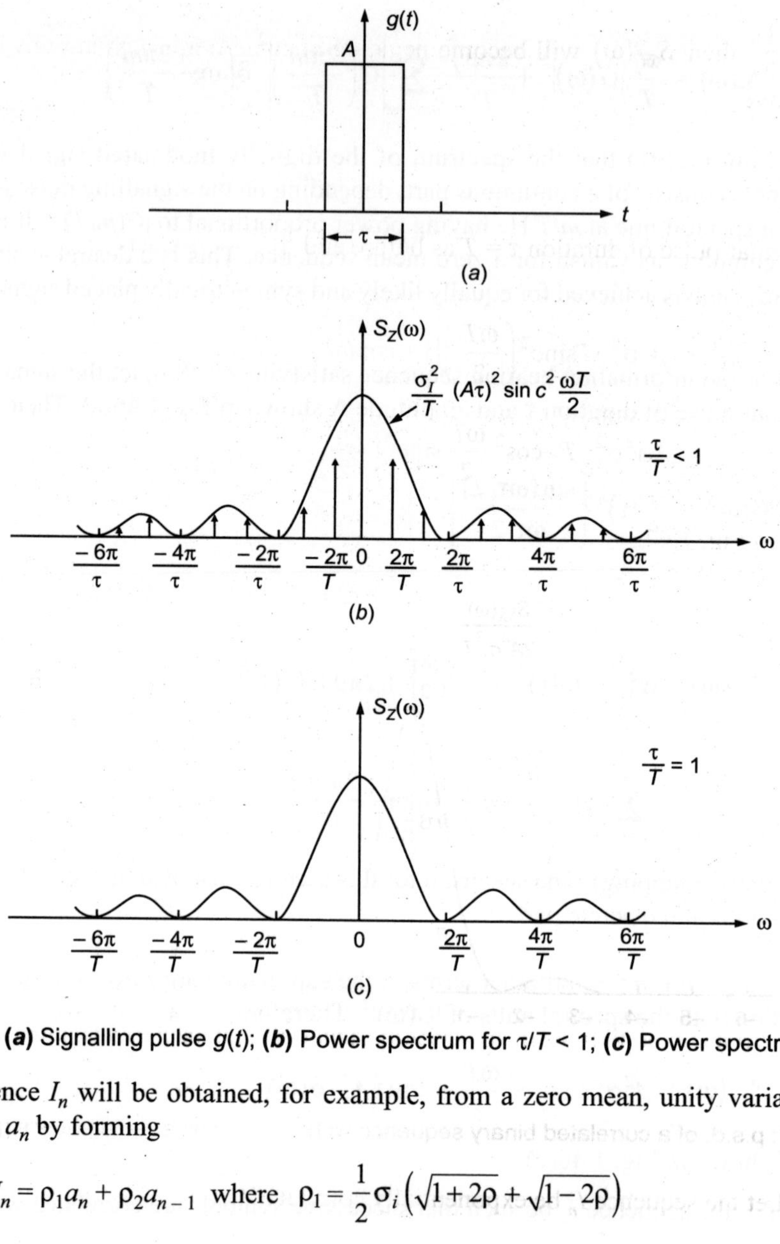

Figure 1.46: (*a*) Signalling pulse $g(t)$; (*b*) Power spectrum for $\tau/T < 1$; (*c*) Power spectrum for $\tau/T = 1$

Such a sequence I_n will be obtained, for example, from a zero mean, unity variance uncorrelated binary sequence a_n by forming

$$I_n = \rho_1 a_n + \rho_2 a_{n-1} \text{ where } \rho_1 = \frac{1}{2}\sigma_i\left(\sqrt{1+2\rho} + \sqrt{1-2\rho}\right)$$

$$\text{and } \rho_2 = \frac{1}{2}\sigma_i\left[\left(\sqrt{1+2\rho} + \sqrt{1-2\rho}\right)\right] \tag{1.296}$$

From (1.286) and (1.287) we get

$$S_{\tilde{x}\tilde{x}}(\omega) = \frac{1}{T}\left|G(\omega)\right|^2 \cdot \sigma_I^2\left[1 + \rho e^{-j\omega T} + \rho e^{j\omega T}\right]$$

$$= \frac{1}{T}\left|G(\omega)\right|^2 \cdot \sigma_I^2\left[1 + 2\rho\cos\omega T\right] \tag{1.297}$$

Note that if $|\rho| > \frac{1}{2}$, then $S_{\tilde{x}\tilde{x}}(\omega)$ will become negative at some frequency and this is not allowable. Hence we must have

$$|\rho| \leq \frac{1}{2}$$

For $g(t)$, a rectangular pulse of duration $\tau = T$ as before and $|\rho| = \frac{1}{2}$, we get

$$\begin{aligned} S_{\tilde{x}\tilde{x}}(\omega) &= A^2\sigma_I^2 \cdot T\text{sinc}^2\left(\frac{\omega T}{2}\right)[1+\cos\omega T] \\ &= 2A^2\sigma_I^2 \cdot T \cdot \cos^2\frac{\omega T}{2}\text{sinc}^2\left(\frac{\omega T}{2}\right) \end{aligned} \tag{1.298}$$

This is shown in Fig. 1.47.

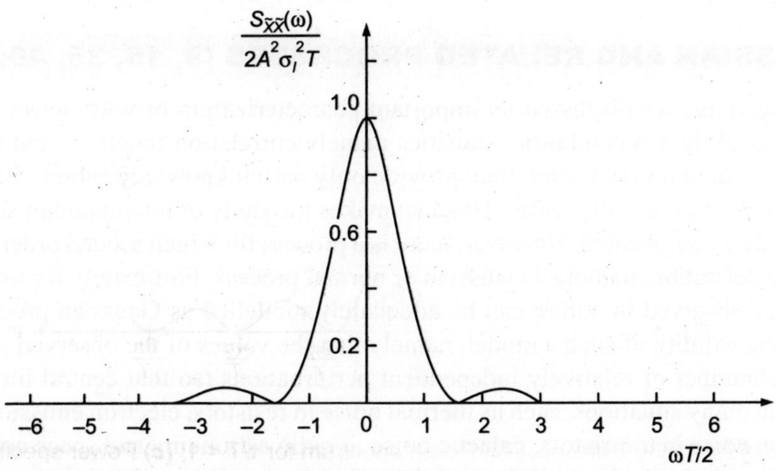

Figure 1.47: p.s.d. of a correlated binary sequence with $\rho = \frac{1}{2}$ and a rectangular signalling pulse

Example 1.16: Let the sequence I_n be exponentially correlated i.e.

$$R_I(m) = \sigma_I^2\rho^{|m|}, \quad |\rho| \leq 1 \tag{1.299}$$

Again from (1.286) and (1.287), we obtain

$$S_{\tilde{x}\tilde{x}}(\omega) = \frac{1}{T}|G(\omega)|^2\,\sigma_I^2 \sum_{m=-\infty}^{\infty} \rho^{|m|}e^{-j\omega mT}$$

$$= \frac{1}{T}|G(\omega)|^2 \cdot \sigma_I^2 \left\{ 2\,\text{Re}\left[\sum_{m=0}^{\infty} (\rho e^{j\omega m T}) \right] - 1 \right\}$$

$$= \frac{1}{T}|G(\omega)|^2\, \sigma_I^2 \left\{ 2\,\text{Re}\left[\frac{1}{1-\rho e^{j\omega T}} \right] - 1 \right\}$$

$$= \frac{1}{T}|G(\omega)|^2\, \sigma_I^2 \left\{ 2\,\text{Re}\left[\frac{1-\cos\omega T + j\rho\sin\omega T}{1+\rho^2 - 2\rho\cos\omega T} \right] - 1 \right\}$$

$$= \frac{1}{T}|G(\omega)|^2\, \sigma_I^2 \frac{1-\rho^2}{1+\rho^2 - 2\rho\cos\omega T}, \quad \rho < 1 \tag{1.300}$$

For ρ very small, it is seen that $S_{\tilde{x}\tilde{x}}(\omega)$ is primarily governed by the spectrum of the signalling pulse except for a small wiggle, which is absent when $\rho = 0$. For ρ close to 1, the term due to the information sequence has a uniformly small value except for values of ω for which $\cos\omega T \simeq 1$.

1.11 GAUSSIAN AND RELATED PROCESSES [9, 15, 35, 49, 50]

In the last few sections, we discussed an important characterization of wide sense stationary random processes in terms of their second order statistics namely correlation functions and the power spectral density functions. Though very useful, they provide only partial knowledge about the process and their complete characterization still eludes us. This fact makes the study of interaction of such processes with systems, enormously complicated. However, there is a process for which second order statistics provides a complete characterisation, namely a Gaussian or normal process. Fortunately for us a large number of random processes observed in nature can be adequately modelled as Gaussian processes. The crucial assumption for the validity of such a model, namely that the values of the observed samples depend on the sum of large number of relatively independent perturbations (so that central limit theorem can be invoked) is met in many situations, such as thermal noise in resistors, electron emission in the thermonic devices, diffusion noise in transistors, galactic noise in radio astronomy and space communications etc. This fact together with the mathematical tractibility of such processes and their interaction with systems, makes their study of special concern to communication systems. In the following sections, we define the Gaussian process and then study some of its important properties and their application to the interaction of such processes with linear and simple nonlinear systems.

Definition and Specification of Gaussian Processes

We have seen earlier that specification of an arbitrary random process $x(t)$ implies stating a rule for determining the joint density function of the samples taken at any finite set of time instant t_i; $i = 1, 2, \ldots n$, i.e., $p_{\mathbf{x}(t)}$ where $\mathbf{x}(t)$ denotes the random vector defined in (1.155) i.e.

$$\mathbf{x}(t) \triangleq [x(t_1), x(t_2) \ldots x(t_n)] = [x_1, x_2, \ldots, x_n]$$

We say that a set of n random variables $x_1, x_2, \ldots, x_n$ are *jointly normally distributed* if their joint characteristic function is given by

$$\phi_{\mathbf{x}}(\mathbf{u}) = \exp\left[j\sum_{i=1}^{n} u_i m_i - \frac{1}{2}\sum_{i=1}^{n}\sum_{k=1}^{n} u_k K_{ki} u_i \right] \tag{1.301}$$

$$= \exp\left[j(\mathbf{u}' \cdot \mathbf{m}) - \frac{1}{2}(\mathbf{u}' K \mathbf{u}) \right]$$

where

$\mathbf{u} \triangleq [u_1, u_2, \ldots, u_n]'$; $u_1, u_2, \ldots, u_n$ are any real numbers

$\mathbf{m} \triangleq [m_1, m_2, \ldots, m_n]'$

$m_i \triangleq E[x_i]$

$K_{ik} \triangleq \text{cov}\,[x_i, x_k] = E[(x_i - m_i)(x_k - m_k)]$

$$K = \begin{bmatrix} K_{11} & K_{12} & K_{1n} \\ K_{21} & K_{22} & K_{2n} \\ \vdots & \vdots & \vdots \\ K_{n1} & K_{n2} & K_{nn} \end{bmatrix} \tag{1.302}$$

If K^{-1} exists and K^{ik} denotes the (i, k)th element of K^{-1}, then it can be shown [57] that $x_1, \ldots, x_n$ possess a joint probability density function, called *multivariate Gaussian density function* denoted by $N(\mathbf{m}, K)$ and given by

$$p_{\mathbf{x}}(\alpha) = \frac{1}{(2\pi)^{n/2} \, |K|^{1/2}} \exp\left[-\frac{1}{2}(\boldsymbol{\alpha} - \mathbf{m})' K^{-1}(\boldsymbol{\alpha} - \mathbf{m})\right]$$

$$= \frac{1}{(2\pi)^{n/2} \, |K|^{1/2}} \exp\left[-\frac{1}{2}\sum_{i-1}^{n}\sum_{k=1}^{n} (\alpha_i - m_i) K^{ik} (\alpha_k - m_k) \right] \tag{1.303}$$

$\boldsymbol{\alpha} = [\alpha_1, \alpha_2, \ldots, \alpha_n]$, $\alpha_1, \alpha_2 \ldots, \alpha_n$ are any real numbers.

1.11.1 Properties of Gaussian Random Variables and Gaussian Processes

1. If $x_1, x_2, \ldots, x_n$ are jointly Gaussain, then every linear functional of $x_1, x_2, \ldots, x_n$ is also a Gaussian random variable i.e.

   $y = \sum_{i=1}^{n} c_i x_i$ is Gaussian for all c_i, provided $E[y^2] < \infty$ (The converse is also true).

2. Any linear transformation of a Gaussian random vector is also a Gaussian vector. Thus, if $\mathbf{x}$ is an n-dimensional Gaussian vector, and L is any $k \times n$ matrix, then $\mathbf{y} = L\mathbf{x}$ is also a k-dimensional Gaussian vector. If $\mathbf{x}$ is a zero mean i.e. $\mathbf{m} = 0$ and its covariance matrix is K, then the n-dimensional joint probability density function of $\mathbf{y} = L\mathbf{x}$, L nonsingular, is given by

   $$p_{\mathbf{y}}(\boldsymbol{\alpha}) = \frac{1}{(2\pi)^{n/2} \, |F|^{1/2}} (-\frac{1}{2}\boldsymbol{\alpha}' F^{-1}\boldsymbol{\alpha}) \tag{1.304}$$

   where $F = LKL'$ and $|F| = |L| \cdot |K| \cdot |L'| = |L|^2 \, |K|$

3. The sum of dependent jointly Gaussian r.v.s. is also Gaussian.
4. Gaussian random variables which are mutually uncorrelated, are also statistically independent i.e. $E[x_i, x_j] = 0$ for all $i, j = 1, \ldots n$ for Gaussian random variables, implies that $x_1, \ldots, x_n$ are statistically independent.
5. For a set of jointly distributed Gaussian r.v.'s with a positive definite covariance matrix K, a linear transformation can be found which transforms the dependent variable into a set of statistically independent variables. This is a consequence of a standard result from matrix theory [20] and concerns the characterization of positive definite matrices. The transformation that diagonalizes K is obtained as solution of eigen-vector equation

$$K\mathbf{x}^{(i)} = \lambda_i \mathbf{x}^{(i)},\ i = 1, \ldots, n \tag{1.305}$$

where λ_i are the eigenvalues of the covariance matrix or equivalently the roots of the characteristic equation

$$|K - \lambda I| = 0 \tag{1.306}$$

For distinct eigenvalues, the eigenvectors are orthogonal and may be made orthonormal by scaling. The diagonalizing transformation A is then given by

$$A = \begin{bmatrix} x_1^{(1)} & x_1^{(2)} & \cdots & x_1^{(n)} \\ x_2^{(1)} & x_2^{(2)} & \cdots & x_2^{(n)} \\ \vdots & & & \\ x_n^{(1)} & x_n^{(2)} & \cdots & x_n^{(n)} \end{bmatrix} \tag{1.307}$$

i.e.

$$A' KA = \lambda \tag{1.308}$$

with

$$\lambda = \begin{bmatrix} \lambda_1 & 0 & \cdots & 0 \\ 0 & \lambda_2 & & 0 \\ \vdots & & & \\ 0 & 0 & \cdots & \lambda_n \end{bmatrix}$$

6. For zero-mean Gaussian r.v.'s, $x_1, \ldots, x_n$, the mixed central moments $E\{x_1^{b_1}, x_2^{b_2} \ldots, x_n^{b_n}\}$ where b_i are positive integers, is given by

$$E\{x_1^{b_1}, x_2^{b_2} \ldots, x_n^{b_n}\} = \frac{1}{2^{B/2}(B/2)!} \frac{\partial^B}{\partial u_1^{b_1} \ldots, \partial u_n^{b_n}} \left[\sum_{i=1}^{n}\sum_{k=1}^{n} u_k K_{ki} u_i\right]^{B/2} \Bigg|\ \text{all } u_i = 0 \tag{1.309}$$

where

$$B = \sum_{i=1}^{n} b_i \tag{1.309a}$$

For B odd, the mixed moments are zero. For the important case of $n = 4$, $b_i = 1$,

$$E[x_1 x_2 x_3 x_4] = E[x_1 x_2]\, E[x_3 x_4] + E[x_1 x_3]\, E[x_2 x_4] + E[x_1 x_4]\, E[x_2 x_3] \tag{1.310}$$

7. Let z_n be a sequence of Gaussian distributed random variables, which converge in mean square sense to a random variable z. Then z is also Gaussian distributed.

Definition of a Gaussian Process

A random process $\{x(t), t \in T\}$ is called a *Gaussian* process if for any n and any subset $\{t_1, t_2, \ldots, t_n\}$ of T, $\mathbf{x}(t) = [x(t_1), \ldots, x(t_n)]$ is a Gaussian random vector i.e. $x(t_1), \ldots, x(t_n)$ are jointly Gaussian distributed with characteristic function of eq. (1.301) or p.d.f. of eq. (1.303).

It is obvious from above that for a Gaussian process, the knowledge of the complete probability law is available from the knowledge of mean function $E[x(t)]$ and the covariance function cov $[x(\tau), x(t)]$.

Properties of the Gaussian Process

These follow from the corresponding properties of the jointly distributed Gaussain random variables:

1. The output of a stable linear system excited by a Gaussian ramdom process is also a Gaussian process.
2. Two different linear functionals of a Gaussian process give rise to a pair of jointly Gaussian random variables. Thus if $x(t)$ is Gaussian, then

$$y_1 = \int_0^T g_1(t) x(t)\, dt$$

and

$$y_2 = \int_0^T g_2(t) x(t)\, dt$$

are jointly Gaussian.

3. Linear transformations of the same Gaussian process yield jointly distributed Gaussian processes.
4. If a Gaussian random process is wide-sense stationary, then it is also strict-sense stationary.

1.11.2 Bandpass Gaussian Process [58]

We recall from Section 1.9.2 that a bandpass random process $x(t)$ can be expressed as

$$x(t) = x_c(t) \cos \omega_c t - x_s(t) \sin \omega_c t$$

where $x_c(t)$ and $x_s(t)$ are lowpass inphase and quadrature processes respectively. Alternatively we can represent a bandpass process by a vector process

$$\mathbf{x}(t) = [x_c(t)\; x_s(t)]'$$

Another representation is in terms of lowpass complex envelope process

$$\tilde{x}(t) = x_c(t) + j x_s(t) \tag{1.311}$$

The complex process $\tilde{x}(t)$ is said to be Gaussian if its real and imaginary parts are jointly Gaussian random processes.

Employing the analytic signal representation for the bandpass process $x(t)$, i.e.

$$z_x(t) = x(t) + j\hat{x}(t) \tag{1.312}$$

We see that if $x(t)$ is Gaussian, then its Hilbert transform $\hat{x}(t)$, being a linear transformation of $x(t)$, is also Gaussian. Moreover, from property 4 above, $x(t)$ and $\hat{x}(t)$ are jointly Gaussian. Similarly $x_c(t)$ and $x_s(t)$ are jointly Gaussian. This means that $x(t)$ and $z_x(t)$ are also Gaussian processes. Furthermore if $x(t)$ is stationary zero-mean Gaussian process (SZMGP) then

$$R_{x_c x_s}(0) = R_{x_s x_c}(0) = (0)$$

and hence $x_c(t)$ and $x_s(t)$ are uncorrelated and therefore statistically independent Gaussian random variables at the same time instant. They also have the same variance σ_x^2. Therefore, the joint p.d.f. at any instant $t = t_1$ is

$$p_{x_c x_s}[x_c(t_1), x_s(t_1)] = \frac{1}{2\pi\sigma_x^2}\exp\left\{-\frac{x_c^2(t_1) + x_s^2(t_1)}{2\sigma_x^2}\right\} \tag{1.313}$$

Since $|x(t)|^2 = |z_x(t)|^2 = x_c^2(t) + x_s^2(t)$ we may also write (1.313) as

$$p_x[x(t_1)] = \frac{1}{2\pi\sigma_x^2}\exp\left[-\frac{|x(t_1)|^2}{2\sigma_x^2}\right] \tag{1.314}$$

$$pz_x[z_x(t_1)] = \frac{1}{2\pi\sigma_x^2}\exp\left[-\frac{|z_x(t_1)|^2}{2\sigma_x^2}\right] \tag{1.315}$$

Thus eqs. (1.314), (1.315) become the p.d.f. of an instantaneous value of the envelope of the bandpass random process $x(t)$.

If the bandpass process $x(t)$ is bandlimited with bandwidth $2W$, then its complex envelope is a lowpass process with bandwidth W and applying the sampling theorem, we can write

$$\tilde{x}(t) = \sum_{b=-\infty}^{\infty} \tilde{x}\left(\frac{n}{2W}\right)\text{sinc}\,(2Wt - n) \tag{1.316}$$

For a finite interval $(0, T)$ of the signal,

$$x(t) = \sum_{n=1}^{2WT} \tilde{x}\left(\frac{n}{2W}\right)\text{sinc}\,(2Wt - n) \tag{1.317}$$

and setting $\tilde{x}\left(\frac{n}{2W}\right) = \tilde{x}_n;\ n = 1, 2, \ldots, 2WT,\ \tilde{x}(t)$ is represented as a complex column vector $\tilde{\mathbf{x}}$.

$$\tilde{\mathbf{x}} = [\tilde{x}_1, \tilde{x}_2, \ldots, \tilde{x}_{2TW}]' \tag{1.318}$$

Let the corresponding mean value vector be $\tilde{\mathbf{m}}$ i.e.

$$\tilde{\mathbf{m}} = [\tilde{m}_1, \tilde{m}_2, \ldots, \tilde{m}_{2TW}]' \tag{1.319}$$

Since each of the elements of (1.318) and (1.319) are complex, an alternative way of writing is

$$\tilde{\mathbf{x}} = [\mathbf{x}_c \quad \mathbf{x}_s]' = [x_{c1}, x_{c2}, \ldots, x_{cN} \quad x_{s1}, x_{s2}, \ldots, x_{sN}]'$$

where

$$x_{ci} = x_c(t_i)$$

$$x_{si} = x_s(t_i)$$

and

$$N = 2TW \tag{1.320}$$

Similarly,

$$\tilde{\mathbf{m}} = [\mathbf{m}_c \quad \mathbf{m}_s]' = [m_{c1}, \ldots, m_{cN} \quad m_{s1}, \ldots, m_{sN}]' \tag{1.321}$$

Let us consider the case of zero-mean bandpass process with a flat p.s.d.of height $N_0/2$ and band-limited to $2W$. Let us choose ω_c to be at the centre of the band so that p.s.d. is symmetrical about ω_c. As seen in Sec. 1.9.2, this choice makes $x_c(t)$ and $x_s(t)$ uncorrelated random processes. Since they are jointly Gaussian, they are also statistically independent. This makes the components x_{ci} and x_{si} as statistically independent. Since the common variance is $2N_0W$, the joint p.d.f. can be written in following equivalent forms as

$$\begin{aligned} p_{\tilde{x}}[\tilde{x}(t)] = p_{\tilde{x}}(\tilde{\mathbf{x}}) &= \frac{1}{(4N_0W)^{2WT}} \exp\left[-\frac{1}{4N_0W}\sum_{i=1}^{2TW}\left(x_{ci}^2 + x_{si}^2\right)\right] \\ &= C_0 \exp\left[-\frac{1}{4N_0W}\sum_{i=1}^{2TW}|\tilde{x}_i|^2\right] \\ &= C_0 \exp\left[-\frac{1}{4N_0W}\tilde{\mathbf{x}} * \tilde{\mathbf{x}}\right] \\ &= C_0 \exp\left[-\frac{1}{4N_0W}\|\tilde{\mathbf{x}}\|^2\right] \\ &= C_0 \exp\left[-\frac{1}{2N_0}\int_0^T \tilde{x}^2(t)\,dt\right] \end{aligned} \tag{1.322}$$

When the p.s.d. is not flat, the inphase and the quadrature components are not statistically independent. Denoting the corresponding covariance matrices as K_c and K_s and cross-covariance matrices as K_{cs} and K_{sc} it may be easily seen that

$$K_c = K_s$$

$$K_{sc} = -K_{cs} = K'_{cs} \tag{1.323}$$

and the overall covariance matrix K_y may be written in the partitioned form as

$$K_y = \left[\begin{array}{c|c} K_c & K_{cs} \\ \hline -K_{cs} & K_c \end{array}\right] \tag{1.324}$$

The p.d.f of a (0, T) segment of the bandpass Gaussian process may now be written as

$$p_{\tilde{x}}(\tilde{\mathbf{x}}) = \frac{1}{\left[(2\pi)^{2N}|K|\right]^{1/2}} \exp\left[-\frac{1}{2}(y-m_y)'K_y^{-1}(y-m_y)\right] \tag{1.325}$$

where $y \triangleq [y_1, y_2, ..., y_{4TW}]'$ is obtained from $\tilde{x}$ by renumbering the components as follows

$$\left.\begin{aligned} y_i &= x_{ci} \\ y_{i+2TW} &= x_{si} \end{aligned}\right\}; \quad 1 \le i \ge 2TW \tag{1.326}$$

1.11.3 Fourier Series Representation

In Section 1.13 on harmonic analysis of random process, we have discussed the Fourier series representation of a random process, namely

$$x(t) = \frac{a_{c0}}{2} + \sum_{k=1}^{\infty} a_{ck} \cos k\omega_0 t + a_{sk} \sin k\omega_0 t$$

where

$$a_{ck} = \frac{2}{T}\int_0^T x(t) \cos k\omega_0 t \, dt$$

$$a_{sk} = \frac{2}{T}\int_0^T x(t) \sin k\omega_0 t \, dt \tag{1.327}$$

For a ZMGSP, $a_{c0} = 0$ and since a_{ck} and a_{sk} are linear functionals of $x(t)$, they are Gaussian random variables. Since for large T, they are uncorrelated, hence statistically independent, and have a common variance of $\frac{2}{T}S(kf_0)$. The p.d.f. of the coefficient is given by

$$p_{a_{ck}}(a_{ck}) = \frac{\sqrt{T}}{2\sqrt{\pi S(kf_0)}} \exp\left[-\frac{Ta_{ck}^2}{4S(kf_0)}\right] \tag{1.328}$$

Therefore, the joint p.d.f of the process is obtained as a product and is given by

$$p_x(x) = p_{ac1, \ldots, a_{sN}}(a_{c1}, \ldots, a_{sN})$$

$$\left[\prod_{k=1}^{N} \frac{T}{4\pi S(kf_0)}\right] \exp\left[\left(-\frac{T}{4}\right)\sum_{k=1}^{N}(a_{ck}^2 + a_{sk}^2)/S(kf_0)\right] \tag{1.329}$$

In terms of the Fourier transform $X_T(f)$ of the [0, T] segment of the process $x(t)$,

$$a_{ck}^2 + a_{sk}^2 = \frac{4}{T^2}|X_T(kf_0)|^2 \tag{1.330}$$

For large T, replacing the sum by the integral in (1.329)

$$\sum_{k=1}^{N} \frac{a_{ck}^2 + a_{sk}^2}{S(kf_0)} \rightarrow \frac{4}{T} \int_0^{\infty} \frac{|X_T(f)|^2}{S(f)} \cdot df$$

and, we can write (1.329) in the form

$$p_x(\mathbf{x}) = C \exp\left[-\tfrac{1}{2}\int_{-\infty}^{\infty} \frac{|X_T(f)|^2}{S(f)} \cdot df\right] \tag{1.331}$$

1.11.4 Envelope of a Bandpass Gaussian Process

We have already seen that in the representation

$$x(t) = x_c(t) \cos \omega_c t - x_s(t) \sin \omega_c t$$

of a SZMGP, $x_c(t)$ and $x_s(t)$ are zero-mean jointly Gaussian and for any fixed instant, the random variables x_c and x_s have the joint p.d.f. given by (1.313).

Let $v(t)$ and $\phi(t)$ denote the envelope and phase of the process respectively. Then

$$v(t) = [x_c^2(t) + x_s^2(t)]^{1/2}; \; v(t) \geq 0$$

and

$$\phi(t) = \tan^{-1} \frac{x_s(t)}{x_c(t)} \qquad 0 \leq \phi \leq 2\pi \quad \text{or} \quad -\pi \leq \phi \leq \pi$$

Also,

$$x_c(t) = v(t) \cos \phi(t)$$
$$x_s(t) = v(t) \sin \phi(t)$$

Knowing the joint p.d.f. of x_c and x_s, we obtain the joint p.d.f. of the envelope and phase as

$$p_{v,\phi}(v, \phi) = p_{xc,\,xs}[x_c(v, \phi)]|J| \tag{1.332}$$

where J, the Jacobian of the transformation, is

$$J = \begin{vmatrix} \dfrac{\partial x_c}{\partial v} & \dfrac{\partial x_c}{\partial \phi} \\ \dfrac{\partial x_s}{\partial v} & \dfrac{\partial x_s}{\partial \phi} \end{vmatrix} = \begin{vmatrix} \cos\phi & -v\sin\phi \\ \sin\phi & -v\cos\phi \end{vmatrix} = v$$

Hence

$$p_{v,\phi}(v, \phi) = \frac{v}{2\pi\sigma_x^2} \exp\left[-\frac{v^2}{2\sigma_x^2}\right], \qquad v \geq 0, 0 < \phi < 2\pi \tag{1.333}$$

where

$$\sigma_x^2 = E[x_c^2(t)] = E[x_s^2(t)]$$

The p.d.f. for envelope v above is obtained as

$$p_v(v) = \int_0^{2\pi} p_{v,\phi}(v,\phi)d\phi = \frac{v}{\sigma_x^2}\exp\left[\frac{v^2}{2\sigma_x^2}\right], v \geq 0 \tag{1.334}$$

This is known as *Rayleigh p.d.f.* and is shown in Fig. 1.48 as the curve $u/\sigma_x = 0$. It is also seen from (1.333) and (1.334) that p.d.f of the phase process, at any instant of time, is uniform over $(0, 2\pi)$ and that v and ϕ are statistically independent random variables.

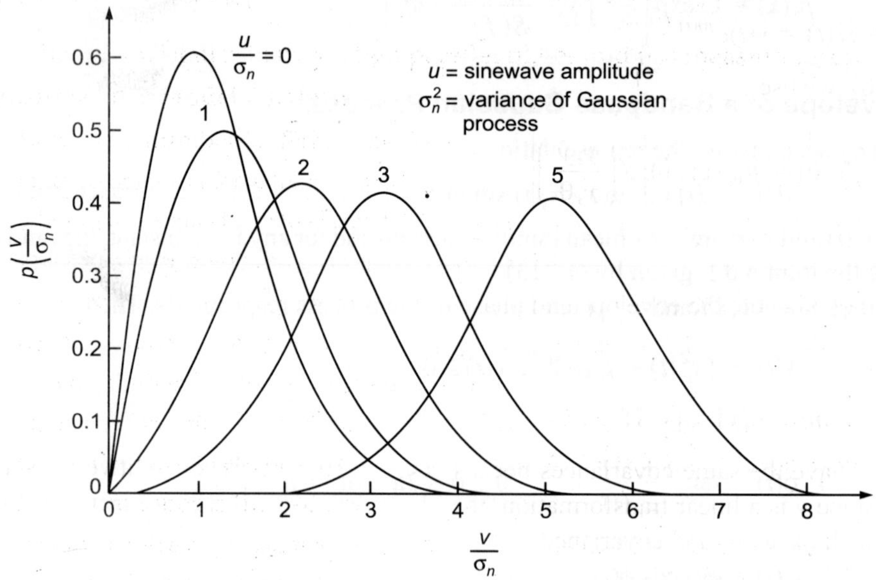

Figure 1.48: p.d.f of envelope v of a sine wave plus a bandpass Gaussian process

The mean and the variance of Rayleigh p.d.f. is

$$E[v(t)] = \sqrt{\pi/2}\cdot\sigma_x$$

$$\text{var}\,[v(t)] = \left(2 - \frac{\pi}{2}\right)\sigma_x^2 \tag{1.335}$$

As we shall see later, we need to know the p.d.f. of the square of the envelope.

Let $w(t) = v^2(t)$, it is easily shown that

$$p_w(w) = \frac{1}{2\sigma_x^2}\left[-\frac{w}{2\sigma_x^2}\right]; \qquad w \geq 0$$

$$E[w(t)] = E[v^2(t)] = 2\sigma_x^2$$

$$\text{var}\,[w(t)] = 4\sigma_x^4 = E^2[w(t)] \tag{1.336}$$

1.11.5 Envelope and Phase Statistics for a Bandpass Deterministic Signal Plus Bandpass Gaussian Process

Let $n(t)$ be ZMSGP, $s(t)$ a known bandpass deterministic function and let

$$y(t) = s(t) + n(t) \tag{1.337}$$

Let the envelope and phase of $s(t)$ be $u(t)$ and $\psi(t)$ and that of $y(t)$ be $v(t)$ and $\theta(t)$ respectively, i.e.,

$$s(t) + j\hat{s}(t) = u(t)e^{j\psi(t)} \tag{1.338}$$

$$y(t) + j\hat{y}(t) = v(t)e^{j\theta(t)} \tag{1.339}$$

To calculate $p(v, \theta)$, we use

$$p_{v,\theta}(v, \theta) = p_{y,\hat{y}}(y, \hat{y})\left| J\left(\frac{y, \hat{y}}{v, \theta}\right)\right| \tag{1.340}$$

Define

$$\mathbf{y} = [y, \hat{y}]' \text{ and } p_{y,\hat{y}}(y, \hat{y}) = p_{\mathbf{y}}(\mathbf{y}) \tag{1.341}$$

Since $E[n(t)] = 0$,

$$m_{\mathbf{y}} = E[\mathbf{y}] = [s \;\; \hat{s}]'$$

Moreover, n and y have the same covariances because they differ only by an additive constant. **y** is a Gaussian vector since **y** is a linear transformation of n. Therefore $p_y(\mathbf{y})$ is a two-dimensional Gaussian density function with mean $m_{\mathbf{y}}$ and covariance matrix

$$K = \begin{bmatrix} \sigma_n^2 & 0 \\ 0 & \sigma_n^2 \end{bmatrix}$$

Since

$$\left| J\left(\frac{y, \hat{y}}{v, \theta}\right)\right| = v$$

from (1.340)

$$p_{v,\theta}(v, \theta) = \frac{v}{2\pi\sigma_n^2} \exp\left\{-\frac{1}{2}[\mathbf{y} - m_{\mathbf{y}}]'K^{-1}[\mathbf{y} - m_{\mathbf{y}}]\right\} \tag{1.342}$$

Substituting (1.341) in (1.342) results in

$$p_{v,\theta}(v, \theta) = \frac{v}{2\pi\sigma_n^2} \exp\left\{-\frac{1}{2\sigma_n^2}(y^2 - 2ns + s^2 + \hat{y}^2 - 2\hat{y}\hat{s} + \hat{s}^2)\right\} \tag{1.343}$$

Alternatively, in terms of u, v, θ and ψ, we can write this as

$$p_{v,\theta}(v, \theta) = \frac{v}{2\pi\sigma_n^2} \exp\left[-\frac{v^2+u^2}{2\sigma_n^2}\right] \exp\left[\frac{uv\cos(\theta-\psi)}{\sigma_n^2}\right] \tag{1.344}$$

and

$$p_v(v) = \frac{1}{2\pi}\int_0^{2\pi} p_{v,\theta}(v,\theta)\, d\theta = \frac{v}{\sigma_n^2}\exp\left[-\frac{v^2+u^2}{2\sigma_n^2}\right] I_0\left(\frac{uv}{\sigma_n^2}\right); u \geq 0 \tag{1.345}$$

where $I_0(q)$ is the modified Bessel function of the first kind and zero order defined as

$$I_0(q) = \frac{1}{2\pi}\int_0^{2\pi} \exp(q\cos\theta)\, d\theta \tag{1.346}$$

Equation (1.345) is known as Rician density function [59].

The p.d.f. of the envelope, $p_v(v)$ is sketched in Fig. 1.48, for various values of u/σ_n. It is seen that when $u \to 0$ the p.d.f. approaches the Rayleigh function (1.334). For small noise i.e. $\frac{u}{\sigma_n} \gg 1$, the p.d.f. approaches the Gaussian function with means u*.

It is to be noted that above derivation does not use a narrowband assumption on $n(t)$. The above results will also be valid for the case of random signal $s(t)$, i.e., for u and ψ, being random, if we substitute $p(v, \theta/u, \psi)$ for $p(v, \theta)$ and $p(v/u, \psi)$ for $p(v)$. From (1.345) it is seen that $p(v/u, \psi)$ is not a function of ψ and we can write the joint p.d.f.

$$p(v, \theta, u, \psi) = p(v, \theta/u, \psi)p(u, \psi)$$

---

*This may be seen from (1.345) also. For $u \to 0$, with $n(t) \gg s(t)$, $I_0\left(\frac{uv}{\sigma_n^2}\right) \to 1$. With $n(t) \ll s(t)$, $\frac{uv}{\sigma_n^2} \gg 1$ or $\frac{u}{\sigma_n} \gg 1$ for $\frac{v}{\sigma_n}$ not too small and noting that

$$\lim_{\alpha\to\infty} I_0(\alpha) \sim \frac{e^\alpha}{\sqrt{2\pi\alpha}}\left[1 + \frac{1}{8\alpha} + 0(\alpha^{-3})\right]$$

$$p(v/u,\psi) \to \frac{v}{v_n^2}\exp\left[-\frac{v^2+u^2}{2\sigma_n^2}\right]\exp\frac{\exp[2uv/2\sigma_n^2]}{\sqrt{2\pi(uv/\sigma_n^2)}}$$

$$= \sqrt{\frac{v}{u}} \cdot \frac{1}{\sqrt{2\pi\sigma_n^2}} \exp\left[-\frac{v^2+u^2}{2\sigma_n^2}\right]$$

since $\sqrt{\frac{v}{u}} \to 1$ as $y \to s$

Hence

$p(v/u,\psi) \sim N(u, \sigma_n^2)$ as $\sqrt{(v/u)} \to 1$ where $N(m, \sigma^2)$ denotes a normal p.d.f. with mean m and variance σ^2.

One minus the cumulative distribution function of the Rician variable is known as the Marcum Q-function [60] given by

$$Q(\alpha, \beta) = 1 - \int_0^{\beta} z \exp\left[-\frac{z^2 + \alpha^2}{2}\right] I_0(\alpha z)\, dz \tag{1.347}$$

$$= \int_{\beta}^{\infty} z \exp\left[-\frac{z^2 + \alpha^2}{2}\right] I_0(\alpha z)\, dz \tag{1.348}$$

where we have used the normalized variables $z = \dfrac{v}{\sigma_n}$ and $\alpha = \dfrac{u}{\sigma_n}$. This function is plotted in Fig. 1.49.

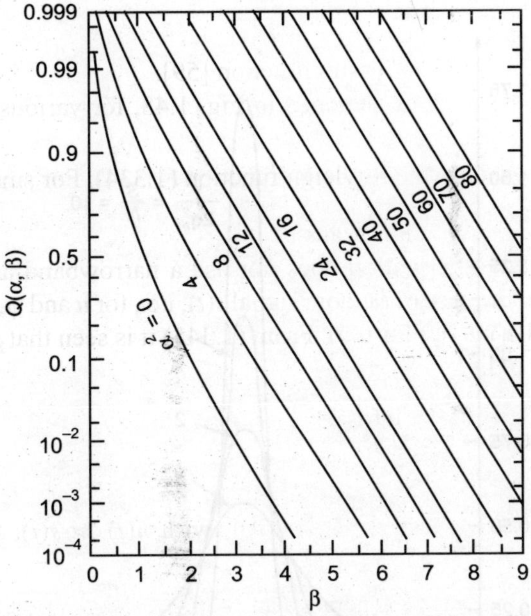

Figure 1.49: Plot of Q-function, given by eq. (1.348) for different values of $\alpha^2 = u^2/\sigma_n^2$

These plots are of great interest in detection problems as discussed in Chapter III. The p.d.f. of the phase process θ conditioned on ψ is obtained as follows. Assuming u to be a known constant,

$$p(\theta/\psi) = \int_0^{\infty} p(v, \theta/\psi)\, dv$$

where $p(v, \theta/\psi)$ is given by (1.344).

Therefore,

$$p(\theta/\psi) = \frac{1}{2\pi\sigma_n^2} \exp\left[-\frac{u^2}{2\sigma_n^2} \sin^2(\psi - \theta)\right]$$

$$\times \int_0^{\infty} v \exp\left\{-\frac{[v - u\cos(\psi - \theta)]^2}{2\sigma_n^2}\right\} dv$$

After some manipulation, it may be shown that

$$p(\theta/\psi) = \frac{\exp(-u^2/2\sigma_n^2)}{2\pi} + \frac{u\cos(\psi-\theta)}{2\sqrt{2\pi\sigma_n^2}}\exp[-u^2/2\sigma_n^2\sin^2(\psi-\theta)] \times\left\{1+\operatorname{erf}\left[\frac{u\cos(\psi-\theta)}{\sqrt{2\sigma_n}}\right]\right\} \quad (1.349)$$

where $\operatorname{erf}(x) = \frac{2}{\sqrt{\pi}}\int_0^x e^{-z^2}\,dz.$

The phase p.d.f. is shown plotted in Fig. 1.50.

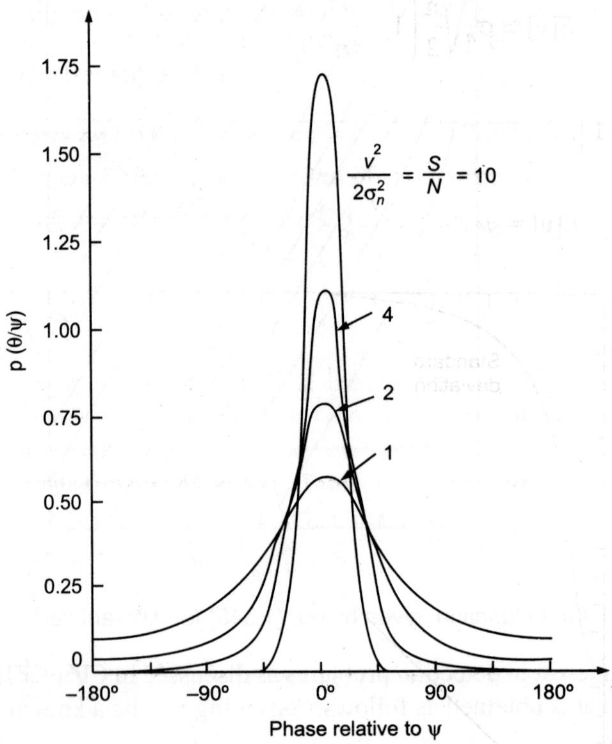

Figure 1.50: p.d.f. of phase of sine wave plus bandpass Gaussian noise

The moments of the envelope of a sine wave plus additive SZMGP are obtained as [59]

$$E[v^k] = \int_0^\infty v^n p_v(v)\,dv$$

$$= \frac{1}{\sigma_n^2}\int_0^\infty v^{n+1}\exp\left[-\frac{v^2+u^2}{2\sigma_n^2}\right] I_0\left(\frac{uv}{\sigma_n^2}\right) dv$$

$$= (2\sigma_n^2)^{k/2}\Gamma\left(\frac{k}{2}+1\right), {}_1F_1\left(-\frac{k}{2};1;-\frac{u^2}{2\sigma_n^2}\right) \quad (1.350)$$

where Γ denotes the gamma function and ${}_1F_1$ is the confluent hypergeometric function defined by the following series:

$${}_1F_1(a;\, b;\, x) = \sum_{r=0}^{\infty}\frac{(a)_r x^r}{(b)_r r!} = 1 + \frac{ax}{b} + \frac{a(a+1)x^2}{2!b(b+1)} + \frac{a(a+1)(a+2)x^3}{3!b(b+1)(b+2)} + \cdots \quad (1.351)$$

The mean and standard deviation of v/σ_n for different u/σ_n are plotted in Fig. 1.51. For small values of SNR $(u^2/2\sigma_n^2 < 0.1)$;

$$E[v] \simeq \sigma_n\sqrt{\frac{\pi}{2}}\left(1-\frac{u^2}{4\sigma_n^2}\right) \quad (1.352)$$

For large SNR $\left(\frac{u^2}{2\sigma_n^2} \gg 1\right)$,

$$E[v] = u \quad (1.353)$$

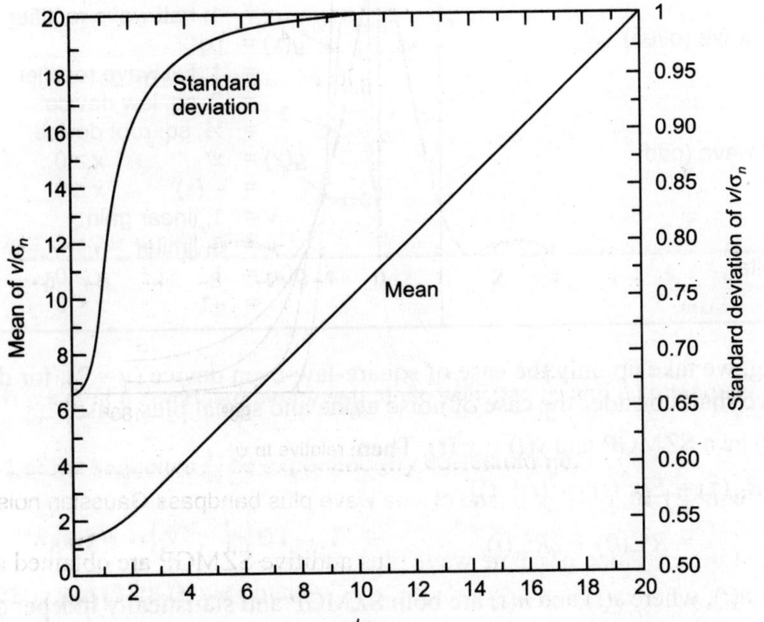

Figure 1.51: Mean and standard deviation for RICIAN distribution

The second and fourth moments of the envelope v are also the mean and second moments of the square of the envelope

$$E(v^2) = E[w] = 2\sigma_n^2\left(1+\frac{u^2}{2\sigma_n^2}\right) \tag{1.354}$$

$$E[v^4] = E[w^2] = 8\sigma_n^2 + 8u^2\sigma_n^2 + u^4 \tag{1.355}$$

$$\text{var }[w] = 4\sigma_n^4 - 4u^2\sigma_n^2 \tag{1.356}$$

1.11.6 Nonlinear Operations on Gaussian Processes

Many signal processing operations in communication systems, such as mixing or heterodyning, detection, limiting etc. involve nonlinear operations on random processes. It is therefore of great interest to study the effect of such nonlinear device implementing these operations on random processes. In our discussion below, we shall only take up the case of zero memory power law devices on stationary zero-mean Gaussian processes. An important class of nonlinear devices is the νth law devices whose transfer characteristic is based on the νth power law. The transfer characteristics of such devices are summarized in Table 1.2.

Table 1.2: The νth power law devices

Device	Transfer characteristic
1. Half wave	$g(x) = ax^\nu \quad x \geq 0$ $= 0 \quad x < 0$ a is a scaling constant $\nu = 1$; half wave rectifier
2. Full wave (even)	$g(x) = \lvert x\rvert^\nu$ $= 1$; full wave rectifier $= 2$; sq. law device $= ½$; sq. root device
3. Full wave (odd)	$g(x) = x^\nu \quad x \geq 0$ $= -(x)^\nu \quad x \leq 0$ $\nu = 1$; linear gain $\nu = 0$; limiter
4. Limiter	$g(x) = 1 \quad x > 0$ $= -1 \quad x < 0$

In the following, we take up only the case of square-law-even device ($\nu = 2$), for detailed study. For the input signal, we shall consider the case of noise alone and signal plus noise.

***Case* 1:** Let $x(t)$ be a SZMGP and $y(t) = x^2(t)$. Then

$$R_{yy}(\tau) = E[x^2(t)\cdot x^2(t+\tau)]$$

$$= R_{xx}^2(0) + 2R_{xx}^2(t) \tag{1.357}$$

Let $x(t) = s(t) + n(t)$, where $s(t)$ and $n(t)$ are both SZMGP and statistically independent. Then,

$$R_{xx}(\tau) = R_{ss}(\tau) + R_{nn}(\tau)$$

$$R_{xx}(0) = R_{ss}(0) + R_{nn}(0) = P_s + P_n$$

$$R_{xx}^2(0) = (P_s + P_n)^2$$

Therefore,

$$R_{yy}(\tau) = (P_s + P_n)^2 + 2R^2_{ss}(\tau) + 2R^2_{nn}(\tau) + 4R_{ss}(\tau)\, R_{nn}(\tau) \tag{1.358}$$

where P_s and P_n are respectively signal and noise powers.

***Case* 2:** Let $x(t) = s(t) + n(t)$ but $s(t)$, now is non-Gaussian. In particular we consider

$$s(t) = A \cos (\omega_0 t + \theta)$$

where A is a constant amplitude and θ is uniformly distributed over $(0, 2\pi)$ and $n(t)$, as before, is SZMGP,

$$\begin{aligned} R_{yy}(\tau) &= E[y(t)y(t+\tau)] = E[x^2(t)x^2(t+\tau)] \\ &= E\{[A^2 \cos^2 (\omega_0 t + \theta) + n^2(t) + 2A \cos (\omega_0 t + \theta)n(t)] \\ &\quad \times [A^2 \cos^2(\omega_0(t+\tau) + \theta) + n^2(t+\tau) + 2A \cos (\omega_0(t+\tau) + \theta)\, n(t+\tau)]\} \end{aligned}$$

Writing

$$R_{yy}(\tau) = \underset{\text{(signal terms)}}{R_{s^2s^2}(\tau)} + \underset{\text{(noise terms)}}{R_{n^2n^2}(\tau)} + \underset{\text{(interaction terms)}}{R_{snsn}(\tau)} \tag{1.359}$$

We have on simplification,

$$\begin{aligned} R_{s^2s^2}(\tau) &= P_s^2\left(1 + \frac{1}{2}\cos 2\omega_0\tau\right) \\ R_{snsn}(\tau) &= P_sP_n + 2P_sR_{nn}(\tau) \cos \omega_0\tau \\ R_{n^2n^2}(\tau) &= P_n^2 + 2R^2_{nn}(\tau) \end{aligned} \tag{1.360}$$

We calculate below the SNR performance of a square law detector. For the output, we call the signal part as the part of $R_{yy}(\tau)$ due to signal alone and noise part as that due to noise alone or due to interaction of noise and signal.*

For Case I, where $s(t)$ and $n(t)$ are both Gaussian and independent,

$$\left(\frac{S}{N}\right)_{\text{out}} = \frac{3(P_s/P_n)^2}{6P_s/P_n + 3} = \frac{(S/N)^2_{\text{in}}}{2(S/N)_{\text{in}} + 1} \tag{1.361}$$

Therefore for high $(S/N)_{\text{in}}$ i.e., $(S/N)_{\text{in}} \gg 1.0$,

$$(S/N)_{\text{out}} \approx \frac{1}{2}(S/N)_{\text{in}}$$

and for low $(S/N)_{\text{in}}$ i.e., $(S/N)_{\text{in}} \ll 1.0$, $(S/N)_{\text{out}} \approx (S/N)^2_{\text{in}}$.

This behaviour of the square law detector is shown in Fig. 1.52.

For large $(S/N)_{\text{in}}$, the detector behaviour is linear whereas for small $(S/N)_{\text{in}}$, the detector shows a *signal suppression effect*, or a threshold effect.

---

*This is a reasonable assumption for most analog communication situations. However, in case of radar detection, it is unreasonable to allocate noise and signal interaction to noise part alone since here we are concerned not with the intelligibility of the received information but rather with the presence or absence of the signal.

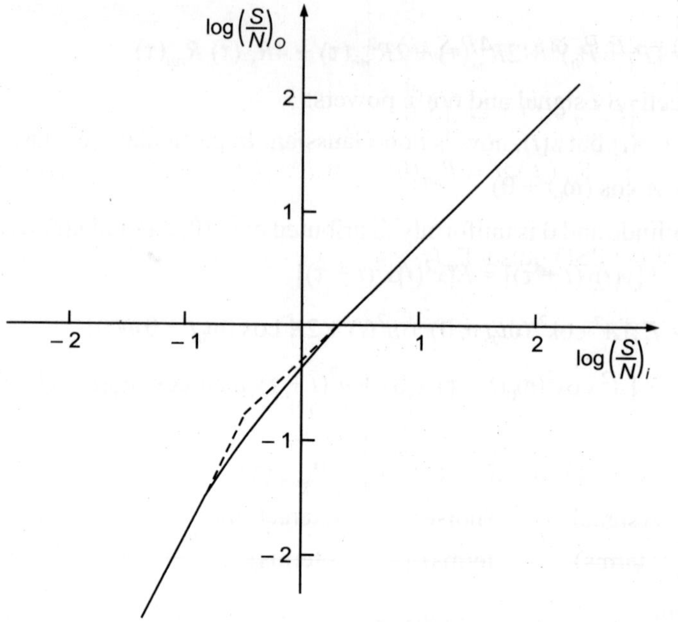

Figure 1.52: SNR characteristic for square law detector

For Case II, i.e., for sine wave with Gaussian noise,

$$(S/N)_{\text{out}} = \frac{3P_s^2/2}{6P_sP_n + 3P_n^2}$$

$$= \frac{1}{2}\frac{(S/N)_{\text{in}}^2}{2(S/N)_{\text{in}} + 1} \tag{1.362}$$

In this case, therefore, $(S/N)_{\text{out}}$ is 3 dB inferior as compared to Case I. However, with amplitude-modulated sine wave plus Gaussian noise (as in AM systems),

$$(S/N)_{\text{out}} = \frac{(S/N)_{\text{in}}^2}{2(S/N)_{\text{in}} + \frac{1}{2}},$$

with the modulation index $m = 1$.

Let us calculate the power spectral density (PSD) of the output process $y(t)$ in case II. Referring to $R_{yy}(\tau)$ of eq. (1.359) and individual terms in (1.360), we obtain the overall PSD by obtaining the PSD corresponding to each of the three terms. Thus

$$S_{ss}(f) = P_s^2\delta(f) + \frac{P_s^2}{4}[\delta(f - 2f_0) + \delta(f + 2f_0)] \tag{1.363}$$

This is sketched in Fig. 1.53(a).

To obtain $S_{sn}(f)$, we shall have to assume some form of $S_{nn}(f)$. Let $S_{nn}(f)$ correspond to narrowband white noise of bandwidth W centered at f_0 as shown in Fig. 1.53(b). The height of the pulse is $P_n/2W$ corresponding to input noise power of P_n. From eq. (1.360),

$$S_{sn}(f) = 2P_sP_n\,\delta(f) + 4P_sS_n(f)*\{\tfrac{1}{2}[\delta(f-f_0)+\delta(f+f_0)]\}$$

$$= 2P_sP_n\,\delta(f) + 2P_s[S_n(f-f_0) + S_n(f+f_0)] \tag{1.364}$$

$$\int_{-\infty}^{\infty} S_{sn}(f)df = R_{sn}(0) = 4P_sP_n = 4hW$$

$h = \dfrac{P_sP_n}{4}$ is the height of the PSD pulse. $S_{sn}(f)$ is shown in Fig. 1.53(c). From eq. (1.360),

$$S_{nn}(f) = P_n^2\delta(f) + 2S_n(f) * S_n(f) \tag{1.365}$$

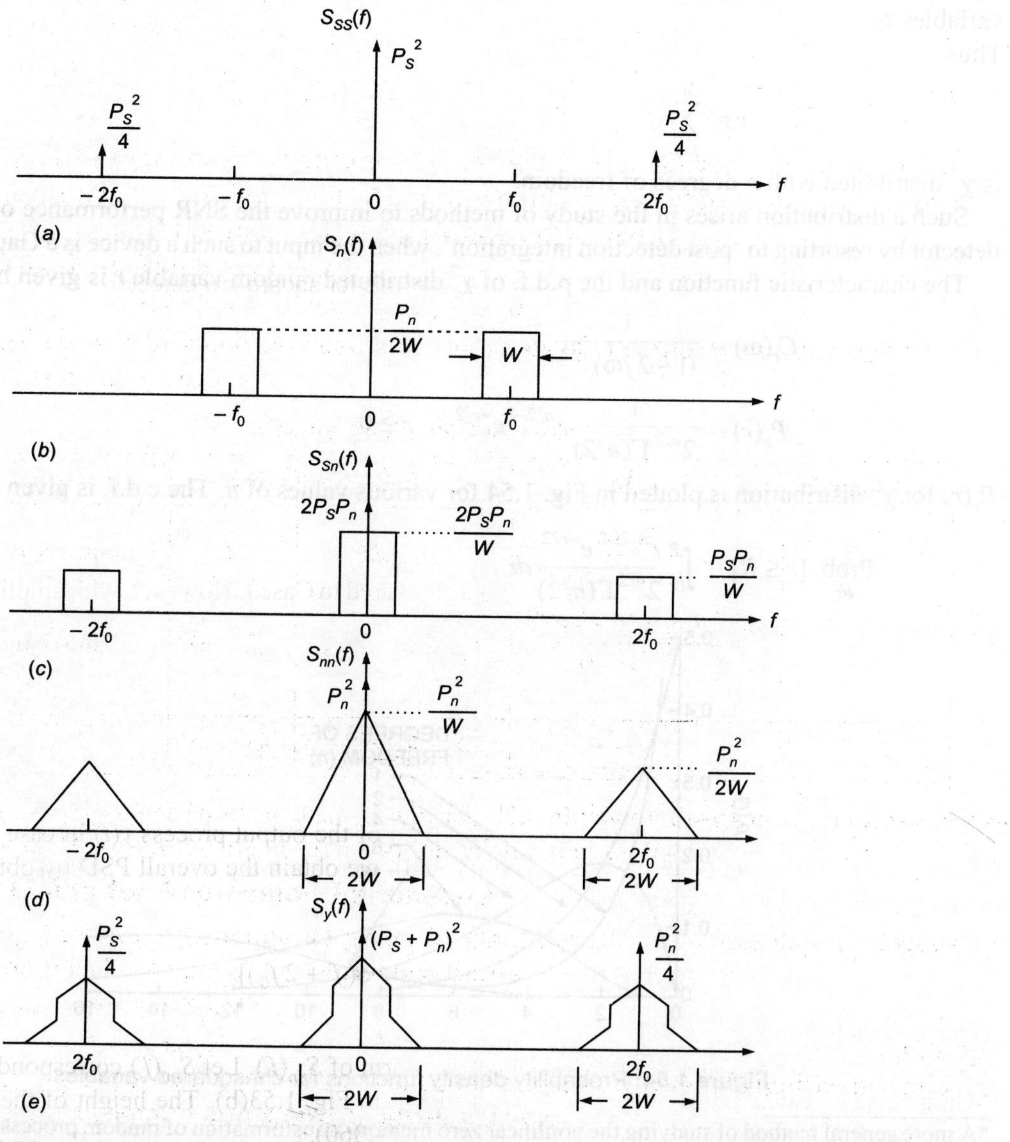

Figure 1.53: p.s.d. of the squared envelope of a sine wave plus bandpass Gaussian process

This is shown in Fig. 1.53(d). Here the height of the triangular spectrum is $P_n^2/2W$ corresponding to total power of $2P_n^2$. Adding up all terms corresponding to eqs. (1.363, 1.364, 1.365), we obtain the overall PSD. This is sketched in Fig. 1.53(e).

We note that the PSD of output $y(t)$ of a square law device is nonzero only in narrowbands centred around zero frequency and $\pm 2f_0$. One recovers the baseband spectrum by passing the output of the square law device through a lowpass filter centred at zero and of bandwidth twice that at the detector input*.

1.11.7 Chi-Squared Distribution [49, 62]

Chi-squared distribution is referred to as the distribution of a random variable r which is a sum of the square of a finite number of statistically independent zero mean, unity variance Gaussian random variables x_i.

Thus

$$r = \sum_{i=1}^{n} x_i^2 \tag{1.366}$$

is χ^2 distributed with n degrees of freedom.

Such a distribution arises in the study of methods to improve the SNR performance of a square law detector by resorting to 'post detection integration', when the input to such a device is a Gaussian process.

The characteristic function and the p.d.f. of χ^2 distributed random variable r is given by

$$C_r(\omega) = \frac{1}{(1-2j\omega)^{n/2}} \tag{1.367}$$

$$P_r(r) = \frac{1}{2^{n/2}\Gamma(n/2)} r^{n/2-1} e^{-r/2}, \quad r \geq 0 \tag{1.368}$$

$P_r(r)$ for χ^2 distribution is plotted in Fig. 1.54 for various values of n. The c.d.f. is given by

$$\text{Prob. } [r \leq R] = \int_0^R \frac{r^{n/2-1}\, e^{-r/2}}{2^{n/2}\Gamma(n/2)} dr \tag{1.369}$$

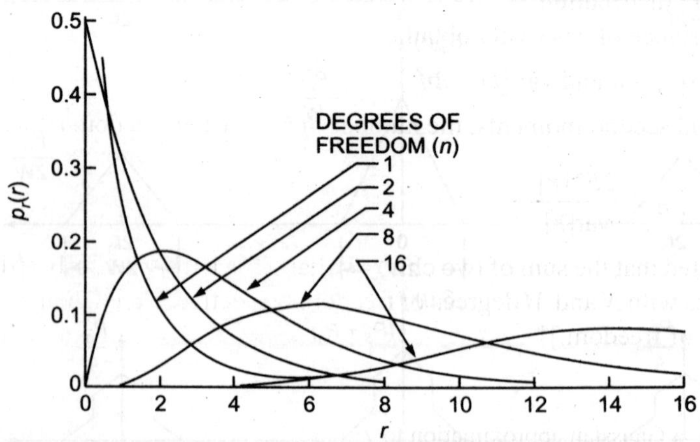

Figure 1.54: Probability density functions for chi-squared variables

*A more general method of studying the nonlinear zero memory transformation of random processes is based on the Fourier series or power series expansion of the output $y(t)$. For details see [9].

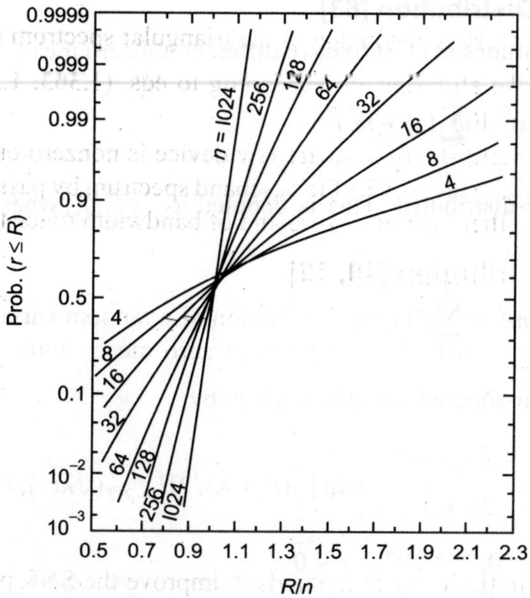

Figure 1.55: Cumulative chi-squared distribution (n = degrees of freedom)

and is plotted in Fig. 1.55. If we make a change of variables

$$p = \frac{n}{2} - 1 \text{ and } u = \frac{R}{2(p+1)^{1/2}},$$

then (1.369) may be written as

$$I(u, p) = \frac{1}{\Gamma(p+1)} \int_0^{u(p+1)^{1/2}} t^p e^{-t}\, dt \tag{1.370}$$

Equation (1.370) is called Pearson's incomplete gamma function and is tabulated in [62] along with tabulated values of χ^2 distribution.

The mean and variance of r is easily obtained as

$$E[r] = n \text{ and var } [r] = 2n \tag{1.371}$$

Knowing the first and second moments, the number of degrees of freedom n is obtained as

$$n = \frac{2E^2[r]}{\text{var}[r]} \tag{1.372}$$

[It should also be noted that the sum of two chi. χ^2 variables is itself χ^2-distributed. In particular if r_N and r_M are chi. χ^2 variates with N and M degrees of freedom respectively, then their sum is also χ^2 distributed with $N + M$ degrees of freedom.]*

---

*For large values of n, a Gaussian approximation to χ^2 p.d.f. may be used. A Gaussian distribution is fitted to χ^2 distribution by equating means and variances. We may therefore write for

$$(n > 100, \text{ say}), \text{Prob } [r > R_0] = \frac{1}{\sqrt{4\pi n}} \int_{R_0}^{\infty} \exp\left[-\frac{(r-n)^2}{4n}\right] dr.$$

Non-central chi-squared Distribution [63]

When the Gaussian random variates in (1.366) have different nonzero means m_i, and unit variance, then

$$r' = \sum_{i=1}^{n} x_i'^2 = \sum_{i=1}^{n} (x_i + m_i)^2 \tag{1.373}$$

is said to be non-centrally χ^2-distributed. This is denoted as $X_n'^2(\lambda)$ where λ is the so-called 'non-centrality parameter' given by

$$\lambda = \sum_{i=1}^{n} m_i^2 = \sum_{i=1}^{n} E[x_i'^2] \tag{1.374}$$

The p.d.f. of non-central chi-squared variable is given by

$$\begin{aligned} p_{r'}(r') &= \frac{1}{2}\left(\frac{r'}{\lambda}\right)^{n-2/4} \exp[-(r'+\lambda)/2] I_{n-2/2}(\sqrt{\lambda r'});\ r' \geq 0 \\ &= 0, \qquad r < 0 \end{aligned} \tag{1.375}$$

where $I_k(\cdot)$ is the modified Bessel function (64) of the first kind and order k. Some plots of this p.d.f. are shown in Fig. 1.56. The mean and variance of r' are obtained as

$$E[r'] = \lambda + n, \text{ and var } [r'] = 4\lambda + 2n \tag{1.376}$$

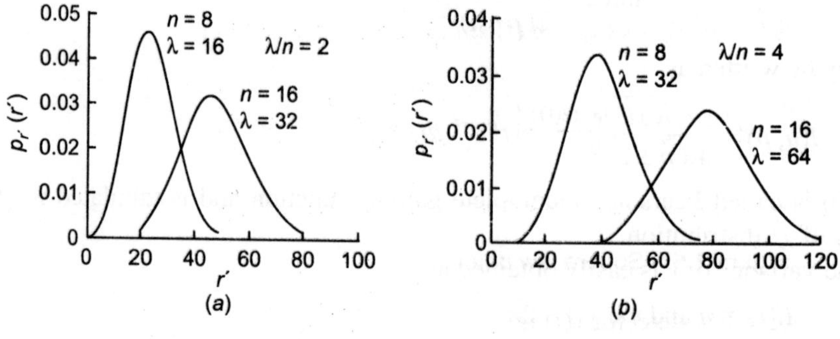

Figure 1.56: Non-central χ^2 density functions
(a) $\lambda/2 = 2$, **(b)** $\lambda/2 = 4$

Gamma and Non-central Gamma Distribution

If y_i's are zero mean Gaussian r.v.'s with common variance σ^2, then

$$q = \sum_{i=1}^{n} y_i^2 \tag{1.377}$$

is said to be Gamma distributed with n degrees of freedom and parameter σ. The p.d.f. for gamma distribution is

$$p_q(q) = \frac{q^{n/2-1} \exp(-q/2\sigma^2)}{2^{n/2}\Gamma(n/2)\sigma^n},\ q > 0, \sigma > 0 \cdot q < 0 \tag{1.378}$$

On comparison of (1.377) and (1.366) it is seen that gamma variable q is related to χ^2 variable r through the transformation

$$q = \sigma^2 r \quad \text{or} \quad y_i = \sigma x_i$$

Similarly,

$$q' = \sum_{i=1}^{n} y_i'^2 = \sum_{i=1}^{n} (y_i + m_i)^2 \tag{1.379}$$

is called a 'non-central gamma variable'. It is easy to see that

$$r' = q'/\sigma^2 = \sum_{i=1}^{n} (y_i/\sigma + m_i/\sigma)^2 \tag{1.380}$$

has non-central χ^2 distribution with non-centrality parameter given by

$$\lambda = \sum_{i=1}^{n} \left(\frac{m_i}{\sigma}\right)^2 \tag{1.381}$$

Example 1.17: Consider the quadratic detector with post detection summation as shown in Fig. 1.57. The input to the system is a sine wave plus a narrowband SZMGP with variance σ_n^2.

$$x(t) = A \cos(\omega_0 t + \theta) + n(t)$$

$$n(t) = n_c(t) \cos \omega_0 t - n_s(t) \sin \omega_0 t$$

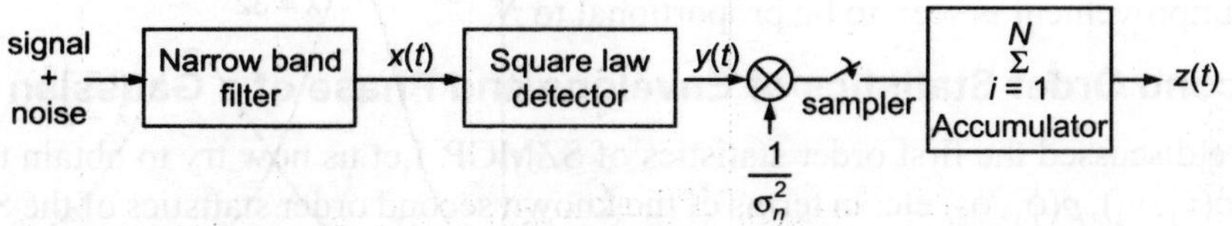

Figure 1.57: Square law detection with post detection summation

The output of the square law detector $y(t)$ is

$$y(t) = |x(t)|^2 = [A \cos \theta + n_c]^2 + [A \sin \theta + n_s]^2$$

The output $z(t)$ after scaling with $1/\sigma_n^2$, sampling and summation is

$$z = \sum_{i=1}^{N} \left\{ \left(\frac{A \cos \theta}{\sigma} + \frac{n_c}{\sigma_n}\right)^2 + \left(\frac{A \sin \theta}{\sigma_n} + \frac{n_i}{\sigma_n}\right)^2 \right\} \tag{1.382}$$

We immediately see that the two summation terms are both noncentral χ^2 with N degrees of freedom and noncentral parameters

$$\lambda_1 = \frac{NA^2}{\sigma_n^2} \cos^2 \theta; \lambda_2 = \frac{NA^2}{\sigma_n^2} \sin^2 \theta \tag{1.383}$$

Since the two summations are also s.i., it follows that z is also noncentral χ^2 with $2N$ degrees of freedom with

$$\lambda = \lambda_1 + \lambda_2 = \frac{NA^2}{\sigma_n^2} \tag{1.384}$$

It is also seen that the distribution is independent of θ and that each sample of the output has two degrees of freedom.

If the SNR at the output is defined as $E^2[z]/\text{var}[z]$, then we see from (1.376) that

$$(\text{SNR})_{\text{out}} = \frac{(\lambda + N)^2}{2[2\lambda + N]} = \frac{\left(\dfrac{NA}{\sigma_n^2} + N^2\right)}{2\left[\dfrac{2NA^2}{\sigma_n^2} + N\right]}$$

$$= \frac{N}{2}\left\{\frac{\left(\dfrac{A^2}{\sigma_n^2} + 1\right)^2}{\dfrac{2A^2}{\sigma_n^2} + 1}\right\}$$

$$= \frac{N}{2}\frac{[2(\text{SNR})_{\text{in}} + 1]^2}{[4(\text{SNR})_{\text{in}} + 1]} \tag{1.385}$$

and the SNR improvement is seen to be proportional to N.

1.11.8 Second Order Statistics of Envelope and Phase of a Gaussian Process

So far we have discussed the first order statistics of SZMGP. Let us now try to obtain the second order statistics like $p(v_1, v_2), p(\phi_1, \phi_2)$ etc. in terms of the known second order statistics of the SZMGP $x(t)$. We use the following results from previous sections.

analytical signal: $z(t) = x(t) + j\hat{x}(t)$

envelope: $v(t) = |z(t)| = \sqrt{x^2 + \hat{x}^2}$

phase: $\phi(t) = z(t) = \tan^{-1}\dfrac{\hat{x}(t)}{x(t)}$

$x(t) = v\cos\phi,\ \hat{x}(t) = v\sin\phi$

If $R_{xx}(\tau)$ is a.c.f. of $x(t)$, then

$R_{\hat{x}x}(\tau) = \hat{R}_{xx}(\tau) = -R_{x\hat{x}}(\tau)$ is an odd function of τ.

$\hat{R}_{xx}(0) = 0, \qquad R_{xx}(0) = \sigma_x^2$

Let us assume that $\sigma_x^2 = 1$ and let $\rho_x(\tau)$ be the normalized a.c.f.of $x(t)$. Since v_i and ϕ_i depend on both x_i and $\hat{x}_i, i = 1, 2, \ldots,$ the joint p.d.f.

$$p(v_1, v_2; \phi_1, \phi_2) = p(x_1, x_2, \hat{x}_1, \hat{x}_2)\left| J\left(\frac{x_1, x_2, \hat{x}_1, \hat{x}_2}{v_1, v_2, \phi_1, \phi_2}\right)\right| \tag{1.386}$$

As the Jacobian $|J|$ involves two independent transformations viz. $x_1,\ \hat{x}_1 \to v_1, \phi_1$ and $x_2,\ \hat{x}_2 \to v_2, \phi_2$ it is simply the product of the Jacobians of these two transformations. Therefore,

$$\left| J\left(\frac{x_1, x_2, \hat{x}_1, \hat{x}_2}{v_1, v_2, \phi_1, \phi_2}\right)\right| = v_1 v_2 \tag{1.387}$$

Let us try to evaluate $p(x_1, x_2, \hat{x}_1, \hat{x}_2)$.

Since $x(t)$ is a SZMGP, $x_1, x_2, \hat{x}_1$ and $\hat{x}_2$ are also zero-mean Gaussian variates. Hence $p(x_1, x_2, \hat{x}_1, \hat{x}_2)$ is a zero-mean multivariate Gaussian p.d.f. with zero-mean vector and a 4 × 4 covariance matrix A, where

$$\Lambda = \begin{bmatrix} 1 & \rho_x(\tau) & 0 & -\hat{\rho}_x(\tau) \\ \rho_x(\tau) & 1 & \hat{\rho}_x(\tau) & 0 \\ 0 & \hat{\rho}_x(\tau) & 1 & \rho_x(\tau) \\ -\hat{\rho}_x(\tau) & 0 & \rho_x(\tau) & 1 \end{bmatrix} \tag{1.388}$$

Thus, in principle, we may write

$$p(x_1, x_2, \hat{x}_1, \hat{x}_2) = \left[(2\pi)^4 \,|\,\Lambda\,|^{-1/2} \exp\left\{ -\frac{1}{2} x' \Lambda x \right\} \right]$$

where

$$x = [x_1, x_2, \hat{x}_1, \hat{x}_2]', \tag{1.389}$$

It is seen that Λ can be written in the partitioned form.

$$\Lambda = \left[\begin{array}{c|c} A & B \\ \hline B & A \end{array}\right]$$

where

$$A = \begin{bmatrix} 1 & \rho_x \\ \rho_x & 1 \end{bmatrix} = A'$$

and

$$B = \begin{bmatrix} 0 & -\hat{\rho}_x \\ \hat{\rho}_x & 0 \end{bmatrix} = -B'$$

Using some results from linear algebra for matrices of above kind, it can be shown that p.d.f. of (1.389) may be written in the following explicit form

$$p(v_1, v_2, \phi_1, \phi_2) = \frac{v_1\, v_2}{(2\pi)^2 (1-R^2)} \exp\left\{-\frac{v_1^2+v_2^2}{2(1-R^2)}\right\} \times \exp\left\{\frac{Rv_1v_2 \cos(\phi_1 - \phi_2 - \psi)}{1-R^2}\right\} \quad (1.390)$$

where $Re^{i\psi} \triangleq (\rho_x + j\hat{\rho}_x)$; $R = |\rho_x + j\hat{\rho}_x|$, $\tan\psi = \frac{\hat{\rho}_x}{\rho_x}$.

A number of other interesting results can now be derived from this p.d.f. For example,

$$p(v_1, v_2, \phi_1) = \int_0^{2\pi} p(v_1, v_2, \phi_1, \phi_2)\, d\phi_2$$

$$= \frac{v_1 v_2}{2\pi(1-R^2)} \exp\left[-\frac{v_1^2+v_2^2}{2(1-R^2)}\right] I_0\left(\frac{Rv_1v_2}{1-R^2}\right)$$

$$= p(v_1, v_2)\, p(\phi_1) \quad (1.391)$$

Therefore,

$$p(v_1, v_2) = \frac{v_1 v_2}{1-R^2} \exp\left[-\frac{v_1^2+v_2^2}{2(1-R^2)}\right] I_0\left(\frac{Rv_1v_2}{1-R^2}\right) \quad (1.392)$$

Since ϕ_2 is chi of v_2 from results on first order statistics, we see here that ϕ_2 is also chi of v_1 and v_2 jointly and therefore we infer that the envelope $v(t)$ and phase $\theta(t)$ are independent random processes. Moreover, $p(\phi_1)$ is uniformly distributed. We also note that, if for two instants t_1, t_2 such that

$$R_{xx}(\tau) = R_{xx}(t_1 - t_2) = 0 \quad \text{or} \quad \rho_x(\tau) = \hat{\rho}_x(\tau) = 0$$

giving $R = 0$, then

$$p(v_1, v_2, \phi_1, \phi_2) = \frac{v_1 v_2}{(2\pi)^2} \exp\left[-\frac{v_1^2+v_2^2}{2}\right]$$

$$= p(v_1)p(v_2)p(\phi_1)(\phi_2) \quad (1.393)$$

where

$$p(\phi_1) = p(\phi_2) = 1/2\pi; \text{ uniform p.d.f.}$$

and

$$p(v) = v \exp\left[-\frac{v^2}{2}\right]; \text{ Rayleigh p.d.f.}$$

This shows that for such a value of time difference v_1, v_2, ϕ_1 and ϕ_2 are all s.i. random variables. (It does not imply that corresponding random processes are s.i.)

If we consider

$$p(\phi_1, \phi_2/v_1, v_2) = \frac{p(\phi_1, \phi_2, v_1, v_2)}{p(v_1, v_2)} = \frac{\exp\left\{\frac{v_1v_2R\cos(\phi_1-\phi_2-\psi)}{1-R^2}\right\}}{(2\pi)^2 I_0\left(\frac{v_1v_2R}{1-R^2}\right)} \quad (1.394)$$

then,

$$p(\phi_2/v_1, v_2) = \int_0^{2\pi} p(\phi_1, \phi_2 / v_1, v_2)\, d\phi_1 = \frac{1}{2\pi}$$

and

$$p(\phi_1/\phi_2, v_1, v_2) = \frac{p(\phi_1, \phi_2/v_1, v_2)}{p(\phi_2/v_1, v_2)} \frac{\exp\left\{\dfrac{v_1 v_2 R \cos(\phi_1 - \phi_2 - \psi)}{1-R^2}\right\}}{2\pi I_0\left[\dfrac{v_1 v_2 R}{1-R^2}\right]} \tag{1.395}$$

This is known as the Viterbi p.d.f. [52, 65] and is of importance in the theory of synchronization. The plot of

$$p(x) = \frac{\exp[\alpha \cos x]}{2\pi I_0(\alpha)}; \quad 0 \le x \le 2\pi \tag{1.396}$$

is shown in Fig. 1.58. It approaches the Gaussian p.d.f. $N\left(\pi, \frac{1}{\alpha}\right)$ for $\alpha \gg 1$.

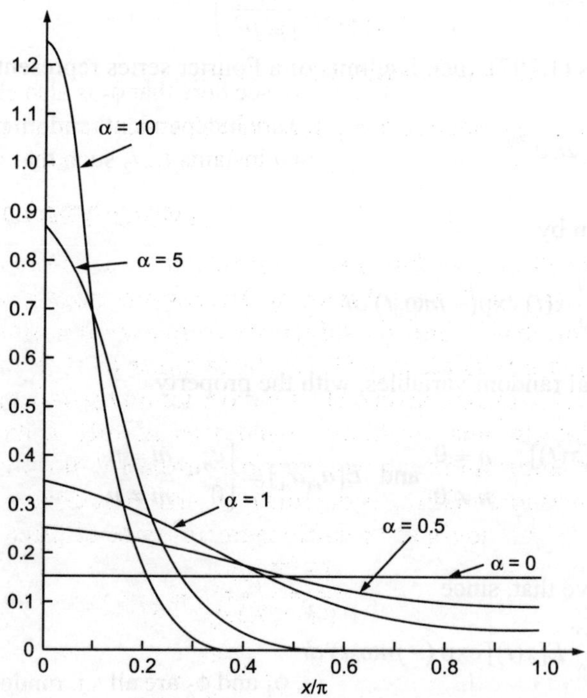

Figure 1.58: Conditional p.d.f. of the phase of a SZMGP (Viterbi density)

1.12 HARMONIC ANALYSIS OF RANDOM PROCESSES [50, 54, 66]

In Section 1.7 we discussed the representation of random processes in terms of power spectral density function which was defined to be the Fourier transform of the correlation function. We did not critically examine the possibility of direct harmonic representation of the process. We do this briefly in this section.

Fourier Series Representation

A random process $x(t)$ is said to be periodic in the mean square sense if its a.c.f. $R_{xx}(\tau)$ is periodic, i.e.,

$$R_{xx}(\tau + T) = R_{xx}(\tau) \tag{1.397}$$

The smallest number T satisfying (1.397) is said to be the period of the process.

Since $R_{xx}(\tau)$ is a deterministic periodic signal in this case, it admits of the Fourier series representation

$$R_{xx}(\tau) = \sum_{n=-\infty}^{\infty} c_n \exp(jn\omega_0\tau); \quad \omega_0 = \frac{2\pi}{T} \tag{1.398}$$

and

$$c_n = \frac{1}{T}\int_0^T R_{xx}(\tau)\exp(-jn\omega_0\tau)\,d\tau \tag{1.399}$$

The corresponding power spectrum is given by

$$S_{xx}(\omega) = 2\pi \sum_{n=-\infty}^{\infty} c_n \delta(\omega - n\omega_0) \tag{1.400}$$

We now show that if $x(t)$ obeys (1.397), then it admits of a Fourier series representation

$$x(t) = \sum_{n=-\infty}^{\infty} a_n e^{jn\omega_0 t} \tag{1.401}$$

where the coefficients a_n, given by

$$a_n = \frac{1}{T}\int_0^T x(t)\exp(-jn\omega_0 t)\,dt \tag{1.402}$$

are uncorrelated and orthogonal random variables, with the property

$$E[a_n] = \begin{cases} E[x(t)]; & n = 0 \\ 0 & n \neq 0 \end{cases} \quad \text{and} \quad E[a_m a_n^*] = \begin{cases} c_n & m = n \\ 0 & m \neq n \end{cases} \tag{1.403}$$

with c_n as in (1.398).

To prove (1.403), we observe that, since

$$E[a_n] = \frac{1}{T}\int_0^T E[x(t)]\exp(-jn\omega_0 t)\,dt$$

and $E[x(t)]$ is constant, $E[a_n] = 0$ except for $n = 0$.

Also,

$$E[a_n^* x(u)] = \frac{1}{T}\int_0^T E[x*(t)x*(u)]\exp(jn\omega_0 t)\,dt$$

$$= \frac{1}{T}\int_0^T R_{xx}(u - t)\exp(jn\omega_0 t)\,dt$$

Letting $u - t = \tau$, and using the periodicity of $R_{xx}(\tau)$, we see that

$$E[a_n^* x(u)] = c_n \exp(jn\omega_0 u) \tag{1.404}$$

Multiplying (1.402) by a_n^* and taking expectation, we get

$$E[a_m a_n^*] = \frac{1}{T}\int_0^T E[a_n^* x(u)] \exp(-jn\omega_0 u)\, du$$

$$= \frac{c_n}{T}\int_0^T \exp(-j(m-n)\omega_0 u)\, du$$

$$= 0 \text{ for } m \neq n$$

To establish the Fourier series representation (1.401), we need to show that

$$E[|x(t) - \sum_{n=-\infty}^{\infty} a_n \exp(jn\omega_0 t)|^2] = 0 \tag{1.405}$$

This is easily done by writing L.H.S. as

$$E[|x(t)|^2] - \sum_{n=-\infty}^{\infty} E[a_n^* x(t)] \exp(-jn\omega_0 t) - \sum_{n=-\infty}^{\infty} E[a_n x^*(t)] \exp(jn\omega_0 t)$$

$$+ \sum_{m=-\infty}^{\infty} \sum_{n=-\infty}^{\infty} E[a_m a_n^*] \exp[j(m-n)\omega_0 t]$$

Using (1.403), (1.404) and (1.398), we get

$$R_{xx}(0) - \sum_{n=-\infty}^{\infty} c_n - \sum_{n=-\infty}^{\infty} c_n^* + \sum_{n=-\infty}^{\infty} c_n = 0,$$

thus establishing the representation (1.401) with property (1.403). When the random process $x(t)$ is not periodic in the mean square sense, i.e., $R_{xx}(\tau)$ is not periodic, then the coefficients (1.402) in the representation (1.401), for a given T, are no longer orthogonal (and uncorrelated) and representation (1.401) is not valid for every t. However, one can easily show that it still holds for $|t| < T/2$ if we consider the periodic extension of $R_{xx}(\tau)$ with basic period of T. In other words, the expansion

$$x_p(t) = \sum_{n=-\infty}^{\infty} a_n^{jn\omega_0 t} \tag{1.406}$$

converges to $x(t)$ in the interval $(-T/2, T/2)$ in mean square sense, i.e.,

$$E[|x(t) - x_p(t)|^2] = 0 \text{ for } |t| < T/2 \tag{1.407}$$

Furthermore, it may be shown that in the limit $T \to \infty$,

$$\lim_{T\to\infty} T \cdot E[a_n a_m^*] = \begin{cases} S_{xx}(n\omega_0) & n = m \\ 0 & n \neq m \end{cases} \tag{1.408}$$

where $S_{xx}(\omega)$ is the p.s.d. of the nonperiodic random process $x(t)$.

Thus $E[|a_n|^2]$, i.e., variance of the Fourier coefficients tends to zero as $1/T$ for $T \to \infty$ whereas the covariance $E[a_n a_m^*]$ goes to zero faster than $1/T$. Thus we may say that the Fourier coefficients are approximately uncorrelated for large T.

Karhunen-Loeve Expansion

It is possible to have a harmonic representation of the Fourier series kind in terms of orthogonal functions $\phi_n(t)$ such that

$$x(t) \sim \sum_{n=1}^{\infty} b_n \phi_n(t), \qquad |t| < T/2 \tag{1.409}$$

where the series converges to $x(t)$ in some sense and b_n are a set of orthogonal random variables (and hence uncorrelated for the zero mean process). Here $\phi_n(t)$'s are a set of complete orthonormal (C.O.N.) functions over the interval $(-T/2, T/2)$, i.e.,

$$\int_{-T/2}^{T/2} \phi_n(t)\phi_n^*(t)\, dt = \delta_{mn} \tag{1.410}$$

and

$$b_n = \int_{-T/2}^{T/2} x(t)\phi_n^*(t)\, dt \qquad n = 1, 2 \ldots \tag{1.411}$$

How do we obtain the functions $\phi_n(t)$ that guarantee the orthogonality of b_n? For this purpose, we assume that the a.c.f. of $x(t)$, namely, $R_{xx}(t_1, t_2)$ is continuous. Then, to ensure the orthogonality of b_n, i.e.,

$$E[b_n b_m^*] = 0 \tag{1.412}$$

it can be shown that $\phi_n(t)$ must satisfy the integral equation

$$\int_{-T/2}^{T/2} R_{xx}(t_1, t_2)\phi(t_2)\, dt = \lambda\phi(t_1), \qquad |t_1| < T/2 \tag{1.413}$$

for some value $\lambda = \lambda_n$, such that

$$E[|b_n|^2] = \lambda_n \tag{1.414}$$

This is easily seen by multiplying (1.409) by b_m^*, taking the expectation and using (1.412),

$$\begin{aligned} E[x(t_1)b_m^*] &= E[\sum_{n=1}^{\infty} b_n b_m^* \phi_n(t_1)] \\ &= E[|b_m|^2]\phi_m(t_1) \end{aligned} \tag{1.415}$$

Also, from (1.411),

$$\begin{aligned} E[x(t_1)b_m^*] &= \int_{-T/2}^{T/2} E[x(t_1)x^*(t_2)]\phi_m(t_2)\, dt_2 \\ &= \int_{-T/2}^{T/2} R_{xx}(t_1, t_2)\phi_m(t_2)\, dt_2 \end{aligned}$$

and hence (1.413) is established.

The converse is also easily proved, i.e., if $\phi_n(t)$ are orthonormal functions satisfying (1.413) and $x(t)$ is represented as in (1.409), then the coefficients b_n are orthogonal. To see this, we again multiply (1.409), by b_m^* and use (1.411) along with (1.413) in taking the expectation

$$E[x(t_1)\cdot b_m^*] = \int_{-T/2}^{T/2} R(t_1, t_2)\phi_m(t_2)\, dt_2$$

$$= \lambda_m \phi_m(t_1)$$

Again, using (1.411),

$$E[b_n b_m^*] = \int_{-T/2}^{T/2} E[x(t_1)b_m^*]\phi_n^*(t_1)\, dt_1$$

$$= \lambda_m \int_{-T/2}^{T/2} \phi_m(t_1)\phi_n^*(t_1)\, dt_1$$

$$= \lambda_m \delta_{mn}$$

The representation

$$x_{KL}(t) = \sum_{n=1}^{\infty} b_n \phi_n(t) \tag{1.416}$$

with b_n as in (1.411) and $\phi_n(t)$ as the eigenfunctions of the autocorrelation function $R_{xx}(t_1, t_2)$ obtained as solutions of the integral equation (1.413) is known as Karhunen-Loeve (KL) representation of random process $x(t)$. It is easily shown that this representation has the property that

$$E[|x(t) - x_{KL}(t)|^2] = R(t,t) - \sum_{n=1}^{\infty} \lambda_n |\phi_n(t)|^2 \text{ for } |t| < T/2 \tag{1.417}$$

and the mean square error in the representation goes to zero if it can be shown that the eigenfunction equation (1.413) has non-zero solutions $\phi_n(t)$ for certain eigenvalues λ_n of λ and that these solutions are orthonormal and have the property that

$$R(t, t) = \sum_{n=1}^{\infty} \lambda_n |\phi_n(t)|^2 \tag{1.418}$$

Example 1.18: Let $x(t)$ be a stationary process with p.s.d. which has the idealized low-pass characteristic with bandlimiting frequency W. Its a.c.f. is given by

$$R_{xx}(\tau) = 2W \frac{\sin 2\pi W\tau}{2\pi W\tau} \tag{1.419}$$

For KL representation of such a process, we must solve the integral equation

$$\int_{-T/2}^{T/2} \frac{\sin W(t-\tau)}{\pi(t-\tau)} \phi(\tau)\, d\tau = \lambda\phi(t) \tag{1.420}$$

Slepian et al. [21] have investigated the solution of this equation and the solutions $\psi_n(c; t)$, $c = WT$ are known as prolate spheroidal wave functions. Except for a scale factor, the solutions of the above integral equation namely the eigenvalues $\lambda_n(c)$ and eigenfunctinos $\psi_n(c; t)$ form a one parameter family of functions depending on product $c = WT$. Substituting $T\tau$ for τ and Tt for t in (1.420), we obtain

$$\int_{-1/2}^{1/2} \psi(\tau) = \frac{\sin c\,(t-\tau)}{\pi(t-\tau)}\, d\tau = \frac{1}{T}\lambda\psi(t) \tag{1.421}$$

where $\psi(t) = \phi(Tt)$

Figure 1.59 shows the plot of $\psi_n(t)$ (for $n = 0, 1, 2, 3$) corresponding to the $c = 4$. The corresponding eigenvalues are 0.996, 0.912, 0.519, 0.110. The prolate spheroidal wave functions $\psi_n(t)$'s have an appearance of damped sinusoids. Figure 1.59(*b*) shows the behaviour of the largest eigenvalue λ_0 with parameter *c*, which is measure of the time-bandwidth product.

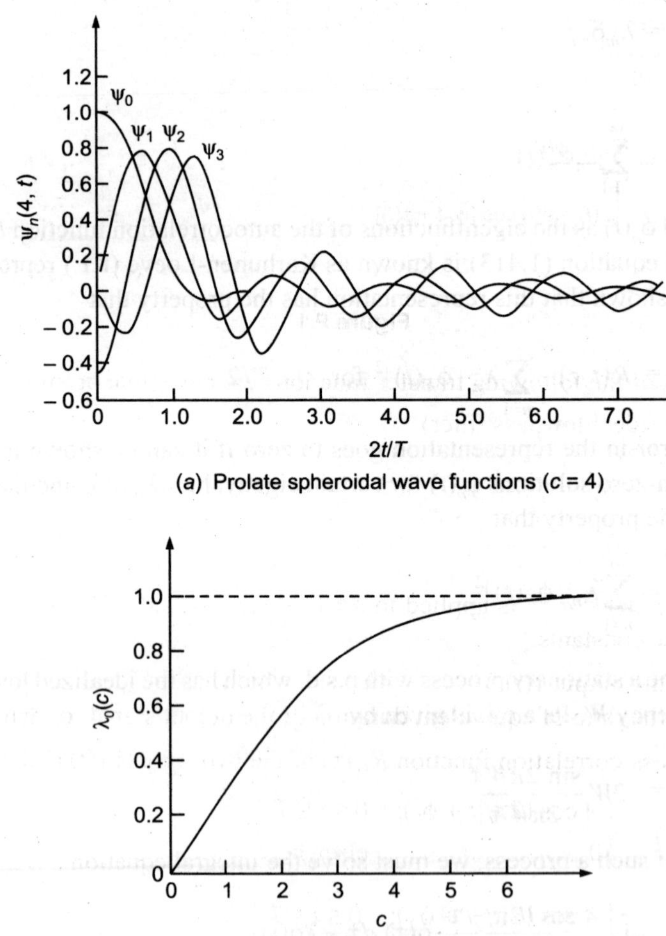

(*a*) Prolate spheroidal wave functions ($c = 4$)

(*b*) Variation of the largest eigenvalue λ_0 with $c = WT$

Figure 1.59

PROBLEMS

1. Obtain the expressions relating the input and output of the following systems (Fig. P.1) and classify them into the categories (i) linear (ii) time invariant (iii) time varying (iv) LTI. For the LTI system, obtain the corresponding transfer function and plot the magnitude and phase response.

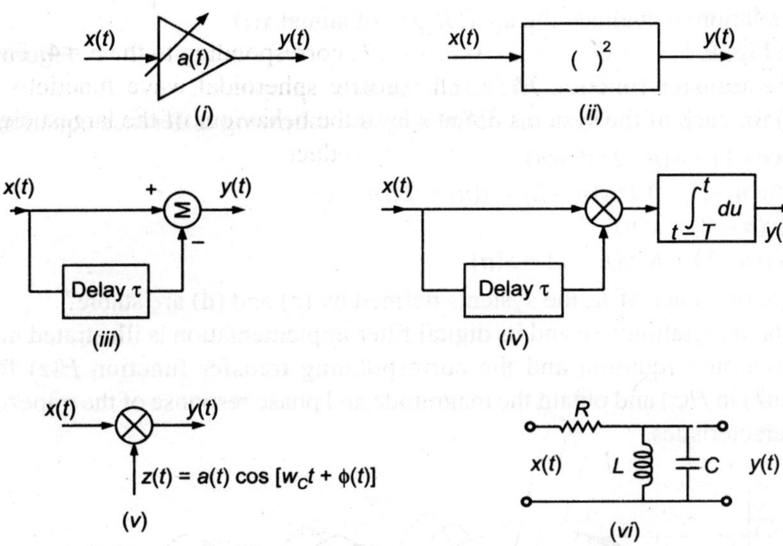

Figure P.1

2. Determine which of the following transfer functions are realizable or unrelizable and why
   (i) rect. $[f/2W]$ (ideal lowpass filter)
   (ii) $\exp[-(\omega t_0)^2/2]$
   (iii) $j\omega/[(2+j\omega)^2+9]$
   (iv) $\dfrac{\pi}{a}\exp(-|\omega|/a)$

3. A Gaussian pulse $x(t) = e^{-\alpha t^2}$ is applied to a LTI system whose transfer function is $H(f) = e^{-\beta f^2}$, where α and β are constants.
   (i) Determine the output $y(t)$.
   (ii) Determine the ratio of equivalent duration of the output signal to input signal.

4. Determine the cross-correlation function $R_{xy}(\tau)$ of the two signals $x(t)$ and $y(t)$ defined by

$$x(t) = \begin{cases} A\cos(2\pi f_1 t + \phi_1); & 0 \le t \le T \\ 0 & \text{elsewhere} \end{cases}$$

$$y(t) = \begin{cases} A\cos(2\pi f_2 t + \phi_2); & 0 \le t \le T \\ 0 & \text{elsewhere} \end{cases}$$

How does changing the frequency difference $|f_1 - f_2|$ affect $R_{xy}(\tau)$? Assume $|f_1 - f_2| \ll \dfrac{f_1 + f_2}{2}$.

5. (i) Show that

$$R_{xy}^{(m+n)}(\tau) = (-1)^n \int_{-\infty}^{\infty} x^{(m)}(t)[y^n(t-\tau)]^* \, dt$$

where $x^m(t)$ denotes the mth derivative of $x(t)$; similarly for $y^n(t)$ and $R_{xy}^{(m+n)}(\tau)$.

(ii) Use this relation to evaluate the a.c.f. $R_{xx}(\tau)$ of signal $x(t)$ shown in Fig. P. 5.

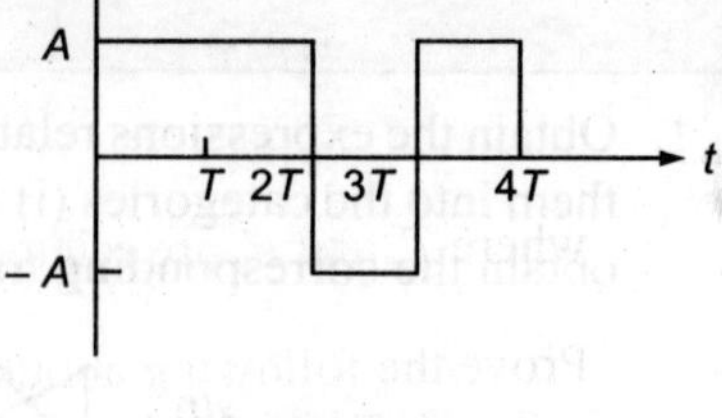

Figure P. 5

6. Determine the transfer function $H(z)$ and the unit sample response $h(n)$ for each of the systems defined by the following difference equations:
   (a) $y(n) - 2y(n-1) + y(n-2) = x(n)$
   (b) $y(n) + 0.6y(n-1) - 0.16y(n-2) = x(n) + 2x(n-1)$
   (c) $y(n) - K^2y(n-2) = x(n)$
   (d) $y(n) - 2Ky(n-1) + K^2y(n-2) = x(n)$

   For what range of values of K, the systems defined by (c) and (d) are stable?
7. The trapezoidal integration rule and its digital filter implementation is illustrated in Fig. P.7 below. Write the difference equation and the corresponding transfer function $H(z)$ for this system. Set $z = \exp(j\omega T)$ in $H(z)$ and obtain the magnitude and phase response of the trapezoidal integrator. Plot these characteristics.

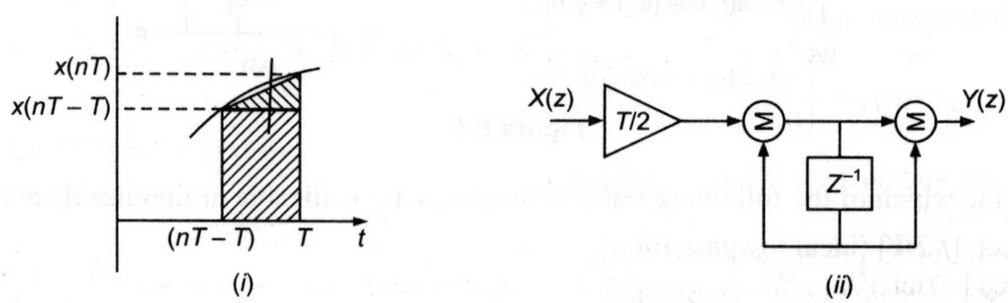

Figure P. 7

8. Determine the direct form realization for each of the following transfer functions:

   (a) $H(z) = \dfrac{3}{1+0.1z^{-1}} + \dfrac{1.5z^{-2}}{(1-z^{-1})(1+0.5z^{-1})}$

   (b) $H(z) = \left[\dfrac{z^{-1}(1-z^{-3})}{(1-z^{-1})(1+0.2z^{-2})}\right]^2$

   (c) $H(z) = \dfrac{1}{(1-0.1z^{-1})(1-0.2z^{-1})}$

9. Consider a periodically sampled sequence $x(kT_s)$, $k = 0, 1, 2, \ldots, N-1$. The discrete Fourier transform (DFT) of this sequence is defined by

$$X(kF_s) = T_s \sum_{n=0}^{N-1} x(nT_s) \exp\left(-j\frac{2\pi kn}{N}\right)$$

and the inverse DFT is defined by

$$x(nT_s) = \frac{1}{NT_s} \sum_{k=0}^{N-1} (kF_s) \exp\left(j\frac{2\pi kn}{N} \right)$$

where T_s and F_s are related by $T_s F_s = \frac{1}{N}$.

Prove the following about DFT:

(i) $x(nT_s)$ and $X(kF_s)$ are periodic with period N.

(ii) If $x_1(nT_s)$ and $x_2(nT_s)$ are two periodic sequences with period N and $X_1(kF_s)$ and $X_2(kF_s)$ are the corresponding DFT's then for the periodic convolution of $x_1(nT_s)$ and $x_2(nT_s)$ defined by

$$y(nT_s) = T_s \sum_{m=0}^{N-1} x_1(mT_s) x_2(nT_s - mT_s)$$

$$Y(kF_s) = X_1(kF_s) \cdot X_2(kF_s)$$

where $Y(kF_s)$ is the DFT of $y(nT_s)$.

10. Find the spectrum of cosine pulse

$$x(t) = \begin{cases} A\cos\omega_0 t & |t| \le \tau/2 \\ 0 & \text{otherwise} \end{cases}$$

and raised cosine pulse

$$x(t) = \begin{cases} (A/2)(1 + \cos 2\omega_0 t); & |t| \le \tau/2 \\ 0 & \text{otherwise} \end{cases}$$

expressing your answer in terms of sine function. In each case sketch the spectrum and estimate the spectral width.

11. A finite duration cosine wave of frequency f_c has precisely N cycles. Sketch its spectrum. How large must N be so that the significant spectral width is less than 1% of f_c?

12. For the double exponential pulse $f(t) = \sqrt{b}\, e^{-b|t|}$, $b > 0$, find the second moment bandwidth $\Delta\Omega_2$, duration ΔT_2 and uncertainty relationship.

13. Obtain the second moment duration-bandwidth relationship for the pulse

$\left[u\left(t + \frac{\tau}{2}\right) - u\left(t - \frac{\tau}{2}\right) \right] \cos\frac{\pi t}{\tau}$. How is it affected by the parameter τ?

14. If a signal is bandlimited to W, show that the magnitude squared of the corresponding analytic signal is bandlimited to $2W$.

15. Show that the time function $(j/\pi t)\exp(-j\omega_c t)$ has the 'shifting' property, i.e.,

$$[(j/\pi t)\exp(-j\omega_c t)] * \tilde{x}(t) = \tilde{x}(t)$$

where $\tilde{x}(t)$ is the complex envelope and asterisk denotes convolution.

16. Consider the RLC filter shown in Fig. P. 16. Let $\omega_0^2 = 1/LC$, $\omega_1 = 1/RC$ and $\omega_1 \ll \omega_0$. Show that the complex impulse response is given approximately by

$$\tilde{h}(t) = \frac{\omega_1}{2} \exp\left(-\frac{\omega_1 t}{2} \right), t \ge 0$$

Obtain the response $y(t)$ of the circuit to an input $x(t)$ which is an RF carrier pulse of amplitude A_1, duration T, and frequency ω_0.

17. For a frequency modulated signal

$$x(t) = \cos\left[\omega_0 t + D_f \int_{-\infty}^{t} m(\tau)\, d\tau\right]$$

Obtain the complex envelope $\tilde{x}(t)$, the inphase and quadrature signals $x_c(t)$ and $x_s(t)$, the envelope $r(t)$ and phase $\theta(t)$. Obtain the same for a single sideband phase modulation (SSB-PM) signal whose complex envelope is given by

$$\tilde{x}(t) = \exp\{jD_p[m(t) + j\hat{m}(t)]\}$$

where $m(t)$ is the modulating signal and $\hat{m}(t)$ its Hilbert transform.

18. (Principle of Stationary Phase [2]). Consider the following integral:

$$\tilde{U}(f) = \int_{-\infty}^{\infty} a(t) \exp[j\mu(t,f)]\, dt$$

where $\mu(t,f) = \theta(t) - 2\pi ft$

The inverse Fourier transform of $\tilde{U}(f)$ is

$$\tilde{u}(t) = a(t) \exp[j\theta(t)]$$

We assume that $a(t)$ varies slowly compared to factors $\cos[\mu(t, f)]$ and $\sin[\mu(t, f)]$, which go through a large number of oscillations within the range of integration. The principle of stationary phase states that $\tilde{U}(f)$ will have its greatest values when $\frac{\partial}{\partial t}[\mu(t, f)] = 0$

(i) Using the first three terms in a Taylor series expansion of $\mu(t, f_k)$, show that, according to this principle, the value of $\tilde{U}(f)$ at $f = f_k$ is approximated as

$$\tilde{U}(f_k) \simeq \sqrt{\frac{2\pi j}{\mu''(t_k, f_k)}}\, a(t_k) \exp[j\mu(t_k, f_k)]$$

where t_k denotes the values of time t for which the first derivative of $\mu(t, f_k)$ is zero and

$$\mu''(t_k, f_k) = \left.\frac{\partial^2 \mu}{\partial t^2}\right|_{t=t_k}$$

(Note: $\int_{-\infty}^{\infty} \exp(jy^2)\, dy = \sqrt{\pi j}$)

(ii) The complex envelope of a pulse $u(t)$ is given as

$$\tilde{u}(t) = \begin{cases} \exp\left(j\pi k t^2\right) & |t| \le T/2 \\ 0 & \text{otherwise} \end{cases}$$

where k is a positive constant. Assuming k is large, use the result for $\tilde{U}(f)$ in (i) to show that the spectrum of $\tilde{u}(t)$ is given by

$$\tilde{U}(f) = \begin{cases} \dfrac{1}{\sqrt{k}} \exp\left(-j\dfrac{\pi f^2}{k} + j\dfrac{\pi}{4}\right), & |f| \le \dfrac{kT}{2} \\ 0 & \text{otherwise} \end{cases}$$

(iii) How does parameter k affect the time-bandwidth product of $u(t)$?

19. If $y(t)$ is the response of an LTI system $H(\omega)$ to a W-bandlimited input $x(t)$ of energy E_x, then show that

$$|y(t)| \le \left|\frac{E_x}{2\pi}\int_{-2\pi W}^{2\pi W} |H(\omega)|^2 \, d\omega\right|^{1/2}$$

$$|x(t)| \le \sqrt{2E_x W}$$

$$|x'(t)| \le 2\pi W\sqrt{\frac{2}{3}E_x W}$$

Furthermore if bandlimited function $x(t)$ is approximated by the finite series

$$x_N(t) = \sum_{n=-N}^{N} x(nT_s)\frac{\sin 2\pi W(t - nT_s)}{2\pi W(t - nT_s)}$$

then show that

$$|x(t) - x_N(t)| \le \sqrt{2We_N}$$

$$\text{where } e_N = \int_{-\infty}^{\infty} |x(t) - x_N(t)|^2 \, dt$$

$$= T_s \sum_{|n|>N} |x(nT_s)|^2$$

20. (Orthogonality principle or projection theorem). An arbitrary real function $x(t)$ is to be approximated by a linear combination

$$\mathbf{x} = \sum_{k=1}^{n} a_k \phi_k(t)$$

of n linearly independent signals $\phi_k(t)$. The constants a_k are to be selected so as to minimize the mean square error

$$e = \int_{-\infty}^{\infty} |x(l) - \sum_{k=1}^{n} a_k \phi_k(t)|^2 \, dt = \langle x - \mathbf{x}, x - \mathbf{x}\rangle$$

where $\langle x, y\rangle$ defines the inner product of two functions $x(t)$ and $y(t)$ as

$$\langle x, y\rangle = \int_{-\infty}^{\infty} x(t) y * (t)\, dt$$

Show that the optimum a_i are such that the error $[x(t) - \mathbf{x}(t)]$ is orthogonal to $\phi_i(t)$, i.e.,

$$\langle x - \mathbf{x}, \phi_i\rangle = 0, \qquad i = 1, \ldots, n$$

Furthermore, if $\{\phi_k(t)\}$ form an orthonormal set, i.e.,

$$\langle \phi_i, \phi_j\rangle = \delta_{ij}$$

then show that the optimum coefficients a_i are given by

$$a_i = \langle x, \phi_i\rangle$$

Also,

$$\langle \mathbf{x}, \mathbf{x}\rangle = \sum_{k=1}^{n} |a_k|^2$$

The optimum mean square approximation $\mathbf{x}(t)$ of $x(t)$ is called the 'projection' of $x(t)$ on the linear space S of functions $\phi_i(t)$, because

$$\langle x - \mathbf{x}, x\rangle = 0 \text{ and } \langle x, \mathbf{x}\rangle = \langle \mathbf{x}, \mathbf{x}\rangle$$

Also,

$$e = \int_{-\infty}^{\infty} |x(t)|^2\, dt - \sum_{i=1}^{n} |a_k|^2$$

21. Given n arbitrary sampling instants t_k and a real W-bandlimited function $x(t)$ with energy E_x, let us determine the constants a_k such that the sum

$$\mathbf{x}(t) = \sum_{k=1}^{n} a_k \phi_k(t); \; \phi_k(t) = \frac{\sin 2\pi W(t - t_k)}{\pi(t - t_k)}$$

is the optimum mean square approximation of $x(t)$. Show that $\mathbf{x}(t)$ is also an interpolator of $x(t)$, i.e., $\mathbf{x}(t_k) = x(k)$.

22. Let a random process $x(t)$ be given by $x(t) = a$ where a is a random variable taking on values 1 and –1 with probabilities p and q respectively. Find the a.c.f. of $x(t)$. Is this process wide sense stationary? Is it ergodic?

23. Let $y(t) = x^2(t)$. Let $x(t)$ be a random process with sample functions of the form $x(t) = a \cos(\omega t + \theta)$ where a and θ are statistically independent random variables and θ is uniformly distributed on $(0, 2\pi)$. Find the a.c.f. of $x(t)$ and $y(t)$. Show that the ensemble average is not equal to time average of the sample function unless a is nonrandom constant.

24. $x(t)$ and $y(t)$ are statistically independent processes which are individually and jointly widesense stationary. Let

$$z(t) = x(t) + y(t)$$

$$\omega(t) = x(t) - y(t)$$

Determine $R_z(\tau)$, $R_\omega(\tau)$, $R_{z\omega}(\tau)$, $R_{\omega z}(\tau)$.

25. Consider the random processes

$$x(t) = \sum_{i=1}^{N} a_i \exp(j\omega_i t)$$

$$x(t) = \sum_{i=1}^{N} a_i \cos \omega_i t + b_i \sin \omega_i t$$

where a_i and b_i are zero-mean uncorrelated random variables with variance σ_i^2. Obtain their a.c.f.

26. Let x_1 and x_2 be two random variables with correlation coefficient ρ and variance σ^2. Find the linear transformation A which transforms them into uncorrelated random variable y_1 and y_2 each with unit variance.

27. For each of random processes $\{x(t), t \geq 0\}$ given below, obtain (i) the mean value function $m(t)$, (ii) covariance kernel $K_x(t_1, t_2)$, (iii) autocorrelation function $R_x(\tau)$ if the process is stationary.
    (a) $x(t) = a \cos \omega t + b \sin \omega t$; ω is a positive constant, a and b are uncorrelated r.v.'s with means 0 and variance σ^2.
    (b) $x(t) = \cos(\omega t + \theta)$; ω is a positive constants and θ uniformly distributed on $(0, 2\pi)$.
    (c) $x(t) = \cos(\omega t + \phi(t))$; ω a positive constant and $\{\phi(t), t > 0\}$ is a random process defined by $\phi(t) = W(t+1) - W(t)$

    where $\{W(t), t \geq 0\}$ is a Wiener process with parameter σ^2.

28. The output of a 'jittery' oscillator is described by

$$x(t) = A \cos(2\pi f t + \theta)$$

where A is a constant, and f and θ are s.i. random variables. Let the p.d.f. of f be $p(f)$ and that of θ is uniformly distributed on $(0, 2\pi)$. Show that the p.s.d. of $x(t)$ is given by

$$S_{xx}(f) = \frac{A^2}{2} p(f).$$

29. Two RC amplifiers with transfer functions $H_\alpha(\omega) = \dfrac{1}{1 + j(\omega/\alpha)}$ and $H_\beta(\omega) = \dfrac{1}{1 + j(\omega/\beta)}$ are cascaded. If the input $x(t)$ to this system is white noise with two-sided p.s.d. $N_0/2$ and the output is $y(t)$, find
    (a) $S_{yy}(\omega)$ and $R_{yy}(\tau)$
    (b) total average noise power $E[Z^2] = \text{var}[Z]$ at the output. Show that it is the same as total average noise power from a single amplifier with transfer function $\dfrac{1}{1 + j(\omega/\omega_0)}$ with $\omega_0 = \dfrac{\alpha \cdot \beta}{\alpha + \beta}$.
    (c) $R_{zz}(\tau)$ when $\alpha = \beta$.

30. We wish to generate a stationary random process $x(t)$, with zero-mean and desired a.c.f. $R_{xx}(\tau)$ or p.s.d. $S_{xx}(f)$ by exciting an LTI filter with impulse response $h(t)$ by an input white noise process of two-sided p.s.d. of $N_0/2$. Determine the condition on $h(t)$ and on the corresponding transfer function $H(f)$ of the filter to achieve this objective. If the filter is to be physically realizable, what is the requirement on $S_{xx}(f)$?

31. In the system shown in Fig. P. 31, $x(t)$ is a Gaussian random process with zero-mean and a.c.f.

$$R_{xx}(\tau) = \frac{2\sin \pi\tau}{\pi\tau}$$

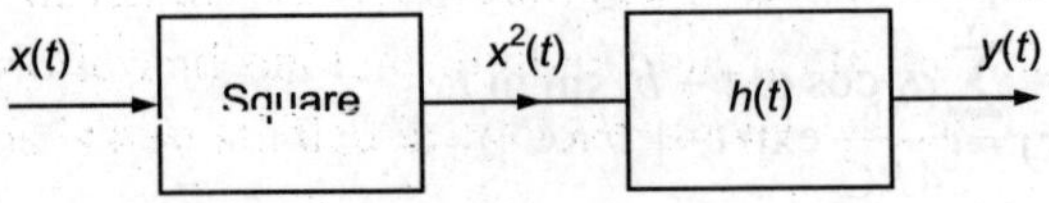

Figure P. 31

Find expressions for $m_y(t)$ and $R_{yy}(t_1, t_2)$. Is $y(t)$ wide sense stationary? (**Hint:** Use the result (1.309)) on mixed moments of jointly distributed Gaussian r.v.'s).

32. Show that the a.c.f. of angle modulated signal $y(t) = \cos(\omega_c t + x(t) + \theta)$ where θ is a uniformly distributed r.v. on $(0, 2\pi)$ is given by

$$R_{yy}(\tau) = \frac{1}{2} \text{ Re } \{\exp(j\omega_c t)\, E[\exp[j(x(t+\tau)) - x(t)]]\}$$

(i) Use Taylor series expansion of $x(t+\tau)$ about t and assume that the second and higher derivatives of $x(t)$ may be neglected. Now show that

$$S_{yy}(\omega) = \frac{1}{4} \cdot 2\pi[p_1(\omega - \omega_c) + p_1(\omega + \omega_c)]$$

where $p_1(x) = p(\dot{x})$ is the p.d.f. of derivative process $\dot{x} = \frac{dx}{dt}$. Assume $p(\dot{x})$ is even. This result is known as 'FM adiabatic theorem' or 'Woodwards theorem'. Interpret the result in terms of 'slowly varying' $x(t)$.

33. A constant level a is to be estimated from its noisy observation $x(t) = a + n(t)$ where the noise $n(t)$ has zero-mean and a.c.f. $R_{nn}(\tau)$. We use time average over an interval as our estimate $\hat{a}$, i.e,

$$\hat{a} = \frac{1}{T}\int_0^T x(t)\, dt$$

(a) Show that $E[\hat{a}] = a$, var $[\hat{a}] = \frac{1}{T}\int_{-T}^{T}\left(1 - \frac{|\tau|}{T}\right) R_{nn}(\tau)\, d\tau$

(b) If $R_{nn}(\tau)$ is of duration much smaller compared to T and its area equals s, then var $[\hat{a}] = \frac{s}{T}$. What should be the value of T so that $p[|\hat{a} - a| \le 0.1a] \ge 0.99$?

(c) What is the value of T if $n(t)$ is $N(0, 1)$, i.e, zero-mean Gaussian process with unity variance?

34. If $x(t)$ is a stationary process with mean m_x and p.s.d. $S_{xx}(\omega)$, show that for the random variable

$$y = \frac{1}{n}[x(T) + x(2T) + \ldots + x(nT)]$$

which represents its sample mean,

$$E(y) = m_x \text{ and var } [y] = \frac{1}{2\pi n^2}\int_{-\infty}^{\infty} S_{xx}(\omega)\frac{\sin^2 n\omega T/2}{\sin^2 \omega T/2}\, d\omega$$

35. Show that the a.c.f of the output $y(t)$ of a simple R - C filter excited by zero-mean white Gaussian noise of two sided p.s.d. of $N_0/2$, is given by

$$R_y(\tau) = \frac{N_0}{4RC}\exp(-|\tau/RC|)$$

Prove that $y(t)$ is a first order Markov process. Such a process has the property that

$$p(y_n/y_{n-1}, y_{n-2} \cdot \cdot, y_1) = p(y_n|y_{n-1})$$

To prove this, define $t_3 - t_2 = t_2 - t_1 = \Delta$, exp, $(-\omega_0|\Delta|) = \rho$, $\omega_0 = \frac{1}{RC}$, $y(t_3) = y_3$, $y(t_2) = y_2$ and $y(t_1) = y_1$, where $t_3 > t_2 > t_1$. Now derive $p(y_1|y_2, y_3)$ and show that $p(y_1|y_2, y_3) = p(y_1|y_2)$.

36. $x(t)$ is a stationary zero-mean Gaussian process with a.c.f. $R_{xx}(\tau) = be^{-k|\tau|}$, in the system of Problem 32 and $h(t)$ is an integrator with impulse response

$$h(t) = \frac{1}{T}\int_{t-T}^{t} du.$$

Show that $y(t)$ provides an unbiased an consistent estimate of the average power in $x(t)$. An *unbiased* estimate of a parameter has a mean value equal to the parameter being measured. A *consistent* estimate is an unbiased estimate whose variance tends to zero as the sample size (in this case, T) tends to infinity.

37. A zero-mean stationary Gaussian random process $x(t)$ with variance σ_x is applied to an ideal limiter whose input-output characteristic is shown in Fig. P. 37. Show that the a.c.f. of the limiter output is given by

$$R_{yy}(\tau) = \frac{2}{\pi}\sin^{-1}\left[\frac{R_x(\tau)}{\sigma_x^2}\right]$$

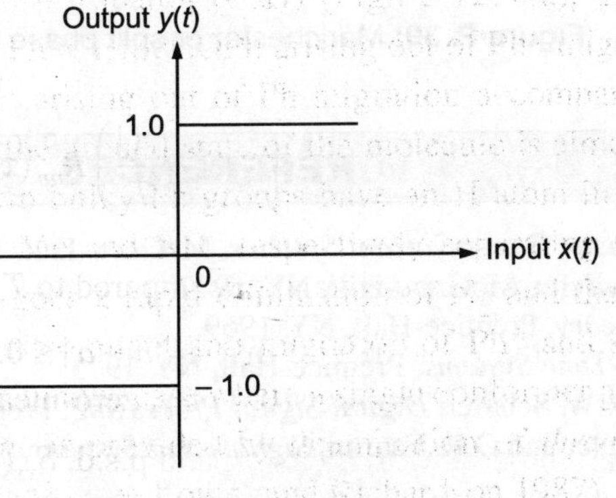

Figure P. 37: Ideal limiter

38. The average power $y = \frac{1}{T}\int_0^T x^2(t)\,dt$ of a zero-mean Gaussian process may be approximated by a central chi-square p.d.f. Using the results on the mean and variance of chi-square p.d.f., show that the number of degrees of freedom of y is approximately

$$\nu \approx R_{xx}^2(0)\left[\frac{1}{T}\int_{-T}^{T}\left(1-\frac{|\tau|}{T}R_{xx}^2(\tau)\right)d\tau\right]^{-1}$$

and for large T,

$$\nu \approx R_{xx}^2(0)\left[\frac{1}{T}\int_{-\infty}^{\infty}R_{xx}^2(\tau)d\tau\right]^{-1}$$

If $x(t)$ is W bandlimited white noise, then show that $\nu = 2WT$.

39. Using the method of Section 1.10, obtain the p.s.d. of a PCM signal which uses Manchester or split phase encoding shown in Fig. P. 39. Here each binary 1 is represented by a positive half-bit period pulse followed by a negative half-bit period pulse. Similarly, a binary 0 is represented by the complement of binary 1 representation. Assume that data have values $a_n = \pm 1$ which are uncorrelated and equally probable.

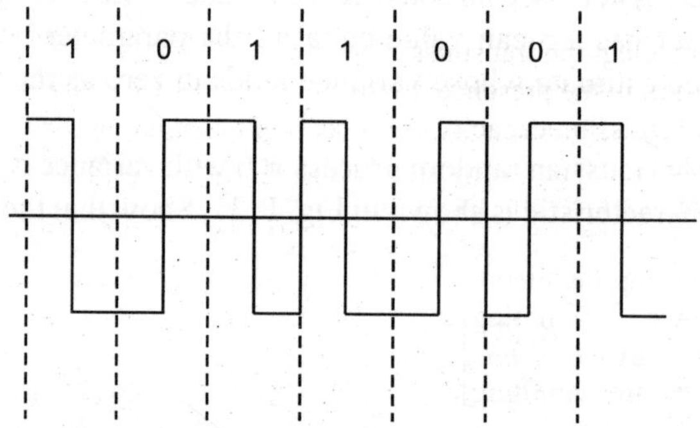

Figure P. 39: Manchester or split phase encoding

REFERENCES

1. Brown, *Analysis of Linear Time-invariant Systems*, McGraw-Hill, 1963.
2. A. Papoulis, *Signal Analysis*, McGraw-Hill, NY, 1977.
3. L.E. Franks, *Signal Theory*, Prentice-Hall, NY, 1969.
4. J.A. Cadzow, *Discrete-Time Systems*, Prentice-Hall, NY, 1973.
5. A.V. Oppenheim and R.W. Schafer, *Digital Signal Processing*, Prentice-Hall, NY, 1975.
6. R.E. Ziemer, W.H. Tranter and D.R. Fannin, *Signals and Systems, Continuous and Discrete*, McMillan, NY, 1983.
7. D. Slepian, 'On bandwidth', Proc. *IEEE*, **64**(3), March 1976, 292–300.
8. I. Gonorovsky, *Radio Circuits and Signals*, Mir Publishers, Moscow, 1974.
9. J.B. Thomas, *An Introduction to Statistical Communication Theory*, John Wiley NY, 1979.
10. I. Kay and R. Silverman, 'On the uncertainty relation for real signals', *Information and Control*, **1**, 1957, 64–76.

11. W. Hilberg and P.G. Rothe, 'The general uncertainty relation for real signals in communication theory', *Information and Control* **18**(2), March 1971, 103–125.
12. D. Gabor, 'Theory of communication', *J. Inst. Elec. Engrs.* (London), **93**, 1946, 429–441.
13. R.C. Dixon, *Spread Spectrum Systems*, John Wiley, NY, 1976.
14. C.E. Cook and M. Bernfeld, *Radar Signals*, Academic Press, NY, 1967.
15. A.D. Whalen, *Detection of Signals in Noise*, Academic Press, NY, 1971.
16. S. Haykin, *Communication Systems*, John Wiley, NY, 1978.
17. S. Stein and J.J. Jones, *Modern Communication Principles*, McGraw-Hill, NY, 1967.
18. W.M. Brown and C. J. Palermo, *Random Processes, Communications and Radar*, McGraw-Hill, NY, 1969, Chapter 5.
19. E. Bedrosian, 'The analytic signal representation of modulated waveforms', *Proc. IRE*, **50**(10), Oct. 1962.
20. A.W. Naylor and G.R. Sell, 'Linear operator theory in engineering and science', *Springer-Verlag Series in Mathematical Sciences (40)*, Springer-Verlag, 1982.
21. D. Slepian, H. O. Pallak and H. J. Landau, 'Prolate spheroidal wavefunctions, Fourier analysis and uncertainty principle, I and II', *Bell Syst. Tech. J.*, **40**(1), 43–84, 1961.
22. A. Papoulis, 'Error analysis in sampling theory', *Proc. IEEE*, 54, July 1966, 947–955.
23. R.M. Lerner, 'Representation of signals', *Lectures on Communication System Theory*, ed. E. J. Baghdady, Wiley, NY, 1961.
24. H.S. Black, *Modulation Theory*, Van Nostrand, 1953.
25. P.F. Panter, *Modulation, Noise and Spectral Analysis*, McGraw-Hill, NY, 1965.
26. Gerlach, *Theory and Application of Statistical Wave Period Processing*, Vol. I, Gorden and Breach, Science Publishers, NY, 1970.
27. E.T. Whittaker, 'On the functions which are represented by the expansion of interpolating theory', *Proc. Roy. Soc. Edinburgh*, **35,** 1915, 181–194.
28. H. Nyquist, 'Certain topics in telegraph transmission theory', *AIEE Trans.*, 47, 1928, 617–644.
29. C.E. Shannon, 'Communication in the presence of noise', *Proc, IRE,* **37**, June 1949, 10–21.
30. V.A. Kotelnikov, 'On the transmission capacity of "ether" and wire in telecommunications', *Izd. Red. Upr. Sryazi RKKA* (Moscow), 1933.
31. A.J. Jerri, 'The Shannon sampling theorem–its various extensions and applications: A tutorial review', *Proc. IEEE*, **65**(11), Nov. 1977, 1565–1596.
32. H.P. Kramer, 'A generalized sampling theorem', *J. Math. and Phys.*, **38**, 1959, 68–72.
33. D.P. Peterson and D. Middleton, 'Sampling and reconstruction of wave number limited functions in *N*-dimensional Euclidean spaces', *Inform. and Contr.*, **5**, Dec. 1962, 279-323.
34. A.V. Balakrishnan, 'A note on the sampling principle of continuous signals', *IRE Trans. Inform. Theory*, **IT-3**, June 1952, 143–146.
35. D. Middleton, *Introduction to Statistical Communication Theory*, McGraw-Hill, NY, 1960.
36. D.P. Peterson and D. Middleton, 'Linear interpolation, extrapolation and prediction of random space time field with limited domains of measurement', *IEEE Trans. on Infor. Theory*, **IT-11**(1), Jan. 1965.
37. D.L. Jagerman and L. Fogel, 'Some general aspects of sampling theorem,' *IRE Trans. On Inform. Theory*, **IT-2**, Dec. 1956, 139–146.
38. D.A. Linden and N.M. Abramson, 'A generalization of sampling theorem', *Inform. and Contr.*, **3**, 1960, 26–31.
39. A. Kohlenberg, 'Exact interpolation of bandlimited functions', *J. App. Phys.*, **24**, 1953, 1432–1436.
40. D.A. Linden, 'A discussion of sampling theorem', *Proc. IRE*, **47**, 1959, 1219–1226.
41. D.A. Gabor, 'Communication theory and cybernetics', *IRE Trans. on Circuit Theory*, **CT-1**(4), Dec. 1954, 19–31.
42. J.L. Yen, 'On the nonuniform sampling of bandwidth limited signals', *IRE Trans. Circuit Theory*, **CT-3**, Dec. 1956 251–257.
43. F.J. Beutler, 'Error-free recovery of signals from irregularly spaced samples', *SIAM Rev.*, **8(3)**, July 1966, 328–335.
44. N. Levinson, *Gap and Density Theorems*, Colloq. Pub. 26, Amer. Math. Soc., New York, 1940.

45. S.Cambanis, N.K. Habib, 'Finite sampling approximations for non-bandlimited signals', *IEEE Trans. Inform. Theory*, **IT-28**, Sept. 1982, 807–809.
46. R. Hulthen, 'Restoring causal signals by analytic continuation: A generalized sampling theorem for causal signals', *IEEE Trans. on Accoust, Speech, Signal Processing*, **ASSP-31**(3), Oct. 1983, 1294–1298.
47. E. Parzen, *Modern Probability Theory and Its Applications*, John Wiley, NY, 1960.
48. W. Feller, *An Introduction to Probability Theory and Its Applications*, John Wiley, 1957.
49. E. Parzen, *Stochastic Processes*, Holden Day, San Francisco, Calif, 1962.
50. A. Papoulis, *Probability, Random Variables and Stochastic Processes*, McGraw-Hill, NY, 1965.
51. K.L. Chung, *Elementary Probability Theory with Stochastic Processes*, Springer-Verlag, 1979.
52. W.C. Lindsay, *Synchronization Systems in Communication and Control*, Prentice-Hall, 1972.
53. J.M. Wozencraft and I.M. Jacobs, *Principles of Communications Engineering*, John Wiley, NY, 1965, pp. 136–143.
54. N. Wiener, *Time Series Extrapolation, Interpolations and Smoothing of Stationary Time Series*, MIT Press, 1964.
55. J.G. Prokias, *Digital Communications*, McGraw-Hill, NY, 1983.
56. J.K. Holms, *Coherent Spread Spectrum Systems*, John Wiley, NY, 1982, Chapter 2.
57. H. Cramer, *Mathematical Methods of Statistics*, Princeton University Press, Princeton, NY, 1964, pp. 118.
58. H. Urkowitz, *Lecture Notes on 'Introduction to Signal Processing'*, Drexel University, Philadelphia, USA, 1968.
59. S.O. Rice, 'Mathematical analysis of random noise', *BSTJ*, **23**, July 1944, 282–332; **24**, Jan. 1945, 46–56. These papers are reprinted in N. Wax, *Selected papers on Noise and Stochastic Processes*, Dover Press, NY, 1954, pp. 133–294.
60. J.L. Marcum, 'A statistical theory of target detection by pulsed radar', *IRE Trans. Inform. Theory*, **IT-6**(2), 259–268.
61. A Wald, *Statistical Theory with Engineering Applications, Wiley*, NY, 1952.
62. M. Abromowitz and I.A. Stegun, *Handbook of Mathematical Functions*, NBS Applied Math. Series, Vol. 55, Supt. of Documents, U.S. Government Printing Office, Washington, DC.
63. P.B Patnaik, 'The noncentral χ^2 and F distribution and their applications', *Biometrika*, **36**, 1949, 202–232.
64. N.W. McLachlan, *Bessel Functions for Engineers*, 2nd Ed., Oxford Univ. Press, London, 1955.
65. A.J. Viterbi, 'Optimum detection and signal selection for partially coherent binary communication', *IEEE Trans. Inform. Theory*, **IT-11**(2), April 1965, 239–246.
66. W.B. Davenport Jr. and W.L. Root, *Random Signals and Noise*, McGraw-Hill, NY, 1958.

CHAPTER 2

Digital Representation of Signals

2.0 INTRODUCTION [1, 2, 3, 4, 5]

The sampling theorem provides an efficient representation of a continuous waveform through discrete samples taken at intervals of $1/2W$, where W is the lowpass bandwidth of the signal $x(t)$. The waveform is reconstructed by multiplying the samples with an interpolating function e.g., $\frac{\sin x}{x}$, $u(t - T)$ etc., as discussed in Chapter 1. The reconstructed signal is given by:

$$\tilde{x}(t) = \sum_{n=-N}^{N} x\left(\frac{n}{2W}\right) \frac{\sin \pi (2Wt - n)}{\pi(2Wt - n)}, N = TW \tag{2.1}$$

where the number of samples = $(2TW + 1)$ in T secs, should be large. In a practical transmission system, the sampled values of $x(t)$ are not correctly received due to noise, distortion, and crosstalk in the medium. If, however, the sample values are digitally represented and then transmitted, the system becomes more robust and noise-resistant.

The representation of an analog waveform in the digital form requires quantization in time as well as in amplitude. The samples $x(n)$ are first quantized and then coded in a binary form to produce the equivalent digital bit stream, and the total signal processing involved to obtain the approximate $\tilde{x}(t)$ is seen as:

$$x(t) \rightarrow \text{sampler} \rightarrow \text{quantizer} \rightarrow \text{coder} \rightarrow \text{Tx path} \rightarrow \text{regenerator}$$
$$\rightarrow \text{decoder} \rightarrow \text{sampler} \rightarrow \text{lowpass filter} \rightarrow \tilde{x}(t)$$

The generation and recovery of the binary equivalent of $x(t)$ is the nowclassical pulse code modulation technique (PCM), which has been used for long distance communication since 1960's. A simple technique of generating PCM signals is to use successive approximation of $x(n)$, as illustrated in Fig. 2.1. Here a sample value of 85 units is successively compared with reference values of 64, 32, 16 etc. and wherever the remainder is greater than the reference, digit 1 is produced, thus giving the binary equivalent of 85 as: 1010101.

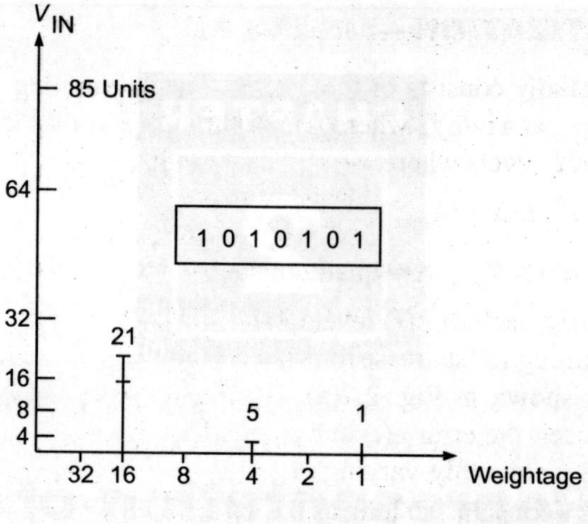

Figure 2.1: Binary encoding using successive approximation

Over the years, various other techniques have been developed for waveform coding, e.g., Delta Modulation (DM) with its various modified forms. While PCM and DM codecs are applicable for any arbitrary waveform, special codecs have been developed for speech waveforms, based on the parametric properties of speech. Such parametric speech coders are also quite a few in number and the Table 2.1 gives a summary of the codecs useful for speech encoding. We shall discuss below some of the important properties of these codecs.

Table 2.1

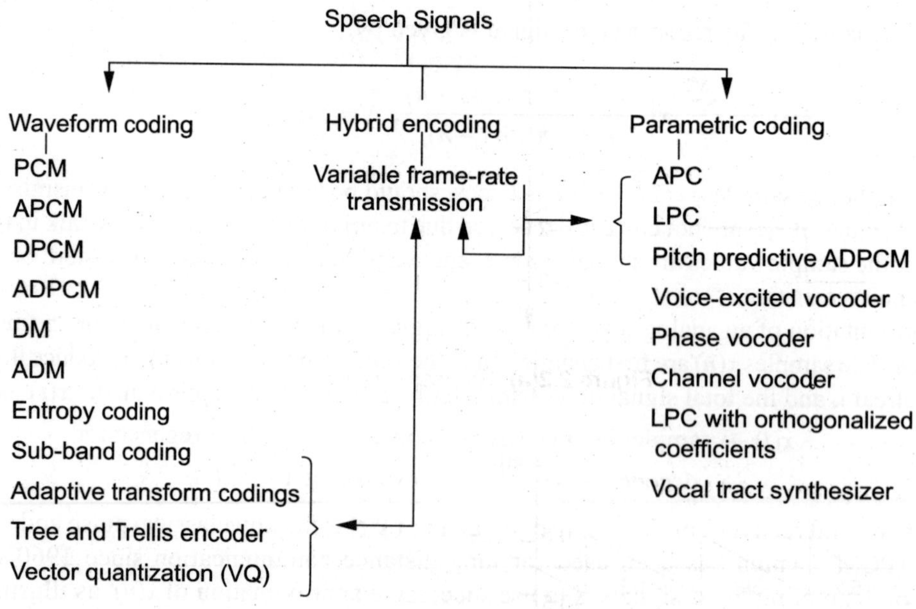

2.1 LINEAR QUANTIZATION—LINEAR PCM [6, 7]

The lincar PCM codec essentially consists of three operations—sampling, quantizing and coding. The message $x(t)$ is sampled as $x(r)$, at a rate f_s, where f_s is slightly greater than the Nyquist rate $2W$, and each sample is quantized to one of L levels where

$$L = 2^B,\ L\cdot\Delta = (X_{\max} - X_{\min}),$$

$$X_{\min} \leq x(t) \leq X_{\max};\ \Delta = \text{quantum step}, \tag{2.2}$$

giving B-bit words representing each of $L(r)$ levels.

Quantization is obtained through a 'staircase transducer' (equivalent to a fixed value quantizer), having the input-output relation as shown in Fig. 2.2(a). If a linear waveform (say a sawtooth voltage) is superimposed on the transducer, the error $e(t)$ due to quantization will be as shown in Fig. 2.2(*b*). The effect of the quantization on a smoothly varying $x(t)$ is shown in Fig. 2.2(*c*), where it is seen that the quantized output $\hat{x}(t)$ remains constant, so long as the signal is confined to a tread and jumps by one full step when the input increases beyond the tread. The quantization error voltage $e(t)$ is somewhat triangular in shape and for an arbitrary input $x(t)$, it is reasonably assumed that all error amplitudes within $\pm\Delta/2$ are equiprobable, giving

$$p(e) = 1/\Delta,$$

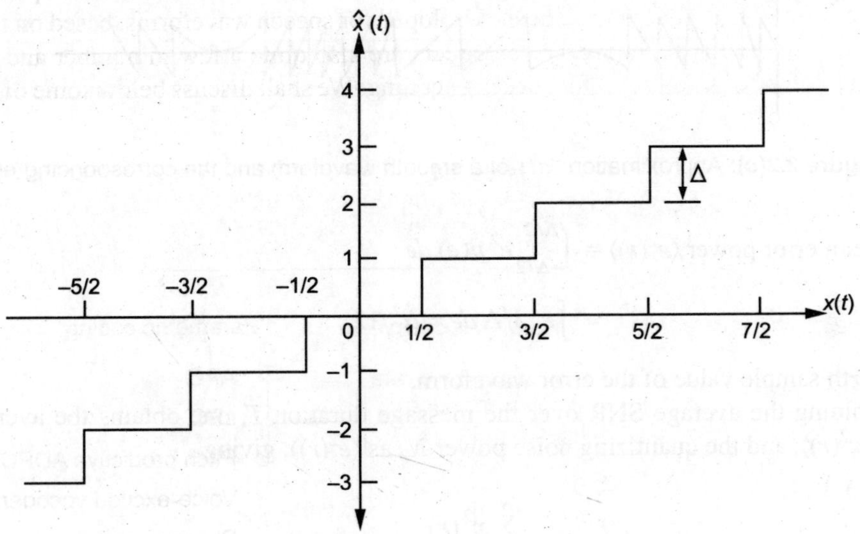

Figure 2.2(*a*): Staircase transducer

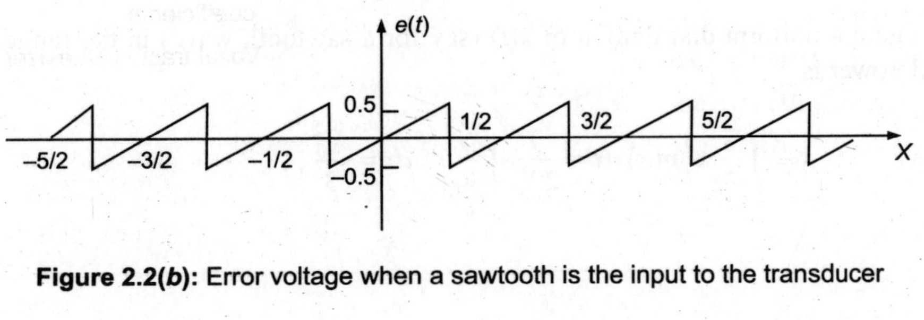

Figure 2.2(*b*): Error voltage when a sawtooth is the input to the transducer

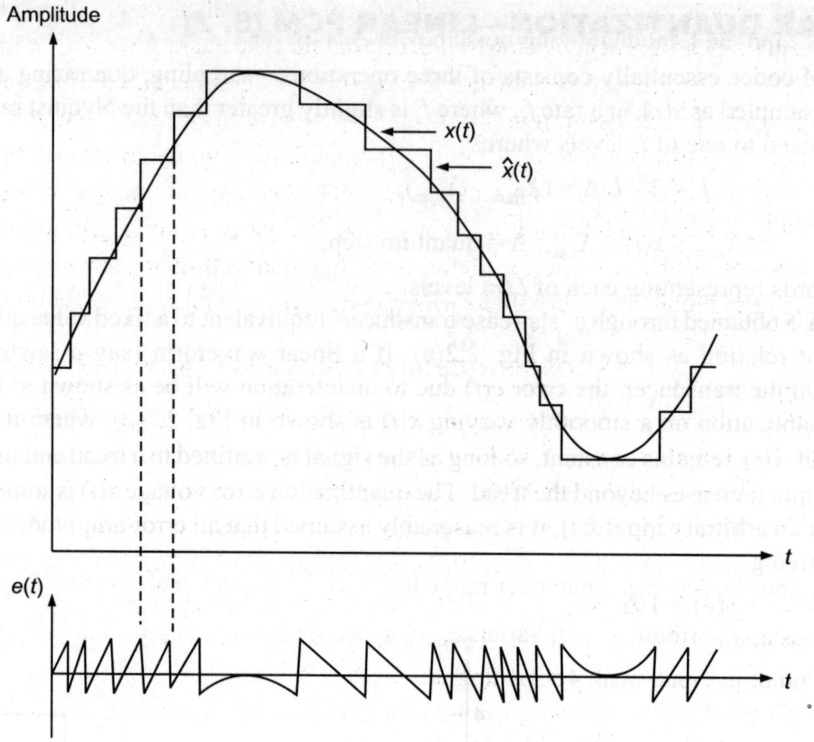

Figure 2.2(c): Approximation $\hat{x}(t)$ of a smooth waveform and the corresponding $e(t)$

Therefore, mean error power $\langle e^2(r)\rangle = \int_{-\Delta/2}^{\Delta/2} e^2 p(e)\, de$

$$= \int e^2 .1/\Delta\, de = \Delta^2/12 \tag{2.3}$$

where $e(r)$ is rth sample value of the error waveform.

For determining the average SNR over the message duration T, one obtains the average message powers S as $\langle x^2(r)\rangle$, and the quantizing noise power N_q as $\langle e^2(r)\rangle$, giving

$$\text{SNR} = \frac{S}{N_q} = \frac{\langle x^2(r)\rangle}{\langle e^2(r)\rangle} = \frac{\sum_{-N}^{N} x^2(r)}{\sum_{-N}^{N} e^2(r)}$$

Assuming again a uniform distribution of $x(t)$ (say for a sawtooth wave) in the range of $\pm X_0$, the average signal power is

$$S = \int_{-X_0}^{X_0} x^2 p(x)\, dx = \frac{1}{2X_0}\int_{-X_0}^{X_0} x^2\, dx = \frac{X_0^2}{3},$$

where $p(x) = \dfrac{1}{2X_0}$ and the total quantizing noise power N_q is

$$N_q = \int_{-X_0}^{X_0} \frac{\Delta^2}{12}\, p(x)\, dx = \frac{\Delta^2}{12} \tag{2.4}*$$

Since the maximum signal level $X_0 = \dfrac{L\Delta}{2} = \Delta \cdot 2^{B-1}$, therefore

$$\text{Peak SNR} = \frac{S}{N_q} = \frac{X_0^2}{3} \Big/ \frac{\Delta^2}{12} = \frac{(2X_0)^2}{\Delta^2} = 2^{2B} = 6B \text{ dB} \tag{2.5}$$

For a sinusoidal signal $a \cos (\omega t + \phi)$,

$$S = a^2/2, \text{ and peak SNR} = \frac{a^2}{2} / (\Delta^2/12)$$

$$= (3/2)2^{2B} = (6B + 2) \text{ dB} \tag{2.6}$$

For a random $x(t)$, the peak-to-peak quantizer range has to be $2X_{max}$ and in all practical cases $X_{max} = 4\sigma_x$ (specially for Gaussian distribution with variance σ_x^2 where $p(|x| > 4\sigma_x) \simeq 10^{-5}$).

Thus SNR for random inputs with 4σ loading is:

$$\text{Peak } S/N_q = \frac{X_{max}^2}{16} \Big/ \frac{\Delta^2}{12} = \frac{3}{16} \cdot 2^{2B}$$

$$= (6B + 5 - 12) = (6B - 7) \text{ dB} \tag{2.7}$$

In practical systems, the sampling rate f_s is made greater than $2W$, say $f_s = \dfrac{8W}{3}$, and in the receiver the lowpass filter bandwidth of W gives a reduction of Q-noise by $\dfrac{4}{3} = 1$dB only.

It should be understood that in deriving the above relations, the following idealised conditions have been assumed:

(a) The quantization noise is a stationary white noise process
(b) The Q-noise is uncorrelated with $x(t)$
(c) The Q-noise has a rectangular distribution over the quantization interval, $p(e) = 1/\Delta$.

All real signals $x(t)$, e.g., speech, are random (although non-stationary in practice) and cover all quantization levels, if $\Delta \ll X_0$, thus satisfying the above assumptions.

---

*For uniform quantization, $N_q = \dfrac{\Delta^2}{12}$ independent of the *pdf* of $x(t)$, since $\int_{-\infty}^{\infty} p(x)\, dx = 1$ for all different *pdf*'s. $N_q = \Sigma e^2(r)$ is obtained from the sampling theorem relation:

$$\text{Signal power} = \frac{1}{T}\int_{-\infty}^{\infty} x^2(t)\, dt = \frac{1}{2TW}\sum_{1}^{2TW} x^2(r).$$

It is seen from eq. (2.5) – (2.7) that S/N_q is dependent on the signal power S, and N_q is dependent only on Δ but not on S. Thus for a random signal with variance σ_x^2, the SNR is:

$$S/N_q = (6B - 7) - \sigma \text{ dB} \tag{2.8}$$

where the signal is σ dB below the reference 0 dB. This variation of S/N_q is shown in Fig. 2.3 for both sinusoidal and random inputs.

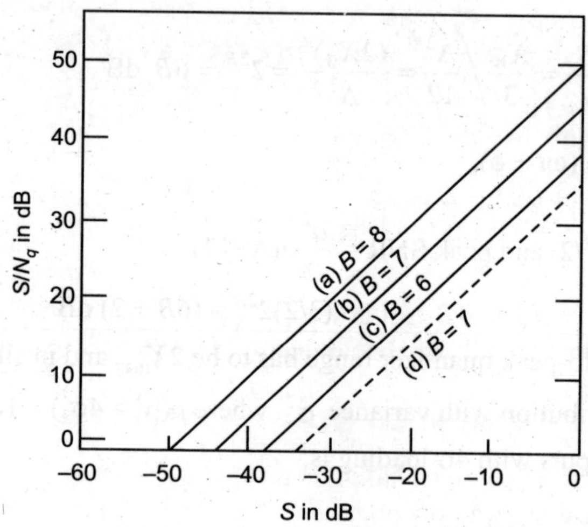

Figure 2.3: *S/N_q* vs *S*; ***(a), (b), (c)***, for sinusoidal input and ***(d)*** for random inputs

Example 2.1: Assume that a repetitive triangular waveform is quantized by 64 levels and the peak-to-peak value of the signal is $2X_0$. Assuming uniform distribution of the signal levels, the mean signal power is

$$S = X_0^2/3$$

and the quantizing noise power

$$N_q = \Delta^2/12, \Delta = X_0/32$$

Therefore,

$$S/N_q = 2^{12} = 36 \text{ dB}.$$

This gives a bandwidth expansion of six times (equivalent to six bits/sample) using Nyquist's bandwidth, but in practice the band-expansion $\simeq$ 9 times.

If, however, the signal is quantized to 81 levels, giving 4 ternary symbols per analog sample, then

$$N_q = \Delta^2/12, \Delta = 2X_0/81$$

Therefore,

$$S/N_q = (81)^2 = 38.2 \text{ dB}.$$

This gives a practical bandwidth expansion $\simeq$ 6 times.

Example 2.2: Assume that the signal amplitude has an exponential density function, given by,

$$p(x) = e^{-2|x|}$$

Then mean signal power $S = \sigma_x^2 = 0.5$.
For $x = 4.5$, $p(x) = 10^{-4}$. As such the effective peak X_0 of the signal $\simeq 4.5 = (4.5)\sqrt{2}\sigma_x = 6.3\sigma_x$.
If now the signal is quantized to 64 levels, then,

$$S/N_q = \frac{12\sigma_x^2}{\Delta^2} = \frac{12 \times 2^{10}}{(6.3)^2} = 25 \text{ dB}$$

For the ternary encoding with 4 symbols/sample,

$$S/N_q = \frac{12\sigma_x^2}{\Delta^2} = 12\left(\frac{27}{4.1}\right)^2 = 27 \text{ dB.}$$

Comparing the results of Ex. 2.1 and 2.2, it is seen that larger values of the crest factor ($= X_0/\sigma_x$) of the signal result into lesser values of S/N_q for linear quantization.

2.2 SPECTRA OF QUANTIZED SIGNALS [8]

The S/N_q relation above has been derived based on the quantization of the sample values of $x(t)$. But it is equally valid if sampling is done after quantization of a smooth $x(t)$, as shown in Fig. 2.2(c). Bennett [8] has derived the spectrum of $e(t)$ of Fig. 2.2(c) as obtained by passing $x(t)$ through a staircase transducer (by using the correlation theorem), and has shown that the normalized spectral density $G_e(\alpha)$ of $e(t)$ for a random flat-band input signal, $0 < f < W$, of power σ_x^2, is given by (with 4σ loading of the quantizer):

$$G_e(\alpha) = G_e\left(\frac{f}{W}\right) = \frac{\Delta^3}{2\pi^3\sigma_x^3}\sqrt{\frac{3}{2\pi}}\sum_{n=1}^{\infty}\frac{1}{n^3}\exp\left(-\frac{3\Delta^2 f^2}{8n^2\pi^2\sigma_x^2 W^2}\right) \quad (2.9)$$

The above error signal spectral density is shown in Fig. 2.4, for various values of B. It is seen that $G_e(\alpha)$ has lesser peak $G_e(0)$ but larger spread for finer quantization (i.e., for larger values of B). It is also shown that

$$\int_0^\infty G_e(\alpha)\, df = \cdot\, \Delta^2/12 = N_q, \text{ as shown earlier.}$$

The PCM coder output is now obtained by sampling the staircase waveform $\hat{x}(t)$ by an ideal sampling function $S(t)$ with $f_s \geq 2W$. The resultant sampled values $\hat{x}(r)$ is coded into binary words for transmission. Thus the sampler output, assuming no correlation between $x(t)$ and $e(t)$,

$$\begin{aligned} y(t) &= \hat{x}(t) * h_s(t),\ h_s(t) = \text{impulse response of the sampler} \\ &= [x(t) + e(t)] * h_s(t) \\ &= [x(t) * h_s(t)] + [e(t) * h_s(t)] \end{aligned} \quad (2.10)$$

The component $[x(t) * h_s(t)]$ gives $x(t)$ after lowpass filtering in the decoder-receiver. To obtain the Q-noise component, one has to multiply $G_e(\alpha)$ by the Fourier series representing the switching function

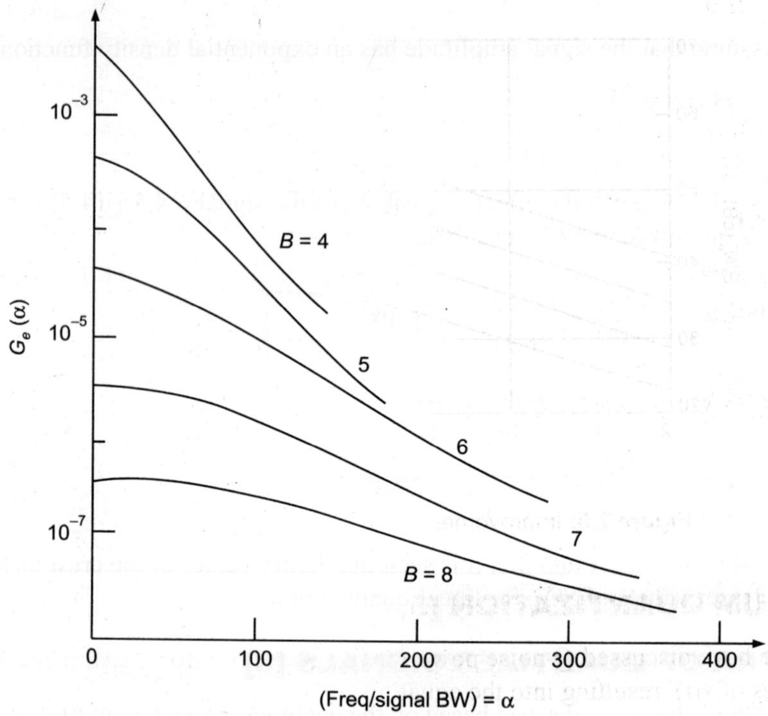

Figure 2.4: Error spectrum for random signals [8]

$S(t)$ and sum up all sideband contributions in the signal band [0, W] Hz. For an ideal switching function, infinite harmonics of f_s with unit amplitude are generated, and each harmonic nf_s beats with a band of error spectrum ($nf_s \pm W$) above and below nf_s, thus translating this error power to [0, W] band. The total error power in the signal band now equals to the area under $G_e(\alpha)$ curve of Fig. 2.4, as the contributions due to individual sidebands around nf_s are added linearly in the signal band. But the total error power accepted by the (0-W) lowpass filter decreases as f_s is increased, because each harmonic of the sampling frequency is now pushed upward into the portion of the error spectrum, with $G_e(\alpha)$ having lesser and lesser values. In the limit, as $f_s \to \infty$, we get the nonsampled case i.e., the output of the staircase transducer only, with the same value of N_q. Thus for $f_s \simeq 2W$,

$$N_q = \int_0^{\infty} G_e(\alpha)\, d\alpha \tag{2.11}$$

But for $f_s > 2W$,

$$N_q = \int_0^{1} G_e(\alpha)\, d\alpha + \sum_{n=2}^{\infty} \int_{(nf_s - W)/W}^{(nf_s + W)/W} G_e(\alpha)\, d\alpha \tag{2.12}$$

Based on eq. (2.12), S/N_q is plotted in Fig. 2.5, showing the decrease of N_q with increase of f_s. The initial points at minimum f_s are determined by eq. (2.11), since the accepted sidebands on the harmonics exactly fill out the entire error spectrum. The asymptotes at large f_s correspond to the area $= \int_0^1 G_e(\alpha) d\alpha$ only and the total improvement in N_q is 15–20 dB only with large f_s. However, this advantage has been utilized in DM and multilevel DM codecs, where $f_s \gg W$.

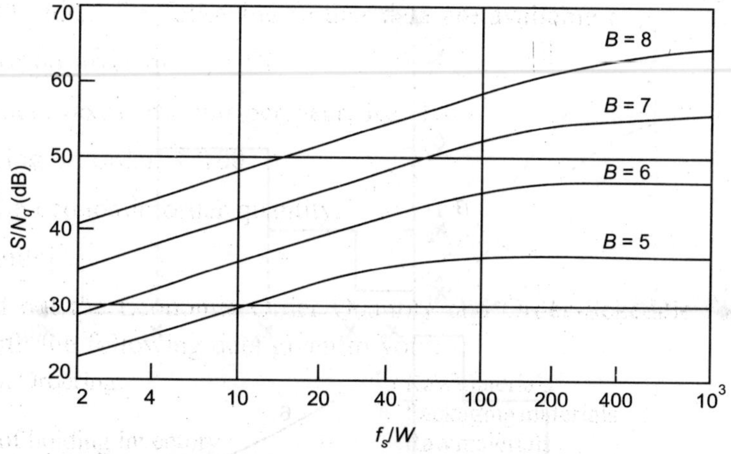

Figure 2.5: Improvement in S/N_q with f_s for random inputs

2.3 OPTIMUM QUANTIZATION [9, 10]

In Section 2.1, we have discussed Q-noise power on the basis of uniform quantization, i.e., Δ being the same for all values of $x(t)$, resulting into the equation

$$N_q = \int \frac{\Delta^2}{12}\, p(x)\, dx, \langle e^2(r) \rangle = \frac{\Delta^2}{12}$$

If, however, $p(x)$ is either Gaussian, Laplace or Gamma density function, then $p(x)$ at $x \leq \sigma_x$ is much larger than $p(x)$ at $x \gg \sigma_x$. As such, $\Delta(x)$, which is now a function of the input, may be made variable and N_q will be minimized if $\Delta(x)$ is small at $x \leq \sigma_x$ and $\Delta(x)$ is large otherwise, resulting into nonlinear (nonuniform) quantization, as shown in Fig. 2.6. This nonlinear quantization may be achieved either through companding (i.e., compressing the signal before quantization and correspondingly expanding the decoded signal at the receiver) or through minimizing N_q by determining the optimum $\Delta(x)$, given the variance and the pdf $p(x)$ of $x(t)$. Such optimum quantizers have been discussed by Max and others [9, 10].

With the nonlinear approximation of Fig. 2.6, Q-noise power is

$$N_q = \langle \{\hat{x}(t) - x(t)\}^2 \rangle = \langle e^2(t) \rangle = \int e^2\, p(e)\, de.$$

Over the piecewise linear portions of $e(t)$ curve, the instantaneous Q-noise is

$$N_q(j) = \int_{x_{j-i}}^{x_j} (\hat{x}_j - x)^2\, p(x)\, dx \tag{2.13}$$

Since

$$p(e) = p(\hat{x} - x) = p(x/\hat{x}) \cong p(x)$$

and the total Q-noise, assuming symmetry in $p(x)$, is

$$N_q = 2 \sum_{j=1}^{L/2} N_q(j), \tag{2.14}$$

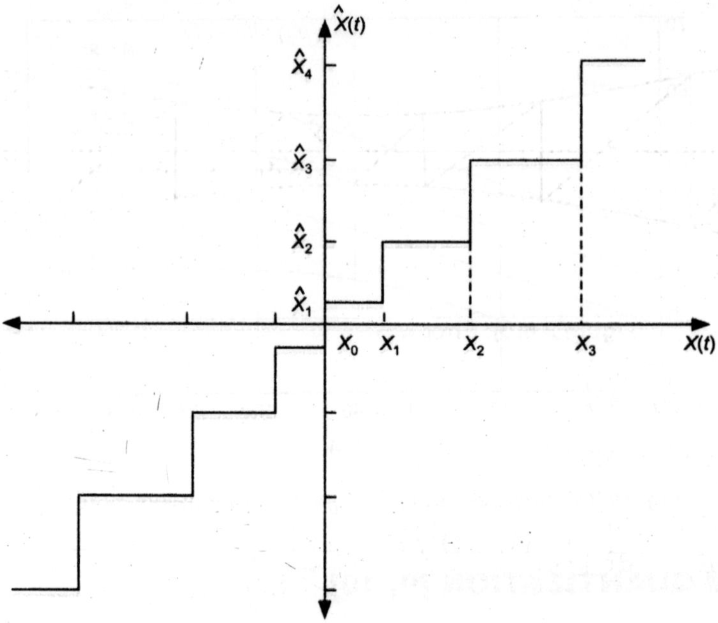

Figure 2.6(a): Nonlinear quantization

where $\hat{x}(L/2)$ represents x in the range $x(L/2-1) < x < \infty$. Since N_q is a function of the two variables $\hat{x}_j$ and x_j, the minimization of N_q requires to partially differentiate N_q with respect to x_j and $\hat{x}_j$ and the results equated to zero. Differentiating with respect to x_j, the eq. (2.14) leads to the condition that*

$$\left|\hat{x}_{j+1} - x_j\right| = \left|\hat{x}_j - x_j\right|$$

and

$$x_j = \frac{(\hat{x}_j + \hat{x}_{j+1})}{2},\ j = 1, 2, \ldots, \left(\frac{L}{2} - 1\right) \tag{2.15}$$

*Proof of Eq. (2.15): Referring to Fig. 2.6(b), it is seen that $e(t)$ is the difference between the step waveform $\hat{x}_j(t)$ and a straight line with a slope of $\pi/4$, and thus

$$\tfrac{1}{2} N_q = \int_0^{\hat{x}_1} (\hat{x}_1 - x)^2 p(x)\,dx + \int_{\hat{x}_1}^{x_1} (\hat{x}_1 - x)^2 p(x)\,dx + \int_{x_1}^{\hat{x}_2} (\hat{x}_2 - x)^2 p(x)\,dx + \int_{\hat{x}_2}^{x_2} \ldots dx + \int_{x_2}^{\hat{x}_3} \ldots dx + \ldots$$

Differentiating the equation with respect to x_j will involve only the two terms with limit x_j and thus,

$$\frac{\partial N_q}{\partial x_j} = \frac{2\partial}{\partial x_j}\left[\int_{\hat{x}_j}^{x_j} (\hat{x}_j - x)^2 p(x)\,dx - \int_{\hat{x}_{j+1}}^{x_j} (\hat{x}_{j+1} - x)^2 p(x)\,dx\right] = 0$$

or

$$(\hat{x}_j - x)^2 p(x)\Big|_{\hat{x}_j}^{x_j} = (\hat{x}_{j+1} - x)^2 p(x)\Big|_{\hat{x}_{j+1}}^{x_j}$$

Assuming $p(x)$ to be constant over the interval $\hat{x}_j$ to $\hat{x}_{j+1}$, $(\hat{x}_j - x_j)^2 = (\hat{x}_{j+1} - x_j)^2$ or $\left|\hat{x}_j - x_j\right| = \left|\hat{x}_{j+1} - x_j\right|$ and hence,

$$x_j = \frac{(\hat{x}_j + \hat{x}_{j+1})}{2} \tag{2.15a}$$

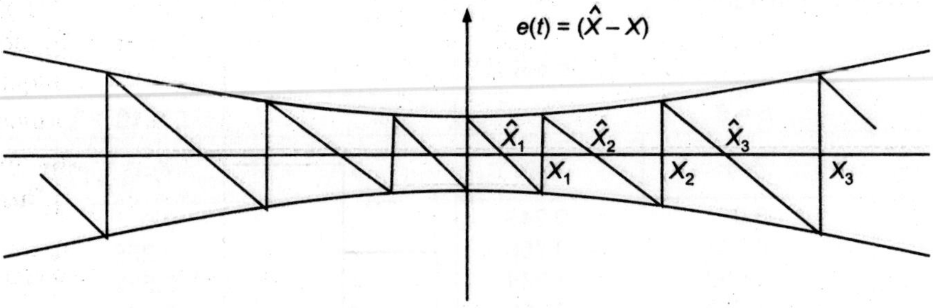

Figure 2.6(*b*): Error waveform due to nonlinear QZR

Eq. (2.15) shows that $\langle e(t)\rangle = 0$

Differentiating eq. (2.14) with $\hat{x}_j$,

$$\frac{\partial N_0}{\partial \hat{x}_j} = \frac{\partial}{\partial \hat{x}_j}\left[\int_{x_{j-1}}^{x_j} (\hat{x}_j - x)^2\, p(x)\, dx\right]$$

$$= 2\int_{x_{j-1}}^{x_j} (\hat{x}_j - x)p(x)\, dx,\ j = 1, 2, \ldots, L/2$$

$$= 0$$

or

$$\hat{x}_j \cdot \int_{x_{j-1}}^{x_j} p(x)\, dx = \int_{x_{j-1}}^{x_j} xp(x)\, dx \tag{2.16}$$

i.e., the mean value of $x(t)$ in the range $x_{j-1} \le x \le x_j$

$$= (\hat{x}_j) \times (\text{probability of } x(t) \text{ being in the range } x_{j-1} \le x \le x_j)$$

Therefore, $\hat{x}_j$ is the centroid of the area $\int p(x)\, dx$ between x_{j-1} and x_j. The two sets of eqns. (2.15) and (2.16) have to be solved simultaneously for $(L - 1)$ unknown parameters of the quantizer and their closed form solutions are not possible, as the eqns. are generally nonlinear. An iterative process has been used by Max, where an initial value of $\hat{x}_1$ is assumed and the subsequent x_j's and $\hat{x}_j$'s are calculated using eqs. (2.15) and (2.16). If $\hat{x}_{L/2}$ satisfies the condition (2.16) for $x_{L/2} < x < \infty$, then the optimum $\hat{x}_j$ and x_j have been obtained. Generally, an iterative process with different initial values of $\hat{x}_j$ has to be followed using numerical methods and a computer, till the optimum values are obtained. Max has given tables of optimum $\hat{x}_j$ and x_j for an $x(t)$ having Gaussian pdf,

$$p(x) = \frac{1}{\sqrt{2\pi}} \exp\left(-x^2/2\right), \qquad \sigma_x = 1 \tag{2.17}$$

and a short list is given in Table 2.2.

Table 2.2: ($\bar{x} = 0, \sigma_{\hat{x}} = 1$) [9]

L = 8			L = 16		
j	x_{j-1}	$\hat{x}_j$	j	x_{j-1}	$\hat{x}_j$
1	0.0	0.245	1	0.0	0.128
2	0.501	0.756	2	0.258	0.388
3	1.050	1.344	3	0.522	0.657
4	1.748	2.152	4	0.799	0.942
			5	1.099	1.256
			6	1.437	1.61
			7	1.844	2.069
			8	2.401	2.733

MSE = 0.03454
S/N_q = 15 dB

MSE = 0.009497
S/N_q = 20 dB

In general, the solution of eqs. (2.15) and (2.16) result in nonuniform distribution of $\hat{x}_j$'s, except in the case where $p(x)$ = constant and the quantization step Δ is

$$\Delta = \hat{x}_j - \hat{x}_{j-1} = x_j - x_{j-1} \tag{2.18}$$

In practical A/D conversion, it is simpler to have equal Δ's, and the optimum Δ for minimum Q-noise for a given $p(x)$ may be calculated by noting that

$$x_j = \Delta \cdot j \text{ and } \hat{x}_j = \Delta \cdot j - \Delta/2 = \left(\frac{2j-1}{2}\right)\Delta \tag{2.19}$$

and

$$N_q = 2\sum_{j=1}^{(L/2-1)} \int_{(j-1)\Delta}^{j\Delta} \left\{\left(\frac{2j-1}{2}\right)\Delta - x\right\}^2 p(x)\,dx + 2\int_{(L/2-1)\Delta}^{\infty} \left\{\frac{(L-1)\Delta}{2} - x\right\}^2 p(x)\,dx$$

For minimum N_q,

$$\frac{dN_q}{d\Delta} = \sum_{j=1}^{(L/2-1)} (2j-1) \int_{(j-1)\Delta}^{j\Delta} \left\{\frac{(2j-1)\Delta}{2} - x\right\} p(x)\,dx + (L-1)\int_{(L/2-1)\Delta}^{\infty} \left\{\left(\frac{L-1}{2}\right)\Delta - x\right\} p(x)\,dx$$

$$= 0 \tag{2.20}$$

Optimum Δ for a uniform quantizer is now given by the solution of the Eq. (2.20). Using $p(x)$ of Eq. (2.17), Max has calculated various Δ's for $L = 2$ to 36 and a few of the results are shown in Table 2.3.

The quantizing levels $(x_j, \hat{x}_j)$ in Table 2.2 and Δ's in Table 2.3 are shown for $\sigma_x = 1$ and hence these values have to be multiplied by σ_x of the signal to get the correct levels.

Further it is shown that for optimum quantizer with $\sigma_x = 1$,

$$N_q = K \cdot L^{-k}, \tag{2.21}$$

where $K = 1.32$, $k = 1.74$ for $L = 4$ and $K = 2.21$, $k = 1.96$ for $L \geq 32$. But for optimum equal $= \Delta$ (uniform) quantizer ($\sigma_x = 1$),

$$N_q = 1.47L^{-1.74} \text{ for } L \text{ large.} \tag{2.22}$$

Table 2.3: ($\bar{x} = 0, \sigma_{\hat{x}} = 1$) [9]

$L = 2^B$	Δ	MSE	S/N_q(dB)
2	1.596	0.3634	4.5
4	0.9957	0.1188	9.3
8	0.586	0.03744	14.3
16	0.3352	0.01154	19.4
32	0.1881	0.00349	24.6

For speech signals, a more appropriate density function is the gamma distribution given by

$$p(x) = \left[\frac{\sqrt{3}}{8\pi\sigma_x |x|}\right]^{1/2} \exp\left(-\frac{\sqrt{3}|x|}{2\sigma_x}\right) \tag{2.23}$$

and a simpler approximation is the Laplacian density:

$$p(x) = \frac{1}{\sqrt{2}\sigma_x} \exp\left(-\frac{\sqrt{2}|x|}{\sigma_x}\right) \tag{2.24}$$

Paez and Glisson [10] have calculated the values of x_j, $\hat{x}_j$ and Δ's for optimum and uniform quantizers using Max's technique and the pdf's of eqs. (2.23) and (2.24). A few of their results are shown in Table 2.4 for optimum quantization.

Table 2.4: ($\bar{x} = 0, \sigma_{\hat{x}} = 1$) [10]

	Gamma density				Laplacian density			
$L \rightarrow$	8		16		8		16	
$j \downarrow$	x_{j-1}	$\hat{x}_j$	x_{j-1}	$\hat{x}_j$	x_{j-1}	$\hat{x}_j$	x_{j-1}	$\hat{x}_j$
1	0.0	0.149	0.0	0.072	0.0	0.222	0.0	0.126
2	0.504	0.859	0.229	0.368	0.504	0.785	0.266	0.407
3	1.401	1.944	0.588	0.791	1.181	1.576	0.566	0.726
4	2.872	3.799	1.045	1.300	2.285	2.994	0.910	1.095
5			1.623	1.945			1.317	1.540
6			2.372	3.798			1.821	2.103
7			3.407	4.015			2.499	2.895
8			5.050	6.085			3.605	4.316
MSE	0.0712		0.0196		0.0548		0.0154	
S/N_q (dB)	11.47		17.07		12.61		18.12	

The values of Δ's for uniform quantizers for the same density functions are shown in Table 2.5.

Table 2.5: $(\bar{x}=0, \sigma_x^2=1)$ [10]

$L = 2^B$	Gamma density			Laplacian density		
	Δ	MSE	S/N_q(dB)	Δ	MSE	S/N_q(dB)
2	1.154	0.668	1.77	1.414	0.500	3.01
4	1.060	0.320	4.95	1.087	0.1963	7.07
8	0.796	0.1323	8.78	0.731	0.0717	11.44
16	0.540	0.05	13.00	0.456	0.0254	15.96
32	0.346	0.0178	17.49	0.281	0.0087	20.60

The optimum uniform and nonuniform quantizers, as discussed above, are particularly sensitive to the pdf of the input to the coder. The nonuniform quantizers are complex in hardware and both the types require step adjustments with the variance of the signal. Some of the natural signals, e.g., speech, are of non-stationary nature and the idle channel noise (during no signal condition) is an important consideration for the coder design. Moreover, speech signals normally have a large dynamic range, say, 40 to 50 dB, and hence the above theoretically optimum codecs do not give satisfactory S/N_q results. Companded and adaptive codecs, as discussed below, are shown to be better practical solutions for quantizing speech and television signals.

2.4 NONUNIFORM QUANTIZATION [11]

It has been shown in eq. (2.8) and Fig. 2.3, that for uniform quantization, the S/N_q decreases with the decrease of the input to the coder. It is, therefore, necessary to quantize the smaller signals with finer steps and larger signals with coarser steps. Keeping the quantizers step Δ constant, this may be achieved if a compressor with a nonlinear characteristic modifies the signal level instantaneously, as is used in the long distance companded telephone circuits. Consider a nonlinear amplifier characteristic, shown in Fig. 2.7 where

$$y = f(x)$$

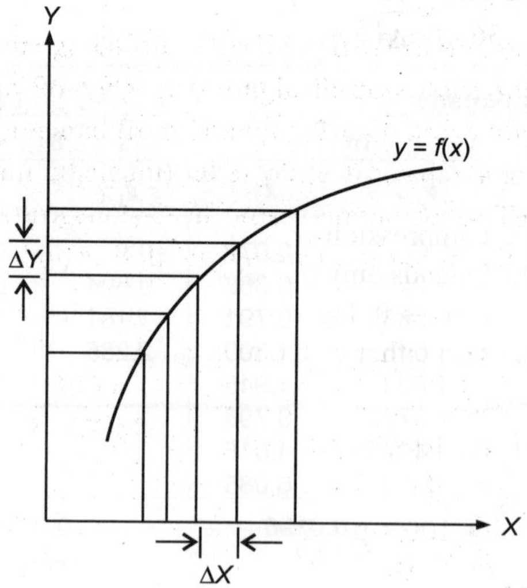

Figure 2.7: Nonlinear amplifier characteristics

and for larger L,

$$\Delta y/\Delta x = dy/dx = f'(x) \tag{2.25}$$

and from eq. (2.3) the Q-noise power for the interval Δx,

$$N_q(\Delta x) = \frac{(\Delta x)^2}{12}$$

and from eqs. (2.4) and (2.25),

$$N_q = \int_0^{X_0} \frac{(\Delta x)^2}{12} p(x)\, dx = \frac{(\Delta y)^2}{12} \int_0^{X_0} \frac{p(x)}{[f'(x)]^2} dx \tag{2.26}$$

If $f(x)$ is assumed to be a logarithmic function, then

$$y = \log_e x \ldots \qquad 0 < x < \infty$$

$$\frac{dy}{dx} = \frac{1}{x} = f'(x)$$

and for $p(x) = \dfrac{1}{X_0}$, $0 < x < X_0$, eq. (2.26) reduces to

$$N_q = \frac{\Delta^2}{12X_0} \int_0^{X_0} x^2\, dx = \frac{\Delta^2}{12X_0} \frac{X^3}{3}\bigg|_0^{X_0} = \frac{\Delta^2 X_0^2}{36},$$

where Δy = equal quantum step Δ and

$$\frac{S}{N_q} = \frac{X_0^2}{3} \bigg/ \frac{\Delta^2 X_0}{36} = \frac{12}{\Delta^2}, \tag{2.27}$$

where $S = \dfrac{X_0^2}{3}$.

It is seen that with logarithmic compression of $x(t)$, and uniform quantization of $y(t)$, the S/N_q is independent of the signal level and depends only on the step size Δ. Although an ideal log-characteristic is not practically obtainable, since for $x \to 0$, $\log x \to -\infty$, a suitable approximation of the compressor characteristic may be obtained by using either of the relations [11]

$$\text{(a)}\quad y = \frac{\log_e (1 + \mu\,|x|)}{\log_e (1 + \mu)}, \quad \mu > 0, \text{ and } |x| \le 1 \tag{2.28}$$

which is known as the μ-law (usually, $100 < \mu < 200$), and

$$\text{(b)}\ y = \begin{cases} \dfrac{A|x|}{1+\log_e A}, & 0 < |x| \le \dfrac{1}{A} \\ \dfrac{1+\log_e A|x|}{1+\log_e A}, & \dfrac{1}{A} < |x| \le 1 \end{cases} \tag{2.29}$$

which is known as the A-law, with $75 < A < 150$.

The overall block diagram for the coder-decoder now includes compressor/expander and quantizer, as shown in Fig. 2.8(a). Figure 2.8(b) shows the compression and expansion characteristics giving the equivalent nonlinear quantization of Fig. 2.6. to calculate S/N_q for the μ-law compressor-quantizer, one notes that for eq. (2.28),

$$\frac{dy}{dx} = f'(x) = \frac{\mu}{\ln(1+\mu)}\left(\frac{1}{1+\mu x}\right), \quad x \ge 0$$

$$= \frac{k_1}{1+\mu x}, \text{ where } k_1 = \frac{\mu}{\ln(1+\mu)} \tag{2.30}$$

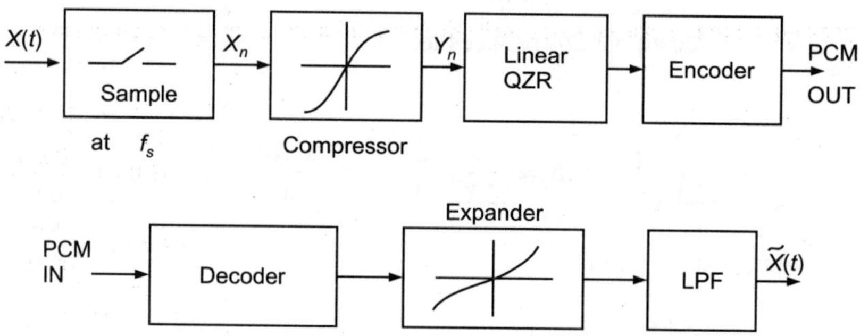

Figure 2.8(*a*): Block diagram of a companded PCM codec

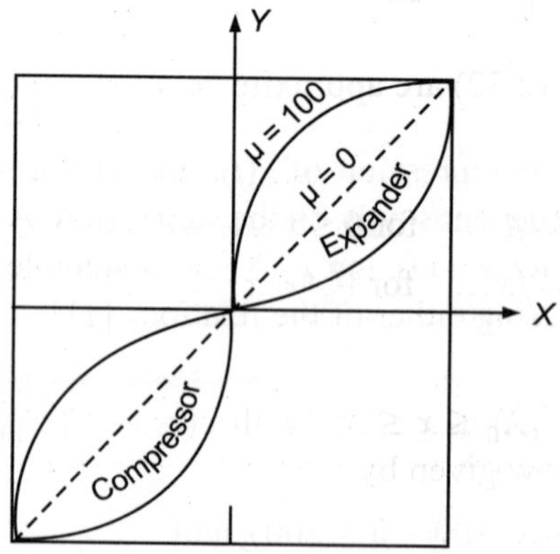

Figure 2.8(*b*): Compressor expander characteristics

For uniform distribution of x, $-X_0 \le x \le X_0$, with $p(x) = 1/(2X_0)$ and using the symmetry of the compressor characteristic, we have from eq. (2.26),

$$N_q = 2\int_0^{X_0}\left(\frac{\Delta^2}{12}\right)\left(\frac{1}{2X_0}\right)\left(\frac{1+\mu x}{k_1}\right)^2 dx$$

$$= \frac{\Delta^2}{12k_1^2}\left(1+\frac{\mu^2 X_0^2}{3}+\mu X_0\right) X_0 \le 1 \text{ and } X_{max} = 1$$

If $L = 2X_{max}/\Delta = 2/\Delta$, then with $S = X_0^2/3$, the SNR is

$$\frac{S}{N_q} = k_1^2 X_0^2 L^2\left[1+\frac{(\mu X_0)^2}{3}+\mu X_0\right]^{-1} \tag{2.31}$$

For a sinewave input, $x(t) = a \sin \theta$,

$$p(x) = \frac{1}{\pi}\frac{1}{\sqrt{a^2-x^2}} \text{ and } S = \frac{a^2}{2}$$

and

$$N_q = \frac{1}{3k_1^2 L^2}\left[1+\frac{(\mu a)^2}{2}+\frac{4\mu a}{\pi}\right].$$

Thus,

$$\frac{S}{N_q} = \frac{3k_1^2 L^2 a^2}{2}\left[1+\frac{(\mu a)^2}{2}+\frac{4\mu a}{\pi}\right]^{-1} \tag{2.32}$$

The two results of eqs. (2.31) and (2.32) are approximately the same.

For the A-law characteristics,

$$f'(x) = \frac{dy}{dx} = \begin{cases} k_2 A \ldots & \text{for } 0 \le x \le 1/A \\ k_2/x \ldots & \text{for } 1/A < x \le 1 \end{cases} \tag{2.33}$$

where $k_2 = 1/(1 + \ln A)$.

For uniform distribution of x, $-X_0 \le x \le X_0$, with $p(x) = 1/(2X_0)$ and using the symmetry of the compressor characteristics, N_q is now given by

$$N_q = \begin{cases} 2\int_0^{X_0}\left(\frac{\Delta^2}{12}\right)\left(\frac{1}{2X_0}\right)\left(\frac{1}{k_2 A}\right)^2 dx \cdots \text{ for } 0 \le X_0 \le 1/A \\ \left[2\int_0^{1/A}\left(\frac{\Delta^2}{12}\right)\left(\frac{1}{2X_0}\right)\left(\frac{1}{k_2 A}\right)^2 dx\right. \\ \left. + 2\int_{1/A}^{X_0}\frac{\Delta^2}{12}\left(\frac{1}{2X_0}\right)\left(\frac{X}{k_2}\right)^2 dx\right] \quad \text{for } 1/A < X_0 \le 1 \end{cases} \tag{2.34}$$

Integrating for $X_0 \le 1/A$,

$$N_q = \frac{\Delta^2}{12}\frac{1}{(k_2 A)^2} = \frac{1}{3L^2 k_2^2 A^2}$$

and

$$\frac{S}{N_q} = \frac{X_0^2}{3N_q} = (X_0 L k_2 A)^2 \tag{2.35}$$

For $1/A < X_0 \le 1$,

$$N_q = \frac{1}{3L^2 X_0 k_2^2 A^3} + \frac{1}{3L^2 X_0 k_2^2}\left[\frac{X_0^3}{0} - \frac{1}{3A^3}\right]$$

and

$$\frac{S}{N_q} = (X_0^3 L^2 k_2^2)\left[\frac{X_0^3}{3} - \frac{2}{3A^3}\right]^{-1} \tag{2.36}$$

Referring to eqs. (2.5) and (2.35), it is seen that the companding advantage for $X_0 \le 1/A$, is equal to $[A/(1 + \ln A)]^2$. For sinewave and other pdf's for $x(t)$, S/N_q can be calculated based on the above equations. Various pdf's and the resultant S/N_q values are also given by Cattermole [11]. Smith [7] has derived the S/N_q equation for random speech-like signals (with approximate exponential pdf) with μ-law companding and is given as

$$\frac{S}{N_q} = \frac{3L^2 k_1^2}{\mu^2}\left[1 + \frac{1}{\mu^2\sigma_x^2} + \frac{\sqrt{2}}{\mu\sigma_x}\right]^{-1}, X_{\text{max}} = 1 = L \cdot \Delta \text{ and } S = \sigma_x^2 \tag{2.37}$$

which is similar to eqs. (2.31) and (2.32). The variation of SNR's with signal amplitude X_0 for μ-law and A-law companding is shown in Fig. 2.9(a) and (b), along with the plot of S/N_q for uniform quantizers. It is observed that the dynamic range of the codecs is now in the range of 30 to 40 dB with flat values of $\frac{S}{N_q} \simeq \frac{3L^2 k_1^2}{\mu^2}$ (or $3L^2 K_2^2$) and a companding advantage of 26 dB. Normally 4σ loading, i.e., $X_{\text{max}} = 4\sigma$, is

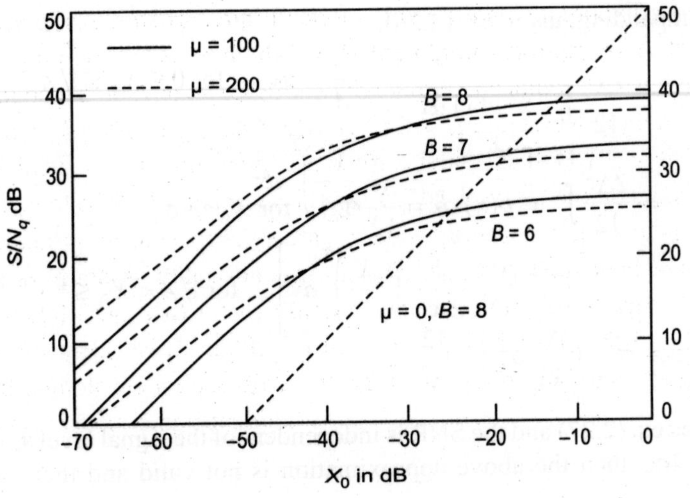

Figure 2.9(*a*): S/N_q vs. X_0 for μ-law PCM

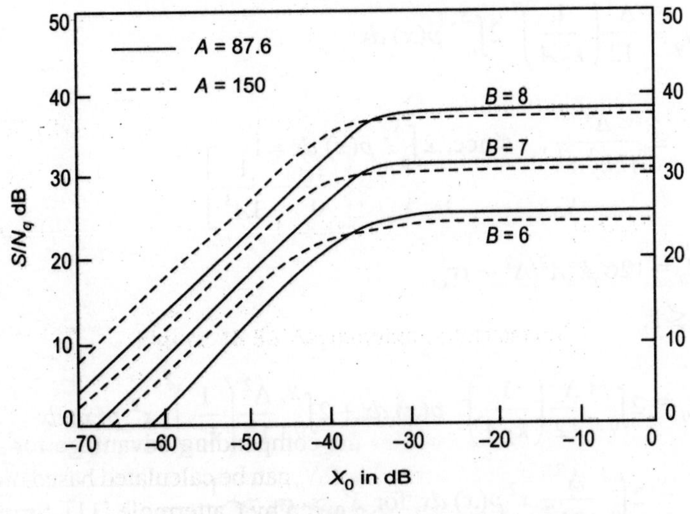

Figure 2.9(*b*): S/N_q vs. X_0 for *A*-law PCM

a satisfactory condition to avoid overload distortion in the codecs; however, the dynamic ranges in the two cases differ to some extent.

Example 2.3: It is instructive to repeate the problem of Example 2.2 with logarithmic and *A*-law companding. Assume the normalised pdf for the signal, given by,

$$p(x) = \frac{1}{\sigma^2} \exp\left(-\frac{2}{\sigma^2} |x|\right),$$

where $\sigma_x^2 = \frac{\sigma^4}{2}$, $X_0 = 6.3\sigma_x$ as before.

For logarithmic companding, using eq. (2.26),

$$N_q = \frac{2\Delta^2}{12}\int_0^{X_0} x^2 p(x)\,dx$$

$$\simeq \frac{2\Delta^2}{12}\int_0^{\infty} x^2 p(x)\,dx = \frac{\Delta^2}{12}\sigma_x^2, \quad \text{for } X_0 \gg \sigma_x$$

Therefore,

$$\frac{S}{N_q} = \sigma_x^2\bigg/\left(\frac{\Delta^2}{12}\sigma_x^2\right) = \frac{12}{\Delta^2}$$

The result is the same as eq. (2.27) and the SNR is independent of the signal level as desired. If, however, $X_0 < 6.3\sigma_x$, say $X_0 = 2\sigma_x$, then the above approximation is not valid and there will be considerable overload distortion.

For A-law companding, using eq. (2.34) and (2.35), for the range $0 \le X_0 \le 1/A$,

$$N_q = \frac{\Delta^2}{12}\left(\frac{1}{k_2 A}\right)^2 2\int_0^{X_0} p(x)\,dx$$

$$\simeq \frac{\Delta^2}{12k_2^2A^2}, \quad \text{since, } 2\int_0^{X_0} p(x)\,dx \simeq 1$$

Therefore,

$$S/N_q = 12\sigma_x^2 k_2^2 A^2/\Delta^2 \propto \sigma_x^2.$$

For the range $1/A < X_0 \le 1$,

$$N_q = 2\int_0^{1/A}\frac{\Delta^2}{12}\left(\frac{1}{k_2 A}\right)^2 p(x)\,dx + 2\int_{1/A}^{X_0}\frac{\Delta^2}{12}\left(\frac{1}{k_2}\right)^2 x^2 p(x)\,dx$$

$$\simeq 2\int_0^{\infty}\frac{\Delta^2}{12k_2^2}x^2 p(x)\,dx, \text{ for } X_0 \gg \sigma_x \gg \frac{1}{A}$$

$$= \frac{\Delta^2\sigma_x^2}{12k_2^2}.$$

Therefore, $S/N_q = \sigma_x^2/N_q = 12k_2^2/\Delta^2$ which is independent of σ_x. For $A \simeq 100$, $k_2 = 1/5.6$ and $\Delta = 2X_0/L$, $X_0 = 1$.

Therefore,

$$\frac{S}{N_q} = 3L^2k_2^2 = 26 \text{ dB}$$

which is approximately the same as in eq. (2.36).

2.4.1 Segmented Approximations [12, 13]

The nonlinear μ-law and A-law characteristics are obtained by using precision temperature-controlled diode networks and the compressor-expander matching is rather critical. It is possible to digitally implement the compression-expansion characteristics, if the nonlinear X-Y curves are approximated by linear segments. In practice, the A-law curve with $A = 87.6$ is approximated by 13-segments and the μ-law curve with $\mu = 255$, by 15-segments. A 13-segment approximation of the A-law curve with $L = 128$, is shown in Fig. 2.10(a), where it is seen that initial 16 quantized $\hat{x}_j$ levels are reserved for $0 < x \le 1/64$, $8\,\hat{x}_j$ levels for $1/64 < x \le 1/32$ and so on.

The slopes of the successive segments are:

For 1st segment,

$$\frac{d\hat{x}_j}{dx} = \frac{A}{1 + \ln A}, \quad 0 < x \le \frac{1}{A},$$

and by inspection,

$$\text{the slope} = \frac{16/64}{1/64} = 16 = \frac{A}{1 + \ln A} \text{ which, gives } A = 87.6. \tag{2.38}$$

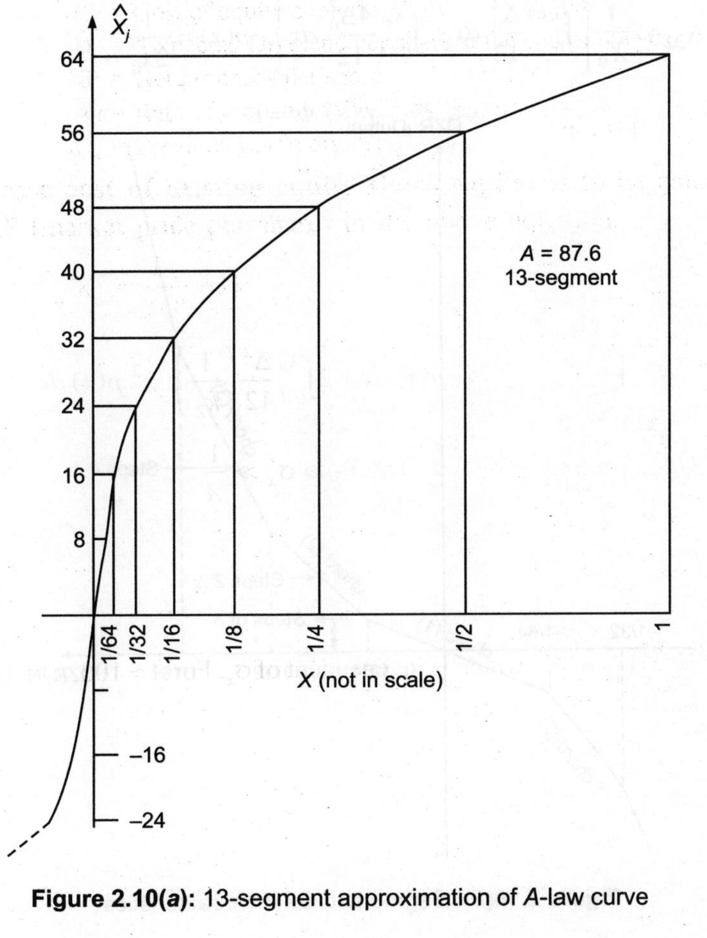

Figure 2.10(*a*): 13-segment approximation of *A*-law curve

For the 2nd segment, the slope = $\dfrac{8/64}{1/64}$ = 8, and subsequent slopes are: 4, 2, 1, 1/2 and 1/4 for the respective chords. This means that the equivalent nonuniform quantizing steps are of heights Δ_0 for $|x| \le 1/64$, $2\Delta_0$ for $1/64 < |x| \le 1/32$, $4\Delta_0$ for $1/32 < |x| \le 1/16$, and $8\Delta_0$ $16\Delta_0$, $32\Delta_0$, $128\Delta_0$ for next successive intervals of x, as shown in Fig. 2.10(b). For $X_{max} = 1$, $\Delta_0 = 2^{-10}$, for 128 level nonuniform quantization, and thus the quantization for $|x| \le 1/64$ is really fine and better by an order of 16 as compared to the 128 level uniform quantization. This gives the S/N_q advantage of 24 dB [$=(16)^2$] at low signal levels. Using eq. (2.5) for uniform quantization with $\Delta_0 = 2^{-10}$, S/N_q for the first interval is

$$\left(\frac{S}{N_q}\right)_1 = \frac{X_0^2/3}{\Delta_0^2/12} = 2^{22} \cdot X_0^2,\ |X_0| \le \frac{1}{64}$$

$$= 2^{10}\ (= 30\text{ dB})\text{ for } |X_0| = \frac{1}{64} \tag{2.39}$$

For the second interval $\dfrac{1}{64} < |X_0| \le \dfrac{1}{32}$,

$$N_{q2} = \frac{1}{X_0}\left[\int_0^{1/64} \frac{\Delta_0^2}{12}\,dx + \int_{1/64}^{X_0} \frac{4\Delta_0^2}{12}\,dx\right],\quad p(x) = \frac{1}{2x_0} \tag{2.40}$$

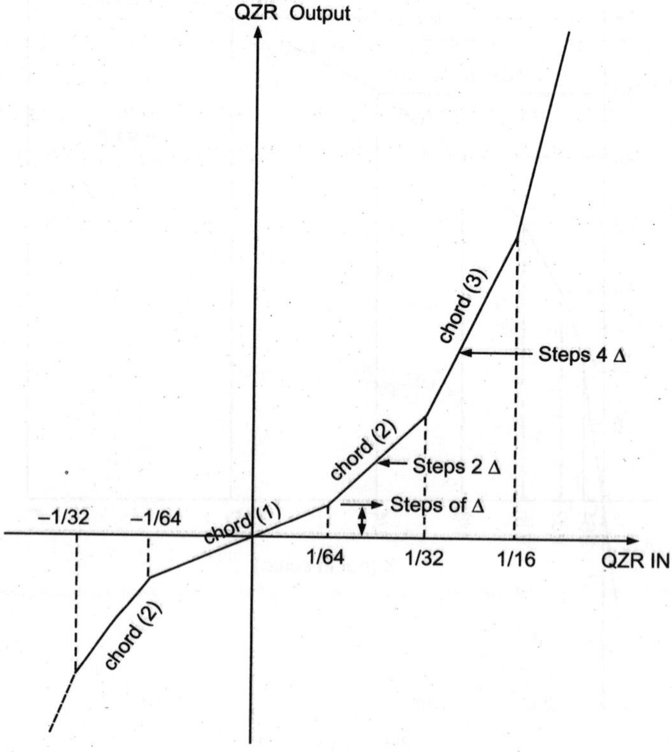

Figure 2.10(*b*): Segmented QZR characteristics

and the limiting value of S/N_{q2} at $|X_0| = \dfrac{1}{32}$ is

$$\frac{S}{N_{q2}} = \frac{2^{13}}{5} (= 32 \text{ dB})$$

The Q-noise for other intevals may be calculated following the above principles, and in general, N_{qj} is given by (assuming jth being the last interval),

$$N_{qj} = \frac{1}{X_0}\left[\int_0^{1/64} \frac{\Delta_0^2}{12} dx + \int_{1/64}^{1/32} \frac{4\Delta_0^2}{12} dx + \cdots + \int_{1/2}^{X_0} \frac{(2^6 \Delta_0)^2}{12} dx\right] \tag{2.41}$$

The SNR values at the corner points of the linear segments for 7-bit segmented A-law companding is given in Table 2.6.

Table 2.6

Segment No.	Limits of $\|X_0\|$	Step-size (normalised)	S/N_q (dB)
1	1/64	2^{-10}	30
2	1/32	2^{-9}	32
3	1/16	2^{-8}	33
4	1/8	2^{-7}	33
5	1/4	2^{-6}	33
6	1/2	2^{-5}	33
7	1	2^{-4}	33

The overall SNR characteristic for smooth or sinusoidal waveforms with segmented companding is shown in Fig. 2.11. As the signals input increases from one segment to the next, the S/N_q values momentarily decrease because of the larger step size used in the quantization and S/N_q reaches the peak

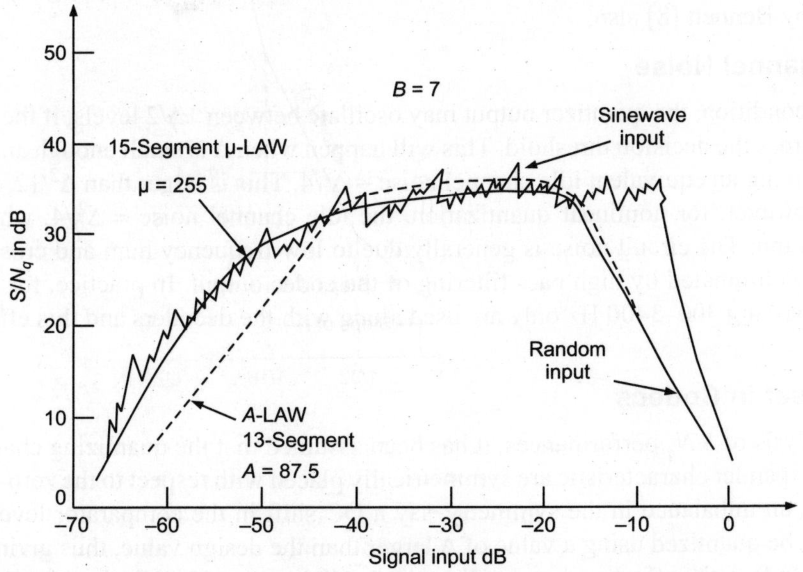

Figure 2.11: SNR characteristics of segmented QZR

value again at the end of the segment only. This gives the ripples in the SNR curve, but with random input, the SNR curve is smooth, as shown in Fig. 2.11. The SNR curve obtained for the 15-segment μ-law companding is also shown in Fig. 2.11.

2.5 PRACTICAL CONSIDERATIONS [14]

In practical codecs, S/N_q further deteriorates due to overload of the quantizer, and dc offsets in comparator circuits. There are also the problems of idle channel noise and SNR deterioration due to noise in the transmission path, resulting in false detection of the transmitted digits. These are briefly discussed below.

2.5.1 Overload Distortion

For random signals, there is always a finite probability that the signal peaks will exceed the quantizer limits (± 1 assumed here) and produce overload distortion. As such the crest factor γ of the signal should match the dynamic range of the quantizer, where

$$\gamma = \frac{\text{Effective } X_{\max}}{\sigma_x}$$

For Gaussian signals, $\gamma \simeq 4$, and the peak S/N_q is reduced by 12 dB as shown in eq. (2.7). For speech signals, a Laplacian density function of eq. (2.24) may be assumed and the probability of the peak exceeding $X_{\max}$ is 10^{-3}, if $\gamma \simeq 5$. This indicates that σ_x for speech should be maintained at 14 dB below $X_{\max}$ of the quantizer. Cattermole [11] has shown that the clipping noise for speech is given by

$$N_c = \sigma_x^2 \exp(-\sqrt{2}/\sigma_x) \tag{2.42}$$

and the total noise $N_T = N_q + N_c$. The overload characteristics shown in Fig. 2.11 corresponds approximately to eq. (2.42). For an increase of $\sigma_x/X_{\max}$ by 5 dB, the S/N_c deteriorates by about 12 dB for 7-bit uniform quantization. Listening tests, however, have confirmed that small peak clipping does not impair the speech quality significantly and as such $\gamma = 4$ is a reasonable value for practical speech codecs, which has been confirmed by Bennett [8] also.

2.5.2 Idle Channel Noise

Under no signal condition, the quantizer output may oscillate between $\pm\Delta/2$ levels, if the circuit noise is large enough to cross the decision threshold. This will happen when Δ is small enough and the rms noise voltage $\sigma_n \simeq \Delta$, giving an equivalent idle channel noise $= \Delta^2/4$. This is larger than $\Delta^2/12$, the N_q value of the quantizer. However, for nonlinear quantization, the idle channel noise $\simeq \Delta_0^2/4$, where $\Delta_0 \ll \Delta$ of uniform quantization. The circuit noise is generally due to low frequency hum and cross-talk; and this can be effectively eliminated by high pass filtering of the codec output. In practice, for speech codecs, bandpass filters passing 300–3400 Hz only are used along with the decoders and this effectively solves the problem.

2.5.3 DC Offset in Codecs

In the earlier analysis of S/N_q performances, it has been assumed that the quantizing characteristics and the compressor-expander characteristic are symmetrically placed with respect to the zero-signal level. If, however, there is an unbalance in the symmetry, say a DC shift in the comparator level, then the low level signals may be quantized using a value of Δ larger than the design value, thus giving a lower S/N_q for small signals. In fact, the effective companding curve will be somewhat displaced and the result is the introduction of some harmonic distortion in the codec output. It may be shown that for A-law companding, the codec output y is related to the input x by:

$$y = x \exp [\pm D(1 + \log A)] \quad \text{for } 1/A < x \leq 1 \tag{2.43}$$

where D is the dc offset in the quantizer. Since compressor-expander circuits are made of diodes, special care is to be taken in matching their characteristics and also to stabilize their performance through temperature control. Theoretical calculations have shown that the SNR impairment could be as large as 6 dB for practical circuits. However, measurements on practical PCM systems have shown only 2 dB loss in S/N_q, mainly because the compression curve is linear at low inputs. In segmented companding using digital hardware, the problem is somewhat eliminated.

2.5.4 Effect of Channel Errors [15]

During transmission through noisy channels, it is possible that one or two bits of a PCM-word would be received wrongly and as such, will produce sharp clicks in the decoded output. If the channel error probability is P_e, then the probability of error in a B-bit word is:

$$P_w = 1 - (1 - P_e)^B \tag{2.44}$$

However the weightage of individual bits are not same, and the effect of an error in the MSB is more significant (about 50 per cent) than an error in LSB (about 1 per cent). The mean square change in the decoded amplitude is therefore given by, (assuming a single error/word only):

$$\frac{1}{B}\sum_{i=1}^{B}(2^{i-1})^2 = \frac{(4^B - 1)}{3B} \tag{2.45}$$

This results in the equivalent noise power at the output given by

$$N_e = \Delta^2[1 - (1 - P_e)^B]\frac{(4^B - 1)}{3B} \simeq P_e \frac{\Delta^2 \cdot 4^B}{3}$$

and for equiprobable signal, $-X_0 < x < X_0$,

$$S = \frac{X_0^2}{3} = \frac{\Delta^2 \cdot 4^B}{12}$$

Thus, the equivalent

$$\frac{S}{N_e} = \left(\frac{\Delta^2 \cdot 4^B}{12}\right)\Bigg/\left(\frac{P_e \Delta^2 4^B}{3}\right) = \frac{1}{4P_e} \tag{2.46}$$

The overall SNR in the system will be now due to both N_q and N_e, and if $N_T = N_q + N_e$, then

$$\frac{S}{N_T} = \frac{S}{N_q + N_e},$$

given by eqs. (2.31), (2.34) and (2.46).

In practical systems, equal weightage is given to both types of noise, giving $N_q = N_e$. Further in a companded system, the effect of the error in MSB and higher order bits is further magnified by the expander characteristic. Listening tests have shown that $P_e \leq 10^{-4}$ has to be maintained in the channel for satisfactory quality of decoded speech. For a 7-bit codec, $P_e = 10^{-4}$, means that there will be, on an average, one click per second, which would be really disturbing in an otherwise good quality decoded speech.

Example 2.4: In practical codecs, 7 bit log-PCM is usually used. This gives $S/N_q \simeq 32$ dB over a 40 dB dynamic range. Assume that the CNR in the channel is 10 dB: then with BPSK transmission, the bit error rate $P_e = 10^{-5}$, giving $S/N_e = 44$ dB. Thus the effective $S/N_T \simeq 32$ dB and is mainly controlled by N_q. But for lower CNR's , S/N_T is controlled by P_e in the channel. As for example,

(a) If CNR = 8.5 dB, then $P_e = 10^{-4}$ and $S/N_e = 34$ dB. This gives $S/N_T = 30$ dB.
(b) If CNR = 7 dB, then $P_e = 10^{-3}$ and $S/N_e = 24$ dB. This gives $S/N_T \simeq 23.4$ dB.

It is thus seen that for achieving an overall $S/N_T = 30$ dB (say), the values of S/N_q and $S/N_e \geq 33$ dB. In a PCM codec, $S/N_e \geq 34$ dB is only acceptable, and CNR ≥ 9 dB for BPSK transmission.

2.6 COMPANDED CODECS [14]

The basic techniques for analog-to-digital conversion of signals are:

(a) Counter method:
  (i) Compensation method using a D-to-A converter in the feedback,
  (ii) Single-slope and Dual-slope method by initially converting the analog signal to a PPM or a PWM signal.
(b) Electron-beam coder.
(c) Parallel-comparison method, using $(2^B - 1)$ parallel comparators.
(d) Successive approximation method, using a D/A converter in the feedback path.

The counter methods, although simple in hardware, are slow in operation and hence, less accurate for communication codecs. They are mostly used in instrumentation systems. Electron beam coder is too costly and the parallel-comparator method, although fast in operation, requires extensive circuitry and hence, are not cost effective. For digital communication systems, the successive approximation method is widely used and has been produced in a chip form for sample rate $f_s \leq 10$ MHz.

The successive approximation principle, discussed in Sec. 2.1, is illustrated in Fig. 2.12 (a) for a 7-bit coder (the 8th bit is used for signalling). It is seen that a sampled pulse X_j is compared successively with reference currents of units 64, 32, …, 2, 1 through the coder-control logic and whenever the difference value is positive, the comparator gives '1' as output and at the same time the weighted value is subtracted from X_j. Thus through $B(= \log_2 L)$ successive operations, a complete code word is generated. The decoder, shown in Fig. 2.12(b), is complementary to the coder. The coder is equivalent to a quantizer (QZR) with feedback as shown in Fig. 2.12(c). The *DA* and *AD* converters are now available in an LSI form and one such 10-bit *ADC* is shown in Fig. 2.13, where the successive approximation technique is used through a *DAC* in the feedback.

In more recent IC's operations of companding, coding and decoding have been combined in a single chip. Examples of this are INTEL 2910A and 2911A codec chips. 2910A incorporates 15-segment μ-law companding with $\mu = 255$, and 2911 A incorporates 13-segment *A*-law companding with $A = 87.6$.

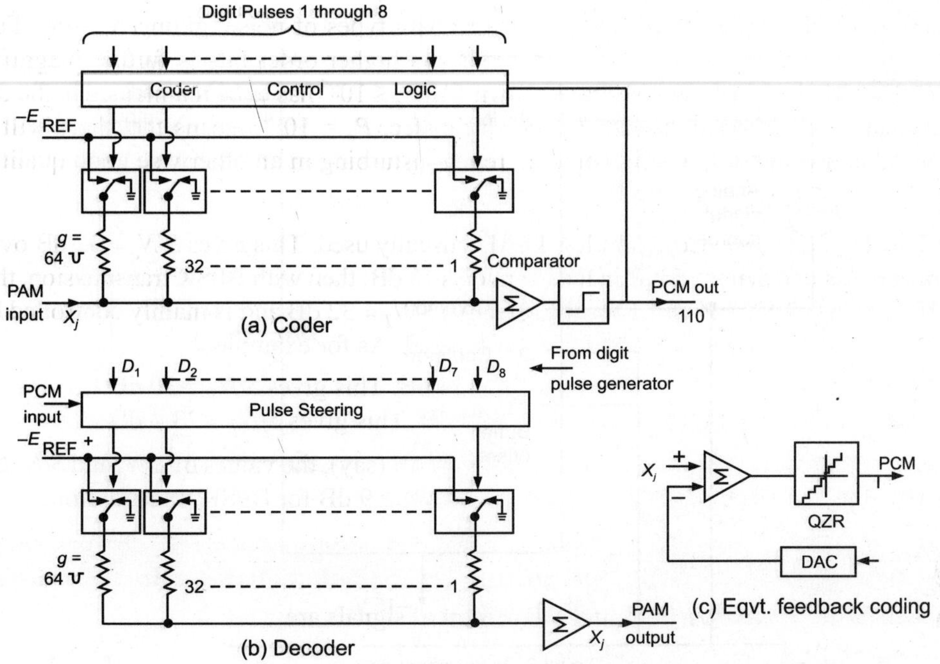

Figure 2.12: 7-bit coder and decoder: **(*a*)** Coder, **(*b*)** Decoder, **(*c*)** Eqvt. feedback coding

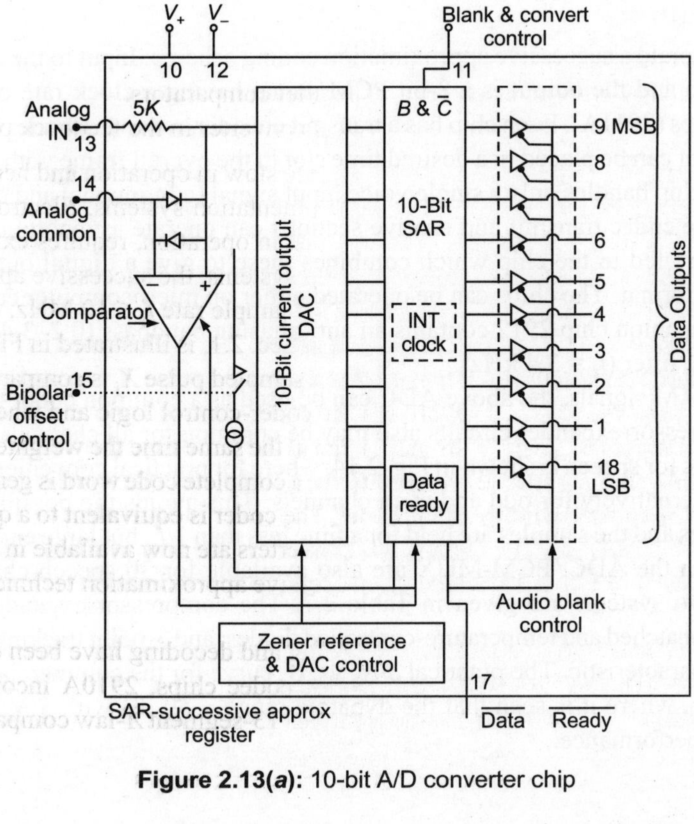

Figure 2.13(*a*): 10-bit A/D converter chip

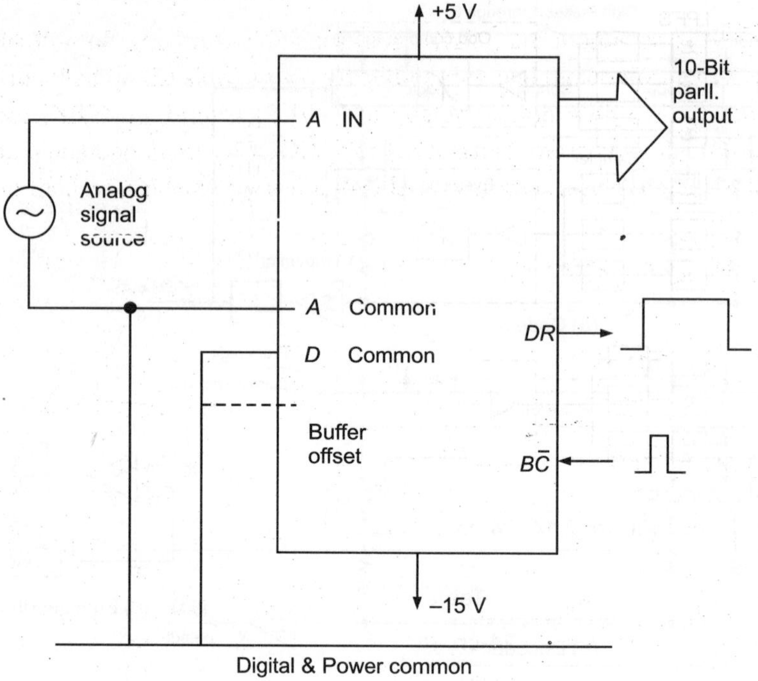

Figure 2.13(*b*): Typical AD chip application

Both the chips incorporate a successive approximation coding scheme. Input to the coder is band-limited (0.2–3.4 kHz) speech and the output is a 8-bit PCM signal having a clock rate of either 1.544 Mbps (2910A) or 2.048 Mbps (2911A). Each chip has a transmit encoder, receive decoder and a control section. The 8-bit coded output can be placed in a desired time slot in the overall frame with the help of a suitable control signal. Each chip handles only a single voice input signal (continuous) and outputs another voice signal (received). The codec transmit and receive sections can operate asynchronously. Signalling bits and frame sync bit are fed to the chip which combines these to give a digital output compatible with either T-1 or CCITT format. The chips can be operated either on microcomputer control mode or direct control mode. A companion chip 2912 contains an anti-aliasing bandpass (0.2–3.4 kHz) transmit filter, and a lowpass receive filter ($f_c = 3.4$ kHz).

For multiplexed PAM signals, the above ADC can be used as a common coder for all the samples in the frame. The compressor-expander circuits also may be common for all channels. Such a multiplexed 24-ch. log-PCM codec for speech is shown in Fig. 2.14, where for practical convenience, two compressor-encoders are used alternatively for odd and even channels. This avoids the possible crosstalk between the successive samples and the samples are held for a time less than $1/f_s$, but large enough for B successive operations required in the ADC. PCM-MUX are also available for 30 speech channels. The relevant parameters of the two systems are given in Table 2.7. The compressor-expander circuits shown in Fig. 2.15 are made of matched and temperature-controlled diodes; and 3-point tracking is used for matching the required μ-law characteristic. The practical S/N_q vs. σ_x curve for the commercial 24-ch. PCM-MUX is shown in Fig. 2.16, where it is seen that the dynamic range ≃ 50 dB with $S/N_q \simeq 32$ dB, giving the required toll quality performance.

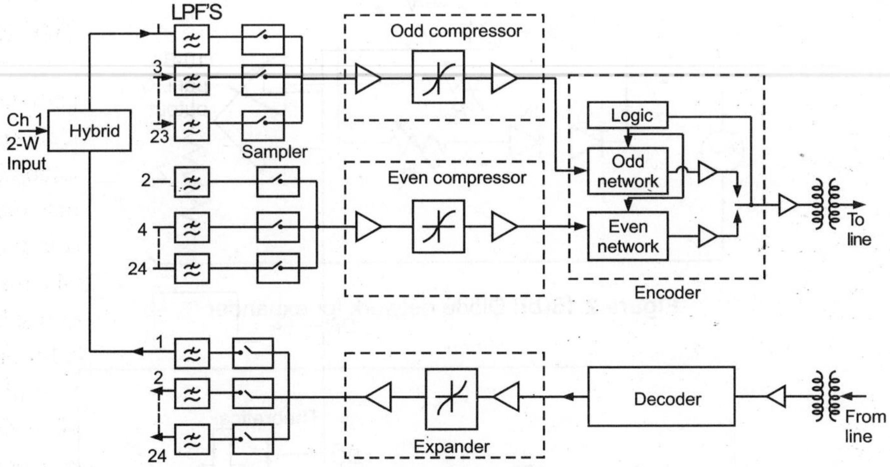

Figure 2.14: 24-channel log-PCM system

Table 2.7

System parameters	24-ch. PCM*	30-ch. PCM
f_s	8 kHz	8 kHz
Frame time	125 µs	125 µs
B	7 + 1	8
Signalling	8th bit	31st channel
Synchronisation	193rd bit in the frame	32nd channel
Companding	µ-law (µ = 100)	A-law (A = 87.6)
Companding gain	26 dB	24 dB
Clock rate	1.544 Mb/s	2.048 Mb/s

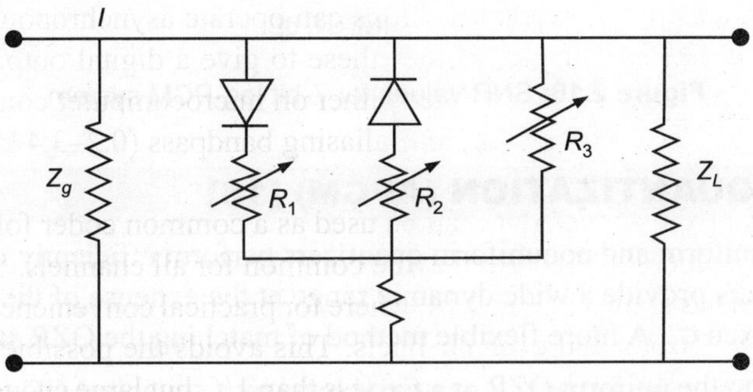

Figure 2.15(*a*): Diode network for compressor

*This is the original T-1 format, which has since been modified. In the modified T_i-carrier, B is usually 8, except that one information bit (LSB) of every sixth sample of a signal (e.g., 1, 7, 13, ... , samples) is used to transmit the signalling information. Every sixth frame, therefore, has 7 × 24 = 168 information bits, 24 signalling bits and 1 framing bit. In all other frames, there are 8 × 24 = 192 information bits and 1 framing bit.

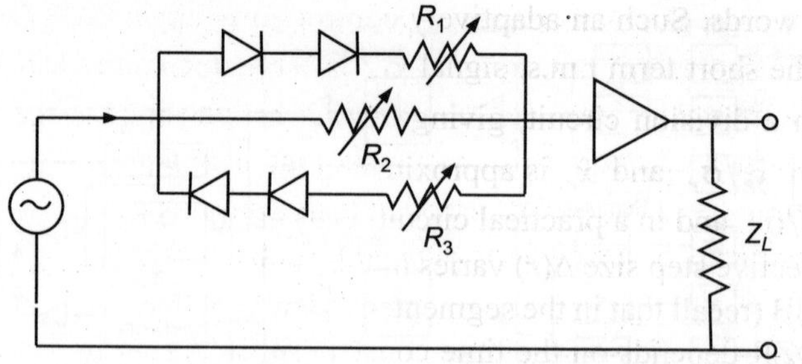

Figure 2.15(*b*): Diode network for expander

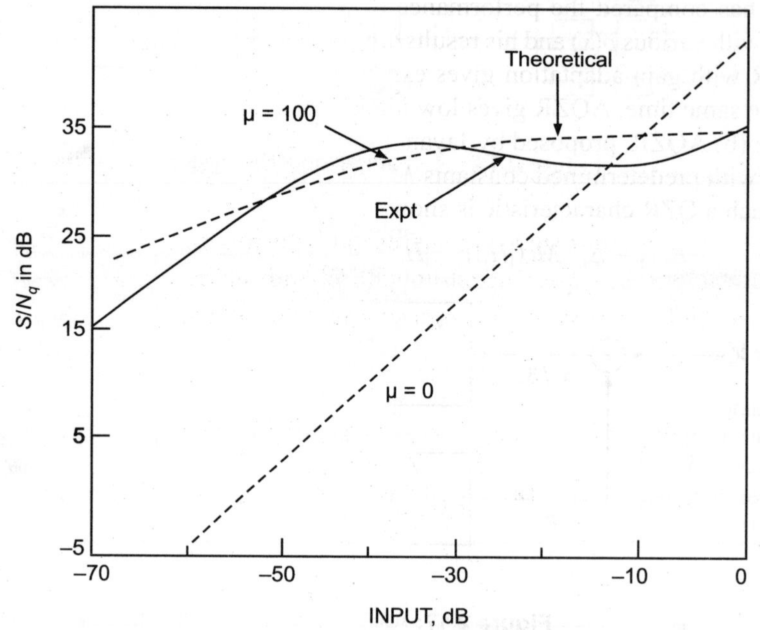

Figure 2.16: SNR values for 7-bit log-PCM system

2.7 ADAPTIVE QUANTIZATION (APCM) [16]

It has been shown that uniform and nonuniform quantizers perform efficiently if the σ_x is held constant, and the companded coders provide a wide dynamic range at the expense of the peak S/N_q obtainable in an uniform QZR with fixed σ_x. A more flexible method of matching the QZR step-size would be either:

(a) to hold the input to the uniform QZR at a fixed level by a suitable AGC circuit, and both the code word and the AGC control signal made available at the decoder, or

(b) to adapt the step size of the QZR instantaneously depending on the magnitude of the previous code word, i.e., QZR should have one/two-word memory and the control signal is predetermined based on the statistics of the signal.

The AGC-controlled adaptation may be either feed-forward or feedback type; but in practice, the feedback type of gain control is preferred, as the control signal may be regenerated at the receiver from the received code words. Such an adaptively controlled uniform QZR (known as APCM) is shown in Fig. 2.17, where the short term r.m.s. signal $\hat{\sigma}_r$ is estimated from QZR output and input sample x_r is controlled through a division circuit, giving $x_r/\hat{\sigma}_r$ as the input to the QZR. At the receiver, $\hat{\sigma}_r$ is reconstructed from $\hat{x}_r/\hat{\sigma}_r$ and $\hat{x}_r$ is approximated through multiplication by $\hat{\sigma}_r$. The overall gain of the QZR, $G(r) = 1/\hat{\sigma}_r$ and in a practical circuit, $G(r)$ has to be limited in the range of $G_{max}/G_{min} \simeq 100$. As a result, the effective step size $\Delta(r)$ varies in the range of $\Delta_{max}/\Delta_{min} = 100$, giving a dynamic range of the QZR to be 40 dB (recall that in the segmented A-law QZR, $\Delta_{max}/\Delta_{min} = 2^6$). It has been shown that the success of the AQZR depends on the time constant of the signal-power averager, and for speech, this should be of the order of 10–20 ms. Such an APCM is also known as syllabically-adapted PCM, as against the instantaneously-adapted PCM of type (b), discussed later.

Noll [16] has compared the performance of 8-level QZR's with μ-law companding and adaptive quantization with various $p(x)$ and his results are shown in Table 2.8. It is seen from the Table 2.8 that the uniform QZR with gain adaptation gives excellent results as compared to those of more complicated QZR's. At the same time, AQZR gives low idle channel noise in the system.

In the type (b) AQZR, proposed by Jayant [4], the QZR has a one-word memory and the step-size Δ_r is multiplied with predetermined constants M_1, M_2, etc., depending upon the value of the previous code word X_{r-1}. Such a QZR characteristic is shown in Fig. 2.18, where the step-size algorithm is

$$\Delta_{r+1} = \Delta_r \cdot M_i(|H_r|), \quad |H_r| \text{ is the previous code word magnitude} \tag{2.47}$$

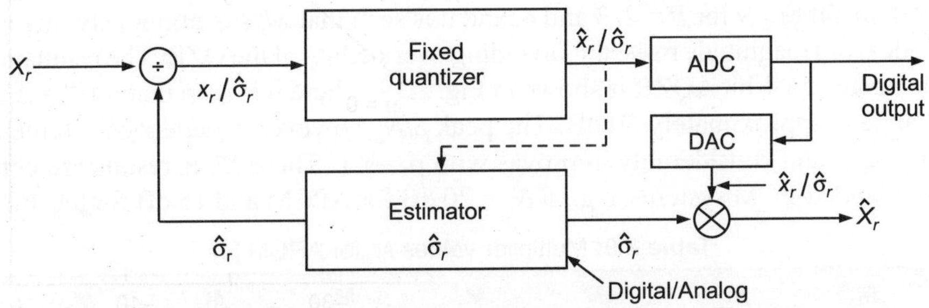

Figure 2.17: AGC-controlled adaptive QZR

Table 2.8: Comparison of S/N_q for 3-bit adaptive and nonadaptive QZR's. τ = time constant of the signal power averager [16]

QZR and $p(x)$	Non-adaptive S/N_q(dB)	Adaptive S/N_q (dB)	
		τ = 16 ms	τ = 128 ms
Non-uniform QZR			
μ-law (μ = 100, $X_{max} = 8\sigma_x$)	9.5	–	–
Gaussian	7.3	15.0	12.1
Laplace	9.9	13.3	12.8
Uniform QZR			
Gaussian	6.7	14.7	11.3
Laplace	7.4	13.4	11.5

It has been shown by computer simulation that the variable step-size Δ_r is mostly centred around the optimum Δ_{opt}, when optimum multipliers M_i are used, thus making the QZR robust. The optimum values

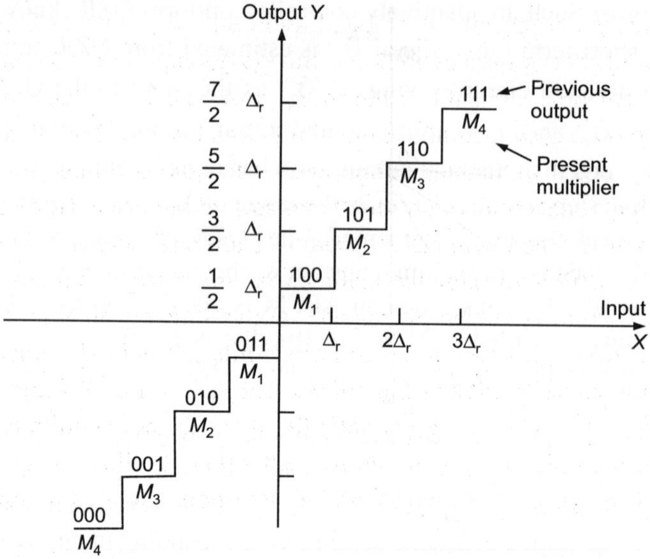

Figure 2.18: Adaptive QZR with one-word memory [4]

of M_i's are given in Table 2.9 for $B = 2$, 3 and 4, and it is seen that M_i's continuously increase in value as the previous code word magnitude rises, thus avoiding overloading of the QZR. The resultant improvement in the dynamic range of a 4-bit AQZR is shown in Fig. 2.19, where it is seen that for $R = \Delta_{max}/\Delta_{min} = 100$, the dynamic range is approximately 50 dB. The peak S/N_q however is dependent on the correlation ρ_1 between X_r and X_{r-1}, and considerably improves with $\rho_1 \to 1$. These SNR results are better than those obtained in companded-PCM systems, e.g., $S/N_q = 20$ dB for APCM and 15 dB for log-PCM with $B = 4$.

Table 2.9: Multiplier values M_i for APCM [4]

$M_i \to$ / $B \downarrow$	M_1	M_2	M_3	M_4	M_5	M_6	M_7	M_8
2	0.60	2.20	–	–	–	–	–	–
3	0.85	1.00	1.00	1.50	–	–	–	–
4	0.80	0.80	0.85	0.90	1.2	1.6	2.0	2.4

Example 2.5: Assume that a multistep waveform is to be quantized with an 8-level AQZR, having the adaptation logic given in Table 2.9. The sample values of $x(t)$ at the sampling instants, T_r, are given as

$$\{X_r\} = \{0.4, 0.4, 0.8, 0.8, 1, 1, 5, 5, 5, 5, -1, -1, -1, -3, -5, -1, 0, 0, \ldots\}$$

It is observed from Fig. 2.18 that for $X_r < \Delta_r$, $\hat{X}_r = \Delta_r/2$ for $X_r \geq \Delta_r$ $\hat{X}_r = 3\Delta_r/2$, for $X_r \geq 2\Delta_r$, $\hat{X}_r = 5\Delta_r/2$ and so on. These values have been used in the table, where C_r is the digital value of $\hat{X}_r$ and $|e_r| = |X_r - C_r \cdot \Delta_r|$. The code words have been assigned on the basis of the codes of Fig. 2.18. It is seen from the table that for abrupt changes of X_r at T_6, T_{10} and T_{15}, the error $|e_r| = \{3.7, 0.08 \text{ and } 0.13\}$, showing that the error is more for the abrupt change in positive slopes of $x(t)$ than for negative slopes. Without adaptation, and taking $\Delta_0 = 1.5$, the approximation is given by

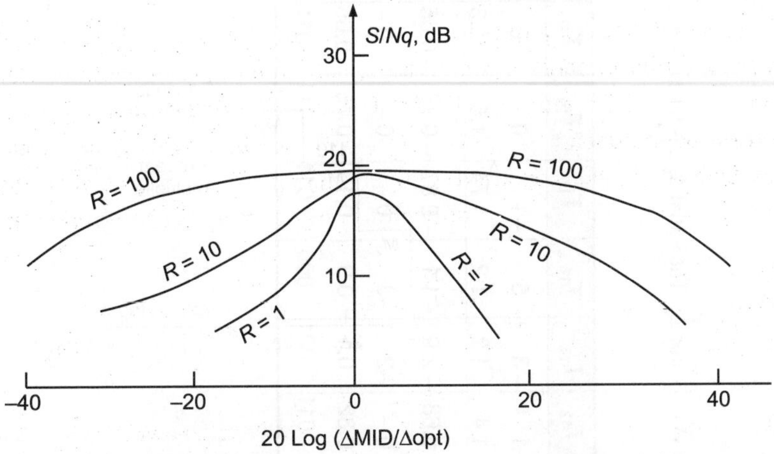

Figure 2.19: Companding characteristics of adaptive QZR [4], with $B = 4$, $p = 0.5$ (correlation), $R = \Delta_{max}/\Delta_{min}$; $\Delta_{mid} = |\Delta_{min} \cdot \Delta_{max}|^{1/2}$

$$\{\hat{X}_r\} = \{0.75, 0.75, 0.75, 0.75, 0.75, 0.75, 5.25, 5.25, 5.25, 5.25, -0.75, -0.75, -0.75, -3.75, -5.75, -0.75, -0.75, ...\}$$

and

$$\{|e_r|\} = \{0.35, 0.35, 0.05, 0.05, 0.25, 0.25, 0.25, 0.25, 0.25, 0.25, 0.25, 0.75, 0.25, 0.25, 0.75, \ldots\}$$

Apparently, the overload distortion (as indicated by the rectangle in $|e_r|$) is more prominent in APCM coding than in linear-PCM, but the quantization is more coarse in L-PCM than in APCM. For smooth waveforms, the approximation with APCM will be much better.

2.8 DIFFERENTIAL PCM (DPCM AND ADPCM) [17]

Many of the live information signals e.g., speech and TV, exhibit significant correlation between successive samples, x_r and x_{r-1}, at Nyquist rate, and the variance of the first difference,

$$\langle d_r^2(1)\rangle = \langle (x_r - x_{r-1})^2\rangle$$

is smaller than σ_x^2 of the signal. Specifically,

$$\begin{aligned}\langle d_r^2(1)\rangle &= \langle x_r^2\rangle + \langle x_{r-1}^2\rangle - 2\langle x_r \cdot x_{r-1}\rangle \\ &= \langle x_r^2\rangle + [2(1-\rho_1)]\end{aligned} \tag{2.48}$$

where ρ_1 is the correlation between x_r and x_{r-1}. In speech signal, $\rho_1 > 0.5$, and $\langle d_r^2(1)\rangle < \sigma_x^2$, and it will be advantageous to use $d_r(1)$ as the input to the QZR and reconstruct x_r through an integrator. Instead of using a straightforward QZR as above, a better approximation to x_r is provided through a feedback predictor, as shown in Fig. 2.20, and the difference $\langle d_r^2\rangle$ is further minimized. With a single-tap (a_1) predictor, the difference signal is given by

The following table shows the values of $\hat{X}_r$, the digital values of the PCM code $\{C_r\}$, the error $\{|e_r|\} = \{| X_r - \hat{X}_r |\}$, and Δ_r. The initial Δ_{min} is taken as 0.5.

Table for Example 2.5

$T_r \rightarrow$	T_0	T_1	T_2	T_3	T_4	T_5	T_6	T_7	T_8	T_9	T_{10}	T_{11}	T_{12}	T_{13}	T_{14}	T_{15}	T_{16}	T_{17}
$\{X_r\}$	0.4	0.4	0.8	0.8	1	1	5	5	5	5	–1	–1	–1	– 3	– 5	–1	0	0
$\{\Delta_r\}$	0.5	0.43	0.36	0.36	0.36	0.36	0.36	0.54	0.81	1.2	1.8	1.6	1.4	1.2	1.2	1.74	1.5	1.3
$\{\hat{X}_r\}$	0.25	0.22	0.54	0.54	0.9	0.9	1.3	1.9	2.80	4.3	– 0.92	– 0.8	– 0.68	– 2.9	– 4.1	– 0.87	– 0.75	– 0.65
$\{C_r\}$	0 +	0 +	1	1	2	2	3	3	3	3	0 –	0 –	0 –	–2	–3	0 –	0 –	0 –
$\{\lvert e_r \rvert\}$	0.15	0.18	0.26	0.26	0.1	0.1	3.7	3.1	2.2	0.7	0.08	0.2	0.32	0.1	0.9	0.13	0.75	0.65
Code words	100	100	101	101	110	110	111	111	111	111	011	011	011	001	000	011	011	011

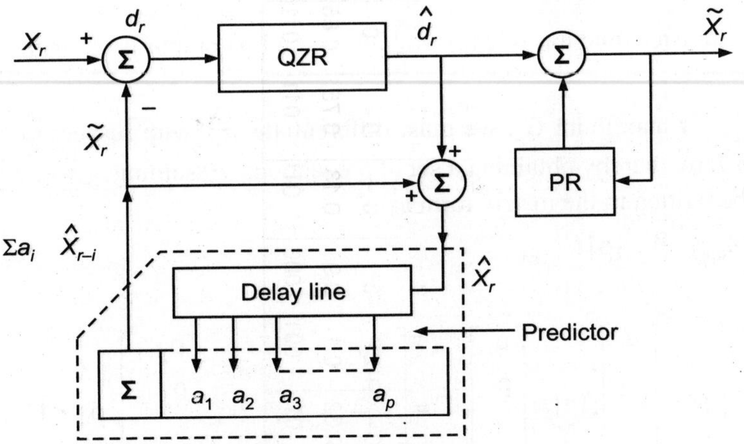

Figure 2.20: Block schematic of DPCM

$$\begin{aligned}\langle d_r^2(a_1)\rangle &= \langle (x_r - a_1 x_{r-1})^2\rangle \\ &= \langle x_r^2\rangle(1 + a_1^2 - 2a_1\rho_1) \\ &= \langle x_r^2\rangle(1-\rho_1^2) \text{ if } a_1 = \rho_1\end{aligned}$$

and

$$\begin{aligned}\frac{S}{N_q} &= \frac{\langle x_r^2\rangle}{\langle e_r^2\rangle} = \frac{\langle x_r^2\rangle}{\langle(\hat{x}_r - x_r)^2\rangle}, \\ &= \frac{\sigma_x^2}{\sigma_e^2} = \left(\frac{\sigma_x^2}{\sigma_d^2}\right)\left(\frac{\sigma_d^2}{\sigma_e^2}\right), \quad \sigma_d^2 = \langle d_r^2\rangle\end{aligned} \tag{2.49}$$

where $e_r = (\hat{d}_r - d_r) = \hat{d}_r - x_r + \tilde{x}_r = (\hat{x}_r - x_r)$.

Now for the QZR, $(S/N)_Q = \sigma_d^2/\sigma_e^2$, and hence the gain due to DPCM,

$$G_p = \frac{\sigma_x^2}{\sigma_d^2} = \frac{\langle x_r^2\rangle}{\langle(x_r - \tilde{x}_r)^2\rangle} \tag{2.50}$$

To maximize G_p, it is necessary to minimize σ_d, with a given σ_x; and this may be obtained through a multi-tap predictor, where the delayed-weighted sum of previous quantized samples forms the linearly predicted value $\tilde{x}_r$ for the incoming sample x_r. Thus,

$$\tilde{x}_r = \sum_{i=1}^{p} a_i \hat{x}_{r-i} \tag{2.51}$$

where the predictor transfer function is: $P(z) = \sum_{i=1}^{p} a_i z^{-i}$, and its output is $\tilde{x}_r$ with input $\hat{x}_r$, as shown in Fig. 2.20.

To minimize σ_d^2, for maximum G_p, we must differentiate σ_d^2 with respect to each a_i and set the derivatives equal to zero, thereby obtaining a set of p equations. Assuming, $\sigma_e^2 \ll \sigma_x^2$ the equations for optimum a_i's may be written in the matrix form as

$$[A_{\text{opt}}] = \Gamma^{-1}[\rho] \tag{2.52)*}$$

where

$$[A] = \begin{bmatrix} a_1 \\ a_2 \\ \vdots \\ a_p \end{bmatrix}; [\rho] = \begin{bmatrix} \rho_1 \\ \rho_2 \\ \vdots \\ \rho_p \end{bmatrix}; \Gamma = \begin{bmatrix} \rho_0 & & \cdots & \rho_{p-1} \\ \rho_1 & & \cdots & \rho_{p-2} \\ \vdots & & & \\ \rho_{p-1} & \rho_{p-2} & \cdots & \rho_0 \end{bmatrix}, \rho_0 = 1$$

where $[\rho]$ is the normalised autocorrelation matrix.

The resulting optimum gain over PCM is

$$G_{\text{opt}} = \left[1 - \sum_{i=1}^{p} a_i \rho_i\right]^{-1} \tag{2.53}$$

---

*Proof of eq. (2.52)

$$\sigma_d^2 = \langle (x_r - \tilde{x}_r)^2 \rangle = \langle [x_r - \sum_{i=1}^{p} a_i \tilde{x}_{r-i}]^2 \rangle$$

$$\simeq \langle [x_r - \sum_{i=1}^{p} a_i x_{r-i}]^2 \rangle, \text{since } \tilde{x}_r = x_r + e_r \simeq x_r \text{ if } \sigma_x^2 \gg \sigma_e^2.$$

Now,

$$\frac{\partial \sigma_d^2}{\partial a_i} = -2\langle [(x_r - \Sigma a_i x_{r-i})\,(x_{r-j})] \rangle,$$

$$= 0, \quad j = 1, 2, \ldots, p;$$

since for all terms in $\Sigma a_i x_{r-i}$, except for $(a_j x_{r-j})$ term, $\frac{\partial}{\partial a_j} = 0$

Therefore,

$$\langle x_r x_{r-j} \rangle = \sum_{i=1}^{p} a_i \langle (x_{r-i} x_{r-j}) \rangle$$

or

$$\phi(j) = \sum_{i=1}^{p} a_i \phi(j-i)$$

which can be written in the matrix form as

$$[\rho] = [\Gamma][A]$$

by defining the normalised autocorrelation as $\rho(j) = \frac{\phi(j)}{\sigma_x^2}$.

For a multitap predictor, the variation of SNR gain with p is shown in Fig. 2.21, where the central curve is the mean value and the shaded area gives the possible deviation for different speakers. It is seen that the gain saturates with 2 or 3 coefficients only and does not reach the ideal value of 13 dB [16]. It is also seen that G_{opt} is independent of the number of bits B per word, given that B is large enough so that $\sigma^2 e \ll \sigma_x^2$.

More sophistication in the DPCM codec can be introduced by adding the step size control as in APCM. Such an ADPCM block diagram is shown in Fig. 2.22, where the step size control is obtained through the short term average power $(\hat{\sigma}_q)^2$ of the QZR output, as was done in APCM of Fig. 2.17. For instantaneous adaptation using one-word memory, as in Fig. 2.18, the step size multiplier constants have been recalculated for DPCM and is given in Table 2.10, where it is seen that the values are somewhat similar to those of Table 2.9. Since on-line calculation of A_{opt} is not generally realizable, an adaptation technique for updating a_i's is used. Speech records of 4–32 msec are stored and the correlation matrix calculated; then optimum a_i's are updated after every 4–32 msec. Along with step size adaptation, the block diagram of Fig. 2.22 shows simultaneous control of a_i's as well. In most ADPCM codecs, the step size control is adaptive but the optimum tap-weights are maintained constant for a particular input signal whose statistics have been pre-determined. With such a fixed-coefficient predictor, the SNR gain of 8–10 dB is easily obtained. The overall performances of DPCM and other systems are shown in Fig. 2.23(a), and Table 2.11 gives SNR gains over PCM. It is seen that ADPCM with frame time of 4 msec only ($\simeq$ 32 samples of speech) gives an improvement of 13 dB and further increase of frame time does not help. In ADPCM with first-order prediction, the coefficients a_i's are derived at the decoder from the received bit stream and are not separately transmitted. However, in adaptive-predictive coding (APC), discussed later in Sec. 2.12, the predictor coefficients have to be transmitted along with QZR output. It has been shown that the advantages of adaptive quantization and adaptive prediction in ADPCM are additive and only second order interactions exist between them. As the prediction improves with $p > 2$, the QZR input tends to become Gaussian and small variations in a_i's do not change the Gaussian nature of d_r-statistics. The theoretical bounds on SNR for PCM, DPCM and DM are shown in Fig. 2.23(b), as given by O'neal [53].

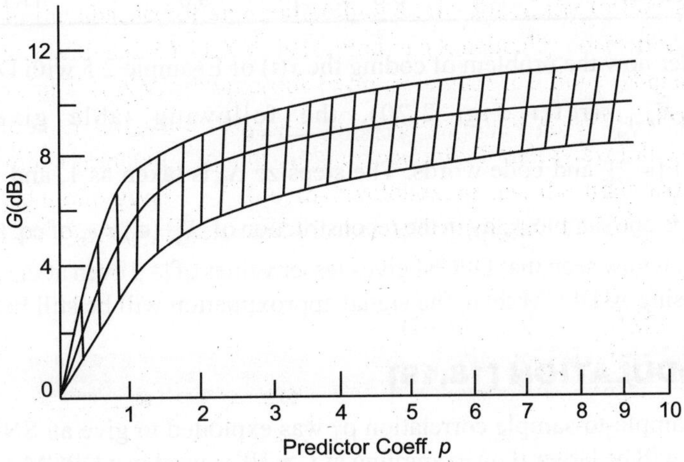

Figure 2.21: SNR gain in DPCM over PCM [4]

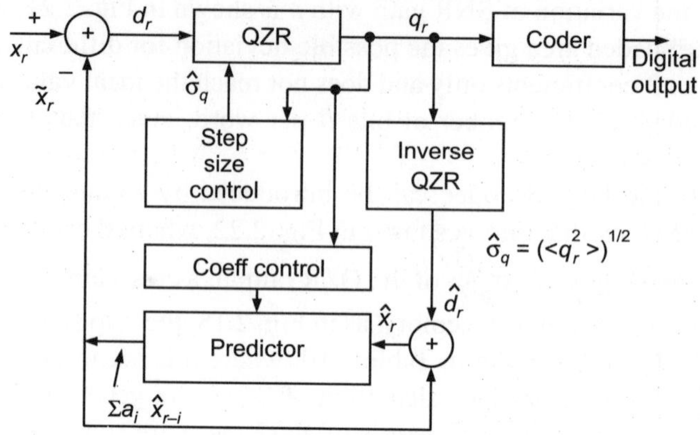

Figure 2.22: Block schematic for ADPCM

Table 2.10: Step size multiplier values M_i for ADPCM [4]

$M_i \rightarrow$ $B \downarrow$	M_1	M_2	M_3	M_4	M_5	M_6	M_7	M_8
2	0.8	1.6						
3	0.9	0.9	1.25	1.75				
4	0.9	0.9	0.90	0.90	1.20	1.60	2.0	2.4

Table 2.11: DPCM performances [16]

Types of codec	SNR gain over PCM in dB with no. of taps (p)			
	1	3	5	10
Nonadaptive DPCM	5.4	8.4	8.6	9.0
ADPCM, T_{AD} = 4 ms	5.6	10.0	11.5	13.0
ADPCM, T_{AD} = 32 ms	5.6	9.6	11.1	12.6

Example 2.6: Consider now the problem of coding the $x(t)$ of Example 2.5 with DPCM. Using $a_1 = \rho_1 = 0.8$ and $\tilde{X}_r = a_1\hat{X}_{r-1}$ (refer Fig. 2.20), the following table gives the values of $\{\tilde{X}_r\}, \{D_r\}, \{D_r\}, \{\hat{C}_r\}\ \{|\in_r|\}$ and code words. The stepsize Δ_0 is taken as 1, and it is noted that $|e_r| = |X_r - \tilde{X}_{r+1}|$ i.e., there is one sample delay in the reconstruction of $\tilde{X}_r$ ($\in_r \neq e_r$ of eq. (2.49)). As compared to the results of Ex. 5, it it now seen that DPCM gives lesser values of $|\in_r|$ even at the sharp discontinuities of the signal waves. Using A-DPCM then, the signal approximation will be still better.

2.9 DELTA MODULATION [18,19]

In DPCM codec, the sample-to-sample correlation ρ_1 was exploited to give an SNR better than that of PCM. This correlation will be larger if oversampling at $f_s \gg 2W$ is used in a DPCM codec and at the same time, the number of bits used in the QZR may be reduced for the same SNR. In the limit $B = 1$ and using

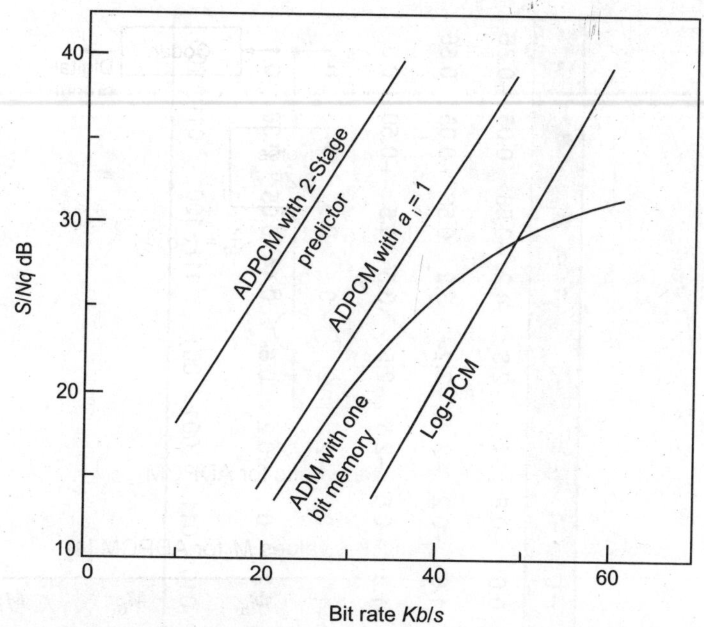

Figure 2.23(*a*): SNR for different codes (with speech input)

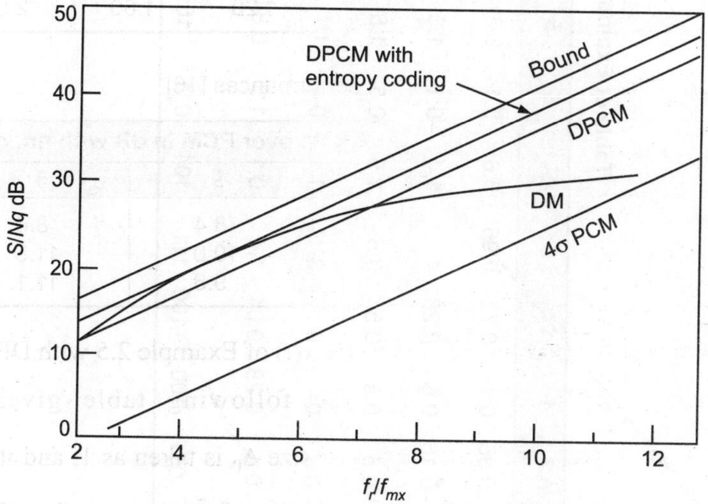

Figure 2.23(*b*): SNR bounds (Refer O'neal [53])

a simple integrator as the predictor, the DPCM codec of Fig. 2.20 is reduced to a Linear Delta-Modulator (LDM), shown in Fig. 2.24(a), which may also be called a unibit DPCM codec. The approximating waveform $\tilde{x}(t)$, the error waveform $e(t)$ and the pulse output $P(t)$ of the DM are shown in Fig. 2.24(b). It is seen that the error signal $e(t)$ is quantized to ± values and then sampled at $f_s \gg 2W$, giving $P(t)$ as the output of the coder. The approximating staircase waveform is $\tilde{x}(t)$ generated by $P(t)$ through the integrator in the feedback. In practice, however, a leaky (RC) integrator is used as the predictor, to avoid accumulation

Table of Example 2.6

$\{X_r\}$	0.4	0.4	0.8	0.8	1	1	5	5	5	5	−1	−1	−1	−3	−5	−1	0	0	0
$\{\tilde{X}_r\}$	0	0.4	0.72	0.96	0.4	0.72	0.96	3.6	5.1	3.7	4.16	0.6	−0.8	−1	−2.8	−4.2	−0.56	0.05	−0.36
$\{D_r\}$	0.4	0	0.08	−0.16	0.6	0.28	4	1.4	−0.1	1.3	−5.2	−1.6	−0.2	−2	−2.2	3.2	0.56	−0.05	0.36
$\{\hat{D}_r\}$	0.5	0.5	0.5	−0.5	0.5	0.5	3.5	1.5	−0.5	1.5	−3.5	−1.5	−0.5	−2.5	−2.5	3.5	0.5	−0.50	0.5
$\{C_r\}$	0	0	0	0	0–	0	3	1	0–	1	−3	−1	0–	−2	−2	3	0	0–	0
$\{\lvert\epsilon_r\rvert\}$	0	0.32	0.16	0.4	0.28	0.04	1.4	0.1	1.3	0.84	1.6	0.2	0	0.2	0.8	0.44	0.05	0.36	0.1
Code words	100	100	100	011	100	100	111	101	011	101	000	010	011	001	001	111	100	011	100

of channel errors at the decoder. Further, the sampler may be replaced by a clocked flip-flop to give NRZ pulses at the coder output.

Consider that DM is approximating a sinewave signal $x(t) = A \sin(\omega_m t + \theta)$. Then the staircase step-height Δ and f_s have to be such that $\tilde{x}(t)$ follows the largest slope of $x(t)$ and the condition to be satisfied is:

$$\text{Max. slope of } x(t) = \omega_m A \le \Delta f_s = \text{slope of } \tilde{x}(t)$$

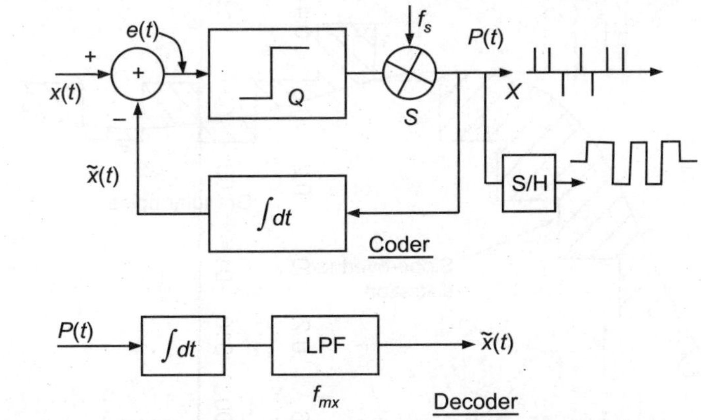

Figure 2.24(*a*): DELTA Modulator

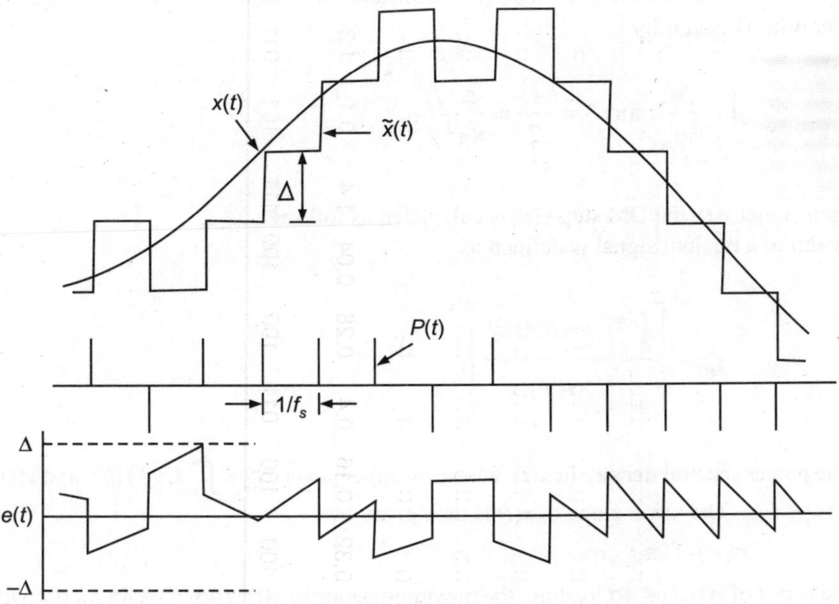

Figure 2.24(*b*): Waveform approximation and *e*(*t*)

or

$$\Delta \geq \omega_m A/f_s \tag{2.54}*$$

If, however, the slope $x'(t) > \Delta f_s$, then overload distortion occurs, and this is shown in Fig. 2.25. One may avoid this by increasing Δ, but at the same time, the granular noise during the period of small slopes of $x(t)$ will increase, as is seen in the Fig. 2.25. Thus the total distortion at the DM receiver output consists of both granular (= quantization noise) and slope overload distortion. One then has to use an adaptive scheme to improve the SNR both at low and high input levels, leading to Adaptive DM (ADM).

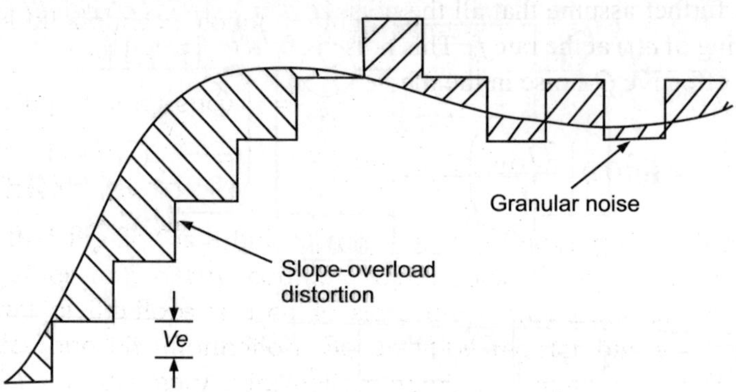

Figure 2.25: Delta modulator and its characteristics

2.9.1 Quantizing Noise in LDM

From the slope overload condition of eq. (2.45) for sinusoidal inputs, it is seen that the maximum signal the coder will handle is given by

$$A = \frac{\Delta f_s}{\omega_m} \text{ and } S = \frac{A^2}{2} = \frac{\Delta^2 f_s^2}{8\pi^2 f_m^2} \tag{2.55}$$

---

*For random input signal $x(t)$, the DM step size is calculated as follows:
Effective bandwidth of a random signal is defined as:

$$f_{\text{eff}} = \left[\frac{\int_{-f_{\max}}^{f_{\max}} f^2 G(f)\, df}{\int_{-f_{\max}}^{f_{\max}} G(f)\, df} \right]^{1/2},$$

where $G(f)$ is the power spectral density of $x(t)$, whose average power $\sigma_x^2 = \int_{-\infty}^{\infty} G(f)\, df$; and $x(t)$ is limited to the frequency band $|f| \leq f_{\max}$. The r.m.s. slope of $x(t)$ is then given by:

$$\sigma(x') = 2\pi f_{\text{eff}}\, \sigma x.$$

Assuming Gaussian pdf of $x(t)$, and 4σ loading, the maximum slope of $x(t) = 4\sigma(x') = 8\pi f_{\text{eff}}\, \sigma x$, which should obey eq. (2.54), i.e.

$$\Delta \geq 8\pi\sigma x f_{\text{eff}}/f_s = 8\pi\sigma x f_{\max}/(\sqrt{3} f_s), \text{ for } G(f) = \text{constant}, \left(\text{since } f_{\text{eff}} = \frac{f_{\max}}{\sqrt{3}}, \text{ for } G(f) = \text{constant} \right). \tag{2.54(a)}$$

and the error waveform of Fig. 2.24(b) is random with peaks of $\pm\Delta$. However, the nature of the waveform is somewhat different than that obtained in a PCM-QZR, and as such, the uniform density of $e(t)$ cannot now be assumed. Further the Q-noise power $\langle e^2(t)\rangle = \Delta^2/3$ with uniform density and $\langle e^2(t)\rangle = \Delta^2/16$ with Gaussian density. From detailed analysis, it is estimated that the actual density is somewhat in between the two and conveniently, we assume that

$$\langle e^2(t)\rangle = \Delta^2/9$$

As in PCM, we may further assume that all the noise power $\langle e^2(t)\rangle$ is contained in the bandwidth $f_s/2$, because of the sampling of $e(t)$ at the rate f_s. This noise is further filtered through the message filter with cutoff at $f_{\max}$ and the effective Q-noise in the message bandwidth is given by

$$N_q = \langle e^2(t)\rangle\left(\frac{2f_{\max}}{f_s}\right) = 2\Delta^2\frac{f_{\max}}{9f_s} \tag{2.56}$$

Hence, the SNR for sinusoidal input,

$$\frac{S}{N_q} = \left(\frac{\Delta^2 f_s^2}{8\pi^2 f_m^2}\right)\Big/\left(\frac{2\Delta^2 f_{\max}}{9f_s}\right) = \left(\frac{9}{16\pi^2}\right)\cdot\left(\frac{f_s^3}{f_{\max}f_m^2}\right) \tag{2.57*}$$

Others [19] have shown that the SNR follows the relation

$$S/N_q \simeq 0.05\,K^2\cdot(f_s^3/f_{\max}f_m^2),\ K \le 1; \tag{2.58}$$

It is now seen that the SNR improves with f_s at the rate of 9 dB/Oct, whereas in PCM, the improvement is exponential. Moreover, at higher signal frequencies f_m, the SNR deteriorates at the rate of 6 dB/Oct (as shown in Fig. 2.29(c)) but fortunately this matches more or less with the falling spectrum of speech. DM is therefore a matching codec for speech and is being used widely.

2.9.2 Improvements in LDM Codecs [19]

The slope-overload distortion in the single-integrator LDM may be minimised by using a second order prediction (by incorporating a double integrator in the feedback path of the coder) in the codec. This circuit approximates $x(t)$ with linear segments, as is commonly used in network approximations, and the

---

*The S/N_q for random input signal has to be modified according as the derivation of eq. (2.54a). Using eq. (2.56) for N_q, and eq. (2.54a) for Δ;

$$\frac{S}{N_q} = \frac{\sigma_x^2}{N} = \frac{9\sigma_x^2 f_s}{2\Delta^2 f_{\max}} = \frac{9\sigma_x^2 f_s}{2f_{\max}}\left(\frac{\sqrt{3}f_s}{8\pi\sigma_x f_{\max}}\right)^2 = \frac{27}{128\,\pi^2}\left(\frac{f_s^3}{f_{\max}^3}\right) \tag{2.57(a)}$$

If, however, uniform spectral density of $\langle e^2(t)\rangle (= \Delta^2/3)$ is assumed, then

$$\frac{S}{N_q} = \frac{9}{128\,\pi^2}\left(\frac{f_s^3}{f_{\max}^3}\right) \tag{2.57(b)}$$

Comparing eqns. (2.57) and (2.57a), it is observed that S/N_q for random while Gaussian signal is approximately 4 dB worse than that for the sinusoidal signal with $f_m = f_{\max}$. However, for the mean speech frequencies ($\simeq f_{\max}/3$), S/N_q is better by 10 dB showing again that the DM codec matches very well for signals with falling spectrum, e.g., speech.

approximation is shown in Fig. 2.26(a). The signal is built up from $P(t)$ by changing the slope of $\tilde{x}(t)$ instead of the height of $\tilde{x}(t)$ as was done in case of Fig. 2.24(a). For a sinewave input, the second differential of $x(t)$ now matches the step-heights Δf_s, and the slope-overload condition is

$$|x''(t)|_{\max} = A\omega_m^2 \le \Delta f_s \tag{2.59}$$

The corresponding SNR has been shown to be

$$S/N_q \simeq 6 \times 10^{-4} K^2 f_s^5 (f_m^4 f_{\max}),\ K \le 1 \tag{2.60)*}$$

The SNR now improves with f_s at the rate of 15 dB/octave, but at the same time SNR falls at the rate of 12 dB/oct at higher signal frequencies f_m. Moreover, with the second order network in the feedback, the circuit tends to become unstable with possible large overshoots, as shown by dotted lines in Fig 2.26(a). As a result, a modified double integrator with a zero in the transfer function $H(s)$, as shown in Fig. 2.26(b), is used in practical systems. Alternatively, the outputs of a double integrator and a single integrator is added in the ratio of $a_1 : a_2$ to obtain $\tilde{x}(t)$ and to provide the necessary stability of the circuit. This circuit shown in Fig. 2.26(c), has been widely used in later developments.

However, the problem of poor SNR at higher f_m still remains and this has been somewhat solved in the following two improved codecs. In one coder, called Delta-Sigma Modulator (D-ΣM) [18], the input $x(t)$ is compared directly with NRZ output $\tilde{x}(t)$ of the clocked comparator, and the difference $e(t)$ is integrated and fed to the comparator, as shown in Fig. 2.27. This is equivalent to integrating $x(t)$ in the earlier codec of Fig. 2.24 and $\int x(t)\,dt$ compared with $\int P(t)\,dt$. Thus the higher slopes at higher frequencies are reduced by 6 dB/Oct, and avoid the slope overload. The decoder is a simple lowpass-filter amplifier. The coder is specially suitable for low frequency telemetry signals including d.c., but may also be used for speech frequencies with a simple 12 dB/Oct filter in the feedback path. The SNR of the codec has been calculated to be

$$\frac{S}{N_q} \simeq 0.02K^2 \left(\frac{f_s}{f_{\max}}\right)^3, \qquad K \le 1 \tag{2.61}$$

---

*Approximate proof of eq. (2.60) is as follows:

The noise waveform $e(t)$ has the peaks $\pm E_0$, and E_0 is the incremental height achieved by $\tilde{x}(t)$ in the interval $1/f_s$, giving

$$E_0 = \int_0^{\tau} \frac{\Delta}{\tau} t\, dt = \frac{\Delta\tau}{2} = \frac{\Delta}{2f_s}, \quad \tau = \frac{1}{f_s}$$

Therefore, $N_q \propto \left(\frac{\Delta^2}{4f_s^2}\right)\left(\frac{f_{\max}}{f_s}\right)$, in the signal bandwidth.

From eq. (2.59), $S = \frac{A^2}{2} = \frac{\Delta^2 f_s^2}{8\pi^2 f_m^4}$ and, therefore,

$$\frac{S}{N_q} \propto f_s^5 \Big/ (f_m^4 f_{\max}) \tag{2.60a}$$

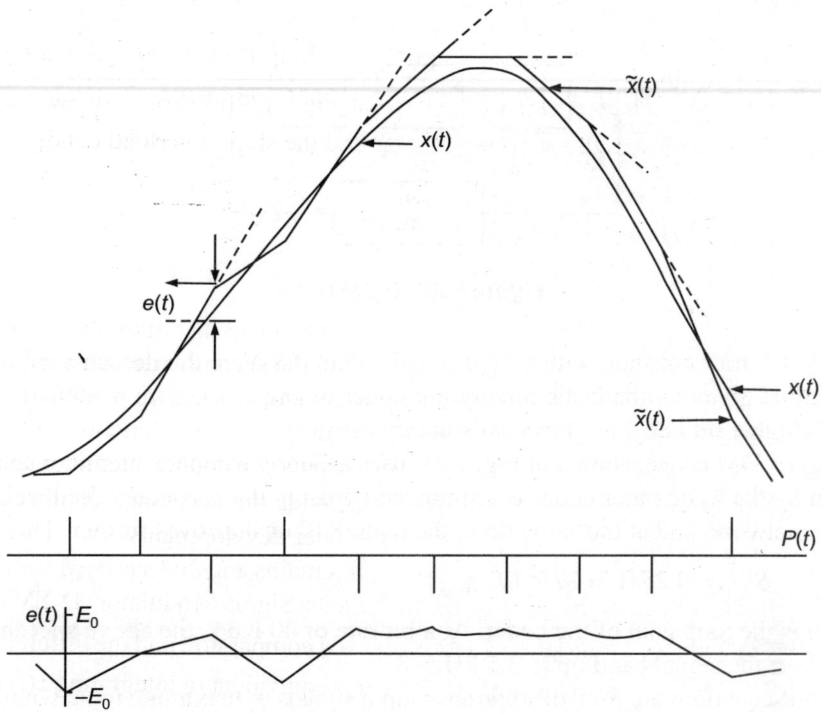

Figure 2.26(*a*): Linear segment approximation of $x(t)$

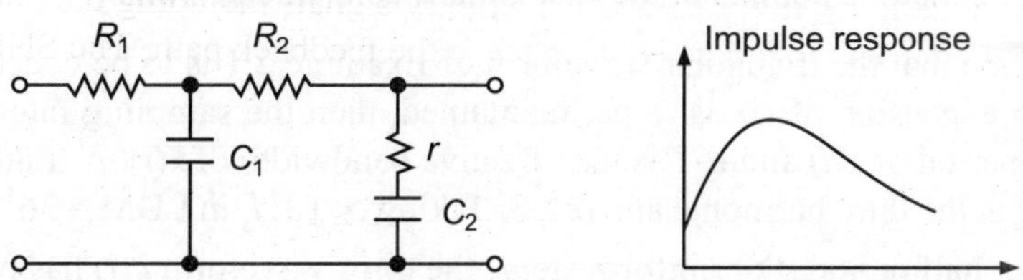

$$H(s) = \frac{S + \gamma}{(S + \alpha)\,(S + \beta),\ \alpha < \beta < \gamma}$$

Figure 2.26(*b*): Modified double integrator and its impulse response

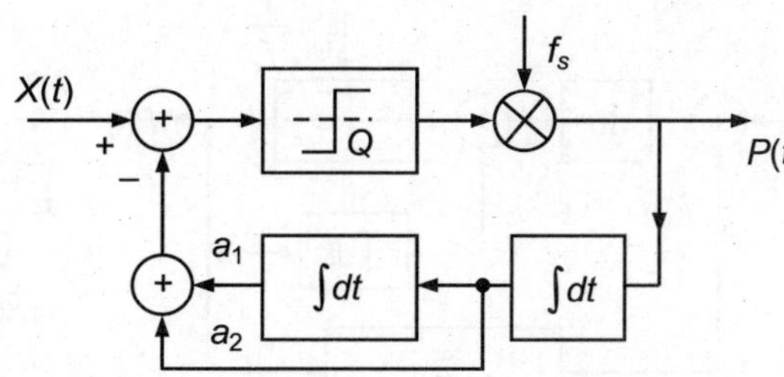

Figure 2.26(*c*): Coder with partial double integrator in the feedback

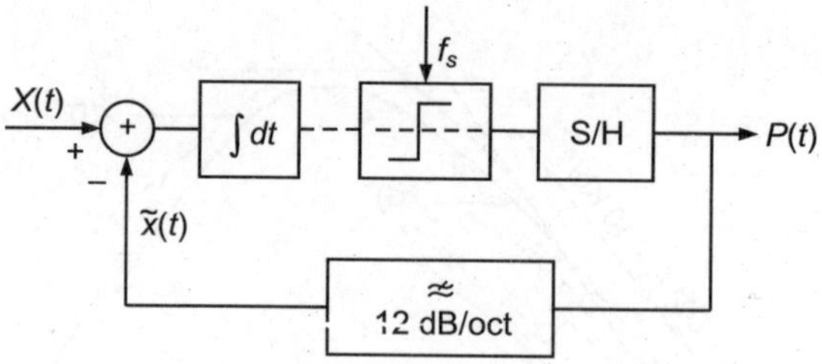

Figure 2.27: *D-ΣM* coder

Although the SNR is now constant with f_m, the actual value of S/N_q in the speech band is approximately 10 dB less than that given by the basic integrating coder of Fig. 2.24, at $f_m = 800$ Hz. The codec then requires a much higher bit rate f_s to give a satisfactory S/N_q.

A more complex DM codec, shown in Fig. 2.28, uses a 2-loop feedback circuit, where the frequency distortion given by the basic inner coder is eliminated by using the secondary feedback through the 12 dB/Oct lowpass network, and at the same time, the overall SNR improves further. The SNR now is

$$S/N_q \simeq 0.25\,|1 + GH|^2 \cdot (f_s/f_{\max})^3 \tag{2.62}$$

where GH ($\simeq 5$) is the loop gain of the coder. At a bit rate of 40 Kb/s, the above speech codec gives an SNR of 40 dB over the signal band up to 3.5 kHz.

The above SNR equations are for 0 dB reference input signals (≡ maximum input without overloading) and the variation of peak SNR with bit rates for different codecs is shown in Fig. 2.29(a). The variations of SNR with signal power S and with f_m are shown in Figs. 2.29(b) and (c). It is observed that the dynamic range of the codecs is poor as in the case of non-companded PCM.

Example 2.7: Assume that the triangular waveform of Example 2.1 is to be coded with linear DM. If the same bandwidth expansion of six is to be maintained, then the sampling rate in LDM is $(12\alpha/T)$, where T is the time period of $x(t)$ and α/T is the effective bandwidth of $x(t)$. A triangular waveform has considerable energy in the third harmonic and $\alpha \geq 3$. Taking $\alpha = 3, f_s$ in LDM $= 36/T$, and the number of steps in $\tilde{x}(t) = 18$ per half period. For uniform steps, the error waveform $e(t)$ has peaks of $\pm\Delta/2$, where the step size $\Delta = 2X_0/18 = X_0/9$.

Therefore, $N_q = \dfrac{\Delta^2}{12} = \dfrac{(X_0/9)^2}{12}$.

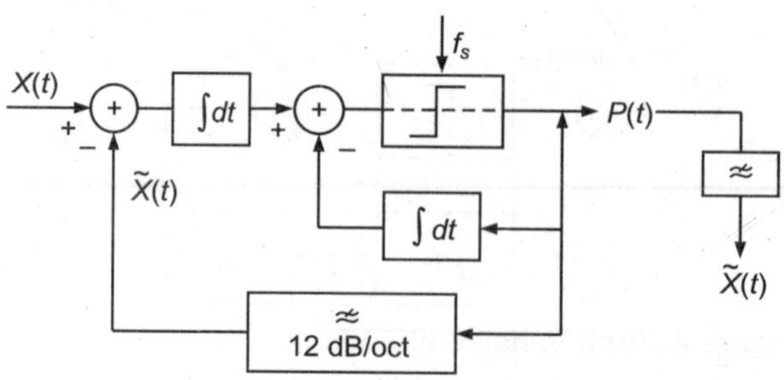

Figure 2.28: 2-loop DM codec

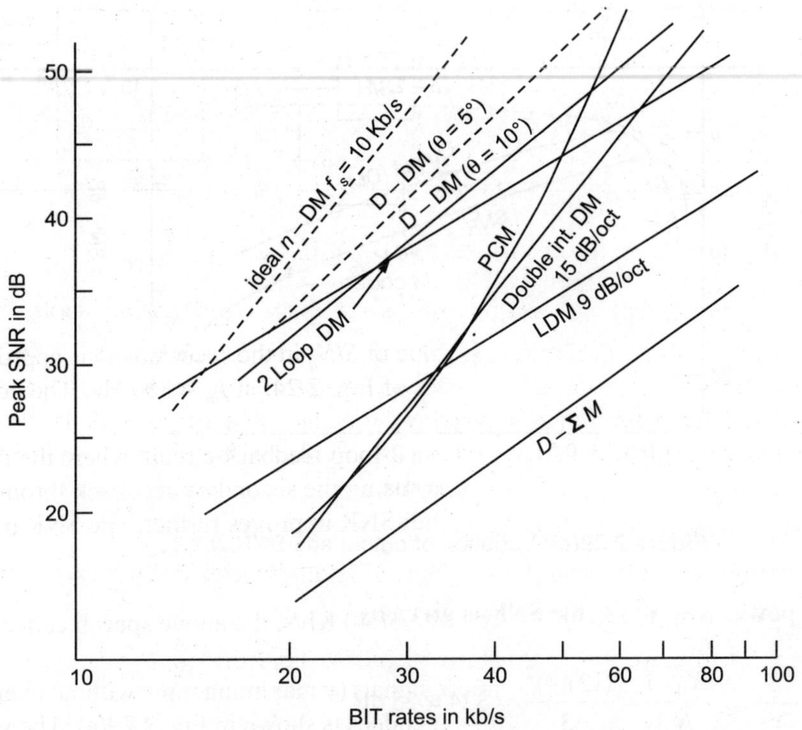

Figure 2.29(*a*): Variation of SNR with PRF for LDM and D-DM codecs (PCM shown for reference)

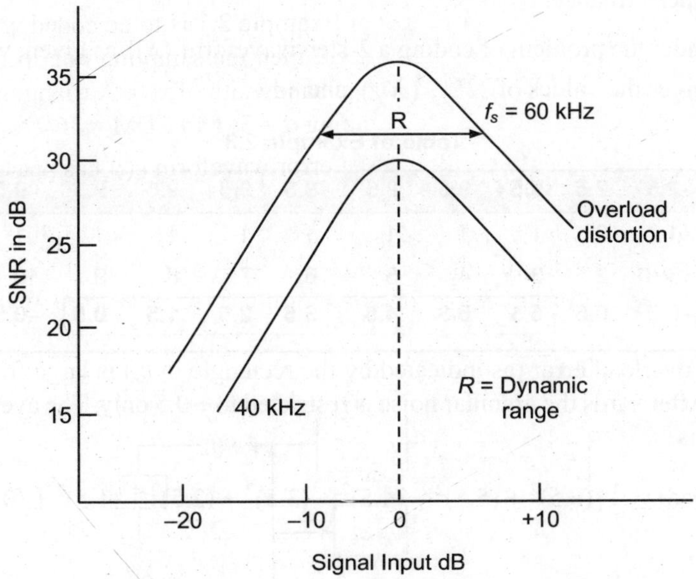

Figure 2.29(*b*): Variation of SNR with *S* in LDM

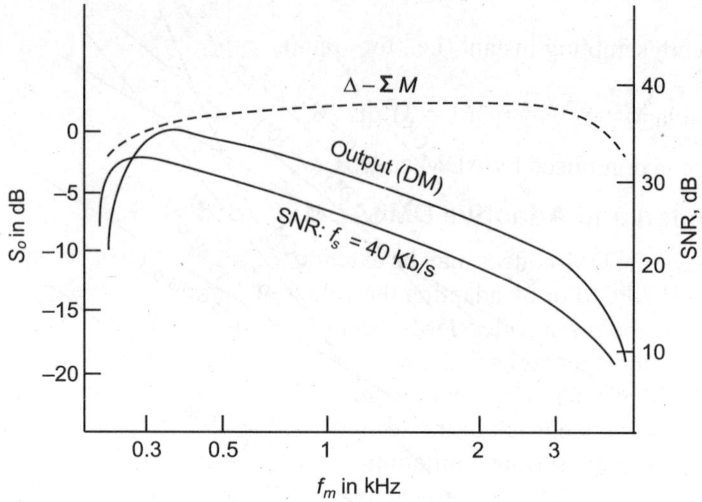

Figure 2.29(c): Variation of output and SNR with f_m

Since the signal power $S = X_0^2/3$, the SNR is given by

$$\frac{S}{N_q} = \frac{X_0^2/3}{N_q} = \frac{(12)(9)^2}{3} = 324 \simeq 25 \text{ dB}.$$

The result is of course worse than that of Ex. 2.1, showing that LDM requires higher bit rate f_s than in PCM for equivalent performance.

Example 2.8: Consider the problem of coding a 2-step waveform $\{X_r\}$ as given, with LDM having $\Delta = 1$. The table below gives the values of $\{P_r\}$, $(\tilde{X}_r)$ and $\{e_r\}$.

Table of Example 2.8

$\{X_r\}$	2.5	2.5	2.5	2.5	9.5	9.5	9.5	9.5	9.5	9.5	9.5	9.5	9.5	9.5	9.5
$\{P_r\}$	1	1	1	0	1	1	1	1	1	1	1	1	0	1	0
$(\tilde{X}_r)$	1	2	3	2	3	4	5	6	7	8	9	10	9	10	9
$\{e_r\}$	1.5	0.5	−0.5	0.5	**6.5**	**5.5**	**4.5**	**3.5**	**2.5**	**1.5**	**0.5**	−0.5	0.5	−0.5	0.5

It is seen that the overload error (as indicated by the rectangle in e_r) is large from the fifth sampling instant to the tenth. Afterwards the granular noise is restricted to ± 0.5 only. The average quantizing noise during the overload is

$$\overline{e_r^2} \text{ (overload)} = \tfrac{1}{7}[(6.5)^2 + (5.5)^2 + (4.5)^2 + (3.5)^2 + (2.5)^2 + (1.5)^2 + (0.5)^2] = 16.23$$

Therefore,

$$S/N_q \text{ (overload)} = \overline{X_r^2}/\overline{e_r^2} = 90.26/16.23 \simeq 8 \text{ dB}$$

But $\overline{e_r^2}$ after the tenth sampling instant (i.e., for granular noise) = 0.25. Therefore,

$$S/N_q \text{ (granular)} = \overline{X_r^2}/\overline{e_r^2} = 361 = 31 \text{ dB}$$

This overload noise is minimised by ADM codecs.

2.9.3 Companded and Adaptive DM (ADM) Codecs [19, 20, 21]

The dynamic range of the DM-codecs may be extended either by using a compressor-expander circuit in the loop, as used in log-PCM or by adapting the values of Δ with the varying values of σ_x, as in APCM. The earlier model of the companded DM-codec, as shown in the Fig. 2.30, used an instantaneous compressor made of diode networks and gave an overall SNR of 30 dB over a dynamic range of 40 dB at the bit rate of 50 Kb/s. The harmonic distortion produced in the compressor-expander circuits is minimised by the negative feedback of the secondary loop [20].

Abate [21] recognised the serious limitations of DM for coding signals having large dynamic range, e.g., speech and TV, and gave two improved adaptive schemes, as shown in Fig. 2.31, where the DM codec is forced to respond adaptively to the changes in the input signal by changing the step size of $\tilde{x}(t)$. The step size Δ is controlled either in a discrete or in a continuous way by the signal slope information. In the discrete scheme of Fig. 2.31(a), when the slope overload occurs, the QZR output is a series of 1's or 0's and the switch control selects a gain K_i greater than K_{i-1}, such that the new larger step is $K_i\Delta$. If the pulse polarity remains unchanged, then higher gains K_{i+1} etc. are chosen successively until the largest K_n is reached. When the polarity of QZR output reverses, the step size decreases in steps. In the decoder, a complementary operation is performed after sensing the input pulses. Since in this discrete adaptive system, the step size is changed at the rate of f_s, the system may be viewed as a DM with instantaneous companding incorporated. In Fig. 2.31(b), the control signal is obtained from the continuous derivatives of $x(t)$ and to conserve channel bandwidth, this must be a very low frequency signal. (In speech, σ_d varies at the syllabic rate only). As such Δ is varied at a slow rate and the resultant DM may be viewed as one with syllabic companding incorporated. In all later developments in ADM codecs, some modifications of the above principles only have been used.

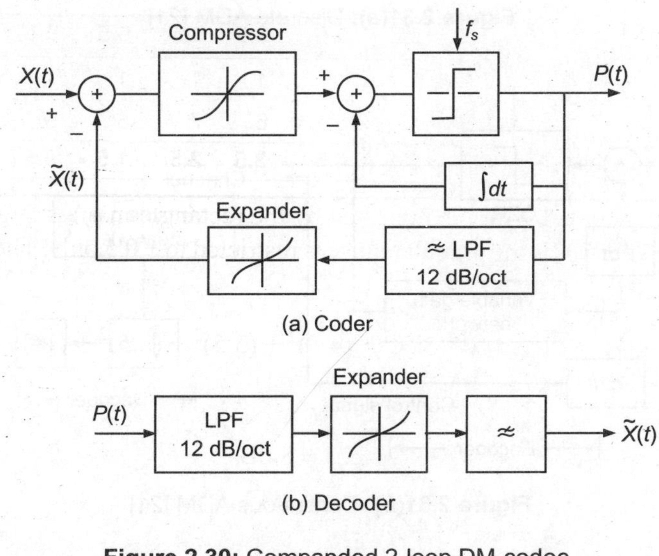

Figure 2.30: Companded 2-loop DM-codec

Abate has defined the slope loading factor F as

$$F = \frac{\Delta f_s}{\sigma'_x} = \frac{\Delta}{\sigma_d} \tag{2.63}$$

where σ'_x = rms value of $x'(t)$, and $\sigma_d = (\langle(x_r - x_{r-1})^2\rangle)^{1/2} = \sigma'_x / f_s$. For LDM, minimum N_q occurs for

$$F = \ln(f_s/f_{max}) \text{ and } \Delta_{opt} = \sigma_d \cdot \ln(f_s/f_{max}) \tag{2.64}$$

since $\Delta f_s \geq |x'(t)|_{max}$. In discrete ADM, the slope overload occurs if $|x'(t)| > K_n \Delta f_s$, and the maximum slope loading factor is

$$F_{max}(ADM) = K_n \Delta / \sigma_d$$

It was found by computer simulation that the minimum N_q did not increase even for large K_n, and thus a companding advantage of $(K_n)^2$ is obtained. The value of K_n is

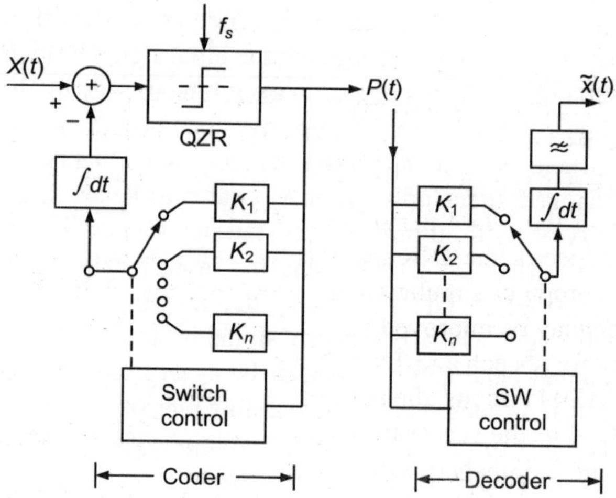

Figure 2.31(*a*): Discrete ADM [21]

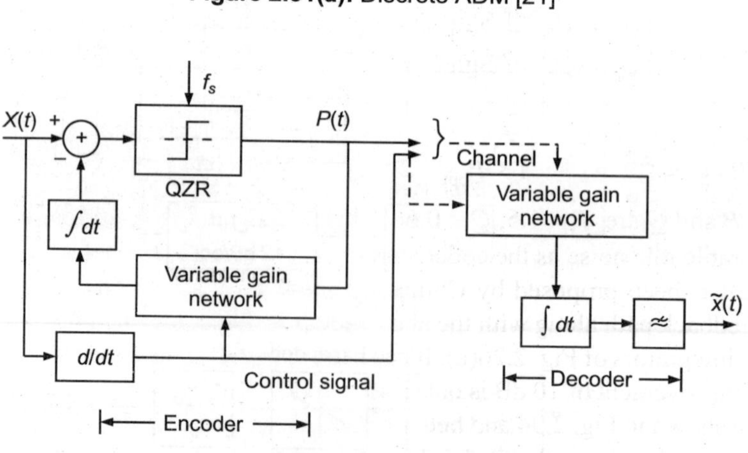

Figure 2.31(*b*): Continuous ADM [21]

$$K_n = \frac{\sigma_x(\max)}{\sigma_x(\min)}$$

and the smallest step size is found to be

$$\Delta(\min) = \sigma_d \cdot \sigma_x(\min) \ln (f_s/f_{\max})$$

The approximate values of the granular noise power N_q and the overload noise power N_0 have been given as

$$N_q = \begin{cases} \dfrac{\Delta_k^2 f_{\max}}{3 f_s} & \text{for } F < 8 \\ \dfrac{F \Delta_k^2 f_{\max}}{24 f_s} & \text{for } F > 8 \end{cases} \tag{2.65}$$

and

$$N_0 \simeq \frac{8\pi^2}{81}(3F + 1) \exp(-3F_{\max})$$

where $\Delta_k = K_n \Delta$ and $F_{\max}$ as defined above.

It is seen from above that the ADM performances depend specifically on the nature of the signal spectrum and for integrated spectra, the SNR and dynamic range are better than those for uniform spectra. These results obtained by computer simulation are shown in Figs. 2.32(a), (b) and (c), It may also be seen that the dynamic range may be improved further by increasing K_n but a practical limit of $4 < K_n < 8$ has been found satisfactory for speech and TV signals.

In later development of ADM circuits, the use of adaptive QZR has been more popular than nonlinear companding in the loop. One of the successful instantaneously adaptive codecs has been proposed by Jayant [22] and shown in Fig. 2.33, where a one-bit memory is used to indicate the continuous increase or decrease of $x'(t)$ and the step sizes are adapted according to the algorithm

$$\Delta_r = \begin{cases} \Delta_{r-1} \cdot P, & \text{if Sgn } e_r = \text{Sgn } e_{r-1} \\ \Delta_{r-1} \cdot Q, & \text{if Sgn } e_r \neq \text{Sgn } e_{r-1} \end{cases}$$

and

$$P \cdot Q \simeq 1, \qquad 1 < P_{\text{opt}} < 2 \tag{2.66}$$

Probable values of P and Q are: $P = 1.5$, $Q = 0.66$. This adaptation results in improved dynamic range, but there is considerable idle noise as the coder starts hunting when $\tilde{x}(t)$ approaches $x(t)$. A refinement of the above codec has been proposed by Childs [23], who uses a combination of single and double integrators in the feedback path along with the above adaptive logic, i.e., the integrator of Fig. 2.33(a) is replaced by the two integrators of Fig. 2.26(c). It has been shown that the codec has a very large dynamic range and an SNR improvement of 10 dB is obtained at comparatively low f_s. Responses of the codec to triangular inputs are shown in Fig. 2.34 and better results are obtained for $P = 1.2$, $Q = 0.8$, $R = a_1/a_2 = 2$. The main advantage of the codec is that its hardware may be almost all-digital and made of IC's mostly.

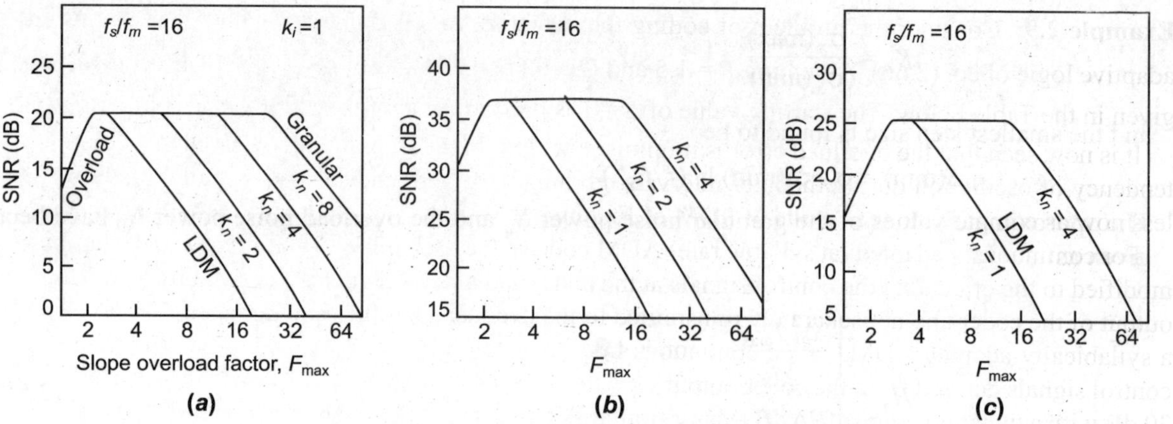

Figure 2.32: SNR and dynamic range in ADM (computer simulation) [21]: **(*a*)** Uniform signal spectrum, **(*b*)** TV signal spectrum, **(*c*)** Speech signal spectrum

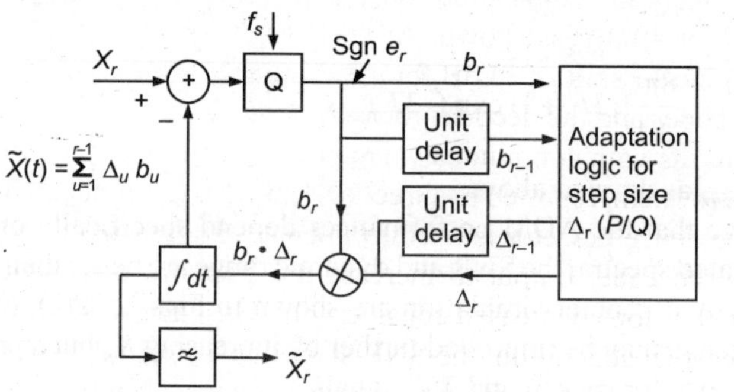

Figure 2.33(*a*): Block diagram of ADM with 1-bit memory

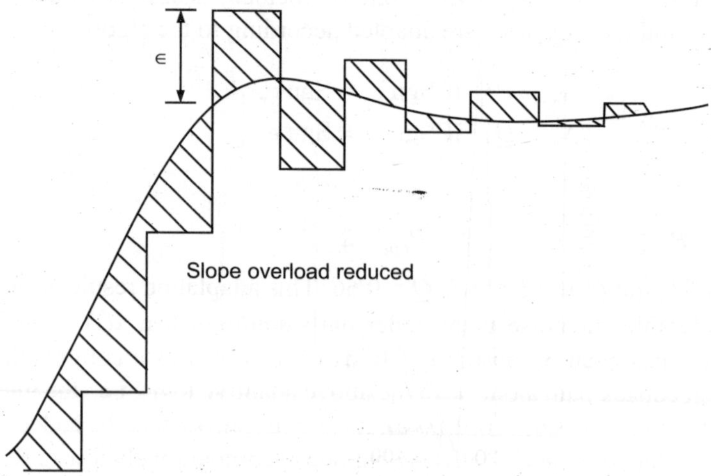

Figure 2.33(*b*): Effect of instantaneous companding

Example 2.9: Consider the problem of coding the waveform of Example 2.8 with ADM having the adaptive logic of eq. (2.66). Assuming $P = 1.5$ and $Q = 0.66$, the values of $\{P_r\}$, $\{\tilde{x}_r\}$, $\{\Delta_r\}$ and $\{e_r\}$ are given in the Table below. The starting value of $\Delta = 1$ is used in the approximation.

It is now seen that the overload error is minimised within three sampling intervals only, but there is a tendency of oscillation during the later intervals, giving a larger idle noise. The overall N_q is however less now as compared to the results of LDM of Example 2.8.

For continuously adapted (at syllabic rate) ADM codecs, the principle shown in Fig. 2.31(b) has been modified to the effect that the control signals at the coder and at the decoder are derived from the digital output of the coder and not separately transmitted to the decoder. Greefkes and de Jager [24] developed a syllabically-adapted ADM, called continuous DM', as shown in Fig. 2.35, where the low frequency control signals derived from the coder output vary the QZR step size. At $f_s = 34$ kHz, the peak SNR was 30 dB with a dynamic range of 25 dB. Many similar systems have been experimented with and one such ADM codec, developed by Cartmale and Steele [25], shown in Fig. 2.36, gave a very wide dynamic range. The basic coder was a D-ΣM circuit, and an nth order power law element was used to control the feedback gain. At $f_s = 192$ Kb/s and with a time constant of 10 ms for the control signal, the SNR was 15 dB with a dynamic range of 40 dB, as shown. At 40 Kb/s and with a partial double integrator in the feedback, the ADM will give an SNR of 35 dB for a dynamic range of 45 dB. However, the matching of the nth law device at the coder and the decoder for varying σ_x will be a critical hardware problem.

The most promising results with continuously adapted codecs have been obtained with the digitally controlled ADM (DCDM) system, reported by Greefkes [26]. In the circuit, shown in Fig. 2.37, a 4-stage shift register with a coincidence detector is used to indicate the sharp rise/fall in $x'(t)$ and the filtered control signal multiplies the coder output to increase the step size continuously till the coincidence output returns zero, when the slope decreased considerably. The control is syllabic and the time constant used is 10–20 ms. With correct adaptation, the SNR should be independent of the input σ_x and be at its peak value. With $f_s = 40$ Kb/s, SNR > 30 dB for a dynamic range of 45 dB and for $f_s = 60$ Kb/s, SNR > 37 dB. SNR values are shown in Fig. 2.37(b) along with the relative amplitudes of the feedback pulses.

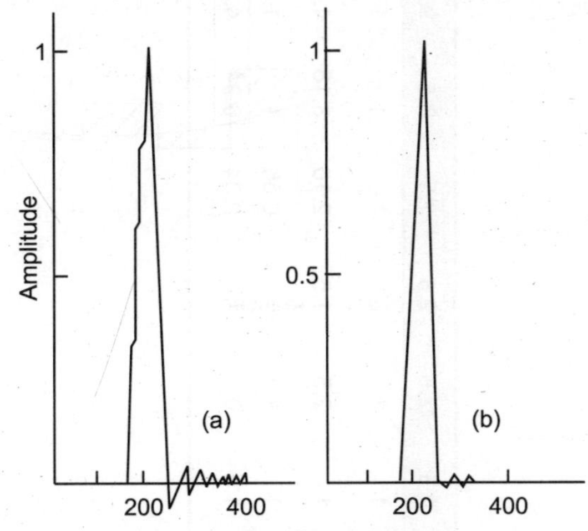

Figure 2.34: Responses to triangular inputs through ADM: **(a)** $P = 1.2$, $Q = 0.6$, $R = 1$, **(b)** $P = 1.2$, $Q = 0.8$, $R = 2$

Table of Example 2.9

$\{X_r\}$	2.5	2.5	2.5	2.5	9.5	9.5	9.5	9.5	9.5	9.5	9.5	9.5	9.5	9.5	9.5	9.5
$\{P_r\}$	1	1	0	1	1	1	1	1	0	1	0	1	1	0	0	1
$\{\tilde{X}_r\}$	1	2.5	1.5	2.16	3.16	4.66	6.91	10.28	8.03	9.53	8.53	9.19	10.19	9.53	8.53	9.19
$\{\Delta_r\}$	1	1.5	1	0.66	1	1.50	2.25	3.38	2.25	1.50	1	0.66	1	0.66	1	0.66
$\{e_r\}$	1.5	0	1	0.34	**6.34**	**4.54**	**2.59**	–0.78	1.47	–0.03	0.97	0.31	–0.69	–0.03	0.97	0.31

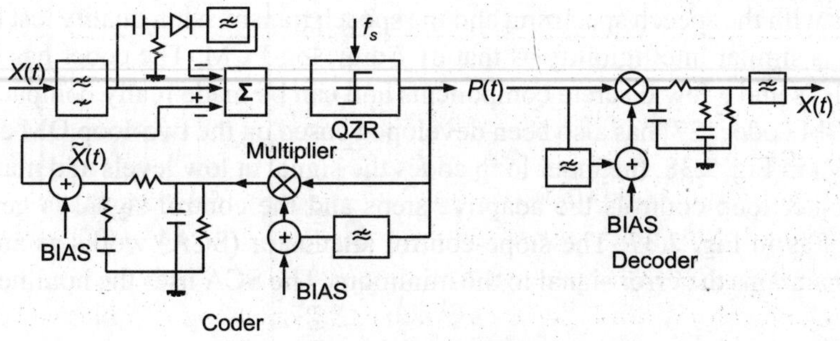

Figure 2.35(*a*): Block schematic of continuous DM

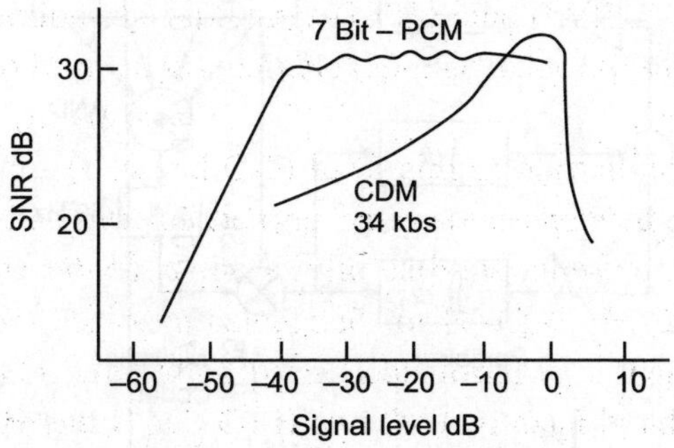

Figure 2.35(*b*): SNR of CDM at 34 Kb/s

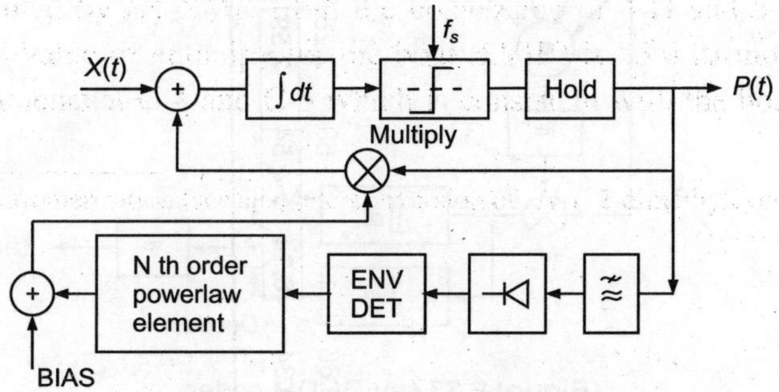

Figure 2.36(*a*): ADM with *D-ΣM*

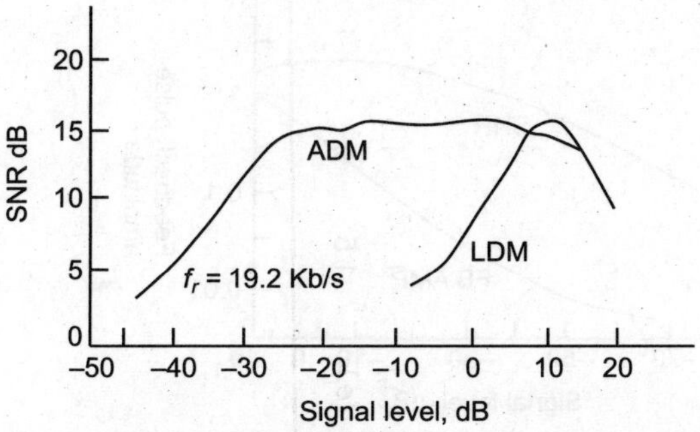

Figure 2.36 (*b*): SNR vs S for companded *D-ΣM*

The coder matches with the speech spectrum, and the speech transmission quality test has shown that the 40 Kb/s DCDM has similar intelligibility as that of 7 digit log-PCM. The coder has been miniaturized around only four IC's and a few discrete components and can be made really compact.

An efficient ADM codec [27] has also been developed based on the two-loop DM circuit of Fig. 2.28. In the circuit shown in Fig. 2.38, the inner loop codes the signal at low levels and maintains the desired 1/0 pattern. The outer loop controls the adaptive steps and the control signal is generated through a coincidence circuit as in Fig. 2.37. The slope-control attenuator (SCA) works as an expander for the digital input and maintains the error signal to the minimum. The SCA uses the nonlinear drain-to-source

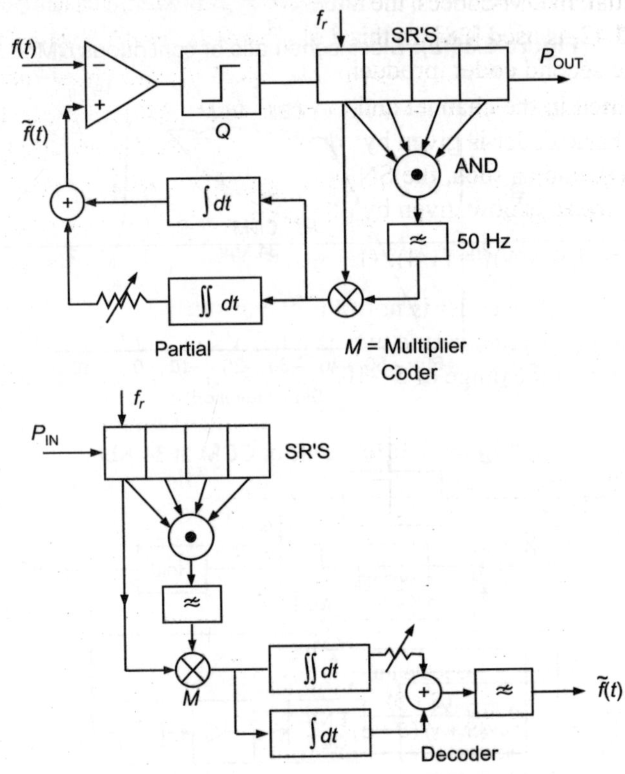

Figure 2.37 (*a*): DCDM codec

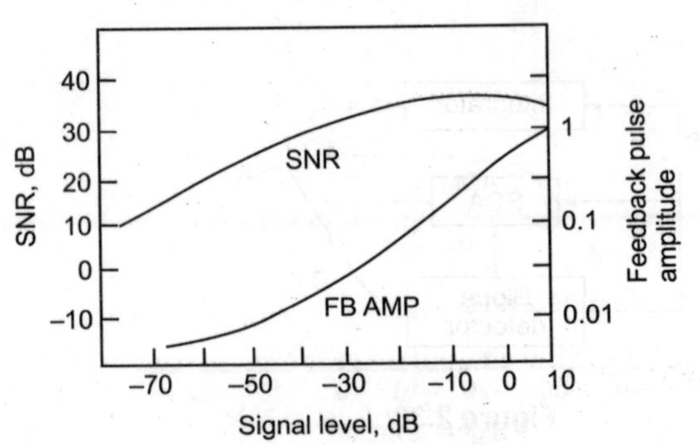

Figure 2.37 (*b*): SNR and feedback characteristics of DCDM; f_s = 40 Kb/s [26]

resistance of a FET and can be used in two alternate configurations as shown in Fig. 2.39. The SNR characteristic of the codec is shown in Fig. 2.40, where it is seen that at $f_s = 32$ Kb/s, the SNR > 28 dB with a dynamic range of 35 dB. The frequency response is flat in the speech band, but the SNR falls by 9 dB at 3.2 kHz; thus the codec is suitable for integrated-spectrum signals. By modifying the circuit, as shown in Fig. 2.41, a flat-spectrum ADM with a constant SNR and constant frequency response over the band, has been developed. This codec has an SNR > 32 dB with a dynamic range of 35 dB at $f_s = 40$ Kb/s; and the overall characteristics are equivalent to those of 7-bit log-PCM.

All the above unidigit DM coders suffer from the defect that the SNR improvement with f_s is only 9 dB/Oct with single integration and 12 dB/Oct with partial double integration, whereas in PCM, the SNR improvement is exponential. In DM-codecs, the rate of SNR improvements can be enhanced if a multistage coder, as shown in Fig. 2.42, is used [28]. In this 2-stage coder, called D-DM, the error $e_1(t)$ of the basic coder is processed in the second coder, producing an error $e_2(t)$. The digital outputs of two coders are multiplexed and transmitted in the channel and the receiver recovers $\tilde{x}(t)$ in a complementary way as shown. The SNR of the basic coder is given by eq (2.62); but the input $e_1(t)$ is now statistically random and uncorrelated with $x(t)$, and as such, the SNR of the second coder is less by 10–12 dB. The overall $(\mathrm{SNR})_{\mathrm{II}}$ of the two-digit codec is now given by [28, 29],

$$(S/N_q)_{II} = 1/4(S/N)_1^2 = (1/4)^3\,(1+\mathrm{GH})^4\,(f_s'/f_{\max})^6 \tag{2.67}$$

where the sampling rate f_s' in each coder is now half the original, i.e., $f_s' = f_s/2$. Equation (2.67) is valid under the assumption that in the receiver the two channel outputs add in phase, but in practice there will be some phase difference θ, in the range of 5°–10°, and the $(S/N_q)_{II}$ is now poorer. The variation of SNR

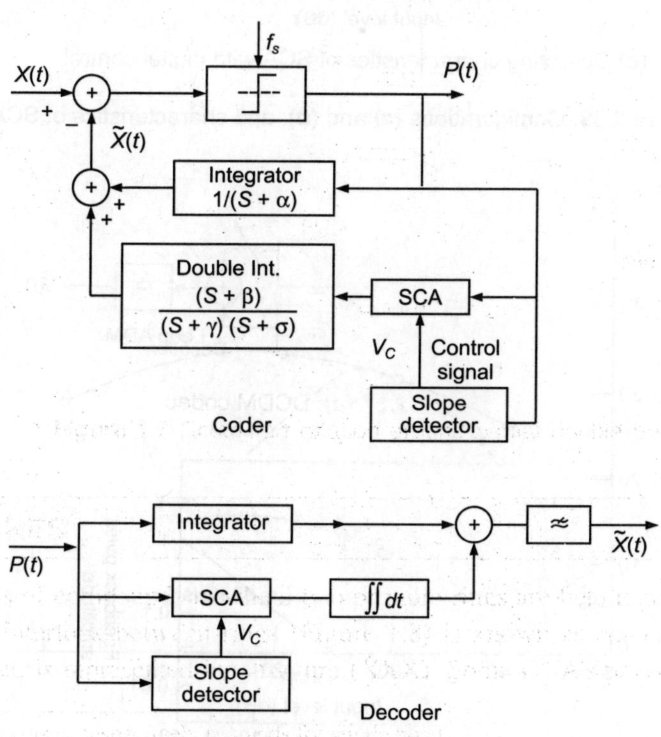

Figure 2.38: 2-loop ADM [27]

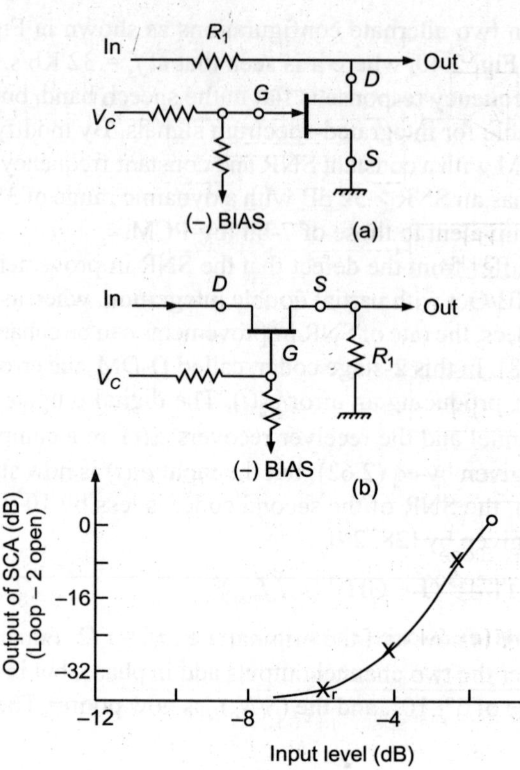

(*c*) Operating characteristics of SCA with digital control

Figure 2.39: Configurations **(*a*)** and **(*b*)**, and characteristics of SCA **(*c*)**

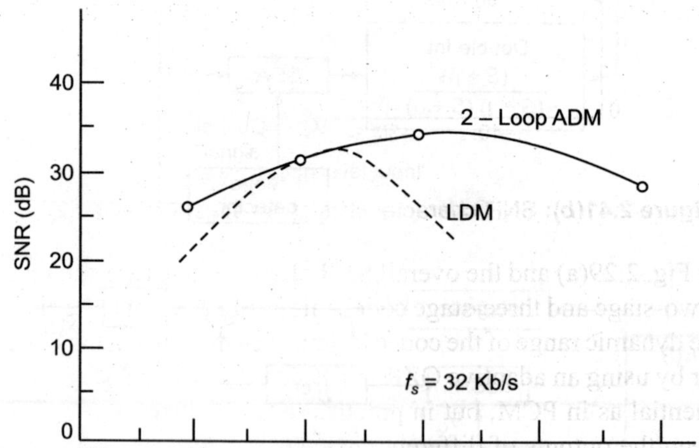

Figure 2.40: SNR characteristics of 2-loop DCDM [27]

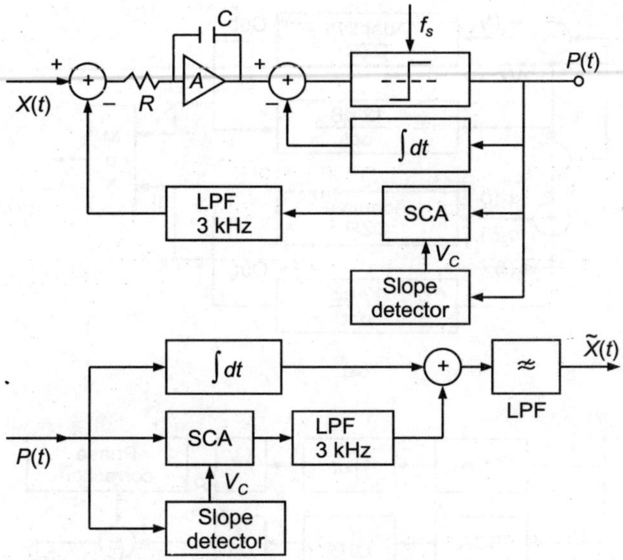

Figure 2.41(*a*): 2-loop DCDM (flat spectrum)

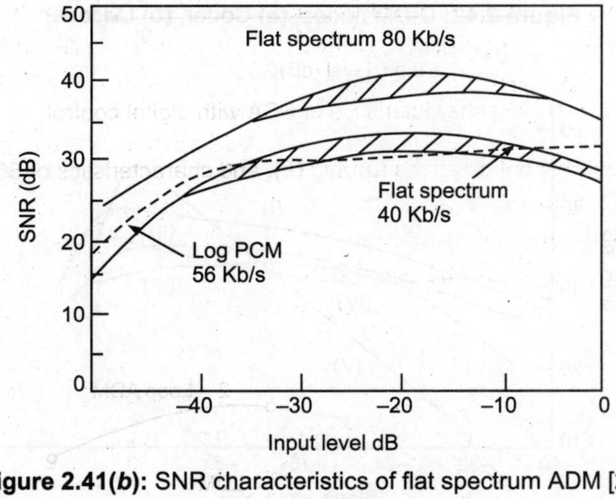

Figure 2.41(*b*): SNR characteristics of flat spectrum ADM [27]

with bit rate is shown in Fig. 2.29(a) and the overall SNR characteristics are shown in Fig. 2.43, where it is seen that the SNR of two-stage and three-stage codecs are better than that of a single-stage codec at the same overall bit rate. The dynamic range of the codec is improved by either using closed loop companding, as shown in Fig. 2.30, or by using an adaptive QZR in the successive codecs. Ideally, the SNR gain in an *n*-stage D-DM is exponential as in PCM, but in practical circuits, there will be some SNR loss due to phase differences between the outputs of different channels at the receiver. In an experimental D-DM system, an SNR $\simeq$ 38 dB with a dynamic range of 45 dB, has been obtained at a bit rate of 60 Kb/s.

2.9.4 Effect of Channel Errors [30]

Considering an uniformly distributed $x(t)$, e.g., a triangular waveform, being coded by LDM or D-ΣM, the signal build up at the receiver output is given by

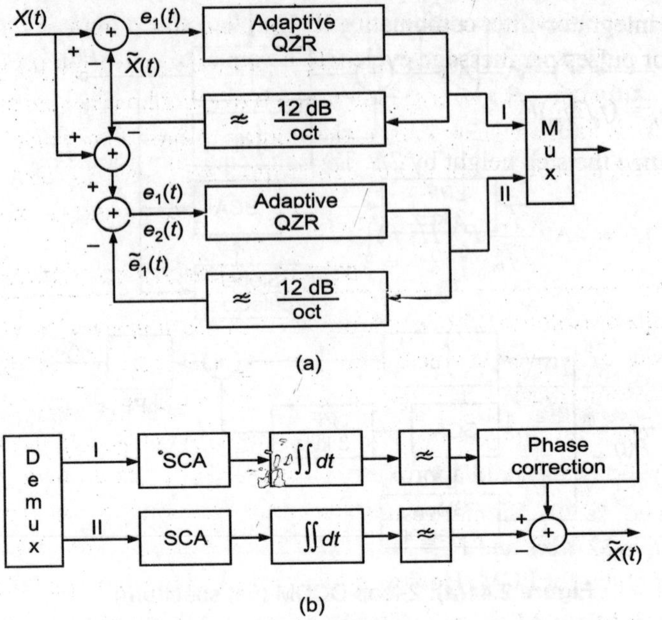

Figure 2.42: D-DM codec, **(*a*)** Coder, **(*b*)** Decoder

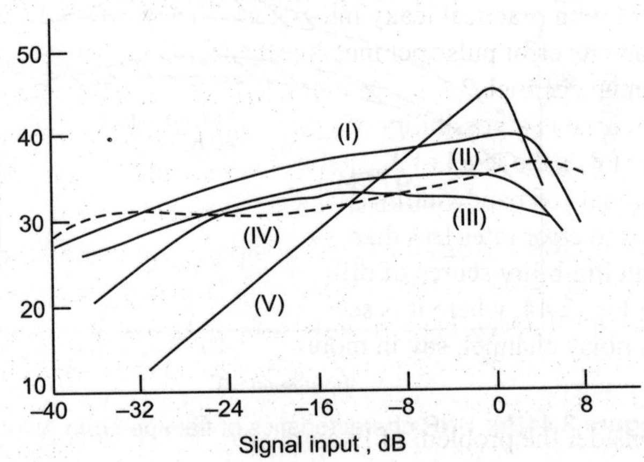

Figure 2.43: Variation of SNR with input level, (i) D-DM, n = 2, 60 Kb/s, θ = 10° (ii) D-DM, n = 2, 40 Kb/s, θ = 10° (iii) 2-loop ADM, 50 Kb/s (iv) Log-PCM, 7-bit (v) D-DM using LDM, 60 Kb/s, θ = 10°

$$\tilde{x}(t) \propto \Delta(f_s/2f_m)|H(f)|$$

and

$$S_0 = \sigma_{\tilde{x}}^2 \propto \frac{\Delta^2}{12}\left(\frac{f_s}{2f_m}\right)^2 |H(f)|^2 \tag{2.68}$$

where peak-to-peak signal voltage is developed by $(f_s/2f_m)$ steps per message cycle and $H(f)$ is the transfer function of the integrator-filter combination. If P_e is the error probability in the channel, then the average number of error pulses per message cycle will be equal to

$$p_n = (f_s/2f_m)P_e$$

and each error will change the step height by 2Δ. Then the equivalent noise power N_e is

$$N_e \propto 4\Delta^2 \frac{f_s}{2f_m} P_e \,|\, H(f) \,|^2 \tag{2.69}$$

and

$$\frac{S_0}{N_e} = \frac{f_s}{f_m} \frac{10^{-2}}{P_e} \tag{2.70}$$

For ADM, eqns. (2.68) and (2.69) hold approximately with $\Delta = \Delta_{mean} = \sqrt{\Delta_{min} \cdot \Delta_{max}}$ and thus (S_0/N_e) remains the same as in eq. (2.70). Similar results have been given by Johnson [30]. For a test signal at 800 Hz in LDM, with $f_s = 32$ Kb/s and $P_e = 10^{-3}$, $(S_0/N_e) \simeq 400 = 26$ dB. Also $(S_0/N_e) \simeq 1/[(2.5)P_e]$, an equation similar to that given for PCM. However, there is a significant perceptual difference in the two codecs having the same P_e. In PCM, an error in MSB gives an amplitude jump in $\tilde{x}(t)$ by X_{max}, whereas in DM, the jump is only by a step of $2\Delta \ll X_{max}$. Moreover, the eq. (2.70) is rather pessimistic in the sense that the energies due to the error pulses in a message cycle have been assumed to be linearly added, which is not true in DM with practical leaky integrators. In any case, with $P_e < 10^{-2}$, there is hardly any possibility of more than one error pulse per message cycle, thus giving a perceptually better score in DM than in PCM for the same channel P_e.

In ADM, however, there is a possibility of larger error steps because of adaptation, but none of the larger steps ($\leq \Delta_{max}$) reaches the level of X_{max} as in PCM. Moreover, the adaptation control voltage is a very low frequency signal (of bandwidth < 100 Hz) in syllabically adapted DM; as such, the control signal does not respond to error rates less than, say, 5 per cent, and the overall reconstructed signal level remains stable. The intelligibility scores of different ADM (DCDM) systems with $f_r = 20$, 14.4, 10 and 7.2 Kb/s are shown in Fig. 2.44, where it is seen that the intelligibility scores do not deteriorate with $P_e < 0.05$ [31]. Thus in a noisy channel, say in mobile digital radio communication, ADM performs much better than PCM.

Example 2.10: Reconsider the problem of Example 2.4. In DM codecs, the channel error rate $P_e \leq 10^{-3}$ is acceptable of speech reconstruction. It is known that for CNR = 7 dB, $P_e = 10^{-3}$ using BPSK transmission. This gives an equivalent $S_0/N_e \simeq 1/(2.5)\, P_e = 26$ dB. But $S/N_q \geq 32$ dB in ADM with $f_s = 40$ Kb/s. Thus effectively $S/N_T = 25.5$ dB; and at the same time the channel CNR of 7 dB is acceptable (whereas in PCM, CNR ≥ 9 dB).

With this background, consider the design of the DCDM codec of Fig. 2.37. From Fig. 2.37 b, the mean step size $\Delta_{mean} = \sqrt{0.01} = 10^{-1}$. Then, $N_e \propto 4\Delta^2_{mean}\, (f_s/2f_m)\, P_e|H(f)|^2 \simeq 8.10^{-5}$ with $|H(f)| \simeq 1$ for low frequencies, $f_s = 40$ kHz, $f_m = 1$kHz and $P_e = 10^{-4}$. This gives $N_e \simeq -41$ dB.

From eq. (2.56), $N_q = 2\Delta^2 \cdot f_{max}/9f_s$.

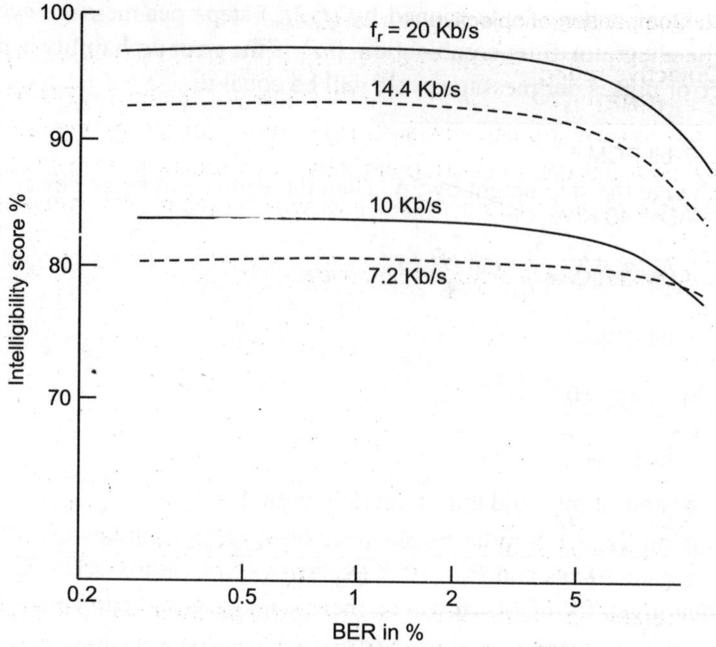

Figure 2.44: Intelligibility scores in ADM (syllabic companding) with different BER [31]

Using the above values and $f_{max} = 4$ kHz, $N_q \simeq -37$ dB, and the overall $N_T \simeq -35.5$ dB. For equal values of N_e and N_q, P_e may be larger, say, $P_e = 2 \times 10^{-4}$. Of course, for speech encoding, a still larger value of P_e may be tolerated.

2.10 SUBJECTIVE RATING [31, 32]

Although SNR characteristics have been the main criteria for evaluating the different codecs, the listening tests and subjective rating do not always conform to the objective rating of codecs based on SNR only, specially at low bit rates. The first observation is that the perceptual quality of the coder varies with the amount of correlation between the signal amplitudes and the error waveforms, and in low-bit-rate ADPCM and ADM codecs, the error waveforms are somewhat signal dependent. As such, the annoyance value of the signal-correlated noise is less for codecs with reasonably good SNR's but the effect is worse when the SNR is below a threshold. It has been further shown that in ADPCM, the error spectrum falls with higher frequencies and is somewhat matched to the long-term speech spectrum itself. Although there is a certain speech-related buzziness in ADPCM noise, the Q-noise is less in ADPCM than in log-PCM during the silent periods of speech (due to smaller Δ during zero-signal periods). In ADM at low bit rates, the slope-overload distortion constitutes the major part of N_q and this is mostly signal-correlated and of less annoyance value. Thus the perceptual quality of ADM is largely controlled by the granular Q-noise and it is recommended that speech quality in low bit-rate ADM may be improved by slow-adaptive strategies (i.e., by operating with some overload) and by making N_q-spectra falling at higher frequencies. On the basis of above reasonings and subjective ratings, the different useful codecs are catalogued in Table 2.12 according to their preferences [4]. It is seen that ADPCM and ADM codecs are subjectively assessed to be better than corresponding PCM codecs having similar SNR's.

Table 2.12: Comparison of objective and subjective performances of A/D codecs

Objective rating (SNR)	Subjective rating (preferences)
7-bit PCM	ADM-40 Kb/s
ADM-40 Kb/s	7 bit PCM
ADM-32 Kb/s	ADM-32 Kb/s
6 bit PCM	4 bit ADPCM
4 bit ADPCM	6 bit PCM
5 bit PCM	3 bit ADPCM
3 bit ADPCM	5 bit PCM
4 bit PCM	4 bit PCM

2.11 LDM-TO-PCM CONVERSION [19, 33]

It has been possible to fabricate LDM coder in a single IC chip and a high bit rate LDM can represent analog $x(t)$ with a high accuracy within $\pm\,\Delta$. The output of LDM can then be applied to an up-down (reversible) binary counter, whose output X_r at any clock instant gives the binary representation of $x(t_r)$. If the counter output is now read at the Nyquist interval $1/(2W)$, then the reading will be equivalent to a PCM word of B bits; and we obtain a linear PCM output, as shown in Fig. 2.45. However, Q-noise spectrum of the LDM output has a bandwidth spread upto $f_s/2$ Hz and there will be excess aliasing noise when the sampling rate is reduced to $2W$ Hz. It is therefore necessary to remove the Q-noise in the band $2W$ to $f_s/2$ Hz and this can be done very efficiently by a digital (FIR) lowpass filter with cutoff at $2W$ Hz. The output of the filter is computed only every M samples ($M = f_s/2W$) and the hardware of the converter is all digital.

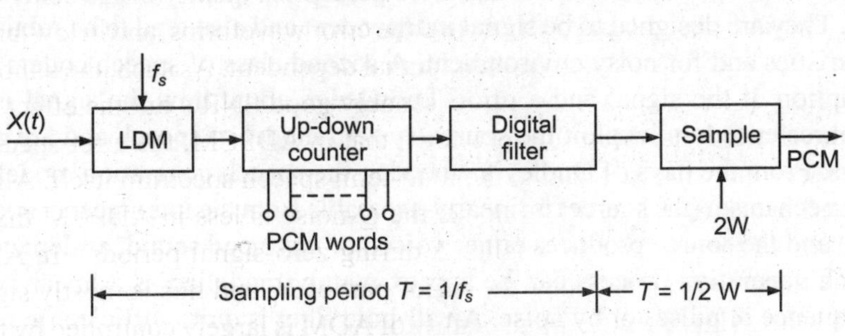

Figure 2.45: LDM-PCM converter

Alternatively, $x(t)$ may be sampled and held at the Nyquist rate and passed through A-law/μ-law compressor, giving compressed samples x_r of width $\tau \simeq 1/2W$. The LDM now approximates the step x_r within τ sec. and the Δ steps counted by the up-down counter as shown in Fig. 2.46.

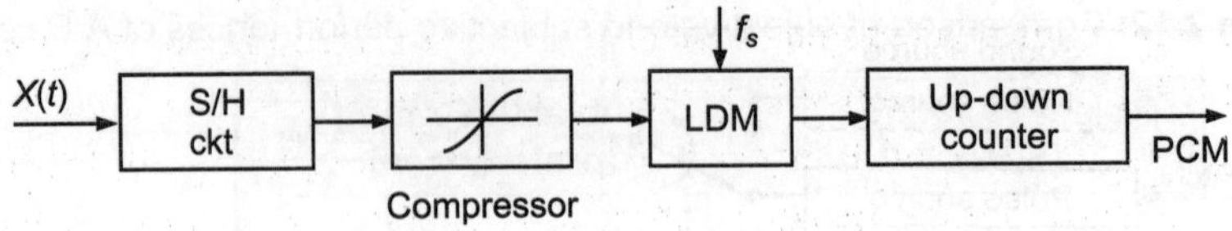

Figure 2.46: LDM-log PCM converter

After the counter is read, the LDM and the counter may be reset to read the next sample x_{r+1} during the next interval. The PCM words at the codec output is now equivalent to the output of a log-PCM codec. For a 7 bit PCM word, the LDM clock has to be:

$$f_s > 128 \times 8000 = 1.024 \text{ Mb/s}.$$

Allowing for the settling time, $f_s \simeq 1.25$ Mb/s.

It is also possible to convert LDM output to ADPCM words at Nyquist rates through a modified circuit shown in Fig. 2.47. The basic approach is to implement ADPCM directly from PCM samples through digital hardware e.g., arithmetic and logical operations only. Because of the two-stage code conversions necessary in the codec, the overall SNR degrades to some extent, but the loss does not exceed 3 dB. It is also possible to convert LDM to ADM and *A*-law PCM through suitable digital hardware [19].

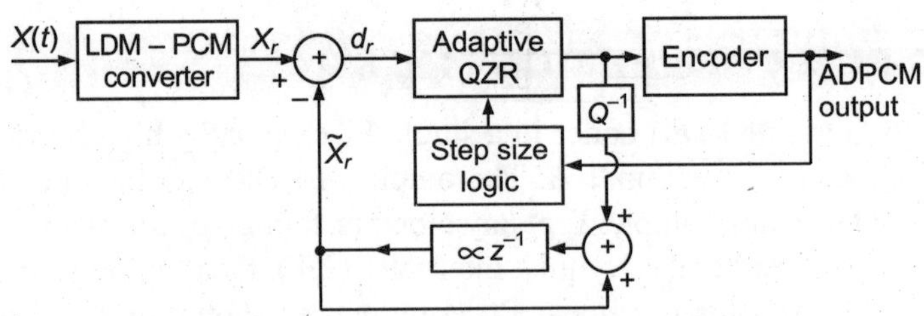

Figure 2.47: LDM to ADPCM converter

2.12 PARAMETRIC SPEECH CODERS [2, 5]

The codecs discussed so far are based on waveform coding and are equally applicable to speech, TV, and telemetry signals. They are designed to be signal independent and also tend to be robust for a wide range of signal characteristics and for noisy environment. A second class of speech coders depends upon the parametric description of the signal and a priori knowledge about how the signal is generated at the source. These 'source encoders' exploit the source-system model of speech and are classified as APC, LPC and vocoders. From the days of Dudley's vocoder, the speech generation model specifies that the sound generating mechanism (the source) is linearly-separable from the intelligence modulating vocaltract filter (the system) and the source produces either voiced or unvoiced sound, as depicted in Fig. 2.48(a).

Thus the speech generation process can be approximately modelled as a linear system driven by a quasi-periodic sequence of pulses or by noise. An all-pole filter is quite sufficient to represent the effect of the vocal tract and subsequent transmission medium of the signal. The equivalent transfer function can be represented by the predictor as shown in Fig. 2.48(b). Since the voiced speech has a long-term correlation and the vocal tract transfer function gives a short-term correlation, the speech model can be further modified and represented by a two-stage predictor with a single pulse as the input as shown in Fig. 2.48(c). As a result, an inverse filter should be able to remove the short-term and long-term redundancy of the speech signal. Such an inverse filter and its equivalent parallel form is shown in Fig. 2.49, where

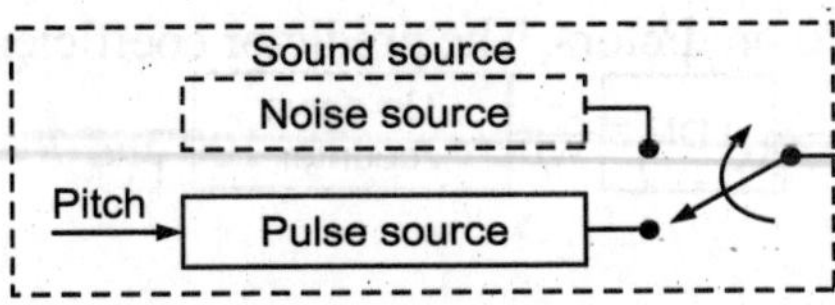

(*a*)

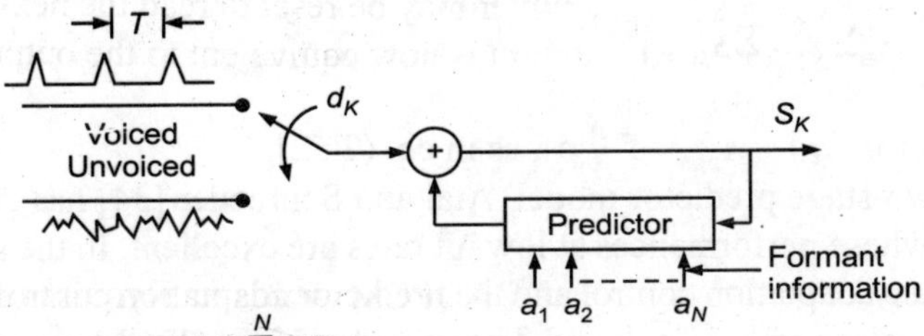

$$S_K = d_k + \sum_{i=1}^{N} a_i S_{k-i}$$

(*b*)

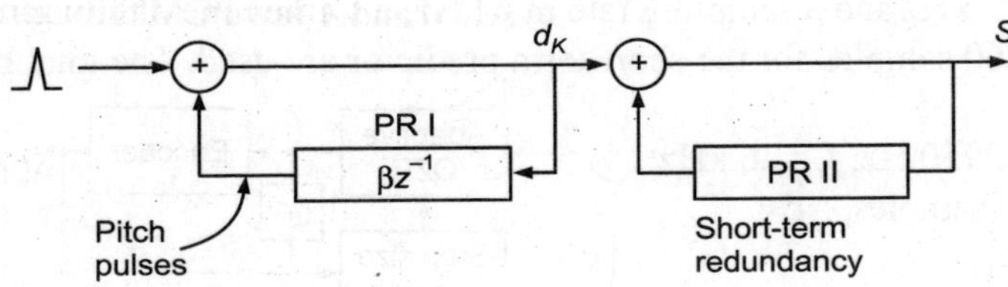

(*c*)

Figure 2.48: Speech models **(*a*)** Source-system model of speech, **(*b*)** Linear predictor model for speech, **(*c*)** Voiced speech model [34]

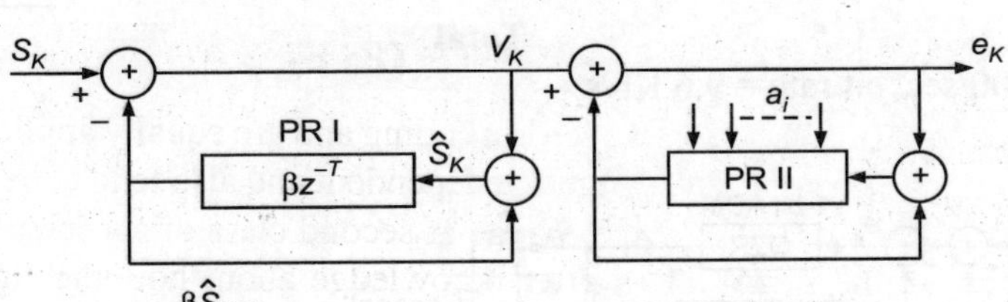

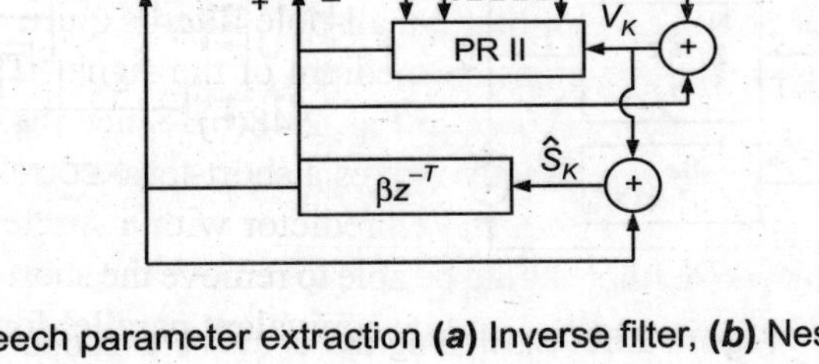

Figure 2.49: Speech parameter extraction **(*a*)** Inverse filter, **(*b*)** Nested inverse filter

the different parameters are extracted through the two predictors. The predictor coefficients β and T are obtained by minimising the function $\sum^{n} V_k^2 = \sum^{n} (s_k - \beta \hat{s}_{k-T})^2$ where $n \simeq 75$–100 and $T \simeq 3$–20 ms. T is selected to maximize the normalised correlation:

$$\rho T = \frac{\sum^{n} s_k s_{k-T}}{[\sum^{n} s_k^2 \cdot \sum^{n} s_{k-T}^2]^{1/2}} \quad \text{and} \quad \beta_{opt} = \frac{\sum^{n} s_k \cdot s_{k-T_{opt}}}{\sum^{n} s_{k-T_{opt}}^2} \tag{2.71}$$

and for the second-predictor, $[a_i] = A_{opt} = \Gamma^{-1}|\rho|$, as in eq. (2.52).

Based on the above two-stage predictor model, Atal and Schroeder [34] has developed an adaptive-predictive codec (APC) whose performances at low bit rates are excellent. In the system block diagram, shown in Fig. 2.50, the step adaptation control and the predictor adaptation control are obtained from the quantizer output and the values of c_r, a_i's, β and T are obtained from the matrix eqns. (2.52) and (2.71). These parameters are transmitted to the receiver along with the quantized residual output and the signal reconstructed in a complementary decoder. This coder is also known as Adaptive Residual coder. For a speech of bandwidth 3 kHz and a sampling rate of 6 kHz, a frame time of 120 samples for the long-term predictor and that of 60 samples for the short-term predictor are used. The total bit rate of the system is calculated as below:

Speech $BW = 2950$ Hz, $f_s = 6$ kHz.

For $P_1(z)$, $N = 120$ samples,	β	...3 bits
	T	... 7 bits
	σ_r	... 4 bits
For $P_2(z)$, $N = 60$ samples,		
Frame = 10 mS, 4 of a_i coeff. 5 × 4		... 20 bits
	q_k (error signal)	...60 bits
	Parameter Normalization	... 2 bits
	Total	96 bits

For a frame rate of 100/sec, bit rate = 9.6 Kb/s.

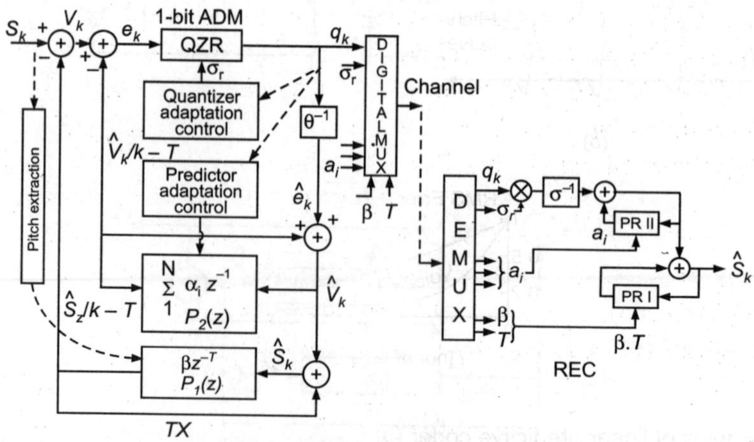

Figure 2.50: Block diagram of the APC codec

The quality of reproduction by this codec is equivalent to 4.5 bit log-PCM at 9600 bits and 4.1 bit log-PCM at 7200 bits. The error rate in the channel may be as high as 10^{-3} for satisfactory reproduction and with an error rate = 10^{-2}, the system performance is marginal. This codec is onc of the best low-bit rate codec for speech. However, calculation of the predictor coefficients and their adaptation cannot be done on-line and it requires a mini-computer to calculate these coefficients.

There is a further modification of the predictive codec developed by Aral and Hanauer [5] and known as Linear Predictive Codec (LPC). In this system shown in Fig. 2.51(a), the speech is again modelled as the sum of weighted delayed samples and the predictor coefficients are obtained in the same manner as in the case of APC by using the correlation matrix of the input signal. The excitation function of the speech is obtained by a peak-picker circuit after removal of the short-term redundancy of the signal, as shown in Fig. 2.51(b). The pitch extraction circuit is very vital to the proper representation of the speech and other methods, such as, Cepstrum technique may be used to determine the pitch frequency. Alternatively, a lowpass filtered speech of 500 Hz BW only is transmitted to the receiver by 1 Kbit DM and is used as the excitation function for the decoder.

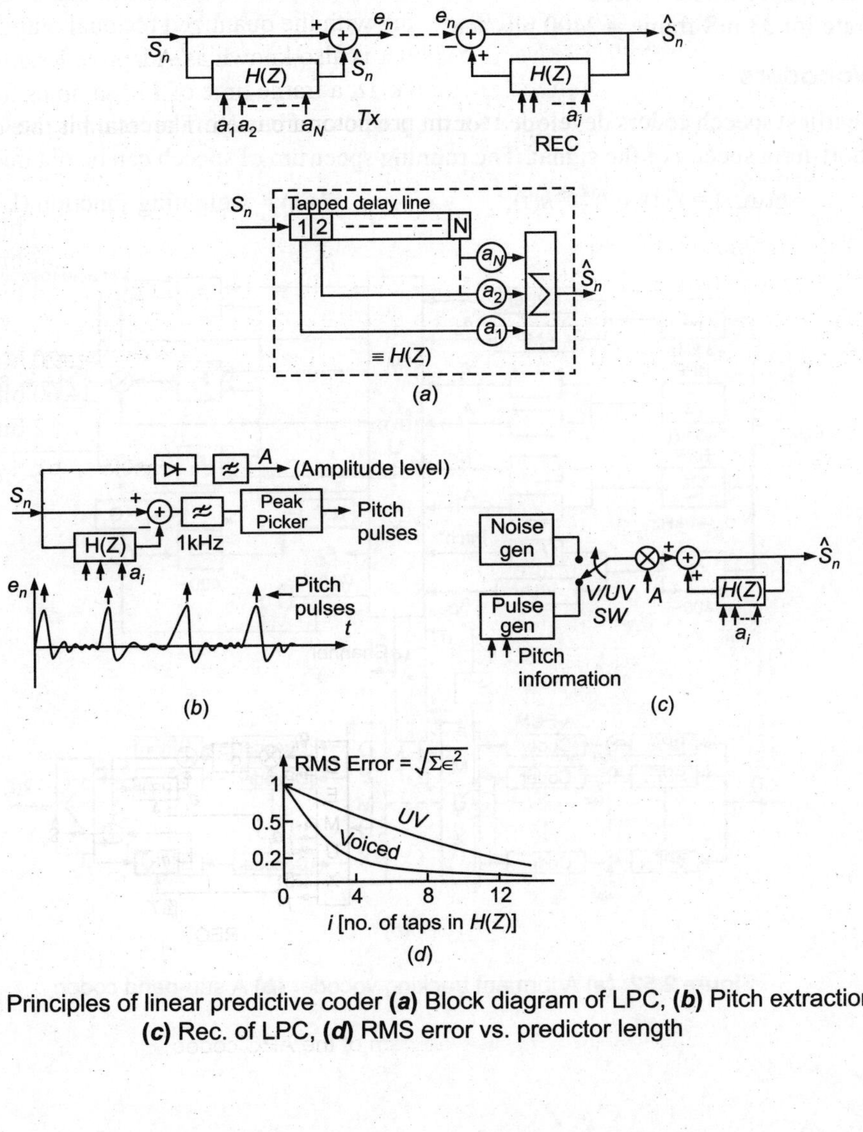

Figure 2.51: Principles of linear predictive coder **(a)** Block diagram of LPC, **(b)** Pitch extraction in LPC, **(c)** Rec. of LPC, **(d)** RMS error vs. predictor length

The receiver structure is shown in Fig. 2.51 (c), where the voiced/unvoiced signal operates pulse generator/noise generator switch and the speech is reconstructed through the predictor $H(z)$ with updated coefficients received from the coder. The RMS error for voiced signal decreases rapidly with the number of taps in the predictor, but the error for unvoiced sound varies slowly with the number of predictor taps, as shown in Fig. 2.51(d). In general, 10 to 12 predictor taps are used. The total number of bits required for a toll quality speech is given as follows:

Parameter	*Bits required*
Pitch	6
V/UV	1
RMS A	5
Freq and BW of poles of $[1 - H(z)]$; $(n = 10–12)$	60
Total	72 bits

Bit rate for 10 mS frame = 7200
and bit rate for 33 mS frame = 2400 bits/sec.

2.12.1 Vocoders

One of the earliest speech coders developed for the purpose of bandcom pression is the Channel vocoder based on short-term spectra of the signal. The running spectrum of speech can be obtained from the eqn.

$$F(\omega, t) = f(t) \cdot e^{-j\omega t} * h(t), \qquad h(\lambda) = \text{weighting function (LP)}$$

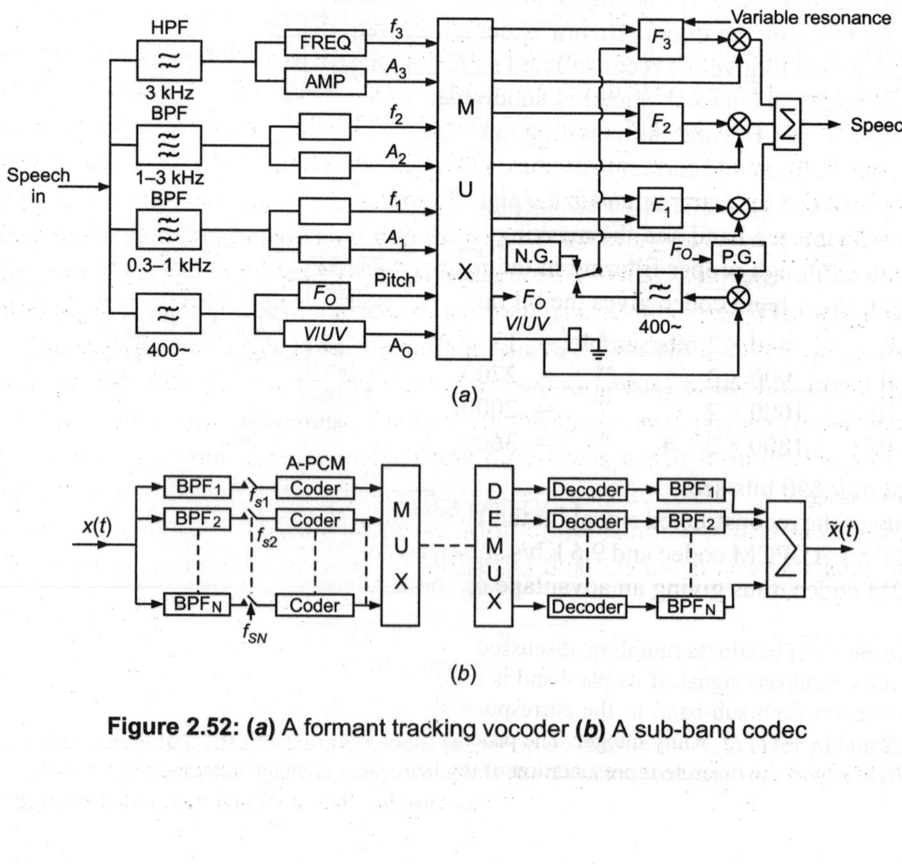

Figure 2.52: (*a*) A formant tracking vocoder **(*b*)** A sub-band codec

Therefore,

$$f(t) = \frac{1}{2\pi h(0)} \int_{-\infty}^{\infty} F(\omega, t) e^{j\omega t}\, d\omega, \; h(0) = 1. \tag{2.72}$$

$F(\omega, t)$ can be measured by a bandpass filter-rectifier combination and the so-called spectrogram is obtained by a bank of filters for speech signal. In spectrograms, it is seen that the voiced sound has distinct formants which vary somewhat with time, but the unvoiced sound does not have any distinct formant structure. Based on the above representation of speech, multi-channel vocoders and formant vocoders have been developed for the last 40 years. The block diagram of a formant tracking vocoder [35] is shown in Fig. 2.52(a). In the particular system, the information regarding the 3 formants and V/UV signals are transmitted to the receiver and the speech is reconstructed through a complimentary circuit. For 16 msec frame time, the number of bits required is 1200 approximately and the articulation score obtained is approximately 80 per cent. In Dudley's vocoders, the bandwidth required is 400/600 Hz approximately for reasonably good articulation.

2.12.2 Sub-band Coding [5]

In contrast with the above parametric vocoders, a frequency domain coder, using waveform coding of narrow bands of the speech signal, gives perceptual advantages over a single-band waveform codec, as discussed in Sections on APCM and ADPCM. In this sub-band coder, the signal band (say, speech) is subdivided into four to eight sub-bands, using appropriate bandpass filters, and then each band is translated to the corresponding low-pass signal*. The resultant lowpass signals are now encoded using APCM coders, as discussed in Sec. 2.7. The digital outputs of the coders are multiplexed and transmitted to the receiver. In the receiver the digital stream is demultiplexed, decoded and filtered, and then the filter outputs are combined to give the reconstructed signal, as shown in Fig. 2.52(b).

In a 4-band codec, the speech signal is subdivided into 4 bands of: (225–450), (450–900), (1000–1500) and (1800–2700) Hz, and the sampling rates are: $f_{s1} = 450, f_{s2} = 900, f_{s3} = 1000$ and $f_{s4} = 1800$. In lower bands, where pitch and formant structure must be accurately preserved, the coding is made more accurate by using 4 or 5 bits/sample, and in the higher bands, fewer bits/sample are used. APCM maintains an optimum S/N_q in each band and the masking of signals in one band by the quantizing noise of other bands is avoided through proper filtering in the receiver. Usually, 16/9.6 Kb/s are used in such codecs. As an example, the 4-band codec gives the bit rates according as:

Band I	450 × 3 bits/sample	= 1350
Band II	900 × 3 "	= 2700
Band III	1000 × 2 "	= 2000
Band IV	1800 × 2 "	= 3600

giving a total of 9,650 bits/sec.

The quality of the reconstructed signal for the 16 Kb/s sub-band codec has been found to be equivalent to that of 22 Kb/s ADPCM codec and 9.6 Kb/s sub-band codec gives a quality similar to that obtained in a 20 Kb-ADM codec, thus giving an advantage of approximately 10 Kb/s in transmission rates.

*Based on the theory of bandpass sampling, discussed in Ch. 1, Sec. 1.5, it is known that the sampling rate $f_s = 2W$ is also valid for a bandpass signal, if its passband is restricted to $[nW$ to $(n + 1)W]$ Hz. Thus in practice, it is not necessary to convert each sub-band to the corresponding lowpass signal, but the bandpass filters should have cutoffs at $nf_s/2$ and $(n + 1) f_s/2$, n any integer. The process is known as 'Integer-band' sampling and the samples at the rate of $2W$/sec. give the discrete representation of the bandpass signal.

2.13 RECENT DEVELOPMENTS IN SPEECH CODING [36]

A general description of various established digitization techniques has been given in earlier sections. Recently some new techniques, e.g., Tree, Trellis and Vector quantization, spectrum shaping of the *Q*-noise, variable frame rate transmission of narrow band LPC parameters etc., have been developed [36]. Some of these are briefly discussed below.

2.13.1 Multipath Search Coding (MSC) [37]

The conventional coding schemes, e.g., PCM, DPCM, ADM, are based on instantaneous decisions of the quantizer, and as such, are called single path coders. In multipath search coding schemes, on the other hand, the future values of input samples $x_{n+1}, x_{n+2}, \ldots$, are considered as well before a (delayed) decision is made about the optimum channel code $\{c_n\}$ to be transmitted. In Fig. 2.53, showing the structure of MSC schemes, the input samples $\{x_n\}$ are fed into the input buffer of length *N*. The encoder compares the buffered vector $\mathbf{X}$ with a collection of possible output sequence $\mathbf{Y}_k$, $k = 1, 2, \ldots, 2^N$. The optimum output sequence is the nearest neighbour sequence based on the squared error criterion

$$E_k = (\mathbf{X} - \mathbf{Y}_k)^T (\mathbf{X} - \mathbf{Y}_k) \tag{2.73}$$

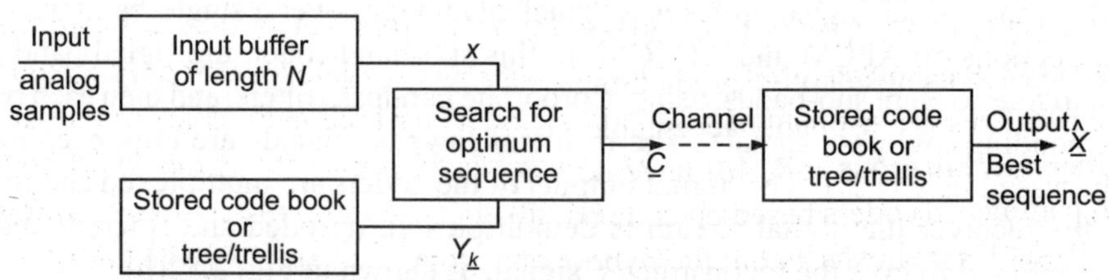

Figure 2.53: Structure of multipath search coding (MSC) scheme

The decoder is informed about the chosen output sequence by a binary channel sequence C_n, and on its basis, the decoder outputs the corresponding output sequence approximating $\{x_n\}$.

MSC coding strategies are based on:

1. Codebook coding, also known as list coding/vector quantization (VQ)
2. Tree coding [38]
3. Trellis coding [39]

In the codebook coding schemes, the set of possible output sequences $\mathbf{Y}_k$, $k = 1, 2, \ldots, 2^N$, is arranged in a codebook whose elements are not restricted in any way. When the optimum output sequence has been found, the corresponding index of that sequence is transmitted as the channel sequence *C* in a binary format using *N* bits, as shown in Fig. 2.54. In Tree and Trellis coding schemes, the output sequences of length *L* are arranged in the form of a tree or trellis of depth *L*, as in Figs. 2.54 (b) and (c). Its branches are populated with reconstruction values, and the different sequences have a number of common elements. Each sequence has a path through the tree or trellis, and the channel sequence then provides information regarding the path map and the equivalent input to the decoder. MSC schemes may work in a block mode, i.e., a complete (B.L.) bit binary sequence is released to the channel after the optimum $\mathbf{Y}_k$ has been found, or in an incremental mode, where only a single symbol is released at a time. The input block is then shifted by one sample and the search is repeated.

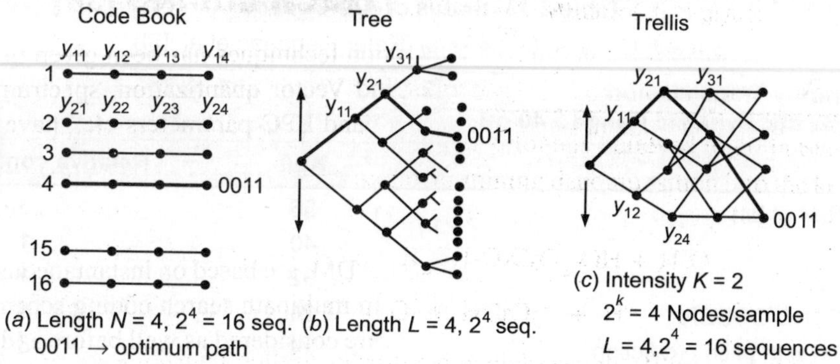

Figure 2.54: MSC coding classes **(*a*)** Code book coding **(*b*)** Tree coding **(*c*)** Trellis coding.

Simulation results show that for code book coding with $N = 8$, SNR improves by 5 dB over DPCM for correlated signals coded at 1 bit/sample (e.g. 8000 bits for speech). In Tree and Trellis encoding, similar SNR advantage is obtained. In Ref. [38], it has been shown that in Tree encoding with (M, L) search algorithm, SNR gain with ADPCM and ADM configurations was 4 to 5 dB at 2 bits/sample rate. The tree-encoded speech using adaptive prediction and quantization is almost of 'toll quality' at 24 Kb/s and the quality at 16 Kb/s is reasonably acceptable.

R. M. Gray and his group [39, 40] have developed Trellis waveform coders, hybrid tree-encoding system, and parametric coders based on code books of inverse filters produced by off-line LPC and VQ techniques. It has been shown that in the hybrid tree-encoding system (a residual excited LPC-RELP), SNR gain over DPCM is 5 to 6 dB for 1 bit and 2 bits/sample code rates, and at the same time, the speech quality was found to be subjectively much improved as compared with other similar systems. Various modifications and improvements are being presently investigated for Inverse Filter matching techniques and multirate voice digitizer using VQ, and their results are given in Ref. [36]. Viswanathan et al. [41] have developed a variable frame rate (VFR) transmission technique for low-bit rate LPC-parameters. In this method, the parameters are transmitted only when their values have changed sufficiently over the interval since their preceding transmission. The application of VFR technique has resulted into narrowband LPC speech coders with average bit rates of 2000–2400 b/s and their subjective speech quality has been found to be equal to or better than that of a 5700 bits fixed frame rate LPC codec.

The present trend in speech coding is to go for more hybridised systems, where various established techniques, e.g., APC/LPC, ADPCM/ADM, Tree/VQ and variable rate transmission, are combined to optimize the performance, thus leading to low bit rate digitized speech in the range 1.5 to 10 Kb/s.

2.13.2 Quality and Complexity

Flanagan et al. have discussed the quality and complexity of different codecs in Ref. [5] and their assessment of the codes is shown in Table 2.13. It is seen that the quality of waveform coders deteriorates significantly at bit rates lower than 16 Kbits, while the vocoder (including LPC) performs optimally at bit rates of 2.4 to 4.8 Kb/s. In the range of 1.2 Kb/s, the codecs are still useful for information storage, as used in voice response systems, although their quality needs to be improved further.

Table 2.13: Status of A/D Codecs

Speech Codecs (0–3.4 kHz) (Refer Flanagan et al. [5])

	Kb/s	Relative complexity
A. Toll-Quality Transmission **$S/N \geq 32$ dB; Dynamic Range ≥ 40 dB**		
Log-PCM (7 bit)	56	3
ADM	40	1
ADPCM	32	1
SUB-BAND CODER	24	5
ADPCM with pitch prediction	24	5
TREE-Encoded-ADPCM	24	10
APC, ATC (Transform), Phase Vocoder (PV), Voice Excited Vocoder (VEV)	16	50–100
B. Communication Quality **$S/N \geq 24$ dB; Dynamic Range ≥ 30 dB**		
Log-PCM	36	
ADM	24	
ADPCM	16	
SUB-BAND	9.6	
APC, ATC, PV, VEV	7.2	
VFR-LPC	2.4	
C. Synthetic Quality		
Channel Vocoder and LPC	2.4	
LPC with orthogonal coeff.	1.2	
Formant vocoder	0.5	

2.14 PICTURE SIGNAL ENCODING [42, 43, 44, 45]

For a high quality A/D conversion of picture signal, it is necessary to have the bit rates of the coder from 100 to 150 Mb/s. Although the standard log-PCM codec may be used at higher speed, it becomes difficult to have the conversion time of the order of 100 ns with the standard circuitry. As such, the serial mode of successive approximation of $x(t)$, as shown in Fig. 2.12, is not feasible and a parallel mode of A/D conversion, as shown in Fig. 2.55, is used to reduced the required conversion time.

In the 3-bit coder shown, seven comparators are used to designate the eight levels, 00 . . . to 111, and the appropriate seven equidistant reference voltages are generated by means of a voltage divider. The outputs of the comparators are converted to binary numbers through a combinatorial logic circuit, whose truth table is shown in Table 2.14. For time synchronisation, the outputs of the comparators $k_1, \ldots k_7$, are initially stored in F/F's and then given to the logic circuit through a command signal synchronised with the bit timing.

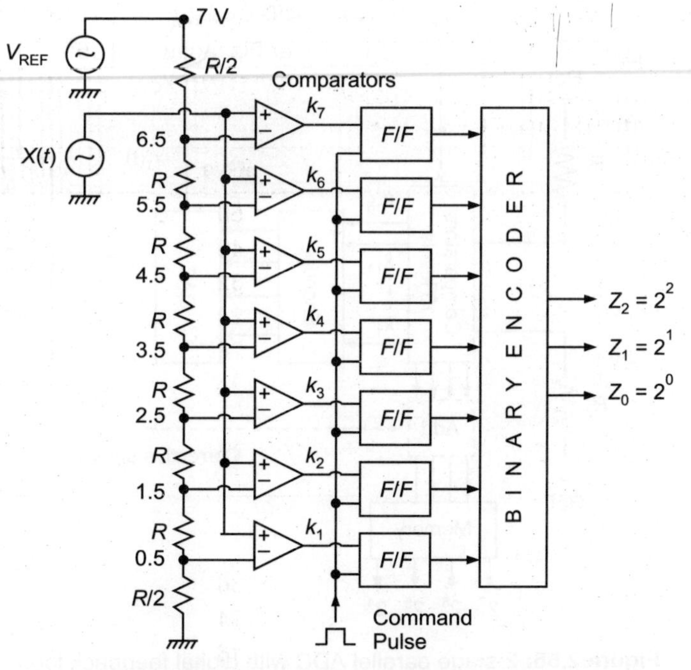

Figure 2.55: Parallel mode A/D converter

Table 2.14: Truth table of the parallel A/D converter

Input $x(t)$ (0–8) V	Comparator states k_i's							Binary output number		
	7	6	5	4	3	2	1	z_2	z_1	z_0
0–0.5	0	0	0	0	0	0	0	0	0	0
0.5–1.5	0	0	0	0	0	0	1	0	0	1
1.5–2.5	0	0	0	0	0	1	1	0	1	0
2.5–3.5	0	0	0	0	1	1	1	0	1	1
3.5–4.5	0	0	0	1	1	1	1	1	0	0
4.5–5.5	0	0	1	1	1	1	1	1	0	1
5.5–6.5	0	1	1	1	1	1	1	1	1	0
6.5–7.5	1	1	1	1	1	1	1	1	1	1

One of the major disadvantages of parallel ADC's is that the number of comparators grows exponentially with the word-length, e.g., for an 8-bit ADC, number of comparators required is 255, which is prohibitive. By using series-parallel configuration, a 2-stage ADC will give 8-bit words with only 30 comparators, as shown in Fig. 2.56. Initially the four MSB's are parallel converted and the result is the coarsely quantized value of the input $x(t)$. The equivalent analog voltage $x_1(t)$ generated through a DAC, is subtracted from $x(t)$ and the remainder Δx is next digitized through a second 4-bit ADC. If the difference $\Delta x = [x(t) - x_1(t)]$ is amplified by a factor of 16, then both the ADCs can operate at the same input voltage range. However, the accuracy requirement in the first ADC is much higher, e.g., in the figure, the first 4-bit ADC must have an accuracy of a 8-bit ADC; otherwise the Δx and hence the output of the second ADC will be in error. Since there is a delay in the first ADC, the $x(t)$ should be passed through a sample and hold circuit, such that $\Delta x(t) = \Delta x(t_0)$ and is constant during the switching time of the second ADC.

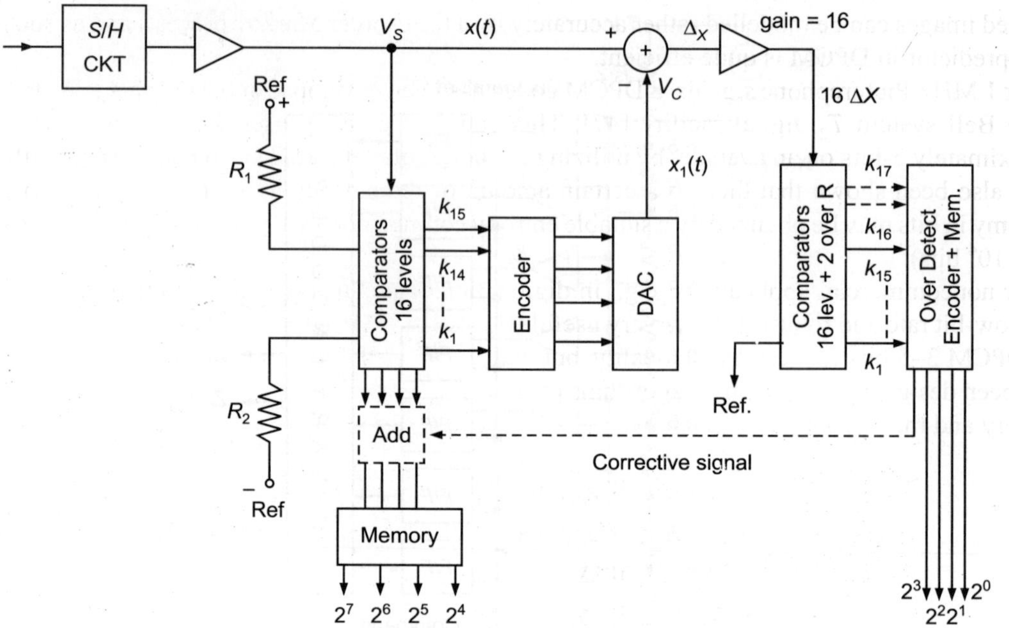

Figure 2.56: 2-stage parallel ADC with digital feedback loop

An efficient high speed 8-bit PCM codec for colour TV has been designed at Comsat Labs. [46], based on the above principles, and the total conversion time is less than 50 ns. The stringent requirement of the accuracy of first ADC is reduced by using a digital feedback circuit which corrects the quantization error in the first ADC. Assume that due to the inaccuracy of one of the reference voltages in the first ADC, $x(t_0)$ = 1101 … (MSBs) is read as 1100, i.e., the reference was higher, then Δx would be larger than 15 units, which will be coded by the second ADC as 1111 only. Thus the correct value of $x(t_0)$ = 11010000 is read as 11001111, i.e., one bit error at LSB position. This can be corrected if the second ADC has more than 15 comparators, say 17; then the comparator outputs will indicate an 'overrange' condition if the input > 15 units. This 'overrange' signal may be fed back to the first ADC as a corrective signal, as shown by the dotted lines in Fig. 2.56 and 0001 may be added to 1100, giving the correct output. The new Δx will then be coded by the second ADC. If on the other hand the reference voltage was lower, then $x(t_0)$ = 1101 … may be read as 1110; but Δx would now be negative (since DAC will produce correct $\hat{x}(t_0)$ and $\Delta x = x(t_0) - \hat{x}(t_0)$). This will indicate an 'underrange' condition and a corrective signal 0001 may be subtracted from 1110, giving the correct $x(t_0)$ = 1101 …. To maintain the high speed required, the DAC is a current-driven ladder network (temperature-compensated), with a settling time of 15 ns only. No OP-AMP is used in the ADC loop; and all voltage adding and subtracting is done directly at the input of the comparators. The 8-bit codec gives an SNR of 45 dB with a full load sinusoidal input.

2.14.1 DPCM and ADM Codecs

It is well known that the bandwidth used for TV signals is much in excess of the total information content in the signal. It has been estimated that TV signals contain only 1.1–1.5 bits of information per pixel. As such it should be possible to encode TV signals with 1.5 to 2 bits per sample. DPCM codecs have been widely used for digitizing TV signals. Experimental results with various images have indicated that sampled images can be modelled rather accurately by a third-order Markov process, and as such a third-order predictor in DPCM is quite efficient.

For 1 MHz-Picturephone signals, a DPCM codec has been developed at 6.312 Mb/s to be transmitted in the Bell system T_2 digital facility [47]. This codec effectively uses 4-bit/sample, but transmits approximately 3-bits on an average, by utilizing the horizontal blanking period and a small buffer store. It has also been shown that there is a certain amount of redundancy in the DPCM output and further economy in bits may be obtained by a suitable entropy coding, along with the necessary buffer store (say, about 10^4 bits).

For noncommercial applications, e.g., in digitization of pictures obtained through satellites, a low-cost, low-bit rate and robust ADM is very useful. Normally, PCM requires 6–8 bits for Nyquist samples and DPCM 3–5 bits for reasonable quality, but with ADM, 1–3 bits/sample are adequate. Such codecs have been designed with adaptive algorithms [48] somewhat similar to that of eqn. (2.26), but uses 2-bit memory and the step sizes are given by:

$$\Delta_r = \begin{cases} \Delta_{r-1} \cdot P, & \text{if } e_{r-1} \text{ and } e_{r-2} \text{ are both of same sign} \\ \Delta_{r-1} \cdot Q, & \text{if } e_{r-1} \text{ and } e_{r-2} \text{ are of opposite sign} \\ \pm 2\Delta_{\min}, & \text{if } |\Delta_{r-1}| = \Delta_{\min} \\ \pm 15\Delta_{\min}, & \text{if } |\Delta_{r-1}| = 15\Delta_{\min} \end{cases} \tag{2.74}$$

Thus, the range of Δ_r is limited to: $\Delta_{\min} \le |\Delta_r| \le 15\Delta_{\min}$ and $P = 1.5$, $Q = 0.5$. Effectively the dynamic range now is 30 (= 30 dB).

It has been shown that such an ADM codec at 16 (approx. 2 bits/pixel) and 22 Mb/s, gives reasonably good picture quality and at the same time can tolerate channel error of 10^{-3} or less. Even at $P_e = 10^{-2}$, the horizontal and vertical synchs operate effectively and the picture does not 'breakup'. The codec may also be used to encode NTSC colour video signals at 24 Mb/s. Further, digitized voice at 32 Kb/s can be transmitted simultaneously during the horizontal synch pulse periods, with no increase in the basic (video) bit rate of 20/24 Mb/s.

The above DPCM and ADM codecs are one-dimensional (1D); but pictures are two dimensional (2D) in nature and as such, correlation exists between neighbouring pixels on the same horizontal line as well as on earlier lines (i.e., in various spatial directions) in the frame, as shown in Fig. 2.57. In a moving

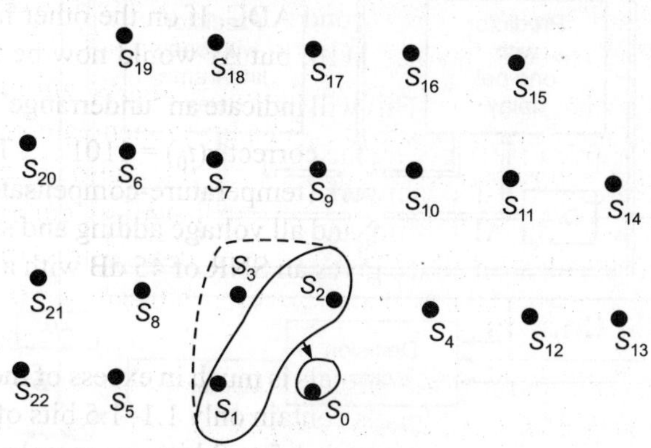

Figure 2.57: Showing pixels adjacent to S_0. Usually S_1, S_2 and S_3 are used to predict S_0. Alternatively S_1 and S_2 are used to predict S_0

picture, frame-to-frame correlation also exists. Thus the predictor in DPCM codecs may make the estimate of the next sample value s_0 as

$$\hat{s}_0 = \Sigma a_i s_i$$

where $\{a_i\}$ are calculated on the basis of the correlation matrix $[\rho_{ij}]$, as discussed in Sec. 2.8. It has been shown that using only s_1, s_2, s_3, the prediction is sufficiently accurate and the differential signal $e_r(t)$ is a set of almost uncorrelated variables, giving near-optimum DPCM coding gain. But it helps if the predictor uses a buffer memory of two line duration and some pixels from the earlier line as well are considered for the approximation of s_0.

In 2D-ADM, using only one-line memory, the weighted average of $\hat{s}_1$, $\hat{s}_2$ and $\hat{s}_3$ in the form of

$$\hat{s}_0 = 0.75\hat{s}_1 + 0.75\,\hat{s}_2 - 0.5\hat{s}_3 \tag{2.75}$$

has been used for prediction the present pixel s_0. The digitized pictures with 2 bits/pixel give reasonably good quality, and a 'look-ahead' algorithm reduces the edge-busyness considerably. Alternatively a dual-mode 2D-ADM, shown in Fig. 2.58 may be operated such that $\hat{s}_0$ estimate is based upon only one of the earlier pixels of the pair $(\hat{s}_1, \hat{x}_2)$, after determining the differences $|s_0 - \hat{s}_1|$, $|s_0 - \hat{s}_2|$ and whichever is smaller. Then $\hat{s}_0 \simeq \hat{s}_1$ or $\hat{s}_2$ whichever is nearest to s_0. Although this 2D-ADM uses a simple approximation of $\hat{s}_0$, as in 1D-ADM, (an additional Marker bit is transmitted to indicate the preference between $\hat{s}_1$ and $\hat{s}_2$) a significant reduction in edge-busyness occurs in 2D-ADM, since at any edge, the movement of the coding path of the DM is adjusted in the direction of the edge. As stated earlier, three dimensional

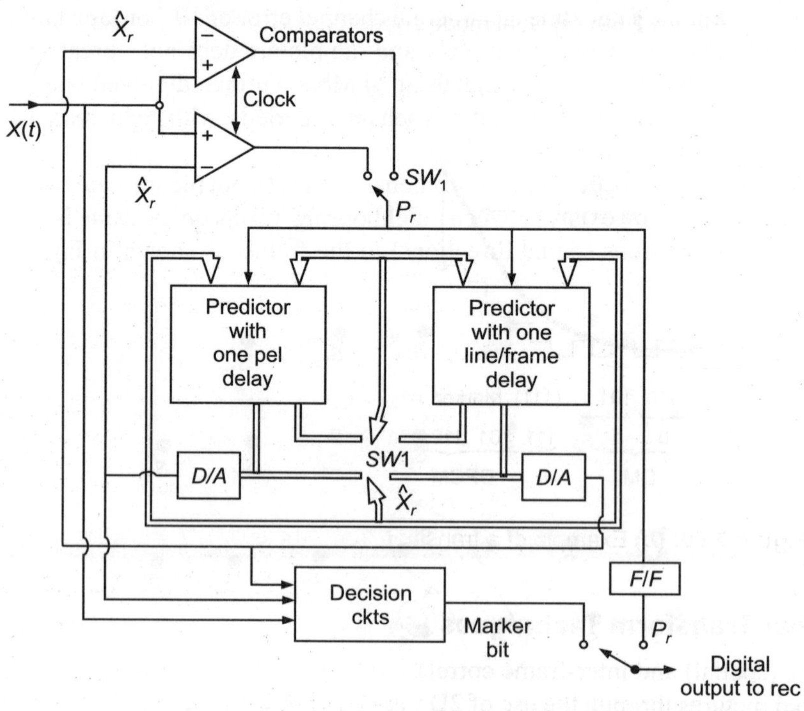

Figure 2.58: A dual-mode 2D-ADM (Intraframe/Interframe prediction)

prediction (including frame-to-frame correlation) may also be used in above coders. Alternatively, the above dual mode 2D-ADM may be modified to select the best predicted $\hat{s}_0$ either from the previous horizontal estimate. i.e., $\hat{s}_0 \sim \hat{s}_1$, or from the previous frame estimate $\hat{s}_0\,(t - \Delta t)$, where Δt = frame time, as indicated in Fig. 2.58. This codec shows good gray level rendering, but also some amount of edge-busyness for moving pictures [49].

In another variation of an adaptive dual-mode codec [50], incorporating one DPCM and one DM coder, the DPCM is switched whenever a sharp transition of the video waveform occurs and DM codes the waveform with small slopes, as shown in Fig. 2.59(a). In Fig. 2.59(b), the transition from the DM to DPCM mode is shown when the large slope occurs, and the two bit streams along with the Marker bits (111) are also shown. The bit rate in DM is three times that of the Nyquist sampling rate f_s used for DPCM, but the rate is reduced to f_s, by using a simple coding table; (001, 010, 100) = 0 and (110, 101, 011) = 1. Whenever DM output is 111/000, coder switches to DPCM. A look-ahead circuit is incorporated to prevent edge-busyness and the output bits are further entropy coded using a flexible buffer memory. Using moderate hardware and at bit rates of 1.5 bits/pixel, the codec performance is equivalent to 3-bit DPCM.

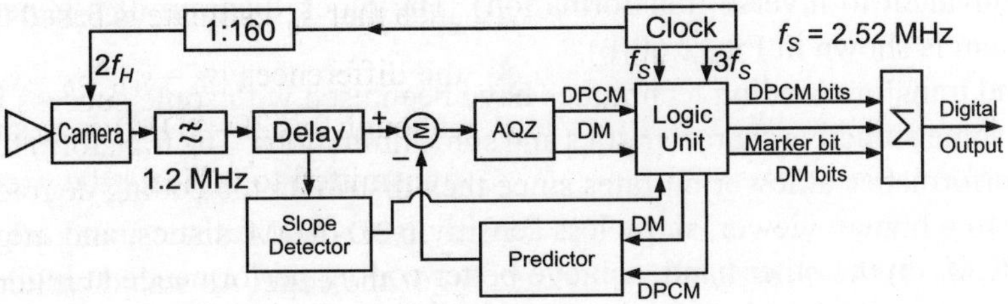

Figure 2.59: (*a*) Dual mode coder for picture-phone signals [50]

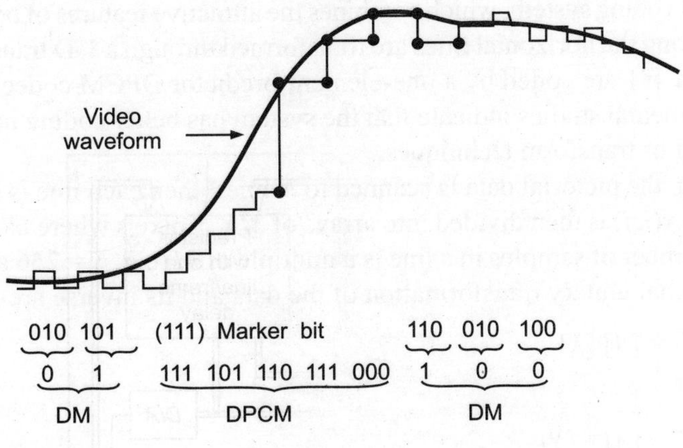

Figure 2.59: (*b*) Example of a transition from DM to DPCM and back to DM [50]

2.14.2 Linear Transform Techniques [44, 51]

The intra-frame (spatial) and inter-frame correlation in picture signals may be utilized to reduce the bit rates of digitized pictures through the use of 2D transform techniques, e.g., K-L, Fourier, Hadamard and others. Transform techniques accomplish both statistical and psychovisual coding by two stage operation:

first operation transforms the set of statistically correlated pels $\{x\}$ into a new set of almost uncorrelated variables $\{y\}$, and the second operation quantizes and codes the variables $\{y\}$ keeping in view the sensitivity of the human vision to the subjective effects of Q-error and channel errors. The transformed variables $\{y\}$ have a varying degree of significance in contributing to both the information content and the subjective quality of the resulting picture. As such, the less significant variables may be discarded without causing any perceptible degradation of the quality of the reconstructed picture, thus leading to bit rate reduction in the transmitted signal. The method of 'principal components' (K-L transform) and such other transforms achieve this quite efficiently.

To understand the application of this coordinate transformation of the pictorial data, consider two pixels x_1, x_2, which are somewhat correlated. Because of their correlation, the sample values will most often fall inside the shaded region of Fig. 2.60(a). Now if one rotates the coordinate system to the positions indicated by y_1 and y_2, two new variables are obtained, one having a larger variance than the other, although the sum of the variances remains the same. Performing this rotation of coordinates for all N^2 pels, one obtains N^2 new variables $\{y\}$ that are uncorrelated and have monotonically decreasing variances. Finally, the estimated picture pels $\{\hat{x}\}$ are obtained through the inverse rotation of the quantized variables $\{\hat{y}\}$ (equivalent to inverse transformation). The block diagram of a generalized Transform Image coding system is shown in Fig. 2.60(b).

Both DPCM and transform coding techniques have been used with some success in coding pictorial data, and each has some attractive characteristics and some limitations. The transform techniques achiever superior coding performance at lower bit rates since they distribute the coding degradation in a manner less objectionable to a human viewer, show less sensitivity to data statistics, and are less vulnerable to channel noise. DPCM, on the other hand, achieve better coding performance at higher bit rate, and the equipment complexity and delay due to coding are minimal. But DPCM is more sensitive to picture statistics, requiring adaptive prediction and the channel errors propagate in the reconstructed picture. Because of the computational complexity of 2-D transform techniques and the comparative simplicity of 1-D transform, a hybrid coding system, which combines the attractive features of both, has been suggested, where the pixels $\{x\}$ along the horizontal lines are transformed through a 1-D transform and each column of the transformed data $\{y\}$ are coded by a one-element-predictor DPCM codec as shown in Fig. 2.61. Theoretical and experimental studies indicate that the system has better coding performances than those of 'stand alone' DPCM or transform techniques.

In the hybrid system, the pictorial data is scanned to N lines, then each line is sampled at the Nyquist rate. This sampled data $x(i, j)$ is then divided into arrays of $M \times N$ pixels where i and j index the rows and the columns and the number of samples in a line is a multiple of M (e.g. $N = 256$ and number of columns $M = 16$). One-dimensional unitary transformation of the data and its inverse are given by:

$$[Y] = [A]\,[X]$$

and

$$[X] = [A]^{-1}\,[Y] \tag{2.76}$$

where $[A]$ is the transformation matrix and $A = A^{-1}$ symmetric orthogonal (orthonormal) matrix. The relation (2.76) can be written in the series form as:

$$y_i(u) = \sum_{j=1}^{M} x_i(j) \cdot A(j,u),$$

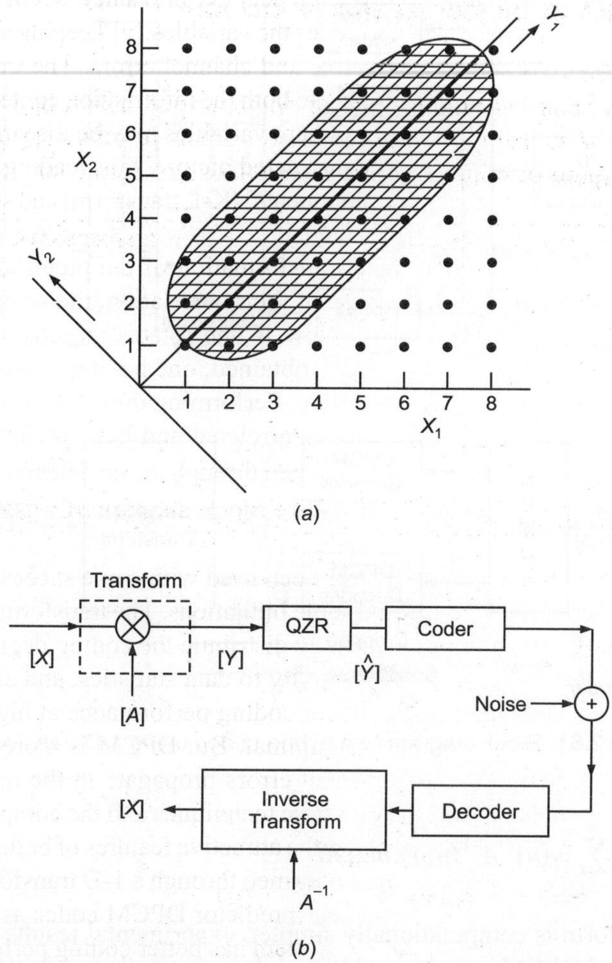

Figure 2.60: (*a*) Rotation of coordinate system for the grey levels of pixels x_1 and x_2. The most likely values are the ones in the shaded area. y_1 and y_2 are the new coordinate system, **(*b*)** Block diagram of the transform coding system

and

$$x_i(j) = \sum_{u=1}^{M} y_i(u) \cdot A^{-1}(u, j), \begin{cases} i = 1, 2, \dots, N \\ j = 1, 2, \dots, M \\ u = 1, 2, \dots, M \end{cases} \tag{2.77}$$

where A (j, u) is the forward transformation kernel.

The column of vectors $y_i(u)$ is now coded by DPCM giving the variables $z_i(u)$, but a number of DPCM codecs, M in this case, as shown in Fig. 2.61, is required since the correlation properties of each column are different from those of others. The picture is reconstructed by the inverse transformation of the estimated variables $\hat{y}_i(u)$ using:

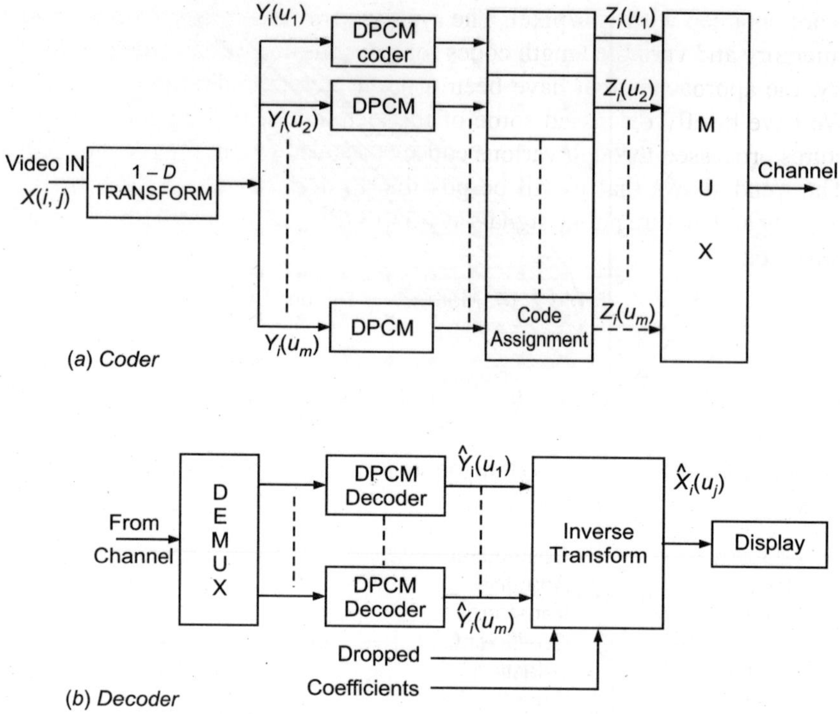

Figure 2.61: Block diagram of a hybrid codec using 1-D transform

$$\hat{x}_i(j) = \sum_{u=1}^{m} \hat{y}_i(u) \cdot A^{-1}(u, j), m \le M \tag{2.78}$$

Since the Hadamard transform is computationally simpler, experimental results with the above hybrid coder using *H*-transform and DPCM have been obtained. It is shown that acceptable quality pictures are obtained by using 1 bit/pixel or less and the degradation due to channel errors with $P_e \le 10^{-2}$ is negligible. Effective SNR's for Markov data have been calculated and they are approximately:

Bit rate/pixel	1	2	3	4
SNR in dB	30	37	42	47

2.14.3 Present Status

The other efficient technique of bandwidth reduction of TV signals is the interframe coding (also known as conditional frame-to frame coder), where the high frame-to-frame correlation of the pixels is used to predict the future values of pixels. One method of coding of video data is simply to transmit the gray levels of the elements that have changed in successive frames with proper addressing information. The receiver then generates the successive frames by replenishing the previous frames with the transmitted information. Naturally less information needs to be transmitted, if only the elements that change beyond a particular threshold level are updated. The use of an appropriate value for the threshold eliminates the granular noise in the background without noticeably degrading the encoded video data. Experiments with picture-phone signals using the 'conditional frame replenishment' technique have indicated that good pictures are obtained with 1bit/pixel. The system generates information at an uneven rate, and as such, a large memory and variable length codes for transmission of data are used [52].

In summary, the approaches that have been used in picture coding may be classified as shown in Table 2.15. We have briefly discussed some of the techniques. In [45], an assessment of the relative quality of pictures processed through various codecs has been given. This is shown in a tabular form in Table 2.16. The trend shows that it will be possible to digitize colour TV signals at 15 Mb/s using Interpolative coding and picture phone signals at 1.5 to 3 Mb/s using interframe coding and incorporating motion compensation.

Table 2.15: Methods of Picture Coding

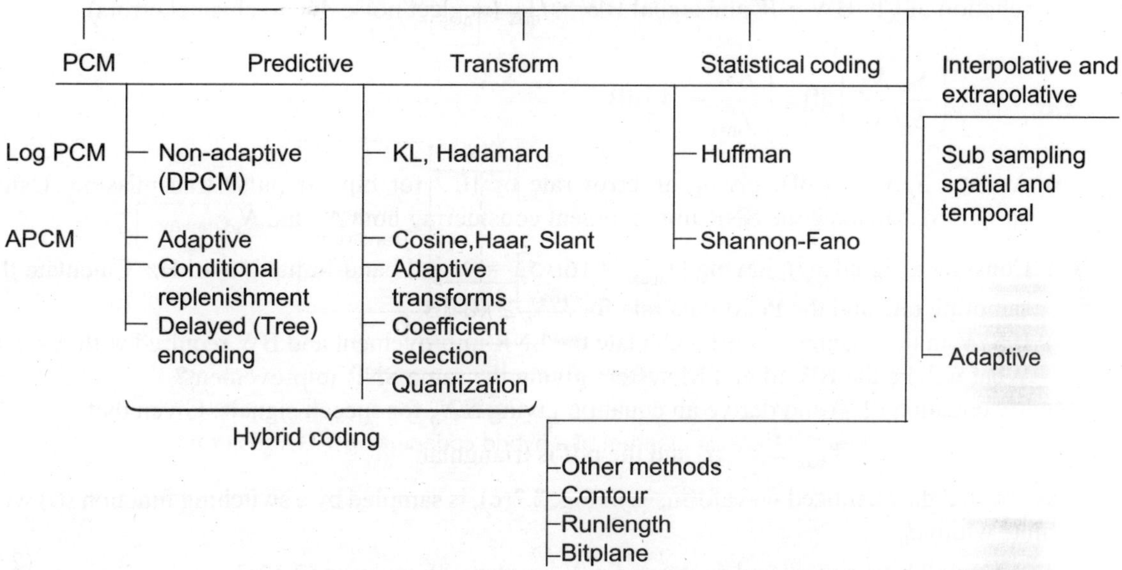

Table 2.16: Picture Codecs: A. Broadcast TV Signals, BW = 4.5 MHZ: B. Picture phone Signals, BW = 1 MHz

Type of Coder	Bit-rate, Mb/s for the Quality					
	Excellent		Good		Fair	
	A	B	A	B	A	B
log PCM	95	20–40	50			
DPCM	45	10	32	6	16	4
ADM	60	10	30	6	16	4
Hybrid Transform/DPCM	40	8	25	6	15	3
Intra-frame coding (adaptive)	50	30	35	20	25	10
F-F conditional replenishment	30	6	25	3	20	1.5
Interpolative coding (futuristic)	20	–	15	–	10	–
F-F coding with motion compensation	–	3	–	1.5	–	1

PROBLEMS

1. Most of the discussions in the chapter has been based on binary-coded PCM. It is possible and spectrally efficient, if 3/4/8/16-ary PCM is used for transmission. Calculate the quantization-noise power N_q and S/N_q in each case. Are these SNR values different from S/N_q for Binary PCM? Assume maximum quantized level to be 243 for ternary and 256 for 4/8/16-ary PCM. Calculate the bandwidth required for transmission in each case, assuming signal BW = 4 kHz.
2. (a) For the transmission of PCM signals in the noisy channel, $S_i/Ni \simeq 10$ dB is to be maintained for low error rate. Based on this, calculate the SNR improvement at the output of the decoder as a function of Ch. BW = W and signal BW = f_{max} (neglect noise due to channel errors).

   Ans: $$\left(\frac{S_0/N_0}{S_i/N_i}\right)\text{dB} = \left(\frac{6\,W}{f_{max}} - 4\right)\text{dB}$$

   (b) Assume $S_i/N_i \simeq 9$ dB, giving an error rate of 10^{-4} for bipolar pulse transmission. Using eq. (2.46), calculate the SNR improvement considering both N_q and N_e.
3. (a) Consider a signal $x(t)$, having $|X_{max}| = 16$, $\sigma_x^2 = 9$, and band-limited to 4 kHz. Calculate the sampling rate and the PCM data rate for $S/N_q \geq 40$ dB.
   (b) If S_i/N_i in the channel = 9 dB, calculate the SNR improvement and BW required with $\Delta = 1/8$. What will be the BW of an FM system giving the same SNR improvement?
4. Modify equation (2.7) and derive an equation giving S/N_q for speech signals. Given that:

   $X_{max} \simeq 2.5\,\sigma_x$ and the pdf is triangular.
5. Assume that the quantized waveform, as in Fig. 2.2(c), is sampled by a switching function $s(t)$ with a finite width d_0.
   (a) What will be the modified equation for N_q in place of equation (2.12)?
   (b) For a random signal with uniform pdf, having $W = 4$ kHz, $f_s = 10$ kHz, and $d_0 = 10$ μs, calculate N_q assuming $\Delta = X_{max}/32$.
6. It has been shown that the quality of PCM-coded speech for $3 \leq B \leq 6$, improves considerably if a dither signal is added to $x(t)$ before encoding and the same is subtracted from the decoded speech [4]. Give a block diagram of the proposed system, and show that N_q at the decoder output is larger than that for the decoder without dither. But after subtracting the dither from the decoder output, N_q is the same as that of the undithered case. [Dither signal is normally generated by a PN sequence generator and the same generator is used at the receiver.]
7. It is sometimes necessary to connect PCM/ADM codecs in tandem. It has been shown [4] that the resultant N_q is increased by 3 dB for each PCM codec-pair in tandem. N_q in ADM codecs in tandem increases by approximately 8 dB per pair.
   (a) Prove the statement regarding PCM codec in tandem.
   (b) Show heuristically the approximate result for ADM codecs in tandem.
8. (a) In testing a PCM codec, a sine wave signal is used as the input. Discuss qualitatively what will be the S vs S/N_q plot as the signal frequency and the signal amplitude are varied over the operating range of $(0 - f_{max})$ in Hz and $|x| \leq X_0$ respectively. Plot such a curve.
   (b) Because of large variations in S vs. S/N_q curve obtained with a sinewave input, a complex $x(t)$ is used for testing PCM codecs. Two types of signals are used: (i) A narrowband Gaussian

signal with $\Delta f \leq f_{max}/20$, (ii) a broad-band noise signal with a slot of Δf. Describe the measuring technique in detail and justify that the experimental results are now realistic.

9. A signal $x(t)$ has $|x| \leq X_0$, $\sigma_x = 1$, and $p(x)$ is symmetrically triangular.
   (a) Calculate S/N_q for a log-PCM codec using $\mu = 200$ and $B = 5$ (refer to eqns. (2.31) and (2.32)).
   (b) Repeat the problem (a) with $p(x)$ Laplacian and $X_0/\sigma_x = 4$.
10. Approximate the μ-law curve with $\mu = 100$ by 2 linear segments, one for $0 < x \leq 1/10$ and the other for $1/10 < x \leq 1$. Calculate S/N_q for $S = -40, -30, -20, -10$ and 0 dB.
11. Discussions of Sec. 2.4.1 on segmented μ-law and A-law, and the calculations shown for S/N_q are based on uniform density function for $x(t)$. Supposing now that $p(x)$ is Gaussian or Laplacian, calculate the new values of S/N_q for different ranges of the input signal. Give Tables, similar to Table 2.6, for both the pdf's.
12. Calculate a few points of the (S/N_q vs. S) curve for the 15-segment μ-law companding shown in Fig. 2.11.
13. Determine the PCM companding law which maximizes the information content of the encoded signal if the probability density function of the input is given by

$$p(x) = \frac{\sqrt{2}}{2\sigma^2} \exp\left(-\frac{\sqrt{2}}{\sigma^2}|x|\right)$$

   Also determine the optimized signal-to-noise ratio in the case.
14. Overload distortions in PCM codecs have been discussed in Sec. 2.5.1. Assuming that the large signal is first passed through a soft limiter, such that the peak of the signal is now limited to the dynamic range of the codec, the total distortion at the output is approximately equal to the sum of the harmonic distortion due to the limiter and the N_q due to the quantizer. Based on this, calculate $S/(N_c + N_q)$ for 5 and 10 dB overload and check with eq. (2.42).
15. (a) A PCM codec is to be designed for giving an overall $S_0/N_0 = 40$ dB with a dynamic range of 35 dB.The PCM words are transmitted in a channel with an error rate $P_e = 10^{-4}$. Calculate the minimum bit rate required in the codec for a random message with $f_{max} = 4$ kHz. Use log-PCM.
   (b) Calculate the bit rate required for the same problem, if an ADM codec is used, instead of the log-PCM codec.
16. Assume that for speech signals, 10^{-2} error rate in ADM is perceptually equivalent to 10^{-4} error rate in PCM. Compare the bit rates required in ADM and PCM for an equivalent $S/N_T = 32$ dB for speech coding.
17. (a) A signal has the sample values given by $x(r) = 2^r$, $r = 0, 1, 2, \ldots, 7$ and repeats itself after 8 samples. The signal is coded by a 4-bit A-PCM codec, whose step adaptation logic is given by eq. (2.47) and the successive multiplier constants M_i's are given in Table 2.9. Make a Table giving $x(r)$, $\Delta(r)$, $\hat{x}(r)$ and the corresponding PCM words. Assume that $R = \Delta_{max}/\Delta_{min} = 10$, and $\Delta_{min} = 1$. Calculate approximately the value of N_q.
   (b) Repeat the problem with an ADM codec using the adaptation logic of eq. (2.66), where $P = 1.5$ and $Q = 0.66$. Calculate approximately the value of N_q.
18. Consider a random signal whose tow-sided power spectral density is 0.5, and bandwidth unity. The signal is of lowpass nature. It is fed into an LDM. Assume a 4σ loading of the coder. If the sampling frequency $f_s = 13$, determine the step size. Also determine the S/N_q at the output of the decoder.

19. The LDM of problem (18) is made adaptive using the following adaptation rule:

L_r	L_{r-1}	L_{r-2}	Group No.	Adaptation constant
0	1	0	1	$M_1 = 0.4$
1	0	1		
0	1	1	2	$M_2 = 0.9$
1	0	0		
0	0	1	3	$M_3 = 1.5$
1	1	0		
0	0	0	4	$M_4 = 2.0$
1	1	1		

In above, L_r is the current digital output, L_{r-1} the previous, and so on. Also M_1 is the step size multiplier when one of the group 1 patterns occurs, and so on.
The maximum and minimum step sizes are limited to $10\,\Delta_0$ and $\Delta_0/10$ respectively, where Δ_0 is the step size of LDM of Problem 18.
Consider the following input to the ADM:

$$x(t) = \begin{cases} 1/2, & 0 \le t \le 10T \\ 25.0, & 10T \le t \end{cases}$$

where $T = 1/f_s$
Construct a Table as shown below and plot the output of ADM.

Sample No.	Input	Integrator output	Sign of error	Multiplier	Step size	Digital output
1						
⋮						
30						

20. Use Jayant's one-bit memory ADM coder with $P = 1.5$ and $Q = 0.5$ and repeat problem 19.
21. *A* 2-bit uniform quantizer is used in the feed-forward path of a DPCM coder, and a one-tap predictor is used in the feedback path. The input signal variance is unity. The auto-correlation of the input for one sampling period delay is 0.95. Assume a 4σ loading of the quantizer. Obtain the optimum value of the predictor gain a_1. Also determine the prediction gains with and without the quantizer.
22. (a) The speech statistics show that the normalized correlation ρ_i between neighbouring samples are given as:

i	0	1	2	3	4	5	6	7	8	9	10
ρ_i	0.1	0.8	0.6	0.4	0.22	0.18	0.1	0	–0.1	–0.15	–0.2

Using only ρ_i's for $i = 1, 2, 3$ calculate a_i's from eqn. (2.52). Then calculate the S/N_q-gain of the DPCM using a 3-tap predictor and check with the value given in Table 2.9.
(b) Check the result with a 5-tap predictor.
(c) Show that for optimum prediction, the prediction error d_r is uncorrelated with the predicted signal $\hat{x}_r$.

23. (a) Television signals are generally modelled as a 3rd-order Markov process. Assuming the values of $\rho_i = 0.7, 0.5, 0.3$ for $i = 1, 2, 3$, calculate the S/N_q gain of a DPCM codec over a PCM codec using $B = 3$, 4 and 6.
    (b) Assume that for TV signal, the ρ_i's along the horizontal direction and along the vertical direction are the same and the corresponding values are as given in problem (a). Calculate the S/N_q-gain in 2D-DPCM using a 4-tap predictor (2 in horizontal direction and 2 in vertical direction).
    (c) Give a physical explanation of S/N_q-gain obtained by using eqn. (2.75). Estimate the approximate S/N_q-gain obtained in the 2D-ADM codec.
24. Considering the 2-loop ADM of Fig. 2.28, derive the eq. (2.62) from basic considerations. Assume that the inner loop is equivalent to a nonlinear amplifier giving distortion $\simeq N_q$ and the outer loop provides the necessary negative feedback to correct the distortion. Why is there a maximum value of $|GH|$?
25. (a) Referring to Fig. 2.36, develop the basic equations to give S/N_q for ADM codecs using the nth-order power-law element in the feedback path. Relate the dynamic range with the order of the control element in the feedback path.
    (b) Using a basic ADM (speech) codec, calculate the values of S/N_q and dynamic range ($\simeq R^2$, where $R = \Delta_{max}/\Delta_{min}$) for a 3rd order power-law element in the feedback path for $f_s = 60$ Kb/s. Repeat the problem for a 4th order power-law element.
26. (a) Derive the eqn. (2.67) for S/N_q of a 2–bit D-DM codec. It has been estimated that the minimum f_s for each codec $\simeq$ 12 Kb/s. Justify the result.
    (b) If the maximum bit rate allowable is 48 Kb/s, calculate the S/N_q obtainable for 2/3/4-stage D-DM codec.
27. Write a brief review of the present status of the Transform coding techniques for picture signals.

REFERENCES

1. Jayant, N.S., Editor, *Waveform Quantization and Coding*, IEEE Press Issue, 1976.
2. Flanagan, J.L., *Speech Analysis, Synthesis and Perception,* Academic Press, 1965.
3. Flanagan, J.L., 'Focal points in speech communication research', *IEEE Trans. Com. Tech.*, Vol. Com-19, p. 1006, Dec. 1971.
4. Jayant, N.S., 'Digital coding of speech waveforms: PCM, DPCM and ADM quantizers', *Proc. IEEE*, Vol. 62, p. 611, May 1974.
5. Flanagan, J.L. et al., 'Speech coding', *IEEE Trans. Comm.*, vol. Com-27, No. 4, pp. 710–736, 1979.
6. Oliver, B.M., Pierce, J. R. and Shannon, C. E., 'The Philosophy of PCM'. *Proc. IRE*, Vol. 36, pp. 1324–1331, 1948.
7. Rabiner, L.R. and Schafer, R.W., *Digital Processing of Speech Signals*, Prentice-Hall Inc., 1978.
8. Bennett, W.R., 'Spectra of quantized signals', *BSTJ*, Vol. 27, pp. 446–472, 1948.
9. Max, J., 'Quantizing for minimum distortion', *IRE Trans. Information Th.*, Vol. IT-6, pp. 7–12, 1960.
10. Paez, M.D. and Glisson, T.H., 'Minimum mean-squared-error quantization in speech PCM and DPCM systems', *IEEE Trans. Comm.*, Vol. Com-20, pp. 225–230, 1972.
11. Cattermole, K.W., *Principles of Pulse-code Modulation*, Iliffe, London, 1970.
12. Candy, J.C., and Wooley, B. A., 'An A/D converter with segmented companding', *IEEE Nat. Telecommun. Conf. Record*, pp. 388–391, Dec. 2–4, 1974.
13. Kaneko, H., 'A unified formulation of segment companding laws and synthesis of coders and digital compandors', *BSTJ*, Vol. 49, p. 1555, Sept. 1970.
14. Owen, F.F.E., *PCM and digital Transmission Systems*, McGraw-Hill, 1982.
15. Noll, P., 'The performance of PCM and DPCM speech coders in the presence of independent and correlated errors', *IEEE Int. conf. on Commun.*, Vol. 11, pp. 30(1)–30(5), 1975.

16. Noll, P., 'Adaptive quantizing in speech coding systems', *Proc. IEEE Zurich Seminar on Digital Communication*, pp. B3(1)–B3(6), March 12–15, 1974.
17. Cummiskey, P., 'Jayant' N. S. and Flanagan, J. L., 'Adaptive quantization in DPCM coding of speech', *BSTJ*, Vol. 52, pp. 1105–1118, 1973.
18. Das, J., 'A critical review of delta modulation systems', *Electro-Technology*, Vol. 16, p. 41, March 1972.
19. Steele, R., *Delta Modulation Systems*, Pentech Press, London, 1975.
20. Das, J. et al., 'An estimation of performances of unidigit and multidigit communication system' *JIETE*, Vol. 15, pp. 95–110, 1969.
21. Abate, J.E., 'Linear and adaptive delta modulation', *Proc. IEEE*, Vol. 55, pp. 298–308, 1967.
22. Jayant, N.S., 'Adaptive delta modulation with a 1-bit memory' *BSTJ*, Vol. 49, pp. 321–342, 1970.
23. Childs, I., *Electronics Letters*, Vol. 7, p. 295, 1971.
24. Greefkes, J.A. and de Jager, F., *Philips Res. Report*, Vol. 23, p. 233,1968.
25. Cartmale, A.A. and Steele, R., *Proc. IEEE*, Vol.117, No. 1915, 1970.
26. Greefkes, J.A. 'A digitally companded delta modulation modem for speech transmission', *Proc. IEEE Int. Conf. Communications*, pp. 7–33, June 1970.
27. Chakravarty, C.V. and Faruqi, M. N.,'Two-loop adaptive delta modulation', *Trans. IEEE Comm.*, Vol. Com-22, pp. 1710–1713, Oct. 1974.
28. Das, J. and Chatterjee, P.K., 'Optimized Δ-Δ modulation systems', *Electronics Letters*, Vol. 3, pp. 286–287, 1967.
29. Steele, R., 'Peak SNR formulas for multistage DM with RL-shaped Gaussian input signals', *BSTJ*, Vol. 61, pp. 347–362, 1982.
30. Johnson, F.P., 'Calculating delta modulator performance', *IEEE Trans. Audio Electroacoustics*, Vol. AU-16, pp. 121–129, 1968.
31. Melnick, M., 'Intelligibility performance of a variable slope delta modulation', *Proc. IEEE* Vol. 59, p. 1382, 1971.
32. Jayant, N.S. and Rosenburg, A. E., 'The preference of slope-overload to granularity in the delta-modulation of speech', *BSTJ*, Vol. 50, p. 3117, Dec. 1971.
33. Goodman, D.J., 'The application of delta modulation to analog-to-PCM encoding', *BSTJ*. Vol. 48, pp. 321-343,1969.
34. Atal, B.S. and Schroeder. M.R., 'Adaptive predictive coding of speech signals', *BSTJ*, Vol. 49, p. 1973, 1970.
35. Olson, H.F., 'Speech processing systems', *IEEE Spectrum*, Vol. I, p. 90, 1964.
36. Special Issue, 'Bit rate reduction and speech interpolation', *IEEE Trans. Comm.*, Vol. Com-30, No. 4, April 1982.
37. Fehn, H.G. and Noll, P., 'Multipath search coding of stationary signals with applications to speech', *IEEE Trans. Comm.*, Vol. com-30, No. 4, April 1982.
38. Jayant, N.S. and Christensen. S. A., 'Tree-encoding of speech using the (*M*, *L*) algorithm and adaptive quantization', *IEEE Trans. Comm.*, Vol. Com-26, p. 1376, 1978.
39. Stewart, L.C., Gray, R.H. and Linde, Y., 'The design of Trellis waveform coders', *IEEE Trans. Comm.*, Vol. Com-30, p. 702, April 1982.
40. Matsuyama, Y., and Gray, R.M., 'Voice coding and tree encoding speech compression systems based upon inverse filter matching', *IEEE Trans. Comm.*, Vol. Com-30, p. 711, April 1982.
41. Viswanathan. V.R. et al., 'Variable frame rate transmission: A review of methodology and application to narrowband LPC speech coding', *IEEE Trans. Comm.*, Vol. Com-30, p. 674, April 1982.
42. Schreiber, W.F., 'Picture coding', *Proc. IEEE*, Vol. 55, p. 320, March 1967.
43. Limb, J.O., 'Source-receiver encoding for TV signals', *Proc. IEEE*, Vol. 55, p. 364, 1967.
44. Habibi, A. and Robinson, G.S., 'A survey of digital picture coding', *Computer*, p. 22, May 1974.
45. Netraveli, A.N. and Limb, J.O., 'Picture coding: A review', *Proc. IEEE*, Vol. 68, p. 366, 1980.
46. Horna, O.A., 'A 150 MBPS *A*/*D* and *D*/*A* conversion system', *Comsat Tech. Review*, Vol. 2, No. 1, p. 39, 1972.
47. Abbott, R.P., 'A differential pulsecode modulation codec for videotelephony using four bits per sample', *IEEE Trans. Comm. Tech.*, Vol. Com-19, p. 907, Dec. 1971.

48. Schilling D.L. et al., 'Video encoding using adaptive delta modulation', *IEEE Trans. Comm.*, Vol. Com-26, p. 1682, 1978.
49. Lei, T.R., Scheinberg, N., Schilling, D. L., 'Adaptivc dclta modulation for video encoding', *IEEE Trans. Comm.*, Vol. Com-25, pp. 1302–1314, 1977.
50. Frei, A.H., Schindler, H. R. and Vettiger, P., 'An adaptive dual-mode coder/decoder for television signals', *Trans. IEEE Comm. Tech.*, Vol. Com-19, p. 933, Dec. 1971.
51. Faruqi, M.N., 'Bandwidth compression of image signals', *JIETE*, Vol. 26, p. 29. 1980.
52. Haskell, B.G., et al., 'Interframe coding of video telephone pictures', *Proc. IEEE*, Vol. 60, p. 792, 1972.
53. J.B. O'neal, Jr. 'A bound on signal-to-quantizing noise ratios for digital encoding systems', *Proc. IEEE*, Vol. 55, pp. 287–292, 1967.

CHAPTER

3

Optimum Detection and Parameter Estimation

3.0 INTRODUCTION

One of the central problems in communication, control and radar systems is to design an optimum receiver which will maximize the information content of the detected signal or minimize the error between the transmitted and the detected signals. The overall design of the optimum receiver-matched filters, detectors, etc., now depends on the statistics of the signal and the disturbing noise. This calls for a knowledge of statistical communication theory, decision theory and related topics. The definitions and classification of optimum receivers are based on the concepts of information theory and decision theory. The receivers may be designed to give the maximum signal-to-noise power ratio (SNR) or the minimum error rate (equivalent to maximum information rate) at the output. A broad classification of the optimum receivers may be based on the following:

(a) Maximum average SNR criterion—as applied to analog signals, e.g., AM, FM, PM, PPM etc. The optimality of the receiver is now characterized by the gain in the output SNR as compared to the carrier-to-noise ratio (CNR) at the receiver input.

(b) Maximum peak-SNR criterion—as applied to radar signals, using the matched-filter receiver, $(SNR)_{pk} = 2E/n_0$ gives the maximum visibility of the radar echoes in the background of noise. This also gives the minimum error probability P_e at the detector output, as P_e decreases with the increase of $(SNR)_{pk}$.

(c) Inverse probability criterion (also known as MAP criterion)—as applied to communication and radar signals. Maximum information is contained in the processed signal where the inverse probability $p(X/Y)$ is maximum. The detector is in the form of an inverse probability computer giving at the output of a posteriori pdf whose maximum determines the correct signal.

(d) Maximum likelihood (ML) criterion—as applied to radar and communication signals, specially for parameter estimation, where the likelihood function $p(Y/X)$ is maximized for optimum solution.

(e) Maximum information transmission criterion—as discussed in Chapter 5. The criterion is equivalent to minimum error rate criterion and the signal received by an ideal observer/receiver through a proper detector using an optimum threshold.

(f) Mean-square error (MSE) criterion—as applied to filtering, prediction and analog transmission-reception.

Based on the above requirements, the basic principles involved in the design of optimum detectors and estimators are discussed in this chapter. In the following sections, discussions start with an introduction to the decision theory followed by the optimum detection of binary, *M*-ary and radar signals, using matched filters (MF) and cross-correlators. The theory of parameter estimation is then applied to measure the size, range and velocity of radar targets, and also for the reception of analog-modulated signals, leading to phase-locked loop (PLL) applications. The problem of optimal signal design for radar signal resolution and *M*-ary digital communication is then discussed and specific solutions are given. The similarity between radar and communication problems has been highlighted, leading to the conclusion that: the basic problems in radar reception, e.g., in clutter noise, and in wideband digital communication, e.g., in spread-spectrum system, have to be solved from the point of view of optimum signal design.

3.1 WEIGHTED PROBABILITIES AND HYPOTHESIS TESTING [2, 3, 4]

In a signal detection problem, either in radar systems or in digital communication, the received signal is always distorted due to the non-ideal characteristics of the channel and the random noise added to the signal at the receiver input. The observer or the receiver has to decide if a desired signal has been received or not in the presence of above noise and distortion. Even if the desired signal is known precisely regarding its shape, amplitude and arrival time, the observable signal at the receiver becomes a random process, and thus, the overall problem of signal detection is basically a statistical decision problem.

Consider that a signal $s(t)$ is intermittently transmitted from a source and the receiver/observer is required to decide if the signal is present or not, when the actual voltage received is $y(t)=s(t) + n(t)$, $n(t)$ = noise voltage. The two situations/events of 'signal not present' and 'signal present' are called:

H_0 : null hypothesis, i.e. 'no signal present' and $y(t) = n(t)$

H_1 : 'signal present' i.e. $y(t) = s(t) + n(t)$

The decision of H_0/H_1 now has to be based on some probabilistic criterion, and one possible criterion is that: 'choose that hypothesis which is most likely to have occurred based on single/multiple observations.' Thus the decision rule is:

$$\text{Decide on } H_1 \text{ if } p(H_1/y) > p(H_0/y) \tag{3.1}$$

and

Decide on H_0 otherwise

This decision rule is known as the 'Maximum a posteriori probability' (MAP) criterion, where $p(H_1/y)$ and $p(H_0/y)$ are a posteriori probabilities. Using Bayes theorem

$$p(H_1/y) = \frac{P_1 p(y/1)}{p(y)}$$

and

$$p(H_0/y) = \frac{P_0 p(y/0)}{p(y)}$$

where a priori probabilities $P_1 = p(1) = p(H_1)$, $P_0 = p(0) = p(H_0)$, $(P_0 + P_1) = 1$; and the conditional probabilities $p(y/1) = p(y/H_1)$, $p(y/0) = p(y/H_0)$. Then, the decision rule may be restated as:

If $\dfrac{P_1 \cdot p(y/1)}{P_0 \cdot p(y/0)} > 1$, then hypothesis H_1 is true and hypothesis H_0 is true otherwise. Equivalently,

$$\lambda_1 = \frac{p(y/1)}{p(y/0)} > P_0/P_1 \text{ for hypothesis } H_1$$

and (3.2)

$$\lambda_0 = \frac{p(y/0)}{p(y/1)} \geq P_1/P_0 \text{ for hypothesis } H_0$$

The ratios λ_1 and λ_0 are known as Likelihood ratios (LR) and the test based on these is called the Likelihood ratio test (LRT). The receiver performing the test based on eq. (3.2) is called an Ideal observer; also the method is equivalent to MAP detection as seen from eq.(3.1).

An alternative decision rule could be to choose a decision threshold y_0 such that:

$$y > y_0 \ldots H_1 \text{ is true}$$

and (3.3)

$$y \leq y_0 \ldots H_0 \text{ is true}$$

At the same time, y_0 is so chosen that the overall error probability P_e is minimum. Suppose that the conditional probabilities $p(y/0)$ and $p(y/1)$ are as shown in Fig. 3.1, where y_0 divides the range of y into the regions $R(0)$ and $R(1)$. Then the error probabilities P_f (= false alarm probability) and P_m (= probability of miss) are given by:

$$P_f = P_0 \int_{R_{(1)}} p(y/0)\, dy = P_0 \int_{y_0}^{\infty} p(y/0)\, dy$$

and

$$P_m = P_1 \int_{R_{(0)}} p(y/1)\, dy = P_1 \int_{-\infty}^{y_0} p(y/1)\, dy \tag{3.4}$$

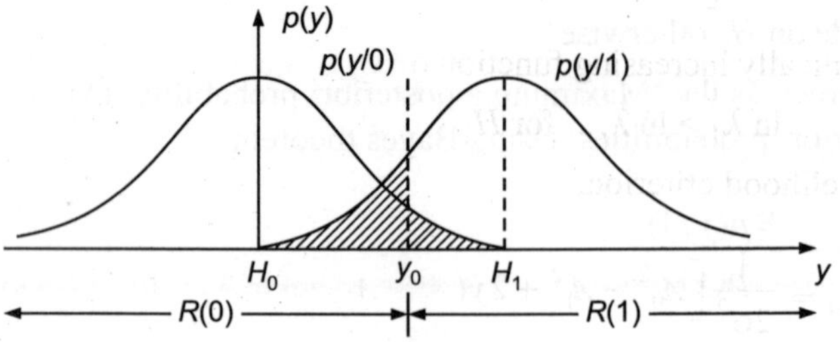

Figure 3.1: Conditional probabilities and decision regions

The total error probability, $P_e = (P_f + P_m)$ is now minimized by choosing a proper y_0. The minimum of P_e is obtained by setting $\partial P_e/\partial y_0 = 0$, which gives

$$P_0 \cdot p(y_0/0) = P_1 \cdot p(y_0/1)$$

or

$$\frac{p(y_0/1)}{p(y_0/0)} = \frac{P_0}{P_1} = \lambda_t$$

then the decision rule is:

$$\lambda_1 > \lambda_t \ldots \text{ for } H_1$$

and (3.5)

$$\lambda_0 \geq 1/\lambda_t \ldots \text{ for } H_0$$

which is the same as eq. (3.2). Thus the minimum Error criterion, MAP criterion and the Ideal observer, all lead to the same likelihood ratio test of eq. (3.5), for unweighted probabilities of H_1 and H_0. The equivalence between eqs. (3.3) and (3.5) is calculated knowing the actual values of H_0, H_1 and the pdf of the noise. The sample y is now called the 'test statistic' for the optimum receiver.

Example 3.1: Consider the reception of $1/O$ binary signals, designated by A_1/A_0 volts and distorted by Gaussian noise in the channel. Referring to Fig. 3.1, $H_0 = A_0$, $H_1 = A_1$ and the pdf's $p(y/0)$ and $p(y/1)$ are now Gaussian. For a single sample or for a single pulse transmission, the conditional probabilities at the receiver, assuming noise with zero mean and variance $= \sigma^2$, are given by:

$$p(y/1) = \frac{1}{\sqrt{2\pi}\sigma} \exp\left[\frac{-(y - A_1)^2}{2\sigma^2}\right]$$

and

$$p(y/0) = \frac{1}{\sqrt{2\pi}\sigma} \exp\left[\frac{-(y - A_0)^2}{2\sigma^2}\right]$$

With source probabilities P_1 and P_0, the likelihood ratio λ_1 is

$$\lambda_1 = \frac{p(y/1)}{p(y/0)} = \exp\left[\frac{-(y - A_1)^2}{2\sigma^2} + \frac{(y - A_0)^2}{2\sigma^2}\right] > \lambda_t \text{ for hypothesis } H_1.$$

Since $\ln \lambda_t$ is a monotonically increasing function of λ_t, it is equally valid to make the decision based on

$$\ln \lambda_1 > \ln \lambda_t \ldots \text{ for } H_1$$

Then, using the log-likelihood criterion,

$$\ln \lambda_1 = \frac{1}{2\sigma^2}\left[A_0^2 - A_1^2 + 2y(A_1 - A_0)\right] > \ln \lambda_t = \ln \frac{P_0}{P_1} \ldots \text{ for } H_1$$

or

$$y > y_0 = \frac{A_1 + A_0}{2} + \frac{\sigma^2}{A_1 - A_0} \ln \frac{P_0}{P_1} \ldots \text{ for } H_1 \qquad (3.6)$$

Alternatively from eq. (3.5),

$$\lambda_t = \frac{P_0}{P_1} = \exp\left[\frac{1}{2\sigma^2}\left\{A_0^2 - A_1^2 + 2y_0(A_1 - A_0)\right\}\right],$$

which leads to the value of the threshold y_0 as in eq. (3.6). The total error P_e is now given by the shaded area in Fig. 3.1. For $P_0 = P_1 = 0.5$, $y_0 = (A_1 + A_0)/2$ and $P_f = P_m$, giving $P_e = 2P_f$. For binary on-off keying using $H_0 = 0$, $H_1 = A$ volts, and $P_1 = P_0 = 0.5$, the threshold $y_0 = A/2$. The error probabilities are now given by

$$P_f = P_m = P_0 \int_{A/2}^{\infty} \frac{1}{\sqrt{2\pi}\sigma} \exp\left(\frac{-y^2}{2\sigma^2}\right) dy$$

Therefore,

$$P_e = \int_{A/2\sigma}^{\infty} \frac{1}{\sqrt{2\pi}} e^{-z^2/2} dZ, Z = y/\sigma = \text{erfc}\,(A/2\sigma), \tag{3.7}$$

where

$$\text{erf}\,(x) \triangleq \int_{-\infty}^{x} \frac{1}{\sqrt{2\pi}} e^{-z^2/2}\, dZ \tag{3.8*}$$

and erfc $(x) = [1 - \text{erf}\,(x)]$.

For bipolar transmission using $+\,A/2$ volt signals, and $P_0 = P_1 = 0.5$, $y_0 = 0$, and

$$P_e = \int_0^{\infty} \frac{1}{\sqrt{2\pi}\sigma} \exp\left[\frac{-(y + A/2)^2}{2\sigma^2}\right] dy$$

$$= \int_{A/2\sigma}^{\infty} \frac{1}{\sqrt{2\pi}} e^{-z^2/2}\, dZ$$

$$= \text{erfc}\,(A/2\sigma) \tag{3.9}$$

Thus it is seen that P_e is the same for both unipolar and bipolar transmissions, but the signal power ($= A^2/2$) in the unipolar case is 3 dB more than that in the bipolar case ($= A^2/4$), giving a 3 dB advantage in the signal-to-noise ratio (SNR) required for bipolar signalling.

---

*An alternative definition of erf (x) is

$$\text{erf}_*\,(x) \triangleq \frac{2}{\sqrt{\pi}} \int_0^x \exp\,(-Z^2/2)\, dZ$$

and

$$\text{erf}_*\,(x) = 2\ \text{erfc}\,(\sqrt{2}x)$$

Using this value of erf$_*(x)$, $P_e = \frac{1}{2}$ erfc $*$ $[A/(2\sqrt{2}\sigma)]$. However, eq. (3.8) will be used in the present text.

3.1.1 Bayes Criterion

The decision rules for the ideal observer, as given by eqs. (3.5) and (3.6), have been derived for the unweighted probabilities P_0, P_1, P_f and P_m. In actual practice, however, the decisions will have to be biased when different decisions are of different importance or have different risk values, e.g., in Radar, the risk of a 'miss' is definitely more than that of a 'false alarm'. In such cases, the probabilities are weighted according to the costs involved, based on a cost matrix and the hypothesis H_1/H_0 is decided on the basis of this cost matrix. In general, the cost matrix [C] associated with the decision matrix [H] may be written as

$$[C] = \begin{bmatrix} C_{11} & C_{10} \\ C_{01} & C_{00} \end{bmatrix} = \begin{bmatrix} C_{11} & C_m \\ C_f & C_{00} \end{bmatrix} \tag{3.10}$$

for

$$[H] = \begin{array}{c|cc} Y/X & 1 & 0 \\ \hline 1 & H_1 & H_0 \\ 0 & H_1 & H_0 \end{array}$$

where C_{ij} is the cost of choosing H_j when actually H_i is ture. In practical situations, the costs C_{11} and C_{00} for correct decisions H_1 and H_0 may be taken as zero (except when some credit is given also for correct decisions) and [C] reduces to

$$[C] = \begin{bmatrix} 0 & C_m \\ C_f & 0 \end{bmatrix} \text{ for the error probabilities } [P] = \begin{bmatrix} 0 & P_m \\ P_f & 0 \end{bmatrix}$$

Using the cost matrix and eq. (3.4), the average risk per decision is now given as

$$\overline{C}(y_0) = P_0\left[C_{00}\int_{R(0)} p(y/0)\,dy + C_{01}\int_{R(1)} p(y/0)\,dy\right]$$

$$+ P_1\left[C_{10}\int_{R(0)} p(y/1)\,dy + C_{11}\int_{R(1)} p(y/1)\,dy\right]$$

$$= P_0 C_f \int_{y_0}^{\infty} p(y/0)\,dy + P_1 C_m \int_{-\infty}^{y_0} p(y/1)\,dy$$

if $$C_{11} = C_{00} = 0 \tag{3.11}$$

The average risk $\overline{C}(y_0)$ is now minimised by chossing the optimum value of the threshold y_0, which is obtained by setting

$$\frac{\partial \overline{C}(y_0)}{\partial y_0} = 0$$

and calculating the value of y_0. The result is:

$$P_0 C_f \cdot p(y_0/0) = P_1 C_m p\,(y_0/1)$$

Therefore,

$$\lambda_t = P_0 C_f / P_1 C_m, \tag{3.12}$$

and the decision rule is the same as eq. (3.5) with the weighted value of λ_t. The value of $\overline{C}\,(\lambda_t)$ is now minimum.

For the problem in Ex. 3.1, the value of the threshold λ_t is

$$\lambda_t = \frac{\exp[-(y_0 - A_1)^2/2\sigma^2]}{\exp[-(y_0 - A_0)^2/2\sigma^2]} = \frac{P_0 C_f}{P_1 C_m}$$

The corresponding decision threshold for the 'test statistic' y is

$$y_0 = \frac{A_1 + A_0}{2} + \frac{\sigma^2}{A_1 - A_0} \ln \frac{P_0 C_f}{P_1 C_m} \tag{3.13}$$

which is the same as eq. (3.6) for $C_f = C_m = 1$. Because of the relative costs, the threshold is biased toward the decision costing more, i.e., for $C_f > C_m$ and $P_0 = P_1, y_0 > (A_1 + A_0)/2$ = the mean value of the signal.

The decision strategy, which minimizes the average risk $\overline{C}$, is known as the Bayes criterion and the minimum average risk $\overline{C}_{\min}$ obtained by using eq. (3.13) for y_0 in eq. (3.11), is known as Bayes risk. When the relative costs $(C_{01} - C_{00}) = (C_{10} - C_{11})$, the Bayes criterion is equivalent to the ideal observer of eq. (3.5). The criterion, defined by eq. (3.12), is also true when the decision is based on measurements of more than one variable. It is necessary, however, to have a priori knowledge of the relative costs and the source probabilities P_0 and P_1. The concept of Bayes strategy requires that a large number of observations or trials are made to ascertain P_0 and P_1 correctly.

3.1.2 Minimax Criterion

In many applications, as in the theory of games, the a priori probabilities P_0 and P_1 are not known and the Bayes risk cannot be calculated. In such cases, Bayes minimum risks $\overline{C}_{\min}\,(y_0)$ for the various values of P_0 (or P_1) are calculated and the maximum of the $\overline{C}_{\min}\,(P_0)$ vs. P_0 curve is considered as the minimax risk $\overline{C}_{\text{mnx}}$, as shown in Fig. 3.2. The average risk for any arbitrary value of P_0 or P_1 can never exceed this $\overline{C}_{\text{mnx}}$, i.e. $\overline{C}_{\text{mnx}}$ gives the worst value of risk for $P_0 = P_{\text{mx}}$. The corresponding decision thresholds λ_t and y_{mnx} are calculated from eq. (3.13).

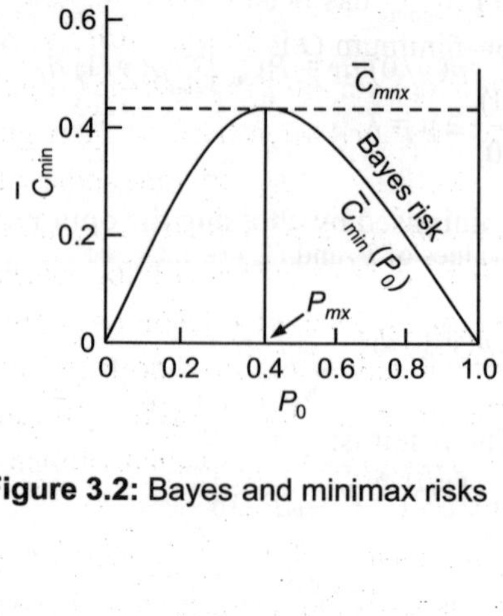

Figure 3.2: Bayes and minimax risks

Consider again the problem of Ex. 1. Assuming that $A_1 = +1, A_0 = -1, \sigma^2 = 2$ and $C_{00} = C_{11} = 0, C_f = 2$, $C_m = 1$, the decision threshold y_0 is given by

$$y_0 = \ln (2P_0/P_1)$$

and

$$\begin{aligned}\overline{C}_{\min}(y_0) &= P_0 C_f \cdot \text{erfc}\,[(y_0 - A_0)/\sigma] + P_1 C_m\, \text{erf}\,[(y_0 - A_1)/\sigma] \\ &= 2P_0\, \text{erfc}\,[(y_0 + 1)/\sigma] + (1 - P_0)\, \text{erf}\,[(y_0 - 1)/\sigma] \end{aligned} \tag{3.14}$$

Equation (3.14) is now plotted in Fig. 3.2 for $0 < P_0 < 1$. It is seen that the maximum of $\overline{C}_{\min}$ occurs for $P_0 = 0.4$ and $y_0 = y_{\text{mnx}} = 0.6$ and the corresponding $\overline{C}_{\text{mnx}} = 0.42$.

Further it may be shown that for the minimax value of P_{mx} and the threshold y_{mnx}, the conditional risks associated with two hypotheses are equal, i.e.,

Risk for decision $H_1 = C_m P_m = C_f P_f =$ risk for decision H_0. On this basis, $\overline{C}_{\text{mnx}}$ and y_{mnx} may be calculated.

3.1.3 Neyman-Pearson Criterion

In some situations, as in radar applications, not only the a priori probabilities but also the cost matrix are not normally known. In such cases, the observer uses a preassigned value of the false-alarm probability P_f which he can afford, and seeks a decision strategy that attains this value and at the same time yields the minimum value of the miss-probability P_m. This strategy is known as the Neyman-Pearson criterion. It corresponds to maximizing the detection probability $P_d = (1 - P_m)$ of a target (in Radar problems) for a given false-alarm probability P_f. Equivalently, one can minimize P_m for a given P_f.

The above maximization problem is generally solved by using the method of Lagrangian (undetermined) multipliers and the equation

$$Q = P_m + \mu P_f, \quad \mu = \text{Lagrangian multiplier,} \tag{3.15}^*$$

is minimized for a given value of P_f. Substituting $C_{00} = C_{11} = 0, P_1 C_m = 1$, and $P_0 C_f = \mu$, in eq. (3.11) the average cost

$$\begin{aligned}\overline{C} &= \int_{-\infty}^{y_0} p(y/1)\, dy + \mu \int_{y_0}^{\infty} p(y/0)\, dy \\ &= P_m + \mu P_f \end{aligned} \tag{3.16}$$

Comparing eqs. (3.15) and (3.16), it is seen that the Neyman-Pearson criterion is a special case of the Bayes criterion. The condition for $\overline{C}_{\min}$ has been obtained in eq. (3.12) and substituting the assumed values as above, the condition for minimum Q is

$$\lambda_1 = \frac{p(y/1)}{p(y/0)} \geq \mu = \lambda_t \tag{3.17}$$

---

*Since P_0 and P_1 are unknown, the values of P_f and P_m in eq. (3.15) are taken to be

$$P_f = \int_{y_0}^{\infty} p(y/0)\, dy$$

and (3.18)

$$P_m = \int_{-\infty}^{y_0} p(y/1)\, dy$$

Thus an LRT minimizes P_m, and equivalently maximizes P_d for the given P_f. The decision rule is then

$$\lambda_1(y) > \lambda_t \text{ ... for } H_1$$

and

$$\lambda_1(y) \leq \lambda_t \text{ ... for } H_0$$

where λ_t and y_0 are calculated from the given P_f.

In most of radar applications $\lambda_1(y)$ is a monotone function of y and there is no minimization of P_m. In case of composite hypothesis testing, however, this is not so and the criterion assumes significance.

Example 3.2: Consider that a radar echo of 1 volt is received along with the corrupting noise of zero mean and variance $\sigma^2 = 1$. On the basis of a single sample y, it is required to find the optimum decision rule for choosing $H_0 = 0$ and $H_1 = 1$, given that $P_f = 0.1$. Then

$$p(y/1) = \frac{1}{\sqrt{2\pi}} \exp\left[-(y-1)^2/2\right]$$

and

$$p(y/0) = \frac{1}{\sqrt{2\pi}} \exp\left[-y^2/2\right]$$

The LR is

$$\lambda_1(y) = \frac{p(y/1)}{p(y/0)} = \exp(y - 1/2)$$

and LRT is

Choose $\exp(y - 1/2) > \lambda_t$ for H_1.

The corresponding threshold y_0 is calculated from

$$P_f = \int_{y_0}^{\infty} p(y/0)\, dy = \int_{y_0}^{\infty} \frac{1}{\sqrt{2\pi}} \exp(-y^2/2)\, dy = \text{erfc}(y_0) = 0.1$$

From Tables of Error Functions,

$$y_0 = 1.28 \text{ for } P_f = 0.1$$

Therefore, the decision rule is

Choose H_1 if $y > 1.28$; and H_0 otherwise.

P_d, based on the single observation y, is

$$P_d = \int_{1.28}^{\infty} p(y/1)\, dy = \int_{1.28}^{\infty} \frac{1}{\sqrt{2\pi}} \exp\left[-(y-1)^2/2\right] dy$$

$$= 0.39$$

The value of λ_t is given by

$$\lambda_t = \lambda(y_0) = \exp(1.28 - 0.5)$$

$$= 2.18$$

and the corresponding decision rule is

Choose H_1 if $\lambda_1(y) > 2.18$ and choose H_0 otherwise.

If we assign a priori probabilities to the signal, say, $P_0 = P_1 = 1/2$, then, from eqn. (3.12), $\lambda_t = 1$

i.e., $\lambda_1(y) = \exp(y - 1/2) > 1$ ... for H_1

Taking logarithm of both sides,

$$y > 1/2 \text{ ... for } H_1.$$

Therefore,

$$P_f = \text{erfc}(1/2) = 0.31 = P_m$$

and

$$P_d = (1 - 0.31) = 0.69$$

It is now seen that P_d is higher in this case, but P_f is also greater than the assigned value of 0.1.

Receiver Operating Characteristics (ROC)

The solutions for the Neyman-Pearson criterion contain all the information required to use Bayes and Minimax criteria. In the theory of hypothesis testing, P_f is called the size of the test and P_d is the power of the test. A plot of P_d vs. P_f, as shown in Fig. 3.3, is often called the 'Receiver Operating Characteristic' (ROC) or the operating characteristic of the test. The curve depends only on the pdf's of the measured quantities under the two hypotheses and not on any costs or a priori probabilities.

In Fig. 3.3, ROC for the problem of Ex. 2 has been drawn. The slope of the ROC at any point gives the value of the corresponding λ_t, since

$$\frac{dP_d}{dP_f} = \lambda_t \tag{3.19}$$

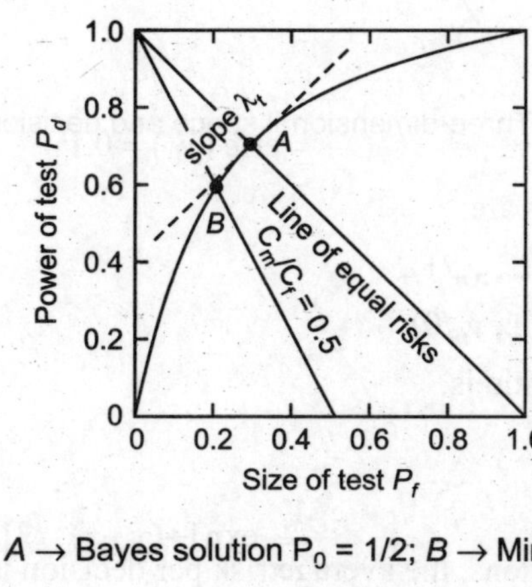

Figure 3.3: ROC of Ex. 3.2; $A \rightarrow$ Bayes solution $P_0 = 1/2$; $B \rightarrow$ Minimax solution with $C_m/C_f = 0.5$

as obtained by differentiating the eq. (3.18). While calculating the Bayes risk, one determines the value of λ_t from eq. (3.12), but this can be determined graphically from the slope of the ROC at a given P_f. Then $\overline{C}_{\min}$ is calculated using eq. (3.11). The dashed line in the figure represents the solution: $\lambda_t = 1$ for Ex. 2 with $P_0 = P_1 = 1/2$. To find the minimax solution, one uses the relation of equal risks: $C_m P_m = C_f P_f$. This gives a straight line in the (P_f, P_d) plane as shown in Fig. 3.3, as lines of equal risks, and the straight lines intersect the ROC at minimax values of P_f and P_d. The slopes of the ROC at these points give the corresponding values of λ_t from which the probability P_0 is calculated using eq. (3.12). The minimax risk is now given by

$$\overline{C}_{mnx}(P_0) = C_m P_m = C_f P_f$$

Thus the ROC of Fig. 3.3 may be used to determine both Bayes and minimax risks.

3.1.4 Hypothesis Testing by Multiple Observations

To improve the visibility of radar echoes and to decrease the error probability in digital communication, the signal pulses are sometimes repeated and the decision H_1/H_0 is made on the basis of the joint pdf of signal plus noise exceeding the optimum threshold. The n observations or measurements, $Y = \{y_1, y_2, ..., y_n\}$ may be regarded as representing a point in an n-dimensional vector space, where the randomly varying received signals will normally cluster together around the point corresponding to $Y(0)$ in the vector space when '0' is being sent and around the point $Y(1)$ when '1' is sent. Assuming that there is a certain distance between $Y(0)$ and $Y(1)$, the decision problem is then to choose an optimum decision surface D which will divide the vector space into the regions $R(0)$ and $R(1)$ as shown in Fig. 3.4, such that the average risks are minimized.

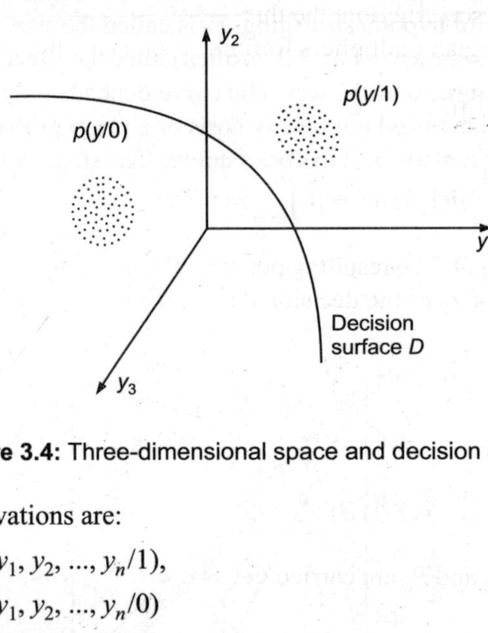

Figure 3.4: Three-dimensional space and decision surface

The joint pdf's of n observations are:

$$p(Y/1) = p(y_1, y_2, ..., y_n/1),$$
$$p(Y/0) = p(y_1, y_2, ..., y_n/0)$$

and the corresponding likelihood ratio is

$$\lambda_1(Y) = \frac{p(Y/1)}{p(Y/0)}$$

Extending eq. (3.11) for n observations, the average risk per decision is given as

$$\overline{C}(D) = P_0\left[C_{00}\int_{R(0)} p(Y/0)\,dY + C_{01}\int_{R(1)} p(Y/0)\,dY\right]$$
$$+ P_1\left[C_{10}\int_{R(0)} p(Y/1)\,dY + C_{11}\int_{R(1)} p(Y/1)\,dY\right]$$

$$= P_0 C_f \int_{R(1)} p(Y/0)\,dY + P_1 C_m \int_{R(0)} p(Y/1)\,dY \ldots$$

$$\text{if } C_{11} = C_{00} = 0; \tag{3.20}$$

where $dY = (dy_1, dy_2, \ldots, dy_n)$ = the volume element in the vector space of n observations. The average risk is now minimized according to Bayes criterion and the decision surface D is obtained such that

$$H_1 \text{ is true if } \lambda_1(Y) > P_0 C_f / P_1 C_m$$

$$= \lambda_t(Y) \tag{3.21}$$

and H_0 otherwise.

The value of $\overline{C}_{\min}$ is obtained by substituting λ_t as the limit of the integrals in eq. (3.20), and $\overline{C}_{\min} = C(\lambda_t)$.

To use the minimax test, eq. (3.20) is used instead of eq. (3.11) for plotting the curve for $\overline{C}_{\min}(P_0)$ in Fig. 3.2. Again we have equal risks, $C_m P_m = C_f P_f$, and $\overline{C}_{\text{mnx}}$ and Y_{mnx} may be calculated from this relation. For the Neyman-Pearson criterion, the threshold $\lambda_t(Y)$ and the decision surface D are obtained by using the method of Lagrangian multipliers forming the linear equation.

$$\overline{C}(D) = P_m + \mu P_f$$

$$= \int_{R(0)} p(Y/1)\,dY + \mu \int_{R(1)} p(Y/0)\,dY \tag{3.22}$$

which is minimized by varying D. The resulting position of D is a function of μ and the corresponding $\lambda_t = k(P_f)$ determines the value of Y_0 of the decision device. Then,

$$P_f = \int_{Y_0}^{\infty} p(Y/0)\,dY$$

and

$$P_d = 1 - \int_{-\infty}^{Y_0} p(Y/1)\,dY \tag{3.23}$$

The detailed calculations of Y_0 and P_d are carried out as in Ex. 2 by simply substituting the joint pdf's in place of $p(y/1)$ and $p(y/0)$.

While solving the problems in Exs. 1 and 2, it has been found convenient to use the logarithm of the likelihood ratio, ln λ, as the test statistic. This is so also for the multiple observations. When the pdf's are jointly Gaussian, this statistics takes a particularly simple form, e.g., y, Σy_i, Σy_i^2 etc., and the numerical solution of the problem is easy. If the observations $\{y_1, y_2, \ldots, y_n\}$ are statistically independent, then ln λ equals the sum of the individual statistics, whatever may be their joint pdf's. Thus ln λ contains all relevant informations regarding the variables and simplified measurements may be used in the statistical experiments. The modified statistic using logarithm, is also known as the 'sufficient statistic', which is a simple function of the received data, but at the same time, contains all the required information for making statistical decisions.

Example 3.3: Consider the extension of the problem in Ex. 3.1, where the transmission time T for the A_1/A_0 pulses is made large as compared to the Nyquist interval $1/B$, B = RF bandwidth of the receiver. Then after coherent detection, the baseband signal $Y(t)$ is sampled n times per T seconds, giving n independent samples per A_1/A_0 transmission. Assuming the disturbing noise to be Gaussian with zero mean and variance σ^2, the joint pdf's for n samples are

$$p(Y/0) = \left(\frac{1}{2\pi\sigma^2}\right)^{n/2} \exp\left[\sum_{j=1}^{n} \frac{-(y_j - A_0)^2}{2\sigma^2}\right]$$

and

$$p(Y/1) = \left(\frac{1}{2\pi\sigma^2}\right)^{n/2} \exp\left[\sum_{j=1}^{n} \frac{-(y_j - A_1)^2}{2\sigma^2}\right]$$

where $Y_j = j$th sample.

The likelihood ratio is now given by

$$\lambda_1(Y) = \exp\left[\frac{A_1 - A_0}{\sigma^2} \Sigma y_j - \frac{n(A_1^2 - A_0^2)}{2\sigma^2}\right]$$

and the decision rule using Bayes criterion is

$$\lambda_1(Y) > \lambda_t = P_0 C_f / P_1 C_m \ldots \text{ for } H_1 \tag{3.24}$$

and H_0 otherwise.

Taking logarithm of both sides, the optimum decision threshold is given by ($A_1 > A_0$)

$$Y_0 = \left[\frac{n(A_1 + A_0)}{2} + \frac{\sigma^2}{A_1 - A_0} \ln \lambda_t\right]$$

and

$$\Sigma y_j > Y_0 \ldots \text{ for } H_1\text{; and } H_0 \text{ otherwise}$$

or

$$\frac{1}{n} \Sigma y_j = \overline{m}_y > \frac{Y_0}{n} \ldots \text{ for } H_1 \tag{3.25*}$$

where $\overline{m}_y$ is the so-called 'sample mean' of the observations. The 'sample mean' is the 'sufficient statistic' for the problem, and P_f, P_m etc. may be calculated from its statistical properties. Since $\overline{m}_y$ is a linear sum of Gaussian random variables, then $\overline{m}_y$ itself is a Gaussian random variable with its means as A_1 for H_1 and A_0 for H_0, and its variance is σ^2/n. The error probabilities are now calculated as:

*It is observed that by using $\ln \lambda_t$ as the decision threshold, the n-dimensional problem of eq. (3.24) is reduced to a one-dimensional problem of eq. (3.25).

$$P_f = \text{erfc}\left(\frac{Y_0/n - A_0}{\sigma/\sqrt{n}}\right)$$

and

$$P_m = \text{erfc}\left(\frac{A_1 - Y_0/n}{\sigma/\sqrt{n}}\right) \tag{3.26}$$

For using Neyman-Pearson criterion, the value of the optimum threshold Y_0/n is calculated from the given P_f, and P_d is then calculated with this value of the threshold.

Example 3.4: Consider that the transmitter sends out bursts of Gaussian noise of power σ_1^2 and σ_0^2 as H_1 and H_0, instead of sending A_1/A_0 volts as in Ex. 3.3. The receiver now has the problem of deciding on H_1/H_0 by making measurements on the n-samples for T seconds. The simplest receiver is a square-law detector-cum-integrator, where the quantity Σy_j^2 is taken as the 'test statistic', and the values of P_f and P_m are calculated accordingly. This may be shown by the analysis as below.

Assuming H_1 and H_0 to be zero-mean Gaussian noise with variances σ_1^2 and σ_0^2 respectively, the joint pdf's for n samples are:

$$p(Y/k) = \left(\frac{1}{2\pi\sigma_k^2}\right)^{n/2} \exp\left[\sum_{j=1}^{n} -\frac{y_j^2}{2\sigma_k^2}\right], \quad k = H_1, H_0$$

and the LR is

$$\lambda_1(Y) = \left(\frac{\sigma_0}{\lambda_1}\right)^n \exp\left[-\sum_{j=1}^{N} y_j^2 \left(\frac{\sigma_0^2 - \sigma_1^2}{2\sigma_0^2\sigma_1^2}\right)\right] > \lambda_t \ldots \text{for } H_1 \tag{3.27}$$

Taking logarithms, the decision rule is

$$\Sigma y_j^2 > \left(\frac{2\sigma_0^2\sigma_1^2}{\sigma_1^2 - \sigma_0^2}\right) \ln\left[\left(\frac{\sigma_1}{\sigma_0}\right)^n \lambda_t\right] = \lambda_t' \quad \ldots \quad \text{for } H_1 \tag{3.28}$$

Thus the test statistic (or sufficient statistic) is Σy_j^2 = sum of the sample squares = Z, and $Z > \lambda_t'$ for the hypothesis H_1. The pdf for Z is the Gamma distribution given by [3]:

$$p(Z) = \frac{1}{\sigma^n 2^{n/2} \cdot \Gamma(n/2)} (Z)^{(n/2)-1} \cdot \exp\left(-\frac{Z}{2\sigma^2}\right) \tag{3.29*}$$

---

*For $\sigma = 1$, the pdf of Z is the χ^2 distribution with n degrees of freedom, given by

$$p(Z) = \frac{1}{2^{n/2} \cdot \Gamma(n/2)} (Z)^{(n/2)-1} \cdot e^{-z/2}, Z \geq 0$$

and

$$P_f = \int_{\lambda'_t}^{\infty} p(Z_0)\, dZ_0, \; Z_0 = \Sigma y_j^2 \;\; \text{with } \sigma = \sigma_0$$

$$P_m = \int_{-\infty}^{\lambda'_t} p(Z_1)\, dZ_1, \; Z_1 = \Sigma y_j^2 \;\; \text{with } \sigma = \sigma_1 \tag{3.30}$$

The value of λ_t will depend on the particular decision criteria used and the numerical values of P_f, P_m and P_d are obtained from tables.

For the special case of $n = 2$,

$$p(Z) = \frac{1}{2\sigma^2} \exp\left(-\frac{Z}{2\sigma^2}\right)$$

and

$$P_f = \exp\left(-\frac{\lambda'_t}{2\sigma_0^2}\right), P_m = 1 - \exp\left(-\frac{\lambda'_t}{2\sigma_1^2}\right)$$

3.1.5 Selection among *M* Hypotheses

So far we have discussed only binary type of decisions. However, in many communication systems, the source transmits M signals corresponding to M hypotheses, and the receiver has to decide on one of the M hypotheses as transmitted, depending on the criteria used. In general, only Bayes criterion is used for such problems, since minimax and Neyman-Pearson criteria are applicable to radar signals with binary hypothesis only. Assuming that cost functions and a priori probabilities are given, the average Bayes risk per decision is obtained as an extension of eq. (3.11), and is given by

$$\bar{C} = \sum_{i=0}^{M-1} P_i \sum_{j=0}^{M-1} C_{ij} \int_{R(j)} p(Y/i)\, dY \tag{3.31}$$

where P_i = a priori probability of the hypothesis H_i, C_{ij} = elements of the cost matrix and $p(Y/i)$ = the joint pdf of n observations, given H_i transmitted. The Bayes risk is now minimized by varying $R(j)$, the decision regions in the vector space for hypothesis H_j (equivalently the decision surfaces D_j are varied for minimum $\bar{C}$). From eq. (3.31), it is noted that the cost associated with the choice of H_j is given by

$$C_j = \sum_{i=0}^{M-1} P_i C_{ij} \int_{R(j)} p(Y/i)\, dY$$

Thus, Bayes decision rule for multiple hypotheses is to choose that hypothesis H_j for which C_j is minimum, i.e.,

$$\alpha_j = \sum_{i=0}^{M-1} P_i C_{ij}\, p(Y/i) \tag{3.32}$$

is a minimum.

As a special case, assume $C_{ii} = 0$ and $C_{ij} = 1$, $i \neq j$ (as is the case in M-ary communication systems discussed later). Then, with equal weighting for all errors,

$$\alpha_j = \sum_{\substack{i=0 \\ i \neq j}}^{M-1} P_i\, p(Y/i) \quad \ldots \text{for all } j$$

Then it is easily shown that

$$\alpha_k - \alpha_j = P_j p(Y/j) - P_k p(Y/k)$$

since in each α_j, the term $[P_j p(Y/j)] = 0$, and all others $(i \neq j, i \neq k)$ are the same in α_k and α_j. Therefore, choosing the minimum α_j corresponds to choosing the maximum of $(\alpha_k - \alpha_j)$, which is equivalent to choosing the largest value of $P_j p(Y/j)$. The decision rule then is

Choose H_j for which $P_j p(Y/j)$ is the largest of all $P_i p(Y/i)$ (3.33)

Using Bayes theorem of inverse probability, the maximizing of $P_j p(Y/j)$ is equivalent to maximizing $p(H_j/Y)$ = the a posteriori probability of H_j given the sample values $\{y\}$. Thus the Bayes criterion reduces to the MAP criterion, as in the case of simple binary detection with unweighted probabilities discussed along with eqs. (3.4) and (3.5). If all P_i are equal, then the decision rule reduces to

Choose H_j for which $p(Y/j)$ is a maximum (3.34)

This is the criterion usually used in the detection of equi-probable M-ary signals. This is also known as the Maximum likelihood criterion.

Example 3.5: Assume that the transmitted alphabets are: –1, 0, 1 and the disturbing noise in the channel is zero-mean Gaussian with variance σ^2. Assume further that all alphabets are equi-probable and $C_{ij} = 1$, $i \neq j$ and $C_{ii} = 0$. It is required to obtain the optimum thresholds for the receiver.

The likelihood functions are given by (see Ex. 3)

$$p(Y/j) = \left(\frac{1}{2\pi\sigma^2}\right)^{n/2} \exp\left[\sum_{i=1}^{n} -\frac{(y_i - j)^2}{2\sigma^2}\right]$$

According to MAP criterion of eq. (3.34), $p(Y/j)$ is to be maximized. Since Σy_i^2 is common in all exp [ ] terms, it is sufficient to maximize the expression

$$\left[\left(\frac{2}{n}\sum_{i=1}^{n} y_i j\right) - j^2\right] \quad \text{for the hypothesis } H_j,$$

where $H_1 = -1$, $H_2 = 0$, $H_3 = 1$. Explicitly, one has to choose the maximum of:

$$\left(-\frac{2}{n}\Sigma y_i - 1\right); \quad 0; \quad \left(\frac{2}{n}\Sigma y_i - 1\right)$$

corresponding to H_1, H_2 and H_3 respectively.

Then for the likelihood function for H_1 to be maximum, the condition is

$$\left(-\frac{2}{n}\Sigma y_i - 1\right) > 0$$

$$> \left(\frac{2}{n}\Sigma y_i - 1\right)$$

i.e., the sample mean $(1/n)\,\Sigma y_i = \bar{m}_y < -1/2$ to satisfy the first inequality, and the second inequality is then automatically satisfied. Similarly, for the likelihood function for H_3 to be maximum, the condition is

$$\left(\frac{2}{n}\Sigma y_i - 1\right) > 0$$

$$> \left(-\frac{2}{n}\Sigma y_i - 1\right)$$

which reduces to

$$\bar{m}_y = \frac{1}{n}\Sigma y_i > 1/2.$$

Thus the threshold to be used are

$$\left.\begin{aligned} &\text{Choose } H_1 \text{ if } \bar{m}_y \le -1/2 \\ &\qquad H_2 \text{ if } -1/2 < \bar{m}_y \le 1/2 \\ &\qquad H_3 \text{ if } \bar{m}_y > 1/2. \end{aligned}\right\}$$

The regions in which each hypothesis is choosen is shown in Fig. 3.5, which is a realistic situation in communication using ternary alphabets. The error probabilities may now be calculated from the pdf of the test statistic $\bar{m}_y$ (refer to eq. (3.26)) and are equal to the sum of the shaded areas in the figure.

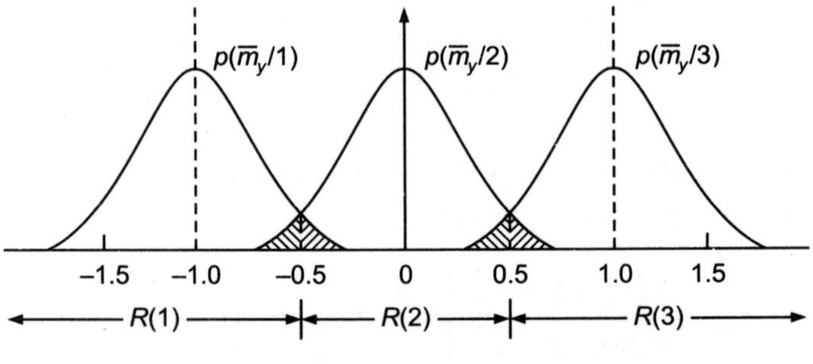

Figure 3.5: pdf of $\bar{m}_y$ and the decision regions

3.1.6 Composite Hypotheses

The hypotheses, so far considered, were simple, i.e., all the necessary pdf's were assumed to be completely known. In practical cases, however, one or both of the pdf's $p(Y/1)$ and $p(Y/0)$ may depend on some parameters $\theta_1, \theta_2, \ldots, \theta_m$ whose values are unknown at the receiver. Specifically, a signal $f(A, \omega, \phi, \tau)$, where A = amplitude, ω = frequency, ϕ = phase and τ = time of arrival, may have one or more of the parameters varying with some unknown pdf's due to signal fading or moving targets. The conditional pdf $p(Y/1)$ is now modified as $p(Y, \theta/1)$, where the components $\theta_1, \theta_2, \ldots, \theta_m$ of θ may be considered as random variables. The hypothesis H_1 is then known as 'composite', where the hypothesis H_0 for noise only is simple. The decision surface D, which divides the regions $R(1)$ and $R(0)$, is now selected on the basis of the average likelihood ratio

$$\lambda_{av}(Y) = \frac{\int \ldots \int p(Y,\theta/1) p(\theta_1) p(\theta_2) \ldots p(\theta_m)\, d\theta_1\, d\theta_2 \ldots d\theta_m}{p(Y/0)} \tag{3.35}$$

If the cost matrix also depends on θ, then the Bayes criterion for hypothesis H_1 is given by

$$\frac{\int \ldots \int p(Y,\theta/1) \ldots p(\theta_1) p(\theta_2) \ldots p(\theta_m) C_m(\theta)\, d\theta_1\, d\theta_2 \ldots d\theta_m}{p(Y/0) C_f} > \frac{P_0}{P_1}$$

But with cost functions independent of θ, the decision rule is reduced to

$$\lambda_{av}(Y) > \lambda_t = P_0 C_f / P_1 C_m \ldots \text{ for } H_1$$

$$\text{and } H_0 \text{ otherwise} \tag{3.36}$$

If the pdf's $p(\theta_1), p(\theta_2), \ldots, p(\theta_m)$ are not known, then the least favourable distribution which maximizes the Bayes minimum $\bar{C}_{min}$ is usually used (somewhat similar to minimax criterion). In using Neyman-Pearson criterion, for a given $P_f = \int_{R(1)} p(Y/0)\, dY$, the miss probability $P_m(\theta) = \int_{R(0)} p(Y, \theta/1)\, dY$ is minimized. In some simple cases, $P_m(\theta)$ is independent of θ and the same decision surface D is optimum for all θ_m's and their prior distributions $p(\theta_m)$. The test is now said to be 'uniformly most powerful'.

Example 3.6: Consider the extension of Example 3.3, where $A_0 = 0$ and $A_1 = A(\theta)$, uniformly distributed over 0 to A_{mx}. It is required to obtain the optimum threshold for the receiver. Assuming the cost functions $C_{11} = C_{00} = 0$, $C_{10} = C_{01} = 1$, and independent of θ, the average LR is, from eq. (3.35):

$$\lambda_{1,av}(Y) = \int_0^{A_{mx}} \exp\left[\frac{A(\theta)}{\sigma^2}\sum_{j=1}^{n} y_j - \frac{nA^2(\theta)}{2\sigma^2}\right] p(\theta)\, d\theta,\; p(\theta) = \frac{1}{A_{mx}}$$

$$> \lambda_t = P_0/P_1 \text{ for } H_1 \tag{3.37}$$

and

$$p(Y/0) = \left(\frac{1}{2\pi\sigma^2}\right)^{n/2} \exp\left[-\sum_{j=1}^{n} \frac{y_j}{2\sigma^2}\right]$$

The decision rule according to Bayes criterion is

$$\bar{m}_y = \frac{1}{n}\Sigma y_j > y_0 \text{ for } H_1$$

and H_0 otherwise, where y_0 is related to λ_t through eq. (3.37). Then, the error probabilities are:

$$P_f = \int_{y_0}^{\infty} p(Y/0)\, dY$$

and

$$P_m = \int_{-\infty}^{y_0} p(Y/1)\, dY$$

where both P_f and P_m are functions of $A(\theta)$.

For the Neyman-Pearson test, the given value of P_f is independent of $A(\theta)$ and the threshold y_0 is calculated from $p(Y/0)$ only. As such the decision threshold is independent of the true mean $\overline{A}(\theta)$ under the hypothesis H_1. The test is therefore 'uniformly most powerful' and can be used without the knowledge of the actual value of the signal $A(\theta)$.

3.2 MF RECEIVER USING MAXIMUM SNR CRITERION [7, 8]

In radar and digital communication systems, the basis problem is to optimally detect known signals, i.e., deterministic time functions of finite duration T, corrupted by random noise. One useful criterion for the optimum receiver is to maximize the output signal-to-noise ratio (SNR) at time T. The maximum-SNR criterion has been used for all analog communication systems and is directly related to the various criteria discussed in Sec. 3.1, since the visibility of all signals improves with the increased SNR at the receiver output. The optimum receiver for the above criterion is shown to be a Matched Filter (MF) or equivalently a correlator, as discussed below.

Consider that the received signal and the output of a linear filter (used as the MF) are given by

$$y(t) = s(t) + n(t)$$

$$y_0(t) = y_s(t) + y_n(t) \tag{3.38}$$

where $y_s(t)$ is the signal component and $y_n(t)$ is the noise component of the output. Then the output SNR at time $t = t_0$ is given by

$$(S_0/N_0) = y_s^2(t_0)\Big/\overline{y_n^2(t_0)} \tag{3.39}$$

which is to be maximized. Since $y_n(t)$ is a random variable, the mean noise power $\overline{y_n^2(t_0)}$ has been used in eq. (3.39). Assuming the impulse response of the linear filter to be $h(t)$ and the Fourier transforms of $s(t)$ and $h(t)$ to be $S(j\omega)$ and $H(j\omega)$ respectively, the following relations can be obtained:

$$\begin{aligned} y_s(t) &= s(t) * h(t) \text{ [convolution integral]} \\ &= F^{-1}[S(j\omega)\cdot H(j\omega)],\ F^{-1}[\ ] = \text{inverse Fourier Transform} \\ &= \frac{1}{2\pi}\int_{-\infty}^{\infty} H(j\omega)\cdot S(j\omega)e^{j\omega t}\, d\omega \end{aligned}$$

Therefore,

$$y_s(t_0) = \frac{1}{2\pi}\int_{-\infty}^{\infty} H(j\omega)\cdot S(j\omega)e^{j\omega t_0}\, d\omega \tag{3.40}$$

The mean square value of $y_n(t)$ is given by

$$\overline{y_n^2(t)} = \frac{1}{2\pi}\int_{-\infty}^{\infty} N(\omega)\left|H(j\omega)\right|^2 d\omega = \overline{y_n^2(t_0)}, \tag{3.41}$$

where $\overline{y_n^2(t)}$ is independent of t, and $N(\omega)$ = power spectral density of $n(t)$.

Assuming that $N(\omega) = n_0/2$ = two-sided spectral density of white noise, and the signal energy $E_s = \int_{-\infty}^{\infty} s^2(t)\,dt = \frac{1}{2\pi}\int_{-\infty}^{\infty}|S(j\omega)|^2\,d\omega$, the output SNR is now given by

$$\frac{S_0}{N_0} = \frac{|y_s(t_0)|^2}{y_n^2(t_0)} = \frac{\left|\int_{-\infty}^{\infty} H(j\omega)S(j\omega)e^{j\omega t_0}\,d\omega\right|^2}{\pi n_0 \int_{-\infty}^{\infty}|H(j\omega)|^2\,d\omega} \tag{3.42}$$

where $y_s^2(t_0) = |y_s(t_0)|^2$, since $y_s(t)$ is a real number. By applying Schwartz inequality*, the eq. (3.42) reduces to

$$\frac{S_0}{N_0} \le \frac{1}{\pi n_0}\int_{-\infty}^{\infty}|S(j\omega)|^2\,d\omega$$

and the maximum S_0/N_0 occurs when the inequality above becomes equality. Thus $(S_0/N_0)_{\max}$ is

$$\left(\frac{S_0}{N_0}\right)_{\max} = \frac{1}{\pi n_0}\int_{-\infty}^{\infty}|S(j\omega)|^2\,d\omega = \frac{2E_s}{n_0} \tag{3.43}$$

and from eq. (3.42), this is true only when

$$H(j\omega) = k \cdot S^*(j\omega)e^{-j\omega t_0},\ k = \text{constant} \tag{3.44}$$

and in the time domain, the impulse response of the filter

$$h(t) = k \cdot s(t_0 - t) \tag{3.45}$$

For signals of duration $(0, T)$, $h(t) = k \cdot s(T - t)$, as the filter output will be maximum at $t_0 = T$ and the filter is physically realizable only if $t_0 \ge T$. It is thus seen that the impulse response of the optimum filter giving maximum SNR at the output is the mirror image of $s(t)$ shifted in time by T seconds. Such a filter is called a Matched filter and is the core of all optimum detectors. Further in the time domain, the filter output

$$y_0(t) = y(t) * h(t) = y(t) * s(t_0 - t),\ k = 1$$

$$= \int_{-\infty}^{\infty} y(\tau)s(\tau + t_0 - t)\,d\tau$$

---

*Schwartz inequality states that if $F_1(j\omega)$ and $F_2(j\omega)$ are complex functions, then

$$\left|\int_{-\infty}^{\infty} F_1(j\omega)\cdot F_2(j\omega)\,d\omega\right|^2 \le \int_{-\infty}^{\infty}|F_1(j\omega)|^2\,d\omega \int_{-\infty}^{\infty}|F_2(j\omega)|^2\,d\omega$$

and the equality holds only if

$$F(j\omega) = kF_2^*(j\omega),\ k = \text{arbitrary constant.}$$

and

$$y_0(t_0) = \int_{-\infty}^{\infty} y(\tau)s(\tau)\,d\tau = \int_0^{t_0} y(t)s(t)\,dt \qquad (3.46)\dagger$$

when decisions are taken at $t = t_0$. Equation (3.46) clearly indicates that the MF output $y_0(t_0)$ is also the output of a 'time correlator', as is often used in optimum receivers.

The signal and noise components of $y_0(T)$ at the filter output are obtained by substituting eq. (3.44) in eqs. (3.40) and (3.41), giving

$$y_s(T) = \frac{1}{2\pi}\int_{-\infty}^{\infty} |S(j\omega)|^2\,d\omega = E_s$$

and

$$\overline{y_n^2(T)} = \frac{n_0}{2}\frac{1}{2\pi}\int_{-\infty}^{\infty} |S(j\omega)|^2\,d\omega = \frac{n_0 E_s}{2} \qquad (3.47)$$

and $(S_0/N_0)_{max}$ is given by eq. (3.43).

Since $y_0(T)$ is a random variable, the mean and variance of $y_0(T)$ are given by

$$E\{y_0(T)\} = E_s$$

---

†Solutions (3.45) and (3.46) may also be derived through the time-domain relations given by

$$\frac{S_0}{N_0} = |y_s(t_0)|^2 / \overline{y_n^2(t_0)} = \frac{\left|\int_0^{\infty} h(\tau)s(t_0-\tau)\,d\tau\right|^2}{\frac{n_0}{2}\int_0^{\infty} h^2(\tau)\,d\tau} \qquad (3.42a)$$

where

$$\overline{y_n^2(t_0)} = E\left\{\int_0^{\infty}\int_0^{\infty} h(\tau)h(z)R_{nn}(z-\tau)\,dz\,d\tau\right\} \qquad (3.41a)$$

$$= \frac{n_0}{2}\int_0^{\infty} h^2(\tau)\,d\tau, \text{ since } R_{nn}(\tau) \text{ for white noise} = \frac{n_0}{2}\delta(\tau)$$

Equation (3.42a) is now maximized by minimizing $\overline{y_n^2(t_0)}$, subject to the constraint that $y_s(t_0)$ is constant. Using the variational method and forming the function $Q = \overline{y_n^2(t)} + \mu y_s(t_0)$, where μ is the Lagrangian multiplier, one obtains the minimum Q when

$$h(\tau) = -\frac{2\mu}{n_0}s(t_0-\tau) = ks(t_0-\tau) \qquad (3.45a)$$

which gives the same result for the impulse response of the MF as eq. (3.45). Equation (3.46) follows immediately. Equation (3.41a) may be used to obtain the impulse response of the MF for non-white random noise, and the result is

$$\int_0^{t_0} h(z)R_{nn}(\tau - z)\,dz = ks(t_0-\tau) \qquad (3.45b)$$

where $R_{nn}(\tau)$ = autocorrelation function of the non-white noise [3].

and

$$\sigma_{y_0}^2 = \text{VAR}\{y_0(T)\} = \frac{n_0 E_s}{2} \tag{3.48}$$

and if $n(t)$ is Gaussian, then $y_0(T)$ is also Gaussian.

To understand the basic principles of the MF receiver, consider Fig. 3.6, where the signal generating filter has the impulse response $s(t)$ and the MF has the impulse response $h(t)$ given by eq. (3.45) and shown in Fig. 3.6(b). Then the signal component at the output is symmetrical around t_0, since with the normalised variable $t' = (t - t_0)$,

$$y_s(-t') = \int_{-\infty}^{\infty} s(-\tau)\cdot s(-t'-\tau)\,d\tau = \int_{-\infty}^{\infty} s(t'-\tau')\cdot s(-\tau')\,d\tau' = y_s(t')$$

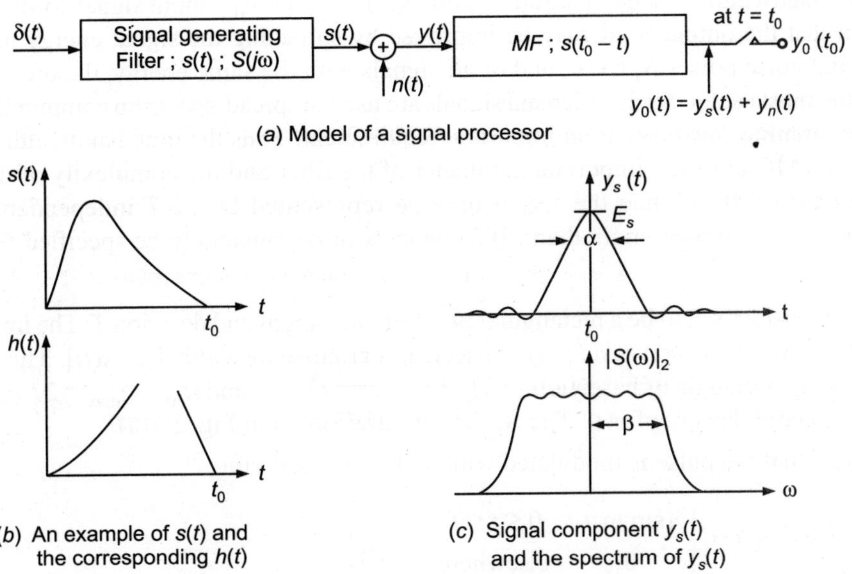

Figure 3.6: Properties of an MF illustrated

The output $y_s(t)$ with its peak E_s at $t = t_0$, and the spectrum of $y_s(t) = |S(j\omega)|^2$ are shown in Fig. 3.6(c). The effective widths of $y_s(t)$ and $|S(j\omega)|^2$, denoted by α and β, are bounded by the relation

$$\alpha\beta \geq \pi \tag{3.49}$$

But one would expect that the MF output should be an impulse of large height so that the signal detection and its parameter estimation (say the delay t_0) are possible with minimum error; thus, in general, $\alpha \geq \pi/\beta$ has to be very small, and β very large. To obtain an impulse at the filter output, the filter could be an inverse with $H(j\omega) = 1/S(j\omega)$, but the gain of such a filter would be very large as $\omega \to \infty$, since $s(t)$ would always be a bandlimited signal. Such an inverse characteristic would then increase $y_n(t)$ infinitely, and hence, the signal $y_s(t)$ would be lost in the noise. The compromise solution is then used, where the signal output is made as close to an impulse as possible, but the noise is suppressed outside the signal band. The transfer function $H(j\omega)$ of the MF should now have $|H(j\omega)| = |S(j\omega)|$; but the phase of $H(j\omega)$ should be inverse of that in $S(j\omega)$, as required by the inverse filter. Thus, $H(j\omega) = S^*(j\omega)$. The output peak

SNR $= 2E_s/n_0$ is now independent of peak power, time duration, bandwidth and waveshape, but depends only on the signal energy E_s. For the purpose of radar signal detection and binary communication, all waveforms having the same energy are equally effective, given that the noise is white and Gaussian and the MF receiver is used. For estimation of signal parameters and for optimum signal resolution, however, the nature of the waveform is also an important variable for optimum results, and various pulse-compression techniques and coded waveforms are used for such purposes.

The output SNR $= 2E_s/n_0$ has been calculated on the basis of white Gaussian noise of infinite bandwidth and power. But in cases where the disturbing noise is bandlimited and of fixed total power, the output SNR of an MF is shown to be

$$\left(\frac{S_0}{N_0}\right)_{max} = 2WT\left(\frac{S}{N}\right)_i = \frac{2E_sW}{N_i} \tag{3.50}$$

where W = MF bandwidth, T = signal duration, and $(S/N)_i = E_s/TWn_0$ = input signal-to-noise power ratio. It is then seen that the output SNR may be improved by spreading the signal energy more and more, keeping the total noise power N_i fixed; and of all signals with the same energy, the one with the largest bandwidth is the most useful. Such wideband signals are used in spread-spectrum communication systems, where the antijamming capability is an important requirement. Thus the time bandwidth product WT of the signal and its MF, is a very important parameter of the filter and the complexity of the filter is also given by the product WT. Since the signal may be represented by $2WT$ independent samples, the corresponding MF requires on more than $2WT$ elements or parameters to be specified for its complete synthesis.

Example 3.7: Consider $s(t)$ to be a rectangular pulse of unit height and duration T. The impulse response of the corresponding MF is $h(t) = s(T - t)$ = a rectangular pulse of width T [$= s(t)$]. The filter response $y_s(t)$ is a symmetrical triangle of basewidth = $2T$ and $y_s(T) = E_s = T$; and $(S_0/N_0)_{max} = 2T/n_0$, where $n_0/2$ is the two-sided spectral density of $n(t)$. The waveforms are shown in Fig. 3.7(a).

Consider now that the pulse is modulated with an RF signal giving,

$$s(t) = \begin{cases} A\sin \omega_0 t, & 0 \le t \le T \\ 0 & \text{elsewhere} \end{cases}$$

The corresponding MF has the impulse responses $h(t)$ as

$$h(t) = \begin{cases} A\sin [\omega_0 (T - t)], & 0 \le t \le T \\ 0 & \text{elsewhere} \end{cases}$$

The filter response $y_s(t)$ is given by (neglecting second harmonics for large ω_0),

$$y_s(t) = \begin{cases} \dfrac{A^2 t}{2} \cos [\omega_0 (T - t)], & 0 \le t \le T \\ \dfrac{A^2}{2}(2T - t) \cos [\omega_0 (T - t)], & T \le t \le 2T \\ 0 & \text{elsewhere} \end{cases} \tag{3.51}$$

and

$$y_s(T) = \frac{A^2T}{2} \text{ and } \left(\frac{S_0}{N_0}\right)_{max} = \frac{A^2T}{n_0} \tag{3.52}$$

Equation (3.51) along with $s(t)$ is plotted in Fig. 3.7(b) and it is seen that the contour of the filter output is the same as that of Fig. 3.7(a). It should be noted that the output of a correlator for the RF pulse would be approximately the same as that of the Fig. 3.7(a) and $\cos \omega_0 t$ term of eq. (3.51) will not be present. The outputs of the MF and the correlator are the same only at $t = T$, and not at other values of t.

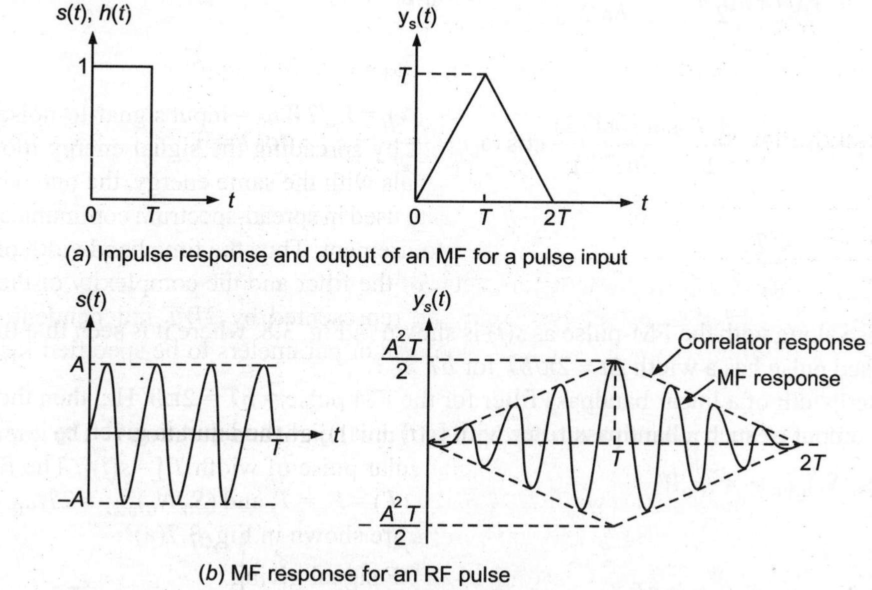

Figure 3.7: Two examples of MF responses: **(a)** Impulse response and output of an MF for a pulse input, **(b)** MF response for an RF pulse

Example 3.8: To obtain better resolution of radar echoes, FM-pulse compression technique has been widely used. The transmitted pulse is of the form,

$$s(t) = \begin{cases} A\cos(\omega_0 t + \frac{1}{2}bt^2), & |t| \le T/2 \\ 0, & \text{elsewhere} \end{cases}$$

The corresponding MF has the impulse response (neglecting the delay),

$$h(t) = A\cos[\omega_0(-t) + \tfrac{1}{2}b(-t)^2], |t| \le T/2$$

$$= A\cos(\tfrac{1}{2}bt^2 - \omega_0 t)$$

An approximate linear MF is given by the transfer function,

$$H(j\omega) = A(\omega)e^{j\beta(\omega)}$$

where $A(\omega)$ is rectangular bandpass and of width = bT,

$$\beta(\omega) = -\omega T_d + \frac{(\omega_i - \omega)^2}{2b},$$

T_d = filter delay, and the quadratic term compensates for the phase-frequency variation of $s(t)$.

The MF response $y_s(t)$ is given by (neglecting the second harmonics)

$$y_s(t) = \frac{A^2}{2} \frac{\sin[bt(T-|t|)]/2}{bt/2} \cos \omega_0 t, \quad |t| \le T$$

and for $t \to 0$,

$$y_s(t \to 0) = \frac{A^2 T}{2} \frac{\sin(btT/2)}{(btT/2)} \cos \omega_0 t \simeq \frac{A^2 T}{2} = E_s \quad \text{for } t = 0$$

and $$\left(\frac{S_0}{N_0}\right)_{\max} = \frac{A^2 T}{n_0} \tag{3.53}$$

The MF response along with the FM-pulse as $s(t)$ is shown in Fig. 3.8, where it is seen that the envelope of the compressed pulse has a width $\Delta t \simeq 2\pi/bT$ for $bT \gg 1$.

Since the bandwidth of a linear bandpass filter for the FM pulse = $bT = 2\pi W$ Hz, then the peak SNR obtained at the output of such a bandpass filter with $[s(t) + n(t)]$ as the input is given by

$$\max (S_0/N_0)_{\text{BPF}} = A^2/n_0 W$$

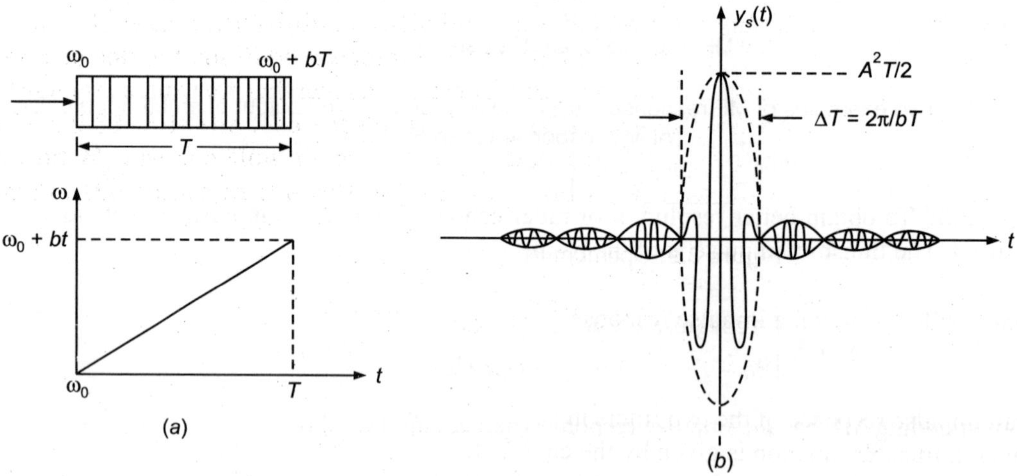

Figure 3.8: *(a)* FM-pulse, ***(b)*** MF-response for the FM pulse

whereas, max $(S_0/N_0)_{\text{MF}} = A^2T/n_0$. Thus the MF has an SNR gain of WT, with reference to a simple bandpass filter (cf. Eq. (3.50)). This gain is called the 'Processing gain of an MF' for coded input signals, and depends only on the time-bandwidth product of the MF.

3.2.1 MF for Non-white Noise

For non-white noise $n(t)$ disturbing the channel, the solution for the appropriate MF is obtained through the integral eqs. (3.41a) and (3.45b) (these are Fredholm integral equations of the first kind). An approximation to the optimum solution of the equations may be obtained by allowing the limits of eq. (3.45b) to extend upto $\pm\infty$ (assuming that a large delay is permissible in the computation) and using convolution. Then, eq. (3.45b) is modified as

$$\int_{-\infty}^{\infty} h(\tau)\, R_{nn}(t-\tau)\, d\tau = s(t_0 - t), -\infty < t < \infty,$$

Taking Fourier transform of both sides, the equation reduces to

$$H(j\omega) \cdot N(j\omega) = S^*(j\omega) \exp(-j\omega t_0)$$

Therefore,

$$H(j\omega) = \frac{S^*(j\omega)\exp(-j\omega t_0)}{N(j\omega)} \tag{3.54}$$

Based on eq. (3.54), the optimum MF is now realized in two parts, the first one is a pre-whitening filter whose output has the modified signal component $s'(t)$ and the whitened noise $n'(t)$, and the second filter is the usual MF matched to the signal $s'(t)$ and the white noise $n'(t)$ as shwon in Fig. 3.9. From the results of the network theory, it is known that $N(j\omega)$ has to be a real function of ω and $N(\omega)$ can be factored as a product of two terms $N_1^+(j\omega)$ and $N_2^-(j\omega)$, such that $N_1^+(j\omega)$ has all its poles and zeros in the left half of the complex s-plane and $N_2^-(j\omega)$ has its poles and zeros in the right half of the plane. Then, $N_1^+(j\omega)$, is a realizable filter function; and the whitening filter has the transfer function of $1/N_1^+(j\omega)$ and its impulse response is $h'_\omega(t)$. The modified signal $s'(t)$ at the output of this filter is

$$s'(t) = s(t) * h'_\omega(t) \tag{3.55}$$

and the whitened noise has $N_\omega(j\omega) = N(j\omega)/N_1^+(j\omega)$.

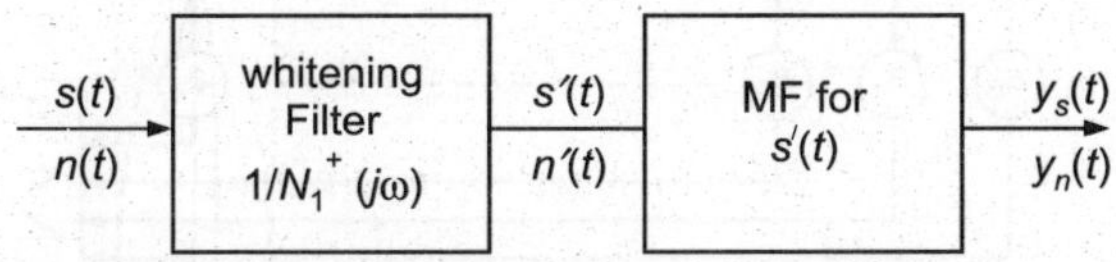

Figure 3.9: Optimum filter for non-white noise

The optimum MF has now the impulse response

$$h(t) = k \cdot s'(t_0 - t) \tag{3.56}$$

The overall impulse response of the two filters in cascade is now given by the convolution $[h'_\omega(t) * h(t)]$, and the overall transfer function is given by the eq. (3.54).

Example 3.9: Consider that the spectrum of the channel noise is RC-shaped, giving $N(\omega) = \dfrac{K}{\omega^2 + \omega_0^2}$.

Then the whitening filter has the transfer function $\dfrac{1}{N_1^+(j\omega)} = \dfrac{(j\omega + \omega_0)}{\sqrt{K}}$. The signal output for this filter $s'(t) = \dfrac{1}{\sqrt{K}}\left\{\omega_0 s(t) + \dfrac{d}{dt}s(t)\right\}$. The impulse response of the MF is then

$$h(t) = \frac{1}{\sqrt{K}}\left\{\omega_0 s(t_0 - t) + \frac{d}{dt}s(t_0 - t)\right\}$$

A corresponding filter network is then designed to match the above $h(t)$.

3.2.2 Practical MF

The simplest MF required in a radar receiver is the IF filter used for the reception of isolated rectangular RF pulses. Since a rectangular RF pulse has sin x/x type of spectral distribution and it is difficult to design a practical filter with such $H(j\omega)$, the practical IF filters generally have either Gaussian, approximately rectangular or tuned-circuit type of $H(j\omega)$. The loss in the peak output SNR of these filters are within 1 dB only as compared to the ideal MF, and a few (5 to 6) cascaded single-tuned IF stages give the best performance. For the pulse-compression filter, using linear FM pulses as the signalling waveform (cf. Ex. 8), a number of bridged-T sections of all pass networks are connected in cascade to form the overall MF.

In consideration of the synthesis of more elaborate MF's as required to optimize the accuracy, resolution and ambiguity of the detected signals (discussed later), the constraints of physical realizability and practical hardware have to be kept in view. As such, the usual approach in the MF design has been to synthesize the signalling waveforms, as required to meet the design specifications, through a cascaded network, e.g., delay lines and all pass networks; and to use the same in a complimentary way to receive the signals as well. Because of the complementary functions of signal generating and signal reception being done by the same network, the problems of realizability and practicality are automatically solved.

A waveform synthesizer using a tapped-delay line (TDL) is shown in Fig. 3.10, where an impulse excites the TDL and the tap outputs are weighted by factors, $\{a_0, a_1, \ldots, a_n\}$ and the summed output is a series of impulses given by

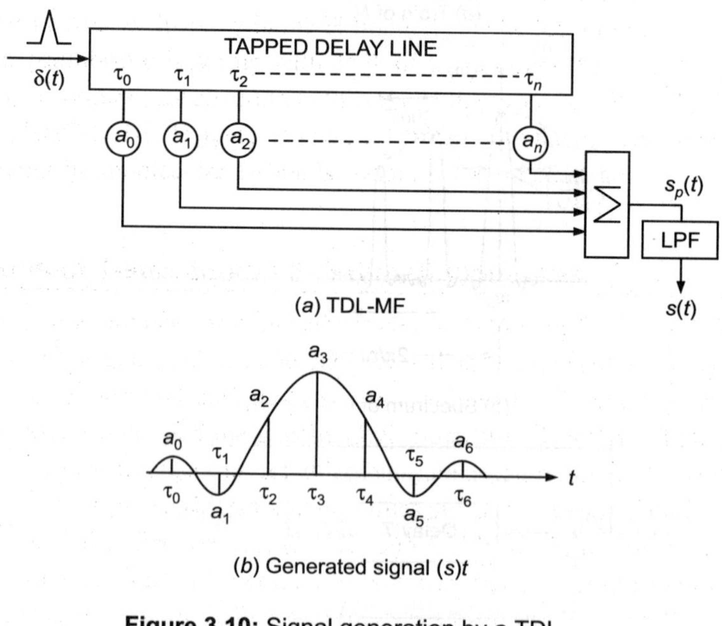

Figure 3.10: Signal generation by a TDL

$$s_p(t) = a_0\delta(t - \tau_0) + a_1\delta(t - \tau_1) + \ldots + a_n\delta(t - \tau_n) \tag{3.57}$$

where $\tau_0, \tau_1, \ldots, \tau_n$ are tap delays. The lowpass filter smooths the output and $s(t)$ is a lowpass signal as shown in Fig. 3.10(b). For the purpose of matched-filtering, the signal $s(t)$ is fed at the right hand side of the TDL, and the summed output will be a sharp pulse, giving the cross-correlated output $y_s(t_0)$ of eq. (3.40). A bandpass signal, e.g., a series of RF pulses, may be generated by modulating a carrier by $s(t)$, but the corresponding MF will have a demodulator plus the TDL as discussed above. A biphase modulated signal may also be generated by using a TDL and a multiplier at the output. However, a TDL, either in the form of an LC-delay line or a charge-coupled device (CCD), is a lowpass MF/waveform generator, and a coherent demodulator is necessary to down convert the RF/IF signals for necessary processing. A surface-acoustic wave (SAW)-TDL, however, can simulate an MF at IF/RF frequencies. For processing of digital signals, it is more appropriate and economical to use digital MF's using shift-registers as the delay units [9].

Example 3.10: Consider that a radar signal consists of N-RF pulses as shown in Fig. 3.11(a). These pulses could be generated by a TDL of Fig. 3.10, where the input is a dc pulse of width Δt and the delays $\tau_0, \tau_1, \ldots, \tau_N$ are multiples of T. The dc pulses now switch an RF source giving the desired signal $s(t)$. The MF for this signal could be of several types.

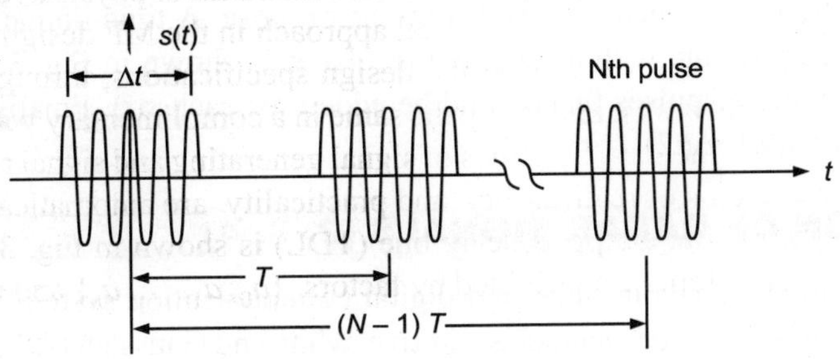

(a) Train of N-RF pulses

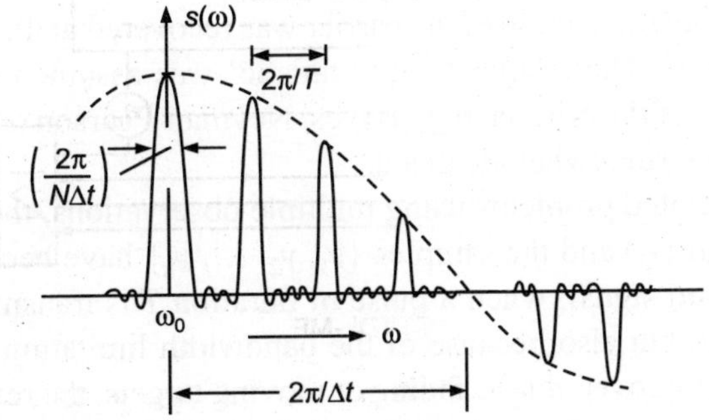

(b) Spectrum of the pulses

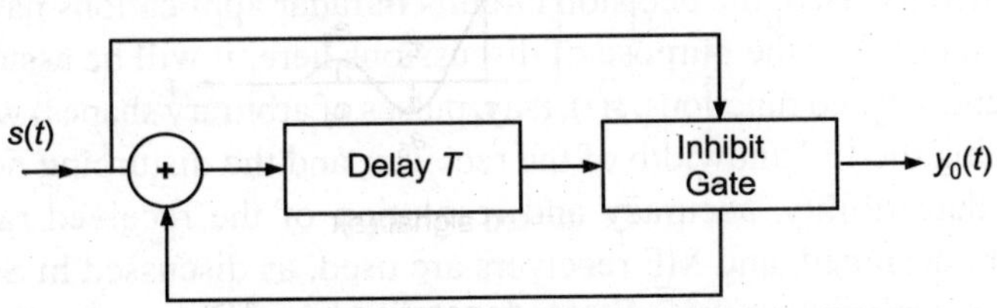

(c) An active comb filter

Figure 3.11: Processing of a train of RF pulses

(a) The same TDL may be used in a complementary way to generate the MF output $y_s(t_0)$.

(b) A two section MF may be used where the first MF_1 matches each of the RF pulses (cf. Ex. 7) to give a sharp spike at its output (assuming that $\Delta t < T/2$), and the second MF_2 now sums up all the N spikes to give a very large spike at the final output.

Assuming for each of the pulses,

$$s_1(t) = \begin{cases} A\cos\omega_0 t, & |t| \le \Delta t/2 \\ 0, & \text{elsewhere,} \end{cases}$$

the output of MF_1 has the peak value of $A^2 \cdot \Delta t/2$; and the final output from MF_2 will have a peak value of $(NA^2 \cdot \Delta t/2)$.

$$\text{The corresponding } (S_0/N_0)_{max} = (NA^2 \cdot \Delta t/n_0) \tag{3.58}$$

(c) The spectrum of the train of pulses is a series of narrow bands around ω_0 in the frequency domain as shown in Fig. 3.11(b). A sin x/x-weighted comb-filter is now a suitable signal processor, and various such filters have been designed from the purpose. A time-domain approximation of the comb-filter is a delay line with as positive feedback as shown in Fig. 3.11(c). The inhibit gate controls the number of pulses that should be added for each $s(t)$. Finally, one obtains the peak pulse of height = $(NA^2 \cdot \Delta t/2)$.

3.3 DETECTION OF RADAR SIGNALS [2, 3, 4]

In general, the transmitted signals in radar and digital communication systems are in the form of RF pulses and the disturbing noise in the channel is additive, white and Gaussian (AWGN). A simple receiver in such cases consists of RF/IF amplifier, detector, sampler, summer (when multiple samples are used for improving the detectibility of signals), and a decision circuit. In Ex. 1, 2 and 3, it was assumed that coherent detection was possible (i.e., phase of the carrier was recovered at the receiver) and the baseband signal had Gaussian properties. The resultant 'test statistic' was assumed Gaussian and the decision threshold arrived at using one of the criteria, e.g., Bayes, Neyman-Pearson or Minimax. Thus the results of eq. (3.25) and eq. (3.26) are somewhat idealized.

In Section 3.1.4 and the related problems using multiple observations, the 'test statistics' have been based on the sample values of $y(t)$ and the samples $\{y_1, y_2, \ldots, y_n\}$ have been assumed independent. In practice, however, the received signal, when a pulse of duration T is transmitted, will be distorted not only due to the noise present, but also because of the bandwidth limitation in the channel, RF/IF and other stages of the receiver. Moreover, due to fading or moving targets, the reflected radar echoes will be received with varying amplitude A, phase ϕ and time delay τ. Unlike in digital communication, where coherent detection is generally used, the decision circuits in radar applications have to be optimised for unknown values of A, ϕ and τ. For the purpose of discussions here, it will be assumed that the received pulse signals are some band-limited functions, $s(t)$, (say, pulses of arbitrary shapes) which are representable by $2WT$ samples, $2W$ being the IF bandwidth of the receiver, and the disturbing noise is AWGN.

Further, to improve detectibility, accuracy and resolution of the received radar echoes, complex signalling waveforms are designed, and MF receivers are used, as discussed in Sec. 3.2. The optimum 'test statistics' for such waveforms, using coherent/incoherent detection and multiple observations will be discussed below.

3.3.1 Envelope Detection

Since coherent detection, using carrier synchronization, is difficult to achieve in many practical radar systems, it is necessary to develop the structures of optimum receivers assuming that the carrier phase is unknown. Reconsider the problem of Example 3.2, and assume that the received signal is of the form:

$$y(t) = \begin{cases} A\cos(\omega_0 t + \phi) + n(t) & \text{for } H_1 \\ n(t) & \text{for } H_0 \end{cases}$$

and the disturbing noise is Gaussian, zero mean with variance σ^2. Due to the uncertainty of phase ϕ, a linear envelope detector is used before forming the test statistic. Referring to Chapter 1, Section 1.7, it is known that the conditional pdf's of the envelope R after detection of a narrrowband IF signal are:

$$p(R/1) = \frac{R}{\sigma^2}\exp(-R^2/2\sigma^2)e^{-\gamma}I_0\left(\frac{AR}{\sigma^2}\right)$$

and (3.59)*

$$p(R/0) = \frac{R}{\sigma^2}\exp(-R^2/2\sigma^2)$$

where $I_0(x)$ is the modified Bessel function of the first kind and zeroth order with imaginary argument, and $\gamma = A^2/2\sigma^2$ = CNR at the receiver input. The likelihood ratio test is now given as

$$\lambda_1(R) = e^{-\gamma}I_0\left(\frac{AR}{\sigma^2}\right) > \lambda_t \quad \text{for } H_1 \tag{3.60}$$

Using Neyman-Pearson criterion, with a given value of P_f, the corresponding decision threshold R_0 is calculated from

$$P_f = \int_{R_0}^{\infty}\frac{R}{\sigma^2}\exp(-R^2/2\sigma^2)\,dR = \exp(-R_0^2/2\sigma^2) \tag{3.61}$$

The value of λ_t is then,

$$\lambda_t = \lambda_1(R_0) = e^{-\gamma}I_0\left(\frac{AR_0}{\sigma^2}\right) \tag{3.62}$$

and the corresponding decision rule is

Choose H_1 if $\lambda_1(R) > \lambda_1(R_0)$

Equivalently, choose H_1, if $R > R_0$; and H_0 otherwise.

Using R_0 as the detection threshold, the detection probability is given by

---

$$^*I_0(x) = \int_0^{2\pi}\exp[x\cos(\theta-\alpha)]\frac{d\theta}{2\pi} \tag{3.59a}$$

$$P_d = \int_{R_0}^{\infty} \frac{R}{\sigma^2} \exp(-R^2/2\sigma^2) e^{-\gamma} I_0\left(\frac{AR}{\sigma^2}\right) dR \tag{3.63}$$

This integral may now be evaluated in terms of the so called Q-function (of Marcum [10]), and P_d calculated for various values of γ. For large values of γ, $I_0(x)$ may be approximated as

$$I_0(x) \simeq \frac{e^x}{\sqrt{2\pi x}}, \quad x \gg 1,$$

then,

$$p(R/1) \simeq \frac{1}{\sqrt{2\pi}\sigma} \exp\left[\frac{-(R-A)^2}{2\sigma^2}\right], \gamma \gg 1$$

which is Gaussian. Then for $\gamma \gg 1$, the detection probability is approximated as

$$P_d \simeq \text{erf}\,[(A - R_0)/\sigma] \tag{3.64}$$

If coherent detection is used, then the 'test statistics' under both H_0 and H_1 are Gaussian (refer to Ex. 2). The decision threshold y_0, for Neyman-Pearson criterion, is now obtained from

$$P_f = \int_{y_0}^{\infty} \frac{1}{\sqrt{2\pi}\sigma} \exp[-y^2/2\sigma^2]\, dy = \text{erfc}\,(y_0/\sigma)$$

and the probability of detection P_d is

$$P_d = \text{erf}\,[(A - y_0)/\sigma] \tag{3.65}$$

For a given P_f, it is now possible to compare the results of coherent/envelope detection using eqs. (3.63) and (3.65). Actual calculations show that for the useful ranges of γ and P_d, the coherent detection gives the same P_d with a CNR approximately 1 dB less than that required for envelope detection [3]. When multiple observations are used, i.e., when a number of incoherent radar echoes are integrated to give better visibility, the pdf's of the sampled envelope are the same as in eq. (3.59) where R_j replaces R for the jth sample. Assuming independence of the samples of signal plus noise, the L.R. is obtained from the joint pdf of n-samples. Thus eq. (3.60) is modified as

$$\lambda_1(R) = \frac{\prod^{n} p(R_j/1)}{\prod^{n} p(R_j/0)} = e^{-n\gamma} \prod^{n} I_0(R_j A/\sigma^2),\ \gamma_j = \gamma \tag{3.66}$$

and $\lambda_1(R) > \lambda_t$ ... for hypothesis H_1.

The actual value of λ_t is obtained from the Neyman-Pearson criterion using the modified eq. (3.62). It is more appropriate to consider $\log_e [\lambda_1(R)]$ and the decision will be based on

$$\sum_{j=1}^{n} \log_e I_0(R_j A/\sigma^2) \geq n_r + \ln \lambda_t \tag{3.67*}$$

Moreover the log I_0-detector may be replaced by a simple diode-envelope detector which approximates the log I_0 characteristics (equivalent to a square law response) at low CNR values. (For multiple observations, it may be assumed that the CNR per pulse is rather small.) The optimum receiver is now of the form, shown in Fig. 3.12, where a diode-envelope detector and a summer/integrator are used to simulate the operation of eq. (3.67). The difference in performances of coherent vs. incoherent receivers with multiple-pulse integration will now be significant and will be further discussed later.

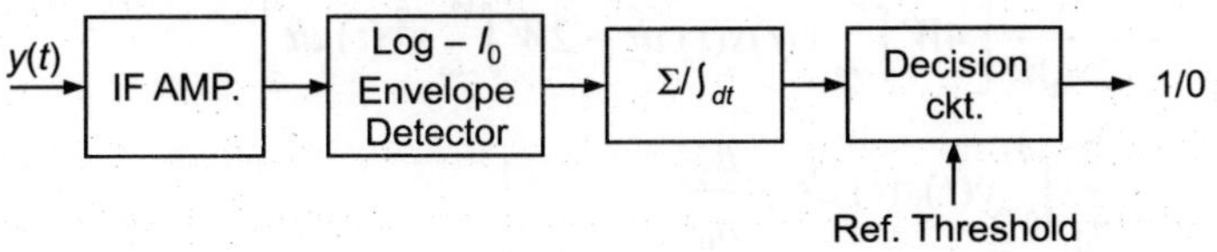

Figure 3.12: Optimum receiver for a train of incoherent pulses

3.3.2 Detection of Exactly Known Signals of Any Shape

Consider that a known complex waveform $s(t)$ is being transmitted and the received signal is:

$$y(t) = \begin{cases} s(t) + n(t) & \text{with signal present, } 0 < t < T \\ n(t) & \text{with no signal} \end{cases} \tag{3.68}$$

where $n(t)$ is zero-mean Gaussian with variance σ^2.

The $s(t)$ is now represented by its $2TW$ samples $\{s_1, s_2, \ldots, s_{2TW}\}$, and $y(t)$ by the samples $\{y_1, y_2, \ldots, y_{2TW}\}$. The conditional pdf's $p(Y/s)$ and $p(Y/N)$ are also represented in the same $2TW$ dimensional vector space. Assuming further that the samples are independent, the L.R. $\lambda_s(Y)$ is given by (similar to eq. (3.24)):

$$\lambda_s(Y) = \frac{p(Y/s)}{p(Y/N)} = \exp\left[\sum_{i=1}^{2TW} \frac{y_i^2}{2\sigma^2} - \sum_{i=1}^{2TW} \frac{(y_i - s_i)^2}{2\sigma^2}\right] \tag{3.69}$$

---

*A comparison of SNR characteristics of linear/quadratic envelope detectors shows that [12]:

$$\text{For } \gamma \gg 1, \text{ the output } S_0/N_0 \simeq \begin{cases} 2(C_i/N_i) & \text{for linear detector} \\ \frac{1}{2}(C_i/N_i) & \text{for quadratic detector} \end{cases}$$

and

$$\text{For } \gamma < 1,\ S_0/N_0 \simeq \begin{cases} 0.92(C_i/N_i)^2 & \text{for linear detector} \\ (C_i/N_i)^2 & \text{for quadratic detector} \end{cases}$$

Thus a linear detector is preferable for the detection of single pulses. However, for post-detection integration of multiple pulses, $\log_e I_0(x) \simeq x^2/4$, $x \ll 1$, and a quadratic detector with an adder simulates the fn. $\Sigma \log_e I_0(x)$. Also, for low CNR, both the detectors perform equally badly.

and the decisions are taken as:

$$\lambda_s(Y) > \lambda_t \text{ for hypothesis } H_s \tag{3.70}$$

and H_N otherwise.

For $TW \gg 1$, the summations may be replaced by integrals, and eqns. (3.69) and (3.70) are further simplified as:

$$\begin{aligned}\ln [\lambda_s(Y)] &= \frac{1}{2\sigma^2}[\Sigma 2 y_i s_i - \Sigma s_i^2] \\ &= \frac{1}{2n_0 W}\left[4W\int_{-\infty}^{\infty} y(t)s(t)\,dt - 2W\int_{-\infty}^{\infty} s^2(t)\,dt\right] \\ &= \frac{2}{n_0}\int_0^T y(t)s(t)\,dt - \frac{E_s}{n_0} \\ &> \ln \lambda_t \ldots \text{ for } H_s \end{aligned} \tag{3.71}*$$

Therefore,

$$U \triangleq \frac{2}{n_0}\int_0^T y(t)\,s(t)\,dt > \beta \triangleq \left(\frac{E_s}{n_0} + \ln \lambda_t\right), \text{ for } H_s, \tag{3.72}$$

where $E_s = \int_{-\infty}^{\infty} s^2(t)\,dt, \sigma^2 = n_0 W, \frac{n_0}{2}$ = two-sided noise spectral density.

The optimum receiver is then an IF cross correlator (or MF) followed by a sampler-comparator, as shown in Fig. 3.13, given that the signal is completely known in A, ϕ and τ. The values of λ_t and β depend on the criteria to be used, and for the Neyman-Pearson test, they are calculated from the given value of P_f. The test statistic U defined by eq. (3.72) is the MF output for $y(t)$ multiplied by $(2/n_0)$ and using eq. (3.48), the mean and variance of U are given by

$$E\{U\} = \frac{2E_s}{n_0}; \text{ and VAR } \{U\} = \frac{2E_s}{n_0}$$

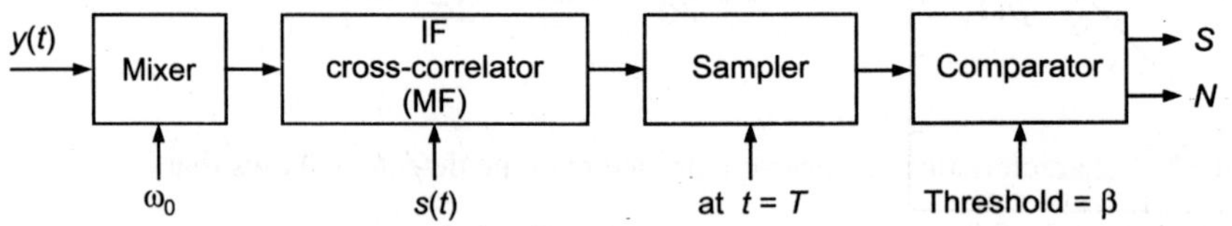

Figure 3.13: Optimum receivers for an exactly known signal

*The following two theorems of sampled signal space have been utilised in deriving eq. (3.71)

$$\left.\begin{aligned} &\text{(i) the signal energy } E = \int_{-\infty}^{\infty} x^2(t)\,dt = \frac{1}{2W}\sum_{1}^{2TW} x_n^2, TW \gg 1 \\ &\text{(ii) } \frac{1}{2W}\sum_{1}^{2TW} x_n \cdot y_n = \frac{1}{2W} X\cdot Y = \int_{-\infty}^{\infty} x(t)y(t)\,dt, TW \gg 1 \end{aligned}\right\} \tag{3.71a}$$

Then the conditional pdf's $p(U/N)$ and $p(U/s)$ are obtained as

$$p(U/N) = \sqrt{\frac{n_0}{4\pi E_s}} \exp\left[-\frac{n_0 U^2}{4E_s}\right]$$

and

$$p(U/s) = \sqrt{\frac{n_0}{4\pi E_s}} \exp\left[-\frac{n_0 (U - 2E_s/n_0)^2}{4E_s}\right] \tag{3.73}$$

The pdf's are Gaussian and the error probabilities are:

$$P_f = \int_{\beta}^{\infty} p(U/N)\, dU = \text{erfc}\left[\beta \cdot \sqrt{\frac{n_0}{2E_s}}\right]$$

and

$$P_m = \int_{-\infty}^{\beta} p(U/s)\, dU = \text{erfc}\left[\left(\frac{2E_s}{n_0} - \beta\right)\sqrt{\frac{n_0}{2E_s}}\right]$$

with

$$P_d = (1 - P_m) \tag{3.74}$$

The above pair of eqs. for P_f and P_d represents the operating characteristics of the receiver of Fig. 3.13, and the ROC is plotted in Fig. 3.14 for various values of $\mu = \sqrt{2E_s/n_0}$. Using Neyman-Pearson criterion,

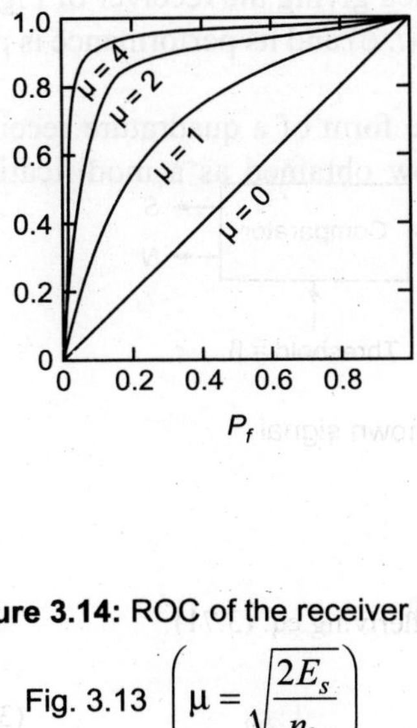

Figure 3.14: ROC of the receiver of Fig. 3.13 $\left(\mu = \sqrt{\frac{2E_s}{n_0}}\right)$

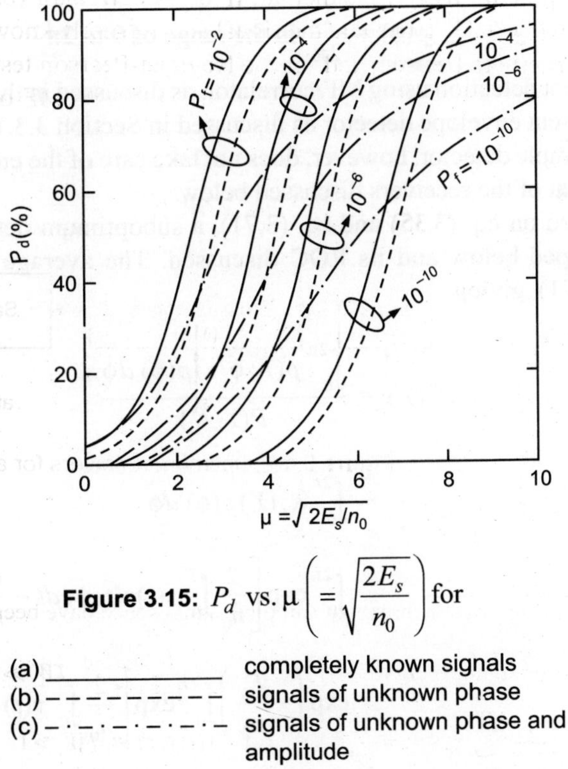

Figure 3.15: P_d vs. $\mu \left(= \sqrt{\frac{2E_s}{n_0}}\right)$ for

(a) ——————— completely known signals
(b) – – – – – – – signals of unknown phase
(c) – · – · – · – signals of unknown phase and amplitude

the variation of P_d with μ for a given P_f is shown in Fig. 3.15, along with other curves for signals with unknown parameters.

If, however, the received signals are of varying amplitudes, but of constant shape, i.e., $s_j(t) = A_j s_0(t)$, where $s_0(t)$ is of constant energy and shape, then the average L.R. $\lambda_s^{av}(Y)$ is to be evaluated, using eq. (3.35). This results into an equation similar to that of eq. (3.37) and the decision rule is

$$\lambda_s^{av}(Y) > \lambda_t$$

For Neyman-Pearson criterion, λ_t is calculated from P_f and is independent of the signal amplitude A_j, but P_d will be dependent on A_j.

3.3.3 Detection of Signals with Unknown Phase

Consider now that the transmitted signal $s(t)$ is known at the receiver, except for its phase ϕ. Then the received signal is of the form:

$$y(t) = \begin{cases} s(t, \phi) + n(t) & \text{with signal present, } 0 < t < T \\ n(t) & \text{with no signal} \end{cases} \tag{3.75}$$

and

$$s(t, \phi) = A \cos(\omega_0 t + \phi).$$

The RF phase ϕ randomly varies in radar echoes, because of the random movement of the target and the least favourable distribution of ϕ is given by

$$p(\phi) = \frac{1}{2\pi} \quad \text{in the range of 0 to } 2\pi \tag{3.76}$$

Coherent detection using MF/correlator, as discussed in the previous section, is not possible now, but the incoherent envelope detector as discussed in Section 3.3.1 may be used giving the receiver of Fig 3.12. This simple detector, however, does not take care of the envelope of $s(t, \phi)$ and its performance is poorer than that of the receivers discussed below.

Based on eq. (3.35) and eq. (3.71), a suboptimum detector in the form of a quadrature receiver is developed below and its ROC discussed. The average L.R. is now obtained as a modification of eq. (3.71), giving

$$\begin{aligned}
\lambda_s^{av}(Y) &= \frac{\int_0^{2\pi} p(Y; \phi/s) p(\phi)\, d\phi}{P(Y/N)} \\
&= \int_0^{2\pi} \lambda_s(Y) p(\phi)\, d\phi \\
&= \int_0^{2\pi} \exp\left[\frac{2}{n_0}\int_0^T y(t) s(t, \phi)\, dt - \frac{E_s}{n_0}\right] p(\phi)\, d\phi \\
&= \exp\left(-\frac{E_s}{n_0}\right) \int_0^{2\pi} \exp\left[\frac{2}{n_0}\int_0^T y(t)\, A\cos(\omega_0 t + \phi)\, dt\right] \frac{d\phi}{2\pi}
\end{aligned} \tag{3.77}$$

Now,

$$\int_0^T y(t)\, A \cos(\omega_0 t + \phi)\, dt$$

$$= \int_0^T y(t) A[\cos \omega_0 t \cos \phi - \sin \omega_0 t \sin \phi]\, dt$$

$$= A \cos \phi\, R \cos \theta_0 - A \sin \phi\, R \sin \theta_0,$$

$$= AR \cos(\phi + \theta_0), \tag{3.78}$$

where

$$R \cos \theta_0 = \int_0^T y(t) \cos \omega_0 t\, dt;$$

$$R \sin \theta_0 = \int_0^T y(t) \sin \omega_0 t\, dt$$

and

$$R^2 = (R^2 \cos^2 \theta_0 + R^2 \sin^2 \theta_0) \tag{3.79}$$

Substituting eq. (3.78) in eq. (3.77), the average L.R. is

$$\lambda_s^{av}(Y) = \exp\left(-\frac{E_s}{n_0}\right) \int_0^{2\pi} \exp\left[\frac{2AR}{n_0} \cos(\phi + \theta_0)\right] \frac{d\theta}{2\pi}$$

$$= \exp\left(-\frac{E_s}{n_0}\right) I_0\left(\frac{2AR}{n_0}\right), \text{ using eq. (3.59a)}$$

$$\geq \lambda_t \text{ for } H_s. \tag{3.80}$$

Since $I_0(x)$ is a monotonically increasing function of x, the decision rule may be based on the envelope R or the squared envelope R^2 as well, and the comparator evaluates

$$R^2 \geq \beta^2 \text{ for } H_s$$

Correspondingly,

$$\lambda_t = \exp\left(-\frac{E_s}{n_0}\right) I_0\left(\frac{2A\beta}{n_0}\right) \tag{3.81}$$

The receiver to obtain the test statistic R^2 follows from eq. (3.79) and is shown in Fig. 3.16, which is a quadrature receiver, using *I-Q* channels, a quadratic detector and a summer. The cross-correlators in the *I* and *Q* channels may be replaced by equivalent MF's matched to $\cos \omega_0 t$ and $\sin \omega_0 t$ respectively; and the resultant optimum receiver is shown in Fig. 3.17, where the same squared envelope R^2 is the test statistic.

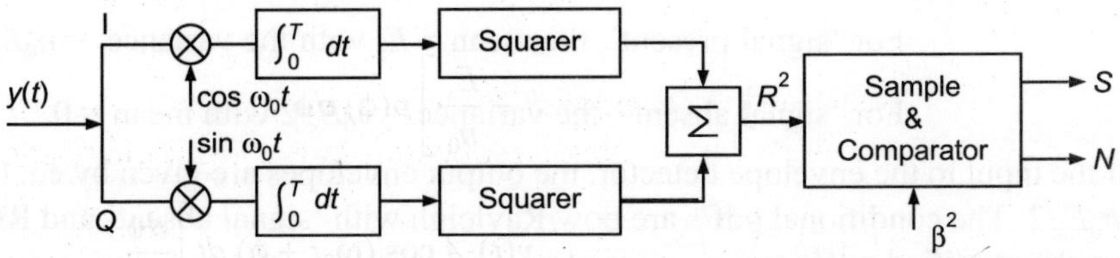

Figure 3.16: Optimum quadrature receiver for signals with unknown phase

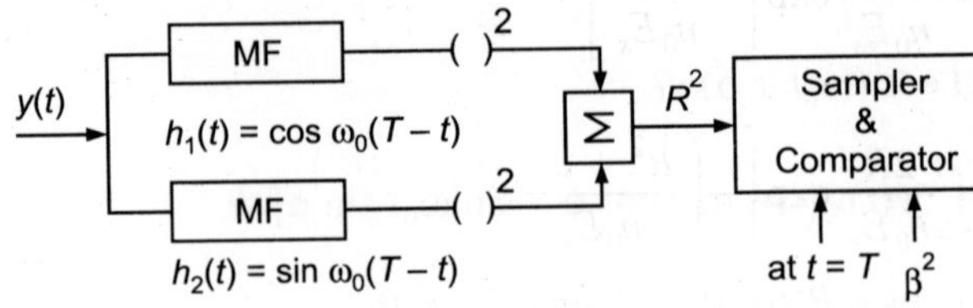

Figure 3.17: Equivalent receiver with MF

The receiver may be further simplified by using an incoherent MF and an envelope detector as shown in Fig. 3.18, where the MF is incoherent in the sense that it is matched to $s(t)$ except for phase ϕ. Choosing arbitrarily that $\phi = 0$, and using eq. (3.46) the MF output with $h(t) = \cos \omega_0 (T - t), 0 < t < T$, is

$$y_0(t) = \int_{-\infty}^{\infty} y(\tau) \cos \omega_0 (T - t + \tau)\, d\tau$$

$$= \cos \omega_0 (T - t) \int_0^T y(\tau) \cos \omega_0 \tau\, d\tau - \sin \omega_0 (T - t) \int_0^T y(\tau) \sin \omega_0 \tau\, d\tau \quad (3.82)$$

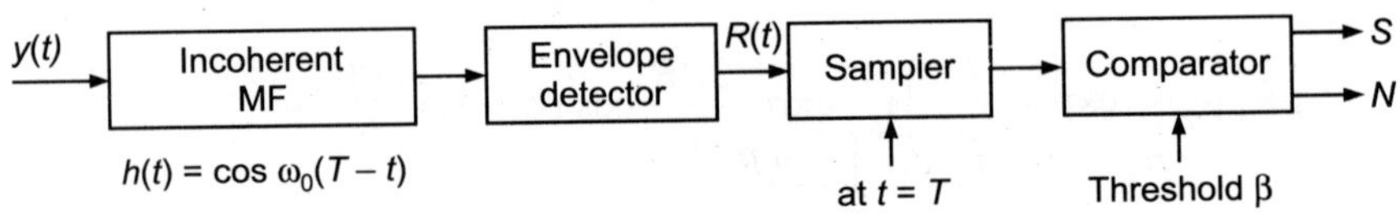

Figure 3.18: Receiver using incoherent MF and envelope detector

Therefore, the squared envelope of $y_0(t)$ at $t = T$ is

$$R^2 = \left[\int_0^T y(\tau) \cos \omega_0 \tau\, d\tau\right]^2 + \left[\int_0^T y(\tau) \sin \omega_0 \tau\, d\tau\right]^2 \quad (3.79)$$

which is the same as eq. (3.79). Since the phase is arbitrary, $h(t) = \sin \omega_0 (T - t)$ is also equally valid for the MF impulse response. Because of the phase uncertainty, an envelope detector is necessary after the incoherent MF and the envelopes obtained in the receivers of Figs 3.16, 3.17 and 3.18 are now the same.

If the envelope $A(t)$ of $s(t)$ is not a constant, but a function of t by design or distorted due to the channel characteristics, then this can be taken care of by suitably designing the MF impulse response as

$$h(t) = A(T - t) \cos \omega_0 (T - t), 0 < t < T, \quad (3.83)$$

and using the same receiver of Fig. 3.18 for optimum detection.

The conditional pdf's $p(R/N)$ and $p(R/s)$ of the test statistic R are now obtained by noting that the MF output $y_0(T)$ of Fig. 3.18 has the mean and variance given by eq. (3.48) and they are:

For 'signal present', the mean = E_s with the variance $= n_0E_s/2$

For 'signal absent', the variance = $n_0E_s/2$ with mean = 0.

With these at the input to the envelope detector, the output envelopes are given by eq. (3.59), where $A = E_s$ and $\sigma^2 = n_0E_s/2$. The conditional pdf's are now Rayleigh with 'signal absent' and Rician with 'signal present'. Thus the modified pdf's are:

$$p(R/N) = \frac{2R}{n_0 E_s} \exp\left[-\frac{R^2}{n_0 E_s}\right]$$

$$p(R/s) = \frac{2R}{n_0 E_s} \exp\left[-\left(\frac{R^2 + E_s^2}{n_0 E_s}\right)\right] I_0\left(\frac{2R}{n_0}\right) \tag{3.84}$$

It may be checked that for large CNR, i.e., for $(2R/n_0) \gg 1$, the Rician pdf $p(R/s)$ approximates to a Gaussian distribution as for eq. (3.59); and thus for $E_s/n_0 \gg 1$,

$$p(R/s) \simeq \frac{1}{\sqrt{\pi n_0 E_s}} \exp\left[-\frac{(R - E_s)^2}{n_0 E_s}\right] \tag{3.85}$$

The error probabilities are then calculated as:

$$P_f = \int_{\beta}^{\infty} p(R/N)\, dR = \exp\left[-\frac{\beta^2}{n_0 E_s}\right]$$

and (3.86)

$$P_d = (1 - P_m) = \int_{\beta}^{\infty} p(R/s)\, dR$$

The threshold value β is obtained from P_f for the Neyman-Pearson test and P_d is then evaluated using the so-called Q-function (of Marcum [10]). Equation (3.86) now completely specifies the ROC of the optimum receiver; and the variation of P_d with $\mu = \sqrt{2E_s/n_0}$ for given values of P_f is plotted in Fig. 3.15. It is seen that the receiver now requires higher signal energy E_s to achieve the same P_d as is obtained by coherent detection [Ref. eq. (3.74)]. The performance of the quadrature receiver or its equivalent is superior to that of the envelope detector of Sec. 3.3.1. The reason is that: the MF before the envelope detector has optimised the input CNR to the detector and thus the detector ouput has improved considerably and gives less P_f and P_m for the given E_s and n_0. However, the statistics given by eqs. (3.59) and (3.84) are of the same form.

3.3.4 Detection of Signals with Unknown Phase and Amplitude

For randomly moving targets, the received radar echoes will have both phase and amplitude varying with time. The signal to be processed is now given by

$$y(t) = \begin{cases} s(t; h, \phi) + n(t); & 0 < t < T \text{ for signal present} \\ n(t), & \text{for signal absent} \end{cases} \tag{3.87}$$

where, $s(t; h, \phi) = hA(t) \cos(\omega_0 t + \phi)$; and ϕ varies from sample to sample, but the observation time T is small compared to the fluctuation time of the amplitude variable $h(t)$, $h > 0$. It is assumed that $A(t)$, ω_0 and τ are known at the receiver. If the variation of h is small and the bias β of the comparator is adjusted according to the variation of h, then the optimum receiver structure is that of Fig. 3.18 with β adaptive. The test is uniformly powerful for all h and the detection system is similar to that used in a search radar system [2] .

In the general case with varying h, the average L. R. of eq. (3.80) is further modified as:

$$\lambda_s^{av}(Y) = \iint \lambda_s(Y) p(h) p(\theta)\, dh\, d\theta$$

$$= \int_h \exp\left(-\frac{h^2 E_s}{n_0}\right) I_0\left(\frac{2hAR}{n_0}\right) p(h)\, dh \tag{3.88}$$

Assuming a particular distribution $p(h)$, the integral of eq. (3.88) may be evaluated by using the Q-function. A reasonably good approximation of $p(h)$ is a Rayleigh distribution, given by

$$p(h) = \frac{h}{h_0^2} \exp\left(-\frac{h^2}{2h_0^2}\right), \quad h \geq 0 \tag{3.89}$$

where h_0 is the most probable value of h.
Using this $p(h)$ and normalising with $h_0 = 1$, $A = 1$, the eq. (3.88) is evaluated as:

$$\lambda_s^{av}(Y) = \int_0^{\infty} h \exp\left[-\frac{h^2}{2}\left(1 + \frac{2E_s}{n_0}\right)\right] I_0\left(\frac{2Rh}{n_0}\right) dh \tag{3.90}^*$$

$$= \frac{n_0}{(n_0 + 2E_s)} \exp\left[\frac{2R^2}{n_0(n_0 + 2E_s)}\right]$$

$$\geq \lambda_t \quad \text{for } H_s \tag{3.91}$$

*Introducing the variable h', given by:

$$\frac{h'^2}{2} = \frac{h^2}{2}\left(1 + \frac{2E_s}{n_0}\right) = \frac{h^2 \alpha^2}{2}$$

$$dh' = \alpha\, dh; \text{ and } I_0\left(\frac{2Rh}{n_0}\right) = I_0\left(\frac{h'}{\alpha}\frac{2R}{n_0}\right)$$

Therefore, eq. (3.90)

$$= \int_0^{\infty} \frac{h'}{\alpha} \exp\left[-\frac{h'^2}{2}\right] I_0\left(\frac{h'}{\alpha}\frac{2R}{n_0}\right) \frac{dh'}{\alpha}$$

$$= \frac{1}{\alpha^2} \exp\left[\frac{2R^2}{\alpha^2 n_0^2}\right] \int_0^{\infty} h' \exp\left[-\frac{h'^2}{2}\right] \exp\left[-\frac{2R^2}{\alpha^2 n_0^2}\right] I_0\left(\frac{h'}{\alpha}\frac{2R}{n_0}\right) dh'$$

$$= \frac{1}{\alpha^2} \exp\left[\frac{2R^2}{\alpha^2 n_0^2}\right] = \text{eq. (3.91)}$$

In obtaining the final result, the following relation has been used,

$$\int_0^{\infty} Z \exp\left(-\frac{\rho + Z^2}{2}\right) I_0\left(\sqrt{\rho} \cdot Z\right) dZ = 1, \tag{3.93}$$

where $Z = h'$, $\sqrt{\rho} = 2R/\alpha n_0$.

The decision rule now is

$$\lambda_s^{\text{av}}(Y) \geq \lambda_t \text{ for } H_s.$$

Equivalently, choose H_s if

$$R^2 \geq \frac{n_0(n_0+2E_s)}{2} \ln\left[\frac{\lambda_t(n_0+2E_s)}{n_0}\right] = \beta^2 \tag{3.92}$$

The optimum receiver is now seen to be the same as that for the signals with unknown phase, given in Figs. 3.16 and 3.18, since the receiver has to obtain the envelope R first for deciding on H_s or H_N.

For the Neyman-Pearson test, P_f is calculated from eq. (3.84) and is the same as given in eq. (3.86a)

$$P_f = \int_\beta^\infty p(R/N)\, dR = \exp\left[\frac{-\beta^2}{n_0 E_s}\right] \tag{3.86}$$

Based on the value of β for the given P_f, the average probability of detection is obtained by averaging the P_d of eq. (3.86) over the range of values of h, $h \geq 0$. Thus from eq. (3.84), the average P_d is:

$$\overline{P}_d = \int_h P_d \cdot p(h)\, dh$$

$$= \int_0^\infty \left[\int_\beta^\infty \frac{2R}{n_0 E_s} \exp\left\{-\frac{(R^2+E_s^2)}{n_0 E_s}\right\} I_0\left(\frac{2R}{n_0}\right) dR\right] h \cdot e^{-h^2/2}\, dh$$

$$= \exp\left[-\frac{\beta^2}{E_s(n_0+2E_s)}\right] \tag{3.94}$$

Further the average signal energy over many intervals of T is calculated as

$$E_{\text{av}} = \int_0^\infty \left(\frac{A^2 T}{2}\right) p(h)\, dh = 2E_s \text{ (normalised)}$$

Therefore,

$$\overline{P}_d = \left[\exp - \left(\frac{\beta^2}{n_0 E_s}\right)\right]^{(n_0/n_0+2E_s)} = (P_f)^{1/(1+\varepsilon)} \tag{3.95}$$

where $\varepsilon = E_{\text{av}}/n_0$. The ROC of the system is plotted in Fig. 3.15 and it is seen that P_d's now further deteriorate because of the amplitude fluctuation.

In a typical radar search problem, the arrival time τ of the echoes are unknown and the received signal is of the form:

$$h(t) = \begin{cases} s(t; h, \phi, \tau) + n(t), \tau_1 < t < (\tau_2 + T) & \text{for signal present} \\ n(t) & \text{for signal absent} \end{cases} \tag{3.96}$$

where $s(t; t, \phi, \tau) = h \cdot A(t-\tau) \cos [\omega_0(t-\tau)+\phi]$, and the observation interval, $\tau_1 < t < (\tau_2 + T)$, is large. The average L.R. is now calculated by further averaging eq. (3.88) or eq. (3.19) over the interval $\tau_1 < t < (\tau_2 + T)$. Thus the test statistic is obtained, using eq. (3.19), as:

$$\lambda_s^{av}(Y) = \int_{\tau_1}^{\tau_2+T} \left[\frac{n_0}{n_0 + 2E_s} \cdot \exp\left(\frac{2R^2}{n_0(n_0+2E_s)} \right) \right] p(\tau)\, d\tau \tag{3.97}$$

The optimum receiver now requires a nonlinear filter with a decision circuit; at the same time, P_d's obtained are small even for large E_s/n_0 because of the large observation interval. It is more appropriate to use 'maximum likelihood' detection technique for signals with unknown arrival time and this gives better values of P_d for a given CNR. This is discussed in Sec. 3.5.5.

3.3.5 Detection of Incoherent Signals by Multiple Observations

In radar applications, sometimes many pulses are transmitted towards the target and the received echoes are observed over a number of intervals, say n, the actual value of n depending on the rate of antenna scan and the repetition rate of the transmitted pulses. If the received echoes can be detected coherently then the results of Example 3.3 and eqs. (3.25) and (3.28) may be used for determining the ROC. But in practice, the echoes will be of varying phase and amplitude, as discussed in earlier sections; but they may be assumed to be statistically independent with narrow-band Gaussian noise disturbing the signals. The received echoes are then, of the form (for known τ):

$$s_j(t_j, \phi_j) = h_j A(t) \cos(\omega_0 t + \phi_j),\ 0 < t < T,\ 1 \le j \le n \tag{3.98}$$

Because of the statistical independence of s_j's, the average L.R. of the echoes is obtained from eqs. (3.66) and (3.88) as:

$$\lambda_s^{av}(Y) = \prod_{j=1}^{n} \exp\left(-\frac{h_j^2 E_s}{n_0} \right) I_0\left(\frac{2h_j A R_j}{n_0} \right) \tag{3.99}$$

Assuming h_j to be small in practical systems and using the approximation $I_0(x) \simeq \exp(x^2/4)$, the average L.R. is appproximated as

$$\ln \lambda_s^{av}(Y) = \sum_1^n \left[\frac{A^2 R_j^2}{n_0^2} - \frac{E_s}{n_0} \right] \cdot \int_0^\infty h_j^2 p(h_j)\, dh_j$$

$$= \overline{h^2} \sum_1^n \left[\frac{A^2 R_j^2}{n_0^2} - \frac{E_s}{n_0} \right] \ge \ln \lambda_t \text{ for } H_s \tag{3.100}$$

where $\overline{h^2} = \int h_j^2 p(h_j)\, dh_j$ = mean square expected signal amplitude. The decision rule is then,

$$Z \triangleq \sum_1^n R_j^2 > \frac{n_0^2}{A^2} \left[\frac{\ln \lambda_t}{\overline{h^2}} + \frac{E_s}{n_0} \right] = \beta_M \text{ for } H_s \tag{3.101}$$

The receiver now has to calculate the sum of the squared envelopes and its block diagram is shown in Fig. 3.19, which is similar to that of Fig. 3.12. The threshold β_M is calculated from the given P_f, using Neyman-Pearson test.

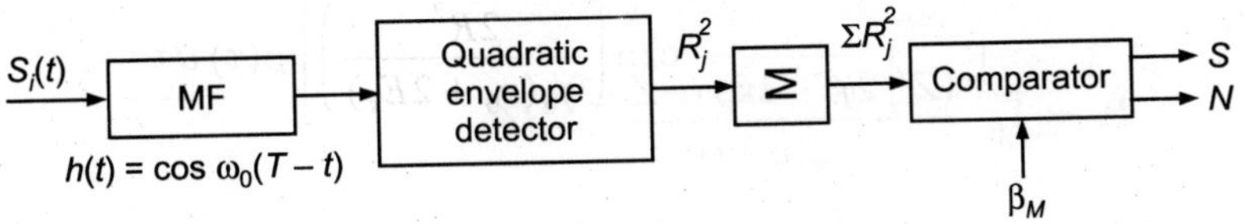

Figure 3.19: Optimum receiver for multiple signals with unknown h and ϕ for low CNR

For simplicity, the test statistic $Z = \sum_1^n R_j^2$ is normalised by dividing R_j^2 by the sample variance, giving

$$Z' = \sum_1^n R_j^2 \Big/ \left(\frac{n_0 E_s}{2}\right)$$

The pdf of Z′, with no signal present is the central χ^2-distribution with $2n$ degrees of freedom (since Z^2 is formed by I and Q channels) and is given by [3] (also refer to eq. (3.29)):

$$p(Z'/N) = \frac{1}{2^n \Gamma(n)} (Z')^{n-1} \exp\left(-\frac{Z'}{2}\right) \tag{3.102}$$

and the false-alarm probability is

$$P_f = \int_{\beta'_M}^{\infty} p(Z'/N)\, dZ', \quad \beta'_M = \beta_M \Big/ \left(\frac{n_0 E_s}{2}\right)$$

$$= 1 - I\left(\frac{\beta'_M}{2\sqrt{n}}, (n-1)\right) \tag{3.103}$$

where $I(u, P)$ is Pearson's form of the incomplete Gamma function and is tabulated in [11].

The pdf of Z' with signal present is the non-central χ^2-distribution with $2n$ degrees of freedom and has the non-central parameter

$$\nu = \frac{2nE_s}{n_0} = \text{sum of SNR's at the MF output.}$$

Then the pdf of (Z'/s) is [3]

$$p(Z'/s) = \tfrac{1}{2}\left(\frac{Z'}{\nu}\right)^{(n-1)/2} \cdot \exp\left[-\frac{Z'+\nu}{2}\right] I_{n-1}\{\sqrt{Z'\nu}\},\ Z' > 0$$

and the probability of detection is

$$P_d = \int_{\beta'_M}^{\infty} p(Z'/s)\, dZ' = Q_M\left[\left(\frac{2nE_s}{n_0}\right)^{1/2}, (\beta'_M)^{1/2}\right] \tag{3.104}$$

where $Q_M(\alpha, \beta)$ is the generalised Marcum Q-function, I_{n-1} is the modified Bessel function of order $(n-1)$.

The eqs. (3.103) and (3.104) have been further approximated for large values of n, and the following results have been obtained [2]:

$$P_f \simeq \int_{\beta'_M}^{\infty} \frac{(Z'/2)^{n-1} \exp(-Z'/2)}{2(n-1)!} dZ'$$

$$\simeq \frac{(\beta'_M/2)^n \exp(-\beta'_M/2)}{(n-1)!(\beta'_M/2 - n + 1)} \tag{3.105}$$

and

$$P_d \simeq \frac{1}{2} - \psi^{(-1)}(x) + \frac{2n + 3v}{6(n+v)^{3/2}} \psi^{(2)}(x) - \frac{n + 2v}{4(n+v)^2} \psi^{(3)}(x) \tag{3.106}$$

where $v = 2n\frac{E_s}{n_0}$, $x = \frac{(\beta'_M - 2n - v)}{\sqrt{2(n+v)}}$, and $\psi^{(j)}(x)$ is defined as (from the definition of Hermite polynomials)

$$\frac{d^j}{dx^j} \exp(-x^2/2) = \sqrt{2}\,\pi\psi^{(j)}(x)$$

but, $\psi^{-1}(x) = \frac{1}{\sqrt{2\pi}} \int_0^x \exp(-y^2/2)\, dy = [\tfrac{1}{2} - \text{erfc}(x)]$.

Using the above equations, the variation of P_d with the overall CNR $= \sqrt{v} = \sqrt{2nE_s/n_0}$ has been plotted in Fig. 3.20, for various values of n and $P_f = 10^{-6}$ and 10^{-10} [2]. It is seen that the total energy required to attain a given P_d increases with n, and it is more difficult to detect the signal when the total energy nE_s is divided among n incoherent pulses than when the energy is concentrated into a single pulse. But for a given average energy E_s, P_d increases with n.

For large values of n, say $n \geq 100$, eq. (3.106) for P_d can be further approximated by taking only the first two terms and the same eq. (3.106) may be used for evaluating P_f as well by setting $v = 0$. Thus for $n \geq 100$, P_f and P_d are given by

$$P_f \simeq \text{erfc}(x'), \text{ where } x' = \frac{(\beta'_M - 2n)}{2\sqrt{n}}$$

and

$$P_d \simeq \text{erfc}(x) \tag{3.107}$$

From these eqs., it may be shown that the minimum detectable signal $E_{\min}$ is inversely proportional to $\sqrt{n}$ and to reduce the value of $E_{\min}$ per pulse to 1/10th of its value for $n = 1$, it is necessary to take about 100 ($= n$) observations. It has also been evaluated that a linear detector is slightly better than the quadratic one for $1 \leq n \leq 70$ and the quadratic one is better only for $n > 70$ (due to the approximation of $\ln I_0(z)$ for small and large values of argument in eq. (3.100)). It has been assumed here that the arrival time τ is known for the echo signal. If, however, τ is unknown, then the 'maximum likelihood' detection will have to be used.

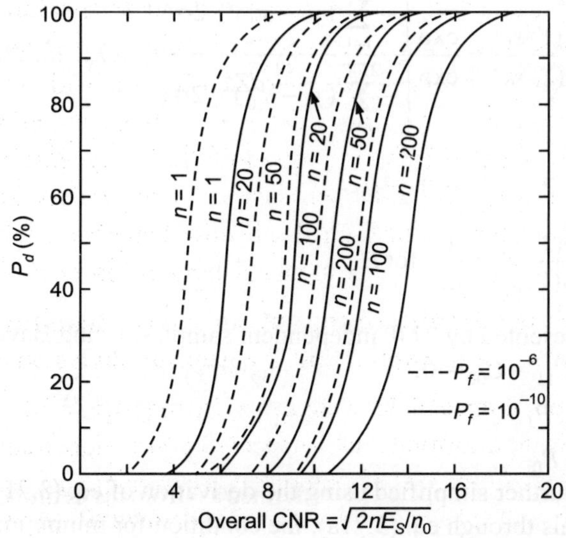

Figure 3.20: Variation of P_d with overall CNR for multiple obervations. Two sets of curves with $P_f = 10^{-6}$ and 10^{-10}

3.4 DETECTION OF SIGNALS IN DIGITAL COMMUNICATION

A simple example of binary communication has been discussed in Ex. 1 and 3, using idealised pulses. MF receivers with maximum SNR criterion has been discussed in Sec. 3.2, where the signalling waveform of the transmitter was optimally detected. In binary and M-ary communication systems, the signalling with two or more distinct waveforms has to be done, and it is necessary to optimize these waveforms for minimum error rate. The waveform design techniques are also relevant to the radar-signal design discussed later.

Most of the efficient digital communication systems utilize coherent detection using MF correlators but in some cases, incoherent detection is also used. As such binary ASK, FSK and PSK systems are widely used. M-ary ASK and PSK systems are used where the source has a large alphabet, but the channel bandwidth is restricted. When the available channel bandwidth is large, the more efficient technique of signalling is to use M-ary orthogonal codes, e.g., M-FSK and binary coded orthogonal signals. In this section, the performances of binary and M-ary signalling techniques are discussed and the M-ary orthogonal signals are shown to perform the best.

3.4.1 Detection of Two Exactly Known Signals

Consider that the two known waveforms $s_1(t)$ and $s_0(t)$ representing '1' and '0' are being transmitted, and the received signal is:

$$y(t) = \begin{cases} s_1(t) + n(t), & 0 < t < T \quad \text{for (1)} \\ s_0(t) + n(t), & 0 < t < T \quad \text{for (0)} \end{cases} \tag{3.108}$$

As an extension of the result of Sec. 3.3.2, for exactly known signals, the likelihood ratios for the bandlimited signals $s_1(t)$ and $s_0(t)$ corrupted by white Gaussian noise, are:

$$\lambda_{s1} = \frac{p(Y/s_1)}{p(Y/s_0)} = \frac{\exp\left[-\sum_{i=1}^{2TW} (y_i - s_{i1})^2/2\sigma^2\right]}{\exp\left[-\sum_{i=1}^{2TW} (y_i - s_{i0})^2/2\sigma^2\right]} > \lambda_t \quad \text{for } H_1$$

and

$$\lambda_{s2} = \frac{p(Y/s_0)}{p(Y/s_1)} > \frac{1}{\lambda_t} \quad \text{for } H_0 \tag{3.109}$$

where $s_1(t)$ and $s_0(t)$ are represented by $2TW$ independent samples. Using Bayes criterion for minimum error, with $C_f = C_m = 1$ and $C_{11} = C_{00} = 0$,

$$\lambda_t = P_0/P_1$$

where $p(s_1) = P_1$ and $p(s_0) = P_0$.

Equation (3.109) is now further simplified using the derivation of eq. (3.71) and converting the sum of sampled signals to integrals through eq. (3.71a), the condition for minimum error is now

$$\ln(\lambda_{s1}) > \ln(P_0/P_1) \quad \text{for } H_1$$

which further reduces to

$$\int_0^T y(t)(S_1 - S_0)\, dt > \left(\frac{n_0}{2}\ln\frac{P_0}{P_1} + \frac{E_1}{2} - \frac{E_0}{2}\right) = \beta \quad \text{for } H_1 \tag{3.110}$$

where, $y(t) = s_1(t) + n(t)$, $S_1 = s_1(t)$, $S_0 = s_0(t)$; $\Sigma s_{i1}^2 = 2WE_1$, $\Sigma s_{i0}^2 = 2WE_0$; and $\Sigma 2y_i(s_{i1} - s_{i0}) = 4W\int_0^T y(t)\,(S_1 - S_0)\, dt$.

The optimum receiver given by eq. (3.110) is a correlation receiver which may be implemented by either a pair of multiplier-integrator circuits or one/two MF's matched to $[s_1(t) - s_0(t)]$ signal, as shown in Fig. 3.21. The MF/correlator output is compared with β, and the decision is made in favour of the larger output. If $P_1 = P_0 = 0.5$ and $E_1 = E_0$, as is generally the case for binary signalling, the threshold β in eq. (3.110) would be zero.

To obtain the ROC of the optimum receiver, eq. (3.110) is rewritten as

$$(u/s_1) = \int_0^T n(t)(S_1 - S_0)\, dt > \left[\frac{n_0}{2}\ln\frac{P_0}{P_1} - \frac{1}{2}\int_0^T (S_1 - S_0)^2\, dt\right] \triangleq u_1$$

i.e.,

$$(u/s_1) > u_1 \text{ for } H_1 \tag{3.111}$$

where $\int_0^T s_1(t)(S_1 - S_0)\, dt = E_1 - \int_0^T s_1(t)s_0(t)\, dt$ has been shifted to the right of eq. (3.110) resulting in eq. (3.111). Since $n(t)$ is Gaussian with zero-mean, then u/s_1 is also Gaussina with zero-mean, and the variance σ_u^2 is obtained as the output of an MF matched to the signal $(S_1 - S_0)$ with $n(t)$ as the input. Referring to eq. (3.48),

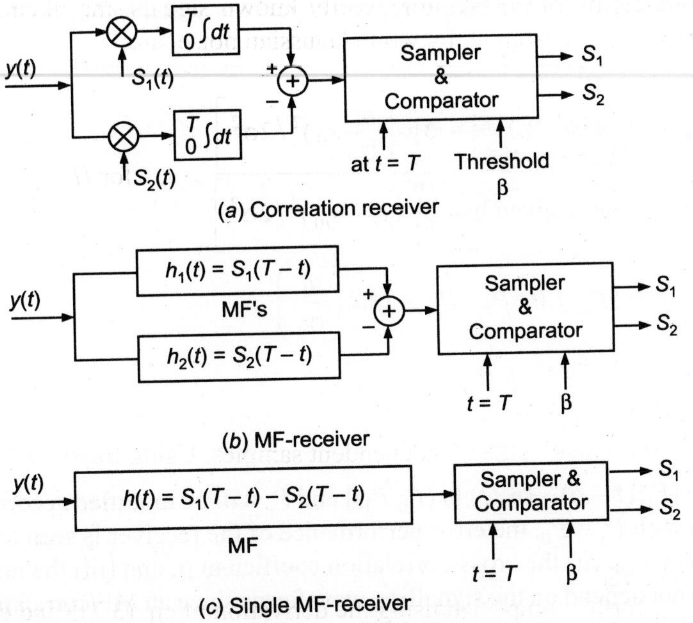

Figure 3.21: Alternative optimum receivers for two known signals: **(a)** Correlation receiver, **(b)** MF-receiver, **(c)** Single MF-receiver

$$\sigma_u^2 = \frac{n_0}{2}\int_0^T (S_1 - S_0)^2\, dt$$
$$= n_0[E_{av}(1-\rho)] \tag{3.112}$$

where $E_{av} = (E_1 + E_0)/2$ = average signal energy; and

$$\rho = \frac{1}{E_{av}}\int_0^T s_1(t)s_0(t)\, dt$$ = normalised correlation coefficient between the signals.

Then the pdf $p(u/S_1)$ is given by:

$$p(u/S_1) = \frac{1}{\sqrt{2\pi\sigma_u}} \exp\left(\frac{-u^2}{2\sigma_u^2}\right) \tag{3.113}$$

The error of deciding H_0 for H_1 occurs when $u/S_1 < u_1$ and this missprobability is given by

$$P_m = \int_{-\infty}^{u_1} p(u/S_1)\, du$$

In general for binary communication, $P_0 = P_1 = 0.5$, and

$$u_1 = -\tfrac{1}{2}\int_0^T (S_1 - S_0)^2\, dt = -[E_{av}(1-\rho)] = -u_1^*$$

Thus,

$$P_m = \int_{-\infty}^{-u_1^*} p(u/S_1)\, du = \int_{u_1^*}^{\infty} p(u/S_1)\, du$$
$$= \text{erfc}\left(\frac{u_1^*}{\sigma_u}\right) \tag{3.114}$$

From symmetry, the probability of mistaking H_1 for H_0 occurs when $(u/S_0) > u_1^*$, and the false alarm probability is:

$$P_f = \int_{u_1^*}^{\infty} p(u/S_0)\,du = \text{erfc}\left(\frac{u_1^*}{\sigma_u}\right) \tag{3.115}$$

The total error probability is now given by

$$P_e = 0.5P_m + 0.5P_f = P_m = \text{erfc}\left(\frac{u_1^*}{\sigma_u}\right) \tag{3.116}$$

where $\left(\frac{u_1^*}{\sigma_u}\right) = \sqrt{(1-\rho)\frac{E_{av}}{n_0}}$.

If $P_1 \neq P_0$, then $u_1^* = [E_{av}(1-\rho) - (n_0/2)\ln(P_0/P_1)]$ and P_e will be modified accordingly. For practical digital communication with $P_1 = P_0$, the error performance of the receiver is seen to depend only on (i) the average signal energy E_{av}, (ii) the cross-correlation coefficient ρ, and (iii) the noise spectral density n_0. Apparently P_e does not depend on the signalling waveforms since an MF/correlation receiver is used. But for minimum P_e, u_1^* is to be maximum and this is possible only for $\rho = -1$ (i.e., $s_1(t) = -s_0(t)$ with n_0, E_{av} constant). Thus the optimum signalling waveforms are also necessary for minimum P_e.

3.4.2 Coherent Binary Communication Systems

Binary signals may be generated using baseband 1/0 and ± 1 pulses, or DSB-AM (ASK), FSK and PSK-RF pulses. The generated signals are of the form:

(a) ASK: $\left.\begin{array}{ll} s_1(t) = A\cos\omega_0 t & 0 \le t \le T \\ s_0(t) = 0 & \end{array}\right\}$

(b) FSK: $\left.\begin{array}{ll} s_1(t) = A\cos\omega_1 t, & 0 \le t \le T \\ s_0(t) = A\cos\omega_0 t, & 0 \le t \le T \end{array}\right\}$

where $(\omega_1 - \omega_0)$ and $(\omega_1 + \omega_0)$ are multiples of π/T and the two signals are orthogonal

(c) CPSK: $\left.\begin{array}{l} s_1(t) = A\cos\omega_0 t,\ 0 \le t \le T \\ s_0(t) = A\cos(\omega_0 t + \pi) = -A\cos\omega_0 t,\ 0 \le t \le T \end{array}\right\}$ (3.117)

The optimum receivers for each of the above signals are the same as shown in Fig. 3.21. The error rate in each of the cases is now calculated using eq. (3.116) and the appropriate values of ρ and E_{av}. The values are

(a) ASK: $E_{av} = E_1/2,\ \rho = 0,\ P_e = \text{erfc}\left(\sqrt{E_1/2n_0}\right)$ and for large CNR,

$$P_e \simeq \sqrt{n_0/\pi E_1}\,\exp(-E_1/4n_0) \tag{3.118}^*$$

---

*For large x, the function erfc (x) is approximated as

$$\text{erfc}(x) \simeq \frac{1}{\sqrt{2\pi}x}\exp(-x^2/2) \tag{3.118a}$$

(b) FSK: $E_{av} = (E_1 + E_0)/2 = E_1 = E_0 = E$ (say), $\rho = 0$;

$$P_e = \text{erfc}\,(\sqrt{E/n_0}) \simeq \sqrt{n_0/2\pi E}\,\exp(-E/2n_0) \text{ for } E/n_0 \gg 1 \tag{3.119}$$

(c) CPSK: $E_{av} = E$, $\rho = -1$;

$$P_e = \text{erfc}\,(\sqrt{2E/n_0}) \simeq \tfrac{1}{2}\sqrt{n_0/\pi E}\,\exp(-E/n_0) \text{ for } E/n_0 \gg 1 \tag{3.120}$$

The various error rates given by eqs. (3.118), (3.119) and (3.120) are shown in Fig. 3.22, where it is seen that for a given E/n_0 (equivalent to CNR), the coherent PSK (CPSK) gives the minimum P_e. This is of course achieved with perfect phase synchronisation of the carrier (with phase error $\Delta\phi \leq 25°$, the CNR loss $\leq$ 1dB) or by using an ideal MF. In general, the P_e vs. E/n_0 curves are 3 dB apart, and $P_e = 10^{-5}$ for E/n_0 = 9.6 dB in PSK, for E/n_0 = 12.6 dB in FSK and for E/n_0 = 15.6 dB in Ask. However, the error performances of incoherent detectors, as discussed later, are within a few dB's of each other, and these detectors are used when carrier synchronisation is difficult to achieve.

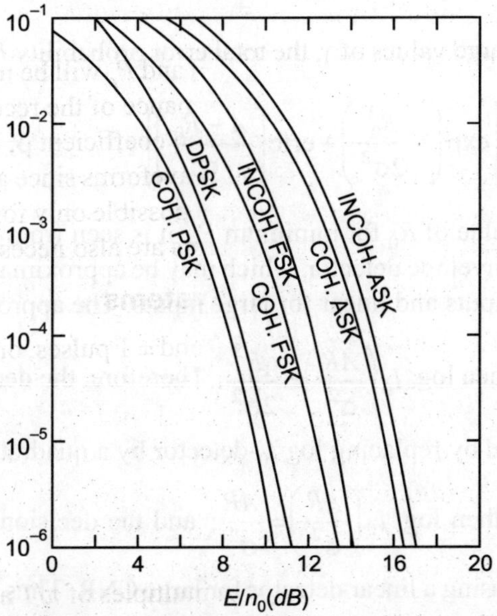

Figure 3.22: Error rates in binary communication systems

3.4.3 Incoherent Binary Communication Systems

The signal sets for ASK, FSK and PSK, given by eq. (3.117), may also be detected incoherently, when the phase synchronisation is not possible at the receiver. The simple envelope detector for ASK has been discussed in Sec. 3.3.1, and the results may be extended for binary communication. Assigning $P_0 = P_1 =$ 0.5; $C_f = C_m = 1$, eq. (3.60) is now modified as

$$\lambda_1(R) = e^{-\gamma} I_0\left(\frac{AR}{\sigma^2}\right) > \lambda_t = \frac{P_0 C_f}{P_1 C_m} = 1 \quad \text{for } H_1 \tag{3.121}$$

Therefore

$$\log_e I_0\left(\frac{AR_0}{\sigma^2}\right) = \gamma,$$

where R_0 is the decision threshold. The total error probability P_e is now given by (using eq. (3.61))

For ASK: $P_e = \frac{1}{2}(P_f + P_m)$

$$= \frac{1}{2}\exp\left(-\frac{R_0^2}{2\sigma^2}\right) + \frac{1}{2}\int_0^{R_0} \frac{R_0}{\sigma^2}\exp\left(-\frac{R^2}{2\sigma^2}\right) e^{-\gamma} I_0\left(\frac{AR}{\sigma^2}\right) dR \tag{3.122}$$

The integral above may be evaluated in terms of the Q-function [10] and P_e calculated for various values of γ. Using the approximation $I_0(x) \simeq (e^x/\sqrt{2\pi x})$ for $x \gg 1$, $p(R/1)$ of eq. (3.59) is approximated as

$$p(R/1) \simeq \frac{1}{\sqrt{2\pi}\sigma}\exp\left[-\frac{(R-A)^2}{2\sigma^2}\right], \gamma \gg 1$$

which is Gaussian. Then for large values of γ, the total error probability P_e of eq. (3.122) is now given by

$$P_e \simeq \frac{1}{2}\left\{\exp\left(-\frac{R_0^2}{2\sigma^2}\right) + \text{erfc}\left[\frac{A-R_0}{\sigma}\right]\right\} \tag{3.123}$$

The obtain an approximate value of R_0 for minimum P_e, it is seen from eqs. (3.60) and (3.121), that the decision is based on a log I_0-envelope detector, which may be approximated by diode-envelope detectors that are nonlinear for small inputs and linear for large inputs. The approximations necessary are:

(i) For $\gamma \ll 1$, $\frac{AR}{\sigma^2} \ll 1$, then $\log I_0\left(\frac{AR}{\sigma^2}\right) \simeq \frac{\gamma R^2}{2\sigma^2}$. Therefore, the decision rule is: Choose H_1 if $R^2 \geq 2\sigma^2$, which is obtained by replacing log I_0-detector by a quadratic envelope detector.

(ii) For $\gamma \gg 1$, $\frac{AR}{\sigma^2} \gg 1$, then $\log I_0\left(\frac{AR}{\sigma^2}\right) \simeq \frac{AR}{\sigma^2}$; and the decision rule is: Choose H_1 if $R \geq A/2$, which is obtained by using a linear detector for large CNR. Thus it is seen that the two thresholds $\sqrt{2\sigma}$ and $A/2$ are different. But a compromise threshold of

$$\frac{R_0}{\sigma} \geq \sqrt{2 + \frac{\gamma}{2}} \text{ for hypothesis } H_1 \tag{3.124}$$

approximately satisfies the two limiting values. Using this compromise threshold R_0, P_e of eq. (3.122) is now calculated for different values of γ, and is shown in Fig. 3.22.

For large CNR, the eq. (3.123) may be further approximated by using $R_0/\sigma = \sqrt{\gamma/2}$ and eq. (3.118a) as

For ASK: $P_e \simeq \frac{1}{2}\left[\exp\left(-\frac{\gamma}{4}\right) + \frac{1}{\sqrt{\pi\gamma}}\exp\left(-\frac{\gamma}{4}\right)\right]$

$$= \frac{1}{2}\exp\left(-\frac{\gamma}{4}\right); \gamma \gg 1$$

$$= \frac{1}{2}\exp\left(-\frac{E}{4n_0}\right) \text{(using transmission in Nyquist bandwidth)} \tag{3.125}$$

Use of an MF before the detector, as in Fig. 3.18, will improve the error performance of the incoherent ASK. However, comparing eqs. (3.118) and (3.125), it is seen that at high CNR, both the detection schemes perform almost equally well, the coherent ASK being superior only at low CNR.

Incoherent FSK

Since the phase is unknown at the receiver, the two received FSK signals will be of the form:

$$y(t) = \begin{cases} s_1(t) + n_1(t) = A\cos(\omega_1 t + \theta_1) + n_1(t) & \text{for } H_1 \\ s_0(t) + n_2(t) = A\cos(\omega_0 t + \theta_0) + n_2(t) & \text{for } H_0 \end{cases} \tag{3.126}$$

where θ_1 and θ_0 are independent random variables, uniformly distributed over the interval $(0, 2\pi)$. The optimum incoherent receiver for signals with unknown phase has been discussed in Sec. 3.3.3 and Fig. 3.18 gives the block diagram of a suitable receiver based on eq. (3.79a). For the two FSK signals, two parallel MF-envelope detectors, shown in Fig. 3.23, are used to obtain the two envelopes R_1 and R_0, which are then compared to decide on H_1 or H_0. In practice, the MF's may be replaced by suitable bandpass filters tuned to ω_1 and ω_0 respectively and with no overlap.

To obtain the ROC of the receiver, assume that H_1 is being transmitted; then, the ω_1-channel receives $[s_1(t) + n_1(t)]$, but the ω_0-channel receives $n_2(t)$ only. Using eq. (3.80), eq. (3.109) and Bayes criterion, the L.R. for $s_1(t)$ may be written as:

$$\lambda s_1 = \frac{\overline{p(Y/s_1)/p(Y/N)}}{\overline{p(Y/s_0)/p(Y/N)}} = \frac{\lambda_{s_1}^{\text{av}}(Y)}{\lambda_{s_0}^{\text{av}}(Y)} \quad \text{(as defined in eq. (3.80))}$$

$$= \frac{\exp(-E_s/n_0)I_0(2AR_1/n_0)}{\exp(-E_s/n_0)I_0(2AR_0/n_0)} \tag{3.127}$$

$$> \lambda_t = P_0/P_1 = 1 \quad \text{for } H_1$$

$$\text{Therefore, } I_0\left(\frac{2AR_1}{n_0}\right) > I_0\left(\frac{2AR_0}{n_0}\right); \text{ or } R_1 > R_0 \quad \text{for } H_1 \tag{3.128}$$

The receiver of Fig. 3.23 decides on H_1 if $R_1 > R_0$ and on H_0 if $R_0 > R_1$. Since the two MF's or the BPF's are tuned to two separate frequency bands around ω_1 and ω_0, the envelopes R_1, R_0 and $n_1(t)$ and $n_2(t)$ are statistically independent, but σ_1^2 and σ_2^2 are equal. The pdf's of envelopes R_1 and R_0 are now given by (eq. (3.59)):

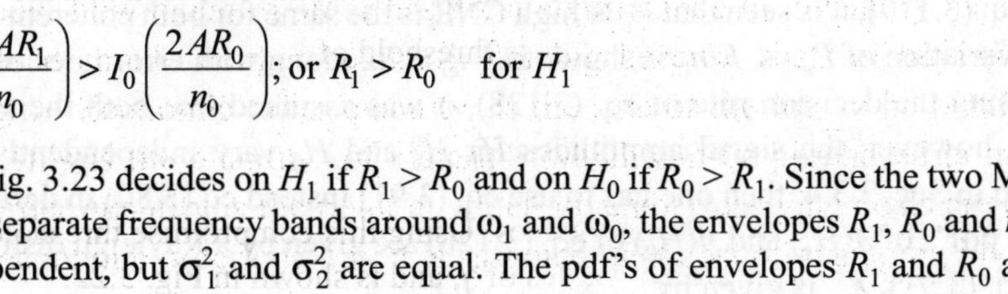

Figure 3.23: Incoherent FSK receiver

$$p(R_1/s_1) = \frac{R_1}{\sigma^2}\exp\left(\frac{-R_1^2}{2\sigma^2}\right)e^{-\gamma}I_0\left(\frac{AR_1}{\sigma^2}\right)$$

$$p(R_0/n_2) = \frac{R_0}{\sigma^2}\exp\left(\frac{-R_0^2}{2\sigma^2}\right) \tag{3.129}$$

The error of deciding on H_0, when H_1 is transmitted, occurs if $R_0 > R_1$ $(= P_m)$ and the error of deciding on H_1, when H_0 is transmitted, occurs if $R_1 > R_0$ $(= P_f)$. From symmetry, $P_m = P_f$ and the total error probability is given by

$$\begin{aligned} P_e &= 0.5\,P_m + 0.5\,P_f = P_m \\ &= \int_{R_1=0}^{\infty} p(R_1)\left[\int_{R_0=R_1}^{\infty} p(R_0)\,dR_0\right]dR_1 \\ &= \int_0^{\infty}\frac{R_1}{\sigma^2}\exp\left(-\frac{R_1^2}{\sigma^2}\right)\exp(-\gamma)\,I_0\left(\frac{AR_1}{\sigma^2}\right)dR_1, \end{aligned} \tag{3.130}$$

since from eq. (3.61) the inner integral = $\exp(-R_1^2/2\sigma^2)$. Using the integral relation given in eq. (3.93), and substituting $Z/\sqrt{2} = R_1/\sigma$, P_e can be written as:

$$\text{For FSK: } P_e = \frac{\exp(-\gamma/2)}{2}\left[\int_0^{\infty} Z\cdot\exp\left(-\frac{Z^2+\gamma}{2}\right)I_0(\sqrt{\gamma}\cdot Z)\,dZ\right]$$

$$= \frac{1}{2}\exp\left(\frac{-\gamma}{2}\right) = \frac{1}{2}\exp\left(-\frac{E}{2n_0}\right),\quad \gamma = \frac{A^2}{2\sigma^2} = \frac{E}{n_0} \tag{3.131}$$

since $\left[\int_0^{\infty}(\,)\,dZ\right] = 1$, by eq. (3.93)

Comparing with eq. (3.119), it is seen that P_e at high CNR is the same for both coherent and noncoherent binary FSK. The variation of P_e vs. E/n_0 is shown in Fig. 3.22 along with error rates of other receivers.

While determining the decision rule of eq. (3.128), it was assumed that both the signal amplitudes were constant. If, however, the signal amplitudes for H_1 and H_0 vary independently, as in a fading channels discussed in Sec. 3.3.4, then one has to use eq. (3.91) instead eq.(3.80) in deriving eq. (3.128). Assuming that the pdf's of $A(H_1)$ and $A(H_0)$ of eq. (3.126) are given by a Rayleigh distribution, as in eq. (3.89), then, for eq. (3.91), λ_{s_1} is given by

$$\lambda_{s_1} = \frac{\lambda_{s_1}^{\text{av}}(Y)}{\lambda_{s_0}^{\text{av}}(Y)} = \frac{[n_0/(n_0+2E_s)]\exp\left[2R_1^2/\{n_0(n_0+2E_s)\}\right]}{[n_0/(n_0+2E_s)]\exp\left[2R_0^2/\{n_0(n_0+2E_s)\}\right]}$$

$$> \lambda_t = 1 \quad \text{for } H_1 \tag{3.132}$$

The decision rule is then: $R_1^2 > R_0^2$; or $R_1 > R_0$ for H_1. Thus the optimum receiver for minimum P_e is the same as shown in Fig. 3.23, for the non-fading channel.

To obtain the values of P_e, it is observed that P_e is a function of A (also of γ), and the average $\bar{P}_e$ will have to be considered. Using eq. (3.131) for a given value of A (H_1), assuming H_1 is being transmitted, the average P_e is

$$\bar{P}_e = \int_0^\infty p\left(A_{H_1}\right)P_e(A)\,dA_{H_1}$$

$$= \int_0^\infty \frac{A}{h_0^2}\exp\left(-\frac{A^2}{2h_0^2}\right)\frac{1}{2}\exp\left[-\frac{A^2}{4\sigma^2}\right]dA$$

$$= \frac{1}{2+E_{av}/n_0}, \text{ where } E_{av} = 2E_s \text{ (by eq. (3.95))} \tag{3.133}$$

It is thus seen that $\bar{P}_e$ for a fading channel does not improve exponentially with the increase in E_{av}, as in eqs. (3.125) and (3.131). To improve $\bar{P}_e$ in such channels, one has to use diversity reception and suitable coding techniques, instead of increasing the signal energy E_s.

Differential PSK (DPSK)

In case of PSK, incoherent detection using envelope detectors is not possible; but differentially coherent PSK may be used instead. The DPSK receiver uses the phase of the previous digit as the reference and detect the phase of the present digit through a coherent detector as shown in Fig. 3.24. To be able to decide on the correct phase of data digits, it is necessary to precode the data in a complementary way, shown in Fig. 3.24(b). Considering a message of {00101001100 ...}, the encoder output consists of {.. 00011000100 ...} and the transmitted phases would be $\{\theta\} = \{..\ 000\pi\pi000\pi00\ ...\}$. It is observed that the transmitted phase θ changes (0 to π) or (π to 0), whenever the input data is '1'; otherwise θ remains unchanged.

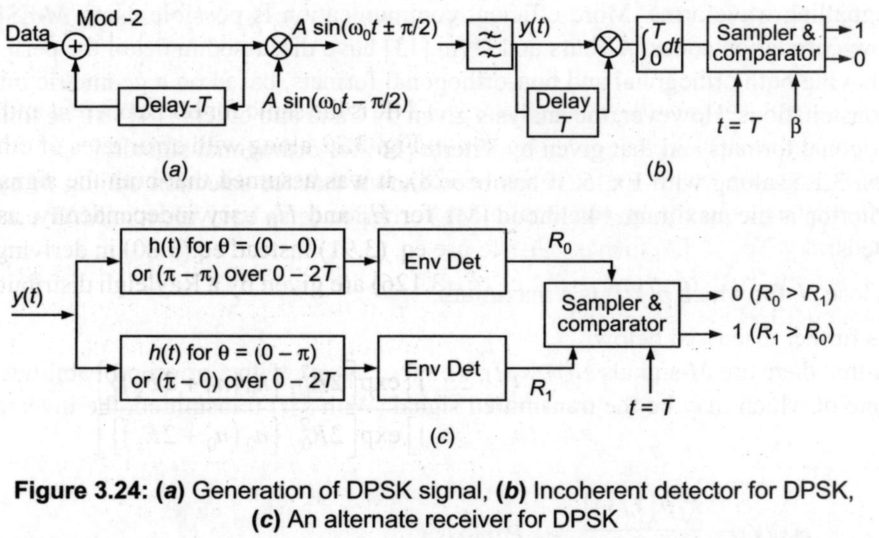

Figure 3.24: (*a*) Generation of DPSK signal, **(*b*)** Incoherent detector for DPSK, **(*c*)** An alternate receiver for DPSK

The differential encoding shown in Fig. 3.24(a) also helps in resolving the phase ambiguity at the receiver output and is sometimes used along with CPSK also. The major difference between CPSK receiver and a DPSK receiver of Fig. 3.24(b) is that: the reference phase used in DPSK is as noisy as the input to the detector and as such, the overall noise at the correlator output is approximately twice that of the CPSK receiver. The general error performance of DPSK is, therefore, expected to be 3 dB worse than that of CPSK; but this is true only for *M*-phase DPSK with $M > 4$. For binary DPSK, an alternative receiver structure is shown in Fig. 3.24(c), which is somewhat similar to the receiver of Fig. 3.23 for incoherent FSK. It is observed that due to the precoding, the transmitted RF pulses are in the form of pairs with phases (0, 0) or (π, π) for '0' message bit and with phases (0, π) or (π, 0) for '1' bit. As such the predetection MF's operate over a two-pulse period (0-2*T*) and distinguishes between the presence of the phase transition (for '1' bit) or its absence. For an input (0, 0) or (π, π), the upper MF and envelop detector will give a large output and the comparator output will be '0'. For an input (0, π) or (π, 0), the comparator output will be '1'. Comparing Figs. 3.23 and 3.24(c), it is evident that the two receivers behave exactly in the same manner, except that the integration time in DPSK is 2*T*, instead of *T* in FSK. The error performance of binary DPSK is then obtained by replacing E/n_0 with $2E/n_0$ in eq. (3.131) giving:

For Binary DPSK

$$P_e = \frac{1}{2}\exp(-\gamma) = \frac{1}{2}\exp(-E/n_0) \tag{3.134}$$

The MF's for Fig. 3.24(c) may be realised using integrate-and-dump filters (BPF followed by an integrate and dump circuit), one running continuously over (0-2*T*) period and the other having a phase inversion at $t = T$. This results into two MF's, matched to two orthogonal waveforms (0, 0)/(1, 1) and (1, 0)/(0, 1), and the detection strategy is then optimum.

3.4.4 Detection of M-ray Signals [14, 15, 16, 17]

For sources generating alphabets $M>2$ (and also for extended sources, as discussed in Chapter 5, Sections 5.1.3 and 5.7.1), it is necessary to use multilevel signalling using multi-amplitude (*M*-ASK) or multiphase (*M*-PSK) signalling waveforms. More efficient communication is possible using *M*-FSK, orthogonal and other low-correlation codes. Arthurs and Dym [13] have discussed in detail the performance of *M*-ary signals having both orthogonal and non-orthogonal formats, based on a geometric interpretation of the signal constellations. However, the analysis given by Cahn and others [14] will be followed here for the nonorthogonal formats and that given by Viterbi [16] for orthogonal signals.

In Secion 3.1.5, along with Ex. 5, it has been shown that for selection among *M* hypotheses, the optimum criterion is the maximum likelihood (ML) criterion (also equivalent to MAP criterion) and the decision rule is:

Choose H_j for which $p(Y/j)$ is a maximum.

The result is further discussed below.

Consider that there are *M*-signals $s_1(t), s_2(t), ..., s_j(t), ...s_M(t)$, with a priori probabilities $P_1, P_2, ..., P_j, ... P_M$, anyone of which may be the transmitted signal. With $s_j(t)$ transmitted, the inverse probabilities are:

$$p(s_j/Y) = \frac{P_j p(Y/s_j)}{p(Y)} = k \cdot P_j \cdot p(Y/s_j) \tag{3.135}$$

since $p(Y)$ is measured at the receiver and is known.

For maximum information, (i.e., for minimum error), the inverse probabilities (or the likelihood functions) are to be maximised, giving the ML-criterion stated above. For band-limited signals $s_j(t)$ and white Gaussian noise, the inverse probabilities may be written as

$$p(s_j/Y) = k{\cdot}P_j{\cdot}p(Y/s_j) = k \cdot P_j\left(\frac{1}{2\pi\sigma^2}\right)^{TW} .\exp\left[-\sum_{i=1}^{2TW}\frac{\left(y_i - s_{ij}\right)^2}{2\sigma^2}\right]$$

and for hypothesis H_1,

$$\log P_j + \frac{1}{2\sigma^2}\Sigma\left[2y_i s_{ij} - s_{ij}^2\right] \underset{j\neq 1}{<} \log P_1 + \frac{1}{2\sigma^2}\Sigma\left[2y_i s_{i1} - s_{i1}^2\right],$$

where the common terms have been neglected. Using eq. (3.71) and eq. (3.71a), the inequalities are written in the integral form as

$$\int_0^T y(t)s_1(t)dt > \int_0^T y(t)s_j(t) + b_j \tag{3.136}$$

where $b_j = (n_0/2)\log P_j - E_j/2$ and $j \neq 1$. For hypothesis H_2, the LHS of the inequality will have $s_2(t)$ in place of $s_1(t)$ and $j \neq 2$; and so on for H_3 etc. For all P_j's $= P$, the ML-criterion reduces to

$$\int_0^T y(t)s_1(t)dt - E_1/2 > \int_0^T y(t)s_j(t) - E_j/2;\ j \neq 1 \text{ for } H_1 \tag{3.137}$$

The optimum receiver is, then, given by either M cross correlators or M-MF's with $h_j(t) = s_j(T - t)$ along with a decision circuit to decide the maximum at $t = T_1$ as shown in Fig. 3.25.

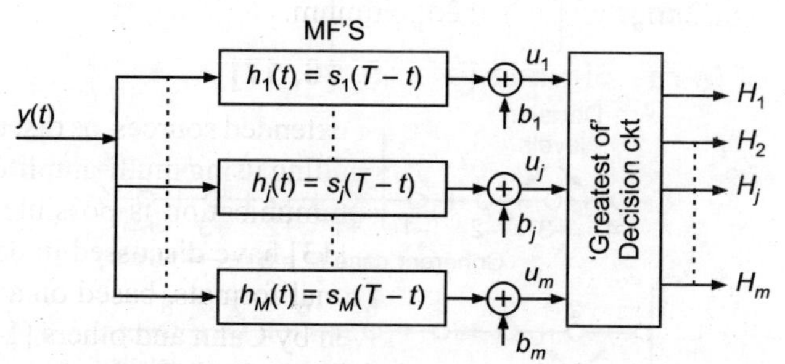

Figure 3.25: Optimum receiver for *M*-ary orthogonal signals

Assuming that $s_1(t)$ is transmitted, an error will occur if any of the $(M-1)$ inequalities of eq. (3.136) is not satisfied. The inequalities may be rewritten in the form (following eq. (3.111))

$$\int_0^T n(t)\left(s_1 - s_j\right)dt > \frac{n_0}{2}\ln\frac{P_j}{P_1} - \frac{1}{2}\int_0^T\left(s_1 - s_j\right)^2 dt$$

$$= \frac{n_0}{2}\ln\frac{P_j}{P_1} - \frac{1}{2}\left(E_1 + E_j\right) + \int_0^T s_1 s_j dt \tag{3.138}$$

and the RHS is to be minimised for minimum error. For equal-energy signals with $E_1 = E_j$ and $P_1 = P_j$, the only way to do this is to make $\int_0^T s_1 s_j dt = 0$, thus making the signals to be orthogonal with respect to each other. This is the optimum solution for M-ary signalling, and orthogonal, bi-orthogonal and low-correlation codes (e.g., *PN* sequences, Gold codes) are used for efficient M-ary digital communication.

However, non-orthogonal M-ary signals, e.g., multilevel ASK (M-ASK) and multiphase PSK (M-PSK) are commonly used where the available channel bandwidth is restricted. The orthogonal codes using multifrequency (M-FSK) signals and binary-coded PSK signals require large bandwidths for efficient transmission, and are useful in wideband space channels. The performance of the above signalling techniques are discussed below.

M-ASK

Consider the signalling waveforms:

$$s_j(t) = A_j(a_0 \cos \omega_0 t),\ 0 \le t \le T, \tag{3.139}$$

and

$$j = 1, 2, \ldots M$$

where $a_0 \cos \omega_0 t$ is the basic signalling waveform and A_j is the level indicating the source alphabet as shown in Fig. 3.26(a). As discussed above, the optimum coherent receiver is an MF a receiver, shown in Fig. 3.26(b), which is similar to Fig. 3.21, but the MF is matched to $a_0 \cos \omega_0 t$. Thus for a transmitted signal $s_j(t)$, the MF output u_j is a Gaussian variable with the mean $= A_jE_0$ and the variance $\sigma_u^2 = n_0E_0/2$, where $E_0 = \int_0^T a_0^2 \cos^2 \omega_0 t\, dt$. The pdf of U/j is then Gaussian and is given by

$$p(U/j) = \frac{1}{\sqrt{2\pi\sigma_u}} \exp\left[-\frac{\left(U - A_jE_0\right)^2}{2\sigma_u^2}\right] \tag{3.140}$$

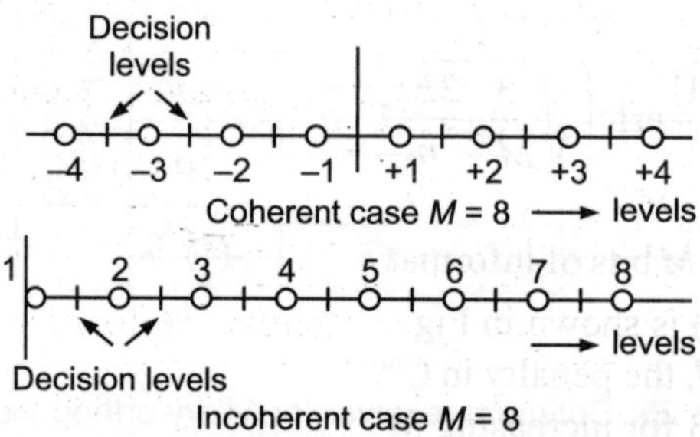

Figure 3.26(*a*): Signal constellations for M-PAM, M = 8

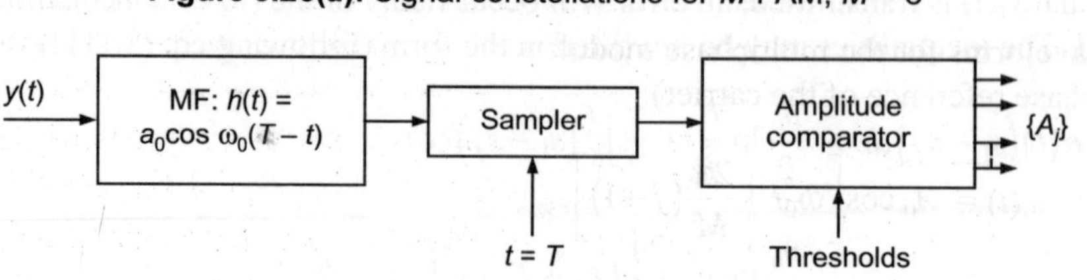

Figure 3.26(*b*): Coherent receiver for M-ASK

From the discussions of Ex. 5, the decision thresholds are at the midlevels $(A_{j-1} + A_j)E_0/2$, and the separation between levels is: $d = E_0$. Now an error occurs when the noise voltage exceeds $\pm E_0/2$ at all levels, except at the two extreme levels where the error can occur in one direction only (resulting into the factor $(M-1)/M$ in the equation for P_e). Assuming that all levels are equiprobable, the average error probability per symbol is given by

$$Pe = \frac{M-1}{M}\left\{\text{Pr. of } \left|u_j - A_j E_0\right| > \frac{E_0}{2}\right\}$$

$$= 2\left(\frac{M-1}{M}\right)\int_{\left(A_j+\frac{1}{2}\right)E_0}^{\infty} p(U/j)\,dU$$

$$= \frac{2(M-1)}{M}\text{erfc}\left(\frac{E_0}{2\sigma_u}\right)$$

$$= \frac{2(M-1)}{M}\text{erfc}\sqrt{\frac{E_0}{2n_0}} \tag{3.141}$$

For $M = 2$, the eq. (3.141) reduces to eq. (3.118) for binary ASK signals. For a level j, the signal energy $E_j = A_j^2 E_0$ and the average energy E_{av} for equiprobable $\pm M/2$ levels, assuming $M \gg 1$, is

$$E_{av} = \frac{E_0}{M}\int_{-M/2}^{M/2} A_j^2\,dA = \frac{M^2 E_0}{12}$$

In terms of E_{av},

$$P_e = \frac{2(M-1)}{M}\text{erfc}\left(\sqrt{\frac{3}{M^2}\frac{2E_{av}}{n_0}}\right) \tag{3.142}$$

and since each symbol carries $\log_2 M$ bits of information, the bit error rate $P_b = P_e/\log_2 M$. The variation of P_b with E_b/n_0 $(= E_{av}/(n_0 \log M))$ is shown in Fig. 3.27 for some values of M.

It is seen that with increasing M, the penalty in CNR for same P_b increases; e.g., for $M = 4$, the loss $\simeq$ 4 dB, but for $M \geq 8$, the loss = 6dB for increasing M by a factor of 2.

M-PSK

The signalling waveforms for the multiphase modulation may be written, as an extension of BPSK, as (neglecting any phase reference of the carrier):

$$s_j(t) = A_0 \cos\left[\omega_0 t + \frac{2\pi}{M}(j-1)\right], j = \{1, 2, \ldots, M\};$$

$$0 \leq t \leq T \tag{3.143}$$

and the signal energy E_j are the same for all j.

The corresponding signal constellations are shown in Fig. 3.28. The ML-criterion for deciding on H_1, for $P_j = P$, is now given by eq. (3.137)

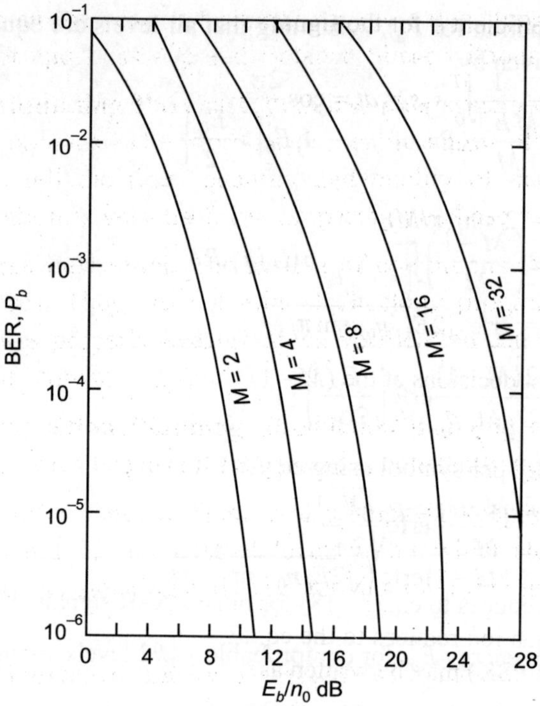

Figure 3.27: P_s vs. E_b/n_0 for M-ASK

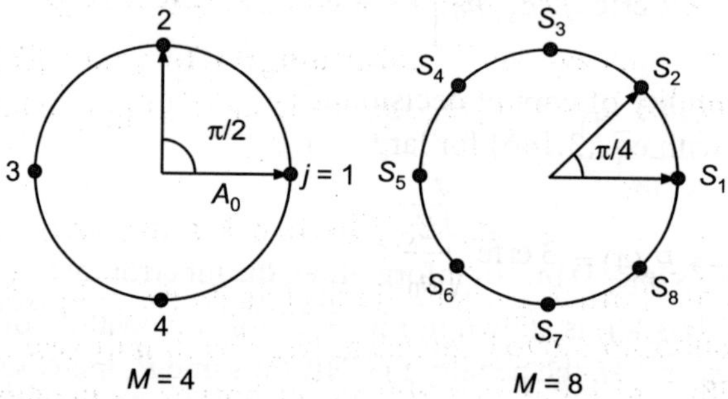

Figure 3.28: Signal constellations for M-PSK, M = 4, 8

modified as

$$\int_0^T y(t)s_1(t)\,dt > \int_0^T y(t)s_j(t)\,dt,\ j \neq 1 \tag{3.144}$$

The corresponding optimum receiver is that given in Fig. 3.25 or its equivalent version with M-cross correlators. The 'Greatest of' decision circuit may be considered equivalent to $(M-1)$ 2-input comparators where the decision for the largest input is obtained through successive comparision of the correlator outputs. Now using the results of Sec. 3.4.1 and eq. (3.116), the error probability at the output of the first comparator is

$$P_{e1} = \text{erfc}\sqrt{(1-\rho)E_s/n_0},\ E_s = E_j$$

where the correlation coefficient ρ for the signal set of eq. (3.143) is calculated as

$$\rho(j, j-1) = \frac{1}{E_s}\int_0^T s_j s_{j-1} dt = \cos\frac{2\pi}{M}$$

and

$$(1-\rho) = 2\sin^2(\pi/M)$$

Therefore,

$$P_{e1} = \text{erfc}\sqrt{2E_s/n_0}\cdot\sin\pi/M \tag{3.145}$$

The probability of correct decisions at the $(M-1)$ comparator outputs is

$$P_c = (1-P_{e1})^{M-1}$$

Then, the error probability per symbol using eq. (3.145) is given by

$$P_e(M) = \{1-(1-P_{e1})^{M-1}\}$$

$$\leq (M-1)\text{erfc}\left(\sqrt{2E/n_0}\cdot\sin\pi/M\right) \tag{3.146}$$

For $M = 2$, the above equation reduces to the eq. (3.120) for binary PSK. The value of $P_e(4)$ for the 4-phase PSK (4-PSK or Q-PSK) may be written as

$$P_e(4) = [1-P_c(4)] = [1-(1-P_e(2))^2]$$

$$= 2\,\text{erfc}\sqrt{2E_s/n_0}\cdot\left[1-\frac{1}{2}\text{erfc}\sqrt{2E_s/n_0}\right] \tag{3.147}$$

where $P_c(4)$ is the probability of correct decision for a 2-bit symbol and $P_e(2)$ is given by eq. (3.120). Comparing eq. (3.147) with eq. (3.146) for large E/n_0, it is observed that

$$\left.\begin{aligned} &\text{Eq. (3.146)} \rightarrow P_e(4) = 3\,\text{erfc}\sqrt{\frac{E_s}{n_0}} \simeq \frac{3}{\sqrt{2\pi}}\sqrt{\frac{n_0}{E_s}}\exp\left[-\frac{E_s}{2n_0}\right] \\ &\text{and} \\ &\text{Eq. (3.147)} \rightarrow P_e(4) \simeq \frac{1}{\sqrt{\pi}}\sqrt{\frac{n_0}{E_s}}\exp\left(-\frac{E_s}{n_0}\right) \end{aligned}\right\} \tag{3.148}$$

Thus eq. (3.146) gives a weaker upper bond for $m > 2$. The reason for that is: Eq. (3.146) assumes equal values of ρ's and equal probability of the error event at all H_j outputs of the decision circuit, independent of the signal transmitted. This is only true when orthogonal codes represent $s_j(t)$ and all codes have the same distance from each other. In M-PSK and M-ASK, however, the signal codes are not equidistant from each other as seen in Figs. 3.26 and 3.28 and the probability of H_j being erroneously decided upon, when either $s_{j-1}(t)$ or $s_{j+1}(t)$ is transmitted, is much more than that of the decision on, say, $H_{j\pm4}$ for the same signal transmission. A tighter bound on P_e is obtained from the following discussions.

In practice, a coherent digital phase detector (with a stored/acquired phase reference at the receiver) is used to measure the phase of the received signal over (0-T). The measured vector v_j will be displaced by

θ from its mean position due to the channel noise, as shown in Fig. 3.29 for the vector v_2. Based on the ML-criterion and the discussions of Ex. 5, the optimum decision thresholds will be along the phasors bisecting the angle $2\pi/M$ between the successive signal vectors. The boundaries of the decision region R_j for hypothesis H_j are, then, the above bisectors, and one such boundary for H_2 is shown in Fig. 3.29. The error will occur whenever the received v_j, given s_j, has a phase error: $|\theta| > \pi/M$. The pdf of the phase θ of a signal consisting of a sinewave plus Gaussian noise (having a Rician envelope distribution) has been shown to be (refer to Sec. 1.11 of Ch. 1)

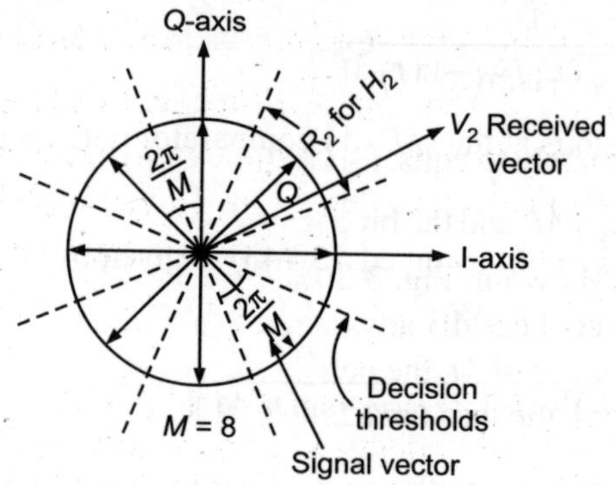

Figure 3.29: Signal vectors and decision thresholds ($M = 8$)

$$p(\theta) = \frac{1}{2\pi} e^{-\gamma}\left[1+\sqrt{4\pi\gamma}\cdot\cos\theta\cdot\exp\left(\gamma\cos^2\theta\right)\text{erf}\left(\sqrt{2\gamma}\cos\theta\right)\right] \tag{3.149}$$

The symbol error probability is then given by

$$P_e(M) = 1-\int_{-\pi/M}^{\pi/M} p(\theta)\,d\theta \tag{3.150}$$

which may be calculated by numerical integration for various values of $\gamma = (E_s/n_0)$ and M.

For $M > 4$ and $E_s/n_0 \gg 1$, a closed form solution of $P_e(M)$ is obtained by noting that

$$\text{erfc}\left(\sqrt{2\gamma}\cos\theta\right) \simeq 1-\frac{e^{-\gamma}\cos^2\theta}{\sqrt{4\pi\gamma}\cos\theta}$$

and

$$p(\theta) \simeq \begin{cases}\sqrt{\dfrac{\gamma}{\pi}}\cos\theta\cdot\exp\left(-\gamma\sin^2\theta\right) & \text{for } |\theta| < \pi/2\\ 0 & \text{for } |\theta| > \pi/2\end{cases}$$

Using these approximations in eq. (3.150),

$$P_e(M) \simeq 1-\int_{-\pi/M}^{\pi/M}\sqrt{\gamma/\pi}\cos\theta\cdot\exp\left(-\gamma\sin^2\theta\right)d\theta$$

$$= 1-\int_{-\sqrt{2\gamma}\sin\pi/M}^{\sqrt{2\gamma}\sin\pi/M} \frac{1}{\sqrt{2\pi}}\exp\left(-\frac{Z^2}{2}\right)dZ,\ Z=\sqrt{2\gamma}\sin\theta$$

$$= 2\,\text{erfc}\left[\sqrt{\frac{2E_s}{n_0}}\cdot\sin\left(\frac{\pi}{M}\right)\right]$$

$$\simeq \frac{1}{\sqrt{\pi E_s/n_0}\,.\sin(\pi/M)}\exp\left[-\left(\frac{E_s}{n_0}\right)\sin^2\left(\frac{\pi}{M}\right)\right] \tag{3.151}$$

The eq. (3.151) reduces approximately to eqns. (3.120) and (3.147) for $M=2$ and 4 respectively. The bit error probability $P_b \simeq P_e(M)/\log_2 M$ and the bit energy $E_b = E_s/\log_2 M$. Using eqns. (3.147) and (3.151), the variation of P_b with E_b/n_0 is shown in Fig. 3.30, where it is seen that the error rates for BPS *K* and *Q*-PSK are almost the same, since the 3 dB advantage in signalling rate compensates for the excess γ required in *Q*-PSK. For higher values of *M*, the penalty in γ for the same values of P_b increases; e.g., for $M=8$, the loss in $E_b/n_0 \simeq 4$ dB and the loss tend 2*s* to 6 dB/for $M > 16$.

Differential M-PSK [14, 15]

As in the case of binary PSK, the phase-comparision detection is also possible for *M*-PSK. At the transmitter, the weightage of the symbol is conveyed by the phase transitions between successive pulses rather than as the absolute phases of the pulses. In the receiver, the differential phase between the successive pulses is measured and the corresponding symbol is decided upon. Since the detection scheme of Fig. 3.24(c) is not applicable to *M*-PSK, the DPSK receiver for $M > 2$ will be an extension of Fig. 3.24(b). As such for $\gamma \gg 1$ and $M \gg 1$, the difference phase has approximately a Gaussian distribution, but because of the noisy reference its variance is twice of that obtained in the CPSK receiver with an undisturbed reference. Therefore, the DPSK receiver introduces a 3 dB degradation in the required E/n_0 for a given P_e. (It has been shown earlier that in binary DPSK, the degradation is very small).

Using a geometric interpretation, Cahn [15] has shown that $P_e(M)$ for DPSK is given by:

$$P_e(M) \simeq 2\,\text{erfc}\left[2\sqrt{E_s/n_0}\cdot\sin\pi/2M\right]$$

$$\simeq \frac{1}{\sqrt{2\pi E_s/n_0}\cdot\sin\pi/2M}\exp\left[-\left(\frac{2E_s}{n_0}\right)\cdot\sin^2\left(\frac{\pi}{2M}\right)\right] \tag{3.152}$$

The corresponding values of P_b vs. E_b/n_0 have been shown in Fig. 3.30 and it is seen that DPSK results are worse by 2.3 dB for $M=4$ and by approximately 3 dB for $M \geq 8$.

Quadrature-AM (QAM)

It is observed from eq. (3.151) and Fig. 3.30 that for $M \to$ large, PSK signalling is rather inefficient. For both *M*-ASK and *M*-PSK, the penalty in CNR for increasing the data rate by one bit (i.e., for increasing *M* by a factor of 2) is approximately 6 dB for $M \to$ large. Alternate signal constellations, using a combination of multiple phases and multiple amplitudes, (APK), have been suggested to reduce this penalty.

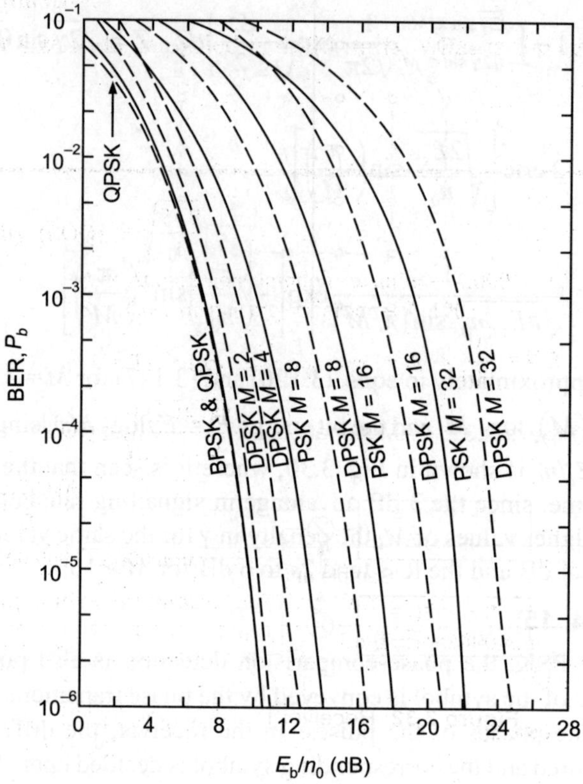

Figure 3.30: BER vs. E_b/n_0 for M-PSK and D-MPSK

Among the possible constellations, the quadrature amplitude modulation (QAM) has been found to be the most efficient and its signalling wave-forms are:

$$s_j(t) = A_j(a_0 \cos \omega_0 t) + B_j(a_0 \sin \omega_0 t), \quad 0 \le t \le T$$

and

$$j = \{1, 2, ..., m\}, \quad m = \sqrt{M} \tag{3.153}$$

Two examples of QAM have been shown in Fig. 3.31, where the I-Q vectors are amplitude modulated by $\sqrt{M}$ levels, as in M-ASK and the dots represent the tips of signal vectors. For $M = 2^k$, k even, the amplitude levels are symmetrical for both I and Q vectors, but for odd vlaues of k, some of the resulting signal points may be omitted as shown in Fig. 3.31, for $M = 32$, $k = 5$. The appropriate receiver is a combination of Q-PSK and M-ASK receivers, as shown in Fig. 3.32. A joint decision on the output of the two comparators gives the desired data word (= symbol).

To obtain the error rate, it is observed that the probability of a correct decision is the probability that both the I-Q vectors are correctly received, since the QAM signal is equivalent to two separate orthogonal PAM signals on the quadrature carriers, and coherent detection is used in the receiver. Then using eq. (3.141), the probability of correct decision is:

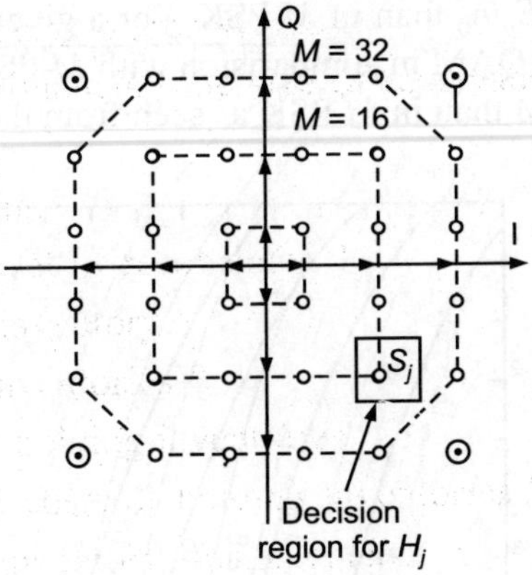

Figure 3.31: QAM signal constellation M = 16, 32

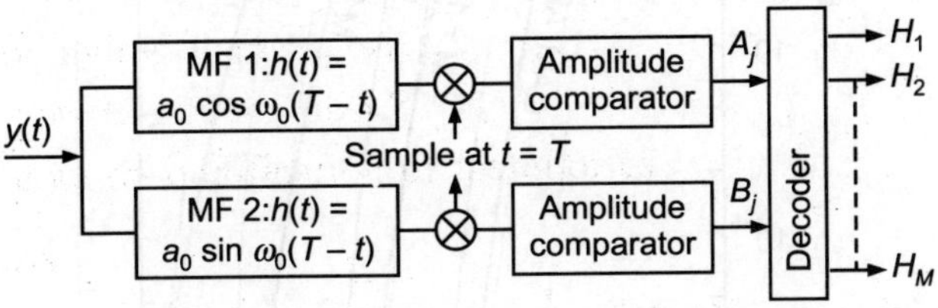

Figure 3.32: Receiver for QAM signals

$$P_c = (1 - P_e(m))^2,$$

and

$$P_e(M) = 1 - [1 - P_e(m)]^2 = 2P_e(m) - P_e^2(m)$$

$$\simeq 2P_e(m) \text{ for } \gamma \gg 1 \triangleright \tag{3.154}$$

where $P_e(m)$, by eq. (3.141), $= \dfrac{2(m-1)}{m}\,\text{erfc}\sqrt{\dfrac{E_0}{2n_0}}$, and $m = \sqrt{M}$, with k even. This result is valid for M = 16, 64, 256. For k odd, i.e., for M = 8, 32, 128, etc., a bound on P_e has been obtained as [17].

$$P_e(M) \le 4\,\text{erfc}\sqrt{E_0/2n_0}, \text{ for } k \ge 3 \tag{3.155}$$

Since the average power on each vector is $m^2 E_0/12, m \gg 1$, the total power transmitted is the sum of the vector powers and the error rate is given by

$$P_e(M) \le 4\,\text{erfc}\sqrt{\frac{3}{M}\frac{E_{av}}{n_0}} \simeq \frac{4\sqrt{Mn_0}}{\sqrt{6\pi E_{av}}}.\exp\left[-\frac{3E_{av}}{2Mn_0}\right] \tag{3.156}$$

Using the relations $E_{av} = kE_b$ and $P_e(M) \simeq kP_b$, the variation of P_b with E_b/n_0 is shown in Fig. 3.33 of M = 4, 16, 64 and 256. By comparing the results of Figs. 3.30 and 3.33, it is observed that the same P_b is obtained in M-ary QAM with less E_b/n_0 than in M-PSK. For a given value of $P_b = 10^{-5}$, the Table 3.1 indicates the power advantage in M-QAM in comparision with M-PSK. Further, the penalty in E_b/n_0 for increasing k is much less in M-QAM than in M-PSK as seen from the Table.

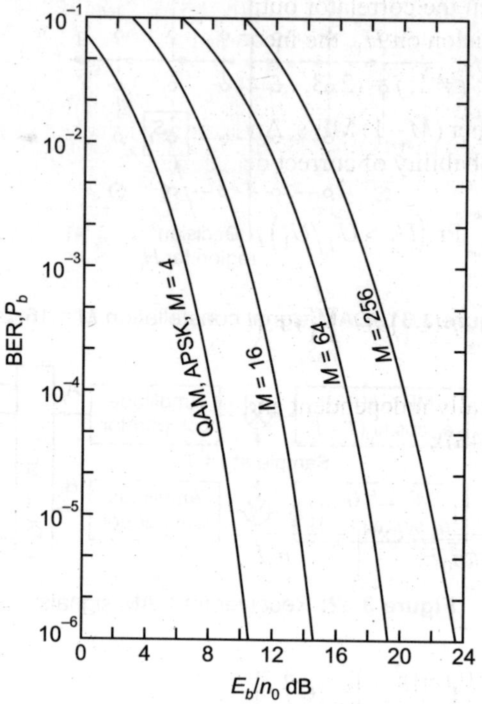

Figure 3.33: BER vs E_b/n_0 for M-ary QAM

Table 3.1: Power efficiency of QAM at $P_b = 10^{-5}$

Signalling method	E_b/n_0 in dB required for		
	$M = 16$	$M = 32$	$M = 64$
M-QAM	13	15	18
M-PSK	17	22	28

3.4.5 Detection of Orthogonal *M*-ary Signals

In the introduction of the earlier section, it has been shown that for equal-energy and equiprobable signals, $\int_0^T s_1(t).s_j(t)\,dt$ should be minimised for minimum error at the output of the multi-MF receiver of Fig. 3.25. Eq. (3.138) is now simplified as:

$$\int_0^T n(t)\left[s_1(t) - s_j(t)\right]dt > \int_0^T s_1(t)s_j(t)\,dt - E_1;\ j \neq 1,\ \text{for } H_1 \tag{3.157}$$

Minimizing the RHS leads to the condition that the signal set $\{s_j(t)\}$ should be orthogonal with respect to each other. Assume that a set of equal-energy and equiprobable signals are used at the transmitter and the receiver of Fig. 3.25 performs the correlation required by eq. (3.136). Assume further that the signal $s_1(t)$ is being transmitted, then the correlator output U_1, corresponding to the MF with $h_1(t) = s_1(T-t)$, has to satisfy, for correct decision on H_1, the inequalities:

$$U_1 > U_j; j \neq 1, j = (2, 3, ..., M) \tag{3.158}$$

and U_j's are the outputs of other $(M-1)$ MF's. An error occurs when any of the $(M-1)$ above conditions is not satisfied. Thus, the probability of correct decision is

$$\begin{aligned} P_c &= \int_{-\infty}^{\infty} \text{Pr.}\left(U_1 > U_j / U_1\right) p\left(U_1\right) dU_1, \text{ for all } j, \quad j \neq 1 \\ &= \int_{-\infty}^{\infty} p\left(U_1\right) dU_1 \prod_{j=2}^{M} \text{Pr.}\left(U_1 > U_j / U_1\right) \end{aligned} \tag{3.159}$$

where U_j's, $j \neq 1$, are statistically independent and identically distributed. The pdf's of the test statistics U_j are given by (from eq. (3.48)):

$$p(U_1) = \frac{1}{\sqrt{\pi n_0 E_s}} \exp\left[-\frac{(U_1 - E_s)^2}{n_0 E_s}\right]$$

and

$$p(U_j) = \frac{1}{\sqrt{\pi n_0 E_s}} \exp\left[-\frac{U_j^2}{n_0 E_s}\right], j \neq 1 \tag{3.160}$$

Further,

$$\begin{aligned} \text{Pr.}(U_j < U_1) &= \int_{-\infty}^{U_1} p\left(U_j\right) dU_j, j \neq 1 \\ &= \text{erfc}\left(\frac{\sqrt{2} U_1}{\sqrt{n_0 E_s}}\right) \end{aligned}$$

Therefore, the symbol error probability is:

$$P_e(M) = 1 - \int_{-\infty}^{\infty} \frac{1}{\sqrt{\pi n_0 E_s}} \exp\left[-\frac{(U_1 - E_s)^2}{n_0 E_s}\right] \left[\text{erf}\left(\frac{\sqrt{2} U_1}{\sqrt{n_0 E_s}}\right)\right]^{M-1} dU$$

and by change of variables

$$y = \frac{(U_1 - E_s)}{\sqrt{n_0 E_s / 2}}; \frac{\sqrt{2} U_1}{\sqrt{n_0 E_s}} = y + \sqrt{\frac{2 E_s}{n_0}},$$

$$P_e(M) = 1 - \int_{-\infty}^{\infty} \frac{\exp(-y^2/2)}{\sqrt{2\pi}} \left[\text{erfc}\left(y + \sqrt{\frac{2E_s}{n_0}} \right) \right]^{M-1} dy \tag{3.161}$$

Since the same probability of error will occur when any of the other $(M-1)$ signals are transmitted and all signals are equi-probable, the eq. (3.161) gives the average probability of symbol errors for the M-ary orthogonal signalling. The integral is not easily evaluated, but the numerical results are available in [16], [17], [18]. A set of curves giving $P_e(M)$ vs. E_s/n_0 (= symbol CNR) is shown in Fig. 3.34, where it is seen that for a given symbol error, say 10^{-5}, the symbol energy required varies only within 2 dB approximately for $M = 2$ to 64. The corresponding E_b (= energy per bit) then decreases considerably with increasing M.

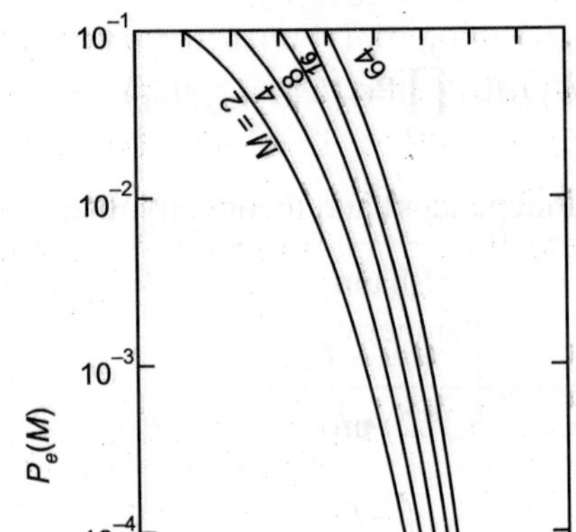

Figure 3.34: Symbol (word) error $P_e(M)$ vs E_s/n_0 for orthogonal codes

The efficiency of the orthogonal signalling becomes evident when the limiting value of $P_e(M)$ is obtained with $M \to \infty$. For k-bit words (alphabets) of the source, $2^k = M$, $E_s = kE_b$ and $E_b/n_0 = E_s/kn_0$ = normalised CNR per bit of information. For $M \to \infty$, $k \to \infty$, $\left(y + \sqrt{2E_s/n_0}\right) \simeq \sqrt{2E_s/n_0}$, and the eq. (3.161) may be approximated by using the asymptotic expression for the error function, as

$$\underset{M\to\infty}{P_e(M)} \simeq 1 - \int_{-\infty}^{\infty} \frac{\exp(-y^2/2)}{\sqrt{2\pi}} dy \left[1 - \frac{\exp-(-kE_b/n_0)}{\sqrt{2\pi}\sqrt{2kE_b/n_0}} \right]^{M-1}$$

$$\leq 1 - \left[1 - \frac{\exp(-kE_b/n_0)}{\sqrt{2\pi}\sqrt{2kE_b/n_0}} \right]^{M}, \text{ since } \int_{-\infty}^{\infty} \frac{\exp(-y^2/2)}{\sqrt{2\pi}} dy = 1.$$

Since $\exp(-kE_b/n_0) \to 0$, for $k \to \infty$, and using the approximation $(1+x)^M \simeq (1+Mx),\ x \ll 1$; $P_e(M)$ is further approximated as

$$P_e(M) \le 1 - \lim_{M\to\infty}\left[1 - \frac{2^k \exp(-kE_b/n_0)}{\sqrt{2\pi}\sqrt{2kE_b/n_0}}\right]$$

$$= \lim_{M\to\infty}\left[\frac{\exp[-(kE_b/n_0 - k\ln 2]}{\sqrt{2\pi}\sqrt{2kE_b/n_0}}\right] \tag{3.162}$$

Therefore, $Pe(M) \to 0$, if $E_b/n_0 > \ln 2$, making the exponent negative. Thus it may be noted that in the limit $k \to \infty$, $M \to \infty$,

$$\left.\begin{aligned} &P_e(M) \to 0 \text{ for } E_b/n_0 > \ln 2 = -1.6 \text{ dB} \\ &\text{and} \\ &P_e(M) \to 1 \text{ for } E_b/n_0 < \ln 2 \end{aligned}\right\} \tag{3.163}$$

This limit of $E_b/n_0 = -1.6$ dB is known as Shannon's limit for error-free communication with orthogonal coding and wideband channels. The channel capacity for such coded communication will be further discussed in Chapter 5.

For large CNR in the channel, eq. (3.162) provides an upper bound on $P_e(M)$, given by

$$P_e(M) \le \sqrt{\frac{n_0}{4\pi E_s}} \cdot M \cdot \exp\left(\frac{-E_s}{n_0}\right),\ \frac{E_s}{n_0} \gg 1 \tag{3.164}$$

A weaker upper bound on $P_e(M)$ may be obtained by considering the multiple-MF receiver of Fig. 3.25 to be one that makes $(M-1)$ *MF*-outputs $U_j, j \neq 1$. Then $P_e(M)$ is upper-bounded by the relation (cf. eq. (3.146)):

$$P_e(M) \le (M-1)P_{e1} \tag{3.165}$$

where, using eq. (3.116), $P_{e1} = \text{erfc}\sqrt{E_s/n_0}$. Therefore,

$$P_e(M) \le (M-1)\,\text{erfc}\sqrt{E_s/n_0}$$

$$= \frac{(M-1)}{\sqrt{2\pi E_s/n_0}}\exp\left[\frac{-E_s}{2n_0}\right],\ \frac{E_s}{n_0} \gg 1 \tag{3.166}$$

It is evident that the lower bound of $P_e(M)$ is simply P_{e1}. Thus,

$$\left.\begin{aligned} &\text{erfc}\sqrt{E_s/n_0} \le P_e(M) \le (M-1)\,\text{erfc}\sqrt{E_s/n_0} \\ &\text{or} \\ &\frac{1}{2}\exp\left(\frac{-E_s}{2n}\right) \le P_e(M) \le \frac{(M-1)}{2}\exp\left(\frac{-E_s}{2n_0}\right) \end{aligned}\right\} \tag{3.167}$$

The limiting value of $P_e(M)$ with $M \to \infty$, using the above bound, leads to the relation

$$P_e(M) \le 2^k \exp(-kE_b/2n_0)$$
$$= \exp[-(kE_b/2n_0 - k\ln 2)]$$

and in the limit $k \to \infty$, $M \to \infty$, $P_e(M) \to 0$, if $E_b/n_0 > 2\ln 2$ (3.168)

This relation, thus, gives the lower limit of E_b/n_0 for error-free communication, 3 dB higher than that of eq. (3.163).

If the signalling waveforms $s_j(t)$ are not orthogonal, but have equal correlation ρ between each other, then, the eq. (3.116) may be directly used in eqs. (3.165) and (3.167), giving the bounds on $P_e(M)$. Thus for equi-correlated signalling waveforms, the error bounds are:

$$\text{erfc}\sqrt{(1-\rho)E_s/n_0} \le P_e(M) \le (M-1)\,\text{erfc}\sqrt{(1-\rho)E_s/n_0} \tag{3.169}$$

It may be shown by detailed analysis that for exact values of $P_e(M)$ for equi-correlated signals the argument of erfc ( ) may be replaced by $\left(y+\sqrt{(1-\rho)2E_s/n_0}\right)$, giving [17],

$$P_e(M) = 1 - \int_{-\infty}^{\infty} \frac{\exp(-y^2/2)}{\sqrt{2\pi}} \left[\text{erfc}\left(y + \frac{(1-\rho)2E_s}{n_0}\right)\right]^{M-1} dy. \tag{3.170}$$

It is interesting to note that an M-ary signal set $s_j(t)$ with cross correlation coefficient ρ and signal energy E_s performs equally well as an orthogonal set with energy equal to $(1-\rho)E_s$. This has been shown to be true for binary signals (of eq. (3.116)) and it is shown to be true for M-ary communication.

Orthogonal Codes

One of the attractive methods of generating orthogonal signals is to phase-modulate a carrier by a binary orthogonal sequence. For a sequence of length M and duration T sec ($T = kT_b$, T_b = bit time), the required bandwidth W is M/T Hz and TW product is M, which is equivalent to the degrees of freedom of the signal space used by the transmitted signals. In such coding, the coder converts the k-information bits into M-element code vectors ($M = 2$) where the chip time T/M sec. The binary sequences are said to be orthogonal if the cross correlation of M-vectors satisfy the conditions:

$$\rho_{ij}(m_i, m_j) = \begin{cases} 1 & \text{for } i = j \\ 0 & \text{for } i \ne j \end{cases} \tag{3.171}$$

Such orthogonal sequences may be generated from the Hadamard matrix using the recurrence formula*,

$$[H_4] = \begin{bmatrix} H_2 & H_2 \\ H_2 & \bar{H}_2 \end{bmatrix}, \text{ where } [H_2] = \begin{bmatrix} 1 & 1 \\ 1 & 0 \end{bmatrix}$$

and Mth order matrix is generated from the $(M/2)$th order one. As an example, an eight-order matrix is given as

*The orthogonal code set may be generated also from the Walsh functions. The codes are also known as the first order Reed-Muller codes.

$$[H_8] \rightarrow \left[\begin{array}{cccc|cccc} 1 & 1 & 1 & 1 & 1 & 1 & 1 & 1 \\ 1 & 0 & 1 & 0 & 1 & 0 & 1 & 0 \\ 1 & 1 & 0 & 0 & 1 & 1 & 0 & 0 \\ 1 & 0 & 0 & 1 & 1 & 0 & 0 & 1 \\ \hline 1 & 1 & 1 & 1 & 0 & 0 & 0 & 0 \\ 1 & 0 & 1 & 0 & 0 & 1 & 0 & 1 \\ 1 & 1 & 0 & 0 & 0 & 0 & 1 & 1 \\ 1 & 0 & 0 & 1 & 0 & 1 & 1 & 0 \end{array}\right] \quad \text{where } [H_8] = \begin{bmatrix} H_4 & H_4 \\ H_4 & H_4 \end{bmatrix}$$

Based on the 1/0 representation of the codevectors, the codes are considered orthogonal if the number of symbol positions in which they are similar equals the number of positions in which they are dissimilar. The test is satisfied by the above set of codes. By interchanging the rows and columns of the derived matrix, the orthogonal codes may be converted to a cyclic one, which is easily generated as the maximal-lenght linear shift-register sequence (also known as PRBS or PNS).

Biorthogonal codes are generated by taking a set of orthogonal code-words and adding to it the complements of each word. Thus $2M$ signals are generated by M-elements only and the total bandwidth requirement is halved. The cross-correlation of these codes is given by

$$\rho_{ij}(m_i, \pm m_j) = \begin{cases} 0 & i \neq j \\ \pm 1 & i = j \end{cases} \tag{3.172}$$

One of the best known biorthogonal codes is the first order Reed-Muller code, an example of which for $M = 16$ is given below:

$$\underset{\text{Orthogonal set}}{\begin{array}{cccccccc} 1 & 1 & 1 & 1 & 1 & 1 & 1 & 1 \\ 1 & 0 & 1 & 0 & 1 & 0 & 1 & 0 \\ 1 & 1 & 0 & 0 & 1 & 1 & 0 & 0 \\ 1 & 0 & 0 & 1 & 1 & 0 & 0 & 1 \\ 1 & 1 & 1 & 1 & 0 & 0 & 0 & 0 \\ 1 & 0 & 1 & 0 & 0 & 1 & 0 & 1 \\ 1 & 1 & 0 & 0 & 0 & 0 & 1 & 1 \\ 1 & 0 & 0 & 1 & 0 & 1 & 1 & 0 \end{array}} \qquad \underset{\text{Complementary set}}{\begin{array}{cccccccc} 0 & 0 & 0 & 0 & 0 & 0 & 0 & 0 \\ 0 & 1 & 0 & 1 & 0 & 1 & 0 & 1 \\ 0 & 0 & 1 & 1 & 0 & 0 & 1 & 1 \\ 0 & 1 & 1 & 0 & 0 & 1 & 1 & 0 \\ 0 & 0 & 0 & 0 & 1 & 1 & 1 & 1 \\ 0 & 1 & 0 & 1 & 1 & 0 & 1 & 0 \\ 0 & 0 & 1 & 1 & 1 & 1 & 0 & 0 \\ 0 & 1 & 1 & 0 & 1 & 0 & 0 & 1 \end{array}}$$

A modified orthogonal code set, known as the simplex code, is generated by deleting the first row and the first column of an orthogonal set. The code set $[H_8]$ is now modified for $M = 7$ as:

$$[H_7] \rightarrow \begin{bmatrix} 0 & 1 & 0 & 1 & 0 & 1 & 0 \\ 1 & 0 & 0 & 1 & 1 & 0 & 0 \\ 0 & 0 & 1 & 1 & 0 & 0 & 1 \\ 1 & 1 & 1 & 0 & 0 & 0 & 0 \\ 0 & 1 & 0 & 0 & 1 & 0 & 1 \\ 1 & 0 & 0 & 0 & 0 & 1 & 1 \\ 0 & 0 & 1 & 0 & 1 & 1 & 0 \end{bmatrix}$$

The codes may be made cyclic by suitably interchanging some rows and columns, and their cross-correlation function is:

$$\rho_{ij}(m_i, m_j) = \begin{cases} 1 & \text{for } i = j \\ -1/(m-1) & \text{for } i \neq j \end{cases} \tag{3.173}$$

The pseudo random binary sequences (PRBS or PNS), generated by feedback shift registers, are also used as semiorthogonal codes with low ρ_{ij}.

It is also observed from the orthogonal code sets that if a codeword is decoded incorrectly, then the number of bits in error is $M/2$, since the distance between the codewords is $M/2$. Thus the bit error probability (BER) is related to the word (symbol) error probability $P_e(M)$ by the equation

$$P_b = \text{BER} = \frac{1}{2} P_e(M) \tag{3.174}$$

and the eqs. (3.161), (3.166) and (3.167) may be modified accordingly to calculate the BER for a given E_b and M. Fig. 3.34 is now redrawn for P_b vs E_b/n_0 (CNR per bit) and is shown in Fig. 3.35. It is now clearly shown that for a given channel with n_0 and P_b specified, the bit energy required decreases considerably with increasing alphabet size, and with $k \to \infty$, $M \to \infty$, Shannon's limit may be reached, giving an ideal coding gain of 11.6 dB at $P_b = 10^{-5}$.

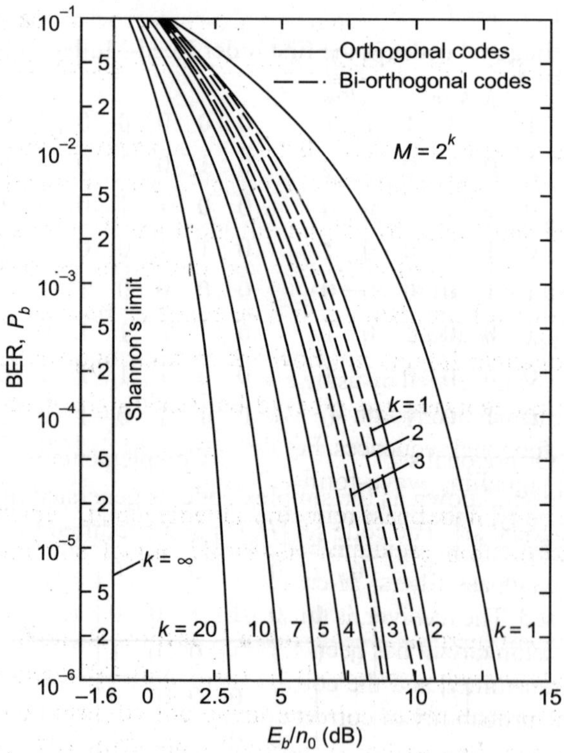

Figure 3.35: BER vs E_b/n_0 for orthogonal and biorthogonal and biorthogonal and biorthogonal M-ary codes

BER for Biorthogonal Codes

It has been observed above that a set of M biorthogonal codes requires only $M/2$ elements in the codeword and the bandwidth required is half of that in the M-element orthogonal set. In the receiver, only $M/2$ cross-correlators/MF's are required, and the decision for, say, H_j is obtained by first comparing the magnitudes of $M/2$ MF-outputs and then by noting if the greatest output is positive or negative. Assuming that the signal $s_1(t)$ is transmitted and the corresponding MF output is positive, the probability of correct decision is given by (cf. eq. (3.159)):

$$\begin{aligned} P_c &= \int_0^{\infty} Pr.\left(|U_1| > |U_j| / U_1\right).p(U_1)\,dU_1,\ j \neq 1 \\ &= \int_0^{\infty} p(U_1)\,dU_1 \prod_{j=2}^{M/2} Pr.\left(|U_1| > |U_j| / U_1\right), \end{aligned} \tag{3.175}$$

where

$$\begin{aligned} Pr.\left(|U_j| < |U_1|\right) &= \int_{-U_1}^{U_1} p(U_j)\,dU_j,\ j \neq 1 \\ &= 1 - 2\ \text{erfc}\left(\sqrt{2}U_1 / \sqrt{n_0 E_s}\right) \end{aligned}$$

Therefore, the symbol error probability is [cf. eq. (3.161)]:

$$P_e(M) = 1 - \int_{-\sqrt{(2E_s/n_0)}}^{\infty} \frac{\exp\left(-y^2/2\right)}{\sqrt{2\pi}} \left[1 - 2\ \text{erfc}\left(y + \sqrt{\frac{2E_s}{n_0}}\right)\right]^{M/2-1} dy \tag{3.176}$$

The BER is $\frac{1}{2}P_e(M)$, for M-large, as in the case of orthogonal codes and a few curves of P_b vs. E_b/n_0 are shown in Fig. 3.35. It is seen that P_b for biorthogonal codes is less than that for orthogonal codes for $k =$ 1, 2, 3. But for large values of k, k ..... 5, P_b for both the cases is the same and the same Shannon's limit of $P_b \to 0$ for $E_b/n_0 > \ln 2$, is obtained for $k \to \infty$, $M \to \infty$.

Orthogonal Signalling with M-FSK

Orthogonal signals may also be generated by using a set of M frequencies, where the frequency difference $\Delta f = 1/T$ Hz ($T = kT_b$). The channel bandwidth $W = M\Delta f = M/T$ Hz and the TW product is M, as in the case of PSK-modulated binary orthogonal sequences. For the M-alphabets of the source, M distinct frequencies over T sec. are used as the signalling wave-forms. Theoretically it should be possible to transmit and detect the wave-forms coherently, if the phase references are available at the receiver. In practice, however, such phase references are difficult to maintain because of the multiplicity of frequencies, and a non-coherent receiver with M-bandpass filters, M-envelope detectors and a decision circuit based on the largest envelope output is used. The receiver, is thus, an extension of Fig. 3.23, and is shown in Fig. 3.36, where the 'Greatest of' decision circuit has been added to decide upon the transmitted symbol H_j.

Assuming that $s_1(t)$ is transmitted and the corresponding envelope-detector output U_1 is compared with all other U_j's, $j \neq 1$, the probability of correct decision is given by (cf. eq. (3,159)).

$$P_c = \int_0^{\infty} p(U_1)\,dU_1 \left[Pr.\left(U_1 > U_j / U_1\right)\right]^{M-1} \tag{3.177}$$

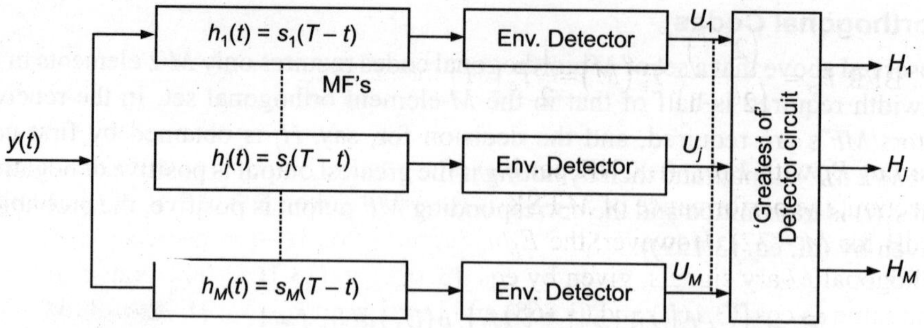

Figure 3.36: Optimum receiver for incoherent *M*-FSK signals

where U_j's, $j \neq 1$, are statistically independent and identically distributed. From eqs. (3.84) and (3.86),

$$p(U_1) = \frac{2U_1}{n_0 E_s} \exp\left[-\left(\frac{U_1^2 + E_s^2}{n_0 E_s}\right)\right] I_0\left(\frac{2U_1}{n_0}\right)$$

$$p(U_j) = \frac{2U_j}{n_0 E_s} \exp\left(-\frac{2U_j^2}{n_0 E_s}\right)$$

and

$$Pr.(U_1 > U_j) = \int_0^{U_1} p(U_j)\, dU_j = 1 - \exp\left(-\frac{U_1^2}{n_0 E_s}\right),\ j \neq 1 \tag{3.178}$$

Further, by using the formula for expansion of $(1 + x)^n$,

$$\left[1 - \exp\left(-\frac{U_1^2}{n_0 E_s}\right)\right]^{M-1} = \sum_{n=0}^{M-1} (-1)^n \binom{M-1}{n} \exp\left(-\frac{nU_1^2}{n_0 E_s}\right) \tag{3.179}$$

Using eq. (3.179) in eq. (3.177), and integrating over U_1,

$$P_c = \sum_{n=0}^{M-1} (-1)^n \binom{M-1}{n} \frac{1}{(n+1)} \exp\left[-\frac{nE_s}{n_0 (n+1)}\right] \tag{3.180}$$

Therefore, the symbol error probability is

$$P_e(M) = 1 - P_c = \sum_{n=1}^{M-1} \frac{(-1)^{n+1}}{n+1} \binom{M-1}{n} \exp\left[-\frac{nE_s}{n_0 (n+1)}\right]$$

and

$$= \frac{\exp(-E_s/n_0)}{M} \sum_{n=2}^{M} (-1)^n \binom{M}{n} \exp(E_s/n_0)$$

$$\text{BER } P_b = \frac{\left(2^{k-1}\right)}{\left(2^k - 1\right)} \cdot P_e(M) \simeq \frac{1}{2} P_e(M) \text{ for } M \gg 1. \quad (3.181)$$

The variation of P_b with E_b/n_0 (= CNR per bit) is shown in Fig. 3.37, where it is seen that, due to incoherent detection, the performance of *M*-FSK is marginally worse than that of *M*-orthogonal signals (by about 0.5 dB for $M = 32$). However, the E_b/n_0 required for $M \to \infty$ decreases in the same way as the case of orthogonal *M*-ary signals, given by eqs. (3.161) and (3.164), The bounds on $P_e(M)$ may be shown to be the same as eqs. (3.167) and (3.168) and in the limit $k \to \infty$, $M \to \infty$, $P_e(M) \to 0$, if $E_b/n_0 > 2 \ln 2$.

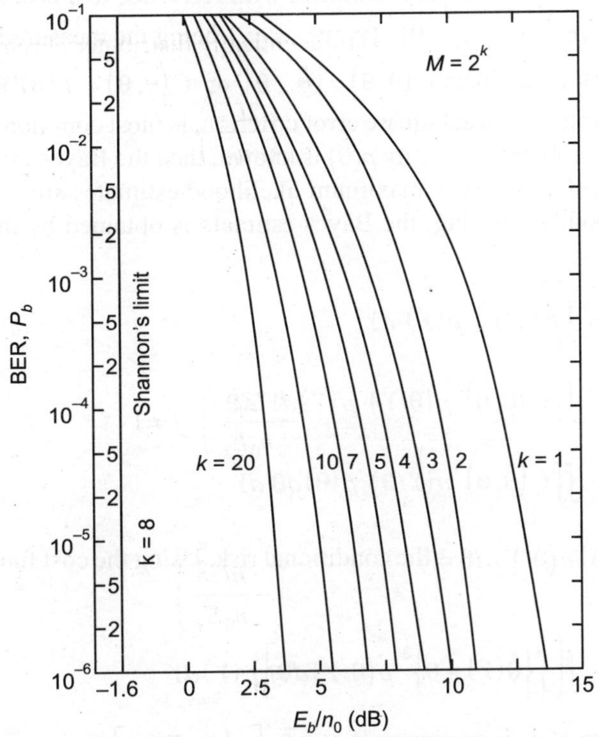

Figure 3.37: BER vs E_b/n_0 for *M*-FSK signals

3.5 ESTIMATION OF SIGNALS PARAMETERS—RADAR PROBLEMS

In radar tracking problems and in analogue communication systems, it becomes necessary to measure the various parameters of the received signals. Specially, in case of radar signals, one determines the target size *h*, range τ, and velocity (in terms of the Doppler shift Δω) whereas in analogue communication, the instantaneous amplitude, phase or frequency is measured to obtain the transmitted information. The theory of parameter estimation discusses methods to find the best estimates of different signal parameters, when both the transmitted parameters and the received signals may vary randomly. Based on the theory of parameter estimation, the techniques of optimum estimation of size, range and velocity of the targets have been developed in this section, both for coherent and incoherent signals. It will be seen that the same matched filters, as are used for optimum detection discussed in earlier sections, are also used here, except that the receivers have to be multiplied when the joint estimation of size, range and velocity is required. The principle of parameter estimation has also been used for the maximum likelihood detection of signals with unknown arrival time and Doppler shift. A particular example of this has been discussed.

Bayes Estimates

In contrast with the binary decision problems where the number of true states are only two and decisions taken for hypotheses H_1 or H_0; in estimation problems, the number of true states and the number of possible observed states are infinite. While finding the best estimate $\hat{\theta}$ of a parameter θ of the transmitted signal, the average risk (cost) $\bar{C}\left[\hat{\theta}/Y\right]$ is minimized with reference to a cost function $C(\hat{\theta},\theta)$, and the conditional PDF $p(Y/\theta) = p(y_1, y_2, ..., y_n/\theta\}$, $\{y_1, y_2, ..., y_n\}$ being the measured quantities corresponding to θ. The cost function may be either $C(\hat{\theta},\theta) = \left|\hat{\theta}-\theta\right|$ or $C(\hat{\theta},\theta) = f(\theta)(\hat{\theta}-\theta)^2$, but the squared-error cost function, similar to the mean-square error criterion, is most commonly used in radar problems. In practice, if the a priori probability density $p(\theta)$ is known, then the Bayes estimate gives the minimum average cost, while with $p(\theta)$ unknown, maximum likelihood estimates are used.

In similarity with hypothesis testing, the Bayes estimate is obtained by minimizing the Bayes risk given by

$$\begin{aligned}\bar{C}\left[\hat{\theta}/Y\right] &= \int R(\hat{\theta}/Y)\cdot p(Y)\cdot dY \\ &= \iint C(\hat{\theta},\theta)\cdot p(\theta/Y)\cdot p(Y)\,dY\,d\theta \\ &= \iint C(\hat{\theta},\theta)\cdot p(Y/\theta)\cdot p(\theta)\,d\theta\,dY \qquad (3.182)\end{aligned}$$

where, $R(\hat{\theta}/Y) = \int C(\hat{\theta},\theta)\,p(\theta/Y)\,d\theta$ = the conditional risk. Using the cost function $C(\hat{\theta},\theta) = (\hat{\theta}-\theta)^2$, the average risk is,

$$\bar{C} = \int\left[\int\{\hat{\theta}(Y)-\theta\}^2\,p(\theta/Y)\,d\theta\right]p(Y)\,dY$$

The mean-square variation of $\hat{\theta}$ is minimum if $\hat{\theta} = \bar{\theta}$ and the average risk $\bar{C}$ is minimised by making $\partial\bar{C}/\partial\hat{\theta} = 0$, giving

$$\begin{aligned}\hat{\theta}(Y) &= \int\theta\,p(\theta/Y)\,d\theta = E[\theta/Y] \qquad (3.183)\\ &= \text{the conditional expected value of } \theta.\end{aligned}$$

If many parameters $(\theta_1, \theta_2, ..., \theta_n)$ are to be simultaneously estimated from the measurements $Y = \{y_1, y_2, ..., y_n\}$, then the average risk of eq. (3.182) is modified as

$$\bar{C}[\hat{\Theta}(Y)] = \int\ldots\int R(\hat{\Theta}/Y)\,p(Y)\,dY \qquad (3.184)$$

If the prior PDF $p(\theta)$ of the parameter values is unknown, then one finds the least favourable distribution, just as in the case of testing of composite hypotheses. The Bayes estimate found by using the least favourable distribution $\overline{p}(\theta)$ is called the Minimax estimate, which gives the maximum value of the Bayes risk $\overline{C}_{\min}$. Since the estimates $\hat{\theta}$ will vary because of the random variations in Y, the best estimate $\hat{\theta}$ should be as close to θ as possible and $p(\hat{\theta}/\theta)$ should be sharply peaked in the neighbourhood of θ. If the mean $\hat{\theta}_j(Y) = \theta_j$, the estimate is said to be unbiased; if not, the difference $(\bar{\hat{\theta}}_j - \theta_j)$ is called the bias of the estimate, The bias of an estimate should be small; also the spread of $\hat{\theta}_j$ about their mean θ_j should be small.

Maximum Likelihood Estimates

If the cost function is not known, then a reasonable method of choosing an estimation strategy is to maximize the a posteriori *pdf* $p(\theta/Y)$ (= MAP estimate), and this is equivalent to maximizing the likelihood function $p(Y/\theta)$ when $p(\theta)$ has a large dispersion. The pdf $p(Y/\theta)$ has a peak in the neighbourhood of the value of θ that maximizes it and the corresponding peaks of $p(Y/\theta)$ and $p(\theta/Y)$ become nearer to each other as $p(\theta)$ becomes more uniform and the initial values of the parameters become more uncertain. One then adopts the estimate of θ as the value which maximize the joint pdf $p(Y/\theta)$ of the observations $\{Y\}$. This procedure gives the so-called maximum likelihood (ML) estimates. The maximizing of $p(Y/\theta)$ is equivalent to maximizing the likelihood ratio $\lambda(Y/\theta)$ and λ_{av} of eq. (3.35) may also be used to obtain the *ML* estimates.

The *ML* estimates of the parameters $(\theta_1, \theta_2, ..., \theta_n)$ may be found by solving the set of equations,

$$\frac{\partial}{\partial\theta_j} p(Y/\Theta) = 0, 1 \le j \le n \tag{3.185}$$

In general, there may be many roots of these equations and one has to choose the solutions $\hat{\Theta}(Y)$ which gives the highest peaks in $p(Y/\Theta)$. It has been shown that the ML estimates are asymptotically normal, unbiased and efficient. When n is large, $p(\hat{\Theta}/\Theta)$ of the estimates have nearly normal distributions with a mean $\hat{\Theta}$ nearly equal to the true values of Θ and with a variance (for the estimation of a signal parameter) given by

$$\sigma^2(\hat{\theta}) \simeq \frac{\left(\partial\langle\hat{\theta}\rangle/\partial\theta\right)^2}{E\left\{\left[\partial/\partial\theta \ln p(Y/\theta)\right]^2\right\}}, \langle\hat{\theta}\rangle = E\{\hat{\theta}\} \tag{3.186}$$

Equation (3.186) is based on the Cramer-Rao bound for $\sigma^2(\hat{\theta})$*.

---

*The Cramer-Rao bound for $\sigma^2(\hat{\theta})$ is given as [3]

$$\sigma^2(\hat{\theta}) \ge \frac{\left(\partial E\{\theta\}/\partial\theta\right)^2}{E\left\{\left[(\partial/\partial\theta)\ln p(Y/\theta)\right]^2\right\}} \tag{3.186a}$$

In the eq. (3.186a), the equality holds only when,

$$(\partial/\partial\theta)\ln p(Y/\theta) = f(\theta)\left[\hat{\theta} - E\{\hat{\theta}\}\right]$$

For signals with additive white Gaussian noise,

$$y(t) = s(t; \Theta) + n(t), \quad 0 \le t \le T$$

Using eqs. (3.69) and (3.71), the likelihood function of $y(t)$ is

$$p(y/\Theta) = K \exp\left[-\frac{1}{n_0}\int_0^T \{y(t) - s(t;\Theta)\}^2 \, dt\right]$$

and

$$\frac{\partial \ln p(y/\Theta)}{\partial \theta_j} = \frac{2}{n_0}\int_0^T [y(t) - s(t;\Theta)]\frac{\partial s(t;\Theta)}{\partial \theta_j} dt \tag{3.187}$$

The *ML*-estimates $\{\hat{\theta}_j\}$ are now obtained by setting the above equation equal to zero and solving for θ_j. For the estimation of a single parameter θ_1, eq. (3.187) reduces to

$$\int_0^T \left[y(t) - s(t;\hat{\theta}_1)\right]\frac{\partial s(t;\hat{\theta}_1)}{\partial \hat{\theta}_1} dt = 0 \tag{3.188}$$

and the variance is given by

$$\sigma^2(\hat{\theta}_1) \simeq \left\{\frac{2}{n_0}\int_0^T \left[\frac{\partial s(t,\theta_1)}{\partial \theta_1}\right]^2 dt\right\}^{-1} \tag{3.189}$$

For coherent signals with coherent detection, one can operate on the envelope function only and consider $y(t)$ as a lowpass signal. The estimates of the amplitude h and delay τ may be obtained by processing the coherently-detected envelope $hB(t - \tau)$ by a lowpass MF along with a suitable multiplier and a sampler. For incoherent signals with complex envelopes due to possible phase or frequency modulation of the carrier (say, for example, due to Doppler shifts), the correlation of the two complex envelopes has to be obtained through an incoherent *MF* followed by an envelope detector, as used in Fig. 3.18 for detection of signals of unknown phase. Some specific problems of estimation in radar signals are discussed below.

Example 3.11: Assume that n independent samples have been taken of a stationary Gaussian noise of variance σ^2. To estimate the mean, say α, one has to maximize the joint pdf $p(Y/\alpha)$, where

$$p(Y/\alpha) = \prod_{i=1}^{n} \frac{1}{\sqrt{2\pi}\sigma}\exp\left(-\frac{(y_i - \alpha)^2}{2\sigma^2}\right)$$

Using log $p(Y/\alpha)$

$$\frac{\partial}{\partial \alpha}\ln p(Y/\alpha) = \sum_{i=1}^{n}\frac{(y_i - \alpha)}{\sigma^2}$$

---

where $f(\theta)$ is not a function of $\{Y\}$ or $\hat{\theta}$, but may be a function of θ. For an unbiased estimator, the bound becomes,

$$\sigma^2(\hat{\theta}) \ge \frac{1}{E\left\{\left[(\partial/\partial\theta)\ln(Y/\theta)\right]^2\right\}} \tag{3.186b}$$

and by equating this to zero, the estimate is

$$\hat{\alpha} = \frac{1}{n}\sum_{i=1}^{n} y_i = \text{sample mean.}$$

The mean $\hat{\alpha} = \alpha$ and the varincee of $\hat{\alpha}$ is:

$$\sigma^2(\hat{\alpha}) = \sigma^2/n$$

3.5.1 Estimation of Target Size and Range (Coherent Signals)

Consider a simple radar problem, where the transmitted signal $s(t) = A(t)\cos(\omega t - \theta)$ and the reflected signal is $h\beta(\tau)\,A(t-\tau)\cos(\omega t - \psi)$, where h is the measure of the target size, τ is the time delay ($= 2d/c$, $d =$ the distance of the target, $c =$ velocity of e.m. waves) corresponding to the target range and $\beta(\tau) = k/\tau^2$ for the target in free space. For coherent detection, only the envelope function $h\beta(\tau)\,A(t-\tau)$ of duration T sec. need be considered for the estimation of target size and range. With band-limited ($\pm$W Hz) Gaussian noise, the received signal may be written as

$$\begin{aligned} y(t) &= \text{envelope of } s(t; h, \tau) + n(t) \\ &= h\beta(\tau)\,A(t-\tau) + n(t) \end{aligned}$$

and the likelihood function for $2TW$ samples,

$$\begin{aligned} p(Y/h;\tau) &= \left(\frac{1}{2\pi\sigma^2}\right)^{TW} \exp\left[-\frac{1}{2\sigma^2}\sum_{1}^{2TW}(y_i - A_i)^2\right] \\ &= K\exp\left\{-\frac{1}{n_0}\int_{\tau_1}^{\tau_m+T}\left[y(t) - A(t; h, \tau)\right]^2 dt\right\} \end{aligned} \tag{3.190}$$

where, the Doppler shift and target velocity are assumed to be small and the range is known to be in the interval $\tau_1 < \tau < \tau_m$.

To determine the target size, assuming that the range τ_0 is known, the above likelihood function is maximized using eq. (3.187), then,

$$p(Y/h) = K\exp-\frac{1}{n_0}\int_{\tau_0}^{\tau_0+T}\left[y(t) - \frac{h}{\tau_0^2}A(t-\tau_0)\right]^2 dt$$

and maximizing $\ln p(Y/h)$ with respect to h, the maximum likelihood estimate $\hat{h}_{ML}$ is

$$\hat{h}_{ML} = \frac{\tau_0^2}{E}\int_{\tau_0}^{\tau_0+T} y(t)\,A(t-\tau_0)\,dt \tag{3.191}$$

where $\int A^2(t-\tau_0)\,dt \simeq E$.

The estimator is, then, a matched filter followed by a multiplier and a sampler, as shown in Fig. 3.38, where the peaky processed pulse as presented to the A-scope is also shown.

The conditional risk (= variance of $\hat{h}$) is given by

$$R_{\text{ML}}(\hat{h}/h) = \overline{(\hat{h}-h)^2} = \left(\frac{1}{2WE}\right)^2 \cdot \overline{\Sigma n_i A_i \Sigma n_j A_j} = \frac{n_0}{2E} = \sigma_{\hat{h}}^2 \tag{3.192}$$

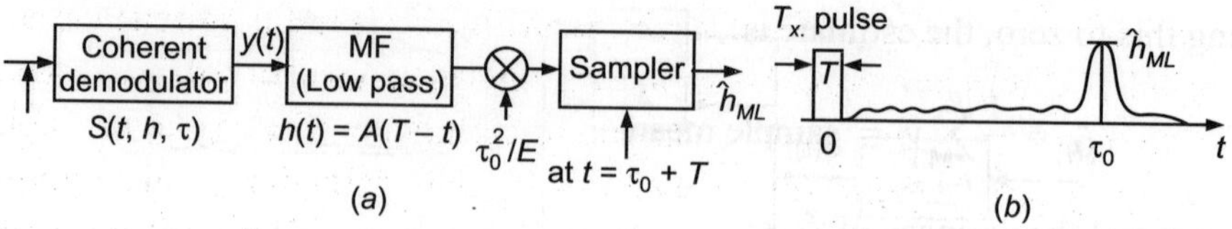

Figure 3.38: (*a*) Estimating receiver for $\hat{h}_{ML}$ using coherent-signal, **(*b*)** Processed pulse as presented to the *A*-scope

Since the PDF's $p(h)$ and $p(\tau)$ are normally unknown, the least favourable distribution $p(h)$ = constant, $p(\tau)$ = constant, may be used to obtain the Bayes estimates. Since now,

$$p(h/Y) = K_1 \cdot p(Y/h)$$

and

$$p(\tau/Y) = K_2 \cdot p(Y/\tau), \quad K_1 \text{ and } K_2 \text{ constants}$$

the Bayes estimates are the same as the ML-estimates. It is however more meaningful to use the ML-estimates in radar problems.

To determine the target range τ, assuming h_0 to be known and $p(\tau)$ = constant, one has to maximize the likelihood function,

$$p(Y/\tau) = K.\exp\left\{-\frac{1}{n_0}\int_{\tau_1}^{\tau_m+T}\left[y(t) - \frac{h_0}{\tau^2}A(t-\tau)\right]^2 dt\right\}$$

and the maximum of ln $p(Y/\tau)$ is obtained by processing the signal in the optimum receiver shown in Fig. 3.39. The maximum of the pulse presented to the *A*-scope is obtained as a maximum of

$$\left[K_1 + \frac{2h_0}{\tau^2}\int_T y(t)\,A(t-\tau)\,dt - \frac{h_0^2}{\tau^4}E\right] \tag{3.193}$$

and the time at which this maximum occurs is the ML-estimate $\hat{\tau}_{ML}$.

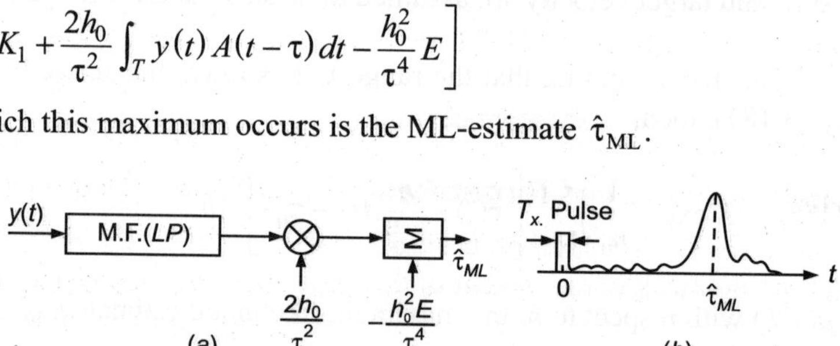

Figure 3.39: (*a*) Estimation for $\hat{\tau}_{ML}$ using coherent signal, **(*b*)** Received pulse on the *A*-scope

3.5.2 Simultaneous Estimation of Size and Range

To obtain the estimates $(\hat{h}, \hat{\tau})$ simultaneously, one has to maximize the function $p(Y/h_j;\,\tau)$ of eq. (3.190) simultaneously with reference to h and τ. One way to implement the dual-function receiver is to use eq. (3.193) and to have an array of matched filter-weighting networks, shown in Fig. 3.40, which would compute $p(Y/\tau)$ for various discrete values of h, the increments of h being chosen as close as necessary. The optimum estimates will, then, be the coordinates of the maximum $p(Y/h_j;\,\tau)$, as will be given from the *A*-scope representation of the various outputs of the array of the function generators.

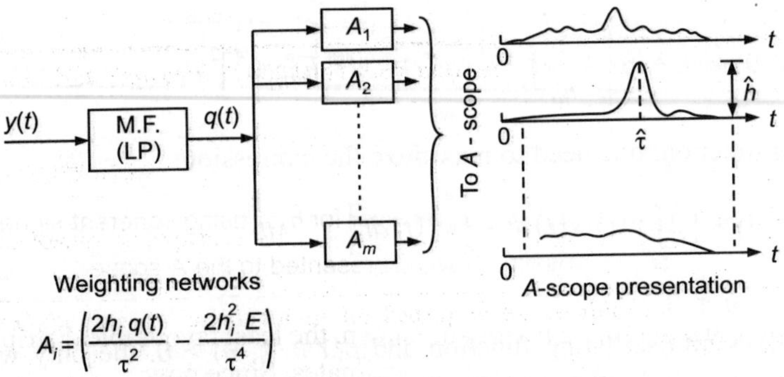

Figure 3.40: Joint estimator using function generators

Alternatively one may use eq. (3.91) and the output of the matched filter may be passed through an array of delay lines whose delays τ_i are incremental in the observation interval $\tau_1 < t < (\tau_m + T)$. If each output is multiplied by the weighting factor τ_i^2/E, the highest peak of all the traces seen on the scope at $t = t(h_{max})$ would give the optimum $(\hat{h}, \hat{\tau})$. This may also be instrumented with a single delay line tapped at $(\tau_1, \tau_2, ..., \tau_m)$, as shown in Fig. 3.41, and each tap-output multiplied by the corresponding weighting factors. The maximum output sampled at $t = t(h_{max})$ gives the ML-estimates $\hat{h}$ and $\hat{\tau}$.

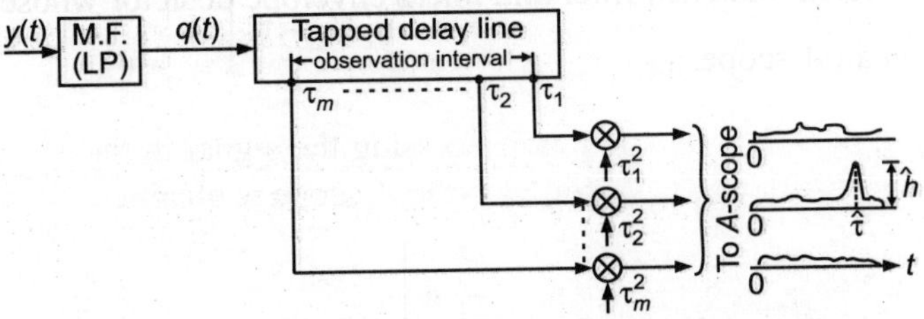

Figure 3.41: Joint estimator using a single tapped delay line

3.5.3 Composite Estimation of Target Range (Incoherent Signals)

Above optimum estimates are based on coherent signals, but the received signals are generally incoherent due to Doppler shift and the random movement of the target. The delayed received signal is now of the form,

$$\begin{aligned} y(t) &= s(t; h, \tau, \theta) + n(t) \\ &= h(\tau)\, A(t - \tau) \cos\,[\omega t - \theta(t)] + n(t) \end{aligned} \tag{3.194}$$

where $\omega t = \omega_0 t + \psi(t)$ including the Doppler shift if any, $h(\tau) = k/\tau^2$, $\theta(t)$ = unknown phase due to the incoherence of the transmitted carrier and varying τ. Since θ is random and gives rise to fine-structure ambiguity function, it is only the arrival time τ of the envelope $A(t - \tau)$ that will be measured to give the optimum estimate $\hat{\tau}$.

Assuming for the present that the Doppler shift is negligible in the interval $t_1 < t < \tau_m + T$ and ω is known, the estimate $\hat{\tau}$ is to be optimized with respect to $h(\tau)$ and $\theta(t)$, meaning thereby that the estimates $\hat{h}$ and $\hat{\theta}$ are to be used to obtain $\hat{\tau}$. The likelihood function to be maximized now is,

$$p(Y/h;\,\tau,\theta) = K\exp\left\{-\frac{1}{n_0}\int_{\tau_1}^{\tau_m+T}\left[y(t)-s(t;h,\tau,\theta)\right]^2 dt\right\} \tag{3.195}$$

Using log-likelihood function, one need to maximize the expression

$$F(h,\tau,\theta) = \left[2\int y(t)\cdot s(t)\,dt - \int s^2(t)\,dt\right] \tag{3.196}$$

to obtain the estimates $\hat{h}, \hat{\tau}, \hat{\theta}$.

Since cos $(\omega t - \theta)$ is an oscillating function and $p(Y/h,\,\tau,\,\theta) > 0$, the only way to maximize the likelihood function is to consider the envelope function $A(t-\tau)$. Referring to Sec. 3.3.3, the envelope function to be maximized is of the form

$$G_1(\tau) = \text{Env. of } F(h,\tau,\theta) = \left|2h\int_{\tau_1}^{\tau_m+T} A^*(t-\tau)\,B(t)\,dt\right| - h^2 E \tag{3.197*}$$

where h = constant, $h^2\int_T A^2(t-\tau)\cos^2(\omega t-\theta)\,dt = h^2E$, and $A^*(t)$ is the conjugate of the complex envelope of $s(t)$; $B(t)$ is the envelope of the received signal $y(t)$. Therefore, for the stationary targets (h = constant), the estimate $\hat{\tau}$ is obtained by maximizing $\left|\int A^*(t-\tau)\,B(t)\,dt\right|$. The optimum estimator, then, consists of an incoherent matched filter and linear envelope detector whose output is maximum at $t_m = (\hat{\tau}+T)$, as seen on an A-scope.

---

*Since with phase or frequency modulation of the carrier, the signal $s(t)$ has a complex envelope, and due to Doppler shift and other nonlinearities, $y(t)$ also has a complex envelope, the envelope correlation of the two signals are calculated as follows:

For tow complex narrow band signals [refer to eq. (1.72) Chapter 1],

$$x(t) = \text{Re}\left\{\tilde{A}_x(t).\exp(j\omega_0 t)\right\}$$

$$y(t) = \text{Re}\left\{\tilde{B}_y(t).\exp(j\omega_0 t)\right\}$$

where $\tilde{A}_x(t)$, $\tilde{B}_y(t)$ are complex envelopes, the cross-correlation function

$$\rho_{xy}(\tau) = \text{Re}\left\{\rho_{\tilde{A}\tilde{B}}(\tau)\exp(j\omega_0 t)\right\}$$

where

$$\rho_{\tilde{A}\tilde{B}}(\tau) = \frac{1}{T}\int_0^T \tilde{A}_x^*(t-\tau)\,\tilde{B}_y(t)\,dt = \text{complex envelope of } \rho_{xy}(\tau)$$

Thus, $\left|\int_T \tilde{A}_x^*(t-\tau)\,B_y(t)\,dt\right|$ is the envelope correlation between $x(t)$ and $y(t)$. This function is maximized to obtain $\hat{\tau}$. If, however, $s(t)$ is simply a rectangular RF pulse, then the envelope $R(t)$ of eq. (3.79), as used for the detection of signals with unknown phase, may be used for obtaining $\hat{\tau}$.

In practice, however, $h(\tau)$ will vary randomly and $G_1(\tau)$ has to be maximized with respect to $h(\tau)$, $h > 0$. Thus making $\partial G_1(\tau)/\partial h = 0$, the estimate

$$\hat{h} = \frac{1}{E}\int_{\tau_1}^{\tau_m+T} A^*(t-\tau)\,B(t)\,dt. \tag{3.198}$$

Using this estimate $\hat{h}$ in eq. (3.197), the *ML*-estimate of τ is obtained by maximizing

$$G(\tau) = \left|\int_{\tau_1}^{\tau_m+T} A^*(t-\tau)\,B(t)\,dt\right|^2 \tag{3.199}$$

The optimum estimator is now a matched filter followed by a quadratic detector, as shown in Fig. 3.42, and the $\hat{\tau}_{ML} = (t_m - T), t_m$ being the time for maximum pulse height. The estimate of the pulse amplitude $\hat{h}$ may be obtained by using $\hat{\tau}$ in eq. (3.198) and it will be seen that the peak value of the detector output is proportional to the square of $\hat{h}_{ML}$.

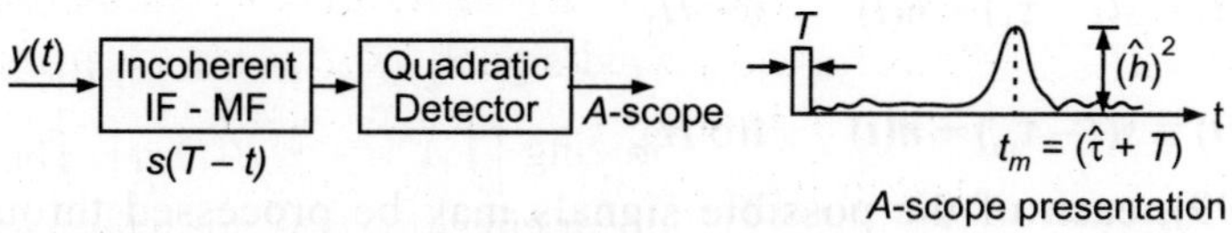

Figure 3.42: Estimator for target range with (h, τ, θ) variable

Since there will be some error in the measurement of $\hat{\tau}$, the variance $\sigma_{\hat{\tau}}^2 = \overline{(\tau - \hat{\tau})^2}$, for large SNR, has been shown to be

$$\sigma_{\hat{\tau}}^2 = \frac{n_0}{2E_r B^2} \tag{3.200}$$

where B = equivalent *IF* bandwidth and E_r = received signal energy. Thus for large SNR, the error depends only on the received energy and receiver bandwidth. For small SNR, the noise peaks will be comparable with the signal peak and the measurement will be very much in error. However, the ML-estimate of τ has been shown to be asymptotically efficient for large SNR, and the above technique is used in conventional radar-range measuring systems.

For multiple observations of the radar echoes, the optimum estimate $\hat{\tau}$ is obtained by following the results of Sec. 3.3.5 and the function to be maximized is:

$$\overline{G(\tau)} \simeq \sum_{1}^{n}\left|\int_{\tau_1}^{\tau_m+T} A^*(t-\tau)\,B(t)\,dt\right|^2 \tag{3.201}$$

where the sum is over nT_0 intervals, T_0 is pulse interval. This multiple estimator is shown in Fig. 3.43, which is similar in principle to that of the optimum detector of Fig. 3.19. The variance is now given by

$$\overline{\sigma_\tau^2} = \frac{n_0}{2B^2}\cdot\frac{1}{\sum_{i=1}^{n} E_i} \tag{3.200a}$$

which is less than that obtained by taking the average of estimates $\{\hat{\tau}_1, \hat{\tau}_2, ..., \hat{\tau}_n\}$.

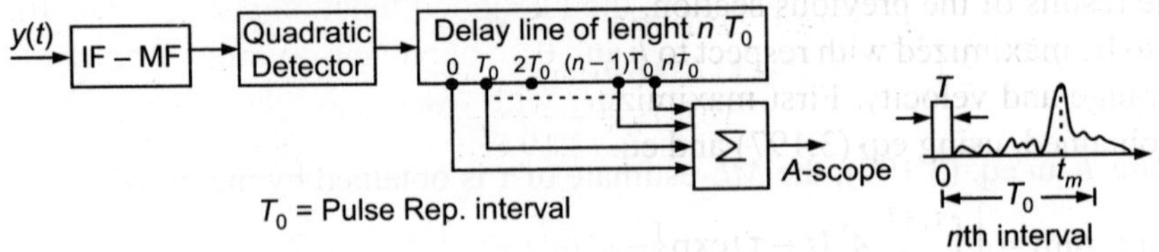

Figure 3.43: Optimum estimator of $\hat{\tau}$ using multiple echoes

An alternative strategy for estimating τ is to take a multiple alternative hypothesis approach to determine if a signal is present, and if so, to determine its time of arrival τ. If the observation interval is finely divided into m subintervals, $\{\tau_1, \tau_2, ..., \tau_m\}$, then assuming that (h, ω) are constant, the required hypotheses are:

$$
\begin{aligned}
y(t) &= n(t) \text{ only} && \text{for } H_0 \\
y(t) &= s(t-\tau_1) + n(t) && \text{for } H_1 \\
&\vdots \\
y(t) &= s(t-\tau_m) + n(t) && \text{fro } H_m
\end{aligned}
\tag{3.202}
$$

Then using eq. (3.199), each of the possible signals may be processed through the receiver of Fig. 3.42 and the output suitably delayed through a TDL as in Fig. 3.41. The delayed outputs, say, $\{u_1, u_2, ..., u_m\}$ are passed through a 'Greatest of' decision circuit to indicate the largest output, say, u_i, as shown in Fig. 3.44. This value is then compared with threshold u_0 and if $u_i < u_0$, H_0 is decided upon. Otherwise, if $u_i > u_0$, H_i is decided upon; and the corresponding $\tau_i = \hat{\tau}$ and the amplitude $|u_i| \propto (\hat{h})^2$. Thus a detection theory approach, in this case, gives an efficient solution to the problem

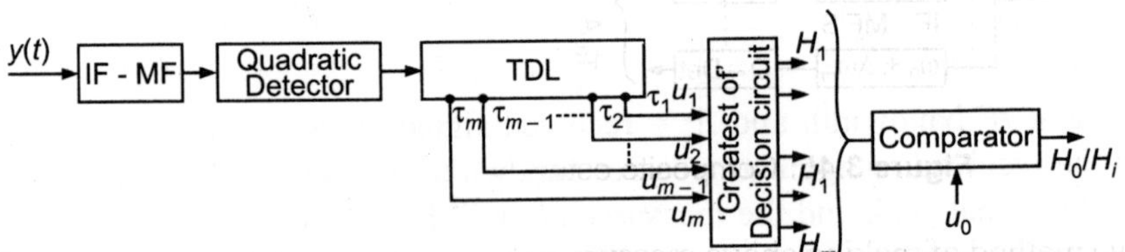

Figure 3.44: Estimation of τ using multiple hypotheses

3.5.4 Composite Estimation of Target Range and Velocity (Incoherent Signals)

It has been assumed in the last section that the Doppler shift is zero and the target has no radial velocity. But in practice, both Doppler shift and range are very useful information to be obtained from the measurements on the received echoes. Modifying the eq. (3.194), the delayed received signal is now of the form

$$
\begin{aligned}
y(t) &= s(t, h, \tau, \Delta\omega, \theta) + n(\tau) \\
&= h(\tau)\, A(t-\tau) \cos\,[\omega_0 t + \Delta\omega(t-\tau) - \theta(t)] + n(t)
\end{aligned}
\tag{3.203}
$$

where, the Doppler shift $\Delta\omega = (2v/c)\omega_0$, v = radial velocity of the target towards the receiver, c = velocity of e.m. waves. Assuming that the bandwidth W of the envelope $A(t)$ is small compared to ω_0

and $2TWv/c \ll 1$, it is found that the envelope distortion is negligible and the received pulse has the same shape as that of the transmitted pulse.

Following the results of the previous section, the likelihood function $p(Y/h, \tau, \Delta\omega, \theta)$, similar to the eq. (3.195), has to be maximized with respect to h and θ, to obtain the optimum estimate $G(\tau, \Delta\omega)$ which determines the range and velocity. First maximizing with respect to θ and then with respect to h, the *ML*-estimate is obtained, using eq. (3.197) and eq. (3.199), by maximizing

$$G(\tau, \Delta\omega) = \left| \int_{\tau_1}^{\tau_m + T} A^*(t-\tau)\exp\{-j\Delta\omega(t-\tau)\}.B(t)\,dt \right|^2 \tag{3.204}$$

If v = constant, then $\Delta\omega$ is also constant; and the effective received RF frequency is $(\omega_0 + \Delta\omega)$. The matched filter of Fig. 3.42 is now matched to the complex envelope $A^*(t-\tau)$ and a carrier frequency $(\omega_0 + \Delta\omega)$; then ML-estimate $\hat{\tau}$ is obtained from the peak pulse seen on the *A*-scope after the quadratic detector.

When the radial velocity is changing, as is the usual case, the carrier is effectively frequency-modulated and many values of $\Delta\omega$ will have to be estimated. Thus to determine the ML-estimates $\widehat{\Delta\omega}$ and $\hat{\tau}$, it is necessary to construct a bank of parallel filters, each matched to $A^*(t).\exp[-j(\Delta\omega_1 + \omega_0)t]$, $A^*(t)\exp[-j(\Delta\omega_2 + \omega_0)t]$, ..., $A^*(t)\exp[-j(\Delta\omega_m + \omega_0)t]$ respectively, as shown in Fig. 3.45. $\{\Delta\omega_1, \Delta\omega_2, ..., \Delta\omega_m\}$ are spaced uniformly over the range of expected frequency shift, the width of the range being limited by the maximum possible target velocity. Each filter is followed by a quadratic detector, whose output is observed on the *A*-scope. The estimate $\widehat{\Delta\omega}$ is the value of $\Delta\omega_p$ associated with that filter whose detector output has the largest peak. The time t_m at which the peak occurs gives the estimate $\hat{\tau}$. The estimate $\widehat{\Delta\omega}$ may be made as accurate as desired by making larger and larger number of filters, but the limit of the resolution is given approximately by $\sqrt{n_0 \cdot 2ET^2}, T$ = pulse-width.

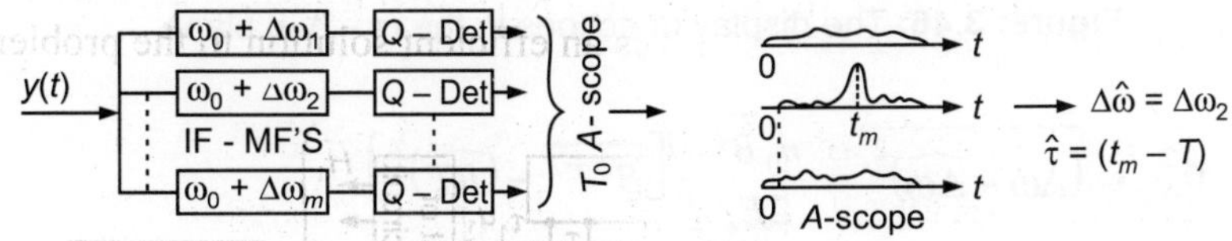

Figure 3.45: Composite estimation of range and velocity

An alternative method of making above measurements is to use a single tapped delay line, whose taps are so weighted through a resistance matrix as to represent the different matched filters of Fig. 3.45, matched to different frequencies $(\omega_0 + \Delta\omega_p)$, as shown in Fig. 3.46. The added outputs of the weighted taps are passed through the quadratic detectors and the resulting time functions are displayed on the *A*-scope. The *m*-outputs of the receiver are used to intensity-modulate *m* horizontally sweeping electron beams impinging on a fluorescent screen. Each beam corresponds to a particular range of Doppler shift (≡ target velocity). A peak on the $(t - \Delta\omega)$ plane shows up as a bright spot on the screen at the appropriate $(\tau - \Delta\omega)$ coordinates giving the estimates $\hat{\tau}$ and $\widehat{\Delta\omega}$.

For approximate rectangular pulses and rectangular bandwidths of the IF filter, the variances of $\hat{\tau}$ and $\widehat{\Delta\omega}$, for large SNR, have been shown approximately to be, similar to eq. (3.200), as

$$\sigma_{\hat{\tau}}^2 = \overline{(\tau - \hat{\tau})^2} = \frac{n_0 T^2}{2E_r} \Big/ \left[\overline{B^2 T^2} - \left(\overline{BT}\right)^2\right]$$

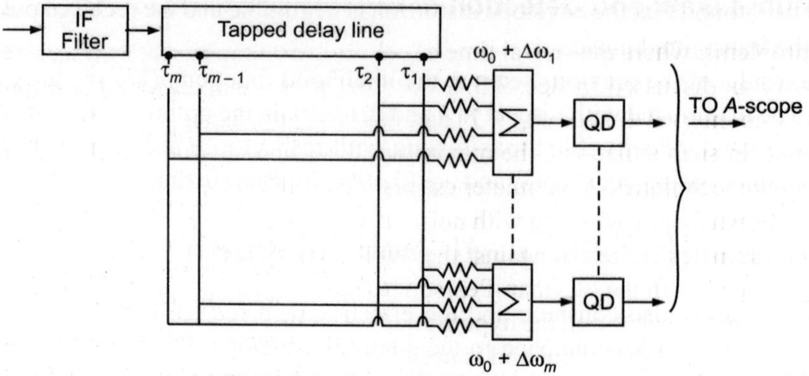

(*a*) Single-delay line estimator of Δ_ω and τ

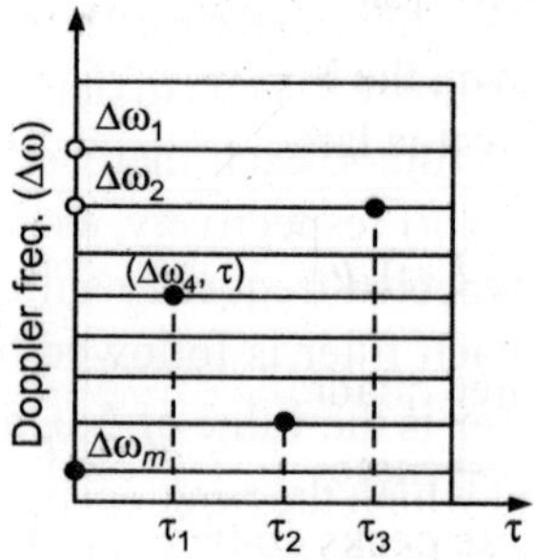

(*b*) The display of echoes in the $(\tau - \Delta\omega)$ plane

Figure: 3.46: The display of echoes in the $(\tau - \Delta\omega)$ plane

$$\sigma^2_{\hat{\Delta}_\omega} = \overline{\left(\Delta\omega - \widehat{\Delta\omega}\right)^2} = \frac{n_0 \overline{B^2}}{2E_r} \Big/ \left[B^2\overline{T^2} - \left(\overline{BT}\right)^2 \right] \tag{3.205}$$

where, $\overline{T^2}$ = mean-square duration of the signal pulse, $\overline{B^2}$ = mean-square width of IF filters, and $\overline{BT}$ = mean cross product of time and frequency. For a given pulse shape, an increase in $\overline{B^2}$ decreases $\overline{T^2}$ and hence, $\sigma^2_{\hat{\tau}}$ cannot be decreased without increasing $\sigma^2_{\hat{\Delta}_\omega}$ at the same time. However, by using sharp rectangular pulses, the most accurate estimates $\hat{\tau}$ and $\widehat{\Delta\omega}$ are possible, but the matched filters have to be very carefully designed.

For the case of multiple observations, it has been shown that the appropriate receiver for a particular $\Delta\omega_p$ is the same as shown in Fig. 3.43. For a range of Doppler shifts, a combination of Figs. 3.43 and 3.45 is used and the receiver is multiplied *m* times, *m* depending upon the frequency resolution required. The error-variances $\sigma^2_{\hat{\tau}}$ and $\sigma^2_{\hat{\Delta}_\omega}$ are now modified by replacing E_r by ΣE_i in eq. (3.205). In making observations over a number of intervals, the target distance and hence τ may change considerably during the time of observation. Suitable compensation for this has to be provided for and this may be done by suitably adjusting the time by which each detector output is delayed as required in Fig. 3.43.

3.5.5 Maximum-Likelihood Detection of Signals of Unknown Arrival Time

In radar search problems, where the arrival time of echoes are unknown, the optimum receiver, based on the decision theory as discussed in Sec. 3.3.4, does not give a satisfactory P_d, as the signal is to be searched over a large interval $\tau_1 < t < (\tau_m + T)$ and the integrated noise power becomes considerable at the receiver output. In such situations, the maximum likelihood measures of the signal parameters are obtained by using the techniques of parameter estimation, as discussed earlier in this section. Assuming that the signal of known form $s_0(t)$ along with noise $n(t)$ is expected at $t = \hat{\tau}$, the signal parameters Θ are estimated and the estimates are tested against the 'Null Hypothesis' H_N, i.e. if they have been possibly generated by $n(t)$ alone. As in the Neyman-Pearson test, a confidence level Θ_0 ($\equiv$ the threshold λ_t) is set so that a given P_f is not exceeded, and the hypothesis H_N is taken as true if $\Theta < \Theta_0$. This method is known as the 'Maximum-likelihood detection' and is used for echoes of unknown τ.

Considering the echoes from the stationary targets with $\Delta\omega = 0$, the received signal is of the form

$$y(t) = h(\tau)\, A(t - \tau) \cos [\omega t - \theta(t)] + n(t) \tag{3.194}$$

Since the values of h, τ and θ are unknown, the ML-estimates of these are made as given by eqs. (3.198) and (3.199). Since the observation interval is large, $G(\tau)$ of eq. (3.199) may be approximated as:

$$G(t) = \left| \int_{-\infty}^{\infty} A^*(t - \tau) B(t)\, dt \right|^2 \tag{3.206}$$

and this is generated by the matched filter-quadratic detector combination shown in Fig. 3.42. The null hypothesis is to be tested now for the estimates $\hat{h}, \hat{\tau}, \hat{\theta}$, and for this, the joint PDF $p(h, \tau, \theta)$ is to be studied for $y(t) = n(t)$ only. Since the noise peaks could occur at any time during the observation interval with uniform probability density, the value of $\hat{\tau}$ is not useful in testing the null hypothesis. Moreover for $y(t) = n(t)$, θ is random in the range $0 < \theta < 2\pi$ with $p(\theta) = 1/2\pi$, and this also is not useful in testing the hypotheses. The amplitude estimate $\hat{h}$ will, however, be small when only noise is present, and the peak $G(\hat{\tau})$ [$\propto (\hat{h})^2$ as given by eq. (3.199)] may be used as the test statistic. If the confidence level G_0 is determined for the given P_f, then,

$$\text{for } G(\tau) < G_0, \qquad H_N \text{ is true}$$

and

$$\text{for } G(\tau) > G_0, \qquad H_s \text{ is true} \tag{3.207}$$

By multiple observations of the echoes, P_d is improved and the $\hat{\tau}_{ML}$ is obtained by maximizing, as in Fig. 3.43, the function

$$\overline{G(\tau)} \equiv \sum_{1}^{n} \left| \int_{-\infty}^{\infty} A^*(t - \tau) \cdot B(t)\, dt \right|^2 \tag{3.208}$$

where the sum is over nT_0 intervals, T_0 = pulse interval. The null hypothesis is now tested by using the statistic

$$Z \triangleq E^2 \sum_{1}^{n} \left(\hat{h}_j\right)^2 = \overline{G(\hat{\tau})} \tag{3.209}$$

and the peak $\overline{G(\hat{\tau})}$ is compared with a bias level G_0 (related to P_f) to decide whether a signal is present or not.

The false-alarm probability P_f is determined from the error rate $e(G_0)$ with which the statistic $G(\hat{\tau})$ crosses the threshold G_0. It has been shown that [2, 12]

$$e(G_0) = \frac{B_s G_0}{\sqrt{4\pi E}} \exp\left(-\frac{G_0^2}{4E}\right) \tag{3.210}$$

and

$$P_f \simeq (\tau_2 + T - \tau_1) e(G_0) \ll 1 \tag{3.211}$$

For multiple observations,

$$P_f(n) = \frac{B_2(\tau_2 + T - \tau_1)}{(n-1)!\sqrt{\pi}} \cdot \left(\frac{G_0}{\sqrt{4E}}\right)^{2n-1} . \exp\left(-\frac{G_0^2}{4E}\right) \tag{3.212}$$

where B_s = effective signal bandwidth. For the given P_f, G_0 is calculated from the above equations and is used as the bias level of the receivers of Figs. 3.42 and 3.43. The detection probability P_d is then calculated from the values of G_0 and $p(G)$ vs $G(\hat{\tau})$ curves. From calculated values of P_d vs. P_f for different B_s and SNR, it may be shown that the changes in the required SNR is small for large changes in the observation interval for a given value of (P_d, P_f). (For an increase of observation interval by a factor of 10^4, the increase in E_b/n_0 is only 2 dB approximately). This is a definite improvement over the results obtained by the detection technique of Sec. 3.3.4.

It has been assumed so far that $\Delta\omega = 0$, but with Doppler shifts, the received signal is of the form given by eq. (3.203). The estimate of $\Delta\omega$ is made by using a bank of filters, as shown in Fig. 3.45, and $\widehat{\Delta\omega}$ and $\hat{\tau}$ are associated with the largest peak of the detected output. The null hypothesis is now tested by comparing the peak of $G(\tau, \Delta\omega)$ with a bias level G_0, corresponding to the allowable P_f. The P_f at each filter output is determined by eq. (3.211) and the total P_f for signal pulses is approximately given by

$$P_f \cong \frac{(W_2 - W_1)(\tau_2 + T - \tau_1) G_0}{\sqrt{4\pi E}} . \exp\left(-\frac{G_0^2}{4E}\right)$$

where $(W_2 - W_1) = m \cdot \Delta\omega$ = total receiver bandwidth. For multiple observations, the value of P_f is modified according as the eq. (3.212).

3.6 PARAMETER ESTIMATION IN COMMUNICATION SIGNALS

In communication systems, the estimation problems relate to:

(i) Acquisition of carrier frequencies and their phases and subsequent tracking of both, as required for coherent detection of digitally-modulated signals.
(ii) Symbol and word synchronisation for digital data.
(iii) Optimum demodulation of AM, FM and Pulse modulated analog signals.
(iv) Optimum filtering.

These problems are discussed in this and next sections using the results of the last section; specifically, the eq. (3.188) gives the required solution in tracking problems. Solutions to the tracking problems finally lead to the well-understood phase-locked loop and its use in FM demodulation gives optimum results. The acquisition problems are better solved by using multiple alternative hypothesis approach, as discussed along with eq. (3.202). For optimum demodulation and filtering, the results include those of Wiener filters, which were originally derived from the minimum-mean-square error criterion. Finally, PLL and FM-FB are discussed as optimum demodulators for FM signals.

3.6.1 Estimation of Signal Amplitudes (Coherent Detection)

Consider a transmitted signal $s(t, h) = hs_0(t)$ where $s_0(t)$ is a known baseband waveform of duration T and band-limited to W Hz, and h is the amplitude parameter to be estimatd. Using coherent detection, any RF signal may be translated to the baseband without any deterioration of the SNR in the baseband. In general, the received signal is then

$$y(t) = (y_1, y_2, ..., y_{2TW}) = hs_0(t) + n(t) \tag{3.213}$$

where the noise power $N = n_0W = \sigma^2$. Assuming the time of arrival of the signal to be known, one may form either the Bayes estimate or the maximum-likelihood estimate of h.

If $p(h)$ is Gaussian with mean h_0 and variance σ_h^2, then using eqs. (3.183) and (3.190), the Bayes estimate is.

$$\hat{h}(y) = \int h \cdot p(h/Y)\,dh$$

where

$$p(h/Y) = p(h) \cdot p(Y/h)/p(Y)$$

$$= K \cdot \exp\left[\left\{\frac{(h-h_0)^2}{2\sigma_h^2} + \frac{1}{2\sigma^2}\sum_{i=1}^{2TW}(y_i - hs_i)^2\right\}\right]$$

$$= K_1 \exp\left[-\frac{1}{2}\left(h - \bar{h}_y\right)^2 \Big/ \sigma_e^2\right] \tag{3.214}$$

where again,

$$\bar{h}_y = \frac{\Sigma s_i y_i + h_0\sigma^2/\sigma_h^2}{2WE + \sigma^2/\sigma_h^2};\ \Sigma s_i^2 = 2WE \text{ and } \sigma_e^2 = \frac{\sigma^2 \cdot \sigma_h^2}{\sigma^2 + 2WE\sigma_h^2}$$

Then, the Bayes estimate, which gives the minimum average risk $\bar{C}_{\min}$ according to eq. (3.183), is (by substituting the integral for Σ)

$$\hat{h}_{\text{Bayes}} = \bar{h}_y = \frac{1}{2WE + \sigma^2/\sigma_h^2}\left[2W\int_T s_0(t)\,y(t)\,dt + \frac{h_0\sigma^2}{\sigma_h^2}\right] \tag{3.215a}$$

and the mean-square error is σ_e^2. The average risk is

$$R_{\text{Bayes}} = \bar{C}_{\min} = \frac{n_0\sigma_h^2}{n_0 + 2E\sigma_h^2}, \tag{3.215b}$$

since $n_0W = \sigma^2$, and σ_e^2 is independent of h. The Bayes estimate here is also the MAP estimate.

The maximum likelihood estimate, on the other hand, is obtained directly from eq. (3.191) as

$$\hat{h}_{ML} = \frac{\Sigma y_i s_i}{\Sigma s_i^2} = \frac{1}{E}\int_T y(t)\, s_0(t)\, dt \tag{3.216}$$

and the average risk is given by (from eq. (3.189)),

$$R_{ML}\left(\hat{h}/h\right) = \overline{\left(\hat{h}-h\right)^2} = \sigma_h^2 = \frac{n_0}{2E} \tag{3.217}$$

Comparing the average risks in the two methods,

$$R_{ML}/R_{Bayes} = 1+\frac{n_0}{2E\sigma_h^2}, \tag{3.218}$$

it is seen that $\hat{h}_{ML}$ would be worse than $\hat{h}_{Bayes}$, if the distribution of h is very narrow (i.e., σ_h^2 small). The receiver in both the cases is implemented by using the coherent detector, cross-correlator and sampler, or a matched filter-detector combination.

3.6.2 Estimation of Amplitude, Phase and Frequency of a Sinewave

Consider now the received signal to be of the form,

$$y(t) = A \sin(\omega_0 t + \theta) + n(t), \quad 0 \le t \le T$$

If the *a priori* probability of A is known, then the maximum inverse probability criterion would be the same as the maximum-likelihood criterion. If now, ω_0 and θ are known, then ML-estimate of A is given from eq. (3.216) as

$$\hat{A}_{ML} = \frac{1}{E}\int_T y(t)\sin\omega_0 t\, dt$$

and the variance of $\hat{A}$ is $\sigma_A^2 = n_0/2E$ (3.219)

The estimator now consists of a cross-correlator and a sampler.

For the estimation of θ, ω_0 and A being known, the eq. (3.188) is used, and the value of $\hat{\theta}$ is calculated by solving

$$\int_0^T \left[y(t) - A\sin\left(\omega_0 t+\hat{\theta}\right)\right]\cos\left(\omega_0 t+\hat{\theta}\right)dt = 0 \tag{3.220}$$

Assuming $\omega_0 T \gg 1$, the equation reduces to (since the integral of the second term = 0),

$$\int_0^T y(t)\cos\left(\omega_0 t+\hat{\theta}\right)dt = 0$$

and expanding,

$$\cos\hat{\theta}\int_0^T y(t)\cos\omega_0 t\, dt = \sin\hat{\theta}\int_0^T y(t)\sin\omega_0 t\, dt$$

Therefore,

$$\hat{\theta}_{ML} = \tan^{-1}\frac{\int_0^T y(t)\cos\omega_0 t\,dt}{\int_0^T y(t)\sin\omega_0 t\,dt} \tag{3.221}$$

The estimator then consists of MF/correlators and a function generator, as shown in Fig. 3.47. For $\theta \ll \pi/2$, the approximation, $\tan\theta \simeq \theta$, may be used. Assuming $p(\theta) = 1/2\pi$, $-\pi \le \theta < \pi$, the MAP estimate is also the same as the ML-estimate.

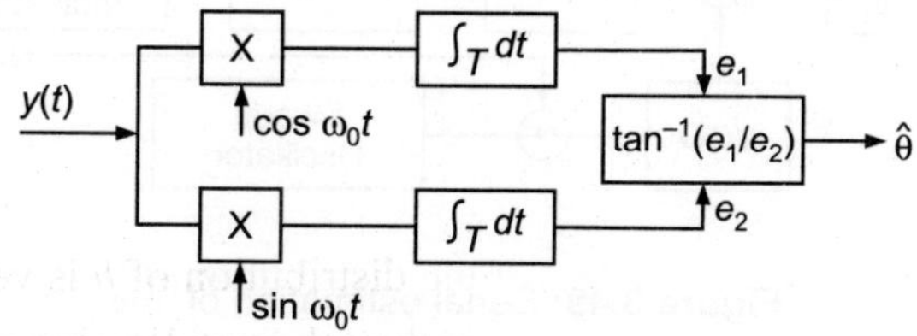

Figure 3.47: Estimator for the phase of sinusoid

To estimate the frequency ω_0 of $y(t)$, the likelihood function $p(Y/\omega, \theta)$ is to be maximized, assuming A to be known and θ uniformly distributed over the range $-\pi \le \theta < \pi$. From the results of Sec. 3.3.3 and 3.5.4, it is seen that the ML-estimate is obtained by maximizing the eq. (3.204), modified as

$$G(\Delta\omega) = \left| \int_T A^*(t-\tau_0)\exp\left[-j\Delta\omega(t-\tau_0)\right]\cdot B(t)\,dt \right| \tag{3.222}$$

where τ_0 is known. With uniform distribution of $\Delta\omega$, this also gives the MAP-estimate. The variance of the estimate is $\sigma^2_{\widehat{\Delta\omega}} = 6n_0/ET^2$. It is now necessary to finely divide the expected frequency range (ω_1, ω_2) into m subintervals $\{\Delta\omega_1, \Delta\omega_2, ..., \Delta\omega_m\}$ and to take the multiple alternative hypothesis approach, as given by eq. (3.202). The estimator now consists of a bank of incoherent MF's followed by envelope detectors as shown in Fig. 3.48 (cf. Figs 3.44 and 3.45) and a 'Greatest of' decision circuit indicates the largest of the detector outputs $\{u_1, u_2, ..., u_m\}$. The largest output, say u_i, corresponds to $\Delta\omega_i$ = the frequency estimate $\widehat{\Delta\omega}$. The same result is obtained if one uses eq. (3.80) and the corresponding Fig. 3.18 as the starting point for the estimation of $\Delta\omega$. Since the signal of TW product = $T(W_2 - W_1)$ is

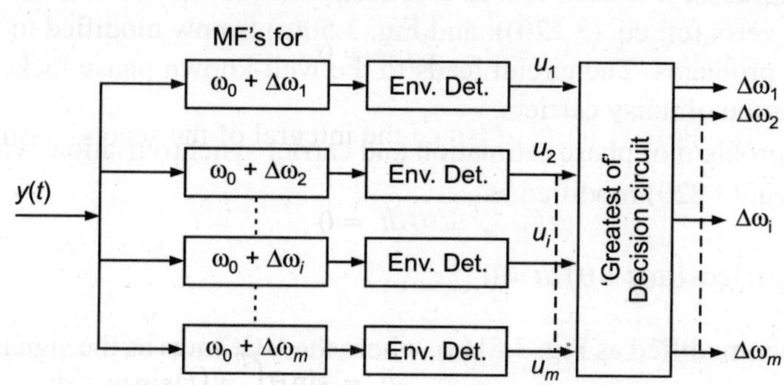

Figure 3.48: Estimation of unknown frequency

completely represented by $2TW$ samples, it is only necessary to have $m = 2TW$ MF's, separated by frequency difference $\Delta\omega = 1/2T$. The number of MF's required may then be quite large, specially when E/n_0 is small. An alternative circuit for frequency estimation is to use a swept oscillator, e.g., a voltage-controlled oscillator (VCO) with a linear sweep, and an incoherent correlator using an MF to obtain the output $\{u_i\}$ serially, as shown in Fig. 3.49. The maximum of $\{u_i\}$ may be indicated by noting that $d\{u_i\}/dt = 0$ for the maximum. Once the maximum is obtained, the corresponding sweep voltage (precalibrated) indicates $\Delta\omega_i = \widehat{\Delta\omega}$ and the tracking of the carrier is now done through a PLL, as discussed below. Other alternative circuits have been discussed in [16].

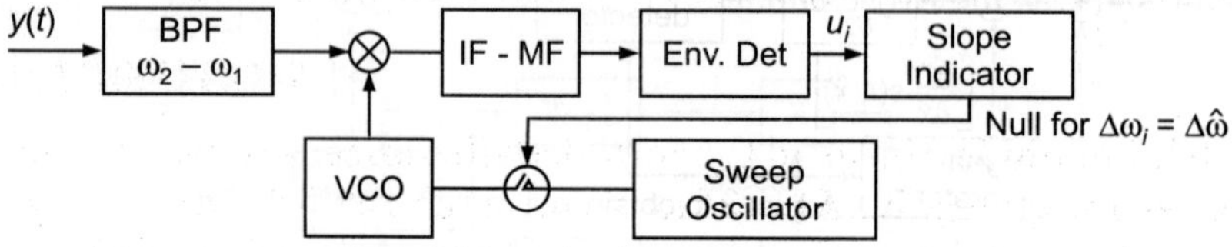

Figure 3.49: Serial estimation of $\widehat{\Delta\omega}$

3.6.3 Phase Locked Loop (PLL) [16]

Consider the eq. (3.188), rewritten as:

$$\int_{-T/2}^{T/2} \left[y(t) - s(t, \hat{x})\right] \frac{\partial s(t, \hat{x})}{\partial \hat{x}} dt = 0 \tag{3.223}$$

where $x(t)$ is the variable to be estimated and the solution of the equation gives $\hat{x}$. The equation can be represented as a feedback network, shown in Fig. 3.50(a), where the signal generator is controlled by the null detector output, and the estimate $\hat{x}$ is obtained from $s(t; \hat{x})$ under the null condition. For the estimation of a single parameter of $s(t; x)$, the correlation function $\rho(x) = \int_T y(t) \cdot s(t; x)\, dt$, passes through a maximum at $x = \hat{x}$, as discussed in earlier sections. Thus a condition for optimum estimator is

$$\left.\frac{\partial_\rho (x)}{\partial x}\right|_{x=\hat{x}} = \int_T y(t) \frac{\partial s(t; \hat{x})\, dt}{\partial \hat{x}} dt = 0 \tag{3.224}$$

Comparing with eq. (3.223), it is seen that as a consequence of eq. (3.224), the integral of the second term of eq. (3.322) is zero; (cf. eq. (3.220)); and Fig. 3.50(a) is now modified to Fig. 3.50(b) for single parameter estimation problems. The circuit leads to the well-known phase locked loop, used specially for phase-synchronisation of noisy carriers.

Consider now the problem of phase estimation and carrier synchronisation, where the solution for $\hat{\theta}$ is obtained from the eq. (3.220) modified as

$$\int_0^T y(t) \cos\left(\omega_0 t + \hat{\theta}\right) dt = 0 \tag{3.225}$$

Fig. 3.50 is now further modified as Fig. 3.51(a), where the VCO acts as the signal generator, generating $B\cos\left(\omega_0 t + \tilde{\theta}\right), \tilde{\theta} \simeq \tilde{\theta}$, and the error voltage $e(t)$ controls the VCO and its phase such that $\in (t) \to 0$. The time differential of the VCO-output phase $\psi(t)$ is given by

$$\frac{d\psi(t)}{dt} = \omega_0 + K_1 e(t)$$

or

$$\psi(t) = \omega_0 t + K_1 \int e(t)\, dt = \omega_0 t + \tilde{\theta} \tag{3.226}$$

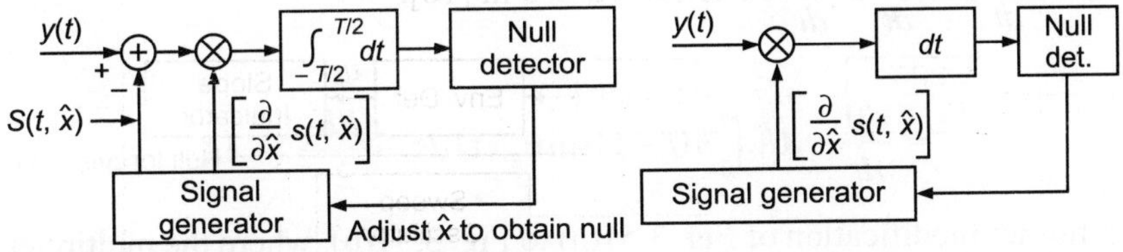

(a) Representation of Eq. (3.223) (b) An equivalent PLL

Figure 3.50

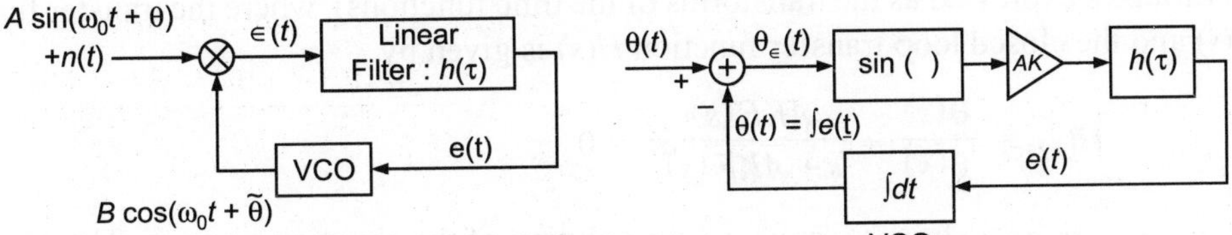

(a) A PLL circuit (b) A modified model of the PLL

Figure 3.51

Thus the integrator required in eq. (3.225) is inbuilt in the VCO, and the linear filter governs the loop bandwidth while eliminating undesired frequencies and noise.

The error voltage $\in(t)$ without noise ($n(t) = 0$) is

$$\in(t) = \frac{AB}{2}\left[\sin(\theta - \tilde{\theta}) + \sin\left(2\omega_0 t + \theta + \tilde{\theta}\right)\right]$$

The linear filter in the loop eliminates the second harmonic and $e(t)$ is

$$e(t) = \frac{AB}{2}\sin(\theta - \tilde{\theta}) \cong \frac{AB}{2}(\theta - \tilde{\theta}), \text{ for } (\theta - \tilde{\theta}) \ll \pi/2 \tag{3.227}$$

According to the principles of negative feedback, the voltage $e(t)$ acts in such a way that the error $\in(t) \to 0$ and $\tilde{\theta} \to \theta$, the true value of the phase.

The PLL may be further modified by considering that filter response is given by

$$e(t) = \int_0^t h(t-\tau) \in(\tau)\, d\tau,$$

$$= \frac{AB}{2}\int_0^t h(t-\tau)\sin\theta_{\in}(\tau)\, d\tau, \tag{3.228}$$

where $h(t)$ = impulse response of the linear filter and $\theta_\in = (\theta - \tilde{\theta})$. Now using eq. (3.226),

$$\theta_\in = \theta - \tilde{\theta} = \theta - K_1 \int e(t)\, dt \tag{3.229}$$

and neglecting the constant phase $\omega_0 t$ in the product terms, the PLL may be described by a differential equation as

$$\frac{d\theta_\in}{dt} = \frac{d\theta}{dt} - \frac{d\tilde{\theta}}{dt}$$

$$= \frac{d\theta}{dt} - AK\int_0^t h(t-\tau)\sin\theta_\in(\tau)\, d\tau, \quad \frac{ABK_1}{2} = AK \tag{3.230}$$

This suggests a further modification of Fig. 3.51(a) to Fig. 3.51(b), where the multiplier is replaced by a difference circuit and a sin ( ) nonlinearity, and the VCO by an integrator. Then the VCO output is proportional to $\int e(t)$ and the loop gain is proportional to the input amplitude A. The PLL may now be linearised by assuming, $\theta_\in \ll \pi/2$, and $\sin\theta_\in \simeq \theta_\in$. The linearised model of the Pll is now shown in Fig. 3.52 (variables expressed as the transforms of the time functions), where the transfer function of the filter is $F(s)$ and the closed loop transfer function $H(s)$ is given by

$$H(s) \triangleq \frac{\tilde{\theta}(s)}{\theta(s)} = \frac{AKF(s)}{s + AKF(s)} \tag{3.231}$$

and

$$\theta_\in(s) = [1 - H(s)]\,\theta(s)$$

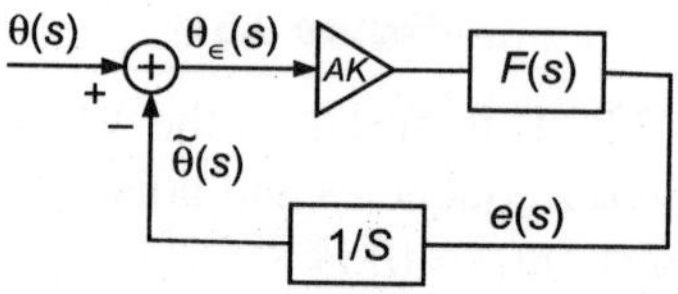

Figure 3.52: The linearised model of the PLL

Based on the transfer function $F(s)$, the PLL may be designated as a first-order, second-order or third-order loop. For $F(s) = 1$, the loop is first-order, for $F(s) = 1 + a/s$, the loop is second-order and for $F(s) = 1 + \frac{a}{s} + \frac{b}{s^2}$, the loop is third-order. The tracking capability of the loops may be calculated as follows:

For the first-order loop, $F(s) = 1$ and $H(s) = AK/(s + AK)$. For an input signal $y(t) = A\sin(\omega t + \theta_0)$, the equivalent input phase θ as defined in eq. (3.229), is $\theta = (\omega - \omega_0)\,t + \theta_0$. Then,

$$\theta(s) = \frac{\omega - \omega_0}{s^2} + \frac{\theta_0}{s}$$

and

$$\theta_\in(s) = [1 - H(s)]\,\theta(s)$$

$$= \frac{\omega - \omega_0}{s(s + AK)} + \frac{\theta_0}{s + AK} \tag{3.232}$$

The steady-state phase error is given by the final-value theorem for Laplace transform as

$$\lim_{t \to \infty} \theta_\in(t) = \lim_{s \to 0} s\theta_\in(s)$$

and from eq. (3.232),

$$\lim_{s \to 0} s\theta_\in(s) = \frac{\omega - \omega_0}{AK} \tag{3.233}$$

thus giving a steady state error. For a fixed input frequency ω, $(\omega - \omega_0)$ is constant and hence the VCO achieves frequency lock, but it can never achieve phase lock.

For the second-order loop, with $F(s) = 1 + a/s$, the closed-loop transfer function is

$$H(s) = \frac{AK(s + a)}{s^2 + AKs + aAK}$$

and

$$\theta_\in(s) = \frac{(\omega - \omega_0) + \theta_0 s}{S^2 + AKs + aAK}$$

which gives the steady-state error,

$$\lim_{s \to 0} s\theta_\in(s) = \lim_{t \to \infty} \theta_\in(t) = 0 \tag{3.234}$$

It is then seen that by introducing an integrator in the loop, the steady-state error is eliminated and the VCO achieves both frequency and phase-lock for a constant frequency input but having random phase, under the condition that $\theta_\in(t) \ll \pi/2$.

For the ease of loop stability, sometimes a zero is introduced in $F(s)$, giving $F(s) = \dfrac{s + a}{s + b}$. This results in a steady-state error,

$$\lim_{t \to \infty} \theta_\in(t) = \lim_{s \to 0} s\theta_\in(s) = \frac{b(\omega - \omega_0)}{aAK} \tag{3.235}$$

This error is similar to that obtained in a first-order loop but decreased by a factor b/a. Thus the loop gain AK should be large for maintaining the steady-state error small.

In case the input frequency varies with time, say for continuous Doppler-shifts, the equivalent phase input $\theta = \alpha t^2 + (\omega - \omega_0)t + \theta_0$. In such cases, both first order loop and the loop with imperfect integrator give unbounded error at $t \to \infty$. For the loop with ideal integrator, the error $\theta_\in(t)_{t \to \infty} = \alpha/aAK$, and the loop gain has to be large to keep the error small. The steady-state error can be made zero only with a third order loop. With $F(s)$ of the form

$$F(s) = 1 + a/s + b/s^2$$

the phase error is,

$$\theta_{\epsilon}(s) = \frac{s^3}{s^3 + AKs^2 + aAKs + bAK}\left(\frac{\alpha}{s^3} + \frac{\omega - \omega_0}{s^2} + \frac{\theta_0}{s}\right) \tag{3.236}$$

which gives $\theta_{\epsilon}(t) \to 0$ for $t \to \infty$. However, the loop has to be properly phase-compensated to avoid instabilitly. All the above results are obtained on the basis of the linear model of the PLL, shown in Fig. 3.52.

The above analysis of the PLL has been done neglectng the input noise in $y(t)$. In the non-linear model of Fig. 3.51, $n(t)$ may be introduced as $n(t)/A$ in the eq. (3.225), giving

$$\int_0^T A\left[\sin(\omega_0 t + \theta) + n(t)\,A\right]\cos(\omega_0 t + \tilde{\theta})\,dt = 0$$

and the Eq. (3.230) is modified as (since A also contributes to the loop gain = AK),

$$\frac{d\theta_{\epsilon}}{dt} = \frac{d\theta}{dt} - K\int_0^t h(t-\tau)\left[A\sin\theta_{\epsilon}(\tau) + n'(\tau)\right]d\tau \tag{3.237}$$

where $n'(\tau)$ has the same σ_n^2 as $n(t)$. For the linear model, superposition holds and the noise power in the loop is the filtered version of the input noise power, as indicated in Fig. 3.53. Assuming $n_0/2$ to be the double-sided noise spectral density of $n(t)$, $n(t)$ white and Gaussian, the variance of $\tilde{\theta}$ is

$$\sigma^2(\tilde{\theta}) = \frac{n_0}{2A^2}\int_{-\infty}^{\infty}|H(j\omega)|^2\frac{d\omega}{2\pi}$$

$$= \frac{n_0}{A^2}B_n \tag{3.238}$$

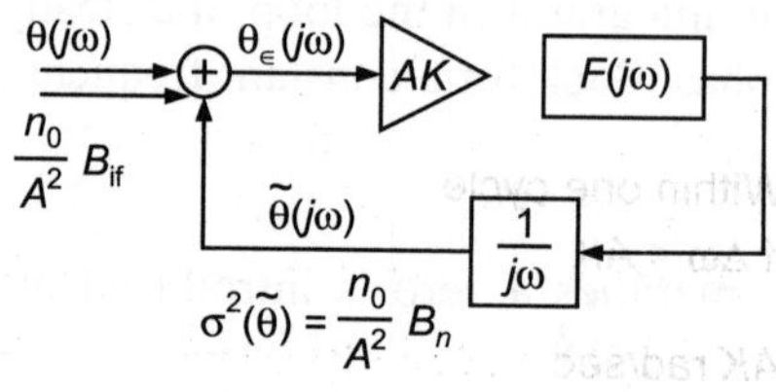

Figure 3.53: Linear model with noise input; input CNR = $\dfrac{A^2}{2n_0B_{if}}$

where, from eq. (3.231), $H(j\omega) = H(s)|_{s=j\omega}$ and the noise bandwidth B_n of the loop is defined as

$$B_n \triangleq \frac{1}{2\pi H(0)}\int_0^{\infty}|H(j\omega)|^2\,d\omega \tag{3.239}$$

In general B_n should be large enough so that the loop is able to track efficiently the variation of the input phase, i.e., $B_n \geq$ the bandwidth of the phase noise of the carrier ω_0, but Bn $\ll$ the system bandwidth which is wide enough because of the modulation of the carrier by digital or analog signals. Thus the loop behaves as a lowpass filter for $n(t)$, thereby improving the loop SNR considerably as compared to the CNR at the RF of IF level of the receiver (SNR gain $\simeq B_{if}/B_n$).

The transient and steady state behaviour of the PLL is generally determined by analysing the nonlinear differential equations of higher orders through phase-plane techniques. Some of the important results of this analysis are given in the Table 3.2 [16].

Table 3.2: Characteristics of PLL

Type of Loop	1st order	2nd order	3rd order
B_n(Hz)	$AK/4$	$(AK + a)/4$; For imperfect integrator: $B_n = \dfrac{AK(AK+a)}{4(AK+b)}$	$\dfrac{AK(aAK+a^2-b)}{4(aAK-b)}$
Pull-in range = $\Delta\omega$(rad/sec); $\Delta\omega_m$ = frequency lock range	$\Delta\omega \le AK$	$\Delta\omega$ very large with T_{acq} large frequency lock range: $^*\Delta\omega_m = 2.6\omega_n$ for $\xi = \dfrac{1}{\sqrt{2}}$; For imperfect integrator: $\Delta\omega_m < 2\omega_n\sqrt{\xi\omega_n b+1}$	Loop unstable for acquisition; $\Delta\omega_n$ same as for 2nd order loops
With linearly varying input frequency $\omega = \omega_0 + Dt$, the search rate $D = 2\Delta\omega/T_0$; $2\Delta\omega$ = frequency range to be searched, T_0 = time period of sweep	. . .	$D \le 2B_n^2, \xi = 1/2$; For imperfect integrator; $D \simeq 2B_n^2\left(1-\dfrac{1}{\sqrt{\text{loop SNR}}}\right)$	$D < \omega_n^2$
Lock-on time = T_{acq}	Within one cycle if $\Delta\omega < AK$	$\dfrac{3.5(\Delta f)^2}{B_n^3}$, Δf = frequency offset	. . .
Hole-on range	AK rad/sec	. . . for all loops	

*Natural frequency of the loop: $\omega_n = \sqrt{aAK}$, and damping constant $\xi = AK/2\omega_n = \omega_n/2a$ (= 0.707 in general).

The most important characteristic of the loop is its pull-in range, i.e., the maximum frequency offset $\Delta\omega$, that may be permitted so that the loop locks on the the input frequency ω_0 and phase θ. This lock-on operation is possible with reasonable $\Delta\omega$, because the difference frequency $\Delta\omega$ produces a d.c. voltage at the phase detector output which pushes the average VCO frequency towards ω_0 in small increments, say, in steps of $\delta\omega$. Thus a major part of the acquisition time T_{acq} is spent in changing the VCO frequency by $\Delta\omega$ in steps of $\delta\omega$, and this results into 'cycle skipping' (or cycle slipping). Once the frequency lock is obtained, the loop stops skipping cycles and begins to pull into phase lock with $\theta_\epsilon \to 0$. The value of T_{acq} is then calculated based on the value of $\delta\omega$ at each step and the time taken for reaching the phase lock after the frequency lock. As an example, for a PLL with $B_n = 5$ Hz, $\Delta f = 10$ Hz and initial $\theta_\epsilon = -\pi/2$, it has been calculated that;

Time required for frequency lock ~ 3 sec

(during which the loop skipped cycles)

and

Time required for phase lock ~ 0.5 sec, giving
Total T_{acq} ~ 3.5 sec.

Although the second-order loop has a large pull-in range, there is a limit of tracking the linearly varying input frequency $\omega = \omega_0 + Dt$. For large D, the value of B_n has to be large, making the phase error $\sigma^2(\tilde{\theta})$ large. Similarly when a VCO with a linear sweep, as in Fig. 3.49, is used to acquire ω_0, the sweep rate is limited by B_n. The Table 3.2 shows the values of different parameters of the PLL obtained by nonlinear analysis, and these may be used for engineering design of systems.

It has been further shown that the frequency of skipping cycles becomes quite large if the loop error $\sigma^2(\tilde{\theta}) = \frac{n_0 B_n}{A^2}$ is larger than 0.25. The frequency of skipping cycles $\simeq \frac{4B_n}{\pi}\exp\left(-\frac{2}{\sigma^2(\tilde{\theta})}\right)$, which gives the value of $10^{-4} B_n$ approximately for $\sigma^2 = 0.25$. When a single cycle is skipped, a transient of 2π occurs in $\tilde{\theta}$ and the loop behaviour deteriorates. This results into the so-called 'threshold' of the input CNR below which the loop (and equivalently PM and FM detectors) SNR decreases rapidly and the linear approximation of the loop does not hold.

3.6.4 Carrier Acquisition and Tracking

The second-order loop is generally used for tracking carrier frequencies in communication systems, since its steady state phase error → 0, if the initial frequency offset is within $\pm\omega_n$. This can also track linearly varying input frequencies within limits, whereas the optimum third-order loop for such inputs has less stable behaviour than the second-order loop. The initial acquisition of unknown input frequencies may be obtained by using a bank of filters separated by $\Delta\omega = 1/2T$, as shown in Fig. 3.48. Alternately a swept VCO may be used as shown in Fig. 3.49, and the maximum sweeping rate is proportional to (aAK) and also to the square of the noise bandwidth B_n of the loop. In case where the initial frequency offset Δf is not too large, the PLL may be operated in both acquisition and tracking modes, by using a large B_n for acquisition and a small B_n for tracking. Since $T_{acq} \simeq 3.5(\Delta f)^2/B_n^3$, the acquisition time is reduced by making B_n large. Once the acquisition is obtained, the loop is switched to the tracking mode by making B_n small, which reduces the tracking error $\sigma^2(\tilde{\theta})$.

To maintain the dynamic range of the input to the PLL within a reasonable limit, it is sometimes necessary to use a bandpass-limiter at the input (in some applications as in multi-access systems, the limiter is essential at the front end). It has been shown that the SNR deterioration due to the limiter is only 1 dB for small CNRs, and the SNR actually improves by upto 3 dB for large CNRs. To simplify the multiplier circuit of the PLL, it is also desirable to use a hard limiter at the output of the VCO, so that the multiplier may be made of switching circuits only. This does not affect the performance of the loop. The overall circuit for carrier acquisition and tracking is now shown in Fig. 3.54, where the input to the PLL has been squared to remove the biphase modulation. It may be shown that the variance of the phase error is increased due to the squaring circuit and as compared to eq. (3.238), the new variance is

$$\sigma^2(\tilde{\theta}) = \frac{n_0 B_n}{A^2}\left(1 + \frac{n_0 B_{if}}{2A^2}\right) \tag{3.240}$$

where B_{if} = bandwidth of the BPF.

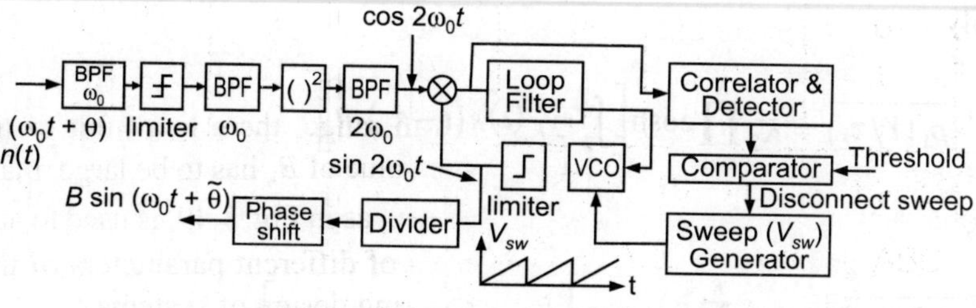

Figure 3.54: Carrier acquisition and tracking circuit

Since $A^2/2n_0B_{if} = \gamma$ = CNR, the phase variance is increased by a factor of $(1 + 1/4\gamma)$; and for γ = 10 dB, there is hardly any deterioration in the loop SNR. Other synchronization circuits, e.g., the Costas loop, also have similar performance.

3.6.5 Symbol (Bit) Synchronization

The problem of bit synchronisation in a digital communication system is somewhat similar to that of the carrier synchronisation. First the acquisition problem is solved by searching for the correct clock time by using multiple alternative hypothesis approach, as discussed along with eq. (3.202) and Fig. 3.44 (also refer to Fig. 3.48 used for acquisition of unknown frequencies). Assume that the bit stream is a low pass rectangular signal (after coherent/incoherent detection) disturbed by white Gaussian noise of zero mean and variance σ^2. Then,

$$y(t) = \sum_{k=1}^{k} a_k s\left[t - kT_b - \tau(t)\right] + n(t)$$

and

$$s(t) = 1, \; 0 < t < T_b, \tag{3.241}$$

where T_b = bit period, $\tau(t)$ = time offset with reference to the local clock, $a_k = \pm 1$, $p(a_k = 1) = 0.5 = p(a_k = -1)$, and the observation is made over a time = KT_b sec. Since $0 \le \tau < T_b$, the interval $(0, T_b)$ is finely divided into $\{\tau_1, \tau_2, \ldots, \tau_i, \ldots, \tau_m\}$ subintervals with $p(\tau) = 1/T_b$. The ML estimate of τ_i for the given $y(t)$ is the value of τ_i which maximizes the likelihood function

$$p(Y/\tau_i) = \frac{p(\tau_i/Y)\,p(Y)}{p(\tau)} = Kp(\tau_i/Y),$$

since $p(\tau)$ is uniform over $(0, T_b)$ and $p(Y)$ is the measured data. Thus the ML estimate is the same as the MAP estimate under the given conditions. Now for the interval k using eq. (3.195).

$$p(Y/\tau_i)k = K_1\left\{p(a_k)\exp\left[-\frac{1}{n_0}\int_{T_b}\left(y_k(t) - a_k s(t - \tau_i)\right)^2 dt\right]\right\} \tag{3.242}$$

and averaging over K intervals,

$$\overline{p(Y/\tau_i)} = \prod_{k=1}^{K} p(Y/\tau_i)_k$$

Since $\int Y_k^2(t)\,dt$ and $\int s^2(t-\tau_i)\,dt$ are constants independent of τ_i and noting that $a_k = \pm 1$, one has to maximize only

$$\overline{p_1(Y/\tau_i)} = K_2 \prod_{k=1}^{K} \cosh\left[\int_0^{T_b} y_k(t)\,s(t-\tau_i)\,dt\right]$$

Equivalently,

$$\log\left[\overline{p_1(Y/\tau_i)}\right] = K_3 \sum_{k=1}^{K} \log\cosh\left[\int_{T_b} y_k(t)\cdot s(t-\tau_i)\,dt\right] \tag{3.243}$$

is to be maximized for estimating $\hat{\tau}$. (K_1, K_2, K_3 are constants).

The acquisition circuit for $\hat{\tau}$ using above principles is shown in Fig. 3.55, which represents eq. (3.243) for $\tau = \{\tau_1, \tau_2, \ldots \tau_i, \ldots, \tau_m\}$. The optimum $\tau_i = \hat{\tau}$ is obtained from the 'Greatest of' decision circuit.

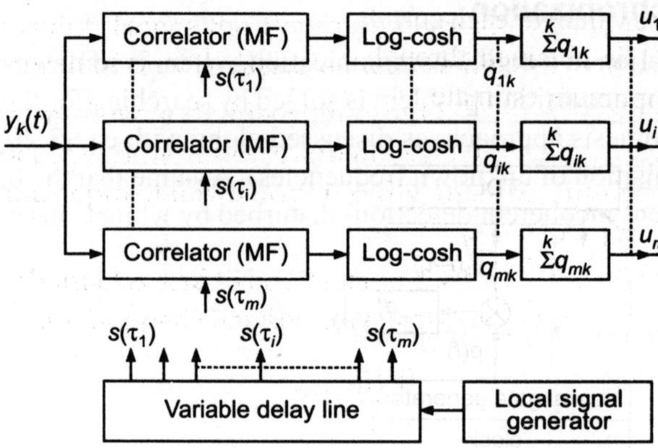

Figure 3.55: MAP/ML estimate of unknown delay for symbols

An alternative circuit for the acquisition may be based on eq. (3.224), which may be rewritten as

$$\int_{T_b} y(t)\frac{\partial s(t-\tau_i)}{\partial \tau_i}\,dt = 0 \tag{3.244}$$

where it is assumed that $s(t-\tau_i)$ is a lowpass signal, e.g., a pulse somewhat symmetrical (due to filtering) and with a peak at the centre. The solution of eq. (3.244) gives $\hat{\tau} = \tau_i$ and the equivalent receiver correlates $y(t)$ with the derivative of the known signal waveform to indicate the null. The derivative $s'(t-\tau)$ may also be modified as a rectangular gating function $g(t-\tau)$, as shown in Fig. 3.56(a), and the relations

$$\left.\begin{aligned} &\int_{T_b} y(t)\,g(t-\tau)\,dt \to \text{minimum} \\ &\text{and} \\ &\int_{T_b} y(t-\tau)\,g(t)\,dt \to \text{minimum} \end{aligned}\right\} \tag{3.245}$$

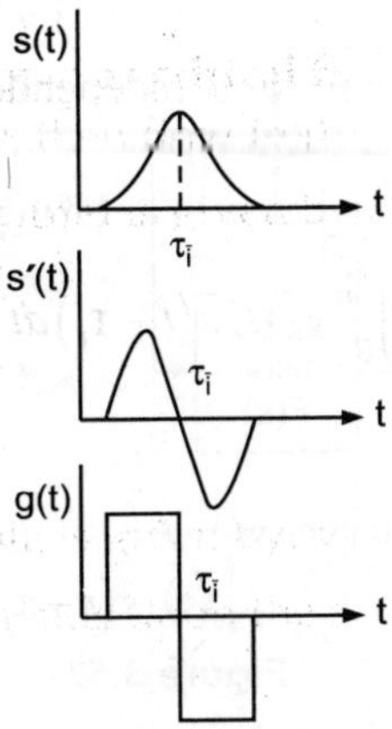

Figure 3.56(*a*): The functions of Eq. (3.245)

hold for $\tau = \hat{\tau}$ and a somewhat symmetrical $y(t)$. The overall acquisition circuit is now shown in Fig. 3.56(b) where the magnitude (absolute value) of each correlator output is summed over KT intervals to give $\{u_i\}$, and the minimum of all $\{u_i\}$ is obtained through a 'Lowest Of' decision circuit. The minimum u_i corresponds to $\tau_i = \hat{\tau}$ = the optimum estimate.

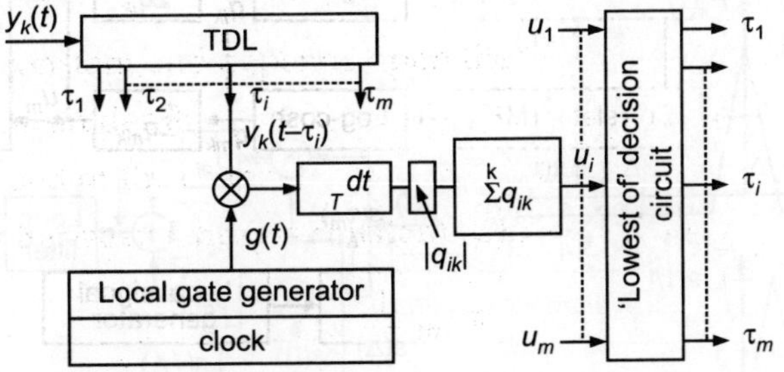

Figure 3.56(*b*): An alternative estimate of τ using gating function

Once the acquisition is obtained, the bit delay τ_i may be tracked by a PLL or by one of its modified versions. The same carrier tracking circuit of Fig. 3.54 may be used along with the squaring circuit to produce line spectra at $2/T_b$ Hz and the clock is phase-synchronized with the input data through suitable phase shifting/delay circuits. Alternatively, the eq. (3.245) provides the basis for the so-called early-late-gate bit synchronizer shown in Fig. 3.57. The input $y(t) = s(t-\tau) + n(t)$ is correlated with locally generated signals $s(t+\Delta)$ and $s(t-\Delta)$ and the difference of the magnitudes (this makes the output bit-independent) of the two outputs control the VCO such that the local signals are shifted to within $\pm\,\Delta$ of $y(t)$, i.e., $s_1(t) = s(t-\tau+\Delta)$ and $s_2(t) = s(t-\tau-\Delta)$. Under this condition, the local signal $s_0(t)$ becomes $s_0(t-\tau)$ and the error voltage $\in(t) \to 0$. The synchronized $s_0(t-\tau)$ is used for the coherent detection of the data. It has been shown that $\Delta = T_b/4$ is optimum and the corresponding mean-square timing error is

$$\frac{\sigma_{\in}^2}{T_b^2} = \frac{1}{8}\frac{B_n T_b}{E_b/n_0} = \frac{1}{8}\frac{n_0 B_n}{A^2} \tag{3.246}$$

where $B_n \simeq 1/(KT_b)$ Hz. Equation (3.246) is seen to be similar to eq. (3.240).

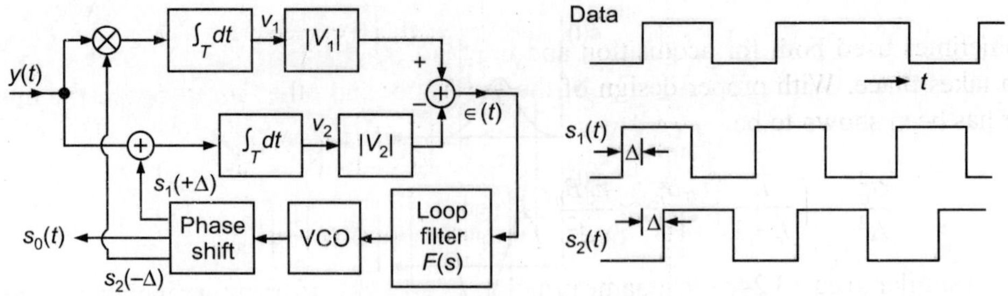

(a) Early-late-gate bit-synchronizer

(b) Waveforms of data, $s_1(t)$ and $s_2(t)$ as used in the cross correlators

Figure 3.57

The above technique of Fig. 3.57 may also be used for Word/Frame synchronisation, e.g., in delay-lock loops (DLL). Assume that an L-lenght ($L = 2^n - 1$) PN-sequence generated by an n stage shift register with feedback, is used as a synchronizing word. The autocorrelation function $\rho_x(\tau)$ of the sequence and the discriminator characteristic that is obtained by combining the $\rho_x(\tau - \Delta)$ and $\rho_x(\tau + \Delta)$ are shown in Fig. 3.58(a). Consider that $y(t) = s(t - \tau) + n(t)$, where $s(t)$ is a PNS of lenght $L = 2^n - 1$. In the receiver shown in Fig. 3.58(b), the incoming $y(t)$ is correlated with the advanced and the delayed sequences,

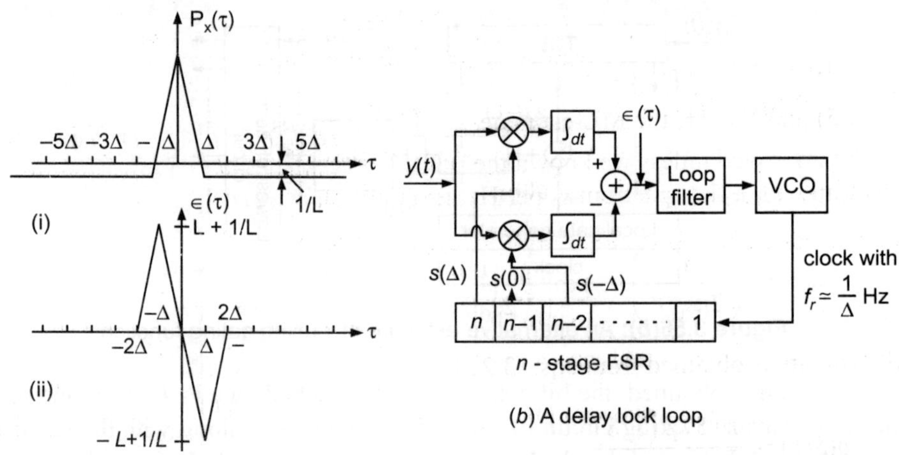

(a) (i) Correlation function $\rho_x(\tau)$
(ii) Discriminator characteristic

(b) A delay lock loop

Figure 3.58

$s(t + \Delta)$ and $s(t - \Delta)$, and their outputs added as shown. Initially, with certain frequency offset of the local VCO, there will be some cycle slipping and as the clock frequency f_r approaches the correct value $1/\Delta$, the sum of the correlator outputs will move into the discriminator characteristic, as shown in (ii) of Fig. 3.58(a). As the sequence further moves towards $\tau > 0$, the error voltage $\in(\tau)$ will be negative, thereby changing f_r in the opposite direction. Thus the VCO clock and its phase will be maintained at the correct value corresponding to $\in(\tau) = 0$ for $\tau = 0$. If the sequence is modulated by ± information bits, then an absolute value-circuit has to be incorporated after the integrators, as used in Fig. 3.57(a). The

DLL is sometimes used both for acquisition and tracking by changing the B_n of the loop filter after acquisition takes place. With proper design of the loop filter and other components, the mean-square delay error has been shown to be

$$\frac{\sigma_\epsilon^2}{\Delta^2} = \left(\frac{L}{L+1}\right)\frac{n_0 B_n}{A^2} \simeq \frac{n_0 B_n}{A^2}, L \gg 1 \tag{3.247}$$

which is again similar to eq. (3.240). The same principle of DLL is also used for ranging, i.e., to determine the distance of moving satellites, using large-length PN sequences as transmitted signals and correlating them with a locally generated PNS in a DLL, somewhat similar to that of Fig. 3.58.

3.7 ESTIMATION OF CONTINUOUS WAVEFORMS

For a random bandlimited analogue signal $s(t)$ corrupted by $n(t)$, the joint pdf of the $2TW$ independent samples is given by eq. (3.24), where the multi-dimensional vector $Y = S + N = \{y_1, ..., y_{2TW}\}$ represents the signal plus noise for $-T/2$ to $T/2$ sec. Then the likelihood function is

$$p(Y/S) = K \exp\left[-\frac{1}{2\sigma_n^2}\sum_{i=1}^{2TW}(y_i - s_i)^2\right] \tag{3.248}$$

where σ_n^2 = variance of $n(t) = n_0 W$.

From eq. (3.185) and Ex. 11, the ML-estimate $\hat{s}_i$ of the samples is simply y_i and it is seen that the estimate cannot be improved unless we known the pdf of $s(t)$.

If $s(t)$ is Gaussian with a flat spectrum s_0 per Hz, then the *a priori* pdf of $s(t)$ is

$$p(S) = \left(\frac{1}{2\pi s_0 W}\right)^{TW} \exp\left[-\frac{1}{2 s_0 W}\sum_{1}^{2TW} s_i^2\right] \tag{3.249}$$

Then the MAP estimate is obtained from eqs. (3.214) and (3.248) as

$$p(S/Y) = \frac{p(S)\,p(Y/S)}{p(Y)}$$

$$= K_y \exp\left[-\frac{1}{2\sigma_n^2}\sum_{1}^{2TW}\left\{y_i^2 - 2s_i y_i + \left(1 + \frac{n_0}{s_0}\right)s_i^2\right\}\right] \tag{3.250}$$

whose maximum occurs for

$$\hat{s}_i = \frac{s_0}{s_0 + n_0} y_i \tag{3.251}$$

Thus, the MAP estimate $\hat{s}_i$ is obtained as a voltage proportional to y_i and the estimate has not improved, as the noise is indistinguishable from the signal.

If, however, the spectral densities of $s(t)$ and $n(t)$ are not uniform, then the signal may be distinguished from noise, as is given by the methods discussed in the well known Wiener's theory of filtering and prediction [20]. Considering now the nonuniform spectral densities $G_s(f)$ and $G_n(f)$ for signal and noise respectively, and working in the frequency domain using the Fourier-series expansion, we have for $T \to \infty$,

$$\left.\begin{aligned} s(t) &= \sum_{m=1}^{TW}\left(a_m(s)\cos m\omega_0 t + b_m(s)\sin m\omega_0 t\right) \\ &\text{and} \\ n(t) &= \sum_{m=1}^{TW}\left(a_m(n)\cos m\omega_0 t + b_m(n)\sin m\omega_0 t\right) \end{aligned}\right\} \tag{3.252}$$

where, $\omega_0 = 2\pi/T$ and $-T/2 < t < T/2$.

The variances $\sigma_m^2(s)$ and $\sigma_m^2(n)$ are

$$\sigma_m^2(s) = \frac{2G_s(mf_0)}{T}; \sigma_m^2(n) = \frac{2G_n(mf_0)}{T} \tag{3.253}$$

Thus the pdf's are obtained by considering the components $a_m(y) = a_m(s) + a_m(n)$, and $b_m(y) = b_m(s) + b_m(n)$, as the Fourier-expansion coefficients of $y(t) = s(t) + n(t)$. Further, $\sigma_s^2 = \sum_m \sigma_m^2(s); \sigma_n^2 = \sum_m \sigma_m^2(n)$; and

$$p(s) = K_s \exp\left[-\frac{1}{2}\sum_1^{TW}\frac{a_m^2(s)+b_m^2(s)}{\sigma_m^2(s)}\right]$$

Then,

$$p(Y/S) = K\exp\left[-\frac{1}{2}\sum_1^{TW}\frac{\left(a_m(y)-a_m(s)\right)^2+\left(b_m(y)-b_m(s)\right)^2}{\sigma_m^2(n)}\right] \tag{3.254}$$

From Bayes theorem, the a posteriori pdf is

$$p(S/Y) = K_y \exp\left\{-\frac{1}{2}\sum_1^{TW}\left[\frac{\left(a_m(y)-a_m(s)\right)^2+\left(b_m(y)-b_m(s)\right)^2}{\sigma_m^2(n)} + \frac{a_m^2(s)+b_m^2(s)}{\sigma_m^2(s)}\right]\right\} \tag{3.255}$$

Maximizing $p(S/Y)$, one obtains the optimum estimates as

$$\left.\begin{aligned} \hat{a}_m(s) &= \frac{\sigma_m^2(s)a_m(y)}{\sigma_m^2(s)+\sigma_m^2(n)} = \frac{G_s(mf_0)a_m(y)}{G_s(mf_0)+G_n(mf_0)} \\ &\text{and} \\ \hat{b}_m(s) &= \frac{G_s(mf_0)b_m(y)}{G_s(mf_0)+G_n(mf_0)} \end{aligned}\right\} \tag{3.256}$$

These estimates are obtained by simply passing the input $y(t)$ through a linear filter having the transfer function,

$$H(f) = \frac{G_s(f)}{G_s(f)+G_n(f)} \tag{3.257}$$

This result is the well-known Wiener's filter, but the filter is unrealizable unless some large delay is introduced. The Bode-Shannon [21] modification of the result gives the physically realizable optimum filter. The variance of the above MAP estimate is

$$\sigma^2_{\text{MAP}} = 2\int_0^W \frac{G_s(f)}{G_s(f)+G_n(f)} G_n(f)\,df \tag{3.258}$$

and the variance of the ML-estimate is

$$\sigma^2_{\text{ML}} = 2\int_0^W G_n(f)\,df \tag{3.259}$$

thus showing the improvement obtained by the optimum filter. The transfer function $H(f)$ of eq. (3.257) may be rewritten as

$$H(f) = \frac{G_s(f)/G_n(f)}{1+G_s(f)/G_n(f)} = \frac{F(f)}{1+F(f)} \tag{3.260}$$

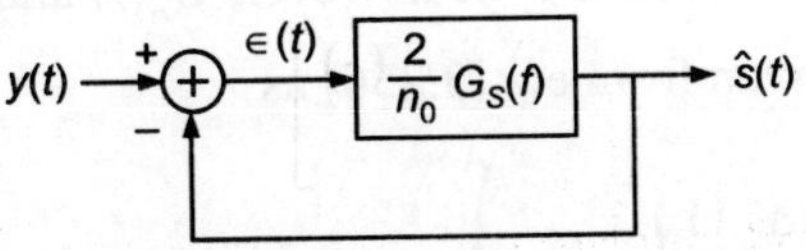

Figure 3.59: Optimum filter equivalent to eq. (3.260)

This has the same form as the closed loop transfer function of eq. (3.231); thus the linear filtering can be performed by a feedback network whose forward gain is $F(f)$, as shown in Fig. 3.59. Since $G_s(f)$ and $G_n(f)$ are rational spectra, and specifically, $G_s(f) = N(\omega^2)/D(\omega^2)$ and $G_n(f) = n_0/2$; their inverse transforms are

$$\rho_{ss}(\tau) = \frac{1}{2\pi}\int_{-\infty}^{\infty} G_s(f)e^{j\omega\tau}\,d\omega$$

and

$$\rho_{nn}(\tau) = \frac{n_0}{2}\delta(\tau) = \frac{1}{2\pi}\int_{-\infty}^{\infty} G_n(f)e^{j\omega\tau}\,d\omega \tag{3.261}$$

where $\rho_{ss}(\tau)$ and $\rho_{nn}(\tau)$ are the autocorrelation functions of $s(t)$ and $n(t)$. Then the equivalent impulse response of the linear filter $F(f)$ is $\rho_{ss}(t)$ and the time domain equation for the filtering process is

$$\hat{s}(t) = \frac{2}{n_0}\int_{-\infty}^{t} [y(\tau)-\hat{s}(\tau)]\rho_{ss}(t-\tau)\,d\tau \tag{3.262}$$

It is then seen that the MAP estimate is the convolution of $[y(t) - \hat{s}(t)]$ with the autocorrelation function of $s(t)$. The equation is somewhat similar to eq. (3.230) modified for the linear model of the PLL. The same result is also obtained from the solution in time domain of the basic MAP-estimate equation [16]

$$\frac{\partial}{\partial \hat{S}} p(S/Y) = \frac{\partial}{\partial \hat{S}} p(S)\,p(Y/S).$$

3.7.1 Estimation of a Modulation Envelope

Consider now the problem of optimum demodulation of AM signals corrupted by noise. For the amplitude-modulated (DSB-AM) transmitted signal, the received signal is

$$y(t) = s(t) + n(t) = [1 + A(t)]\cos\omega_0 t + n(t) \tag{3.263}$$

The problem of obtaining the optimum estimate of $A(t)$ becomes simplified if we note that $A(t)$ produces the double sidebands around ω_0, extending over $(\omega_0 \pm \omega_A)$ and the noise spectrum entering the IF filter also extends over the same band $(\omega_0 \pm \omega_A)$. Since the information is in the sidebands only, the synchronous detection of $y(t)$ by using a local carrier $\cos\omega_0 t$ maintains the low-frequency output SNR equal to the input SNR at the IF, and the translated low frequency noise is Gaussian and flat-spectrum if the IF noise spectrum is Gaussian and Flat-spectrum. The estimate of $A(t)$ is now obtained from the low frequency output of the synchronous detector.

Assuming that $A(t)$ is Gaussian with zero mean, and the spectral density $G_A(f)$ is constant but limited to W Hz, and also assuming $n(t)$ to be white and Gaussian with zero mean, the PDF $p(A/Y)$ is obtained from eq. (3.250) by suitable substitutions. Maximizing $p(A/Y)$, the MAP estimate $\hat{A}$ is of the form of eq. (3.251), giving only a proportionate result. If, however, $G_A(f)$ and $G_n(f)$ are not uniform, then the estimates $\hat{a}_m(A)$ and $\hat{b}_m(A)$ are given from eq. (3.256) as

$$\left.\begin{aligned} \hat{a}_m(A) &= \frac{G_A(f)\,a_m(Y)}{G_A(f)+G_n(f)} \\ &\text{and} \\ \hat{b}_m(A) &= \frac{G_A(f)\,b_m(y)}{G_A(f)+G_n(f)} \end{aligned}\right\} \tag{3.264}$$

where $G_n(f)$ = equivalent I.f. noise spectrum $2G_{n1}(f) = 2G_{n2}(f)$, and $G_{n1}(f) = G_{n2}(f)$ = IF noise spectrum. The optimum receiver now consists of a synchronous detector followed by an optimum linear filter. This receiver performs better than the simple envelope detector at low input SNR, but at high input SNR, the two give almost the same output SNR.

Referring to eq. (3.262), the time domain solution for $\hat{A}(t)$ is obtained as

$$\hat{A}(t) = \frac{2}{n_0}\int_{-\infty}^{t}\left[y(t)\cos\omega_0 t - \hat{A}(t)\right]\rho_{AA}(t-\tau)\,d\tau \tag{3.265}$$

The equation is represented in Fig. 3.60, where the frequencies around $2\omega_0$ are eliminated by the lowpass filter in the loop, and only the low-frequency modulating signals are available at the output.

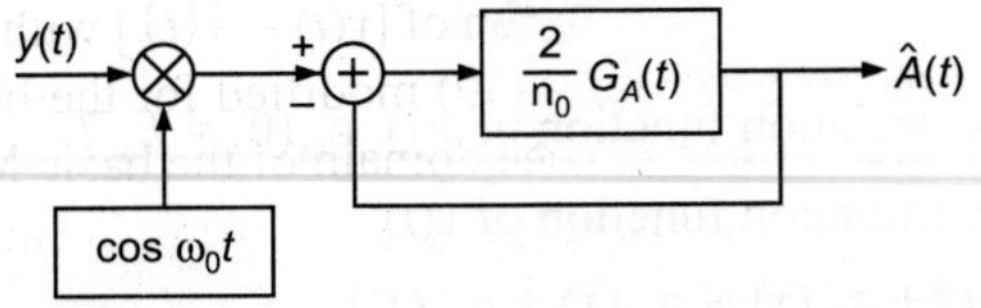

Figure 3.60: Filter representing eq. (3.265) for AM demodulation

3.7.2 Optimum Filters with Mean-Square Error Criterion [20, 21]

The MAP estimates discussed above are not really significant in the evaluation of analog modulation systems; but the mean square error (MSE) criterion is more meaningful and is widely used. The MSE criterion is also equivalent to the Bayes criterion with a quadratic cost function. Thus the estimator for $s(t)$ is so chosen that the average squared error is minimised, giving

$$\overline{\epsilon^2} = E\left\{\left[s(t-\delta) - \hat{s}(t-\delta)\right]^2 / y(t)\right\}, \tag{3.266}$$

as minimum for the optimum estimator. Assume that the filter has the input and the output given by

$$y(t) = s(t) + n(t)$$

and

$$\hat{s}(t-\delta) = \int_0^\infty h(\tau)\, y(t-\tau)\, d\tau,$$

where $h(t)$ is the impulse response, $h(t) = 0$ for $t < 0$; and δ is the delay of the physical filter. The MS error is now given by

$$\overline{\epsilon^2} = \lim_{T\to\infty} \frac{1}{2T} \int_{-T}^{T} dt \left[s(t-\delta) - \int_0^\infty h(\tau)\, y(t-\tau)\, d\tau \right]^2$$

$$= \lim_{T\to\infty} \frac{1}{2T} \int_{-T}^{T} \left[s^2(t-\delta) + \int_0^\infty h(\tau)\, y(t-\tau)\, d\tau \int_0^\infty h(\sigma)\, y(t-\sigma)\, d\sigma \right.$$
$$\left. -2s(t-\delta) \int_0^\infty h(\tau)\, y(t-\tau)\, d\tau\, dt \right]$$

Interchanging the order of the integrals and the limit,

$$\overline{\epsilon^2} = \lim_{T\to\infty} \frac{1}{2T} \int_{-T}^{T} s^2(t-\delta)\, dt$$
$$+ \int_0^\infty h(\tau)\, d\tau \int_0^\infty h(\sigma)\, d\sigma . \lim_{T\to\infty} \frac{1}{2T} \int_{-T}^{T} y(t-\tau)\, y(t-\sigma)\, dt$$
$$-2 \int_0^\infty h(\tau)\, d\tau \lim_{T\to\infty} \frac{1}{2T} \int_{-T}^{T} s(t-\delta)\, y(t-\tau)\, dt$$
$$= \rho_{ss}(0) + \int_0^\infty h(\tau)\, d\tau \int_0^\infty h(\sigma)\, d\sigma\, \rho_{yy}(\tau-\sigma)$$
$$-2 \int_0^\infty h(\tau)\, d\tau \cdot \rho_{ys}(\tau-\delta) \tag{3.267}$$

where

$\rho_{ss}(\tau)$ = autocorrelation function of $s(t)$, $\rho_{ss}(0) = E_s/T$

$\rho_{yy}(\tau)$ = autocorrelation function of $y(t)$

$= [\rho_{ss}(\tau) + \rho_{ns}(\tau) + \rho_{sn}(\tau) + \rho_{nn}(\tau)]$

$\rho_{ys}(\tau)$ = crosscorrelation function between $s(t)$ and $n(t)$

$= [\rho_{ss}(\tau) + \rho_{ns}(\tau)]$

The condition for minimum $\overline{\epsilon^2}$ is obtained by using the calculus of variations. The argument used is: if $h(t)$ actually minimizes $\overline{\epsilon^2}$ and if $h(t)$ is replaced by $[h(t) + ag(t)]$, where a is a real number and $g(t)$ is an arbitrary function of t, then the net effect is to increase $\overline{\epsilon^2}$. Thus for a fixed $g(t)$, $\overline{\epsilon^2}$ is a function of a, and $\overline{\epsilon^2}(a)$ is

$$\overline{\epsilon^2}(a) = \rho_{ss}(0) + \int_0^\infty d\tau \int_0^\infty d\sigma\big[h(\tau)h(\sigma) + ah(\tau)g(\sigma) + ag(\tau)h(\sigma) + a^2 g(\tau)g(\sigma)\big]\rho_{yy}(\tau-\sigma) - 2\int_0^\infty \big[h(\tau) + ag(\tau)\big]\rho_{ys}(\tau-\delta)\,d\tau \tag{3.268}$$

which should be minimum fo $a = 0$. Then setting the differential of $\overline{\epsilon^2}(a)$ equal to zero for $a = 0$, the $\overline{\epsilon^2}$ is minimum for

$$\left.\frac{\partial\overline{\epsilon^2}}{\partial a}\right|_{a=0} = \int_0^\infty g(\tau)d\tau\left[\int_0^\infty h(\sigma)\rho_{yy}(\tau-\sigma)d\sigma - \rho_{ys}(\tau-\delta)\right] = 0$$

Since this must hold for all functions $g(t)$, then the necessary condition is that $\overline{\epsilon^2}$ will be minimum for that $h(t)$ which satisfies,

$$\int_0^\infty h(\sigma)\rho_{yy}(\tau-\sigma)d\sigma = \rho_{ys}(\tau-\delta),\ \tau \ge 0 \tag{3.269}$$

This is the wellknown Wiener-Hopf integral equation of the first kind, and is the basis for the design of optimum filters. The condition is also sufficient since eq. (3.268) may be written as

$$\overline{\epsilon^2}(a) = \overline{\epsilon^2}(a=0) + 2aI_1 + a^2 I_2,$$

where $\overline{\epsilon^2}(a=0)$ is given by eq. (3.267), and

$$I_1 = \int_0^\infty g(t)\left[-\rho_{ys}(t-\delta) + \int_0^\infty h(\tau)\rho_{yy}(t-\tau)d\tau\right]dt = 0, \text{ by eq. (3.269)}$$

$$I_2 = \int_0^\infty\int_0^\infty g(\tau)g(\sigma)\rho_{yy}(\tau-\sigma)\,d\tau\,d\sigma$$

$$= \left\langle\left\{\int_0^\infty g(\tau)\big[s(t-\tau) + n(t-\tau)\big]d\tau\right\}^2\right\rangle$$

$=$ positive for all values of $g(\tau)$.

Therefore, $\overline{\epsilon^2}(a) = \overline{\epsilon^2}(0) = a^2 I_2$ and hence positive, i.e., $\overline{\epsilon^2}$ only increases by perturbing $h(t)$, if $I_1 = 0$, i.e., if eq. (3.269) is satisfied. This proves the sufficiency of the condition given by eq. (3.269).

For the filtering problem discussed in Sec. 3.7, it is observed that $s(t)$ and $n(t)$ are uncorrelated giving $\rho_{ys}(\tau) = \rho_{ss}(\tau)$, and $\rho_{yy}(\tau) = \rho_{ss}(\tau) + \rho_{nn}(\tau)$; the eq. (3.269) is now modified as

$$\int_0^\infty h(\sigma)\left[\rho_{ss}(\tau-\sigma)+\rho_{nn}(\tau-\sigma)\right]d\sigma = \rho_{ss}(\tau-\delta),\ \tau \ge 0 \tag{3.270}$$

Substituting eq. (3.270) in eq. (3.267), the MSE with the optimum filter as the estimator reduces to

$$\begin{aligned}\overline{\epsilon^2_{\min}} &= \rho_{ss}(0) - \int_0^\infty \rho_{ss}(\delta-\tau)h(\tau)\,d\tau \\ &= \rho_{ss}(0) + \int_0^\infty h(\sigma)\rho_{nn}(\tau-\sigma)\,d\sigma - \rho_{ss}(\tau-\delta)\end{aligned} \tag{3.271}$$

It is observed that the optimum estimator is only realizable with some delay δ, and once its impulse response $h(\sigma)$ is obtained from eq. (3.269), the MS error is calculated from eq. (3.271).

The solution of eq. (3.269) requires the knowledge of the correlation functions of the signal and the noise, or equivalently the spectral densities of the signal and the noise. Then the MS error and the transfer function of the estimator/filter may be solved in the frequency domain as well. With a large delay δ in the transfer function, the Wiener-Hopf eq. (3.270) may be rewritten with a change of variables as,

$$\int_{-\delta}^\infty \left[\rho_{ss}(\alpha-\beta)+\rho_{nn}(\alpha-\beta)\right]h(\beta+\delta)\,d\beta = \rho_{ss}(\alpha) \tag{3.272}$$

where $(\tau - \delta) = \alpha$, and $\sigma = (\beta + \delta)$ and $(\tau - \sigma) = (\alpha - \beta)$.

Letting $\delta \to \infty$, the eq. (3.272) becomes the Wiener-Hopf equation with infinite delay. With the limits of the integral $(-\infty, +\infty)$, the equation may be solved by taking transforms of both sides. Using eq. (3.261), the transformed equation is

$$[G_s(\omega) + G_n(\omega)]\, H(j\omega)e^{j\omega\delta} = G_s(\omega)$$

Therefore,

$$H(j\omega) = \lim_{\delta\to\infty} \frac{G_s(\omega)e^{-j\omega\delta}}{G_s(\omega)+G_n(\omega)} \tag{3.273}$$

This is the idealised transfer function of the estimator/filter. It is assumed that the spectra $G_s(\omega)$ and $G_n(\omega)$ are rational function of ω^2 and the filter has infinite delay δ; thus the realizability criterion of the filter is satisfied.

To obtain the MS error, it is noted that the ideal transfer function of the filter in the absence of noise would be simply a delay given by $e^{-j\omega\delta}$. Then the total error spectral density is the sum of two errors given by:

$$G_\epsilon(\omega) = |H(j\omega)|^2 G_n(\omega) + |H(j\omega) - e^{-j\omega\delta}|^2 G_s(\omega) \tag{3.274}$$

Substituting $H(j\omega) = A(\omega)e^{-j\theta}(\omega)$, where $A(\omega)$ and $\theta(\omega)$ are real, the total MSE is:

$$\overline{\epsilon^2} = \frac{1}{2\pi}\int_{-\infty}^\infty G_\epsilon(\omega)\,d\omega$$

$$= \int_{-\infty}^{\infty} A^2(\omega) G_n(\omega) \frac{d\omega}{2\pi} + \int_{-\infty}^{\infty} \left[A^2(\omega) + 1 - 2A\cos(\theta - \omega\delta)\right] G_s(\omega) \frac{d\omega}{2\pi}.$$

$\overline{\epsilon^2}$ is now minimized for the maximum value of cos ( ), i.e., for $\theta = \omega\delta$. Then the MSE is rewritten as:

$$\overline{\epsilon^2} = \int_{-\infty}^{\infty} \left[\left\{ A(\omega)\sqrt{G_s(\omega) + G_n(\omega)} - G_s(\omega) \Big/ \sqrt{G_s(\omega) + G_n(\omega)} \right\}^2 + G_s(\omega) - G_s^2(\omega) \Big/ \left\{ G_s(\omega) + G_n(\omega) \right\} \frac{d\omega}{2\pi} \right] \tag{3.275}$$

Now the MS error is further minimized by making { }^{2} = 0, which give the condition for minimum $\overline{\epsilon^2}$ as:

$$A(\omega) = \frac{G_s(\omega)}{G_s(\omega) + G_n(\omega)}$$

or

$$H(j\omega) = A(\omega)\exp(-j\omega\delta) = \frac{G_s(\omega)\exp(-j\omega\delta)}{G_s(\omega) + G_n(\omega)} \tag{3.276}$$

and from eq. (3.275),

$$\overline{\epsilon^2_{\min}} = \frac{1}{2\pi} \int_{-\infty}^{\infty} \left[G_s(\omega) - \frac{G_s^2(\omega)}{G_s(\omega) + G_n(\omega)} \right] d\omega$$

$$= \frac{1}{2\pi} \int_{-\infty}^{\infty} \frac{G_s(\omega) G_n(\omega)}{G_s(\omega) + G_n(\omega)} d\omega \tag{3.277}$$

This is sometimes known as the irreducible error since the observation interval for the estimation has been made ($-\infty$ to $+\infty$). Thus eq. (3.277) is the lower bound of the MS error, obtained with an infinite-delay filter. It is further observed that the transfer function of the optimum filter obtained from the MAP estimate, given by eq. (3.257) is exactly the same as obtained from the MSE criterion, given by eqs. (3.273) and (3.276). Thus the MAP estimate and the MSE criterion give the same results for optimum filtering problems.

Physically Realizable Filters

It has been observed that the optimum filter of eq. (3.276) is not physically realizable unless $\delta \to \infty$, i.e., the filter delay is very large. However, with a little sacrifice in the MSE, a realizable filter with $h(t) = 0$, for $t < 0$, may be obtained from $H(j\omega)$of eq. (3.276), by using the Bode-Shannon technique [22]. The steps required are as follows:

(1) Using Laplace Transform notations

$$H(s) = \frac{G_s(s)\exp(-s\delta)}{G_s(s) + G_n(s)} = \frac{G_s(s)}{G_y(s)} \exp(-s\delta)$$

Now $G_y(s)$ is factored as:

$$G_y(s) = G_y^+(s)G_y^-(s) \tag{3.278}$$

where $G_y^+(s)$ contains all the poles and zeros in the left half plane and $G_y^-(s)$ contains all the poles and zeros in the right half plane. The input $y(t)$ is passed through a realizable filter with the transfer function $H_1(s) = 1/G_y^+(s)$. The resultant output $y_1(t)$ is now a decorrelated random function with a spectral density of unity.

(2) The residual part of H(s) is now given by

$$H_2(s) = \frac{H(s)}{H_1(s)} = \frac{G_s(s)\exp(-s\delta)}{G_y^-(s)} \tag{3.279}$$

The resultant impulse response is not physical since $h_2(t) \neq 0$, for $t < 0$, and $H_2(s)$ is unrealizable. However, by making a partial fraction expansion of $H_2(s)$ as

$$H_2(s) = \frac{a_1}{(s+p_1)} + \frac{a_2}{(s-p_2)} + \frac{a_3}{(s+p_3)} + \frac{a_4}{(s-p_4)}, \ldots, \text{etc.}$$

one has to choose only those factors with poles in LHP and realize a modified $H_2^+(s)$ with $h_2^+(t) = 0$.

Then

$$H_2^+(s) = \left[\frac{G_s(s)}{G_y^-(s)}\right]_+ \quad \text{with } \delta = 0 \tag{3.280}$$

and the overall transfer function for the zero-delay realizable suboptimum filter is

$$H^+(s) = \frac{1}{G_y^+(s)}\left[\frac{G_s(s)}{G_y^-(s)}\right]_+ \tag{3.281}$$

and the MSE is now given by

$$\overline{\epsilon^2} = \frac{1}{2\pi}\int_{-\infty}^{\infty}\left\{\frac{G_s(\omega)G_n(\omega)}{G_s(\omega)+G_n(\omega)} + \left[\frac{G_s(\omega)}{G_y^+(\omega)}\right] + \left[\frac{G_s(\omega)}{G_y^-(\omega)}\right]_-\right\}d\omega, \tag{3.282}$$

thereby showing an additional term in the MSE.

In the special case of white noise with $G_n(\omega) = n_0/2$, $G_s(\omega) = N(\omega^2)/D(\omega^2)$ and $\delta = 0$, the eq. (2.281) may be written as:

$$H^+(j\omega) = \left\{1 - \frac{\sqrt{n_0/2}}{[n_0/2 + G_s(\omega)]_+}\right\} \tag{3.283}$$

where [$]_+$ indicates the factors having all poles and zeros in the LHP (as in case of $G_y^+(s)$).

$$\text{The resulting MSE is } \overline{\epsilon^2} = \frac{n_0}{2}\int_{-\infty}^{\infty} \log_e\left[1+\frac{G_s(\omega)}{n_0/2}\right]\frac{d\omega}{2\pi} \tag{3.284}$$

These equations are now used in solving the estimation problems in PM and FM systems.

Example 3.12: Consider $G_s(\omega) = \dfrac{1}{1+\omega^2}$ and $G_n(\omega) = 1$

Then the realizable sub-optimum filter with $\delta = 0$ is obtained as:

$$H(s) = \frac{G_s(s)}{G_y(s)},$$

and

$$G_y(s) = \frac{1}{(1-s^2)}+1 = \frac{2-s^2}{1-s^2} = \frac{(\sqrt{2}+s)(\sqrt{2}-s)}{(1+s)(1-s)}$$

then,

$$G_y^+(s) = \frac{\sqrt{2}+s}{1+s} \text{ and } H_1(s) = \frac{1+s}{\sqrt{2}+s}$$

$$H_2(s) = \frac{H(s)}{H_1(s)} = \frac{G_s(s)}{G_y^-(s)} = \frac{1}{(1-s^2)}\frac{(1-s)}{(\sqrt{2}-s)} = \frac{1}{(1+s)(\sqrt{2}-s)} = \frac{\sqrt{2}-1}{1+s}+\frac{\sqrt{2}+1}{\sqrt{2}-s}$$

Therefore,

$$H_2^+(s) = \frac{0.44}{(1+s)}$$

and the overall

$$H^+(s) = \frac{(1+s)}{(\sqrt{2}+s)}\cdot\frac{0.44}{(1+s)} = \frac{0.44}{(\sqrt{2}+s)}$$

The MS error by eq. (3.284) $\simeq$ 0.4, whereas the minimum MS error by eq. (3.277) $\simeq$ 0.36.

3.7.3 Optimum Phase and Frequency Demodulators

It has been shown so far that cross-correlators and optimum linear filters are essential elements of the optimum detection and demodulation systems. In digital communication and in radar signal reception, a delayed replica of the transmitted signal was assumed to be available at the receiver (in the form of a digital store or the matched filter) but in analogue communication systems, this is not so. The best that one can do now, is to form an initial estimate of the transmitted signal and then cross-correlate $y(t)$ with this, thereby improving the estimate. The optimum PM and FM demodulators are, therefore, based on the familiar feedback type of receiver using the principles of signal tracking and control.

Consider first a phase-modulated signal $s(t)$ received along with the noise $n(t)$ as

$$y(t) = A_0 \cos[\omega_0 t + m_p A_m(t)] + n(t) \tag{3.285}$$

where $m_p A_m(t)$ represents the phase modulation of the carrier ω_0. Since PLL has been shown to be an ideal phase tracking circuit and the Wiener filter as an ideal smoothing network, the optimum phase demodulator is obtained by combining both the devices, as shown in Fig. 3.61(a), where the loop bandwidth is larger than the message bandwidth W. Assuming $G_A(f)$ to be band-limited to W Hz and $W \ll f_0$, the optimum estimate $\hat{A}_m(t)$ for the PLL-demodulator is now given by

$$\hat{A}_m(t) = B\int_{-\infty}^{t} y(\tau)\cos\left[\omega_0\tau + m_p A_m(\tau)\right]\cdot f(t-\tau)\,d\tau \tag{3.286}$$

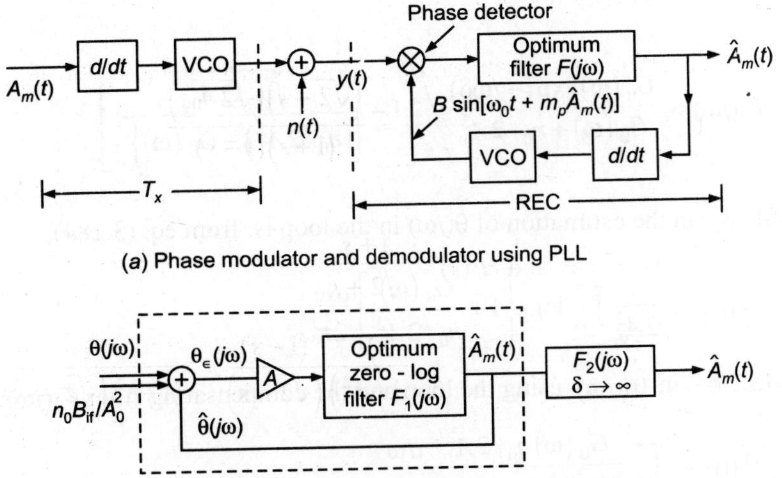

(a) Phase modulator and demodulator using PLL

$\hat{\theta}/\theta_\epsilon(j\omega) = H^+(j\omega)$ of eq. (3.283)

(b) Linear model or the optimum phase demodulator

Figure 3.61

where $f(t)$ is the impulse response of the loop filter $F(j\omega)$, such that the overall loop response is $h(t)$ of eq. (3.269). From the operational point of view, it is seen that the feedback portion of the receiver is the same as that of the modulator and demodulation takes place due to the signal tracking occurring in the loop and the initial estimate is continuously updated due to the continuous comparision in the multiplier which acts as a phase detector.

In conformity with the linear model of the PLL, shown in Fig. 3.53, the linear equivalent of Fig. 3.61(a) is shown in Fig. 3.61(b). It is observed that with a large-delay loop filter, as required by the optimum filter solution, $\delta \to \infty$, of eq. (3.273), the loop will tend to become unstable and the condition that θ should be small for linear approximation cannot be fulfilled; thus it is necessary to replace the ideal filter with a zero-lag filter with the transfer function $H^+(j\omega)$ of eq. (3.283). In the linear loop, $\theta(t)$ is the variation in phase of the carrier due to the modulating signal $A_m(t)$, and from eq. (3.285).

$$\theta(t) = m_p A_m(t) = s(t) \text{ of eq. (3.266)}$$

and

$$G_s(\omega) = G_\theta(\omega) = m_p^2 G_A(\omega) \tag{3.287}$$

With these values substituted in eq. (3.283), the loop filter transfer function is given by

$$F_1(j\omega) = \frac{1}{A_0}\frac{H^+(j\omega)}{1-H^+(j\omega)} = \frac{1}{A_0}\left\{\left[\frac{n_0}{2A_0^2} + G_\theta(\omega)\right]_+ - \sqrt{\frac{n_0}{2A_0^2}}\right\}\Bigg/\sqrt{\frac{n_0}{2A_0^2}} \tag{3.288}$$

where the equivalent $G_n(\omega) = n_0/2A_0^2$. To compensate for the increase in ϵ^2 due to the zero-lag filter, as given by eq. (3.282), one need to use a compensating filter $F_2(j\omega)$ such that the overall transfer function is

$H^+(j\omega)\, F_2(j\omega) = H(j\omega)$ of eq. (3.273)

Therefore,

$$F_2(j\omega) = \frac{G_\theta(\omega)\exp(-j\omega\delta)}{G_\theta(\omega) + n_0/2A_0^2}\Bigg/\left[1 - \frac{\sqrt{n_0/2A_0^2}}{\left\{\left(n_0/2A_0^2\right) + G_\theta(\omega)\right\}_+}\right] \tag{3.289}$$

The resulting MSE $\overline{\epsilon_1^2}$ in the estimation of $\theta(j\omega)$ in the loop is, from eq. (3.284),

$$\overline{\epsilon_1^2}(\theta) = \frac{n_0}{2A_0^2}\int_{-\infty}^{\infty}\log_e\left[1 + \frac{G_\theta(\omega)}{n_0/2A_0^2}\right]\frac{d\omega}{2\pi} \tag{3.290}$$

and the overall MSE $\overline{\epsilon^2}$ in $\hat{\theta}(j\omega)$ using the loop and the compensating filter $F_2(j\omega)$ in tandem is

$$\overline{\epsilon^2}(\theta) = \int_{-\infty}^{\infty}\frac{G_\theta(\omega)\, n_0/2A_0^2}{G_\theta(\omega) + n_0/2A_0^2}\frac{d\omega}{2\pi} \tag{3.291}$$

Example 3.13: Assume that the message spectrum $G_A(f)$ is constant over $\pm W$ Hz, then

$$G_\theta(f) = \frac{m_p^2\overline{A_m^2(t)}}{2W} = \frac{m_p^2}{2W}, \text{ with } \overline{A_m^2(t)} = 1 \text{ and } G_A(f) = \frac{1}{2W}$$

Assume further that $m_p = 5$ (equivalent to the deviation ratio in FM) and input CNR $= (C/N)_i = A_0^2/2n_0B = 10$, $B = B_{if}$. Then,

$$\frac{2A_0^2}{n_0} = 40B_{if};\quad B_{if} \simeq (2m_p + 1)W$$

and

$$\overline{\epsilon_1^2} = \frac{n_0}{2A_0^2}\int_{-W}^{W}\log_e\left(1 + \frac{2A^2 m_p^2\overline{A_m^2(t)}}{2Wn_0}\right)df \simeq 0.04.$$

therefore, with the zero-lag filter the message SNR

$$\frac{S_0}{N_0} = \frac{m_p^2}{\overline{\epsilon_1^2}} = 625\ (= 28\ \text{dB})$$

With the infinite lag filter,

$$\overline{\epsilon^2} = \int_{-W}^{W} \frac{m_p^2/(2W \cdot 40B)}{m_p^2/2W + 1/(40B)} df = 1/220 \simeq 0.005$$

Therefore, the message SNR $S_0/N_0 = m_p^2/\overline{\epsilon^2} = 5500\ (= 37.4\ \text{dB})$. It is now seen that S_0/N_0 with infinite-lag filter is better by 9.4 dB over that of the realizable zero-lag filter. The results of the conventional discriminator-demodulator is similar to that obtained by using the optimum filter.

For an input CNR = 1, with the same parameters as above,

$$\overline{\epsilon_1^2} = 0.29 \text{ and } S_0/N_0 \text{ (zero-lag filter)} = 87.5\ (= 19.4\ \text{dB})$$

and

$$\overline{\epsilon^2} = 1/22 \text{ and } S_0/N_0 \text{ (infinite-lag filter)} = 550\ (= 27.4\ \text{dB}).$$

Threshold Performance

It has been discussed along with Table 3.2 that the frequency of skipping cycles in a PLL becomes large if the loop noise $\frac{n_0 B_n}{A^2} > 0.25$. Thus the value of $\sigma_\epsilon^2 = 0.25$, may be taken as a threshold below which the loop performance deteriorates and the linear equivalence is not maintained. It is seen from the calculations of Ex. 13, that for an input CNR = 0 dB, the loop error $\overline{\epsilon_1^2} = 0.29$, which just exceeds the threshold value. In conventional PM demodulators (using limiter-discriminators) the threshold value of CNR $\simeq$ 10 dB for $m_p = 5$; thus, the PLL as a phase demodulator extends this threshold by approximately 10 dB. This characteristic of the PLL makes it most attractive for PM receivers operating at low CNRs. For large input CNRS, i.e., *C*/*N* above threshold, the SNRs at the output of the PLL demodulators are similar to the values obtained with the conventional demodulators, as shown by calculations of Ex. 12.

It has been shown in Sec. 3.6.3, that the steady state error in a second-order loop tends to zero and the loop is widely used where a compromise between the tracking error and the noise error has to be made. Moreover, a second order loop is easily designed, still maintaining the most important advantage of the loop, i.e., the threshold extension characteristic. As an example, consider a second-order loop shown in Fig. 3.53, where $F(j\omega)$ would be a simple RC integrating filter, whose 3 dB bandwidth $B(3\text{dB}) = \frac{2}{\pi} B_n$. Now making the message bandwith $W = B(3\ \text{dB}) = \frac{2}{\pi} B_n$, the loop noise $\sigma^2(\hat{\theta}) = \frac{n_0 B_n}{A^2} = \frac{n_0}{A^2} \cdot \frac{\pi}{2} W = \frac{\pi}{4}\left(\frac{W}{B}\right)\left(\frac{N}{C}\right)_i$ [using the values of Ex. 13]. Since the threshold vlaue of $\sigma_\epsilon^2 = 0.25$, for stable loop operation,

$$(C/N)_i \geq \pi(W/B) \tag{3.292}$$

For $m_p = 5$, $W/B = 1/11$; therefore, $(C/N)_i \geq \pi/11$
and for $m_p = 1$, $W/B = 1/3$; therefore, $(C/N)_i \geq \pi/3$

Thus it is seen that the loop will behave properly with very low values of CNR. Further filtering of the message through the network $F_2(j\omega)$ is now carried out to obtain the minimum MSE given by eq. (3.291). The overall filter design is also simplified as compared to that of eq. (3.289), and the problem of the stability of the loop with a higher-order filter is greatly minimized.

Frequency Demodulation

An FM signal corrupted with noise is of the form

$$y(t) = A_0 \cos\left[\omega_0 t + m_f \int_{-\infty}^{t} A_m(t)\,dt\right] + n(t) \tag{3.293}$$

where the instantaneous frequency is

$$\omega_i = \frac{d}{dt}\left[\omega_0 t + m_f \int_{-\infty}^{t} A_m(t)\,dt\right] = \omega_0 + m_f A_m(t) \tag{3.294}$$

Based on the principles of the optimum phase-demodulation discussed above, the optimum frequency estimator, using a PLL is shown in Fig. 3.62, which is similar to Fig. 3.61(a), except that the differentiator has been removed. The linear equivalent of the FM demodulator is the same as in Fig. 3.53 with infinite lag optimum filter connected in tandem and $\theta(t) = m_f \int_{-\infty}^{t} A_m(t)\,dt$, $\hat{\theta}(t) = m_f \int_{-\infty}^{t} \tilde{A}_m(t)\,dt$. It is interesting to note that the PLL demodulates the FM signal without using any discriminator in the usual sense. The necessary differentiation of $\theta(t)$ is obtained from the inverse operation of the integration in the VCO.

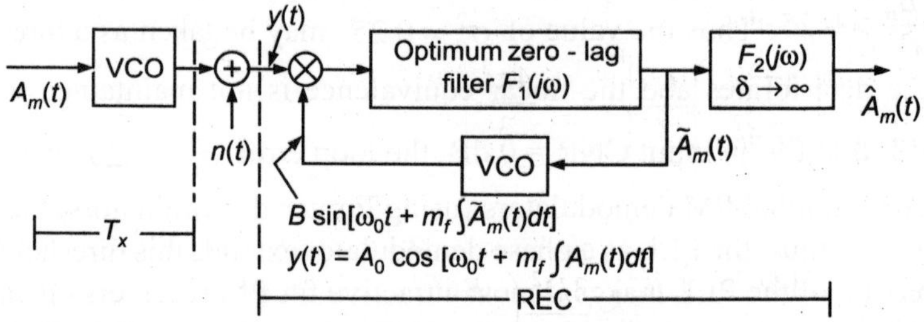

Figure 3.62: FM modulator and optimum demodulator

For the analysis of the linear model, It should be noted that $G_\theta(\omega) = m_f^2 G_A(\omega)/\omega^2$, because of the integrating VCO, and the eqs. (3.288) and (3.289) for $F_1(j\omega)$ and $F_2(j\omega)$ have to be modified accordingly. The transfer functions are then given by

$$F_1(j\omega) = \frac{j\omega}{A_0}\left\{\sqrt{\frac{2A_0^2}{n_0}\left[G_\theta(\omega) + \frac{n_0}{2A_0^2}\right]_+} - 1\right\}$$

(since the differentiator of Fig. 3.61(a) is now absorbed in $F_1(j\omega)$ of Fig. 3.62), and

$$F_2(j\omega) = \frac{\left[m_f^2 G_A(\omega)/\omega^2 \exp(-j\omega\delta)\right]}{\left[m_f^2 G_A(\omega)/\omega^2 + \left(n_0/2A_0^2\right)\right]} \Bigg/ \left\{1 - \frac{\sqrt{n_0/2A_0^2}}{\left[n_0/2A_0^2 + m_f^2 G_A(\omega)/W^2\right]_+}\right\} \tag{3.295}$$

The resulting MSE $\overline{\epsilon_1^2}$ in the loop is

$$\overline{\epsilon_1^2}(\theta) = \frac{n_0}{2A_0^2}\int_{-\infty}^{\infty} \log_e\left[1+\frac{m_f^2 G_A(\omega)/\omega^2}{n_0/2A_0^2}\right]\frac{d\omega}{2\pi} \tag{3.296}$$

and the overall minimum MSE using the compensating filter $F_2(j\omega)$ in tandem is

$$\overline{\epsilon^2(f)} = \int_{-\infty}^{\infty} \frac{n_0 m_f^2 G_A(\omega)/2A_0^2}{m_f^2 G_A(\omega)/\omega^2 + n_0/2A_0^2}\frac{d\omega}{2\pi} \tag{3.297}$$

Both the eqs. (3.288) and (3.296) give the loop MSE $\overline{\epsilon_1^2}$ for the θ-estimate, but the eq. (3.297) is for the MSE in the frequency estimate (equivalent to the MSE in $\hat{A}_m(t)$). As a result $\overline{\epsilon^2(f)}$ is obtained by multiplying the argument in eq. (3.291) by ω^2 after substituting $\left(m_f^2 G_A(\omega)/\omega^2\right)$ for $G_\theta(\omega)$.

Example 3.14: Assume the same parameters as in Example 3.13 for an FM system. In FM, m_f/W = Deviation ratio D = 5 (say); $B_{if} = B = (2D+1)W$; and the message output $S_0 \propto m_f^2$. Further, $\overline{A_m^2(t)} = 1$, $G_A(f) = 1/2W$, and $(C/N)_i = A_0^2/2n_0B = 10$ (say); then using eq. (3.297),

$$\overline{\epsilon^2(f)} \simeq \int_{-W}^{W}\left(\frac{n_0}{2A_0^2}\right)\frac{\omega^2 m_f^2 G_A(\omega)}{m_f^2 G_A(\omega)}df, \text{ neglecting } \frac{n_0}{2A_0^2} \text{ in the denominator}$$

$$= \left(\frac{4\pi^2 n_0}{2A_0^2}\right)\frac{f^3}{3}\Bigg|_{-W}^{W} = \frac{4\pi^2}{3}\left(\frac{n_0}{A_0^2}\right)W^3$$

Therefore,

$$\frac{S_0}{N_0} \simeq \frac{3\left(D^2W^2\right)A_0^2}{4\pi^2 n_0 W^3} \simeq \frac{3D^2}{\pi^2}\left(\frac{B}{2W}\right)\left(\frac{C}{N}\right)_i \tag{3.298}$$

$$\simeq \frac{3}{\pi^2}D^3\left(\frac{C}{N}\right)_i$$

For, $(C/N)_i = 10$, $S_0/N_0 = 375 \simeq 26$ dB.

For small CNR's, the loop error is obtained from eq. (3.296), and for $(C/N)i = 1$,

$$\overline{\epsilon_1^2}(\theta) = \frac{n_0}{2A_0^2}\int_{-2\pi W}^{2\pi W} \log_e\left[1+\frac{m_f^2 1/2W}{n_0\omega^2/2A_0^2}\right]\frac{d\omega}{2\pi}$$

$$= \frac{n_0 W}{A^2}\left[2\sqrt{X}\tan^{-1}1/\sqrt{X} + \log_e(1+X)\right]$$

$$= 0.176 \tag{3.299}$$

where $$X = \frac{m_f^2 A_0^2}{4\pi^2 W^3 n_0} = \frac{D^2 A_0^2}{4\pi^2 n_0 W}; \text{ and } \frac{A_0^2}{n_0 W} = 22$$

then the loop SNR $= m_f^2 \big/ \overline{\epsilon_1^2(\theta)} = 21.5$ dB.

As in the case of the PM demodulator, the PLL-FM demodulator also extends the threshold as seen from the above calculations. In a practical PLL-demodulator shown in Fig. 3.63, the loop filter can be a simple one to suppress the second harmonics and the overall loop may behave as a second-order one so that the steady-state error tends to zero. The optimum filter in tandem improves the final S_0/N_0 in the message band, and an approximately-ideal demodulator is obtained. The overall SNR characteristic of PLL-demodulator is shown in Fig. 3.64, where it is seen that the threshold is approximately at 0 dB.

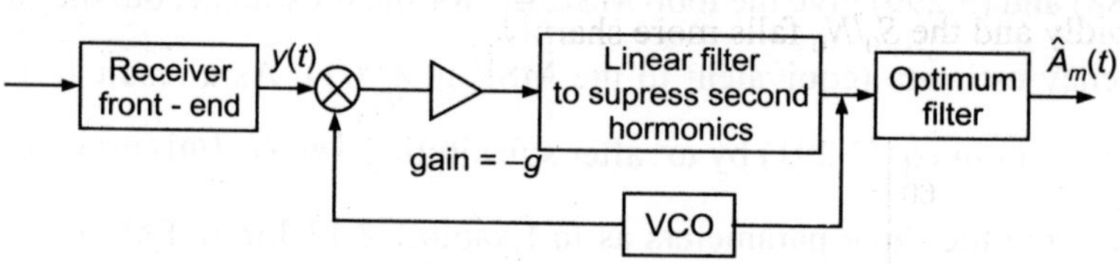

Figure 3.63: A practical PLL as an FM demodulator

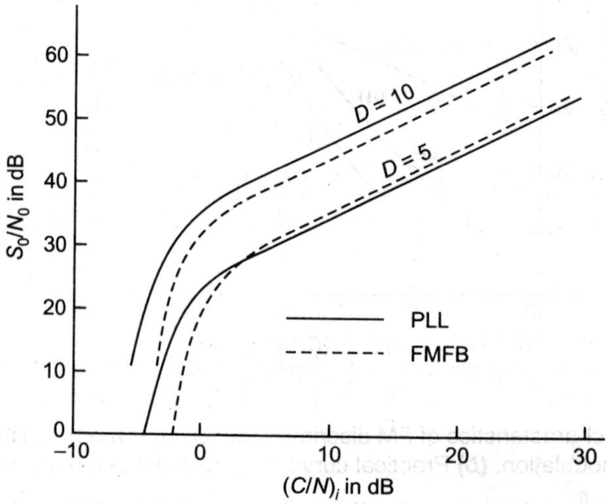

Figure 3.64: SNR characteristics of FMFB and PLL-FM demodulators with sinusoidal modulation

For the conventional FM demodulator using the limiter-discriminator or the zero-crossing counter followed by a lowpass filter, the SNR improvement above threshold is given by

$$\frac{S_0}{N_0} = \frac{3D^2}{2}\frac{B}{W}\left(\frac{C}{N}\right)_i, \left(\frac{C}{N}\right)_i \geq 10 \text{ dB} \tag{3.300}$$

The threshold is normally 10–12 dB for sinusoidal modulation and $D \geq 5$, but is worse for more complex modulating waveforms. It has also been shown that there are two thresholds observed, viz., (a) the first threshold at $(C/N)_i \geq 10$ dB, below which the noise increases suddenly and (b) second threshold at 0 dB approximately below which the noise output levels off but the signal suppression occurs and the so-called 'capture' phenomena takes place.

The noise characteristics below threshold have been derived from Rice's 'clicks'-analysis [6] and it has been shown that the output SNR for rectangular IF is given by,

$$\frac{S_0}{N_0} = \frac{\frac{3}{2}D^2\gamma(B/W)}{1+8\sqrt{3\gamma}(B/2W)^2 \operatorname{erfc}\sqrt{2\gamma}}, \gamma = \left(\frac{C}{N}\right)_i \tag{3.301}$$

These results are shown in Fig. 3.65 and it is seen that the general trend for S_0/N_0 is similar for both unmodulated and modulated cases. For the modulated case, the number of 'clicks' below the threshold increases more rapidly and the S_0/N_0 falls more sharply.

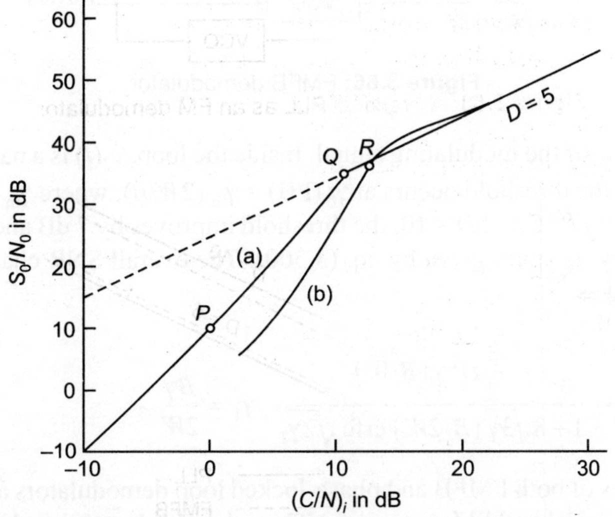

Figure 3.65: SNR characteristics of FM discriminator; *P*, *Q*, *R* are the different thresholds: ***(a)*** Ideal curve for no modulation, ***(b)*** Practical curve for sinusoidal modulation $D = 5$, rectangular IF

The conventional FM-demodulator may also be modified by using a feedback loop, thereby improving its threshold, as shown in Fig. 3.66, and is known as the FMFB detector (also frequency-compressive feedback system). In this circuit, the bandwidth of $H_1(f) = 2W$ and the VCO-output is centred at $(\omega_0 - \omega_1)$. Then after the product demodulator, the output V_2 is

$$v_2(t) = \frac{A_0 B}{2}\cos\left(\omega_1 t + m_f \int A_m(t)\,dt - \tilde{\theta}(t)\right) \tag{3.302a}$$

and

$$v_3(t) = \frac{d}{dt}\left(m_f \int A_m(t)\,dt - \tilde{\theta}(t)\right)$$

$$= m_f A_m(t) - \frac{d\tilde{\theta}(t)}{dt}$$

$$= \tilde{\theta}'(t)/K \tag{3.302b}$$

With proper loop operation,

$$v_3 \to 0 \text{ and } \tilde{\theta}'(t) = \frac{K m_f A_m(t)}{1+K} \simeq m_f A_m(t) \tag{3.303}$$

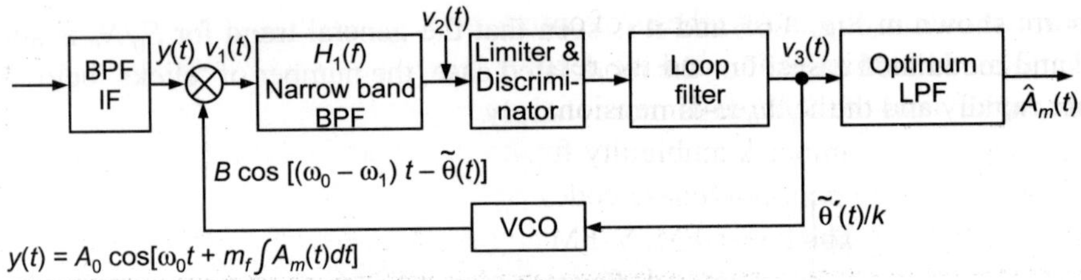

Figure 3.66: FMFB demodulator

Thus $\tilde{\theta}'(t)$ is the estimate of the modulating signal. Inside the loop, $v_2(t)$ is a narrow band FM due to the frequency feedback and the threshold occurs at $\gamma_{th}(FB) = \gamma_{th}\cdot(2W/B)$, where γ_{th} corresponds to the value of D in the FM signal. For $D = 5$ and $D = 10$, the threshold improves by 7 dB and 10 dB respectively. But above threshold, the S_0/N_0 is again given by eq. (3.300). The overall SNR characteristics, given by eq. (3.301), is now modified as

$$\frac{S_0}{N_0} = \frac{\frac{3}{2}D^2\gamma(B/W)}{1 + 8\sqrt{3\gamma}(B/2W)\,\text{erfc}\,\sqrt{2\gamma_1}}, \gamma_1 = \frac{B\gamma}{2W} \tag{3.304}$$

The overall performances of both FMFB and phase-locked loop demodulators are similar in all respects, as shown in Fig. 3.64. In wideband FM systems for TV and multiplexed signals having $D = 5$ and $B = 50$ MHz, as used in the Telstar experiments, the practical threshold improvement has been found to be about 5 dB.

3.8 SIGNAL DESIGN

In problems where the radar echoes are received from multiple targets, it is necessary to design the radar system to have maximum resolving capacity, both in range and velocity. The requirement for maximum signal detectibility is that the transmitted pulse has maximum energy E, which is obtained by increasing T and the peak power of the transmitter. By increasing T, however, the spread of the matched-filter response increases, thereby decreasing the accuracy of the parameter estimation and the resolving capacity. On the other hand, the optimum resolving capacity requires that extremely narrow pulses are transmitted. These conflicting requirements are satisfied by using the pulse-compression technique and the coded waveforms where a large-duration pulse is compressed to a narrow pulse at the receiver with the added advantage of increased immunity to interfering signals.

In a way, the problems are similar to those encountered in the multiplexed communication systems, e.g., in *M*-ary and spread-spectrum (SS) systems, where either synchronous or asynchronous codes are to be identified in the background of many similar codes transmitted and received simultaneously. The problems become worse when the transmission medium is dispersive or the targets have different velocities, thus adding random Doppler shifts to the received signals. The linear FM pulse is the forerunner of all coded waveforms that are used for optimum target resolution under difficult conditions. At present many efficient waveforms, e.g., PN-sequences, Binary phase codes, Polyphase codes, Discrete frequency sequences etc., are used along with the pulse-compression technique realised through MF-receivers. For *M*-ary and SS systems, orthogonal codes are generated using Hadamard matrix, Walsh transform, Gold codes and FT-matrix.

For the purpose of analysis and synthesis of optimum radar waveforms, the concept of the Ambiguity function is introduced in this section and the relative merits and demerits of various coded waveforms are discussed on the basis of the three-dimensional responses of the complimentary matched filters. It is shown that the idealised thumbtack ambiguity function is approximated by some of the binary phase-coded waveforms, but the Frank-polyphase codes have the maximum peak-to-sidelobe ratio when the Doppler shift is negligible. The linear-FM, V-FM and step-FM are the codes most suitable for use when only the range resolution is to be optimised. The requirements of maximum measurement accuracies are, however, incompatible with those of optimum signal resolution, and paradoxically, it is found that some of the waveforms, theoretically optimum from the measurement point of view, are unsuitable for general radar use. Thus, no waveform is optimum from all points of view and the choice of a waveform for a particular application may be subject to many compromises.

3.8.1 Signal Resolution [2]

Consider a simple problem of detecting two signals mixed with Gaussian noise, when the received signal is of the form,

$$y(t) = h_1 s_1(t) + h_2 s_2(t) + n(t) \tag{3.305}$$

In the absence of the specifications of all the prior probabilities and costs, it is convenient to determine the maximum likelihood estimates of h_1 and h_2 and the decision taken by comparing them with some bias level h_0, related to the given false-alarm probability P_f, as was discussed in Sec. 3.5.1. It has been shown that the ML-estimates h_1 and h_2 are those values which minimize the squared error (equivalent to maximizing $p(Y/h)$:

$$\in = \int_{\tau_1}^{\tau_2+T} \left[y(t) - \hat{h}_1 s_1(t) - \hat{h}_2 s_2(t) \right]^2 dt \tag{3.306}$$

The required estimates are then,

$$\left.\begin{aligned} \hat{h}_1 &= \frac{1}{\left(1-\rho_{12}^2\right)} \int_{\tau_1}^{\tau_2+T} \left[s_1(t) - \rho_{12} s_2(t) \right] y(t)\, dt \\ \hat{h}_2 &= \frac{1}{\left(1-\rho_{12}^2\right)} \int_{\tau_1}^{\tau_2+T} \left[s_2(t) - \rho_{12} s_1(t) \right] y(t)\, dt \end{aligned}\right\} \tag{3.307}$$

where the cross-correlation $\rho_{12} = \int_{\tau_1}^{\tau_2+T} s_1(t)\cdot s_2(t)\,dt$, and the signals are normalised as $\int_{\tau_1}^{\tau_2+T}[s_1(t)]^2\,dt = \int_{\tau_1}^{\tau_2+T}[s_2(t)]^2\,dt = 1$. The required estimates $\hat{h}_1$ and $\hat{h}_2$ are obtained by passing the input $y(t)$ through filters matched to $[s_1(t)-\rho_{12}s_2(t)]/(1-\rho_{12}^2)$ and $[s_2(t)-\rho_{12}s_1(t)]/(1-\rho_{12}^2)$ respectively, and by sampling the filter outputs at the end of the observation interval T. The decisions on H_0, H_1, H_2 and H_{12} are obtained on the basis ot comparisons of $\hat{h}_1$ and $\hat{h}_2$ with a bias level h_0 determined by the given P_f.

Under the null-hypothesis that $h_1 = 0$, the p.d.f. of $\hat{h}_1$ is

$$p(h_1/N) = \sqrt{\frac{1-\rho_{12}^2}{\pi n_0}}.\exp\left[-\frac{(1-\rho_{12}^2)\hat{h}_1^2}{n_0}\right]$$

and

$$P_f = \Pr.\{|\hat{h}_1| > h_0\}_{h_1=0} = 2\,\text{erfc}\left[h_0\sqrt{2(1-\rho_{12}^2)/n_0}\right] \tag{3.308}$$

$$P_d = \Pr.\{|\hat{h}_1| > h_0\}_{h_1\neq 0} = 1 - \text{erfc}\,Z_2 + \text{erfc}\,Z_1$$

where

$$Z_1 = h_1\sqrt{2(1-\rho_{12}^2)/n_0} + X;\; Z_2 = h_1\sqrt{2(1-\rho_{12}^2)/n_0} - X$$

and

$$X = h_0\sqrt{2(1-\rho_{12}^2)n_0}\,.$$

For fixed P_f, the probability of detection P_d is an increasing function of X, and is greater for smaller values of ρ_{12}. The more the signals overlap, the nearer is the parameter ρ_{12} to one and the larger is the SNR required to attain a given P_d for either of the signals (cf. eqs. (3.116) and (3.169)). The relations for P_f and P_d for h_2 are similar to eq. (3.308).

In case of narrow-band signals of the form $h_1A_1(t)\cdot\cos(\omega t - \theta_1)$ and $h_2A_2(t)\cos(\omega t - \theta_2)$, having unknown phases and amplitudes, the ML-estimates of h_1 and h_2 are obtained as (for $\Delta\omega = 0$)

$$\hat{h}_1 = \frac{1}{(1-|\rho_{12}|^2)}\left|\int_{\tau_1}^{\tau_2+T}\left[A_1^*(t)-\rho_{12}^*A_2^*(t)\right]B(t)\,dt\right|$$

$$\hat{h}_2 = \frac{1}{(1-|\rho_{12}|^2)}\left|\int_{\tau_1}^{\tau_2+T}\left[A_2^*(t)-\rho_{12}^*A_1^*(t)\right]B(t)\,dt\right| \tag{3.309}$$

where $B(t)$ is the complex envelope of $y(t)$ and

$$\int_T |A_1(t)|^2\,dt = \int_T |A_2(t)|^2\,dt = 1,\ \rho_{12} = \int_T A_1(t)\,A_2^*(t)\,dt$$

The estimates $\hat{h}_1$ and $\hat{h}_2$ are obtained by passing $y(t)$ through filters matched to the narrowband signals with envelopes $[A_1(t) - \rho_{12}A_2(t)]/(1 - |\rho_{12}|^2)$ and $[A_2(t) - \rho_{12}\cdot A_1(t)]/(1 - |\rho_{12}|^2)$ respectively, the filters being followed by linear detectors. The output of each of these detectors at $t = T + \tau_2$ is compared with a bias level h_0 and the decisions H_0, H_1, H_2 and H_{12} are taken accordingly. The false-alarm and detection probabilities are

$$P_f = \exp\left(-h_0^2/2X\right)$$

$$P_d = Q\left(\gamma.\sqrt{1-|\rho_{12}|^2},\, h_0/\sqrt{X}\right) \tag{3.310}$$

where, $\gamma = \sqrt{2E/n_0}$, $X = n_0\Big/\left(1-|\rho_{12}|^2\right)$ and $Q(\alpha, \beta)$ is the Q-function.

It is thus seen that for signal transmitted pulses, the probability of detection depends on the quantity $\{2E(1 - |\rho_{12}|^2/n_0)\}$ and the P_d is reduced when two received signals occur close together in time. Making the observation interval sufficiently large compared to the pulse duration T, the limits of the integrals in eq. (3.309) may be taken from $-\infty$ to $+\infty$, and from Schwarz inequality,

$$|\rho_{12}|^2 \le \int_{-\infty}^{\infty} |A_1(t)|^2\, dt \cdot \int_{-\infty}^{\infty} |A_2(t)|^2\, dt = 1 \tag{3.311}$$

so that $|\rho_{12}| \le 1$; and $|\rho_{12}| = 1$ when $A_1(t) = A_2(t)$. The resolution of the signals now depends on ρ_{12}; the smaller it is, the easier to resolve $A_1(t)$ and $A_2(t)$. ρ_{12} is the measure of the signal-overlap and may be called as the 'Complex cross-correlation' of signals, or the 'Resolution parameter'.

In multi-alphabet digital communication systems, as discussed in Sec. 3.4.5, ρ_{12} is made as small as possible by using orthogonal or biorthogonal codes, so that the alphabets may be detected correctly at the receiver. For radar signals, the waveforms are so designed that ρ_{12} is sufficiently small even for large T and $\Delta\omega$, as is obtained with pulse-compression techniques.

3.8.2 The Ambiguity Function [23]

It has been assumed so far that the medium is coherent and the targets are stationary, so that the Doppler shift $\Delta\omega = 0$. In radar signals, the second signal is usually a replica of the first, but delayed in time by τ and with a relative Doppler shift $\Delta\omega$. Then,

$$A_2(t) = A_1(t - \tau)\cdot \exp\,[j\Delta\omega(t - \tau)]$$

and

$$\rho_{12} = \chi(\tau.\Delta\omega) = \int_{-\infty}^{\infty} A_1(t)\cdot A_1^*(t-\tau)\cdot \exp\left[-j\Delta\omega(t-\tau)\right]dt \tag{3.312}$$

Using Fourier Transform,

$$\chi(\tau, \Delta\omega) = \frac{1}{2\pi}\int_{-\infty}^{\infty} F^*(\omega - \Delta\omega)\cdot F(\omega)\cdot e^{j\omega\tau}\cdot d\omega \tag{3.313}$$

The normalised 'resolution parameter' $|\chi(\tau, \Delta\omega)|$ may also be written as

$$|\chi(\tau, \Delta\omega)| = \left|\int A(t)\cdot A^*(t-\tau)\cos\Delta\omega t\cdot dt\right| \tag{3.314}$$

A three-dimensional plot of χ with τ and $\Delta\omega$ as variables indicate the accuracy and the ambiguity given by the transmitted waveform $s(t)$. In case of matched filters, a similar function $\chi(-t, \Delta\omega)$ gives the response of the filter to a frequency-shifted input signal, where

$$\chi(-t, \Delta\omega) = \int_{-\infty}^{\infty} s(\tau)\cdot s^*(\tau - t)\cdot \exp[j\Delta\omega\tau]\,d\tau$$

$$= \int_{-\infty}^{\infty} F^*(f)\cdot F(f+\Delta f)\cdot \exp(j\omega\tau)\,df \qquad (3.315)$$

In the design and analysis of MF waveforms and also for the coded waveforms, the above resolution parameter determines the usefulness of a particular transmitted signal in the measurement of velocity (Doppler) and range of a moving target or for resolving velocity and range among several moving targets. From this point of view, $\chi(\tau, \Delta\omega)$ is an important concept in certain applications of radar systems.

Based on the resolution parameter, the Ambiguity function is defined as $[\chi(\tau, \Delta\omega)\cdot\chi^*(\tau, \Delta\omega)]$, where

$$\chi(\tau, \Delta\omega) = \int_{-\infty}^{\infty} A(t)\cdot A^*(t+\tau)\cdot \exp(-j\Delta\omega t)\,dt$$

$$= \int_{-\infty}^{\infty} F^*(f)\cdot F(f+\Delta f)\cdot \exp(-j\omega\tau)\,df \qquad (3.316)$$

The ambiguity function can also be derived from the 'mean-square departure' of a waveform and its shifted self, given by

$$\int_{-\infty}^{\infty} \left| s(t) - s(t+\tau, f+\Delta f)\right|^2 dt = \int_{-\infty}^{\infty} \left| F(\omega) - F(\omega+\Delta\omega, t+\tau)\right|^2 d\omega$$

$$= 2E - 2\text{Re}\,\{\chi(\tau, \Delta\omega).\exp[-j(\omega_0 + \Delta\omega)\tau]\} \qquad (3.317)$$

Thus $\chi(\tau, \Delta\omega)$ is the measure of the 'mean-square departure' between two similar waveforms and is generally considered to be the best criterion to be used for the radar signal design.

The general properties of the ambiguity function are:

(a) $|\chi(\tau, \Delta\omega)|_{max} = |\chi(0, 0)| = 2E$ and $|\chi(\tau, \Delta\omega)| \le |\chi(0, 0)|$

(b) $|\chi(\tau, \Delta\omega)| = |\chi(-\tau, -\Delta\omega)|$

$$\left.\begin{aligned} &\text{(c)}\ |\chi(\tau, 0)| = \left|\int A(t)\cdot A(t-\tau)\,dt\right| \\ &\text{and} \\ &|\chi(0, \Delta\omega)| = \left|\int A^2(t)\cdot\cos\Delta\omega t\cdot dt\right| \end{aligned}\right\} \qquad (3.318)$$

(d) $\iint \left|\chi(\tau, \Delta\omega)\right|^2 d\tau\cdot d\Delta\omega = \left|\chi(0,0)\right|^2 = (2E)^2$

= total volume under the ambiguity surface

The volume invariance property of eq. (3.318d) shows that the improvement of resolution by decreasing $|\chi|$ over a large portion of the $(\tau, \Delta\omega)$-plane has to be balanced by increasing its value somewhere else, thereby increasing the ambiguity at some points in the plane. Thus the resolution cannot be improved independently of the ambiguity and vice-versa. Some examples of the ambiguity diagrams are shown in Fig. 3.67, where the nature of the central peaks and their base areas are also given.

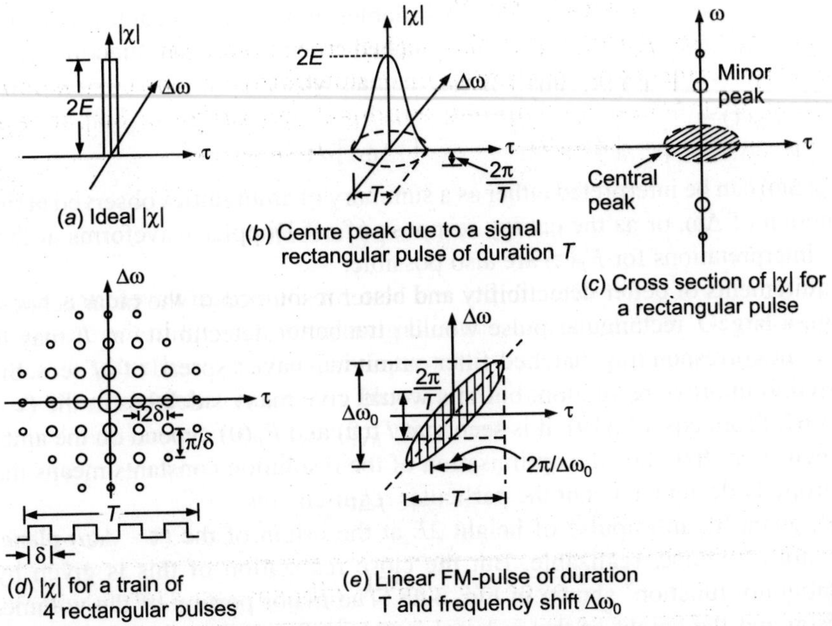

Figure 3.67: Examples of ambiguity diagrams—**(a)** Ideal $|\chi|$; **(b)** Centre peak dut to a single rectangular pulse of duration T; **(c)** Cross section of $|\chi|$ for a rectangular pulse; **(d)** $|\chi|$ for a train of 4 rectangular pulses; **(e)** Linear FM-pulse of duration T and frequency shift $\Delta\omega_0$ [23]

The resolution capability of a radar waveform may now be defined in terms of $|\chi(\tau, \Delta\omega)|$ as:
The time resolution constant for $\Delta\omega = 0$,

$$T_R(0) = \frac{\int_{-\infty}^{\infty} |\chi(\tau, 0)|^2 \, d\tau}{|\chi(0,0)|^2}$$

and the frequency resolution constant for $\tau = 0$,

$$F_R(0) = \frac{\int_{-\infty}^{\infty} |\chi(0, \Delta\omega)|^2 \, d\Delta\omega}{|\chi(0,0)|^2} \tag{3.319}$$

The waveform which furnishes the smallest value for $T_R(0)$ may be considered to have the greatest potential for resolving two signals in time and similarly, $F_R(0)$ has to be the smallest for the maximum resolution in frequency. The product $T_R(0) \cdot F_R(0)$ gives the measure of the combined (time delay and Doppler shift) resolving capability of a waveform. As for example, for a single tone Gaussian pulse, $T_R(0) \cdot F_R(0) = 1$, but for a single rectangular pulse, $T_R(0) \cdot F_R(0) = 2/3$; and the conclusion is made that the rectangular pulse is more suitable for simultaneous measurement of τ and $\Delta\omega$. The resolution constants have been further modified to take into account the changes in τ and $\Delta\omega$, giving

$$T_R(\Delta\omega) = \frac{\int_{-\infty}^{\infty} |\chi(\tau, \Delta\omega)|^2 \cdot d\tau}{|\chi(0,0)|^2}$$

and

$$F_R(\tau) = \frac{\int_{-\infty}^{\infty} |\chi(\tau, \Delta\omega)|^2 \cdot d\Delta\omega}{|\chi(0,0)|^2} \tag{3.320}$$

The quantity $T_R(\Delta\omega)$ can be interpreted either as a summary of ambiguities observed at the matched filter output as a function of $\Delta\omega$, or as the energy response of off-Doppler waveforms at the matched-filter output. Similar interpretations for $F_R(\tau)$ are also possible.

The dual requirements of better detectibility and better resolution of the radar echoes are not always compatible, e.g., a large-T rectangular pulse would give better detectibility as E may be increased by increasing T, but the corresponding matched-filter output will have a spread of $2T$ secs. Similarly, a pulse train may be used to improve resolution, but this would give many sidelobes in the $(\tau - \Delta\omega)$ plane, as shown in Fig. 3.67. From eqs. (3.319), it is seen that $T_R(0)$ and $F_R(0)$ depend on the ambiguity function $|\chi(\tau, \Delta\omega)|$ and hence on $A(t)$, thus the optimisation of the resolution constants means that the envelope $A(t)$ has to be properly designed to suit the particular requirements.

The ideal $|\chi|$, given by an impulse of height $2E$ at the origin of the $(\tau - \Delta\omega)$ plane and shown in Fig. 3.67(a), is not of course, realizable. But the close realization of this is given by the so-called 'Thumbtack ambiguity function' shown in Fig. 3.68. The major portion of the volume, however, lies under the pedestal and the height of the pedestal is $(2E)^2/\alpha\beta$, where α = effective time duration and β = effective bandwidth of the signal waveform. Although many waveforms give similar $|\chi(\tau,\Delta\omega)|$, e.g. by the Binary-Phase codes using PN-sequences, the specific waveform giving the thumbtack function is not known.

Ideally a noise-modulated waveform gives an ambiguity diagram which approaches the thumbtack characteristics of Fig. 3.68. A rotationally symmetric ambiguity surface is obtained by using the Hermite waveforms, given by [23],

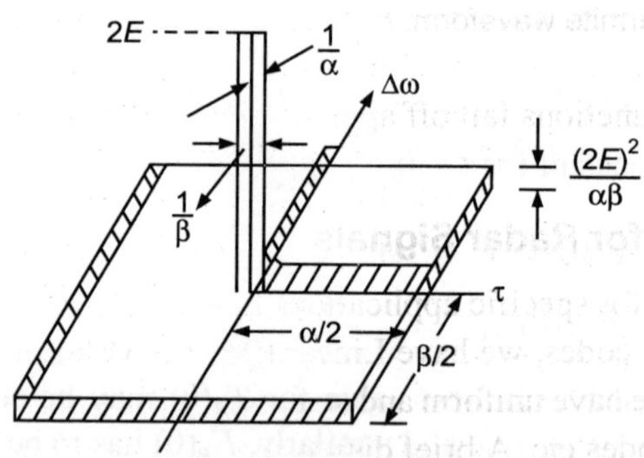

Figure 3.68: The thumbtack ambiguity function

$$A_n(t) = \frac{2^{1/4}}{\sqrt{n!}}.\exp\left(-\pi t^2\right).H_n\left(2.\sqrt{\pi}t\right) \tag{3.321}$$

where the Hermite polynomial $H_n(x)$ is defined as

$$H_n(x) \triangleq (-1)^n.\exp\left(\frac{x^2}{2}\right).\frac{d^n}{dx^n}\left[\exp\left(x^2/2\right)\right]$$

The ambiguity function of these waveforms, as shown in Fig. 3.69, is

$$\chi_n(\tau, \Delta\omega) = \exp[-(\pi/2)(\tau^2 + \Delta\omega^2)] \cdot L_n[\pi(\tau_2 + \Delta\omega^2)] \quad (3.322)$$

where $L_n(x)$ is the nth-order Laguerre polynomial defined by

$$L_n(x) \triangleq \frac{1}{n!} e^x \cdot \frac{d^n}{dx^n}\left(x^n \cdot e^{-x}\right)$$

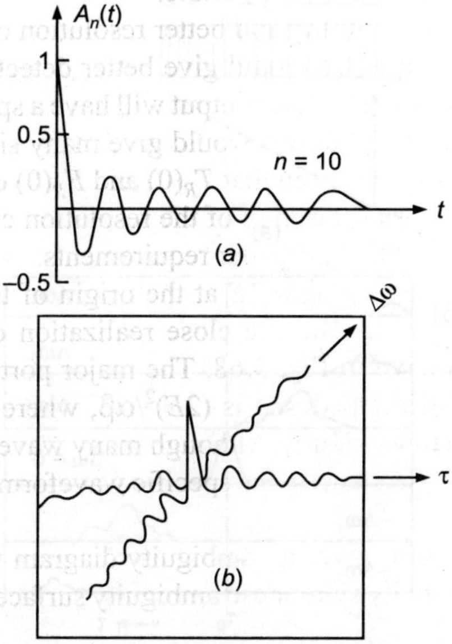

Figure 3.69: ***(a)*** Hermite waveform, n = 10; ***(b)*** Response surface $|x(\tau, \Delta\omega)|$; [23]

The sidelobes of $|\chi_n|$ for these functions fall off approximately as $\sqrt{t}$, whereas in the linear FM pulses, the sidelobes fall off as $1/t$.

3.8.3 Coded Waveforms for Radar Signals

The design of radar waveforms for specific applications has been based on (a) Analogue codes, and (b) Discrete codes. Under analogue codes, we have Linear FM, V-FM, parabolic FM, Hermite waveforms etc., and under discrete codes, we have uniform and staggered pulse trains, binary phase codes, polyphase codes, Frank codes, Huffman codes etc. A brief discussion of these is given below.

Analogue Codes

Historically, the FM-pulse compression technique is the forerunner of all coded waveforms now used in various radar systems. By using a swept carrier for a duration of T sec. in the transmitter, the received signal may be processed through a matched filter having a linear delay vs. frequency characteristic thus resulting into a compressed pulse, giving a good time-resolution (see Ex. 8 and Fig. 3.8). For a linear FM-pulse (also known as Chirp signal) having $\omega = \omega_0 + \mu t$, $0 < t \leq T$, and a rectangular envelope, the ambiguity function is of the form

$$|\chi|(\tau, \Delta\omega) = \begin{cases} \left(1 - \frac{|\tau|}{T}\right) \cdot \left|\frac{\sin x}{x}\right|, & \ldots |\tau| < T \\ 0, & \ldots |\tau| > T \end{cases} \tag{3.323}$$

and the cross-section of $|\chi|$ has been shown in Fig. 3.67(e). The composite surface of $|\chi|$ is shown in Fig. 3.70, where the range-Doppler ambiguity is considerable.

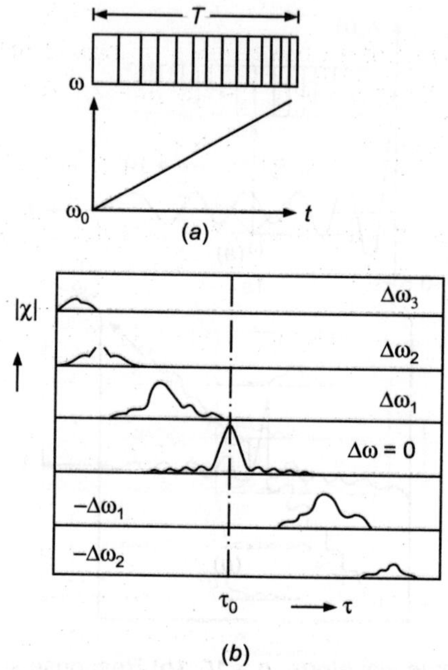

Figure 3.70: (*a*) Linear FM waveform [23]; **(*b*)** Composite response surface with variable τ and $\Delta\omega$

To avoid the ambiguity in the $(\tau - \Delta\omega)$ plane, the linear *V*-FM pulses, where

$$\omega = \begin{cases} \omega_0 + \mu t, & \ldots 0 < t < T/2 \\ \omega_0 - \mu t, & \ldots -T/2 < t < 0 \end{cases} \tag{3.324}$$

have been used. The ambiguity surface is now of the form

$$|\chi(\tau, \Delta\omega)| = |\chi_{11} + \chi_{22}| \tag{3.325}$$

where

$$\chi_{11}(\tau, \Delta\omega) = \left(1 - \frac{2|\tau|}{T}\right) \cdot \frac{\sin x}{x}, -T/2 < \tau < T/2$$

and

$$\chi_{22}(\tau, \Delta\omega) = \chi_{11}^*(-\tau, \Delta\omega) \tag{3.326}$$

Also,

$$|\chi(0, \Delta\omega)| = \frac{\sin T\,\Delta\omega/2}{T\Delta\omega/2}$$

The surface given by eq. (3.325) is shown in Fig. 3.71, whereit is seen that the matched filter output splits into two as the Doppler shift increases, thus giving the estimate $\hat{\tau}$ as the mean of two pulse positions. The importance of the linear V-FM waveforms is that it is the most basic FM pulse compression function that yields uncorrelated measurements of range and velocity (i.e., the error coupling is zero). However, if more than one signal is present, there is an increased amount of uncertainty as to the number of targets and their ranges and velocities. The off-Doppler ambiguity may be partially reduced by using parabolic V-FM waveforms, but this increases the sidelobe responses of the on-Doppler and off-Doppler matched-filter outputs.

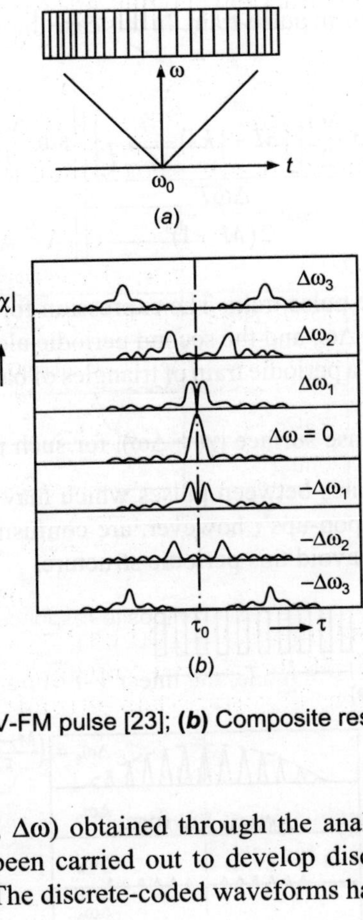

Figure 3.71: (*a*) Linear V-FM pulse [23]; (*b*) Composite response surface $|\chi(\tau, \Delta\omega)|$

Discrete Waveforms

Becuase of the unsatisfactory $\chi(\tau, \Delta\omega)$ obtained through the analogue waveforms discussed above, considerable investigations have been carried out to develop discrete waveforms which give better resolution and lesser ambiguities. The discrete-coded waveforms has the general form,

$$s(t, h, \Delta\omega, \theta) = \begin{cases} \sum_{n=1}^{N} h_n \cdot P_n(t) \cos\left[(\omega_0 + \Delta\omega_n)t + \theta_n\right], 0 < t \leq N\delta \\ 0, \ldots \text{ elsewhere} \end{cases} \tag{3.327}$$

where $P_n(t)$ = unit height pulse of duration δ, $N\delta = T$.

By introducing the modulation on the three parameters h_n, $\Delta\omega_n$ and θ_n, the discrete codes may be divided into three groups, viz.,

(a) Binary codes, where $\{\theta_n\} = \{\Delta\omega_n\} = 0$; $h_n = \{1, 0\}$,

(b) Phase codes, where $\{\Delta\omega_n\} = 0$; $h_n = \{1, 0\}$; $\theta_n = \{\theta_1, \theta_2, ..., \theta_i\}$,

(c) Frequency codes, where $\{\theta_n\} = 0$; $h_n = \{1, 0\}$; $\{\Delta\omega_n\} = \{\Delta\omega_1, \Delta\omega_2, ..., \Delta\omega_i\}$.

Examples of binary codes are the constant carrier pulse trains, where the pulse trains may be uniformly spaced or staggered. The waveforms are now given by

$$s(t) = \begin{cases} \sum_{n=1}^{N} h_n \cdot P_n(t) \cos\omega_0 t, & 0 \le t \le N\delta \\ 0, & \text{elsewhere} \end{cases} \tag{3.328}$$

where $\{h_n\} = \{1 ... 1 ... 1 ...\}$ for uniform pulse-train. In this case the ambiguity function may be described as

$$|\chi(\tau, \Delta\omega)| = \sum_{k=-M}^{M} \left| \frac{\sin\frac{\Delta\omega}{2}\left[(M-|k|)\frac{T}{M-1}\right]}{\sin\frac{\Delta\omega T}{2(M-1)}} \right| \left| \frac{\sin\frac{\Delta\omega}{2}\left[\delta - \left|\tau - \frac{kT}{M-1}\right|\right]}{\Delta\omega/2} \right| \tag{3.329}$$

where M is the number of 1's in the pulse train. The expression for $|\chi|$ consists of two factors, the first independent of τ but periodic along $\Delta\omega$, and the second periodic along τ for constant $\Delta\omega$. The matched filter response for $\Delta\omega = 0$ consists of a periodic train of triangles of basewidth 2δ and the heights weighted by $\left(1 - \frac{|k|}{M}\right)$. The composite response surface $|\chi(\tau, \Delta\omega)|$ for such pulse trains are shown in Fig. 3.72, and it is seen that there are dead zones between pulses which may be used for accurate measurement purposes. The periodic ambiguity 'pop-ups', however, are confusing the case of multiple targets and other waveforms have to be used to avoid this periodic structure.

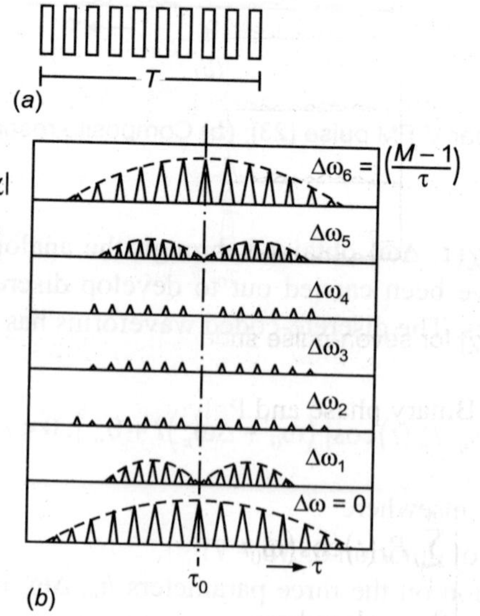

Figure 3.72: Response surface for a uniform pulse train [23]: **(a)** Uniform pulse train, **(b)** Response surface $|\chi(\tau, \Delta\omega)|$

The periodic structure of the pulse trains may be destroyed by staggering the pulses randomly, say for example, by making

$$\{h_n\} = \begin{cases} 1, & \text{for } n = 1, 6, 10, 17, 21, 27, 30, 36 \\ 0, & \text{... elsewhere} \end{cases} \tag{3.330}$$

The ambiguity function for a staggered pulse train is shown to be symmetric about both the τ^- and $\Delta\omega$-axis, and resembles the thumbtack characteristics of Fig. 3.68.

It is also possible to have an 'optimum' staggered pulse train for which $|\chi(\tau, 0)|$ has an uniform sidelobe of magnitude unity in the region $0 \le |t| < T$. In this case, the number of 1's = M is given by

$$M = \sqrt{2T \cdot \Delta f}, \quad \Delta f = 1/\delta \tag{3.331}$$

and the peak-to-sidelobe ratio is equal to $M = \sqrt{2T \cdot \Delta f}$. An example of the optimum staggered pulse train is, 1100101, where M is only 4. For higher values of M, a 'suboptimum' staggered pulse train may be designed, where $|\chi(\tau, 0)| \le 1$ for $\tau \ne 0$, and an example of which is

$$\{h_n\} = \begin{cases} 1, & \text{for } n = 1, 4, 11, 17, 25, 29, 34 \\ 0, & \text{...elsewhere} \end{cases} \tag{3.332}$$

for which the peak-to-sidelobe ratio is 7 : 1, as shown in Fig. 3.73.

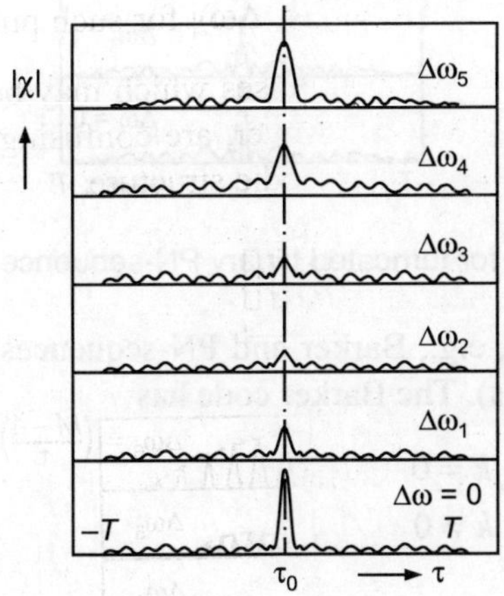

Figure 3.73: Response surface $|\chi|$ for seven-pulse suboptimum staggered pulse train in eq. (3.332) [23]

Phase codes consist mainly of Binary phase and Polyphase codes. The waveforms for Binary phase codes are given by

$$s(t) = \begin{cases} \sum_{1}^{N} P_n(t)\cos(\omega_0 t + \theta_n), & \text{...} 0 \le t \le N\delta = T, \\ 0, & \text{... elsewhere} \end{cases} \tag{3.333}$$

and

$$\sum_{1}^{N} P_n(t)\cos(\omega_0 t+\theta_n) = \sum_{1}^{N} C_n \cos\omega_0 t.$$

where $\theta_n = (0, \pi)$, and $C_n = \exp(j\theta_n) = \pm 1$. The binary sequences $\{\theta_n = 0, \pi\}$ and $\{C_n = \pm 1\}$ are thus interchangeable. The general ambiguity function for the sequence $\{C_n = \pm 1\}$ is similar to that of the staggered pulse trains and is shown in Fig. 3.74. Specifically the autocorrelation function at discrete points on the τ-axis is

$$\chi(k\delta, 0) = \delta \sum_{n=1}^{N-|k|} C_n \cdot C_{n+k} \tag{3.334}$$

which may be evaluated conveniently by the multiplication technique used in the numerical solution of convolution integrals.

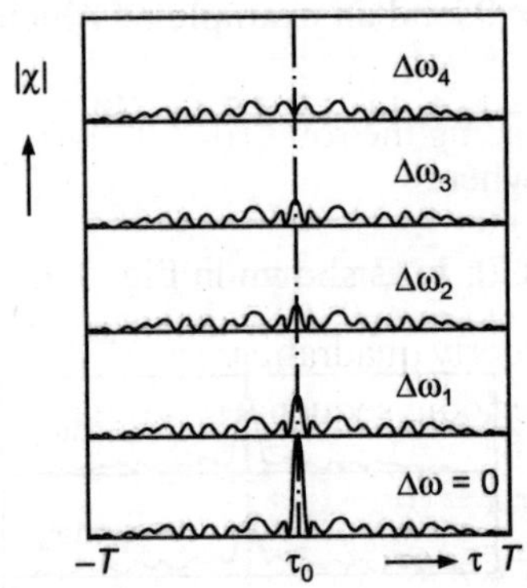

Figure 3.74: Response surface |χ| for turncated binary PN-sequence phase-coded signal, N = 31 [23]

Two important binary sequences, e.g., Barker and PN-sequences, have been used extensively (see Secs. 3.8.5 and 3.8.6 for more details). The Barker code has

$$\chi(k, 0) = \begin{cases} N, & k = 0 \\ \pm 1, 0, & k \neq 0 \end{cases} \tag{3.335}$$

and for the periodic PN-sequences,

$$\chi(k, 0) = \begin{cases} N, & k = 0 \\ -1, & k \neq 0 \end{cases} \tag{3.336}$$

Unfortunately, Barker sequences exist only for $N \leq 13$, e..g., for $N = 13$,

$$C_n = \{+++++--++-+-+\} \tag{3.337}$$

But the periodic PN-sequences generated by canonic shift registers may have N very large, and these periodic pulse trains have been used in high resolution radars with limited range. For general applications, the truncated PN-sequences are used, as it avoids the problem of CW-operation and does not tie the maximum range of the radar to $(2^k - 1)\delta$. The composite response surfaces for both the sequences are similar except that the Barker sequence has some similarities with the V-FM waveforms and shows the crossing-ridge line structure. For N large, the peak-to-side-lobe ratio of the truncated PN-sequences approaches $\sqrt{N}$, but in Barker codes, the ratio is less.

In the group of polyphase codes, special mention may be made of Barker polyphase and Frank polyphase sequences. The waveforms are given by eq. (3.333), except that θ_n is now multivalued, and C_n is a complex coefficient. The general ambiguity function still has the form similar to that of the staggered pulse trains, except that the terms C_{n+k} and C_{n+K+1} are now conjugated. Frank's quantized phase codes are generated from a symmetric matrix $M \times M = N$, where the elements are $2\pi p/M$, $p = 0, 1, 2, ..., M-1$, ... $(M-1)^2$, as given by

$$[p] = \begin{bmatrix} 0 & 0 & 0 & \cdots & 0 \\ 0 & 1 & 2 & \cdots & M-1 \\ 0 & 2 & 4 & \cdots & 2(M-1) \\ \vdots & \vdots & \vdots & \vdots & \vdots \\ 0 & (M-1) & 2(M-1) & \cdots & (M-1)^2 \end{bmatrix} \tag{3.338}$$

The coded sequence is formed by placing the rows (or columns) side by side, yielding a sequence of $M^2 = N$ elements, e.g., for $M = 4$, $N = 16$,

$$\{\theta_n\} = \{0, 0, 0, 0/0, \pi/2, \pi, \tfrac{3}{2}\pi/0, \pi, 0, \pi/0, \tfrac{3}{2}\pi, \pi, \pi/2\}$$

Because the phases progress in an orderly quadratic manner, the Frank-coded waveforms are also known as quantized-phase linear FM. As periodic sequences, these codes have

$$\chi(\tau, 0) = \begin{cases} N, & \text{for } k = 0 \\ 0, & \text{for } k \neq 0 \end{cases} \tag{3.339}$$

but when truncated beyond a single period, this optimum sidelobe performance is not obtained (e.g., peak-to-sidelobe ratios are: 9 : 1 for $N = 9$, 16 : 1.4 for $N = 16$, 36 : 2 for $N = 36$) and 64 : 2.6 for $N = 64$). However, in general, the sidelobes are smaller than in linear FM signals, and for large N, the peak-to-sidelobe ratio approaches $\pi\sqrt{N}$, an improvement over the PN-sequences. For large Doppler shifts, the advantages of these codes are lost as there are large cyclic ambiguity pop-ups (upto 30% of the peak signal output) as in linear FM waveforms. Thus the quantized phase codes are unsuitable when there are large Doppler shifts in an ensemble of received signals, but are excellent codes for signal resolution with $\Delta\omega = 0$,

Huffman sequences are generated by modifying eq. (3.327) as

$$s(t, h, \theta) = \begin{cases} \sum_{n=1}^{N} h_n P_n(t) \cdot \cos(\omega_0 t + \theta_n), & 0 < t < T; \quad \Delta\omega = 0 \\ 0, & \text{elsewhere} \end{cases} \tag{3.340}$$

In general h_n is multivalued and $\theta_n = (0, \pi)$. Noting that $s(t)$ may be written as $s(t) = \sum_{1}^{N} C_n \cos\omega_0 t$, where $C_n = h_n \cos\theta_n$, the values of $\chi(k\delta, 0)$ may be obtained at discrete points on τ-axis by the numerical method. An example of Huffman's sequence with $N = 14$ is given by

$$\{C_n\} = \{-0.57, 0.27, -0.56, 0.55, -0.14, -0.28, -0.31, -1.043, 0.5, 0.037, -0.34, 0.22, 0.34\}$$

The sequence waveform and its $\chi(k\delta, 0)$ is shown in Fig. 3.75, where it is shown that the peak-to-sidelobe ratio is 13 : 1. Modified Huffman sequences with N = 13 have also been developed where both h_n and θ_n are multivalued and these sequences are quasi-orthogonal among themselves [24]. Three such sequences are given in the Table 3.3.

In these sequences, the sidelobe suppression is 32 dB and the cross-correlation between the sequences is 13–20 dB (below 0 dB). They are useful in situations where two or more radar systems have to operate at the same site and the coupling between the systems have to be minimal. Longer sequences can be developed from the product of two short sequences and suitably designed MF's will maintain the quasi-orthogonality between the signals.

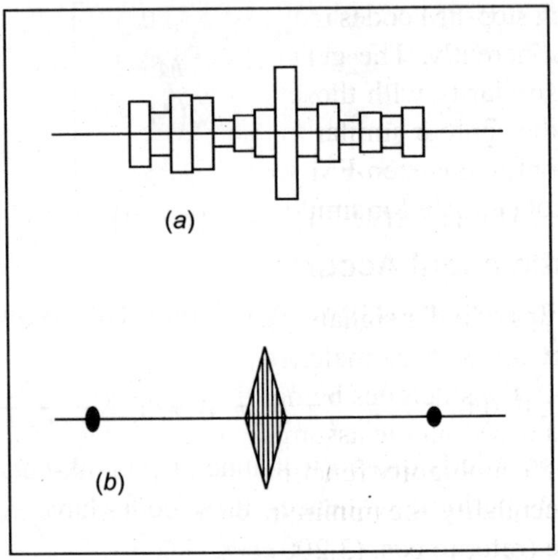

Figure 3.75: An example of Huffman's biphase code [23]. (*a*) Signal waveform. (*b*) Correlation peak at the MF output; peak-to-sidelobe ratio 13 : 1

Table 3.3: Modified hufïman sequences

Sequence I		Sequence II		Sequence III	
h_n	θ_n (degrees)	h_n	θ_n (degrees)	h_n	θ_n (degrees)
1.0	0	1.244	0	0.997	0
1.153	0	1.393	–16.4	1.40	–32.2
0.823	0	0.939	–26.8	1.173	–34.4
0.964	0	0.930	–46.7	0.253	–42.4
0.932	0	0.434	111.7	0.334	155.4
0.957	180	1.398	104.7	1.079	89.0
1.248	180	0.976	–86.1	0.524	–34.0
0.957	0	0.864	–104	1.118	–148.2
0.932	0	1.303	57.3	0.431	35.5
0.964	180	0.617	–134.6	0.823	–40.4
0.823	0	0.622	25.8	1.135	80.0
1.153	180	0.90	–163.6	1.409	–147.0
1.0	0	0.804	0	1.003	0

Frequency codes are derived from the discrete frequency sequences and may be described as

$$s(t) = \sum_{n=1}^{N} h_n \cdot P_n(t)\cos(\omega_0 + \omega_n)t, \; ...0 \le t \le N\delta. \tag{3.341}$$

where, $h_n = 1/0$. The stepped FM-waveforms have their frequencies spaced $1/\delta$ apart and the frequencies may be increasing or decreasing with time. Moreover, the signal may be continuous with all h_n's = 1, or it may be discontinuous with some h_n's = 0. It is found that the spectrum of such FM-waveforms are similar to those of the quantized phase codes (e.g. Frank codes), and the time-bandwidth products of both waveforms are also similar: M^2 in case of phase codes and N^2 in case of step-FM codes. The matched-filter receiver for such step-FM codes may be implemented by a bank of filters matched to $\Delta\omega_n$, whose outputs are summed coherently. The general ambiguity response for the continuous step-FM waveforms has considerable similarity with those of the linear-FM and polyphase-coded signals. The cross-section of $\chi(\tau, \Delta\omega)$ has the contour similar to that of linear FM shown in Fig. 3.67(e). The overall response function of the discontinuous step-FM waveforms has also structural similarity with above. These waveforms are, thus, not suitable for simultaneous measurement of range and velocity.

3.8.4 Measurement (Estimation) Accuracies

The two important characteristics of radar signals, i.e., resolution and ambiguity, have been discussed so far and the general conclusions are that the matched-filter outputs of the waveforms should be sharply peaked and narrow about $\tau = 0$, the sidelobes being of extermely low level. Measurement (estimation) accuracy, on the other hand, is based on the assumption of a single signal being present at the receiver and is related to the SNR and the ambiguity function $|\chi(\tau, \Delta\omega)|$ in the region about $\tau \simeq 0, \Delta\omega \simeq 0$.

Based on the Cramer-Rao inequality, the minimum error variances in the delay time and Doppler shift measurements are shown to be (refer to eqs. (3.200) and (3.205)),

$$\left.\begin{aligned} \left(\sigma_\tau^2\right)_{\min} &= \overline{\left(\tau - \hat{\tau}\right)^2_{\min}} = \frac{n_0}{2E\beta^2} \\ \left(\sigma_{\Delta\omega}^2\right)_{\min} &= \overline{\left(\Delta\omega - \widehat{\Delta\omega}\right)^2_{\min}} = \frac{n_0}{2E\alpha^2} \end{aligned}\right\} \tag{3.342}$$

where $\alpha^2 = \dfrac{2\pi^2}{E}\displaystyle\int_{-\infty}^{\infty} t^2 \left|A(t)\right|^2 dt$ and $\beta^2 = \dfrac{2\pi^2}{E}\displaystyle\int_{-\infty}^{\infty} f^2 \left|F(f)\right|^2 df$.

Then the rms range error $\epsilon_r = \dfrac{1}{2}c\sigma_\tau$ and the rms velocity error $\epsilon_v = \dfrac{1}{2}\lambda\sigma_{\Delta\omega}$, where c = velocity of the transmitted e.m. waves, λ = the transmitted wavelength. The product $\alpha\beta$ of the waveform is a measure of the limit of the accuracy of simultaneous range and velocity estimates and for accurate joint estimates, the product $\alpha\beta$ has to be large. The above minimum error variances hold if one of the parameters τ or $\Delta\omega$ is assumed known and the other is to be estimated. In general, the accuracies also depend on the error (range-Doppler) coupling factor ρ, defined by

$$\rho = \frac{\pi}{E \cdot \alpha\beta}\int_{-\infty}^{\infty} t \cdot \theta'(t) \left|A(t)\right|^2 dt,$$

$\theta(t)$ = phase modulation function, and eq. (3.342) hold for $\rho = 0$. For finite error-coupling, the eqs. (3.342) are modified, as

$$\sigma_\tau^2 = \frac{n_0}{2E\beta^2(1-\rho^2)}; \quad \sigma_{\Delta\omega}^2 = \frac{n_0}{2E\alpha^2(1-\rho^2)} \tag{3.343}$$

The equations hold for $(1-\rho^2) \geq 0$ and this is a useful check for the calculated values of α, β and ρ of any waveform.

From the definitions of α and β, $\alpha^2 = \pi^2T^2/3$, for signals of rectangular envelope having any type of phase modulation function $\theta(t)$, and $\beta^2 = \pi^2B^2/3$ for signals having a rectangular spectrum bandlimited over the interval B. Thus for any waveform having envelope discontinuities, e.g., rectangular RF pulses, $\beta \to \infty$, and for practical purposes, a modified spectrum skirt function is used to get finite values of β.

For linear FM-rectangular envelope pulse-compression signals, $\beta^2 = \pi^2B^2/3$ for large BT-product and the error-coupling factor $\rho = \pi^2TB/3\alpha\beta = 1$, thus giving very large measurement errors σ_τ and $\sigma_{\Delta\omega}$, a result unacceptable for practical purposes. The more realistic values for error variances have been obtained as

$$\begin{aligned} \sigma_\tau^2 &= \frac{(\pi TB)^2}{18}.(\sigma_\tau^2)_{\min} = \frac{T^2 n_0}{12E} \\ \sigma_{\Delta\omega}^2 &= \frac{(\pi TB)^2}{18}.(\sigma_{\Delta\omega}^2)_{\min} = \frac{B^2 n_0}{12E} \end{aligned} \tag{3.344}$$

This shows that error variances increase with time-frequency product and that the requirements of better resolution is not compatible with that of better measurement accuracies. However, for $\Delta\omega \ll B$, the errors will be small. The estimation-error variances are related to the behaviour of the ambiguity function about $\tau = \Delta\omega = 0$, and it is shown that the signals having smaller values of σ^2 are those whose matched-filter response functions fall off more rapidly near the origin. The estimation errors may be further reduced by making measurements over a large number of repeated signals as was discussed in Sec. 3.5.3 and the reduced error is

$$(\sigma^2)N \simeq \sigma^2/N \tag{3.345}$$

where it is assumed that the SNR and the parameter being measured do not change over the interval of N pulses.

For the cases of discrete coded waveforms, the behaviour of range-Doppler coupling is dependent on $\theta'(t)$ which has to be evaluated on the assumption of a finite switching time from state to state. In case of binary codes, the eq. (3.345) may be used for the overall error calculation, but for the binary phase codes, $\theta'(t)$ with finite switching time gives rise to the impulse-like FM function and α, β, ρ, tend to be small as the length N of the code increases, giving $\sigma^2 \simeq \sigma_{\min}^2$. These codes, therefore, meet the requirements of independent range and velocity measurements. The linear step and Frank-coded waveforms, however, behave like the linear FM waveforms, having similar values of ρ, and the values of σ^2 are similar to those of eqs. (3.344).

It is apparent from the above discussions that the requirements of minimum σ^2 and minimum ambiguity responses are not always compatible, examples of which are the linear FM and Frank-coded waveforms. Since $\sigma_{\min}^2$ is based on the behaviour of $|\chi(\tau, \Delta\omega)|$ at the origin only, it conveys little information regarding the undesired ambiguity responses at other points in the $(\tau - \Delta\omega)$ plane. Thus it is found in practice that some of the least favourable signals for general radar use have theoretically optimum measurement accuracies. and many compromises have to be made before the choice of a particular radar signal is arrived at, specially, in multiple and dense target environments.

3.8.5 Optimum Waveforms for M-ary Signalling

Generation and detection of M orthogonal codes have been discussed in Sec. 3.4.5, where it was assumed that an ideal synchronisation at the receiver was achieved through auxiliary signals. It is, however, more desirable to have in-built synchronising capability of the codes used for signalling. Thus the code set $\{C_i\}$ should have the dual properties given by

$$\rho_{ij}(C_i, C_j) = \begin{cases} N, & \text{for } i = j \\ 0, & \text{for } i \neq j \end{cases}$$

and (3.346)

$$\rho_{ii}(C_i, C_{i+\tau}) = \begin{cases} N, & \text{for } \tau = 0 \\ 0, & \text{for } \tau \neq 0 \end{cases}$$

where ρ_{ii} = autocorrelation of the codes (ACF), ρ_{ij} = cross-correlation of the codes (CCF), and N = length of the codes. Such codes are not practically realizable, but by relaxing the second property as

$$\rho_{ii} = \begin{cases} N, & \tau = 0 \\ \in, & \tau \neq 0 \end{cases}$$

where $\in \ll N$, realizable codes may be generated.

Consider the bi-orthogonal code set for $M = 16$ given by the (16 × 8) matrix of eq. (3.172a). Let the code vectors be multiplied (mode 2), element by element, by a relaxed Barker code B, given by (in bipolar form):

$$\{B_{1i}\} = \{1, 1, -1, 1, -1, -1, -1, 1\}$$

whose (3.347)

$$\{\rho_{ii}\} = \{8, -1, 0, -1, 0, -3, 0, 1\}$$

Then multiplying the code set (C_i) by $\{B_{1i}\}$, the modified code set $\{C_i'\}$ is obtained as

$$\begin{bmatrix} C_0' \\ C_2' \\ C_4' \\ C_6' \\ C_8' \\ C_{10}' \\ C_{12}' \\ C_{14}' \end{bmatrix} = \begin{bmatrix} 1 & 1 & 0 & 1 & 0 & 0 & 0 & 1 \\ 1 & 0 & 0 & 0 & 0 & 1 & 0 & 0 \\ 1 & 1 & 1 & 0 & 0 & 0 & 1 & 0 \\ 1 & 0 & 1 & 1 & 0 & 1 & 1 & 1 \\ 1 & 1 & 0 & 1 & 1 & 1 & 1 & 0 \\ 1 & 0 & 0 & 0 & 1 & 0 & 1 & 1 \\ 1 & 1 & 1 & 0 & 1 & 1 & 0 & 1 \\ 1 & 0 & 1 & 1 & 1 & 0 & 0 & 0 \end{bmatrix} \tag{3.348}$$

The even numbered codes are only shown and the odd numbered codes $\{C_1', C_3', ..., C_{15}'\}$ are the complements of the above set. It will be noted that all the code vectors are orthogonal to each other and they have reasonably acceptable autocorrelation ρ_{ii} with small sidelobes. As for example

$$\{\rho_{ii}(C_6')\} = \{8, -1, 0, 3, 0, 1, 0, 1\}$$

and

$$\{\rho_{ii}(C'_{14})\} = \{8, 1, 0, -3, 0, -1, 0, -1\}$$

Thus the correlation peaks obtained at the MF output for each of the M signals transmitted may be used to synchronise the receiver clock and the sampling instants for the detection of codewords.

Similarly, for $M = 32$, bi-orthogonal codes (Reed-Muller codes), one may use the synchronising code $\{B_{2i}\}$ given by

$$\{B_{2i}\} = \{1, -1, -1, 1, -1, 1, -1, -1, 1, 1, -1, -1, -1, -1, -1, 1\}$$

where

$$\{\rho_{ii}\} = \{16, -1, 2, 1, -2, 1, 2, -1, 0, 1, 2, -1, 2, -1, -2, 1\}$$

The resultant codevectors for $M = 32$ are given in Table 3.4. The higher order codes (say for M = 64, 128 etc.) may be generated by combining higher order Hadamard matrices with relaxed Barker codes of required length.

An alternative method of synchronization is to use a suitably modified Barker code in lieu of the all-zero code-vector, and to transmit this code repeatedly at intervals of nT sec. or whenever there is an all-zero message vector at the coder input. Continuous transmission of the sychronising code is also used in some systems, but this requires additional signal power/extra bandwidth for efficient performance.

The functional diagram of M-16 bi-orthogonal code generator is shown in Fig. 3.76, where the binary message vector, say, 1 0 1 1, is fed to the AND gates in parallel, and the gates are controlled by the outputs of the binary dividers as shown. The gate outputs are now summed through MOD-2 adders, resulting into the waveforms shown in Fig. 3.76(b). It can be easily checked that the summed output is the code $C_{11} = \{10100101\}$ and by multiplying with $\{B_{1i}\}$, the desired code $C'_{11} = \{01110100\}$ is obtained. A suitable MF, say, using a tapped delay line (TDL), detects the code in an optimum manner.

Table 3.4: Biorthogonal code set and the modified vectors incorporating the synchronous signal

Code No.	Basic Biorthogonal codes	Modified Reed-Muller codes
C_0	0 0 0 0 0 0 0 0 0 0 0 0 0 0 0 0	1 0 0 1 0 1 0 0 1 1 0 0 0 0 0 1
C_2	0 1 0 1 0 1 0 1 0 1 0 1 0 1 0 1	1 1 0 0 0 0 0 1 1 0 0 1 0 1 0 0
C_4	0 0 1 1 0 0 1 1 0 0 1 1 0 0 1 1	1 0 1 0 0 1 1 1 1 1 1 1 0 0 1 0
C_6	0 0 0 0 1 1 1 1 0 0 0 0 1 1 1 1	1 0 0 1 1 0 1 1 1 1 0 0 1 1 1 0
C_8	0 0 1 1 1 1 0 0 0 0 1 1 1 1 0 0	1 0 1 0 1 0 0 0 1 1 1 1 1 1 0 1
C_{10}	0 1 0 1 1 0 1 0 0 1 0 1 1 0 1 0	1 1 0 0 1 1 1 0 1 0 0 1 1 0 1 1
C_{12}	0 1 1 0 0 1 1 0 0 1 1 0 0 1 1 0	1 1 1 1 0 0 1 0 1 0 1 0 0 1 1 1
C_{14}	0 1 1 0 1 0 0 1 0 1 1 0 1 0 0 1	1 1 1 1 1 1 0 1 1 0 1 0 1 0 0 0
C_{16}	0 0 0 0 0 0 0 0 1 1 1 1 1 1 1 1	1 0 0 1 0 1 0 0 0 0 1 1 1 1 1 0
C_{18}	0 1 0 1 0 1 0 1 1 0 1 0 1 0 1 0	1 1 0 0 0 0 0 1 0 1 1 0 1 0 1 1
C_{20}	0 0 1 1 0 0 1 1 1 1 0 0 1 1 0 0	1 0 1 0 0 1 1 1 0 0 0 0 1 1 0 1
C_{22}	0 0 0 0 1 1 1 1 1 1 1 1 0 0 0 0	1 0 0 1 1 0 1 1 0 0 1 1 0 0 0 1
C_{24}	0 0 1 1 1 1 0 0 1 1 0 0 0 0 1 1	1 0 1 0 1 0 0 0 0 0 0 0 0 0 1 0
C_{26}	0 1 0 1 1 0 1 0 1 0 1 0 0 1 0 1	1 1 0 0 1 1 1 0 0 1 1 0 0 1 0 0
C_{28}	0 1 1 0 0 1 1 0 1 0 0 1 1 0 0 1	1 1 1 1 0 0 1 0 0 1 0 1 1 0 0 0
C_{30}	0 1 1 0 1 0 0 1 1 0 0 1 0 1 1 0	1 1 1 1 1 1 0 1 0 1 0 1 0 1 1 1

The odd numbered codes are formed by complementing the preceding even numbered codes, e.g., C_{15} is the complement of C_{14}. The synchronization signal = 1001010011000001. For PSK transmission, 1/0 in the Table stands for ± 1, i.e. phase reversal of the carrier.

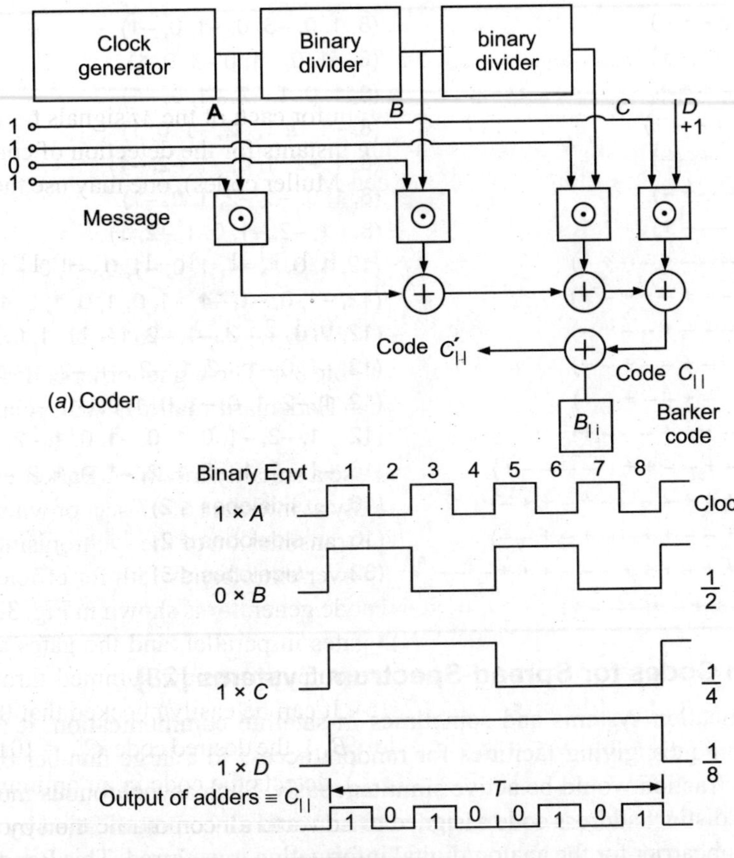

Figure 3.76: Functional diagram of *M*-16 code generator—**(a)** Coder; **(b)** Waveforms of the code generator

Barker Codes [25]

Ideal Barker codes with ACF = (0, ± 1) for $\tau \neq 0$, are known for N = 1, 2, 3, 4, 5, 7, 11 and 13 only. But by relaxing the above condition, it is possible to generate many more quasi-ideal codes. Some of these codes for $N \geq 8$ with their ACF is given in Table 3.5. More details are given in [25].

It is possible to generate longer codes by using dot products of smaller codes, e.g., the code $C_{11} \times C_{11}$, generated by the dot product of Barker code with $N_1 = N_2 = 11$, has an ACF peak of 121 and sidelobes 0 or –11. Thus many suitable combinations of codes may be found for the specific applications.

Table 3.5: Relaxed barker codes

N	Code vectors	ACF $[\rho_{ii}(\tau)]$
11	(+ + + – – – + – – + –)	(11, 0, –1, 0, –1, 0, –1, 0 –1, 0 –1)
	(– + – – + – – – + + +)	(ideal)
	and their complements.	
13	(+ + + + + – – + + – + – +)	(13, 0, 1, 0, 1, 0, 1, 0, 1, 0, 1, 0, 1)
	(+ – + – + + – – + + + + +)	(ideal)
	and complements.	

(Contd.)

8	(+ + + − − − + −)	(8, 1, 0, −3, 0, −1, 0, −1)
	(+ + − + − − − +)	(8, −1, 0, −1, 0 −3, 0, 1)
	(+ + + + − − + +)	(8, 1, 0, 1, −2, −1, 0, −1)
	(+ + + − + − − +)	(8, −1, 0, 1, −2, −1, 0, 1)
	(+ + + − − + − −)	(8, 1, −2, 1, 0, −1, −2, −1)
	(+ + − − − − + −)	(8, 1, 0, −1, −2, 1, 0, −1)
	(+ − + + − − − +)	(8, −1, −2, −1, 0, 1, −2, 1)
12	(+ + + − − − − + − − + −)	(12, 1, 0, 1, −4, 1, 0, −1, 0, −1, 0, −1)
	(+ + + − − − + − + + − +)	(12, −1, 0, −1, −4, −1, 0, 1, 0, 1, 0, 1)
	(+ + + + − − − + − − + −)	(12, 1, 0, 1, −2, −1, −2, 1, −2, −1, 0, −1)
	(+ + + − − − + − − + − +)	(12, −1, 0 −1, −2, 1, −2, −1, −2, 1, −0, 1)
	(+ + + − − − + − − + − −)	(12, 1, −2, 1, 0, −1, 0, 1, 0, −1, −2, −1)
	(+ − + + − + + + − − − +)	(12, −1, −2, −1, 0, 1, 0, −1, 0, 1, −2, 1)
16	(+ − − + − + − − + + − − − − − +)	(16, −1, −2, 1, −2, 1, 2, −1, 0, 1, 2, −1, 2, −1, −2, 1)
	(+ + + − + + + − − − − + − + + −)	(16, ..., sidelobes ≤ 2)
	(+ + + + + − − + + − − + − + − −)	(16, ..., sidelobes ≤ 2)
32	(+ + + + + + − + + + − − − + + + − − − + − + + − + + − + − − − +)	(32, ..., sidelobes ≤ 3)

3.8.6 Optimum Codes for Spread-Spectrum Systems [28]

In mobile communication systems and sometimes in satellite communication, it is necessary to use multiaccessing techniques, giving facilities for random access to a large number of possible users of whom only a small fraction would be active simultaneously. Such asynchronous multiplexing requires that each user has a distinct address code assigned to him and all communication by the user would use this address as the subcarrier for the analog/digital information transferred. This Random access discrete address (RADA) system utilizes a large channel bandwidth B as compared to the information bandwidth W, and B/W is known as the processing gain of the system. The address coding technique is somewhat similar to that used in M-ary signalling through orthogonal codes discussed in Sec. 3.4.5, with the difference that the code transmissions between different pairs of users are now asynchronous (although the transmitter-receiver pair is generally synchronized for the optimum detection of the desired information). Such asynchronous multi-access systems are also known as spread-spectrum (SS) systems, since the system operates by spreading the information spectrum over the total channel bandwidth B.

Consider that a data bit of duration T_0 is multiplied by an address code of length N and chip time τ_c, as shown in Fig. 3.77. Then the BW expansion $= T_0/\tau_e = B/W =$ Processing gain G_p. If the address codes of the K users are distinct and orthogonal, and the receiver uses crosscorrelator/MF detection, then the $(K-1)$ undesired (random) signals in the channel may be considered as random interference to the desired receiver. In the limit of $K \to$ large, the sum of the $(K-1)$ random signals will be equivalent to Gaussian random noise with variance $\sigma^2 = (K-1)S$, where S is the average received power from each of the interfering users. Thus, the CNR at the input of the receiver $= S/(K-1)S = 1/(K-1)$ and the output SNR is, from eq. (3.50),

$$S_0/N_0 = G_p \cdot \text{CNR} = G_p/(K-1) \tag{3.349}$$

where the MF gain $G_p = B/W = N$. It is now seen that for a reasonable $S_0/N_0 \geq 10$ dB with $P_e \leq 10^{-5}$, the code length $N \geq 10K$. Considering the ambient channel noise and other external interferences, N has to be somewhat larger than $10K$.

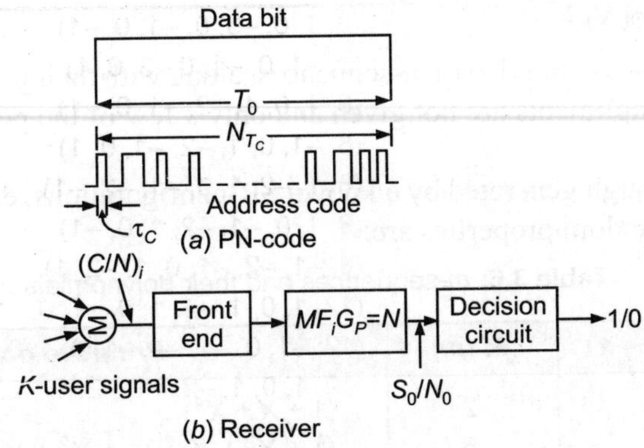

Figure 3.77: SS-code and the MF receiver **(a)** PNH-code; **(b)** Receiver

The address codes are generated mainly by using

(a) Long quasi-orthogonal sequences giving code division multiaccessing (CDMA)
(b) Time-frequency (TF) matrix/frequency hopping (FH) codes giving Pulse address multiaccessing (PAMA)
(c) Hybrid techniques

PNS Codes

For CDMA, maximal length Pseudo-noise sequences (PNS) (also known as PRBS) [26], generated by feedback shift registers (SR), are generally used, since the acyclic ACF of long PNS has small sidelobes and many quasi-orthogonal PNS may be generated if N is sufficiently large. In general, a sequence length of $N = 2^k - 1$, is generated by connecting k-stage SR in tandem and giving a feedback to the input from two or more SR-outputs through a MOD-2 adder, as shown in Fig. 3.78. Maximal-length PNS (or m-sequences) are generated by certain primitive polynomials (refer to Sec. 1, Ch. 2), and the compliments of the polynomials also give valid feedback connections. In the case of the 4-stage SR, shown in Fig. 3.78, the primitive polynomial used is $f(X) = (1 + X + X^4)$ and its compliment is: $\bar{f}(X) = X^4(1 + X^{-1} + X^{-4}) = (X^4 + X^3 + 1)$. Thus the feedback connection may be taken either from $(X_1 \oplus X_4)$ or from $(X_3 \oplus X_4)$, and two distinct 15-length m-sequences are obtained as

$$\{0001 \quad 00110 \quad 101111\} \text{ or } \{000 \quad 1111 \quad 010 \quad 11001\}$$

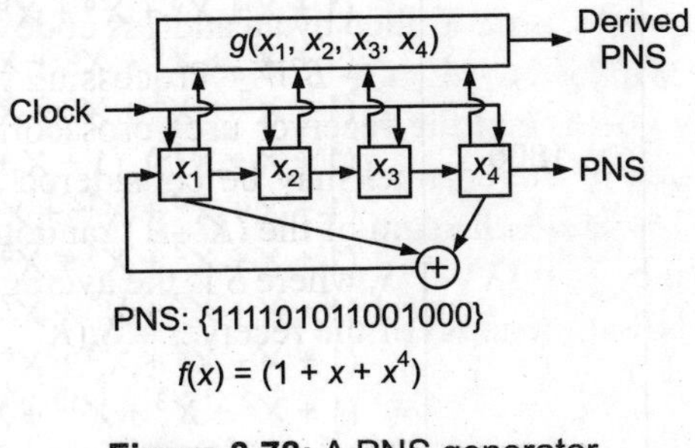

Figure 3.78: A PNS generator

By combining the outputs of the SR's through a linear/nonlinear switching circuit $g(X_1, X_2, X_3, X_4, ...)$, a derived sequence may also be obtained. Maximum number of m-sequences that can be generated by k-SR's is given by

$$M(m) = \phi(N)/k \tag{3.350}$$

where $\phi(N)$ is Euler function. A short list of m-sequences, along with their primitive polynomials is given is Table 3.6, where the compliments are not given and only a few of the polynomials shown. For more details, refer to [19, 26].

The PN sequences, although generated by a known SR-configurations, exhibit random properties for $N \rightarrow$ large. These pseudorandom properties are:

Table 3.6: *m*-sequences and their polynomials

SR-lenght $=k$	$N = 2^k - 1$	$M(m)$	Primitive polynomials $f(x)$
4	15	2	$(1 + X + X^4)$
5	31	6	$(1 + X^2 + X^5)$, $(1 + X^2 + X^3 + X^4 + X^5)$, $(1 + X + X^2 + X^4 + X^5)$
6	63	6	$(1 + X + X^6)$, $(1 + X + X^2 + X^5 + X^6)$, $(1 + X^2 + X^3 + X^5 + X^6)$
7	127	18	$(1 + X^3 + X^7)$, $(1 + X + X^2 + X^3 + X^7)$, $(1 + X + X^2 + X^4 + X^5 + X^6 + X^7)$, $(1 + X^2 + X^3 + X^4 + X^7)$, $(1 + X + X^2 + X^3 + X^4 + X^5 + X^7)$, $(1 + X^2 + X^4 + X^6 + X^7)$, $(1 + X + X^7)$, $(1 + X + X^3 + X^6 + X^7)$, $(1 + X^2 + X^5 + X^6 + X^7)$.
8	255	16	$(1 + X^2 + X^3 + X^4 + X^8)$, $(1 + X^3 + X^5 + X^6 + X^8)$, $(1 + X + X^2 + X^5 + X^6 + X^7 + X^8)$, $(1 + X + X^3 + X^5 + X^8)$, $(1 + X^2 + X^5 + X^6 + X^8)$, $(1 + X + X^5 + X^6 + X^8)$, $(1 + X + X^2 + X^3 + X^4 + X^6 + X^8)$, $(1 + X + X^6 + X^7 + X^8)$
10	1023	60	$(1 + X^3 + X^{10})$, $(1 + X^2 + X^3 + X^8 + X^{10})$, $(1 + X^3 + X^4 + X^5 + X^6 + X^7 + X^8 + X^9 + X^{10})$, $(1 + X^3 + X^4 + X^5 + X^6 + X^7 + X^8 + X^{10})$ etc.
11	2047	176	$(1 + X^2 + X^{11})$, $(1 + X^2 + X^5 + X^8 + X^{11})$, $(1 + X^2 + X^3 + X^7 + X^{11})$, $(1 + X^2 + X^3 + X^5 + X^{11})$, $(1 + X^2 + X^3 + X^{10} + X^{11})$, $(1 + X + X^3 + X^8 + X^9 + X^{10} + X^{11})$, $(1 + X^3 + X^6 + X^8 + X^{11})$, $(1 + X^2 + X^4 + X^6 + X^8 + X^{10} + X^{11})$, etc.
15	32,767	1800	$(1 + X + X^{15})$, $(1 + X + X^5 + X^{10} + X^{15})$, $(1 + X + X^3 + X^{12} + X^{15})$, $(1 + X + X^2 + X^4 + X^5 + X^{10} + X^{15})$, $(1 + X + X^2 + X^6 + X^7 + X^{11} + X^{15})$, $(1 + X + X^2 + X^3 + X^6 + X^7 + X^{15})$, $(1 + X^2 + X^3 + X^{12} + X^{15})$, etc.
20	1,048,575	24,000	$(1 + X^3 + X^{20})$, $(1 + X^3 + X^5 + X^9 + X^{20})$, $(1 + X^2 + X^3 + X^6 + X^8 + X^{11} + X^{20})$, etc.
21	2,097,151	84,672	$(1 + X^2 + X^{21})$, $(1 + X^2 + X^7 + X^{14} + X^{21})$, $(1 + X^2 + X^5 + X^{13} + X^{21})$, etc.

(a) *The balance property:* In each complete period of the sequence, the number of 1's differs from the number of 0's by at most one.
(b) *The run property:* Among the runs of consecutive 1's and 0's in one complete period of the sequence, a half of the runs of each kind are of length 1, a quarter of each kind are of length 2, an eighth are of length 3, and so on, as long as these fractions give meaningful numbers of runs.
(c) *The correlation property:* The cyclic correlation of a PNS is given by eq. (3.336), and in normalised form

$$\rho_{ii}(\tau) = \begin{cases} 1, & \text{for } \tau = 0 \\ -1/N, & \text{for } \tau \neq 0 \end{cases}$$

Based on above, the poser spectrum of a PNS is given by

$$G(\omega) = \frac{N-1}{N}\left(\frac{\sin \omega/2f_r}{\omega/2f_r}\right)^2 \sum_{\substack{-\infty \\ n\neq 0}}^{\infty} \delta\left(\omega - \frac{2\pi n f_r}{N}\right) + \frac{1}{N^2}\delta(\omega) \tag{3.351}$$

where $N = 2^k - 1$ and f_r = clock frequency of the PNS generator. The spectrum thus has a contour given by $(\sin x/x)^2$ and the equivalent noise bandwidth $B_n \simeq f_r$. Although the cyclic ACF's of *m*-sequences are ideal, the acyclic ACF of a PNS has considerable sidelobes and for $N \to$ large, the peak sidelobes have magnitudes $\simeq \sqrt{N}$. The CCF between pairs of *m*-sequences of the same length *N* is then quite large and is given by:

$$\rho_{ij}(C_i, C_j) > \sqrt{N-1}, \tag{3.352}$$

Some values of peak $|\rho_{ij}|$ are given in Table 3.7 and it is seen that all such codes cannot be used as the signalling waveforms for SS systems unless *N* is very large, say, $N \geq 10^4$.

Table 3.7: Maximum cross-correlation values between pairs of *m*-sequences

k	N	Cyclic ρ_{ij}	Max. $\lvert\rho_{ij}\rvert$	Normalised $\rho_{ij} = \lvert\rho_{ij}\rvert/N$
5	31	31, –1	11	0.35
6	63	63, –1	23	0.36
7	127	127, –1	41	0.32
9	511	511, –1	113	0.22
10	1023	1023, –1	383	0.37
11	2047	2047, –1	287	0.14
12	4095	4095, –1	1407	0.34

Gold [27] has, however, shown that certain pairs of *m*-sequences have CCF bounded as

$$\rho_{ij}(C_i, C_j) \leq 2^{(k+1)/2} + 1 \text{ for } k \text{ odd}$$

and

$$\rho_{ij}(C_i, C_j) \leq 2^{(k+2)/2} + 1 \text{ for } k \text{ even and not divisible by 4} \tag{3.353}$$

As for example, using two ($k = 6$) – PNS generators with polynomials,

$$f_1(X) = X^6 + X + 1$$

$$f_2(X) = X^6 + X^5 + X^2 + X + 1,$$

it is calculated that $\rho_{ij} \leq 2^4 + 1 = 17$. Such pairs of PNS are called 'preferred' sequences. From a pair of preferred sequences, say C_i and C_j, a new sequence C_l may be generated by taking MOD-2 sum of C_i with N cyclically shifted versions of C_j or vice versa as shown in Fig. 3.79. Thus N new sequences are generated with the CCF bound given by eq. (3.353), and including C_i and C_j, the total number of such sequences are $(2^k + 1)$.

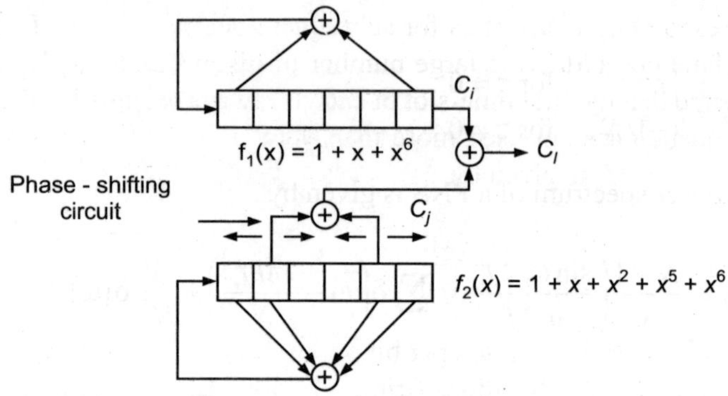

Figure 3.79: Generator for Gold code, cyclic shifts of C_j are controlled by the phase-shifting circuit

These sequences are known as Gold sequences (codes) whose ACF values are the same as the CCF values. Thus the codes have poorer ACF than *m*-sequences, but better CCF values. Further it has been shown that the Gold codes have three-level CCF's and Table 3.8 gives their values and relative frequencies of occurrence. It is seen that even values of *k* (not divisible by 4) give better CCF values for the above Gold codes.

Table 3.8: CCF values and their relative frequencies for Gold codes

k = SR-lenght	*N*	Normalised CCF	Relative frequency
k odd	$2^k - 1$	$-1/N$	~ 0.50
		$-(2^{(k+1)/2} + 1)/N$	~ 0.25
		$(2^{(k+1)/2} - 1)/N$	~ 0.25
k even, but not divisible by 4	$2^k - 1$	$-1/N$	~ 0.75
		$-(2^{(k+2)/2} + 1)/N$	~ 0.125
		$(2^{(k+2)/2} - 1)/N$	~ 0.125

Using Gold codes or *m*-sequences with $N \to$ large, as the addresses for the users in a CDMA(SS) system, the false-address probability for DPSK transmission and MF-reception, is calculated from eq. (3.349), as

$$P_f = \frac{1}{2}\exp(-\gamma) = \frac{1}{2}\exp\left(-N/(K-1)\right) \tag{3.354}$$

where γ = CNR, N = BT for NRZ pulses, and K = number of simultaneous users. The system has strong anti-jamming capability, but suffers from strong-talker interference. However, error-correcting codes may be used to improve the performance further.

TF Matrix Codes

In TF-matrix coding shown in Fig. 3.80, the bit-time T is subdivided into n slots and the available RF bandwidth B is subdivided into m slots, giving rise to $(m \times n)$ slots in the TF-plane, where $T = n\tau$, $B = m\Delta f$, and $\Delta f \geq 1/\tau$. In the transmitter, the information bits control an address matrix, such that a few short pulses (carrier bursts) out of $(m \times n)$ available slots are transmitted per bit. In the receiver, the pulses are received through narrow bandpass filters and the resulting video pulses are either counted or summed through corresponding delay lines for subsequent decision on 1/0. It is seen that using only a few slots in the TF-plane per address, a large number of distinct addresses are possible. To make the addresses unique, assume that the first time-slot of each is always occupied and in general, the same time slot and the same frequency are not used more than once in an address. Using these restrictions, the number of unique addresses M_A is given by:

$$M_A = \frac{m!}{(m-q)!q!} \cdot \frac{(n-1)!}{(n-q)!} \tag{3.355}$$

where q is the number of pulses transmitted per bit of information. The first part of the formula gives the number of available frequency combinations (triple, quadruples etc.) and second part gives the number of addresses for one frequency combination. As for example, using $m = 4$, $n = 6$ and $q = 3$, $M_A = 80$, but using $m = 8$, $n = 10$ and $q = 5$, as in Fig. 3.80, $M_A = 169{,}344$. Thus for a given matrix size, a large number of addresses may be generated, under the control of a PNS generator, whose output at any instant would select the addresses in a random manner.

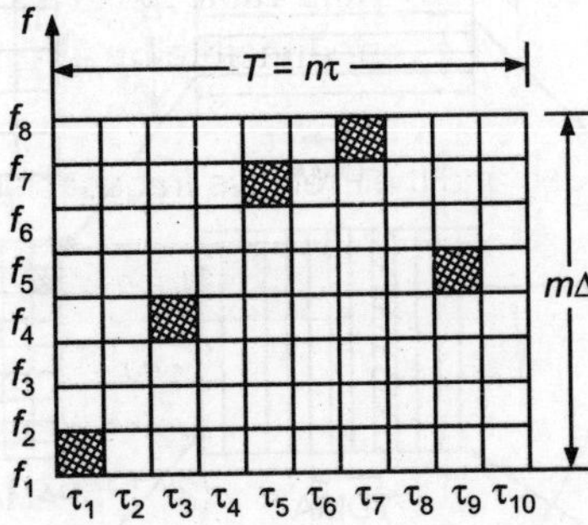

Figure 3.80: TF-matrix code/FFH

To obtain the false-address probability P_f, it is assumed that the receiver recognizes the address through a multiple-coincident circuit and the receiver-transmitter link is synchronized. Further, it may be assumed that the pulses occur obeying the Poisson's distribution, and then, P_f is calculated as

$$P_f = (1 - e^{-\lambda\tau})^q \tag{3.356}$$

where $\lambda\tau = \dfrac{q\tau K}{mT} = \dfrac{qK}{BT} = \dfrac{qK}{m \cdot n}$; K = number of simultaneous users. To obtain an error rate $P_f \simeq 10^{-5}$ with $q = 5$, $\lambda\tau = 0.1$, and for $K = 50$, $(m \times n) = 2500$; but by increasing q, the matrix size may be reduced. Here, also, the error-correcting codes improve the anti-jamming capability of the system to a large extent.

The above signalling technique is generally known as Fast Frequency Hopping (FFH), since q frequencies are transmitted per signalling bit/symbol. In contrast to this, there are slow FH-schemes, where the carrier frequency is randomly hopped in every signalling interval T, under the control of a PNS code generator. The 1/0 signals are transmitted by FSK and the receiver demodulates the signal incoherently after changing the random frequencies to a constant IF by mixing the input with the output of a frequency synthesizer controlled by a synchronised PNS-code generator. Whenever a collision with other transmitter signals occur, then one bit/symbol of information is lost; but this may be retrieved if a suitable error-correcting code is used in tandem.

Many methods of FH signalling have been used and it is also possible to use time hopping (TH) signalling, analogous to FH, to introduce randomness in SS signal [17]. Combining the different techniques of FH, TH and CDMA, many hybrid signalling techniques have been suggested, as shown in Fig. 3.81 [29]. The use of a particular technique depends upon the requirements of anti-jamming capability and the availability of the channel bandwidth required for the link. Analog codes, e.g. multislope linear FM signals, modified Huffman sequences, PPM-PN-FSK etc., have also been suggested as possible SS codes. Cook [30] has suggested that FM signals with different slopes may be used as quasi-orthogonal signals in the bounded TF-plane. For a linear FM pulse, the ith signalling waveform is (see Ex. 8)

$$s_i(t) = A\cos\left(\omega_i t + \frac{1}{2}\alpha_i t\right), \quad 0 < t \le T_i, \tag{3.357}$$

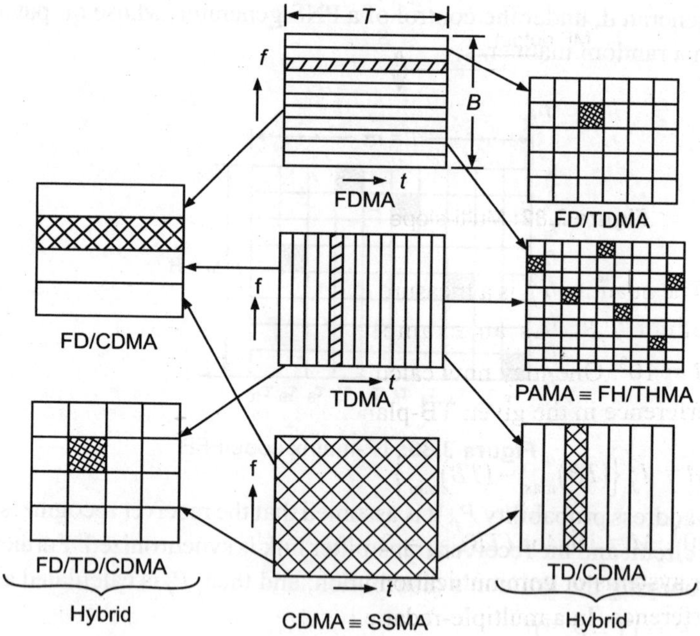

Figure 3.81: Generation of hybrid SS codes [29]

where α_i = time-frequency slope = $2\pi B_i/T_i$, as shown in Fig. 3.82. Suppose that another signal $s_j(t)$ with a slope α_j is received by the MF matched to α_i, then the undesired peak I_j due to $s_j(t)$ is given by

$$I_i^2 = (|\gamma| T_i B_i)^{-1} \tag{3.358}$$

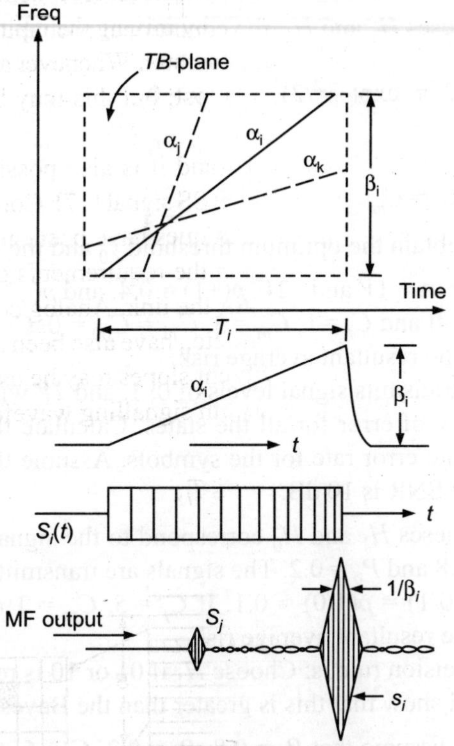

Figure 3.82: Multi-slope FM pulses and MF outputs

where $\gamma = (\alpha_j - \alpha_i)/\alpha_i$. The quantity I_j^2 is a measure of mutual interference and depends on the mismatch factor γ and the product T_iB_i. As an example, for a required signal to interference ratio $S_i/I_j^2 = 30$ dB, $|\gamma|\, T_iB_i = 10^3$. One may now calculate the number of discrete signals that may coexist with the allowable interference in the given TB-plane, and the number is given by

$$M = I_j^2 \left\{(TB)_{\max} - (TB)_{\min}\right\} \tag{3.359}$$

Then for an $S/I_j^2 = 10^3$, $M = 10$, for $(TB)_{\max} = 2 \times 10^4$ and $(TB)_{\min} = 10^4$. But it is possible to have approximately 11 radar systems or communication links in the *TB*-plane having an area of 10^4 with an acceptable mutual interference. In a multiple-radar environment, the mutual interference of the order of –15 to –20 dB may be tolerated and the bandwidth expansion required will be smaller. Similar values of interference are also acceptable for binary decision circuits, e.g., with CNR = 10 dB, $P_e \simeq 10^{-5}$. Further refinement of the coding technique has been suggested and it is possible to introduce identification signals in the form of secondary phase modulation of the linear FM waveform by sinusoidal signals which results into discrete pulse groups (as secondary addresses) at the MF output. Thus the system may be made quite versatile, specially for applications in the multi-radar environment.

PROBLEMS

1. Assume that for hypotheses H_1 and H_0, the conditional probabilities are:

$$H_1: p(y/1) = \sqrt{2/\pi}\ \exp(-y^2/2), \qquad y > 0$$

$$H_0: p(y/0) = e^{-y}, \qquad y > 0$$

and $P_1 = 0.3$, $P_0 = 0.7$, $C_f = C_m$

Using Bayes criterion, obtain the optimum threshold y_0 and the resultant average risk $\bar{C}(y_0)$.

2. In a binary transmission of $+1V$ and $-2V$, $p(+1) = 0.4$, and $p(-2) = 0.6$. The variance of the added Gaussian noise is 0.5 watt and $C_f = 1$, $C_m = 3$, $C_{11} = C_{00} = 0$. Calculate the optimum threshold for Bayes decision rule ad the resultant average risk.

3. A 3-level PCM system transmits signal levels of 0, 1, and $2V$ with equal probability. It is required to have equal probability of error for all the states. Calculate the decision thresholds for such a case, and also the average error rate for the symbols. Assume that the channel noise is Gaussian and the average channel SNR is 10 dB.

4. (a) Assume that hypotheses H_1 and H_0 correspond to the signal 11 and 00 respectively, and the probabilities $P_1 = 0.8$ and $P_0 = 0.2$. The signals are transmitted through a binary channel with error probability $p(0/1) = p(1/0) = 0.1$. If $C_f = 5$, $C_m = 1$ and $C_{11} = C_{00} = 0$, obtain Bayes decision rule and the resultant average risk.

   (b) Suppose that the decision rule is: Choose H_1 if 01 or 10 is received (since $P_1 > P_0$). Calculate the average risk and show that this is greater than the Bayes risk.

5. For Ex. 3 in the chapter, assume that $P_1 = 0.8$, $P_0 = 0.2$, $C_f = 5$, $C_m = 1$, $A_1 = 2$, $A_0 = 0$ and $\sigma^2 = 1$. Calculate Bayes average risk $\bar{C}$ with different values of n, $n = 5$, 10, 15 and 20. How large n should be to decrease $\bar{C}$ from that with $n = 1$ by a factor of 10^{-2}?

6. In Ex. 3, with parameter values as above except that C_f and C_m are unknown, it is required to have $P_f = 10^{-4}$. Calculate the optimum threshold using Neyman-Pearson criterion for $n = 10$,15 and 20. Calculate also the corresponding P_d.

7. (a) The input to an MF is a pulsed sinewave given by

$$s(t) = \begin{cases} 2\cos\left(2\pi 10^3 t + \pi/6\right), & \text{for } 0 < t < T, \quad T = 10^{-3} \text{ sec.} \\ 0, & \text{otherwise} \end{cases}$$

   Calculate the maximum $(S/N)_0$ at the output of the MF, given that $n_0/2 = 10^{-6}$ W/Hz.

   (b) Draw the structure of the MF using TDL, giving clock rate, number of taps and the required weighting functions.

8. For two symbol transmission, assume $s_1(t)$ as shown in Fig. P-8 and $s_2(t) = -s_1(t)$. Calculate $(S/N)_0$ at the receiver output, given that $n_0/2 = 100\ \mu$W/Hz.

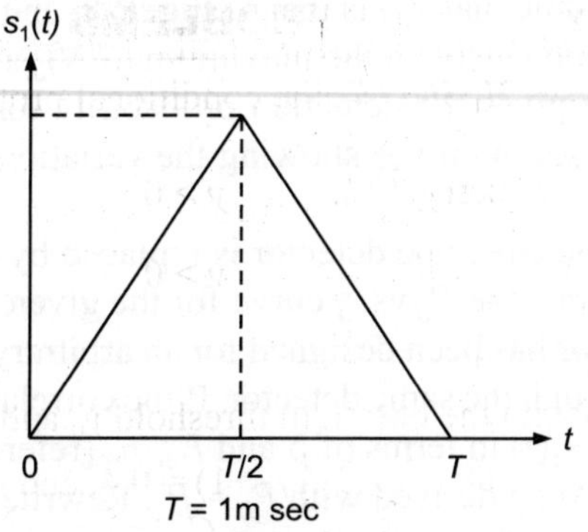

Figure P.8

9. (a) For the given signal waveforms in Figs. P.9(a) and P.9(b), construct the corresponding MF's using TDL. Indicate the clock rate, number of taps and the required weighting functions.

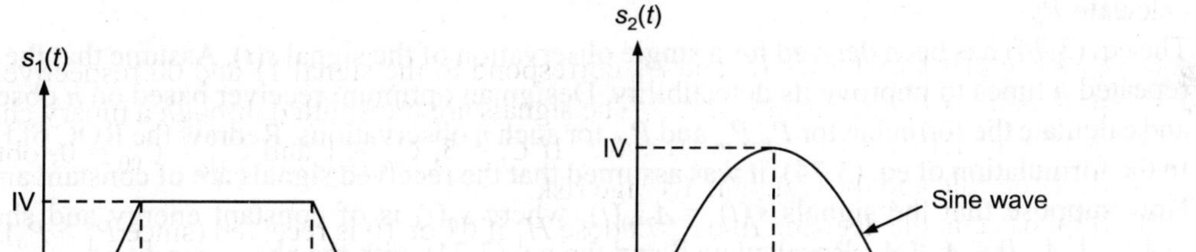

Figure P.9

(b) Use the waveforms for two-symbol transmission and calculate the optimum $(S/N)_0$ at the receiver output in each case if the input $(S/N)_i = 10$ dB.

10. Assume that an MF has the impulse response:

$$h(s) = \begin{cases} A, & 0 \le t \le T \\ 0, & \text{otherwise} \end{cases}$$

A signal $s(t) = e^{-\alpha t}$, $0 \le t \le T$, disturbed by Gaussian noise is received through this MF. What is the peak $(S/N)_0$ at the MF output?

11. A rectangular pulse of height A and width T is received through a Gaussian filter with the impulse response,

$$h(t) = (1/\beta) \exp [-(t - t_0)^2/2\beta^2], -\infty < t < \infty, t_0 > 0, t_0 \gg \beta;$$

Derive an expression for $(S/N)_0$ at the MF output. At what value of t will (S/N_0) be maximum?

12. In Ex. 9 in the chapter, assume that $s(t)$ is that of Fig. P-8, and $\omega_0 = 10^3$, $K = 1$. Obtain the impulse response $h(t)$ for the MF and calculate the maximum $(S/N)_0$ at the MF output.
13. Consider eq. (3.63) which gives the detection probability for incoherent signals using Neyman-Pearson criterion. Draw a set of curves showing the variation of P_d vs. γ for $P_f = 10^{-3}, 10^{-5}, 10^{-7}$ and $\gamma = 1, 2, 3, ..., 10$ dB.
14. If in the above problem, the envelope detector is replaced by a square-law detector, then give the expression for P_d, and redraw the P_d vs. γ curve for the given P_f and γ.
15. Assume that an MF detector has been designed for an arbitrary signal $s(t)$. Now a correlated signal $s_1(t)$ is being received through the same detector. If the correlation coefficients between the signals is ρ, then calculate P_d for $s_1(t)$ in terms of ρ and E_{av}/n_0 [refer to Section 3.4.1].
16. Eqns. (3.114–3.116) have been derived with $P_0 = P_1$. Rewrite the equations with $P_0 \neq P_1$.
17. The sampled representations of two signals $s_1(t)$ and $s_2(t)$ are

$$s_1(t) = \{1, -1, 1, -1, 1\};\ s_2(t) = \{1, -1, -1, -1, -1\}$$

$P_1 = 0.3, P_2 = 0.7$, and the disturbing Gaussian noise is zero-mean and $\sigma_n^2 = 1$. The signals are received through MF detectors. Design the MF and calculate the optimum threshold for minimum P_e. Also calculate P_e.
18. The eq. (3.74) has been derived for a single observation of the signal $s(t)$. Assume that the signal is repeated n times to improve its detectibility. Design an optimum receiver based on n observations and calculate the formulae for P_f, P_m and P_d, for such n observations. Redraw the ROC of Fig. 3.14.
19. In the formulation of eq. (3.74), it was assumed that the received signals are of constant amplitude. Now suppose that the signals $s_j(t) = A_j s_0(t)$, where $s_0(t)$ is of constant energy and shape, and $p(A_j) = 1/A_0$, $0 \le A_j \le A_0$. Recalculate β and the eq. (3.74) with the above condition.
20. For binary transmission, the signals $s_1(t)$ and $s_2(t)$ are

$$s_1(t) = A \sin \omega_0 t + B \cos(\omega_1 t + \theta),\ 0 \le t \le T$$

$$s_2(t) = B \cos(\omega_1 t + \theta), \qquad 0 \le t \le T$$

where A, B, ω_0, ω_1 are known and θ is uniformly distributed over $(0, 2\pi)$. The signals are disturbed by white Gaussian noise in the channel. If $(\omega_0 - \omega_1) = \pi/T$, show that the optimum receiver uses the test statistic $= \int_0^T y(t) \sin \omega_0 t\, dt$. Obtain the equation for P_e.
21. (a) In Sec. 3.4.3, incoherent detection of ASK signals has been discussed using $P_0 = P_1 = 0.5$. Assume that $P_0 \neq P_1$, and $P(A = 0) = p_0$, $p(A = A_0) = p_1$, $p_1 + p_0 = 1$. Calculate P_e.

    (b) Assume that A has the p.d.f., $p(A) = \sum_{i=1}^{m} p_i \delta(A - A_i)$. Obtain the optimum receiver and the values of P_f and P_m.
22. Assume that the signals $s_1(t)$ and $s_2(t)$ for binary transmission are: $s_1(t) = (1 + m \cos \omega_0 t + \phi) s_0(t)$ and $s_2(t) = 0$, where $s_0(t)$ is a band-limited Gaussian process centred around ω_1, $\omega_1 \gg \omega_0$, $m \ll 1$. The channel noise is Gaussian. Design the optimum receiver and calculate P_e.
23. In Section 3.3.5, the test statistic for the decision has been taken as $\sum_1^n R_j^2 > \beta_M$, and a square law detector is used in the receiver. Suppose that an envelope detector was used in the receiver giving the test statistic as: $\sum_1^n R_j > \alpha_M$. Derive the fresh set of equations for P_f and P_d, and compare them with eqs. (3.103) and (3.104).

24. For coherent binary FSK signalling (refer to eq. (3.117)) it was assumed that the signals $s_1(t)$ and $s_0(t)$ were orthogonal. Assume that the two signals are further separated by the conditions:

$$(\omega_1 - \omega_0) \simeq 1.4\pi/T \text{ and } (\omega_1 + \omega_0) \gg 0.$$

Calculate the new P_e and show that this is less than the value given by eq. (3.119).

25. Assume that a ternary communication system uses the alphabets:

$$s_0(t) = 0,\ s_1(t) = A \cos \omega_0 t,\ s_2(t) = -A \cos \omega_0 t;\ 0 \le t \le T$$

If the channel noise is white and Gaussian with spectral density $= n_0/2$, then obtain the optimum receiver for minimum P_e. Calculate the values of P_e for different values of E_b/n_0.

26. The eqs. (3.118)–(3.120) were obtained by assuming perfect phase synchronisation of the carriers. It is now assumed that there is some carrier phase offset $\Delta\theta$ at the receiver. Recalculate the eqs. showing the effect of $\Delta\theta$.

27. The error probability in M-ASK given by eq. (3.142) has been calculated based on coherent detection. What will be the corresponding value of P_e using incoherent detection?

28. Recalculate the eq. (3.151) for $P_e(M)$ of M-PSK systems assuming a carrier phase offset of $\Delta\theta \ll 2\pi/M$.

29. (a) A data transmission system uses repetitive code and PSK to improve P_e at the receiver. Assume that the information bits are repeated 8 times, thus increasing the bandwidth of the channel 8 times as well. Detection is done by using majority logic, i.e., 3 errors per word are allowed for correct decision. Draw the block diagram of a coherent receiver for the system. Calculate P_e at the receiver output for $E_b/n_0 = 5, 7, 10$ dB.
    (b) If the same data bits are now transmitted using biorthogonal codes and approximately the same BW-expansion is used, then what will be P_e for $E_b/n_0 = 5, 7, 10$ dB?

30. Repeat the problems 29(a) and (b), using binary ASK and noncoherent reception.

31. Show a scheme for QPSK transmission of a (32 × 64) biorthogonal code. Calculate the BW-expansion in the channel and P_e for $E_b/n_0 = 3, 5, 7, 10$ dB.

32. Figure 3.30 shows the P_b vs. E_b/n_0 characteristics of M-PSK systems, using Nyquist bandwidth. However, to reduce intersymbol interference, the raised cosine pulses with excess bandwidth given by $\alpha = 0.5$ and $\alpha = 1$, are normally used. Redraw the curves of Fig. 3.30, with $\alpha = 0.5$ and $\alpha = 1$.

33. (a) Eqs. (3.214) and (3.215a) were derived using a Gaussian p.d.f. for the amplitude h. Obtain the Bayes estimate for the same problem using a squared-error cost function and a uniform p.d.f. for h, e.g., $p(h) = 0.5, -1 < h < 1$.
    (b) Repeat the problems with $p(h) = (1/2A) \exp(-|h|A)$.

34. In Ex. 11, assume that the mean of the samples α is known but the variance σ^2 is unknown. Determine the ML-estimate of σ^2 from the given sample values $\{y_i\}$.

35. (a) Derive from fundamental considerations, the eqs. (3.200) and (3.200a) for $\sigma_{\hat{\tau}}^2$ = the variance of $\hat{\tau}$.
    (b) What is the variance of $\hat{\tau}$ obtained through the solution of eq. (3.202)?

36. (a) Assume that the received radar signal is of the form

$$y(t) = A(t - \tau) \cos [\omega_0 t + \alpha t^2 - \theta(t)] + n(t)$$

where α is related to the acceleration of the moving target. Obtain an ML-estimate of α, assuming that $A(t)$, τ and ω_0 are known. What is the value of $\sigma^2(\hat{\alpha})$?

(b) Assuming $A(t)$ and ω_0 to be known, obtain a simultaneous estimate of τ and α and the corresponding variances $\alpha^2(\hat{\alpha})$ and $\sigma^2(\hat{\tau})$.

37. Using linear approximation of the PLL, calculate the transfer function of the optimum loop filter to minimize θ_ϵ^2, given that:
(i) $\theta(t) = a_0 + at + at^2$; (ii) $\theta(t) = a_0 + at^3$.
Are these filters realizable?

38. In a PLL receiver for FM, design the zero-lag loop filter and the infinite-lag tandem filter for a modulating signal given by:

(i) $A(t) = a_0 + a_1 \sin \omega_1 t, \qquad \omega_1 \ll \omega_0$

(ii) $A(t) = a_1 \sin \omega_1 t + a_2 \sin \omega_2 t, \qquad \omega_1, \omega_2 \ll \omega_0.$

Are these filters realizable?

39. In an FMFB receiver, the bandwidth of the loop IF filter is 3 times the audio bandwidth and $D = 10$ in the received signal. What is the threshold value at the receiver? If the modulating signal spectrum is

$$G_s(\omega) = \frac{1}{(1+\omega^4)}$$

design the optimum low pass filter in the loop. Will the audio SNR improve if another lowpass filter in tandem outside the loop is used?

40. The Wiener filter discussed in the text may also be used for pure prediction or prediction plus filtering.

(a) Design an optimum predictor for $G_s(\omega) = \dfrac{1}{(1+\omega^4)}$.

(b) Design a predictor-cum-optimum filter for

$$G_s(\omega) = \frac{1}{(1+\omega^4)} \text{ and } G_n(\omega) = \frac{2\omega^2}{1+\omega^2},$$

REFERENCES

1. Middleton, D., *An Introduction to Statistical Communication Theory,* McGraw-Hill, N.Y., 1960.
2. Helstrom, C.W., *Statistical Theory of Signals Detection,* 2nd Edition, Pergamon Press, Oxford, 1968.
3. Whalen, A.D., *Detection of Signals in Noise,* Academic Press, N.Y., 1971.
4. Das, J., 'Optimum detection and communication in presence of noise', *ITE Students Journal,* Vol. 14, Pt. I, II, III and IV, pp. 27, 106, 156, 212, 1973.
5. Van Trees, H.L., *Detection, Estimation and Modulation Theory,* Pt. 1, John Wiley, N.Y., 1968.
6. Schwartz, M., Bennett, W.R. and Stein, S., *Communication Systems and Techniques,* McGraw-Hill, N.Y., 1966.
7. Special Issue on Matched-Filter, *IRE Trans. on Inf. Theory,* Vol. IT-6, June 1960.
8. Turin, G.L., 'An introduction to matched filters', *IRE Trans. Inf. Theory,* Vol. IT-6, pp. 310–329, 1960.
9. Harmen, W.W., *Principles of Statistical Communication Theory,* McGraw-Hill, N.Y., 1963

10. Marcum, J.I. and Swerling, P., 'Studies of target detection by pulsed radar', Special Monograph Issue, *IRE Trans. Inf. Theory,* Vol. IT-6, April 1960.
11. Pearson, K., *Tables of Incomplete Gamma Function,* Cambridge Univ. Press, London and N.Y., 1965.
12. Rice, S.O., 'Mathematical analysis of random noise', *BSTJ,* Vols. 23 and 24, July 1944 and Jan. 1945.
13. Arthurs, E. and Dym, H., 'On the optimum detection of digital signals in the presence of white Gaussian noise–A geometric interpretation and a study of three basic data transmission systems', *IRE Trans. on Commn. Systems,* Vol. CS-10, pp. 336–372, Dec. 1962.
14. Cahn, C.R., 'Performance of digital phase-modulation communication systems', *IRE Trans. on Comm. Systems,* Vol. CS-7, pp. 3–6, May 1959.
15. Cahn, C.R., 'Combined digital phase and amplitude modulation communication systems', *IRE Trans. on Comm. Systems,* Vol. CS-8, pp. 150–155, Sept. 1960.
16. Viterbi, A.J., *Principles of Coherent Communication,* McGraw-Hill, N.Y., 1966.
17. Proakis, J.G., *Digital Communications,* McGraw-Hill, N.Y., 1983.
18. Lindsay, W.C. and Simon, M.K., *Telecommunication System Engineering,* Prentice-Hall, N.Y., 1973.
19. Holmes, J.K., *Coherent Spread Spectrum Systems,* John Wiley & Sons, N.Y., 1982.
20. Wiener, N., *Extrapolation, Interpolation and Smoothing of Stationary Time Series,* John Wiley, N.Y., 1950.
21. Laning, J.H. and Battin, R.H., *Random Processes in Automatic Control,* McGraw-Hill, N.Y., 1956.
22. Bode, H.W. and Shannon, C.E., 'A simplified derivation of linear least square smoothing and prediction theory', *Porc. IRE,* Vol. 38, pp. 423–424, April 1950.
23. Cook, C.E. and Bernfeld, M., *Radar Signals,* Academic Press, N.Y., 1967.
24. Huffman, D.L., 'Interference blanker for dissimilar modified Huffman sequences', *IEEE Trans.,* Vol. EMC-18, p. 118, 1976.
25. Moharir, P.S. and Selvarajan, A., 'Sequences with good auto-correlation for pulse compression and synchronisation', Tech. Report CIP/01P/1.1, EE Deptt. IIT Kanpur, Jan. 1974.
26. Golomb, S.W., Shift Register Sequences, Holden-Day, San Francisco, 1967.
27. Gold, R., 'Optimal binary sequences for spread spectrum multiplexing', *IEEE Trans. Inf. Th.,* Vol. IT-13, pp. 619–621, Oct. 1967.
28. Dixon, R.C., *Spread Spectrum Systems,* Wiley-Interscience, N.Y. 1976.
29. Das, J., 'A review of the RADA technique', *IETE Students Journal,* Vol. 19, pt. 1, pp. 15–32, 1978.
30. Cook, C.E., 'Linear FM signals formats for beacon and communication systems', *IEEE Trans.,* Vol. AES-10, p. 471, 1974.

CHAPTER 4

Error Correcting Codes

4.0 INTRODUCTION [1, 2, 3, 4]

The optimum detection techniques for signals corrupted by channel noise have been discussed in Chapter 3; however, the rate of transmission of information through a noisy channel requires to be improved through proper signal design and coding. Shannon's Channel Capacity theorem states:

$$C = W \log_2 (1 + S/N) \text{ bits} \tag{4.1}$$

where C = channel capacity, S = signal power, N = noise power and W = channel bandwidth. Since for PSK signalling with no coding, $P_e = 10^{-5}$ for $E_b/n_0 = 9.6$ dB (Chapter 3), and from eqn. (4.1), $P_e \to 0$ for $E_b/n_0 = -1.6$ dB with ideal coding and $W \to \infty$ (as discussed in Chapter 5), the capacity theorem promises a potential coding gain of 11 dB. Since the publication of Shannon's results in 1948, the system designers have been constantly searching for coding methods to improve the rate of transmission in noisy channels. Basically two approaches have been used: (a) one approach has been to expand the information format into larger and larger sets of orthogonal or biorthogonal codes and to use optimum detection techniques at the receiver by using matched filters, correlators etc. as discussed in Chapter 3; (b) the other approach has been to use error-detecting and correcting codes, e.g., block codes, and convolutional codes, where the redundant codes are non-orthogonal, and zero-error rate in the limit is achieved by closely packing the waveforms within a finite bandwidth. In a practical communication system, shown in Fig. 4.1, the ultimate efficiency in transmission may be obtained either by using forward-error-control techniques (FEC) or by using a feedback channel along with error detection and retransmission technique (ARQ). The ARQ method is used when the accuracy requirement is severe and some delay is permissible, e.g., in data transmission systems. If however, the time delay is not permissible and the transmission-reception

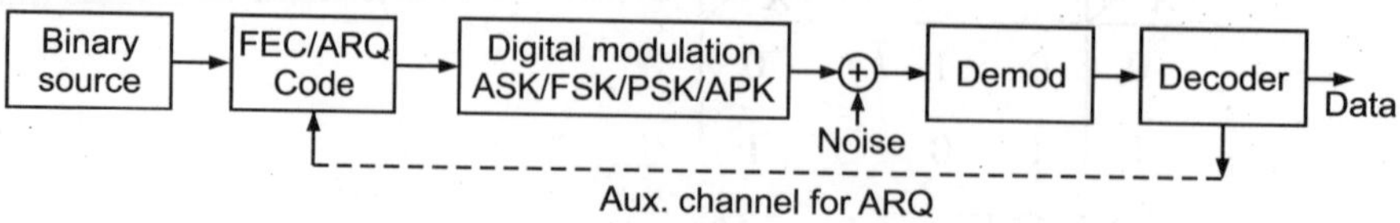

Figure 4.1: A practical communication system

has to be continuous as in, say, satellite links and long distance telephone channels, then FEC methods have to be used.

For the discussions in this chapter, the channels are modelled as either a Binary Symmetric Channel (BSC) or a Binary Erase Channel (BEC), as shown in Fig. 4.2. FEC is generally used in a BSC and ARQ in a BEC.

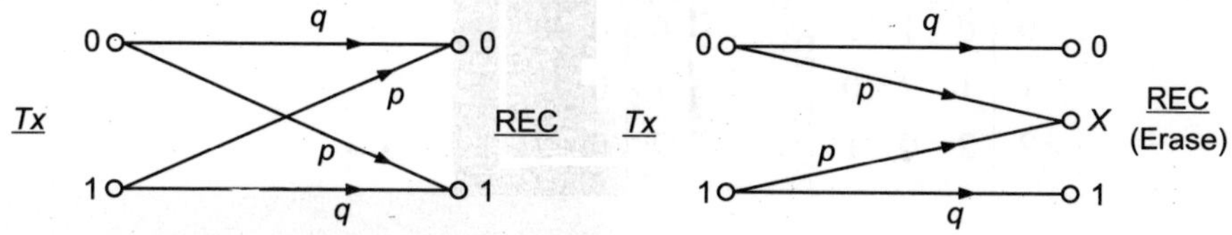

$C = 1 + q \log q + p \log p$

(a) B.S.C.

$C = (1 - 0) = q$

(b) B.E.C.

Figure 4.2: Channel models

Errors occur in the channel either in a random manner or in bursts, e.g., in telephone channels and high frequency radio channels. Specific burst-error correcting codes have been developed for bursty channels. In general, the error correcting/detecting codes are of two types: (a) Block codes, (b) Convolutional codes. Among the block codes, Hamming, Golay and BCH codes have been widely used. But convolutional codes with Viterbi/sequential decoding techniques show better performances in real channels. Shannon's Coding theorem states that for a fixed transmitter power and a transmission rate $R < C$, the error probability P_e in the channel is given by

$$P_e \leq \exp\,[-nE(R)], \quad R < C \tag{4.2}$$

where $E(R)$ is a function of R and n = number of bits in a codeword. Then, $P_e \to 0$, for $n \to \infty$, which requires a long delay for coding and decoding. This exponential bound will be proved later. Historically Hamming's signal-error correcting codes [5] was first developed in 1950 and this was followed by other important developments through the work of Elias (1954), Golay (1954), Slepian (1956), Bose and Chaudhury (1960), and Reed and Solomon (1960) [6, 7, 8, 9, 10]. The chapter discusses the principles of block codes, cyclic codes and convolutional codes. The basic coders and decoders for these codes are explained and their capabilities in improving the error rate in the received data are discussed. Some of the important codes, e.g., Goaly, BCH, Reed-Solomon, Wyner-Ash etc., are also specially discussed. Some emphasis is given for the convolutional codes, as they appear to be more efficient with the present day available hardware.

4.1 GALOIS FIELDS, VECTOR SPACES AND MATRICES

In conformity with switching circuit algebra, the modulo-2 arithmetic is used in the discussions of codes and polynomials. The mod-2 addition and multiplication is defined as

$X \backslash Y$	0	1
0	0	1
1	1	0

(a) $X \oplus Y$

$X \backslash Y$	0	1
0	0	0
1	0	1

(b) $X \odot Y$

(4.3)

The binary alphabet 0 and 1 is called a Field of two elements (or a binary field), which is denoted by GF(2). In binary arithmetic, $-X = X$, and $X - Y = X + Y$. Similarly for three-valued variables, the mod-3 arithmetic is given by

$$\begin{array}{c|ccc} \oplus & 0 & 1 & 2 \\ \hline 0 & 0 & 1 & 2 \\ 1 & 1 & 2 & 0 \\ 2 & 2 & 0 & 1 \end{array} \qquad \begin{array}{c|ccc} \odot & 0 & 1 & 2 \\ \hline 0 & 0 & 0 & 0 \\ 1 & 0 & 1 & 2 \\ 2 & 0 & 2 & 1 \end{array} \tag{4.3a}$$

Polynomials $f(X)$ with 1 or 0 as the coefficients are manipulated using the above relations.

Example 4.1: Consider the product of the polynomials $f_1(X) = X + 1$ and $f_2(X) = X^3 + X + 1$. Then,

$$\begin{aligned} f_1(X) \odot f_2(X) &= X^4 + X^2 + X + X^3 + X + 1 \\ &= X^4 + X^3 + X^2 + 1 \text{ (since } X + X = 0) \end{aligned}$$

Similarly, the division of $f_2(X) = X^6 + X^5 + X^3$ by $f_1(X) = X^3 + X + 1$ is carried out as:

$$\begin{array}{r l}
 & (X^3 + X^2 + X + 1) \quad \leftarrow \text{(Quotient)} \\
X^3 + X + 1 \,\big) & X^6 + X^5 \qquad + X^3 \\
 & X^6 \qquad + X^4 + X^3 \\ \hline
 & \quad X^5 + X^4 \\
 & \quad X^5 \qquad + X^3 + X^2 \\ \hline
 & \qquad X^4 + X^3 + X^2 \\
 & \qquad X^4 \qquad + X^2 + X \\ \hline
 & \qquad\quad X^3 \qquad + X \\
 & \qquad\quad X^3 \qquad + X + 1 \\ \hline
 & \qquad\qquad\qquad 1 \rightarrow \text{Remainder}
\end{array}$$

Example 4.2: Let $f(X) = (X^4 + X + 1)$ Then,

$$\begin{aligned} [f(X)]^2 &= X^8 + X^2 + 1 + 2X^5 + 2X^4 + 2X \\ &= f(X^2), \text{ (since } 2X^5 = 2X^4 = 2X = 0). \end{aligned}$$

Roots of polynomials may also be calculated by using the condition that: If $f_1(1) = 0$, then $(X - 1)$ is a factor of the polynomial $f_1(X)$, and the factors of polynomials may be calculated in the normal way. From Ex. 1, $(X + 1)$ is a factor of $f_1(X) = X^4 + X^3 + X^2 + 1$, since $f_1(1) = 0$. A polynomial $f(X)$ of degree m is said to be irreducible over the binary field GF(2), if $f(X)$ does not have a factor of degree less than m and greater than zero. The polynomials, e.g., $X^3 + X + 1$, $X^4 + X + 1$, $X^5 + X^2 + 1$ etc., are irreducible polynomials.

Fields with 2^m symbols are called Galois fields, GF(2^m), and its arithmetic is derived as follows: Consider a polynomial $p(X)$ of degree m (with binary coefficients) and then using a new variable α, such that $p(\alpha) = 0$, a Table of powers of α is developed. If $p(X)$ is a primitive polynomial, then $\alpha^0, \alpha^1, ..., \alpha^{(q-2)}$, $q = 2^m$, will have distinct values and are called the set of $q(= 2^m)$ field elements. But $\alpha^{q-1} = \alpha^{2^m - 1} = 1$. The element α is called a primitive element of the field GF(2^m) and the polynomial $p(X)$ is called a primitive polynomial.

Example 4.3: GF(2^3) may be formed from GF(2) as the ground field by using the third degree irreducible polynomial $p(X) = X^3 + X^2 + 1$. Then α is the root of $X^3 + X^2 + 1$, i.e., $(X - \alpha) = 0$, and $\alpha^3 + \alpha^2 + 1 = 0$, giving $\alpha^3 = \alpha^2 + 1$. The seven non-zero field elements of GF(2^3) are now given in Table 4.1.

Table 4.1: Representation of GF(2^3)

$\alpha^0 = 1$	001
$\alpha^1 = \alpha$	010
$\alpha^2 = \alpha^2$	100
$\alpha^3 = \alpha^2 + 1$	101
$\alpha^4 = \alpha^2 + \alpha + 1$	111
$\alpha^5 = \alpha + 1$	011
$\alpha^6 = \alpha^2 + \alpha$	110
$\alpha^7 = \alpha^3 + \alpha^2 = 1 = \alpha^0$	001

The representation of GF(2^4) using $p(X) = X^4 + X + 1$ is given in Table 4.12 later in the chapter.

4.1.1 Vectors

The polynomials $f(X)$ with binary coefficients can be represented as an ordered sequence of binary symbols, such as, $(1 + X + X^2 + X^5) \rightleftarrows (1 \quad 1 \quad 1 \quad 0 \quad 0 \quad 1)$, with the highest-power coefficient at the rightmost (since in actual practice, the high-order coefficients are transmitted first). This ordered sequence of binary digits is generally called an 'n-tuple' over GF(2) and may be considered as a vector. The set of all binary n-tuples $\{V_n\}$ is called a Vector space over GF(2). For n binary digits in each vector, the vector space consists of 2^n vectors. As for example, for $n = 3$,

$$\{V_n\} = \{000, 001, 010, 011, 100, 101, 110, 111\}$$

The vectors obey the following postulates:

(a) For every pair of vectors α and β in V_n, there is a vector γ in V_n such that: $\alpha \oplus \beta = \gamma$.

(b) For all α, β, γ in V_n; $(\alpha \oplus \beta) \oplus \gamma = \alpha \oplus (\beta \oplus \gamma)$.

(c) For all α, β in V_n; $\alpha \oplus \beta = \beta \oplus \alpha$.

(d) There is an all-zero vector 0 in V_n such that:

$$\alpha \oplus 0 = 0 \oplus \alpha = \alpha \text{ for every } \alpha \text{ in } V_n.$$

(e) For every α in V_n, there is a vector $-\alpha$, such that:

$$\alpha \oplus (-\alpha) = (-\alpha) \oplus \alpha = 0.$$

The vector $(-\alpha)$ is called the additive inverse of α.

(f) For all scalars C drawn from GF(2), and for all α in V_n, the scalar product $C\alpha$ is a vector in V_n, e.g.,

$$\{u\} = \{c_1v_1 + c_2v_2, ... + c_kv_k\} \in V_n, \tag{4.4}$$

where C_i's are from GF(2) and are called the coefficients of v_i.

Subspaces

A subset S of V_n is called a subspace if:

(a) all-zero vector is in S,

(b) the sum of any two vectors in S is also a vector in S.

Dot Product

The inner product or dot product of two vectors is defined as:

If $v_1 = \{x_1, x_2, x_3\}$ (triple)

$v_2 = \{y_1, y_2, y_3\}$ (triple)

then,

$$v_1 \odot v_2 = \{x_1y_1 \oplus x_2y_2 \oplus x_3y_3\},$$

where $\oplus$ and $\odot$ are mod-2 operations. If $v_1 \odot v_2 = 0$, then v_1 and v_2 are said to be orthogonal.

Example 4.4: For $n = 4$, the vector space consists of the vectors:

$$\{V_4\} = \{0000, 0001, 0010, 0011, 0100, 0101, 0110, 0111, 1000, 1001, 1010, 1011, 1100, 1101, 1110, 1111\}$$

It is easily checked that $v_i + v_j = v_k$, i.e.,

$$(0011) \oplus (0100) = (0111).$$

Similarly, the scalar products:

$$c_1(0011) \oplus c_2(0100) \oplus c_3(0101) = (0110)$$

For $c_1 = 1, c_2 = 0, c_3 = 1$.

Out of the 16 vectors in V_n, consider the subspace

$$S = \{0000, 1100, 0011, 1111\}$$

then this satisfies the requirement that sum of any two vectors is S is also a vector in S.

Dependent and Independent Sets

A set of vectors $v_1, v_2, ..., v_n$ is said to be linearly dependent if and only if there are scalars $c_1, c_2, ..., c_n$ from GF(2), not all zero, such that:

$$c_1v_1 + c_2v_2 + ... + c_nv_n = 0.$$

If the set is not linearly dependent, then they are said to be independent.

A subset of vectors is said to span a vector space if every vector in the vector space is equal to a linear combination of the vectors in the subset and these vectors in the subset are called a generating system of the vector space. In any vector space or subspace, there is at least one set of linearly independent vectors which spans the space; this set is called the basis of the vector space. The number of vectors in a basis is called the dimension of the vector space. Thus, the vector space V_n of all binary n-tuples is an n-dimensional vector space, which is spanned by a basis of n linearly independent vectors.

If $\{v_1, v_2, ..., v_k\}$ are k linearly independent vectors, $k < n$, then all combinations of $\{v_1, v_2, ..., v_k\}$ of the form:

$$\{u\} = \{c_1v_1 + c_2v_2 + ... + c_kv_k\} \tag{4.5}$$

form a k-dimensional subspace S of V_n. There are 2^k possible combination of $\{v_1, v_2, ..., v_k\}$; thus S has 2^k distinct vectors and is a k-dimensional subspace of V_n.

Example 4.5: Consider the set of vectors v_1, v_2, v_3, v_4 from $\{V_4\}$. Then, for,

$$v_1 = 0011,\ v_2 = 0101,\ v_3 = 0100,\ v_4 = 0010$$

the sum $v_1 \oplus v_2 \oplus v_3 \oplus v_4 =$ (0000).
Thus the set of vectors is linearly dependent.

But the set of vectors, $v_1 = 0111$, $v_2 = 1011$, $v_3 = 1101$, $v_4 = 1110$, are seen to be linearly independent. Any vector in $\{V_4\}$ is a linear combination of these vectors. Then v_1, v_2, v_3 and v_4 span the vector space V_4.

4.1.2 Matrices

The code vectors, discussed in Ex. 4 and Ex. 5, may also be arranged in the form of a matrix and the usual matrix manipulation may be carried out. In general, a rectangular array of $m \times n$ elements is called a matrix, but the matrices with binary elements only will be discussed in this chapter.

Consider a matrix $[A]$ with elements a_{ij}, as follows:

$$[A] = \begin{bmatrix} a_{11} & a_{12} & \cdots & a_{1n} \\ a_{21} & a_{22} & \cdots & a_{2n} \\ \vdots & \vdots & \cdots & \vdots \\ a_{m1} & a_{m2} & \cdots & a_{mn} \end{bmatrix} \tag{4.6}$$

where a_{ij} = '0' or '1'. The matrix is over GF(2), where each row is a binary n-tuple and each column is a binary m-tuple. The following are the useful properties of the matrix:

(a) The rows and columns are known as row and column vectors.

(b) The vector space consisting of all linear combinations of the independent row vectors of a matrix is called the row space of the matrix. Similarly, the column space is defined. The row rank of a matrix is the dimension of its row space and is equal to the maximum number of independent rows of the matrix.

(c) The row rank and row space of a matrix are invariant under elementary operations, e.g.,

  (i) Interchange of any two rows,

  (ii) Multiplication of any row by a non-zero field element,

  (iii) Addition of any scalar multiple of one row to another.

(d) Elementary row operations can be used to simplify a matrix to a standard form, say, the Echelon Canonical form or a triangular matrix.

(e) If the inner product of two vectors $v_i \odot v_j$ vanishes (= zero), then v_i and v_j are said to be orthogonal.

(f) The subset of all n-vectors (n-tuples) orthogonal to a subspace S_1 of n-tuples forms a subspace S_2 of V_n. The subspace S_2 is called the null space of S_1. Thus if a vector is orthogonal to every vector of a set which spans S_1, it is in the null space of S_1.

(g) If the dimension of S_2 is k, then the dimension of the null space of S_1 is $(n - k)$.

(h) If $[A_1]$ and $[A_2]$ are two matrices, and if $[\,A_1A_2^T\,]$ has all zero elements, then the row space of A_2 is contained in the null space of A_1 and vice versa. If the row rank of A_1 and row rank of A_2 add to n, then the row space of A_2 is the null space of A_1 and vice versa.

Example 4.6: Consider a matrix $[A]$ as:

$$[A] = \begin{bmatrix} 0 & 1 & 1 & 0 & 1 & 0 \\ 1 & 0 & 1 & 1 & 0 & 0 \\ 1 & 1 & 0 & 0 & 0 & 1 \end{bmatrix} \quad (4.7)$$

This is a 3×6 binary matrix where each row is a 6-tuple and each column is a 3-tuple. Further, the rows are linearly independent. The linear combinations (sum of scalar products) of these rows form the subspace S of V_6, which consists of $2^6 = 64$, 6-tuples. The subspace S has 2^3 elements, as shown and it is a 3-dimensional subspace of V_6.

$$[S] = \begin{bmatrix} 0 & 0 & 0 & 0 & 0 & 0 \\ 0 & 1 & 1 & 0 & 1 & 0 \\ 1 & 0 & 1 & 1 & 0 & 0 \\ 1 & 1 & 0 & 0 & 0 & 1 \\ 1 & 1 & 0 & 1 & 1 & 0 \\ 1 & 0 & 1 & 0 & 1 & 1 \\ 0 & 1 & 1 & 1 & 0 & 1 \\ 0 & 0 & 0 & 1 & 1 & 1 \end{bmatrix}$$

Example 4.7: Consider the matrices A_1 and A_2 as shown

$$[A_1] = \begin{bmatrix} 1 & 0 & 0 & 1 & 1 & 0 & 1 \\ 0 & 1 & 0 & 1 & 1 & 1 & 0 \\ 0 & 0 & 1 & 0 & 1 & 1 & 1 \end{bmatrix}, \; [A_2] = \begin{bmatrix} 1 & 1 & 0 & 1 & 0 & 0 & 0 \\ 1 & 1 & 1 & 0 & 1 & 0 & 0 \\ 0 & 1 & 1 & 0 & 0 & 1 & 0 \\ 1 & 0 & 1 & 0 & 0 & 0 & 1 \end{bmatrix}$$

It is easily verified that

$$[A_1 \cdot A_2^T] = \begin{bmatrix} 0 & 0 & 0 & 0 \\ 0 & 0 & 0 & 0 \\ 0 & 0 & 0 & 0 \end{bmatrix},$$

and hence the row space of A_2 is the null space of A_1.

Proof: $[A_1] = [I_3 \mid P]$ and $[A_2] = [P^T \mid I_4]$, by partitioning,

and I_k = Unit matrix.

Therefore,

$$[A_1]\cdot[A_2^T] = [I_3 \mid P] \times \begin{bmatrix} P \\ \hline I_4 \end{bmatrix} = [P + P] = 0; \text{ since } X + X = 0. \quad (4.8)$$

4.1.3 Linear Switching Circuits [2]

The encoder and decoder circuits are usually constructed with a number of adders (mod-2), memory devices and constant multipliers. For digital circuits, mod-2 adders can be constructed with basic OR/NOR logics, and memory devices may be either delay lines or binary shift registers (SR's). Multiplication of two polynomials $g(X)$ and $f(X)$, and the division of one by the other are also mechanised using mod-2 adders and SR's. As a convention, the high-order coefficients of a polynomial are transmitted first, e.g., for the polynomial,

$A(X) = a_0 + a_1X + a_2X^2 + ... + a_nX^n \Longleftrightarrow (1\ 1\ 1\ ...\ 1)$, a's = 1/0; the rightmost bit is transmitted first in any operation. The product of two polynomials $A(X)$ and $B(X)$ yield:

$$\begin{aligned} C(X) = A(X) \cdot B(X) &= (a_0 + a_1X + ... + a_kX^k)\,(b_0 + b_1X + ... + b_nX^n) \\ &= a_0b_0 + (a_1b_0 + a_0b_1)X + (a_0b_2 + a_1b_1 + a_2b_0)X^2 + ... \\ &\quad + (a_{k-1}\,b_n + a_k\,b_{n-1})\,X^{(k+n-1)} + a_kb_nX^{k+n} \end{aligned} \tag{4.9}$$

This may be realised with the circuits of Fig. 4.3(a) or (b), where $A(X)$ is the input and the coefficients of $B(X)$ are given as the weighting factors connected to the mod-2 adders. Since high-order coefficients are transmitted first, the highest-order coefficient a_kb_n of the product polynomial $C(X)$ is obtained first at the output of Fig. 4.3(a). Then the coefficient of $X^{(k+n-1)}$ is obtained as the sum of $\{a_{k-1}b_n + a_kb_{n-1}\}$, the first term directly and the second term through the storage device SR1. Lower-order coefficients are then generated through successive SR's and mod-2 adders. After $(k + n - 1)$ shifts, the SR's contain $\{0, 0, ..., a_0, a_1\}$ and the output is $(a_1b_0 + a_0b_1)$, which is coefficient of X. After $(k + n)$ shifts, the SR-content is $\{0, 0, ..., a_0\}$ and the output is a_0b_0. The product is now complete and SR content is zero. Figure 4.3(b) performs the multiplication in a similar way, but note the arrangements of SR's and ordering of the coefficients in the two circuits. (Also note that the impulse response of the SR circuits is $h(X) = B(X)$ and hence, the product $C(X) = A(X) * B(X)$).

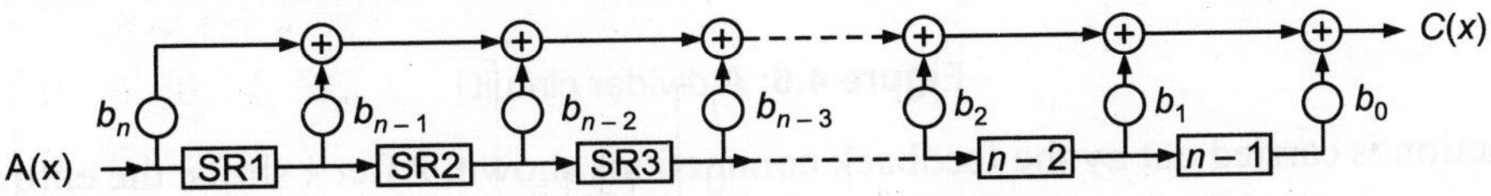

(a) A circuit for multiplying $A(X)$ by $B(X)$

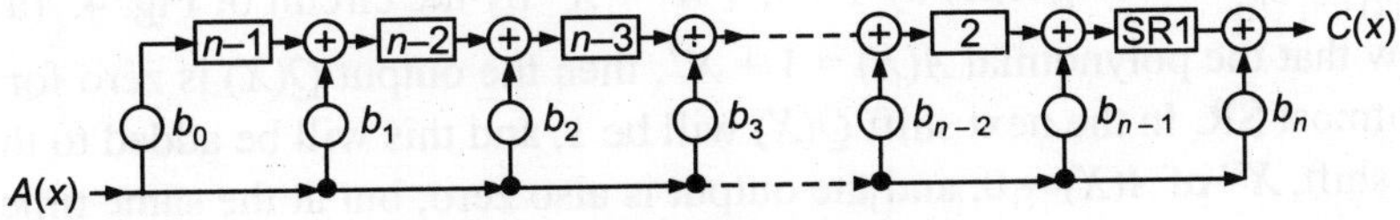

(b) An alternative circuit for multiplication

Figure 4.3

It is possible to combine two multiplication operations into one by using a modification of Fig. 4.3(b) as shown in Fig. 4.4, where the output $C(X)$ is:

$$C(X) = A_1(X) \cdot B_1(X) + A_2(X) \cdot B_2(X)$$

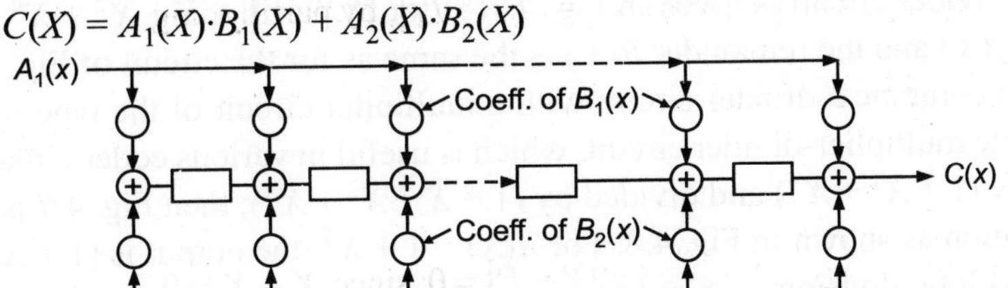

Figure 4.4: A 2-input multiplier

Example 4.8: Consider that a polynomial $A(X)$ is to be multiplied by

$$B(X) = 1 + X^2 + X^4 + X^6$$

Then both the circuits of Fig. 4.5(a) and (b) give the product $C(X)$.

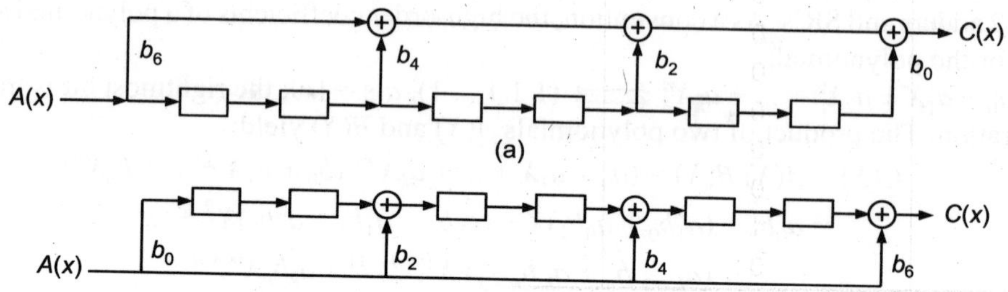

Figure 4.5: Product circuits for $B(X) = 1 + X^2 + X^4 + X^6$

As in the case of multipliers, the division of $A(X)$ by $B(X)$ may be accomplished by using SR's and mod-2 adders, as shown in the circuit of Fig. 4.6. In a division circuit, say $A(X)/B(X)$, the first coefficient of the quotient is $a_k/b_n = q_j$, $k \geq n$, and $q_j \cdot B(X)$ is to be subtracted from $A(X)$.

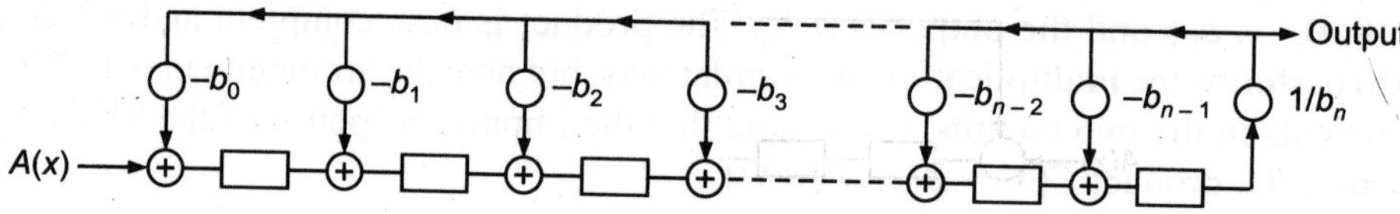

Figure 4.6: A divider circuit

This subtraction is carried out by the feedback connections shown. After k shifts, the entire quotient will appear at the output and the remainder is stored in the SR's.

Example 4.9: Consider dividing $A(X)$ by $1 + X + X^2 + X^4$ by the circuit of Fig. 4.7(a).

Consider now that the polynomial $A(X) = 1 + X^7$; then the output $Q(X)$ is zero for the 4 shifts till X^7 reaches the rightmost SR. In the next shift $Q(X)$ will be 1, and this will be added to the 1st, 2nd and 3rd SR. In the next shift, X^6 of $A(X) = 0$, and the output is also zero; but at the same time the stored 1's are shifted to 2nd, 3rd and 4th SR's. The successive input to the divider, the SR content and the output $Q(X)$ are shown in Table 4.2. Thus the quotient is $(1 + X + X^3)$ and the remainder is zero, as may be checked by long division. If, however, $A(X) = 1 + X + X^2 + X^7$, then, the SR-content after the 8th shift will be 0110; thus $R(X) = X + X^2$, with $Q(X) = 1 + X + X^3$.

An alternative divider circuit is given in Fig. 4.7(b) for $B(X) = 1 + X + X^2 + X^4$. It is easily checked that the quotient $Q(X)$ and the remainder $R(X)$ is the same as for the circuit of Fig. 4.7(a).

It is possible to combine a divider circuit with a multiplier circuit of the type of Fig. 4.3(b), giving thereby a composite multiplier-divider circuit, which is useful in various coder circuits. Suppose $A(X)$ is to be multiplied by $(1 + X^2 + X^4)$ and divided by $(1 + X + X^2 + X^4)$; then Fig. 4.7 is further extended to include multiplication as shown in Fig. 4.8. For $A(X) = 1 + X^7$, the output is $(1 + X + X^2 + X^4 + X^7)$ as may be checked by long division.

Table 4.2: SR content in a Divider Circuit

No. of shifts	Input a_i of $A(X)$	SR content	Output symbol q_j of $[Q(X)]$
0	0	0 0 0 0	0
1	1	1 0 0 0	0
2	0	0 1 0 0	0
3	0	0 0 1 0	0
4	0	0 0 0 1	0
5	0	1 1 1 0	1
6	0	0 1 1 1	0
7	0	1 1 0 1	1
8	1	0 0 0 0	1
9	0	0 0 0 0 $R(X) = 0$	0

(Output symbols for shifts 5–8: $Q(X) = 1 + X + X^3$)

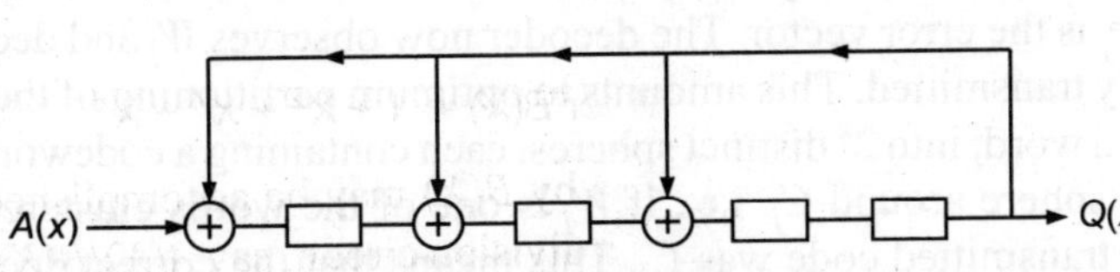

(*a*) Divider circuit

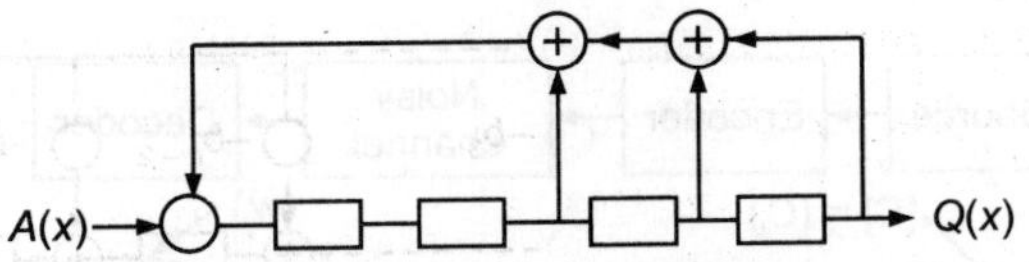

(*b*) Alternative divider circuit

Figure 4.7: Divider circuits for $B(X) = 1 + X + X^2 + X^4$

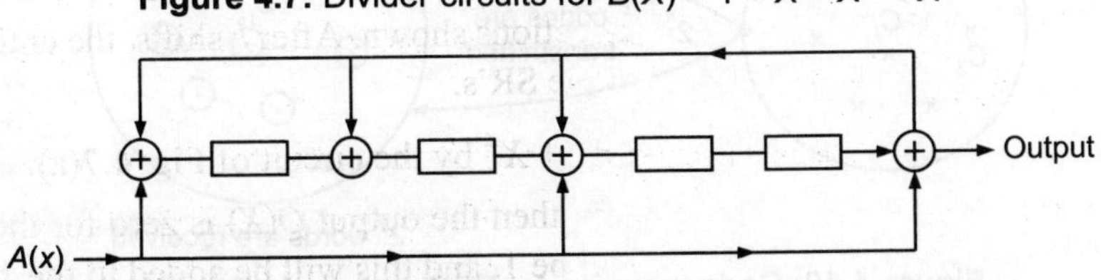

Figure 4.8: Circuit for simultaneously multiplying by $(1 + X^2 + X^4)$ and divided by $(1 + X + X^2 + X^4)$

In certain coders/decoders, a random sequence generator, also known as PN-sequence generator, is utilised. Such a generator is realised by using feedback to a series of SR's through mod-2 adders. With certain feedback conditions, the length of the sequence is maximum and is given by $(2^m - 1)$, where m SR's are used in cascade. Figure 4.9 shows a PNS generator of length $(2^m - 1) = 255$, and its generator polynomial is

$$f(X) = 1 + X^3 + X^5 + X^6 + X^8.$$

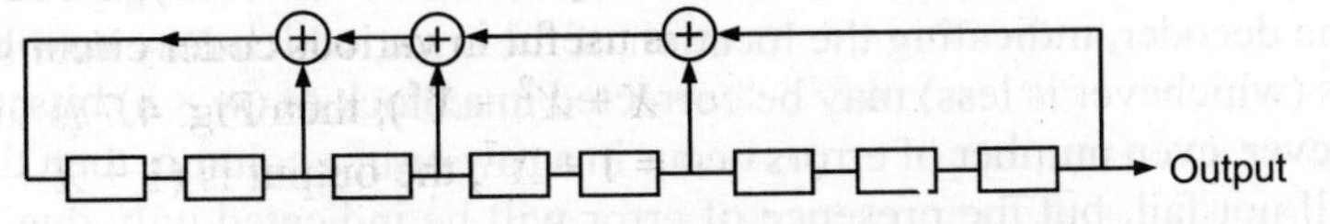

Figure 4.9: A PNS generator of length 255

4.2 BLOCK CODES

For the purpose of encoding messages for error protection, the long messages are broken into message blocks consisting of, say, k bits of information. Then redundant bits (normally known as parity bits) are added to these k bits according to certain rules of coding and the code words of length n bits, inclusive of $(n - k)$ parity bits, are transmitted through the noisy channel. In the receiver, the k message bits are decoded from the erroneous received message blocks using suitable algorithms. With k bits of information per block, there are 2^k possible distinct messages (or words), out of the 2^n words that may be generated with n bits. This set of 2^k words is called a block code, and each codeword is a code vector from the vector space V_n of all n-tuples. Further, a linear code is defined as a set of 2^k n-tuples, which is a subspace of the vector space V_n of all n-tuples.

The process of encoding and decoding data blocks may be looked upon as a problem of mapping the source outputs to match the channel characteristics, as shown in Fig. 4.10. The encoder generates 2^k distinct codewords $\{C_j\}$, which are corrupted by the noise in the channel. Then the received words are: $\{W_j\} = \{C_j + e_j\}$, where e_j is the error vector. The decoder now observes W_j and decides which particular codeword was most likely transmitted. This amounts to optimum partitioning of the n-dimensional space enclosing 2^n points, each a word, into 2^k distinct spheres, each containing a codeword. If now the received word W_j falls within the sphere around C_j, i.e., if W_j is one of the words clustered around C_j, then the decoder decides that the transmitted code was C_j. This means that the correct decision is possible only when the received W_j differs from C_j in a few bit positions only; and this is bounded by Hamming Distance discussed in Sec. 4.2.2.

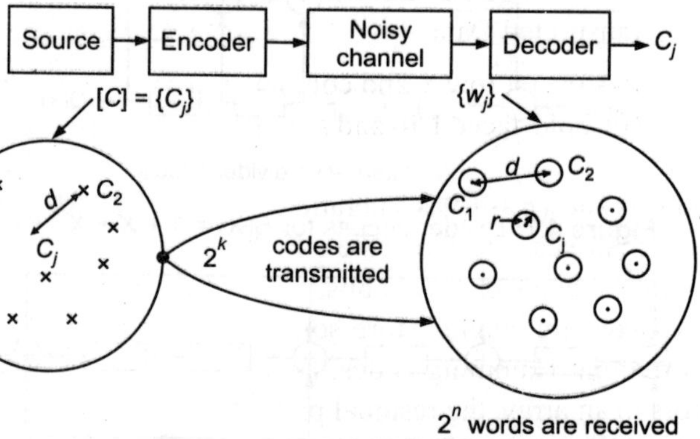

Figure 4.10: Code mapping for SEC-codes, $d \geq 3$, $r = 1$

Based on the above principles, different techniques of generating block codes, e.g., group code, Hamming code etc., and also their decoding techniques are discussed below. The concept of Hamming distance and matrix formulations is also explained.

4.2.1 Parity-check Encoding [5]

Consider that the data bits are arranged in a $(m - 1) \times (n - 1)$ matrix form, as given below; and parity checks (either even or odd) are added for the rows and columns in the nth column and the mth row respectively. If a single error occurs in a row or a column, then the corresponding row and column checks will fail in the decoder, indicating the location of the error, which may now be corrected. Thus a total of m or n errors (whichever is less) may be corrected in a block of $(m \times n)$ bits, transmitted through the channel. If, however, even number of errors occur in a row or in a column, then the parity check over that row/column will not fail, but the presence of error will be indicated only due to check failures in corresponding column/row.

$$\left[\begin{array}{c:c} \text{Data bits} & \text{Check bits} \\ (m-1)\times(n-1) & \text{for rows} \\ \hdashline \text{Check bits} & \text{Check for} \\ \text{for column} & \text{check bits} \end{array}\right] \rightleftarrows \left[\begin{array}{cccccc:c} 1 & 1 & 0 & \cdots & 1 & & 0 \\ 1 & 0 & 1 & \cdots & 0 & & 0 \\ \vdots & & & & & & \\ 1 & 1 & 0 & \cdots & 1 & & 1 \\ \hdashline 1 & 0 & 1 & \cdots & 0 & & 0 \end{array}\right] \begin{array}{c} \uparrow \\ (m-1)\ \text{rows} \\ \downarrow \end{array}$$

$$\leftarrow (n-1)\ \text{columns} \rightarrow$$

Example 4.10: With $n = 5$, $m = 5$, a data block is shown. Even checks are then shown on the 6th row and 6th column. In the decoder, the received data matrix is checked with a 7th row and 7th column indicating the check failures.

$$\left[\begin{array}{ccccc:c} 0 & 0 & 0 & 0 & 0 & 0 \\ 1 & 1 & 1 & 0 & 0 & 1 \\ 1 & 1 & 0 & 1 & 1 & 0 \\ 1 & 0 & 1 & 1 & 0 & 1 \\ 0 & 1 & 1 & 0 & 1 & 1 \\ \hdashline 1 & 1 & 1 & 0 & 0 & 1 \end{array}\right] \rightarrow \left[\begin{array}{ccccc:c} 0 & \mathbf{1} & 0 & 0 & 0 & 0 \\ \mathbf{0} & 1 & 1 & 0 & 0 & 1 \\ 1 & 1 & 0 & 1 & 1 & 0 \\ 1 & 0 & 1 & 1 & \mathbf{1} & 1 \\ 0 & 1 & 1 & 0 & \mathbf{0} & \mathbf{0} \\ \hdashline 1 & 1 & \mathbf{0} & 0 & 0 & 1 \end{array}\right] \begin{array}{c} \mathbf{1} \\ \mathbf{1} \\ 0 \\ \mathbf{1} \\ 0 \\ \mathbf{1} \end{array}$$

$$\begin{array}{cccccc} \mathbf{1} & \mathbf{1} & \mathbf{1} & 0 & 0 & \mathbf{1} \end{array} \leftarrow \text{Check}$$

Transmitted Data — Received Data

It is seen that the checks on 1st row × 2nd column, 2nd row × 1st column and 6th row × 3rd column have failed (as indicated by bold-faced **1/0**) and are easily corrected. But double errors in 5th row and 5th column are not detected.

The above technique is known as Elias's iteration technique, and may be extended to a third dimension by stacking the ($m \times n$) matrices, one above the other, in the vertical (z) axis. Checks are then introduced along the z-axis, say, over a block of ($r - 1$) bits, and two errors along any axis may now be corrected if the errors along other axes are single. More sophisticated three-dimensional codes are generated by using single-error correcting Hamming codes, as discussed in Sec. 4.8. Since the technique corrects almost all single errors in an array, the residual probability of error p_1 after decoding the array, is given by (with p = error probability in the BSC):

$$\begin{aligned} Np_1 &= Np - Np{\cdot}q^{N-1} \\ &\le (Np)^2, \text{ where } N = m \times n \end{aligned} \tag{4.10}$$

or

$$p_1 = Np^2,\ p < \frac{1}{N}$$

But at the same time, the rate of transmission R is decreased as:

$$\begin{aligned} R &= \left(\frac{m-1}{m}\right)\cdot\left(\frac{n-1}{n}\right) \\ &= \left(1 - \frac{1}{m}\right)^2, \text{ for } m = n \end{aligned}$$

$$\simeq \left(1-\frac{2}{m}\right), \quad m \gg 1 \tag{4.10a}$$

An alternative method of improving the performance of the simple parity checks, as above, is to incorporate diagonal checks n the $(m \times m)$ block of digits as shown in the matrix below.

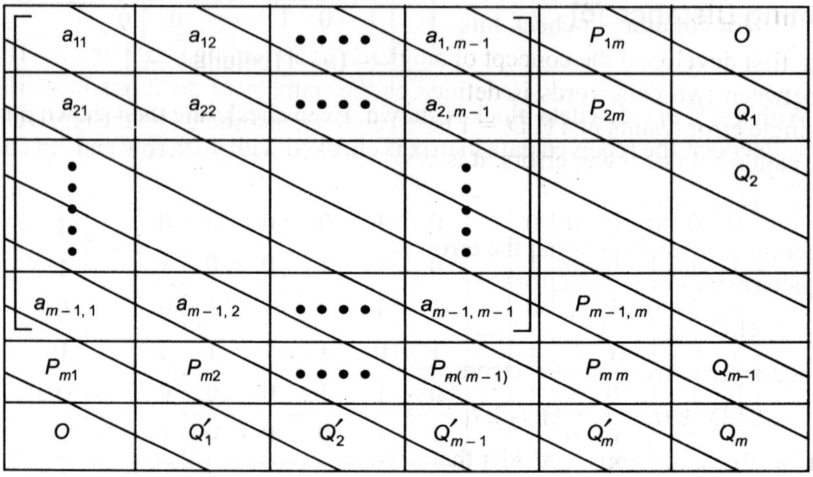

where
$$\begin{cases} a_{11} \; \cdots \; a_{m-1,\,m-1} = \text{information bits,} \\ P_{1m} \; \cdots \; P_{mm} \quad = \text{Mod-2 sum of rows,} \\ P_{m1} \; \cdots \; P_{mm} \quad = \text{Mod-2 sum of columns,} \\ \left.\begin{matrix} Q_1 \; \cdots \; Q_m \\ Q'_1 \; \cdots \; Q'_m \end{matrix}\right\} = \text{Mod-2 sum of diagonal entries.} \end{cases}$$

Whenever two errors occur in a row/column, the ambiguity can be resolved by checking the diagonal parity checks. All single/double errors, except the pattern $\begin{bmatrix} 1 & 1 \\ 1 & 1 \end{bmatrix}$, will now be corrected by the parity checks.

Example 4.11: Consider that the same data block as in Example 4.10 is transmitted and the received data block has errors as shown by (×).

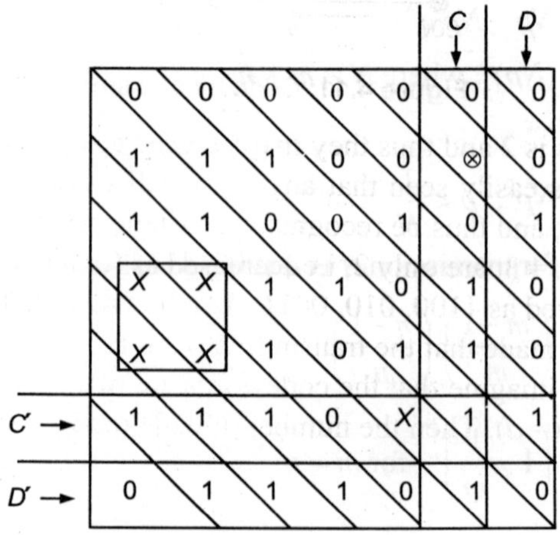

Initially, the row/column checks are made and some errors corrected. Then the diagonal checks are used along with row/column checks and more errors corrected as indicated by ⊗. But the error pattern $\begin{bmatrix} \times & \times \\ \times & \times \end{bmatrix}$ cannot be corrected by these checks. However, this will be a rare event.Checks along the other diagonals may also be used for more error correction.

4.2.2 Hamming Distance [6]

R.W. Hamming first developed the concept of single-error correcting codes and the Hamming distance (H.D.). H.D. between two codewords is defined as the number of positions in which the codewords differ. Then a single error results in a H.D. = 1 between the transmitted and the received words. Thus for detecting error pattern of t or fewer errors, it is necessary and sufficient that

$$\text{H.D.} \geq (t + 1) \tag{4.11}$$

For error correction, it is necessary that the errors should be detected and located as well, and hence for correcting all patterns of t or fewer errors,

$$\text{H.D.} \geq (2t + 1) \tag{4.11a}$$

and for combined error-correction of t_1 errors and error detection of t_2 errors,

$$\text{H.D.} \geq (t_1 + t_2 + 1),\ t_2 \geq t_1 \tag{4.11b}$$

To understand the above relations, consider that 3-bit codewords are plotted at the vertices of a cube, as shown in Fig. 4.11. It is seen that if the codewords used are

{000, 101, 110, and 011},

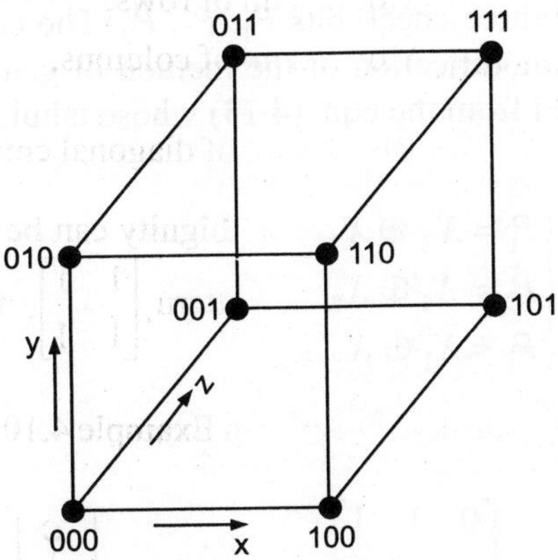

Figure 4.11: Concept of distance

then H.D. between the words is 2 and thus they may be used as codewords with capability of detecting one error per word only. It is easily seen that any error will locate the received words at the vertices which are not the codewords, and thus be recognised as single errors.

The codewords, with H.D. = 3, are only 2, i.e., {000, 111} (or any other combination with H.D. = 3). If a codeword (000) is received as {100, 010, 001}, then it is seen that these are nearer to (000) than to (111). Hence, the decision is made that the transmitted word is (000).

For n-bit words, one may imagine that the codewords are placed at the vertices of an n-dimensional cube with the required H.D. (= d). Then the number of codewords N with $d \geq (2t + 1)$, is given by

$$N(n, d) \le \frac{2^n}{\sum_{i=0}^{t} \binom{n}{i}} \tag{4.12}$$

Based on eqn. (4.12), Table 4.3 gives some typical values of N for different values of n and t.

Table 4.3: Values of N

n \ t	1	2	3	5	7	10	15
3	2						
7	2^4						
15	2^{11}	2^7	2^5				
31	2^{26}	2^{21}	2^{16}	2^{11}	2^6		
63	2^{57}	2^{51}	2^{45}	2^{36}	2^{24}	2^{18}	2^7
127	2^{120}	2^{113}	2^{106}	2^{92}	2^{78}	2^{64}	2^{36}
255	2^{247}	2^{239}	2^{231}	2^{215}	2^{199}	2^{179}	2^{139}
511	2^{502}	2^{493}	2^{484}	2^{466}	2^{448}	2^{421}	2^{376}

4.2.3 Group Codes

Consider that 3-bit messages are to be coded such that single-errors in the received vectors may be corrected. Then from eqn. (4.11a), HD ≥ 3, and from Table 4.3, $n = 6$ to accommodate eight messages consisting of X_1, X_2, X_3 bits and three check bits P_1, P_2, P_3. The coded message vector now reads as $m_j = X_1X_2X_3\ \ P_1P_2P_3$. By slight modification of the method of generating parity checks, discussed in Ex. 10, P_1, P_2 and P_3 are obtained from the eqn. (4.13) whose tabular form is also shown below.

	X_1	X_2	X_3
P_1		√	√
P_2	√		√
P_3	√	√	

$$\equiv \begin{cases} P_1 = X_2 \oplus X_3 \\ P_2 = X_1 \oplus X_3 \\ P_3 = X_1 \oplus X_2 \end{cases} \tag{4.13}$$

In matrix form,

$$\underline{P_1P_2P_3} \equiv \underline{X_1X_2X_3} \cdot \begin{bmatrix} 0 & 1 & 1 \\ 1 & 0 & 1 \\ 1 & 1 & 0 \end{bmatrix}$$

Calling the parity matrix as $[P]$,

$$\underline{P_1P_2P_3} = \underline{X} \cdot [P] \tag{4.13a}$$

For a systematic code, where the message bits are transmitted either at the beginning or at the end of the code C_j, a code-generating matrix may now be specified as

$$[G] = [I_kP], \text{ where } I_k = \text{unit matrix of order } k \tag{4.14}$$

and the resulting codes will be of the form: $\underline{X_1X_2X_3\ P_1P_2P_3}$. In general, the code vector C_j is given as: | message bits ¦ parity bits |, and the following example illustrate this.

Example 4.12: Consider a (6, 3) code, where $n = 6$ and k = number of message bits = 3. The generating matrix $[G]$ is given by:

$$[G] = \left[\begin{array}{ccc|ccc} 1 & 0 & 0 & 0 & 1 & 1 \\ 0 & 1 & 0 & 1 & 0 & 1 \\ 0 & 0 & 1 & 1 & 1 & 0 \end{array}\right]$$

and the code vectors $C_j = \underline{m_1} \cdot [G]$. All the code vectors are obtained from

$$[C] = [m]\,[G]$$

Table 4.4 shows all the code vectors for 3-bit messages.

Table 4.4: Code vectors C_j

Messages m_j X_1, X_2, X_3	Code vectors C_j $X_1\, X_2\, X_3 \quad P_1\, P_2\, P_3$
m_1 0 0 0	0 0 0 0 0 0
m_2 0 0 1	0 0 1 1 1 0
m_3 0 1 0	0 1 0 1 0 1
m_4 1 0 0	1 0 0 0 1 1
m_5 0 1 1	0 1 1 0 1 1
m_6 1 0 1	1 0 1 1 0 1
m_7 1 1 0	1 1 0 1 1 0
m_8 1 1 1	1 1 1 0 0 0

In the receiver, one need to generate fresh parity bits P_1', P_2', P_3' from the received message bits and check with the received parity bits $P_1\, P_2\, P_3$. From eqn. (4.13), it is evident that this cross-checking of parity bits may be obtained from the equation

$$\begin{aligned} P_1 \oplus X_2 \oplus X_3 &= e_1 \\ P_2 \oplus X_1 \oplus X_3 &= e_2 \\ P_3 \oplus X_1 \oplus X_2 &= e_3 \end{aligned} \qquad (4.15)$$

If there is no error, then $e_1 = e_2 = e_3 = 0$. But if there is a single error in the received code, then one or two of the e's will be '1' solution of eqn (4.15) yields the error pattern, as given in Table 4.5. By comparing the results of parity checks by using eq. (4.15) and the entries in Table 4.5, the single errors in the messages are now corrected.

Table 4.5: Error pattern

Error in digits	e_1 e_2 e_3
X_1	0 1 1
X_2	1 0 1
X_3	1 1 0
P_1	1 0 0
P_2	0 1 0
P_3	0 0 1
No error	0 0 0

By slight rearrangement of eqns. (4.13) and (4.15), it is possible to located the error position directly from the binary equivalent of (e_1 e_2 e_3). Suppose that code C_j has the binary locations 1, 2, 3, ..., 6, for $X_1\ X_2\ X_3\ P_1\ P_2\ P_3$ and the [P] is changed to

$$\begin{bmatrix} 1 & 1 & 0 \\ 1 & 0 & 1 \\ 0 & 1 & 1 \end{bmatrix}$$

giving the modified eqn. (4.13) as

$$\begin{aligned} P_1 \oplus X_1 \oplus X_2 &= 0 \\ P_2 \oplus X_1 \oplus X_3 &= 0 \\ P_3 \oplus X_2 \oplus X_3 &= 0 \end{aligned} \tag{4.16}$$

By rearranging eqn. (4.16), one could obtain the error check in the receiver as

$$\begin{aligned} e_1 &= X_1 \oplus X_3 \oplus P_2 \\ e_2 &= X_2 \oplus X_3 \oplus P_3 \\ e_3 &= P_1 \oplus P_2 \oplus P_3 \end{aligned} \tag{4.17}$$

The eqn. (4.17) now gives the error pattern as in Table 4.6, and it is seen that the error location is directly given by the binary equivalent of (e_3, e_2, e_1). This technique of assigning the parity bits was first given by Hamming [6] while formulating the classical Hamming codes.

Table 4.6: Modified error pattern

Location	Error in digits	e_2 e_1
1	X_1	0 0 1
2	X_2	0 1 0
3	X_3	0 1 1
4	P_1	1 0 0
5	P_2	1 0 1
6	P_3	1 1 0

Based on Table 4.3 and above discussions the requirements of check bits for (n, k) single-error correcting (SEC) codes is given by:

$$2^{n-k} \geq (n+1) \tag{4.18}$$

where n = number of total bits in the code, k = number of message bits and $(n - k)$ = number of check bits; and d = 3 for t = 1. Table 4.7 gives the values of n, k and $(n - k)$ for some useful SEC codes. Because of the redundancy introduced, the rate of transmission R is now:

$$R = \frac{k}{n} = \frac{n-r}{n} = 1 - \frac{r}{2^r - 1}$$

and

$$R \to 1 \quad \text{for } r \gg 1, \quad \text{where } r = (n-k).$$

Table 4.7

k	1	4	11	26	57
$n-k$	2	3	4	5	6
n	3	7	15	31	63

Example 4.13: Consider the (15, 11) SEC code with $n = 15$, $k = 11$ and 4 parity bits. Arranging the message and parity bits as:

$$\{P_1P_2P_3X_1X_2X_3X_4P_4X_5X_6X_7X_8X_9X_{10}X_{11}\}$$

the parity bits are generated from the equations

$$P_1 = X_1 \oplus X_2 \oplus X_4 \oplus X_5 \oplus X_7 \oplus X_9 \oplus X_{11}$$
$$P_2 = X_1 \oplus X_3 \oplus X_4 \oplus X_6 \oplus X_7 \oplus X_{10} \oplus X_{11}$$
$$P_3 = X_2 \oplus X_3 \oplus X_4 \oplus X_8 \oplus X_9 \oplus X_{10} \oplus X_{11}$$
$$P_4 = X_5 \oplus X_6 \oplus X_7 \oplus X_8 \oplus X_9 \oplus X_{10} \oplus X_{11}$$

For the message: {11010110011}, the parity bits are: {1000}; and the transmitted vector is {101010100110011}. If the received vector is {101010100110010}, then all the checks will fail giving $\{e_1e_2e_3e_4\}$ as {1111}, indicating that the 15th position is in error.

4.2.4 Generalisation

The above methods of code generation and of obtaining the error pattern at the receiver may be generalised by using the matrix formulation as discussed below.

Form eqn. (4.14), the code generating matrix $[G]$ is written as:

$$[G] = [I_k \vdots P] \tag{4.14)*$$

and the code vector is obtained as:

$$[C] = [m]\cdot[G] \tag{4.19}$$

where $[m]$ = message vector and $[C]$ = code vectors.

If the error vector of the channel is $[E]$, then the received vector $[W]$ is given by:

$$[W] = [C] \oplus [E]$$
$$= [mG] \oplus [E] \tag{4.20}$$

The parity check matrix $[H]$ should be such that in the decoder, the syndrome S obtained by using the relation:

$$[S] = [W]\cdot[H^T] \to 0, \text{ if } [E] \text{ is zero,}$$

i.e., $$[C][H^T] = [m]\cdot[G]\cdot[H^T] = 0 \tag{4.21}$$

* It is equally valid to have

$$[G] = [P \vdots I_k]; \text{ and } [H] = [I_{n-k} \vdots P^T]$$

This is possible if $[G]$ and $[H]$ matrices are orthogonal, i.e., the row space of H is contained in the null space of G and vice versa. $[H]$ then may be written as:

$$[H] = [P^T : I_{n-k}] \tag{4.22}†$$

Then,

$$\begin{aligned} S = [W]\cdot[H^T] &= [C \oplus E]\cdot[H^T] \\ &= [C\cdot H^T] \oplus [E\cdot H^T] = [E \cdot H^T] \end{aligned}$$

and

$$\begin{aligned} [E] &= [S]\cdot[H^T]^{-1} \\ [W+E] &= [C] \end{aligned} \tag{4.23}$$

Example 4.14: Consider the (6, 3) code already discussed in Example 4.12.

The generating matrix was:

$$[G] = \left[\begin{array}{ccc|ccc} 1 & 0 & 0 & 0 & 1 & 1 \\ 0 & 1 & 0 & 1 & 0 & 1 \\ 0 & 0 & 1 & 1 & 1 & 0 \end{array}\right]$$

and the codes C_j generated are given in Table 4.4.

The corresponding $[H]$ is:

$$H = \left[\begin{array}{ccc|ccc} 0 & 1 & 1 & 1 & 0 & 0 \\ 1 & 0 & 1 & 0 & 1 & 0 \\ 1 & 1 & 0 & 0 & 0 & 1 \end{array}\right]$$

It is easily seen that $[G \cdot H^T] = 0$.

If a code vector $C_j = \{1\ 1\ 1\ 0\ 0\ 0\}$ is received without error, then

$$S_j = [W_j]\cdot[H^T] = \underline{0\,0\,0}, \text{ indicating no error.}$$

But if $W_j = \{1\ 1\ 1\ 0\ 0\ 1\}$, then $S_j = \underline{0\,0\,1}$, indicating an error in the last position (by inspection of $[H^T]$).

Standard Array

It is seen that for k-bit messages, we have n-bit codewords and out of 2^n possible n-tuples, we use only 2^k n-tuples as codewords. Thus for each codeword, we have $(2^{n-k} - 1)$ words with H.D. less than $(2t + 1)$. The n-tuples can now be arranged in columns such that the topmost n-tuple is the code vector and other n-tuples in the same column are the allowable received vectors with t-errors. The arrangement of the n-tuples with 2^k columns and 2^{n-k} rows is called a Standard Array for the given (n, k) linear code. The rows are called Cosets and the leftmost n-tuple of each row is called a coset leader. All the rows and columns of the array are disjoint, i.e., every n-tuple appears once and only once in the array. Assume a generating matrix

---

†$[G]\cdot[H^T] = [I_k \vdots P] \cdot \left[\dfrac{P}{I_{n-k}}\right] = [P \oplus P] = 0.$, since $X \oplus X = 0$. This proves that row space of $[G]$ is orthogonal to $[H]$, i.e., the inner product of a vector in the row space of $[G]$ and a row of $[H]$ is zero. Also, $[H]\cdot[C^T] = 0$.

$$[G] = \begin{bmatrix} 0 & 1 & 1 & 1 & 0 & 0 \\ 1 & 0 & 1 & 0 & 1 & 0 \\ 1 & 1 & 0 & 0 & 0 & 1 \end{bmatrix}$$

which generates 8 codevectors, as shown at the top of the columns of the standard array given in Table 4.8. Since the (6, 3) code corrects all single errors, the received vectors with single errors will all lie in the same column. Thus the correct decision is that the top n-tuple is the transmitted vector. The coset leader corresponding to the received vector gives the location of the error. There is also a one-to-one correspondence between the syndrome generated by eqn. (4.21) and the coset leader (which gives the correctable error pattern). Top seven rows correct single errors, and the last row corrects one double-error pattern. The above decoding scheme is called Table-look-up decoding and is equivalent to Maximum likelihood decoding.* Alternatively $S_j = W_jH^T$ may be calculated and the corresponding coset leader added to W_j to obtain the correct C_j.

Table 4.8: Standard Array

Syndrome	Coset leader	*n*-tuples						
000	000000	110001	101010	011100	011011	101101	110110	000111 {code vectors
110	000001	110000	101011	011101	011010	101100	110111	000110
101	000010	110011	101000	011110	011001	101111	110100	000101
011	000100	110101	101110	011000	011111	101001	110010	000011
001	001000	111001	100010	010100	010011	100101	111110	001111
010	010000	100001	111010	001100	001011	111101	100110	010111
100	100000	010001	001010	111100	111011	001101	010110	100111
111	001001	111000	100011	010101	010010	100100	111111	001110

Example 4.15: Consider $W_j = 100010$. The top n-tuple in the 3rd column $= 101010 = C_j$ and the coset leader = 001000.

The corresponding $S_j = \underline{100010} \cdot \begin{bmatrix} 1 & 0 & 0 \\ 0 & 1 & 0 \\ 0 & 0 & 1 \\ 0 & 1 & 1 \\ 1 & 0 & 1 \\ 1 & 1 & 0 \end{bmatrix} = \underline{001}$

and the corresponding coset leader = 001000.

Therefore, $C_j = 100010 + 001000 = 101010$.

Example 4.16: It is possible to use the 6-bit codes to correct some of the double errors also, if the number of messages are less, say, 4 only.

---

*Assuming that all codewords are equally likely to be transmitted, the best decision at the receiver is to decode into a codeword C_j such that H.D. between W_j and C_j is minimum. Such decoding is called Maximum-likelihood decoding.

Consider the (6, 2) code with 4 messages, {00, 01, 10, 11} and a generating matrix

$$G = \begin{bmatrix} 1 & 1 & 1 & 0 & 1 & 0 \\ 1 & 0 & 1 & 1 & 0 & 1 \end{bmatrix}$$

Then the standard array is shown in Table 4.9. It is seen that 6 single errors, 7 double errors and 2 triple errors are correctable by using the Table-lookup technique.

Table 4.9: A standard array for 4 messages

	Coset leader	*n*-tuples			Syndrome
Code words →	000000	111010	101101	010111	0000
Single errors	100000	011010	001101	110111	1000
	010000	101010	111101	000111	0100
	001000	110010	100101	011111	0010
	000100	111110	101001	010011	0001
	000010	111000	101111	010101	1110
	000001	111011	101100	010110	1011
Double errors	110000	001010	011101	100111	1100
	101000	010010	000101	111111	1010
	100100	011110	001001	110011	1001
	100010	011000	001111	110101	0110
	100001	011011	001100	110110	0011
	010100	101110	1110001	000011	0101
	010001	101011	111100	000110	1111
Triple errors	110100	001110	011001	100011	1101
	110001	001011	0111000	100110	0111

Referring to the code mapping shown in Fig. 4.10, assume that the eight code vectors of Table 4.8 (at the top of the columns of the standard array) are at the vertices of a 6-dimensional cube, and the H.D. between the codes ≥ 3. In the receiver, the 6-dimensional space is partitioned into 8 distinct spheres with radius = 1; then all received code vectors with single errors will lie within the spheres and the correct decoding is done. The vectors with more errors will fall outside the spheres or in wrong spheres; thus causing a wrong decision in decoding. This is the general principle of decoding all block codes. It is possible to extend the above arguments for t-error correcting codes, where $d \geq 2t + 1$.

Based on the above concept, a perfect code is defied as a code, where it is possible to form a standard array with all patterns of t or less errors and no others, as the coset leaders; i.e., number of coset leaders $= \sum_{i=0}^{t} \binom{n}{i}$. A code is said to be quasi-perfect, if its coset leaders contain all error patterns with t or less errors and some patterns with $(t + 1)$ errors, but none with $(t + 2)$ or more errors, i.e.,

$$\sum_{i=0}^{t} \binom{n}{i} < \text{number of coset leaders} < \sum_{i=0}^{t+1} \binom{n}{i}.$$

Referring back to Fig. 4.10 it is evident that for a perfect code, the n-dimensional space having 2^n words in it is completely filled up with code spheres of radius t, while the spheres are disjoint and distinct. But in quasi-perfect codes, the code spheres of radius t are disjoint, but do not fill up the total code space. Some code spheres with radius $(t + 1)$ are also allowed to coexist.

The SEC Hamming code and 3-error correcting Golay code are perfect codes. But the (6, 3) group code above is quasi-perfect. Similarly, (15, 5) BCH, (8, 4) Hamming and simplex codes are quasi-perfect.

Linear block codes can be generated by combinatorial switching circuits by using the parity matrix of eqn. (4.13a). As an example, the parity checks of eq. (4.13) are obtained by the circuit of Fig. 4.12, where $P_1P_2P_3$ are generated through mod-2 adders and the coded word is transmitted after suitable multiplexing of the data and parity bits.

A complementary operation is carried out by the decoder circuit of Fig. 4.12(b). The error vectors $e_1e_2e_3$ are generated from the received vector $\{P'_1P'_2P'_3X'_1X'_2X'_3\}$, by using eq. (4.15) and the correction signals are generated from the AND gates according to Table 4.5. The corrected data $X_1X_2X_3$ are now obtained from the mod-2 adders as shown. The SR's are used for storing the data for the duration of processing and if required the corrected data may be stored in another set of SR's for further serial operations.

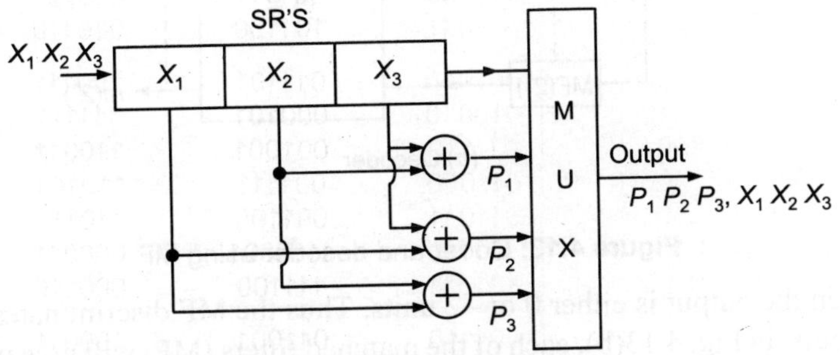

(a) Coder for generating the parity checks

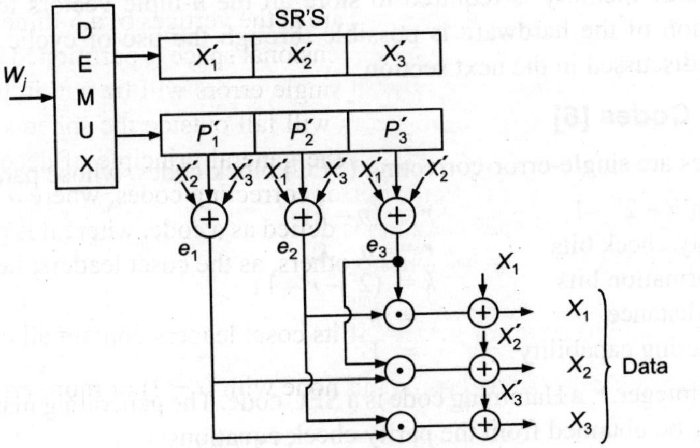

(b) Decoder

Figure 4.12: Coder and decoder for eqn. (4.13)

Alternatively a set of matched filters may be used to decode the received vectors, as shown in Fig. 4.13. Suppose a code vector 001110 of Table 4.4 is stored, in a set of SR's as in Fig. 4.13(a) and the weighted outputs are added to give a value of + 6 units. If now any other vector of Table 4.4 is used as the

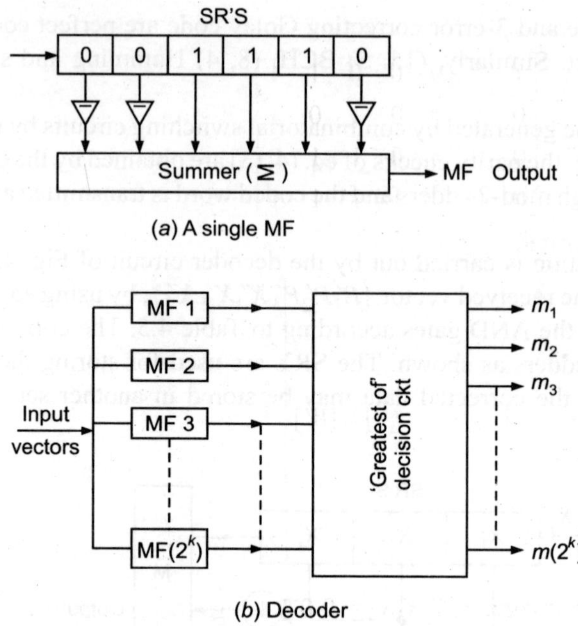

Figure 4.13: Coder and decoder using MF

input to this MF, then the output is either 0 or –2 units. Thus the MF discriminates between the correct and the incorrect inputs. In Fig. 4.13(b), each of the matched filters (MF) will give maximum output only when the correct input is applied, and a comparison of the outputs is done through the 'greatest of' decision circuit. The output of this circuit indicates the correct message. In the Table-look-up method, on the other hand, a large memory is required to store all the n-tuple vectors for comparison purposes. Further simplification of the hardware is possible through the use of cyclic codes and shift-register coder-decoders, as discussed in the next section.

4.2.5 Hamming Codes [6]

The Hamming codes are single-error-correcting (SEC) block codes, whose parameters are

Codelength $n = 2^r - 1$,	$r = (n - k)$
No. of parity check bits	$r = n - k$,
No. of information bits	$k = (2^r - r - 1)$
Hamming distance	$= 3$
Error correcting capability	$= 1$

and for any positive integer, r, a Hamming code is a SEC code. The generating matrix of a (7, 4) systematic Hamming code may be obtained from the parity check equations:

$$\begin{aligned} P_1 &= X_2 \oplus X_3 \oplus X_4 \\ P_2 &= X_1 \oplus X_3 \oplus X_4 \\ P_3 &= X_1 \oplus X_2 \oplus X_4 \end{aligned} \qquad (4.24)$$

and

$$[G] = \begin{bmatrix} 1 & 0 & 0 & 0 & 0 & 1 & 1 \\ 0 & 1 & 0 & 0 & 1 & 0 & 1 \\ 0 & 0 & 1 & 0 & 1 & 1 & 0 \\ 0 & 0 & 0 & 1 & 1 & 1 & 1 \end{bmatrix} \tag{4.25}$$

The syndrome S is generated from:

$$[S] = [W] \cdot [H^T] = [W] \cdot \begin{bmatrix} 0 & 1 & 1 \\ 1 & 0 & 1 \\ 1 & 1 & 0 \\ 1 & 1 & 1 \\ 1 & 0 & 0 \\ 0 & 1 & 0 \\ 0 & 0 & 1 \end{bmatrix} = \underline{S_1 S_2 S_3} \tag{4.26}$$

Example 4.17: For a message, $X_1X_2X_3X_4 = 0\,0\,1\,0$, the code $C_j = 0\,0\,1\,0\,1\,1\,0$. If the received vector $W_j = 0\,0\,1\,0\,0\,1\,0$, then the syndrome $S = 1\,0\,0$, and by inspection of H^T, the error is in the 5th position, giving $E_j = 0000100$. If it is required that the Syndrom pattern should give the error location in binary form, then the error checks in the receiver shoud be similar to that of eqn. (4.17) and is given by:

$$\begin{aligned} s_1 &= X'_4 \oplus P'_1 \oplus P'_2 \oplus P'_3 \\ s_2 &= X'_2 \oplus X'_3 \oplus P'_2 \oplus P'_3 \\ s_3 &= X'_1 \oplus X'_3 \oplus P'_1 \oplus P'_3 \end{aligned} \tag{4.27}$$

The syndrome for the above W_j is now 1 0 1, indicating the error in the 5th position. A table similar to that of Table 4.6 may be written for the (7, 4) code.

From Table 4.7, it is seen that (7, 4), (15, 11), (31, 26), (63, 57) are useful SEC Hamming codes. Example 13 discussed earlier is a (15, 11) Hamming code which was non-systematic. The generating matrix for a systematic (but not cyclic) (15, 11) code is

$$[G] = \begin{bmatrix} 1 & 0 & 0 & 0 & 0 & 0 & 0 & 0 & 0 & 0 & 0 & 1 & 1 & 0 & 0 \\ 0 & 1 & 0 & 0 & 0 & 0 & 0 & 0 & 0 & 0 & 0 & 1 & 0 & 1 & 0 \\ 0 & 0 & 1 & 0 & 0 & 0 & 0 & 0 & 0 & 0 & 0 & 0 & 1 & 1 & 0 \\ 0 & 0 & 0 & 1 & 0 & 0 & 0 & 0 & 0 & 0 & 0 & 1 & 1 & 1 & 0 \\ 0 & 0 & 0 & 0 & 1 & 0 & 0 & 0 & 0 & 0 & 0 & 1 & 0 & 0 & 1 \\ 0 & 0 & 0 & 0 & 0 & 1 & 0 & 0 & 0 & 0 & 0 & 0 & 1 & 0 & 1 \\ 0 & 0 & 0 & 0 & 0 & 0 & 1 & 0 & 0 & 0 & 0 & 1 & 1 & 0 & 1 \\ 0 & 0 & 0 & 0 & 0 & 0 & 0 & 1 & 0 & 0 & 0 & 0 & 0 & 1 & 1 \\ 0 & 0 & 0 & 0 & 0 & 0 & 0 & 0 & 1 & 0 & 0 & 1 & 0 & 1 & 1 \\ 0 & 0 & 0 & 0 & 0 & 0 & 0 & 0 & 0 & 1 & 0 & 0 & 1 & 1 & 1 \\ 0 & 0 & 0 & 0 & 0 & 0 & 0 & 0 & 0 & 0 & 1 & 1 & 1 & 1 & 1 \end{bmatrix} \tag{4.28}$$

The coder and decoder for Hamming codes are designed using the circuits of Figs. 4.12 and 4.13. Of course, they are also implemented by using shift registers and cyclic codes.

The SEC Hamming code may be made to detect double errors as well by adding an extra parity check in its $(n + 1)$th position; thus (8, 4), (16, 11) etc. codes have H.D. = 4 and correct single errors with detection of double errors.

4.3 BINARY CYCLIC CODES

An (n, k) linear code $[C]$ is called a cyclic code if cyclic shifts of the codes are also code vectors of $[C]$. Thus in polynomial form, if $C_1(X)$ is a code vector, given by

$$C_1(X) = C_0 + C_1X + C_2X^2 + \dots + C_{n-1}X^{n-1},\ C\text{'s} \rightarrow 0/1$$

then the shifted vector $C_2(X) = C_{n-1} + C_0X + C_1X^2 + \dots + C_{n-2}X^{n-1}$ is also a code vector. Some of the properties of cyclic codes are:

(a) In an (n, k) cyclic code, there exists one and only one generator polynomial $g(X)$ of degree $(n-k)$, and

$$g(X) = 1 + g_1X + g_2X^2 + \dots + X^{n-k} \tag{4.29}*$$

(b) The generator polynomial $g(X)$ of an (n, k) cyclic code is a factor of (X^n+1); e.g.,

$$(X^n + 1) = g(X)\cdot h(X) \tag{4.30}$$

(c) If $g(X)$ is a polynomial of degree $(n-k)$ and is a factor of $(X^n + 1)$, then $g(X)$ generates an (n, k) cyclic code.

(d) A code polynomial $C(X)$ can be expressed as

$$C(X) = m(X)\cdot g(X)$$

$$= (m_0 + m_1X + m_2X^2 + \dots m_{k-1}X_{k-1})\cdot g(X) \tag{4.31}$$

and there are 2^k code vectors in an (n, k) code. Thus the code polynomials can be generated by multiplying $m(X)$ by $g(X)$, and $g(X)$ is called the Generator polynomial. However, this method generates non-systematic codes.

(e) To generate systematic codes of the form

$$C(X) \rightarrow \{P_0P_1 \dots P_{n-k-1} \mid m_0m_1 \dots m_{k-1}\},$$

the equivalent generator matrix is $G = [P \mid I_k]$.

Multiplying $m(X)$ by X^{n-k} (to shift $m(X)$ to the rightmost entries in the rows), and dividing $X^{n-k}\cdot m(X)$ by $g(X)$, we have

$$X^{n-k}\cdot m(X) = Q(X)\cdot g(X) + R(X) \tag{4.32}$$

or

$$C(X) = R(X) + X^{n-k}\cdot m(X) = Q(X)\cdot g(X)$$

where $Q(X)$ is the quotient and $R(X)$ is the remainder polynomial. Equation (4.32) indicates that $[X^{n-k}\cdot m(X) + R(X)]$ is a polynomial of degree $(n-1)$ or less and is a multiple of $g(X)$ as well, as required

*From this section onwards, all '+' signs indicate modulo sum ⊕.

by the property (d) stated above. Hence, $C_j(X) = [X^{n-k} \cdot m_j(X) + R_j(X)]$ is a valid code corresponding to the message $m_j(X)$. These codes are generated by the SR-circuits similar to that shown in Figs. 4.7 and 4.8.

Example 4.18: For the (7, 4) SEC code, $m(X) = m_0 + m_1X + m_2X^2 + m_3X^3$, and $X^n + 1 = X^7 + 1 = (1 + X + X^3)(1 + X + X^2 + X^4)$. Then, using $g(X) = 1 + X + X^3$,

$$R(X) = X^3 \cdot m(X) + Q(X) \cdot g(X)$$

$R(X)$ is now obtained from a divider circuit, as in Fig. 4.14(a). Initially, S_1 is closed ans S_2 at D. The 4-bit message is sent out in the line and at

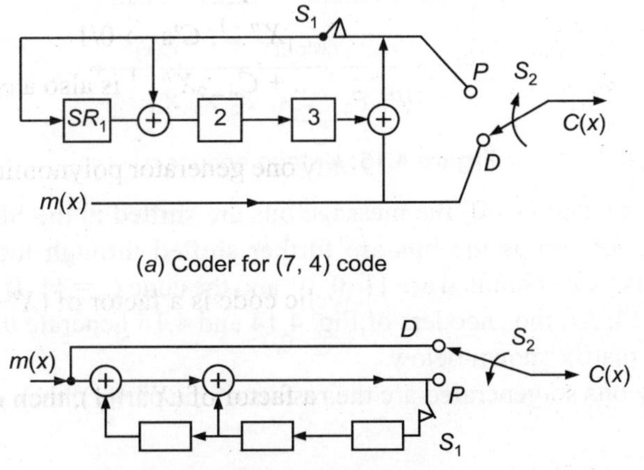

(a) Coder for (7, 4) code

C(x) → {P1 P2 P3 X1 X2 X3 X4}

(b) An alternative encoder circuit

Figure 4.14: ($n - k$) stage encoder for SEC code

the same time $m(X)$ is loaded in the SR's producing $Q(X)$ and $R(X)$ simultaneously. Connecting $m(X)$ at the rightmost point of the divider circuit is equivalent to multiplying $m(X)$ with X^3 (refer to Fig. 4.5(b)). After $m(X)$ is loaded in the SR's, and the division is complete, the content of the SR's is $R(X)$. S_1 now opens and S_2 moves to P. During next shifts, the parity bits are sent out in the line.

Suppose $m(X) = \{1 \quad 0 \quad 1 \quad 1\}$, then SR contents (for Fig. 4.14(a)) are as given below.

No. of shifts	Input $m(X)$	SR content	Output $P_1P_2P_3$
1	1	1 1 0	
2	1	1 0 1	
3	0	1 0 0	
4	1	1 0 0	
		S_1 opens, S_2 moves to P	
5	0	0 1 0	0
6	0	0 0 1	0
7	0	0 0 0	1

Thus the transmitted $C_j = \{1 \quad 0 \quad 0 \quad 1 \quad 0 \quad 1 \quad 1\}$

An alternative encoder is shown in Fig. 4.14(b), whose SR contents are:

$$\{1\,0\,0, 1\,1\,0, 1\,1\,1, 1\,1\,1, 0\,1\,1, 0\,0\,1\} \text{and } P_1\,P_2\,P_3 \rightarrow \{1\,0\,0\}.$$

It is seen that both the above coders use $(n-k)$ stages of SR's. Based on eqn. (4.30), an alternative coder with k-stage SR's may be obtained using the polynomial divider for $h(x) = (X^n + 1)/g(X)$. Such a coder for the (7, 4) code is shown in Fig. 4.15 for $h(X) = 1 + X + X^2 + X^4$.

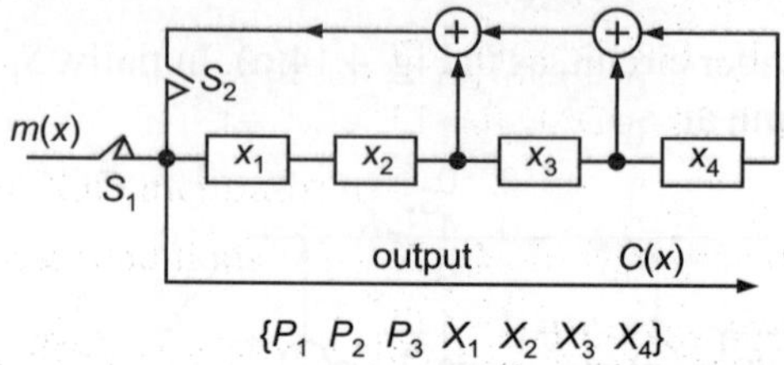

Figure 4.15: *k*-stage encoder

Initially with switch S_1 on and S_2 off, the message bits are shifted in the SR's. Then with S_1 off and S_2 on, the parity bits are formed as the bits are further shifted through the SR's. For the message $m(X) = \{1\ \ 0\ \ 1\ \ 1\}$, the parity bits obtained are $\{1\ \ 0\ \ 0\}$ and the code $C_j = \{1\ \ 0\ \ 0\ \ 1\ \ 0\ \ 1\ \ 1\}$ as obtained by the encoders of Fig. 4.14. All the encoders of Fig. 4.14 and 4.15 generate the same code vectors and have the same generating matrix shown below.

Conceptually the parity bits so generated are the results of the parity check equations (similar to eqn. (4.24)) given as

$$\left.\begin{aligned} P_3 &= X_2 \oplus X_3 \oplus X_4 \\ P_2 &= X_1 \oplus X_2 \oplus X_3 \\ P_1 &= X_1 \oplus X_3 \oplus X_4 \\ &= X_1 \oplus X_2 \oplus P_3 \end{aligned}\right\} \equiv \qquad (4.33)$$

P_1	P_2	P_3	X_1	X_2	X_3	X_4
		√		√	√	√
	√		√	√	√	
√		√	√	√		

The eqn. (4.33) may be written in a tabular form as shown and it is evident that the parity bits are formed by cyclic shifts of the message vector $m(X)$. The generator for the parity bits is then the encoder of Fig. 4.15. From eqn. (4.33), the parity matrix $[P]$ is give as:

$$[P] = \begin{bmatrix} 1 & 1 & 0 \\ 0 & 1 & 1 \\ 1 & 1 & 1 \\ 1 & 0 & 1 \end{bmatrix}$$

and the generating matrix $[G]$ is:

$$[G] = \left[P \mid I_k\right] = \begin{bmatrix} 1 & 1 & 0 & 1 & 0 & 0 & 0 \\ 0 & 1 & 1 & 0 & 1 & 0 & 0 \\ 1 & 1 & 1 & 0 & 0 & 1 & 0 \\ 1 & 0 & 1 & 0 & 0 & 0 & 1 \end{bmatrix} \qquad (4.34)$$

Using the generating matrix of eqn. (4.34), the (7, 4) code vectors are given in Table 4.10.

Table 4.10: Code vectors for $(7^4\ 4)$ SEC code

m_j	message m_j $X_1\ X_2\ X_3\ X_4$	Code vectors $P_1\ P_2\ P_3\ X_1\ X_2\ X_3\ X_4$
1	0 0 0 0	0 0 0 0 0 0 0
2	0 0 0 1	1 0 1 0 0 0 1
3	0 0 1 0	1 1 1 0 0 1 0
4	0 0 1 1	0 1 0 0 0 1 1
5	0 1 0 0	0 1 1 0 1 0 0
6	0 1 0 1	1 1 0 0 1 0 1
7	0 1 1 0	1 0 0 0 1 1 0
8	0 1 1 1	0 0 1 0 1 1 1
9	1 0 0 0	1 1 0 1 0 0 0
10	1 0 0 1	0 1 1 1 0 0 1
11	1 0 1 0	0 0 1 1 0 1 0
12	1 0 1 1	1 0 0 1 0 1 1
13	1 1 0 0	1 0 1 1 1 0 0
14	1 1 0 1	0 0 0 1 1 0 1
15	1 1 1 0	0 1 0 1 1 1 0
16	1 1 1 1	1 1 1 1 1 1 1

It is not apparent that $[G]$ of eqn. (4.34) is cyclic, but this has happened because of the restriction that $[G]$ is of the form $\left[P \mid I_k\right]$. The technique of writing $[G]$ from $g(X)$ is that: assume the first row of $[G]$ to be $g(X)$ and it is seen that polynomial representation of this row has a degree $(n-k)$. In particular, from eqn. (4.34),

$$g(X) = 1 + X + X^3$$

and the following rows are obtained from the right shift of $g(X)$, with the condition that the last four columns of $[G]$ should be of the form $[I_k]$. If there are 1's in the $(n-k+1)$ position of any shifted row, then $g(X)$ is added to the row to form a vector with zero in that position. Since sum of the rows is also a valid vector, this technique produces the desired $[G]$ with the rightshift of $g(X)$. As for example, the 2nd row of eqn. (4.34) is the shifted vector $X{\cdot}g(X)$, but the 3rd row would be $X^2{\cdot}g(X) = 0011010$, violating the form $[I_k]$. Then $(1 + X^2){\cdot}g(X) = 1110010$, which satisfies the condition. Thus $[G]$ is formed by shifting and adding $g(X)$ when required; the overall $[G]$ is now given by

$$[G] = \begin{bmatrix} g(X) \\ X{\cdot}g(X) \\ \left(X^2+1\right)g(X) \\ \left(X^3+X+1\right)g(X) \end{bmatrix} = \begin{bmatrix} 1101000 \\ 0110100 \\ 1110010 \\ 1010001 \end{bmatrix} \tag{4.35}$$

Example 4.19: Consider (15, 11) Hamming code, where

$$(X^{15} + 1) = (1 + X)(1 + X + X^2)(1 + X + X^4)(1 + X + X^2 + X^3 + X^4)(1 + X^3 + X^4).$$

Since $g(X)$ is to be of degree $(n-k) = 4$, choose $g(X) = (1 + X + X^4)$. Then the generator matrix may be written as:

$$[G] = \begin{bmatrix} g(X) \\ X \cdot G(X) \\ X^2 \cdot g(X) \\ (X^3+1)g(X) \\ (X^4+X+1)g(X) \\ (X^5+X^2+X)g(X) \\ (X^6+X^3+X^2+1)g(X) \\ (X^7+X^4+X^3+X)g(X) \\ (X^8+X^5+X^4+X^2+1)g(X) \\ (X^9+X^6+X^5+X^3+X+1)g(X) \\ (X^{10}+X^7+X^6+X^4+X^2+X+1)g(X) \end{bmatrix} = \begin{bmatrix} 1100 & 10000000000 \\ 0110 & 01000000000 \\ 0011 & 00100000000 \\ 1101 & 00010000000 \\ 1010 & 00001000000 \\ 0101 & 00000100000 \\ 1110 & 00000010000 \\ 0111 & 00000001000 \\ 1111 & 00000000100 \\ 1011 & 00000000010 \\ 1001 & 00000000001 \end{bmatrix} \quad (4.36)$$

One could use $g(X) = (1 + X^3 + X^4)$ or $(1 + X + X^2 + X^3 + X^4)$ as well for writing $[G]$. It can be easily checked that all the three polynomials give the same rows of $[P]$ or their equivalent through the combination of rows. The encoder for the (15, 11) code is now given as in Fig. 4.16, where S_1 opens on last 4 bits and S_2 is down for first 11 bits and a total of 15 shifts is required for generating each codeword.

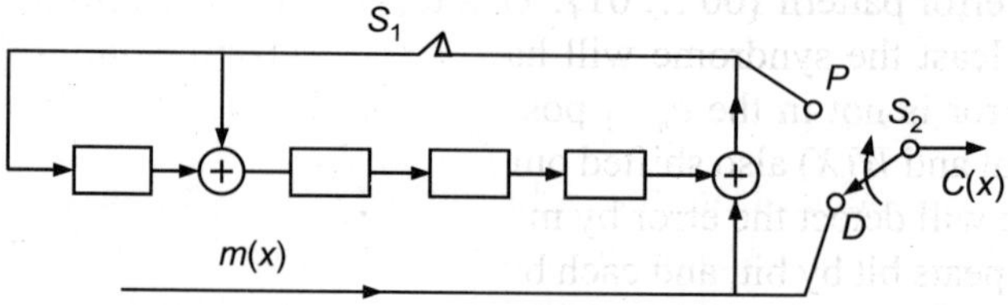

Figure 4.16: Encoder for (15, 11) code with $g(X) = 1 + X + X^4$

4.3.1 Decoders for Cyclic Codes

For the purpose of detection of errors and their subsequent correction, the syndrome of the received word is first calculated. The syndrome polynomial $S(X)$ is the mod-2 sum of the received parity check bits $W_p(X)$ and the parity check bits calculated from the received information bits $W_i(X)$. Then,

$$S(X) = W_p(X) + \text{Remainder of } [W_i(X)/g(X)] \quad (4.37)$$

where the received vector $W(X) = W_p(X) + W_i(X)$. Since $W_p(X)$ has degree less than $g(X)$, Rem. $[W_p(X)/g(X)] = W_p(X)$. Using above two equations,

$$S(X) = \text{Rem. } [W(X)/g(X)] = \text{Rem } [E(X)/g(X)]. \quad (4.38)$$

where $E(X)$ is the error vector and $W(X) = C(X) + E(X)$.

Thus the syndrome of any n-tuple is its residue modulo $g(X)$, and all legitimate codewords $C(X)$ have all-zero syndromes. Since the syndrome of $W(X)$ contains the information about $E(X)$, this information is used for error detection and correction. The syndrome calculation is done by a division circuit which is identical to the encoder at the transmitter and the resultant $S(X)$ is used to retrieve the estimated $E(X)$

from the syndrome evaluator Table. For binary codes, the coefficients of $S(X)$ can be used directly as a memory address; and the corresponding $E(X)$ is stored at the memory location. The calculated $S(X)$ provides the required $E(X)$ through a ROM and $W(X) + E(X) = C(X)$ is calculated. The message bits $m(X)$ are now computed by dividing $C(X)$ by $g(X)$. Such a decoder, as shown in Fig. 4.17, may be used for both non-systematic and systematic codes. For (7, 4) Hamming code, there are 3 syndrome bits and hence a memory with a 3-bit address by 7-bit wordlength is used. This decoder is practical with values of $(n - k) \le 14$ bits, but other techniques, specifically for systematic codes, are more economical for large length codes, as discussed below.

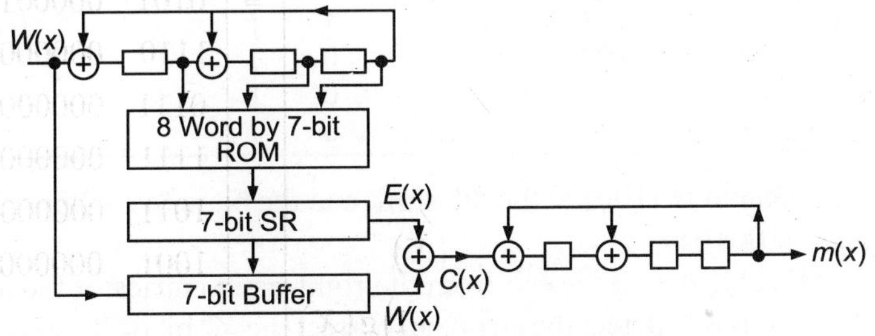

Figure 4.17: Syndrome decoder for a non-systematic (7, 4) cyclic code

For systematic codes, it is easier to test only the syndrome for those error patterns that have '1' only in the high order position, say, the nth position. Errors in other positions are decoded later, using the cyclic structure of the code. It is necessary that the syndrome evaluator table contains only those syndromes which correspond to an error pattern $\{00 \ldots 01\}$. This error is now corrected if the syndrome evaluator can detect the error; at least the syndrome will have some nonzero value if there is any error in the received $W(X)$. If the error is not in the e_{n-1} position, then the syndrome contents will be cyclically shifted with all zero input and $W(X)$ also shifted out from the buffer. Whenever an error bit comes out of the buffer, the syndrome will detect the error by matching the particular stored pattern; and then correct the error. The process repeats bit by bit; and each bit is tested in turn for error and is corrected whenever an error is located. Such a decoder is called the Meggitt decoder.

Consider the decoder circuit, shown in Fig. 4.18, for the (7, 4) cyclic code generated by $g(X) = 1 + X + X^3$. The entire received vector W_j is entered into the SR's bit by bit and at the same time W_j is stored in the buffer memory. After the 3rd shift, the division process will start and after the 7th shift, the syndrome will be stored in the SR's. If $S(X) = \{000\}$, then $E(X) = 0$ and W_j is read out of the buffer. Since we may find $S(X)$ only from $E(X)$ with non-zero coefficients, assume that the received $E(X) = \{0000001\}$. Then the SR-contents are given as: $\{001, 110, 011, 111, 101\}$, showing $S(X) = \{101\}$ after the 7th shift. At the 8th shift, the SR-content is $\{100\}$ and this may be used through a coincidence circuit to correct the error bit coming out of the buffer at the 8th shift. If, however, the error polynomial $E(X) = \{0001000\}$, then the SR content will be $\{100\}$ at the 11th shift and the error will be corrected when the buffer delivers the error bit at the 11th shift.

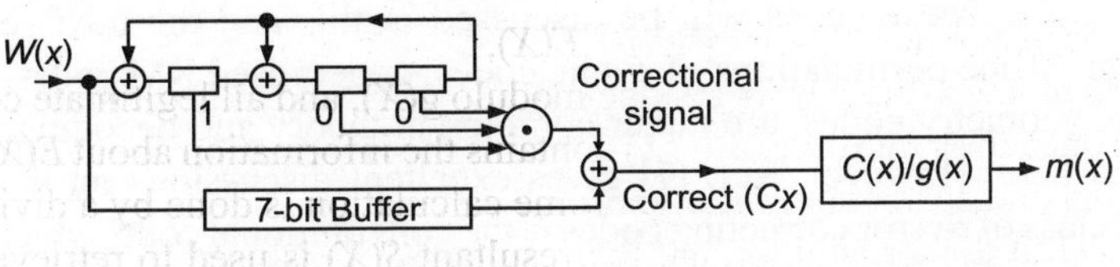

Figure 4.18: The Meggitt decoder for (7, 4) cyclic code

Example 4.20: Suppose for the (7, 4) cyclic code, the received $W_j = 1001010$. Then the SR-contents are as follows:

No. of shift	Input	SR-content for 1001010	Input	SR-content for 1001111
1	0	000	1	100
2	1	100	1	110
3	0	010	1	111
4	1	101	1	001
5	0	100	0	110
6	0	010	0	011
7	1	101	1	011 ← Indicates error
8	0	100	0	111
9	–	–	0	101
10	–	–	0	100

Thus it is seen that SR-content is {100} at the 8th shift, and the bit in the X^6-position of W_j is corrected, giving the correct $C(X) = \{1001011\}$.

If, however, $W_j = \{1001111\}$, then it is seen from the right-hand entries of the above table that at the 10th shift the syndrome content will detect the error and correct the X^4 bit of W_j, giving $C(X) = \{1001011\}$.

The decoder for the (15, 11) cyclic code, using $g(X) = 1 + X + X^4$, is shown in Fig. 4.19, where it may be easily checked that the SR-content at 16th shift is {1000} if the error $E(X) = X^{14}$. Thus, a coincident circuit gives the correction signal to the buffer output, as explained above.

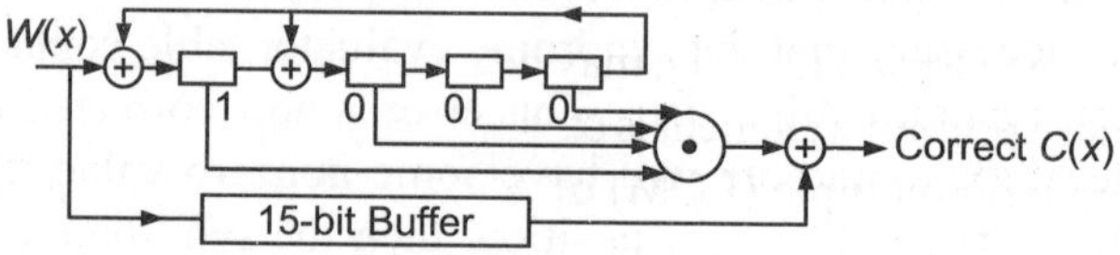

Figure 4.19: Decoder for (15, 11) cyclic code

The above decoders are meant for cyclic SEC codes. However, the Meggitt decoder may be generalised for multiple-error correcting codes, e.g., (15, 7) BCH code. An error-trapping decoder is a modification of a Meggitt decoder that is used for certain cyclic codes.

4.4 MULTIPLE-ERROR CORRECTING CODES

So far SEC codes have been discussed. Multiple-error correcting codes can be developed by the method of forming standard arrays, e.g., the array in Table 4.9, with the required value of H.D. > $2t$, t = number of errors in the received vector. More elegant methods, however, have been developed for t-error correcting codes, and these have resulted in BCH, Reed-Solomon (RS), Golay, Reed-Muller, Simplex and other codes. Most of these codes are cyclic codes and are decoded by Meggitt or Error-trapping decoders. Decoding of BCH and RS codes is rather complex, requiring solution of simultaneous equations and matrix inversion. Fortunately, some of these codes are also majority-logic decodable and the decoders for them are very simple. These codes will be discussed in this and the next sections. Other t-error correcting codes, e.g., Affine permutations of cyclic codes, generalised RM codes, Euclidean geometry codes, and projective geometry codes, are rather complex in theory and decoding techniques, and they will not be discussed here. However, Ref. [7] gives excellent discussions on these codes. One other important and useful class of t-error correcting codes is the convolutional code, which has found extensive use in modern communication systems. This will be discussed in detail in a later section.

4.4.1 BCH Codes [8]

The Bose-Chaudhuri-Hocquenghem (BCH) codes, discovered in 1959–60, are a class of cyclic codes with powerful error-correcting properties and well-known implementation algorithms. BCH (binary) codes may also be considered as a binary group code, where the necessary and sufficient conditions to be satisfied are:

(a) For a t-error correcting (n, k) binary group code, an $(n \times r) = [H^T]$ matrix exists, such that any $2t$ row vectors of the matrix are mutually independent, $(r = n - k)$.

(b) For any positive integers m and t $(t < 2^{m-1})$, a BCH code has the parameters:

$$\begin{aligned} &\text{Block length} && n = (2^m - 1) \\ &\text{No. of parity check bits} && r = n - k \le mt \\ \text{and } &\text{Minimum} && \text{H.D.} \ge 2t + 1. \end{aligned} \tag{4.39}$$

Similar to Table 4.7, the parameters of some of the useful BCH binary codes are given in Table 4.11. The values for larger n are given in Ref. [2] and [3].

Table 4.11: Parameters of BCH codes

n	7	15	15	15	31	31	31	31	31	63	63	63	63	63	63	63	63	63	63	63
k	4	11	7	5	26	21	16	11	6	57	51	45	39	36	30	24	18	16	10	7
t	1	1	2	3	1	2	3	5	7	1	2	3	4	5	6	7	10	11	13	15

The alphabet of a BCH code for $n = (2^m - 1)$, may be represented as the set of elements of an appropriate Galois Field, GF (2^m), whose primitive element is α. Then the generator polynomial of the t-error correcting BCH code is the least common multiple (LCM) of $M_1(X)$, $M_2(X)$, ..., $M_{2t}(X)$, i.e.,

$$g(X) = \text{LCM}\,[M_1(X), M_2(X) \ldots M_{2t}(X)], \tag{4.40}$$

where $M_i(X)$ is the minimum polynomial of α^i, $i = 1, 2, \ldots, 2t$; as given in Table 4.12 for $m = 4$. Since the minimum polynomials for $\alpha^2, \alpha^4, \ldots, \alpha^{2t}$, i.e., for even power of α, are the same as for $\alpha, \alpha^2, \alpha^3 \ldots$ etc., as seen in Table 4.12, the generator polynomial is reduced to:

$$g(X) = \text{LCM}[M_1(X), M_3(X), \ldots, M_{2t-1}(X)] \tag{4.41}*$$

Since the degree of each polynomial is m or less, the degree of $g(X)$ is atmost mt, and $r = n - k \le mt$. For many of the BCH codes, as given Table 4.11, the minimum H.D. = $2t + 1$.

Consider now a (15, 7) double-error correcting BCH binary code, $m = 4$, where the generating polynomial $g(X)$ is given by

$$\begin{aligned} g(X) &= \text{LCM}[M_1(X), M_3(X)] \\ &= M_1(X) \cdot M_3(X) \\ &= (1 + X + X^4)(1 + X + X^2 + X^3 + X^4) \\ &= 1 + X^4 + X^6 + X^7 + X^8 \end{aligned} \tag{4.42}$$

and $\alpha, \alpha^2, \alpha^3$ and α^4 are the roots of $g(X)$. From eqn. (4.39), the code is a (15, 7) cyclic code with minimum $d \ge 5$. Actually, $d = 5$ since $g(X)$ has weight 5.

*BCH codes generated by eqn. (4.41) are called primitive BCH codes.

For triple-error correcting BCH binary code over GF(2^4), with $n = 15$, $k = 5$, $r = 10$, the generating polynomial is:

$$\begin{aligned} g(X) &= \text{LCM}[M_1(X), M_3(X), M_5(X)] \\ &= M_1(X) \cdot M_3(X) \cdot M_5(X) \\ &= (1 + X + X^4)(1 + X + X^2 + X^3 + X^4)(1 + X + X^2) \\ &= 1 + X + X^2 + X^4 + X^5 + X^8 + X^{10} \end{aligned} \quad (4.43)$$

Table 4.12: Elements of GF (2*) Modulo Polynomial = ($X^4 + X + 1$)

Sl. No.	Element as a power of primitive element	Binary representation	Minimal* polynomials
1	0	0 0 0 0	–
2	$\alpha^0 = 1$	0 0 0 1	$X + 1$
3	α^1	0 0 1 0	$X^4 + X + 1$
4	α^2	0 1 0 0	$X^4 + X + 1$
5	α^3	1 0 0 0	$X^4 + X^3 + X^2 + X + 1$
6	$\alpha^4 = \alpha + 1$	0 0 1 1	$X^4 + X + 1$
7	$\alpha^5 = \alpha^2 + \alpha$	0 1 1 0	$X^2 + X + 1$
8	$\alpha^6 = \alpha^3 + \alpha^2$	1 1 0 0	$X^4 + X^3 + X^2 + X + 1$
9	$\alpha^7 = \alpha^3 + \alpha + 1$	1 0 1 1	$X^4 + X^3 + 1$
10	$\alpha^8 = \alpha^2 + 1$	0 1 0 1	$X^4 + X + 1$
11	$\alpha^9 = \alpha^3 + \alpha$	1 0 1 0	$X^4 + X^3 + X^2 + X + 1$
12	$\alpha^{10} = \alpha^2 + \alpha + 1$	0 1 1 1	$X^2 + X + 1$
13	$\alpha^{11} = \alpha^3 + \alpha^2 + \alpha + 1$	1 1 1 0	$X^4 + X^3 + 1$
14	$\alpha^{12} = \alpha^3 + \alpha^2 + \alpha + 1$	1 1 1 1	$X^4 + X^3 + X^2 + X + 1$
15	$\alpha^{13} = \alpha^3 + \alpha^2 + 1$	1 1 0 1	$X^4 + X^3 + 1$
16	$\alpha^{14} = \alpha^3 + 1$	1 0 0 1	$X^4 + X^3 + 1$
17	$\alpha^{15} = 1 = \alpha^0$	0 0 0 1	$X + 1$

*The minimum polynomial or minimum function of $\beta (= \alpha^i)$ is defined as the monic polynomial $M_i(X)$ of the smallest degree with coefficients in the ground field F(i.e. GF(2) in this case), such that $M_i(\beta) = M_i(\alpha^i) = 0$, where α is the primitive element of GF(2^m) and $\beta (= \alpha^i)$ is an element of the extension field GF(2^m). A short list of minimum polynomials is given in Appendix *B*.

This is a (15, 5) cyclic code, with minimum $d = 7$.

For a single-error correcting BCH code, $g(X) = M_1(X) = 1 + X + X^4$, with $n = 15$, $k = 11$, as discussed in Ex. 19. Thus Hamming codes form a subclass of the (primitive) BCH codes.

For other values of m, say $m = 3, 5, 6, 7, \ldots$, etc. one uses $M_1(X), M_3(X), \ldots, M_{2t-1}(X)$ from GF (2^m). These minimum polynomials are listed in Appx. *C* of Ref. [2], and are used to write $G(X)$. Some examples are:

(a) For GF(2^3), the SEC (7, 4) code has $g(X) = M_1(X) = 1 + X + X^3$.

(b) For GF(2^5), $M_1(X) = 1 + X^2 + X^5$, $\quad M_3(X) = 1 + X^2 + X^3 + X^4 + X^5$,

$M_5(X) = 1 + X + X^2 + X^4 + X^5$, $\quad M_7(X) = 1 + X + X^2 + X^3 + X^5$, etc.

Thus $g(X)$ for SEC (31, 26) code has $g(X) = M_1(X) = 1 + X^2 + X^5$,

$$\begin{aligned} g(X) \text{ for } (31, 21), t \le 2, \text{ code } &= M_1(X) \cdot M_3(X) \\ &= 1 + X^3 + X^5 + X^6 + X^8 + X^9 + X^{10} \end{aligned}$$

and $g(X)$ for (31, 16), $t \leq 3$, code $= M_1(X) \cdot M_3(X) \cdot M_5(X)$

$$= 1 + X + X^2 + X^3 + X^5 + X^7 + X^8 + X^9 + X^{10} + X^{11} + X^{15}$$

(c) For GF(2^6), $M_1(X) = 1 + X + X^6$

$$M_3(X) = 1 + X + X^2 + X^4 + X^6$$

$$M_5(X) = 1 + X + X^5 + X^5 + X^6$$

$$M_7(X) = 1 + X^3 + X^6, \text{ etc.}$$

Then for SEC (63, 57) code $g(X) = M_1(X)$,

$g(X)$ for (63, 51), $t \leq 2$ code $= M_1(X) \cdot M_3(X)$,

$g(X)$ for (63, 45), $t \leq 3$ code $= M_1(X) \cdot M_3(X) \cdot M_5(X)$,

and

$g(X)$ for (63, 39), $t \leq 4$ code $= M_1(X) \cdot M_3(X) \cdot M_5(X) \cdot M_7(X)$.

The generating matrix $[G]$ and the parity check matrix $[H]$ of the above codes may be written according to the rules discussed in Section 4.3. As an example, for the (15, 7) cyclic code with $g(X) = 1 + X^4 + X^6 + X^7 + X^8$ with $t \leq 2$, the parity check matrix is (for systematic codes):

$$[H] = \begin{bmatrix} 1 & 0 & 0 & 0 & 0 & 0 & 0 & 0 & 1 & 1 & 0 & 1 & 0 & 0 & 0 \\ 0 & 1 & 0 & 0 & 0 & 0 & 0 & 0 & 0 & 1 & 1 & 0 & 1 & 0 & 0 \\ 0 & 0 & 1 & 0 & 0 & 0 & 0 & 0 & 0 & 0 & 1 & 1 & 0 & 1 & 0 \\ 0 & 0 & 0 & 1 & 0 & 0 & 0 & 0 & 0 & 0 & 0 & 1 & 1 & 0 & 1 \\ 0 & 0 & 0 & 0 & 1 & 0 & 0 & 0 & 1 & 1 & 0 & 1 & 1 & 1 & 0 \\ 0 & 0 & 0 & 0 & 0 & 1 & 0 & 0 & 0 & 1 & 1 & 0 & 1 & 1 & 1 \\ 0 & 0 & 0 & 0 & 0 & 0 & 1 & 0 & 1 & 1 & 1 & 0 & 0 & 1 & 1 \\ 0 & 0 & 0 & 0 & 0 & 0 & 0 & 1 & 1 & 0 & 1 & 0 & 0 & 0 & 1 \end{bmatrix} \tag{4.44}$$

This $[H]$ is used in majority logic decoding techniques where applicable.

For the purpose of decoding BCH codes, consider the received vector $W(X) = C(X) + E(X)$, where $C(X)$ is the code vector and $E(X)$ is the error vector. Since $g(X)$ is a factor of $C(X)$ and $\alpha, \alpha^2, ..., \alpha^{2t}$ are the roots of $g(X)$, α being the primitive element of GF(2^m), then

$$W(\alpha^i) = C(\alpha^i) + E(\alpha^i), \quad i = 1, 2, ..., 2t$$

$$= E(\alpha^i), \quad \text{since } C(\alpha^i) = 0$$

or

$$W(\alpha^i) = \sum_{j=0}^{n-1} e_j \left(\alpha^i\right)^j, \quad e_j = 1 \text{ or } 0 \text{ and } E(X) = \Sigma e_j X^j \tag{4.45}$$

This set of $2t$ equations involves only the components of the error pattern $E(X)$ and $C(X)$ is not involved. Solving this set of equations for e_j, the error pattern can be determined.

The syndrome vectors S_i are now defined as

$$S_i = W(\alpha^i)$$

$$= \omega_0 + \omega_1 \alpha^i + \omega_2 (\alpha^i)^2 + ... + \omega_{n-1} (\alpha^i)^{n-1} \tag{4.46}$$

where $\omega_0, \omega_1, ..., \omega_{n-1}$ are the coefficients of $W(X)$ and $i = 1, 2, ..., 2t$.

Using eqn. (4.45),

$$S_i = E(\alpha^i) = e_0 + e_1\alpha^2 + e_2(\alpha^i)^2 + ... + e_{n-1}(\alpha^i)^{n-1} \tag{4.47}$$

where at most t coefficients are non-zero. The above syndrome vectors are different from the syndrome polynomials $S(x)$ as given in eqn. (4.37), but contain equivalent information. To simplify the notations of eqn. (4.47), assume that p errors have occurred, $0 < p \leq t$, and their locations and magnitudes are given by $X_1, X_2, ..., X_p$ and $Y_1, Y_2, ..., Y_p$ respectively. Then the $2t$ syndrome vectors may be written as

$$\begin{aligned} S_1 &= Y_1X_1 + Y_2X_2 + ... + Y_pX_p \\ S_2 &= Y_1X_1^2 + Y_2X_2^2 + ... + Y_pX_p^2 \\ &\vdots \\ S_{2t} &= Y_1X_1^{2t} + Y_2X_2^{2t} + ... + Y_pX_p^{2t} \end{aligned} \tag{4.48}^*$$

This set of equations has one unique solution, but the equations are non-linear and hence too difficult to solve. To simplify the solution, an Error-locator polynomial is defined as

$$\lambda(Z) = \lambda_pZ^p + \lambda_{p-1}Z^{p-1} + ... + \lambda_1Z + 1 \tag{4.49}$$

which has zeros at the inverse error locations $X_l^{-1}, l = 1, 2, ..., p$. Then

$$\lambda(Z) = (1 - ZX_1)(1 - ZX_2) ... (1 - ZX_p) \tag{4.50}$$

If the coefficients, $\lambda_p, \lambda_{p-1} ... \lambda_1$ of $\lambda(Z)$ can be determined, then the zeros $\lambda(Z)$ can be found to obtain the error locations. These coefficients are related to the syndromes of eqn. (4.48) by the matrix equation:

$$\begin{bmatrix} S_1 & S_2 & S_3 & \cdots & S_p \\ S_2 & S_3 & S_4 & \cdots & S_{p+1} \\ S_3 & S_4 & S_5 & \cdots & S_{p+2} \\ \vdots & \vdots & \vdots & \vdots & \vdots \\ S_p & S_{p+1} & S_{p+2} & \cdots & S_{2p-1} \end{bmatrix} \cdot \begin{bmatrix} \lambda_p \\ \lambda_{p-1} \\ \lambda_{p-2} \\ \vdots \\ \lambda_1 \end{bmatrix} = \begin{bmatrix} -S_{p+1} \\ -S_{p+2} \\ -S_{p+3} \\ \vdots \\ -S_{2p} \end{bmatrix} \tag{4.51}$$

This matrix is non-singular if there are p errors and the equations may be solved by using matrix-inversion. But the matrix is singular if the number of errors is less than p.

The steps for decoding are as follows: [3, 4, 7]

(a) First compute the syndromes using GF(2^m).

(b) As a trial, set $p = t$ and compute the determinant of the matrix $[M]$ of eqn. (4.51). If it is non-zero, then $p = t$ and if it is zero, then $p < t$. Then check with $p = (t - 1)$ and stop when the determinant is not zero.

(c) Solve for $\lambda(Z)$ by inverting the matrix $[M]$ and using eqn. (4.51).

(d) Find the zeros of $\lambda(Z)$ to find the error locations. Use trial and error method, known as Chien search. For Binary codes, locating X_l completes the solution.

---

*Here Y's are 1 or 0 for binary codes, but they are multivalued for a multialphabet code, say, RS code.

Example 4.21: Consider (15, 7), double-error correcting BCH code with

$$g(X) = X^8 + X^7 + X^6 + X^4 + 1$$

If $m(X) = (1 + X)$, the transmitted code $C(X) = X^9 + X^6 + X^5 + X^4 + X + 1$. The received vector $W(X) = X^9 + X^8 + X^6 + X^5 + X^4 + X^3 + X + 1$. Using GF($2^4$) and modulo polynomial $= X^4 + X + 1$, the syndromes are computed as

$$S_1 = \alpha^8 + \alpha^3 = \alpha^{13}$$

$$S_2 = \alpha^{16} + \alpha^6 = \alpha^{11}$$

$$S_3 = \alpha^{24} + \alpha^9 = 0$$

$$S_4 = \alpha^{32} + \alpha^{12} = \alpha^7$$

since $\alpha^9 + \alpha^6 + \alpha^5 + \alpha^4 + \alpha + 1 = 0$.

The determinant of $M = \begin{vmatrix} S_1 & S_2 \\ S_2 & S_3 \end{vmatrix} \neq 0$, hence, $p = 2$.

Then,

$$M = \begin{vmatrix} \alpha^{13} & \alpha^{11} \\ \alpha^{11} & 0 \end{vmatrix}$$

therefore,

$$M^{-1} = \begin{vmatrix} 0 & \alpha^4 \\ \alpha^4 & \alpha^6 \end{vmatrix}$$

and

$$\begin{bmatrix} \lambda_2 \\ \lambda_1 \end{bmatrix} = \begin{bmatrix} M^{-1} \end{bmatrix} \begin{bmatrix} 0 \\ \alpha^7 \end{bmatrix} = \begin{bmatrix} \alpha^{11} \\ \alpha^{13} \end{bmatrix}$$

Thus,

$$\lambda(Z) = \alpha^{11} z^2 + \alpha^{13} z + 1 = (\alpha^8 z + 1)(\alpha^3 z + 1)$$

Therefore, the errors have occurred at the 3rd and 8th positions, giving

$$E(X) = X^8 + X^3.$$

The result was somewhat evident from the value of $S_1 = \alpha^8 + \alpha^3$, as the errors were placed at the zero values of the vector $C(X)$.

If the errors were placed at non-zero values of $C(X)$, say at X^6 and X^4, then, $W(X) = X^9 + X^5 + X + 1$. Using the modulo polynomial $= X^4 + X + 1$, the syndromes are calculated as:

$$\left.\begin{aligned} S_1 &= W(\alpha) = \alpha^{12} \\ S_2 &= W(\alpha^2) = \alpha^9 \\ S_3 &= W(\alpha^3) = \alpha^{10} \\ S_4 &= W(\alpha^4) = \alpha^3 \end{aligned}\right\} \to |M| = \begin{vmatrix} S_1 & S_2 \\ S_2 & S_3 \end{vmatrix} \neq 0; \text{ hence } p = 2.$$

Then,

$$[M] = \begin{bmatrix} \alpha^{12} & \alpha^9 \\ \alpha^9 & \alpha^{10} \end{bmatrix}; \therefore M^{-1} = \begin{bmatrix} \alpha^6 & \alpha^5 \\ \alpha^5 & \alpha^8 \end{bmatrix}$$

and

$$\begin{bmatrix} \lambda_2 \\ \lambda_1 \end{bmatrix} = M^{-1} \begin{bmatrix} \alpha^{10} \\ \alpha^3 \end{bmatrix} = \begin{bmatrix} \alpha^{10} \\ \alpha^{12} \end{bmatrix}$$

Therefore,

$$\lambda(Z) = \alpha^{10}Z^2 + \alpha^{12}Z + 1 = (\alpha^6 Z + 1)(\alpha^4 Z + 1)$$

and the errors have occurred at the 6th and 4th positions of $W(X)$, giving $E(X) = X^6 + X^4$.

Various numerical techniques, e.g., Chien search technique, Berlekamp's iterative scheme etc. have been developed for solving the error-locator polynomials. A general purpose computer or a special hardware is also used for decoding BCH codes with large number of errors. For details, see [3, 4, 7]. The Meggitt decoder, discussed in Sec. 4.3.1 may be used for decoding short length BCH codes [7]. Also the majority-logic decoding techniques, as discussed later, are used for decoding some of the BCH codes.

4.4.2 Reed-Solomon Codes [9]

Reed-Solomon (RS) codes are a subclass of BCH codes, where the code symbols are from GF(q), ($q \neq 2^m$ in general, but usually q is taken as 2^m). A t-error correcting RS code has the following parameters (where symbols are in terms of m tuples):

$$\begin{aligned} &\text{Block length } n = (q-1) \text{ symbols} \\ &\text{No. of parity check symbols } r = n - k = 2t \\ &\text{Minimum H.D.} = d = 2t + 1 \text{ symbols} \end{aligned} \tag{4.52}$$

Consider $q = 2^m$, α as primitive, then the generator polynomial for the RS-code is given by:

$$g(X) = (X + \alpha)(X + \alpha^2) \ldots (X + \alpha^{2t}) \tag{4.53}$$

The polynomial has a degree of $2t$ and is a maximum distance separable (MDS) code, i.e., RS code has maximum H.D. for given values of n and r. Consider a (15, 11) double-error correcting RS-code with $n = 15$, $r = 4$. Then $q = 2^4$, and using eqn. (4.53) and Table 4.12,

$$\begin{aligned} g(X) &= (X + \alpha)(X + \alpha^2)(X + \alpha^3)(X + \alpha^4) \\ &= \alpha^{10} + \alpha^3 X + \alpha^6 X^2 + \alpha^{13} X^3 + X^4 \end{aligned} \tag{4.54}$$

Comparing the binary double-error correcting BCH code of eqn. (4.42), it is seen that the degree of the $g(X)$ of BCH code is 8, but the degree of $g(X)$ of the RS code is 4 only. As such, RS code is a (15, 11) code, whereas the binary BCH is a (15, 7) code. For a message vector $(X + 1)$, the transmitted code $C_1(X)$ for binary BCH is:

$$C_1(X) = 1 + X + X^4 + X^5 + X^6 + X^9$$

and as a 15-tuple binary code $\{C_1\} = \{110011100100000\}$. The RS code for the same message vector is:

$$\begin{aligned} C_2(X) &= (1 + X)(\alpha^{10} + \alpha^3 X + \alpha^6 X^2 + \alpha^{13} X^3 + X^4) \\ &= \alpha^{10} + \alpha^{12} X + \alpha^2 X^2 + X^3 + \alpha^6 X^4 + X^5 \end{aligned}$$

and in the 15-tuple form,

$$\{C_2\} = \{\alpha^{10}, \alpha^{12}, \alpha^2, 1, \alpha^6, 1, 0, 0, 0, 0, 0, 0, 0, 0, 0\}$$

The equivalence of C_2 in the binary form is obtained from Table 4.12 as:

$$\{C_2\} = \{0111,\ 1111,\ 0100,\ 0001,\ 1100,\ 0001, 0000, 0000, 0000, 0000, \\ 0000, 0000, 0000, 0000, 0000\} \tag{4.55}$$

In terms of binary digits, the RS code here is a (60, 44) code. The RS encoder accepts 11 binary 4-tuples of data and forms 15 binary 4-tuples of coded symbols. The binary BCH code, on the other hand, accepts 7 bits and converts these into a 15-bit code. Both codes have a minimum distance of 5; but in RS code, this means that the distance is measured in terms of 4-tuple symbols. Thus the rate of transmission in RS codes is much higher for the same error correcting capability. For channel transmission purposes, one can transmit RS codes either in the binary-word form or as 16-level symbols.

Example 4.22: Using GF(2^3), for $t = 2$, one gets $n = q - 1 = 7$; $k = 3$. Thus (7, 3) RS code is a double-error correcting code. The generator polynomial is: (using Table 4.1):

$$g(X) = (X + \alpha)(X + \alpha^2)(X + \alpha^3)(X + \alpha^4)$$
$$= \alpha^3 + X + \alpha^3X^2 + \alpha^2X^3 + X^4$$

If $m(X)$ is a sequence of 3 octal symbols (= 9 bits), then, say,

$$m(X) = (\alpha^3 + X + \alpha^4X^2),$$

then the code $C(X)$ is:

$$C(X) = m(X)\cdot g(X) = \alpha^6 + \alpha^6X^2 + X^3 + \alpha^4X^5 + \alpha^4X^6 \tag{4.56}$$

which is a sequence of octal symbols whose binary equivalent is:

$$\{110, 000, 110, 001, 000, 111, 111\}$$

For decoding RS codes with errors, the steps given for BCH codes are followed till the error locations are calculated. Now in eqn. (4.48), $X_1, X_2, ..., X_p$ are known, and the set of equations becomes a set of $2t$ linear equations with $Y_1, Y_2, ..., Y_p$ as unknowns. The first p equations can be solved for the error magnitudes $Y_1, ..., Y_p$, if the determinant of the matrix is non-zero. If $p = t$, the determinant is non-zero, and $Y_1, Y_2, ..., Y_p$ are calculated from the relation:

$$\begin{bmatrix} Y_1 \\ Y_2 \\ \vdots \\ Y_p \end{bmatrix} = \begin{bmatrix} X_1 & \cdots & X_p \\ X_1^2 & \cdots & X_p^2 \\ \vdots & & \vdots \\ X_1^p & \cdots & X_p^p \end{bmatrix}^{-1} \times \begin{bmatrix} S_1 \\ S_2 \\ \vdots \\ S_p \end{bmatrix} \tag{4.57}$$

4.4.3 Binary Golay Code

Golay code is a (23, 12) perfect binary code that corrects all errors $t \leq 3$. Based on the relation

$$\left[\binom{23}{0} + \binom{23}{1} + \binom{23}{2} + \binom{23}{3}\right]\cdot 2^{12} = 2^{23} \tag{4.58}$$

the code satisfies eqn. (4.12) and has been shown to be perfect 3-error correcting code and cyclic as well. The code has been used in several practical systems. The generator polynomial of the code is obtained from the relation,

$$(X^{23} + 1) = (X + 1)\ g_1(X)\ g_2(X),$$

where

$$g_1(X) = 1 + X^2 + X^4 + X^5 + X^6 + X^{10} + X^{11}$$
$$g_2(X) = 1 + X + X^5 + X^6 + X^7 + X^9 + X^{11} \tag{4.59}$$

One can use the SR-implementation for the encoder using either $g_1(X)$ or $g_2(X)$ as the divider polynomial. The minimum HD of the code is 7, and hence, $t = 3$. Besides the binary Golay code, there is also a perfect ternary (11, 6) Golay code with $d = 5$.

Error-Trapping Decoder

An error-trapping decoder is a modification of a Meggitt decoder that can be used for certain cyclic codes, e.g., short-length BCH, Golay and burst-error correcting codes. From eqn. (4.38), we have syndrome polynomial as:

$$S(X) = \text{Rem. } [E(X)/g(X)] \tag{4.38}$$

If now the error $E(X)$ is confined to the $(n - k)$ parity check positions $(1, X, ..., X^{n-k-1})$ of $W(X)$, then $E(X) = S(X)$, since the degree of $E(X)$ is less than that of $g(X)$. Thus, correction can be carried out by simply adding $S(X)$ to $W(X)$. Even if $E(X)$ is not confined to the $(n - k)$ parity check positions of $W(X)$, suppose that $E(X)$ has non-zero values clustered together such that the length of the non-zero values is less than the syndrome length, then also the syndrome will exhibit an exact copy of the error pattern after some cyclic shifts of $E(X)$. For each error pattern, the syndrome content $S(X)$ (after the required shifts) is subtracted from the appropriately shifted $W(X)$, and the corrected $W(X)$ recovered.

Sometimes, when error trapping cannot be used for a given code, the test patterns can be modified to include the few troublesome error patterns along with the general test. Such a modified error-trapping decoder is possible for the (23, 12) Golay code, where the error pattern $E(X)$ will be of length 23 and weight 3 or less ($t \leq 3$). The syndrome register has a length 11, and if $E(X)$ has length greater than 11, then the error pattern is not trapped by cyclically shifting $S(X)$. In this case, it is shown that one of the three error bits must have at least five zeros on one side of it and at least six zeros on the other side. Therefore, all $E(X)$ patterns can be cyclically shifted into one of the following three configurations (with bit positions numbered e_0 through e_{22}):

(a) All errors ($t \leq 3$) occur in the 11 high-order bits
(b) One error occurs in position e_5 and other two errors occur in the 11 high order bits
(c) One error occurs in position e_6 and the remaining two errors are in the 11 high order bits.

In the decoder, shown in Fig. 4.20, the received code vector $W(X)$ is fed at the rightmost stage of the syndrome generator, equivalent to multiplying $W(X)$ by X^{11}. Then, the syndromes corresponding to e_5 and e_6 are obtained as (using $g_1(X)$ as the generator polynomial):

$$S(e_5) = R_{em}[X^{16}/g_1(X)] = X + X^2 + X^5 + X^6 + X^8 + X^9$$

and

$$S(e_6) = R_{em}[X^{17}/g_1(X)] = X^2 + X^3 + X^6 + X^7 + X^9 + X^{10} \tag{4.60}$$

The syndrome vectors for the error e_5 or e_6 will be {01100110110} or {00110011011} respectively. Two more errors occurring in the 11 high-order bit positions will cause two 1's in the appropriate positions of the syndrome vectors, thereby complementing the vector for e_5 or e_6. Based on the above relations, the decoder operates as follows:

(a) The entire received vector is shifted in the syndrome generator (with switch G_1 closed) and the syndrome $S(X)$ corresponding to $X^{11} \cdot W(X)$ is formed.

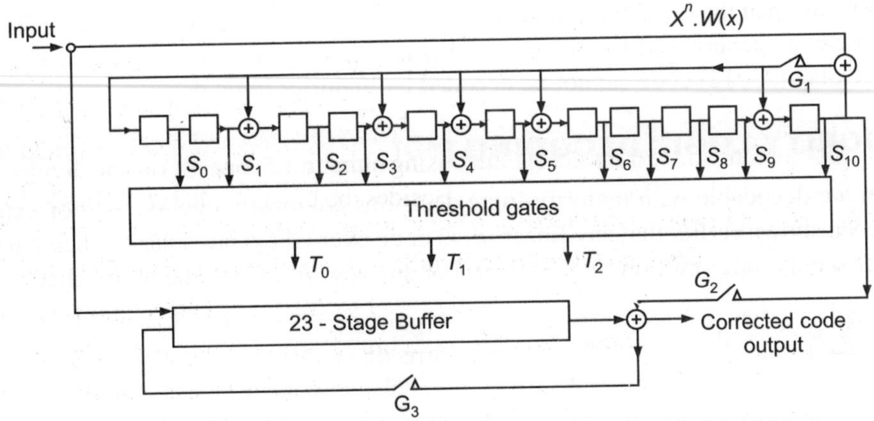

Figure 4.20: Error-trapping decoder for Golay code (23, 12)

(b) If all three (or less) errors are confined to $X^{12}, X^{13}, ..., X^{22}$ of $W(X)$, then the syndrome matches the errors in those positions. The weight of the syndrome is now less or equal to 3. This is checked by a threshold gate and the gate output T_0 switches G_2 on and G_1 off. $W(X)$ is now received from the buffer and corrected by the syndrome bits (as they are clocked bit by bit) through the mod-adder circuit.

(c) If the test in (b) fails, then it is assumed that one error is either at e_5 or at e_6 and the other 2 errors are in the 11 high order bits of $W(X)$. Then if wt. of $S(X) > 3$ (in test (b)), then the weights of $[S(X) + S(e_5)]$ and of $[S(X) + S(e_6)]$ are tested. The decisions are:

  (i) If wt. of $[S(X) + S(e_5)] \leq 2$, then the decision ($T_1 = 1$) is that one error is at position e_5 and 2 errors are at positions where $[S(X) + S(e_5)]$ are non-zero.

  (ii) If wt. of $[S(X) + S(e_6)] \leq 2$, the decision ($T_2 = 1$) is that one error is at position e_6 and 2 errors are at positions where $[S(X) + S(e_5)]$ are non-zero. The above tests are arranged through combinatorial switching circuits and the appropriate corrections in $W(X)$ are made as $W(X)$ is read from the buffer.

(d) If the above tests fail, then with G_1 and G_3 on and G_2 off, the syndrome and buffer contents are shifted by 1 bit. Tests (b) and (c) are now repeated. Bit by bit shifting of $S(X)$ and $W(X)$ is continued till the errors are located, and the corrected. A maximum of 23 shifts will be required to complete the process. After correction of $W(X)$, the corrected $W'(X)$ is further processed through a divider circuit to obtain the message $m(X) = W'(X)/g(X)$.

There is an alternate method of decoding a Golay code, and the method is based on the assumption that when the block of 23 bits with $t \leq 3$ is cyclically shifted, it will be observed that "at most one error will lie out-side the 11 high order bits of $W(X)$" at some shift. This fact is utilized in 'Systematic Search Decoder' where the test (b) is first carried out. If the test is unsuccessful, then 1st bit of $W(X)$ is inverted and the check is made if wt. of $[S(X)] \leq 2$. If the test is successful, then the non-zero positions of $S(X)$ give the two error locations (similar to test (c) above) and the other error is at 1st position. If this fails, then the syndrome content is shifted cyclically, each time testing for wt. $[S(X)] \leq 3$; and if not, invert 2nd, 3rd, ... and 12th bit of $W(X)$ successively and test for wt. $[S(X)] \leq 2$. Since not all errors are in the parity check section, an error must be detected in one of the shifts. Once one error is located and corrected, the other two errors are easily located and corrected by test (b). Sometimes this systematic search coder is simpler in hardware than the error-trapping decoder, but the latter is faster in operation. The systematic search decoder can be generalised for decoding other multiple-error-correcting cyclic codes. It is to be noted that the Golay (23, 12) code cannot be decoded by majority-logic decoders as discussed below.

4.5 MAJORITY-LOGIC DECODING [10]

The codes that are decodable with majority-logic decisions are mostly cyclic codes or extended cyclic codes. It has been shown earlier that any linear (n, k) code over GF(q) has a parity check matrix $[H]$ and the code-words satisfy the condition: $[C][H^T] = 0$. For any row $\underline{h_i}$ of $[H]$, the parity check equations are

$$\sum_{j=1}^{n-1} h_{ij} C_j = 0, \tag{4.61}$$

where j indicates columns, and the summation is over the inner product of the row vector h_i and the code vector. Any linear combination of the rows of $[H]$ also forms another parity check equation and there are a total of q^{n-k} equations that can be formed in this way.

Consider the (7, 3) simplex code, which is dual to the (7, 4) Hamming code. This has $HD = 4$ and corrects one error. This can be generated by the generator matrix $[G]$ and its parity check matrix is $[H]$, as given below:

$$[G] = \begin{bmatrix} 100 & 1011 \\ 010 & 1110 \\ 001 & 0111 \end{bmatrix} \text{ and } [H] = \begin{bmatrix} 110 & 1000 \\ 011 & 0100 \\ 111 & 0010 \\ 101 & 0001 \end{bmatrix} \tag{4.62}$$

Assume that an error vector $\{e_0, e_1, ..., e_6\}$ is being checked by forming the syndromes S_0, S_1, S_2, S_3 as (since $CH^T = 0$):

$$\begin{aligned} S_0 &= e_0 + e_1 + e_3 \\ S_1 &= e_1 + e_2 + e_4 \\ S_2 &= e_0 + e_1 + e_2 + e_5 \\ S_3 &= e_0 + e_2 + e_6 \end{aligned} \tag{4.63}$$

Further forming the parity-check sums as:

$$\begin{aligned} A_1 &= S_0 = \underline{e_0} + e_1 + e_3 \\ A_2 &= S_1 + S_2 = \underline{e_0} + e_4 + e_5 \\ A_3 &= S_3 = \underline{e_0} + e_2 + e_6 \end{aligned} \tag{4.64}$$

it is observed that all of the check sums A_1, A_2, A_3, check the error bit e_0, and no other error bit is checked by more than one check sum. Then a majority decision can be taken that $e_0 = 1$ if two or more of A_j's are non-zero. If $e_0 = 0$, and any other bit is in error, then only one of the A_j's will be non-zero. It is said that the check sums A_j's are orthogonal on the error bit e_0. A circulating SR-memory circuit along with a few logic circuits, as shown in Fig. 4.21, forms the hardware of the decoder.

Initially, the received code vector $W(X)$ is loaded into the SR's and check sums A_1, A_2 and A_3 are formed in the circuit. If e_0 is in error, then the majority-logic output is 1 and e_0 is corrected as it is shifted

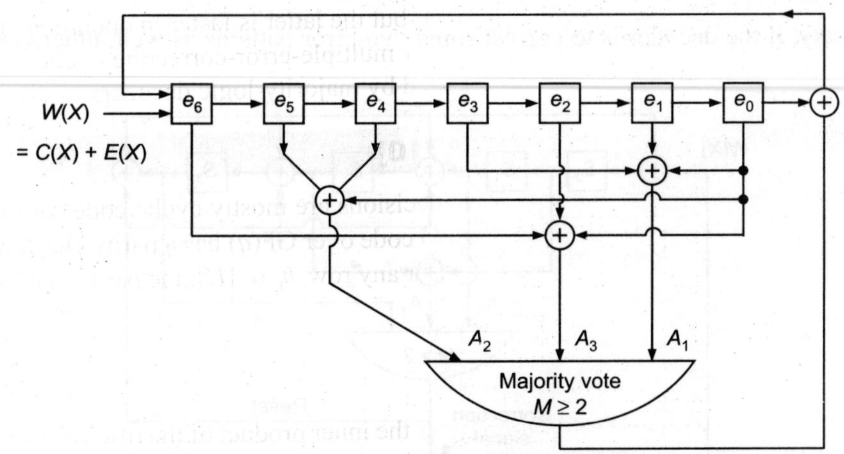

Figure 4.21: A majority-logic decoder for (7, 3) simplex code

out of the buffer. If e_0 is correct, then e_1 is checked after one shift of the SR content. Thus, all the bits are checked by successive shifts and the corrected $C(X)$ is reloaded in the buffer. It is possible to correct single errors only by using two of the check sums. But using three check sums, the decoder also corrects some double error patterns. The decoder will correct all single errors and detect all double error patterns, if the decision is made on the basis of:

(i) $A_1 = 1, A_2 = 1, A_3 = 1$, for single errors
(ii) One or more checks fail for double errors.

The above simplex (7, 3) code is also a cyclic code and its generator polynomial $g(X) = 1 + X^2 + X^3 + X^4$. Then one could generate the syndromes at the decoder by using a divider circuit, as in Fig. 4.22.

Based on the above $g(X)$, the parity check matrix is

$$[H] = \begin{bmatrix} 1000 & 110 \\ 0100 & 011 \\ 0010 & 111 \\ 0001 & 101 \end{bmatrix}$$

and the syndromes are calculated as

$$S_0 = e_0 + e_4 + e_5 \qquad S_1 = e_1 + e_5 + e_6$$

$$S_2 = e_2 + e_4 + e_5 + e_6 \qquad S_3 = e_3 + e_4 + e_6$$

The check sums are then formed as

$$\begin{aligned} A_1 &= S_1 = e_1 + e_5 + \underline{e_6} \\ A_2 &= S_0 + S_2 = e_0 + e_2 + e_6 \qquad (4.65) \\ A_3 &= S_3 = e_3 + e_4 + \underline{e_6} \end{aligned}$$

It is now seen that the check sums are orthogonal on e_6 and this can be corrected through the decision circuit, $M \geq 2$ shown in Fig. 4.22. Successive bits are checked for single error in the block. The feedback signal is necessary, if the decoder is to correct some two-error patterns as well, otherwise this may be omitted.

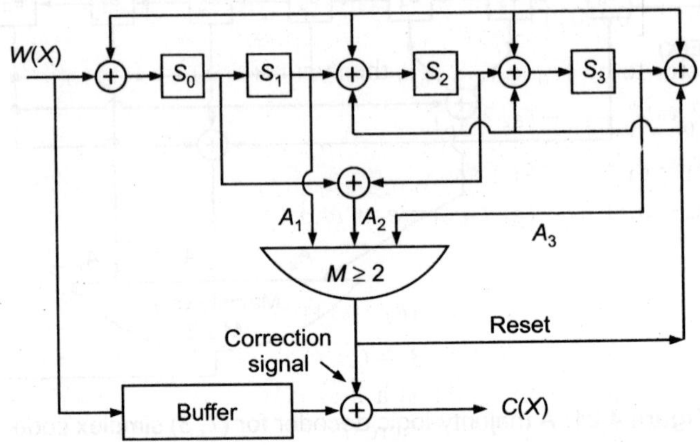

Figure 4.22: Another majority-logic decoder for (7, 3) simplex code

The results of the above example may be generalized in the following:

*(a) A set of parity-check sums $A_1, A_2 \ldots A_j$ is said to be orthogonal on the error bit e_l, if e_l is checked by all check sums A_j in the set, and no other error bit is checked by more than one check sum. The error bit e_l is now detected by a majority vote of the check sums A_j.

(b) Successive bits are tested at each cyclic shift of the syndrome and e_l located. $W(X)$ is corrected as the bits are shifted out of the buffer.

(c) If from the syndrome equations, the number of check sums formed $j = (d - 1)$, where d = minimum H.D. of the code, then the cyclic code is said to be completely orthogonalizable in one step. The errors can be correctly estimated if $t \leq j/2$.

(d) A set of j parity check sums $A_1, A_2, \ldots, A_j$ is said to be orthogonal on a group of error bits $\{e_{il}\}$, if and only if (i) every error bit in $\{e_{il}\}$ is checked by all check sums A_j and (ii) no other error bit is checked by more than one check sum.

A code is said to be L-step orthogonalizable (or L-step majority-logic decodable) if L steps of orthogonalization are required to make a decoding decision on an error bit e_l. If $j = (d - 1)$, then the code is completely L-step orthogonalizable; and the decoder requires L-levels of majority gates for decision making.

(e) The type of decoders, using the check sums based on the syndromes, as shown in Fig. 4.22, is known as type I decoder. The type of decoders, where the received word is directly used to form the check sums (through some buffer) as shown in Fig. 4.21, is known as type II decoder. Any of the decoders may be used in practice. It is evident from eqns. (4.64) and (4.67) that both the decoders may be organised from the H-matrix.

As an example of 2-step orthogonalizable code, consider the (7.4) Hamming code generated by $g(X) = 1 + X + X^3$. Its generator matrix has been given in eqn. (4.34) and its parity check matrix H is:

---

*The term 'orthogonal' as used here directly conflicts with its usage in the study of vector spaces, and as such, some authors [7] prefer the word 'concurring'. So it may be stated that 'A set of check sums, $A_1 \ldots A_j$ is concurring on the coordinate l, if e_l is checked by all check sums in the set. However, here the terms 'orthogonal' is used as is common in many other references.

$$[H] = \begin{bmatrix} 1001011 \\ 0101110 \\ 0010111 \end{bmatrix} \tag{4.66}$$

Assuming the error vector to be $\{e_0, e_1, ..., e_6\}$, the syndromes are:

$$\begin{aligned} S_0 &= e_0 + e_3 + (e_5 + e_6) \\ S_1 &= e_1 + e_3 + e_4 + e_5 \\ S_2 &= e_2 + e_4 + (e_5 + e_6) = e_2 + e_5 + (e_4 + e_6) \end{aligned} \tag{4.67}$$

and check sum

$$A_1 = S_0 + S_1 = e_0 + e_1 + (e_4 + e_6)$$

It is then seen that S_0 and S_2 are orthogonal on $B_1 = (e_5 + e_6)$; and A_1 and S_2 are orthogonal on $B_2 = (e_4 + e_6)$. Further, B_1 and B_2 are orthogonal on e_6. Thus a 2-step majority vote will locate the error on e_6. The corresponding decoder is shown in Fig. 4.23, where the 2nd level majority logic circuit gives the correction signal and the stored $W(X)$ is corrected as the bits are read out from the buffer. Correct decoding is achieved if $t \le j/2 = 1$ error. Comparing this decoder with the decoder of Fig. 4.18, it is seen that the majority logic gives a '1' signal when the syndrome state is {101}, which is the same for error detection in Fig. 4.18. Thus the basic principles of the two decoders are the same.

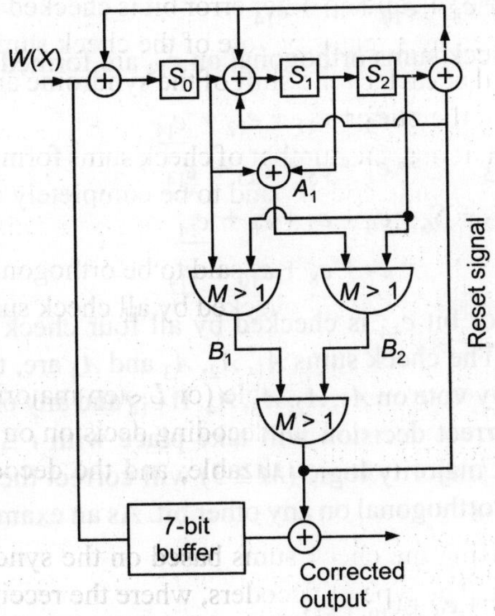

Figure 4.23: 2-step majority-logic decoder for (7, 4) Hamming code

4.5.1 Majority-Logic Decodable Codes

Two examples of majority-logic decoding of cyclic codes have been discussed above. In general, it has been shown that $(2^m - 1, 2^m - m - 1)$ Hamming codes, m any integer, are completely orthogonalizable in $(m - 1)$ steps but the Meggitt decoders discussed earlier are more efficient. Among the BCH codes, the completely orthogonalizable codes are: (a) (15, 7) code with $t \le 2$ is 1-step orthogonalizable, (b) the subclass of $(2^m - 1, m + 1)$ codes for $t \le (2^{m-2} - 1)$, $m \ge 3$ are 2-step orthogonalizable, (c) (31, 16) code with $t \leq 3$ is 3-step orthogonalizable. The Reed-Muller (RM) code is L-step decodable, and has been used in space probe communication systems; however, its H.D. is not very attractive in general, as compared to BCH codes. Other majority-logic decodable codes are: (a) Maximum length (simplex) codes, (b) Difference set codes, (c) Finite projective geometry codes (of which RM code is a subclass), (d) cyclic codes based on Affine permutation (these are l-step decodable) [7], and (e) a subclass of convolutional codes. A few of the more popular codes will be discussed below.

(15, 7) BCH Code

The parity check matrix of the (15, 7) 2-error correcting BCH code has been given in eqn. (4.44), with $g(X) = 1 + X^4 + X^6 + X^7 + X^8$. Assuming an error vector to be $\{e_0, e_1, ..., e_{14}\}$, the syndromes are calculated from $[eH^T]$ as:

$$
\begin{aligned}
S_0 &= e_0 \dots + e_8 + e_9 + e_{11} \\
S_1 &= e_1 \dots + e_9 + e_{10} + e_{12} \\
S_2 &= e_2 \dots + e_{10} + e_{11} + e_{13} \\
S_3 &= e_3 \dots + e_{11} + e_{12} + e_{14} \\
S_4 &= e_4 \dots + e_8 + e_9 + e_{11} + e_{12} + e_{13} \\
S_5 &= e_5 \dots + e_9 + e_{10} + e_{12} + e_{13} + e_{14} \\
S_6 &= e_6 \dots + e_8 + e_9 + e_{10} + e_{13} + e_{14} \\
S_7 &= e_7 \dots + e_8 + e_{10} + \dots + e_{14}
\end{aligned} \tag{4.68}
$$

From the syndromes, 4 parity check sums orthogonal on e_{14} are formed as

$$
\begin{aligned}
A_1 &= S_3 &&= e_3 + e_{11} + e_{12} + \underline{e_{14}} \\
A_2 &= S_1 + S_5 &&= e_1 + e_5 + e_{13} + \underline{e_{14}} \\
A_3 &= S_0 + S_2 + S_6 &&= e_0 + e_2 + e_6 + \underline{e_{14}} \\
A_4 &= S_7 &&= e_7 + e_8 + e_{10} + \underline{e_{14}}
\end{aligned} \tag{4.69}
$$

It is, therefore, seen that the error bit e_{14} is checked by all four check sums and no other error bit is checked by more than one sum. The check sums A_1, A_2, A_3 and A_4 are, then, orthogonal on e_{14} and this may be corrected through majority vote on A_1, A_2, A_3, A_4. If e_{14} and any other bit are in error, then at least three sums will be '1' and a correct decision will take place with $t \leq 2$. A circuit similar to that of Figs. 4.21 and 4.22, with 4-input majority logic ($M \geq 3$) will correct the code is cyclic, it is possible to form another set of 4 check sums orthogonal on any other bit. As an example, the check sums orthogonal on e_{12} are

$$
\begin{aligned}
A_1' &= S_1 = e_1 + e_9 + e_{10} + \underline{e_{12}} \\
A_2' &= S_3 = e_3 + e_{11} + \underline{e_{12}} + e_{14} \\
A_3' &= S_0 + S_4 = e_0 + e_4 + \underline{e_{12}} + e_{13} \\
A_4' &= S_6 + S_5 = e_5 + e_6 + e_8 + \underline{e_{12}}
\end{aligned} \tag{4.70}
$$

It is also evident that with $t \geq 3$, the circuit will fail, except for few special combination of error bits, say for example, $\{e\} = \{e_{11}, e_{12}, e_{14}\}$. This code can also be decoded by a Meggitt decoder, similar to that shown in Fig. 4.19.

(15, 5) BCH Code

The generating matrix for a (15, 5) triple-error correcting BCH code has been given in eqn. (4.43) as

$$g(X) = 1 + X + X^2 + X^4 + X^5 + X^8 + X^{10} \tag{4.43}$$

This code can be decoded by a 2-step majority-logic decoder. The parity check matrix of the code is

$$[H] = \begin{bmatrix} 100000000010101 \\ 010000000011111 \\ 001000000011010 \\ 000100000001101 \\ 000010000010011 \\ 000001000011100 \\ 000000100001110 \\ 000000010000111 \\ 000000001010110 \\ 000000000101011 \end{bmatrix} \tag{4.71}$$

Assuming an error vector $\{e_0, e_1, ..., e_{14}\}$, the syndromes are calculated as

$$\begin{aligned} S_0 &= e_0 + e_{10} + e_{12} + e_{14} \\ S_1 &= e_1 + e_{10} + e_{11} + e_{12} + e_{13} + e_{14} \\ S_2 &= e_2 + e_{10} + e_{11} + e_{13} \\ S_3 &= e_3 + e_{11} + e_{12} + e_{14} \\ S_4 &= e_4 + e_{10} + e_{13} + e_{14} \\ S_5 &= e_5 + e_{10} + e_{11} + e_{12} \\ S_6 &= e_6 + e_{11} + e_{12} + e_{13} \\ S_7 &= e_7 + e_{12} + e_{13} + e_{14} \\ S_8 &= e_8 + e_{10} + e_{12} + e_{13} \\ S_9 &= e_9 + e_{11} + e_{13} + e_{14} \end{aligned} \tag{4.72}$$

It is observed that it is not possible to form 6 check sums orthogonal on e_{14} from the given matrix, as required for $t \leq 3$. It is, therefore, necessary to have 2-step majority logic for the decoder. By manipulating the rows of eqn. (4.72), it is possible to form the following check sums in groups of six:

$$\left.\begin{aligned} A_1 &= S_0 + S_8 &&= e_0 + e_8 + (e_{13} + e_{14}) \\ A_2 &= S_1 + S_5 &&= e_1 + e_5 + (e_{13} + e_{14}) \\ A_3 &= S_3 + S_6 &&= e_3 + e_6 + (e_{13} + e_{14}) \\ A_4 &= S_4 &&= e_4 + e_{10} + (e_{13} + e_{14}) \\ A_5 &= S_7 &&= e_7 + e_{12} + (e_{13} + e_{14}) \\ A_6 &= S_9 &&= e_9 + e_{11} + (e_{13} + e_{14}) \end{aligned}\right\}$$

(Contd.)

$$\left.\begin{aligned}
A_7 &= S_0 &&= e_0 + e_{10} + (e_{12} + e_{14})\\
A_8 &= S_1 + S_2 &&= e_1 + e_2 + (e_{12} + e_{14})\\
A_9 &= S_3 &&= e_3 + e_{11} + (e_{12} + e_{14})\\
A_{10} &= S_4 + S_8 &&= e_4 + e_8 + (e_{12} + e_{14})\\
A_{11} &= S_6 + S_9 &&= e_6 + e_9 + (e_{12} + e_{14})\\
A_{12} &= S_7 &&= e_7 + e_{13} + (e_{12} + e_{14})
\end{aligned}\right\}$$

$$\left.\begin{aligned}
A_{13} &= S_0 + S_5 &&= e_0 + e_5 + (e_{11} + e_{14})\\
A_{14} &= S_1 + S_8 &&= e_1 + e_8 + (e_{11} + e_{14})\\
A_{15} &= S_2 + S_4 &&= e_2 + e_4 + (e_{11} + e_{14})\\
A_{16} &= S_3 &&= e_3 + e_{12} + (e_{11} + e_{14})\\
A_{17} &= S_6 + S_7 &&= e_6 + e_7 + (e_{11} + e_{14})\\
A_{18} &= S_9 &&= e_9 + e_{13} + (e_{11} + e_{14})
\end{aligned}\right\}$$

$$B_1 = S_1 + S_3 + S_4 = e_1 + e_3 + e_4 + e_{14}$$

$$B_2 = S_0 + S_2 + S_6 = e_0 + e_2 + e_6 + \underline{e_{14}}$$

$$B_3 = S_5 + S_8 + S_9 = e_6 + e_8 + e_9 + \underline{e_{14}}$$

It is now seen that the check sums $(A_1 \ldots A_6)$ are orthogonal on $(e_{13} + e_{14}) = B_4$; $(A_7 \ldots A_{12})$ are orthogonal on $(e_{12} + e_{14}) = B_5$; $(A_{13} \ldots A_{18})$ are orthogonal on $(e_{11} + e_{14}) = B_6$; and B_1, B_2, B_3 are orthogonal on e_{14}. By using the 1st step decisions on $(e_{11} + e_{14})$, $(e_{12} + e_{14})$ and $(e_{13} + e_{14})$ and then combining these decisions with the check sums B_1, B_2 and B_3, the final majority vote is taken on the check sums B_1, B_2, B_3, B_4, B_5 and B_6 all of which are orthogonal on e_{14}. The corresponding decoder circuit is shown in Fig. 4.24, as a type-I 2-step decoder; and a similar type-II decoder may also be used for the purpose. It is of course possible to have other sets of check sums to satisfy the requirements of the decoder. In general, it requires considerable ingenuity to form the proper check sums for L-step decoders.

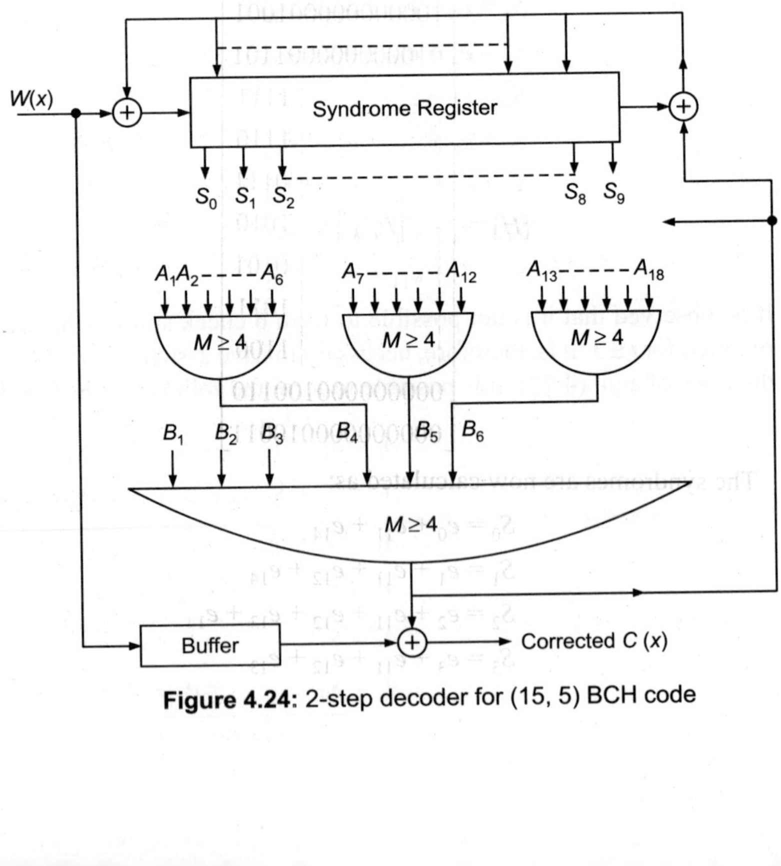

Figure 4.24: 2-step decoder for (15, 5) BCH code

4.5.2 Simplex (Maximum Length) Codes

It has been shown in Chapter 3, that binary orthogonal codes are generated from a Hadamard matrix using a recursive formula:

$$H_1 = \begin{bmatrix} H & H \\ H & -H \end{bmatrix}, \text{ where } H = \begin{bmatrix} 1 & 1 \\ 1 & -1 \end{bmatrix} \tag{4.73}$$

Any higher order orthogonal set may be generated by repeated use of eqn. (4.73). By deleting the last column which is all zero, it is possible to generate simplex codes with the following parameters:

$$\begin{aligned} &\text{Coder length } n = 2^m - 1 \\ &\text{No. of information bits } k = m \\ &\text{Minimum H.D. } d = 2^{m-1} \end{aligned} \tag{4.74}$$

Consider the top four rows of the (16 × 16) Hadamard matrix as shown below:

$$\begin{bmatrix} 1111111100000000 \\ 1111000011110000 \\ 1100110011001100 \\ 1010101010101010 \end{bmatrix} \rightarrow [G] = \begin{bmatrix} 111101011001000 \\ 011110101100100 \\ 001111010110010 \\ 111010110010001 \end{bmatrix} \tag{4.75}$$

By deleting the last column and rearranging the other columns, a generating matrix $[G]$ for a (15, 4) code with HD = 8 is obtained. It may also be checked that the $[G]$ belongs to a cyclic code with

$$g(X) = 1 + X + X^2 + X^3 + X^5 + X^7 + X^8 + X^{11} \tag{4.76}$$

The corresponding parity check matrix is:

$$[H] = \begin{bmatrix} 100000000001001 \\ 010000000001101 \\ \qquad [I_{n-k}] \qquad 1111 \\ \qquad\qquad\qquad 1110 \\ \qquad\qquad\qquad 0111 \\ \qquad\qquad\qquad 1010 \\ \qquad\qquad\qquad 0101 \\ \qquad\qquad\qquad 1011 \\ \qquad\qquad\qquad 1100 \\ 000000000100110 \\ 000000000010011 \end{bmatrix}$$

The syndromes are now calculated as:

$$\begin{aligned} S_0 &= e_0 + e_{11} + e_{14} \\ S_1 &= e_1 + e_{11} + e_{12} + e_{14} \\ S_2 &= e_2 + e_{11} + e_{12} + e_{13} + e_{14} \\ S_3 &= e_3 + e_{11} + e_{12} + e_{13} \end{aligned}$$

$$S_4 = e_4 + e_{12} + e_{13} + e_{14}$$
$$S_5 = e_5 + e_{11} + e_{13}$$
$$S_6 = e_6 + e_{12} + e_{14}$$
$$S_7 = e_7 + e_{11} + e_{13} + e_{14}$$
$$S_8 = e_8 + e_{11} + e_{12}$$
$$S_9 = e_9 + e_{12} + e_{13}$$
$$s_{10} = e_{10} = e_{13} + e_{14}$$

The parity check sums orthogonal on e_{14} are formed as:

$$\begin{aligned}
A_1 &= S_0 &&= e_0 + e_{11} + \underline{e_{14}} \\
A_2 &= S_1 + S_8 &&= e_1 + e_8 + \underline{e_{14}} \\
A_3 &= S_2 + S_3 &&= e_2 + e_3 + \underline{e_{14}} \\
A_4 &= S_4 + S_9 &&= e_4 + e_9 + \underline{e_{14}} \\
A_5 &= S_5 + S_7 &&= e_5 + e_7 + \underline{e_{14}} \\
A_6 &= S_6 &&= e_6 + e_{12} + \underline{e_{14}} \\
A_7 &= S_{10} &&= e_{10} + e_{13} + \underline{e_{14}}
\end{aligned} \tag{4.77}$$

The codes are now decodable with a 1-step majority-logic decoder, as shown in Fig. 4.25, which is a Type-I decoder. Other variations of the decoder are also possible. Since there are seven orthogonal check sums, the decoder corrects 3 errors in 15 bits of the code word, along with some more 4-error patterns as well. In general, the simplex codes can be generated with a generator polynomial

$$g(X) = (X^n + 1)/p(X), \tag{4.78}$$

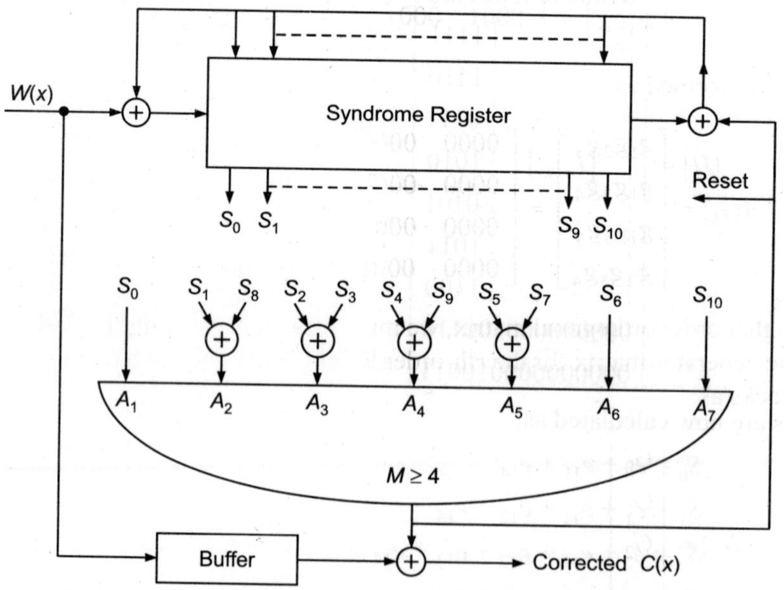

Figure 4.25: 1-step decoder for (15, 4) simplex code

where the parity polynomial $p(X)$ is any primitive polynomial of degree m. The weight of the code is 2^{m-1}. The (15, 4) code above uses $p(X) = 1 + X + X^4$, hence $g(X)$ is as given in eqn. (4.76). Higher order codes, e.g., (31, 5), (63, 6) etc., may be generated in a similar way and they are 1-step majority-logic decodable.

4.5.3 Reed-Muller (RM) Codes

Reed-Muller codes are a class of binary group codes, which are majority-logic decodable and have a wide range of rates and minimum distances. These codes are generated from the top five rows of a Hadamard matrix and the inner product of these rows. Consider the top rows of a (16 × 16) orthogonal matrix given as:

$$[G_0] = [11111111 \quad 1111 \quad 1111]$$

$$[G_1] = \begin{bmatrix} g_1 \\ g_2 \\ g_3 \\ g_4 \end{bmatrix} = \begin{bmatrix} 00000000 & 1111 & 1111 \\ 00001111 & 0000 & 1111 \\ 00110011 & 0011 & 0011 \\ 01010101 & 0101 & 0101 \end{bmatrix} \tag{4.79}$$

Then by using dot products g_1g_2, g_1g_3, g_1g_4, g_2g_3, g_2g_4 and g_3g_4, six more rows may be formed as:

$$[G_2] = \begin{bmatrix} g_1g_2 \\ g_1g_3 \\ g_1g_4 \\ g_2g_3 \\ g_2g_4 \\ g_3g_4 \end{bmatrix} = \begin{bmatrix} 0000 & 0000 & 0000 & 1111 \\ 0000 & 0000 & 0011 & 0011 \\ 0000 & 0000 & 0101 & 0101 \\ 0000 & 0011 & 0000 & 0011 \\ 0000 & 0101 & 0000 & 0101 \\ 0001 & 0001 & 0001 & 0001 \end{bmatrix} \tag{4.80}$$

Further rows may be formed as:

$$[G_3] = \begin{bmatrix} g_1g_2g_3 \\ g_1g_2g_4 \\ g_1g_3g_4 \\ g_2g_3g_4 \end{bmatrix} = \begin{bmatrix} 0000 & 0000 & 0000 & 0011 \\ 0000 & 0000 & 0000 & 0101 \\ 0000 & 0000 & 0001 & 0001 \\ 0000 & 0001 & 0000 & 0001 \end{bmatrix} \tag{4.81}$$

Starting with a higher order orthogonal matrix, we may form more rows similar to the ones above. In the light of above, the generator matrix for the rth-order Reed-Muller code of block length $n = 2^m$ is defined by an array of blocks as:

$$[G] = \begin{bmatrix} G_0 \\ G_1 \\ G_2 \\ \vdots \\ G_r \end{bmatrix} \tag{4.82}$$

where G_0 is a vector of length $n = 2^m$ having all one's; G_1 is an $m \times 2^m$ matrix; and $\{G_2 \ldots G_r\}$ are formed by taking dot products of the rows of G_1 choosing l rows at a time, $l = 2, 3, \ldots (m - 1)$. Since there are $\binom{m}{l}$ ways of choosing the rows of $G_2, G_3, \ldots, G_r$, then the total number of rows, which is equal to the number of message bits k, is given by:

$$k = 1 + \binom{m}{1} + \binom{m}{2} \ldots + \binom{m}{r}$$

$$\text{the number of check bits} = n - k = 1 + \binom{m}{1} + \binom{m}{2} \ldots + \binom{m}{m-r-1} \tag{4.83}$$

and the H.D. of the code $d = 2^{m-r}$ = minimum weight.

Thus by combining eqns. (4.79), (4.80) and (4.81), a third-order RM code has $m = 4$, $n = 16$, $r = 3$, and is generated by a (15×16) matrix given as:

$$[G] = \begin{bmatrix} G_0 \\ G_1 \\ G_2 \\ G_3 \end{bmatrix}$$

This gives a (16, 15) binary code with $d = 2$ (in fact, it is a simple parity-check code). But a (16, 11) binary code with $d = 4$ is generated by the matrix:

$$[G] = \begin{bmatrix} G_0 \\ G_1 \\ G_2 \end{bmatrix}; \quad r = 2 \tag{4.84}$$

This code is a (15, 11) Hamming code with an extra parity check, and corrects single errors and detects double-errors. It is possible to delete the first column, which has a single non-zero entry, and convert the code having $n = (2^m - 1)$ without changing its error-correcting capabilities.

The RM code is detected through a set of majority logic gates which are very easily implemented. Consider the (16, 11) RM code generated by using $[G]$ of eqn. (4.48) and the matrices of eqns. (4.79) and (4.80). The code vectors for a given message vector, $\{m_0, m_1, \ldots, m_{10}\}$ are generated as:

$$\begin{aligned}
C_0 &= m_0 \\
C_1 &= m_0 + m_4 \\
C_2 &= m_0 + m_3 \\
C_3 &= m_0 + m_3 + m_4 + m_{10} \\
C_4 &= m_0 + m_2 \\
C_5 &= m_0 + m_2 + m_4 + m_9 \\
C_6 &= m_0 + m_2 + m_3 + m_8 \\
C_7 &= m_0 + m_2 + m_3 + m_4 + m_8 + m_9 + m_{10} \\
C_8 &= m_0 + m_1
\end{aligned} \tag{4.48a}$$

$$
\begin{aligned}
C_9 &= m_0 + m_1 + m_4 + m_7 \\
C_{10} &= m_0 + m_1 + m_3 + m_6 \\
C_{11} &= m_0 + m_1 + m_3 + m_4 + m_6 + m_7 + m_{10} \\
C_{12} &= m_0 + m_1 + m_2 + m_5 \\
C_{13} &= m_0 + m_1 + m_2 + m_4 + m_5 + m_7 + m_9 \\
C_{14} &= m_0 + m_1 + m_2 + m_3 + m_5 + m_6 + m_8 \\
C_{15} &= m_0 + m_1 + m_2 + m_3 + m_4 + m_5 + m_6 + m_7 + m_8 + m_9 + m_{10}
\end{aligned}
$$

It is easily seen that each of the following four check sums give the message bit m_{10}

$$
\begin{aligned}
C_0 + C_1 + C_2 + C_3 &= A_1 \to m_{10} \\
C_4 + C_5 + C_6 + C_7 &= A_2 \to m_{10} \\
C_8 + C_9 + C_{10} + C_{11} &= A_3 \to m_{10} \\
C_{12} + C_{13} + C_{14} + C_{15} &= A_4 \to m_{10}
\end{aligned}
\tag{4.85}
$$

If now m_{10} is decided upon from the result of the majority-logic gate having A_1, A_2, A_3 and A_4 as inputs, then m_{10} will be correct even if there is an error in the code vectors. If there are two errors, there will be no majority vote and double-error pattern is detected. The above principle may now be used to detect correctly the other message bits (with a single error in the code vector) by forming other groups of check sums, i.e., four distinct check sums for each of the message bits $\{m_5, m_6, ..., m_9\}$. These check sums are:

$$
\begin{aligned}
&\left.\begin{aligned}
A_5 &= C_0 + C_1 + C_4 + C_5 &&= m_9 \\
A_6 &= C_2 + C_3 + C_6 + C_7 &&= m_9 \\
A_7 &= C_8 + C_9 + C_{12} + C_{13} &&= m_9 \\
A_8 &= C_{10} + C_{11} + C_{14} + C_{15} &&= m_9
\end{aligned}\right\} \\
&\left.\begin{aligned}
A_9 &= C_0 + C_2 + C_4 + C_5 &&= m_8 \\
A_{10} &= C_1 + C_3 + C_5 + C_7 &&= m_8 \\
A_{11} &= C_8 + C_{10} + C_{12} + C_{14} &&= m_8 \\
A_{12} &= C_9 + C_{11} + C_{13} + C_{15} &&= m_8
\end{aligned}\right\} \\
&\left.\begin{aligned}
A_{13} &= C_0 + C_1 + C_8 + C_9 &&= m_7 \\
A_{14} &= C_2 + C_3 + C_{10} + C_{11} &&= m_7 \\
A_{15} &= C_4 + C_5 + C_{12} + C_{13} &&= m_7 \\
A_{16} &= C_6 + C_7 + C_{14} + C_{15} &&= m_7
\end{aligned}\right\} \\
&\left.\begin{aligned}
A_{17} &= C_0 + C_2 + C_8 + C_{10} &&= m_6 \\
A_{18} &= C_1 + C_2 + C_9 + C_{11} &&= m_6 \\
A_{19} &= C_4 + C_6 + C_{12} + C_{14} &&= m_6 \\
A_{20} &= C_5 + C_7 + C_{13} + C_{15} &&= m_6
\end{aligned}\right\} \\
&\left.\begin{aligned}
A_{21} &= C_0 + C_4 + C_8 + C_{12} &&= m_5 \\
A_{22} &= C_1 + C_5 + C_{10} + C_{13} &&= m_5 \\
A_{23} &= C_2 + C_6 + C_{10} + C_{14} &&= m_5 \\
A_{24} &= C_3 + C_7 + C_{11} + C_{15} &&= m_5
\end{aligned}\right\}
\end{aligned}
\tag{4.86}
$$

Through the majority votes, $\{m_5, m_6, ..., m_9\}$are estimated by using above check sums. Then the contributions of these message bits to the code vectors are subtracted from $\{C_0, C_1, ..., C_{15}\}$ and the new code vectors are

$$\left.\begin{aligned} &C_0' = m_0, C_1' = m_0 + m_4, C_2' = m_0 + m_3, C_3' = m_0 + m_3 + m_4 \\ &C_4' = m_0 + m_2, C_5' = m_0 + m_2 + m_4, C_6' = m_0 + m_2 + m_3 \\ &C_7' = m_0 + m_2 + m_3 + m_4, C_8' = m_0 + m_1, C_9' = m_0 + m_1 + m_4 \\ &C_{10}' = m_0 + m_1 + m_3, C_{11}' = m_0 + m_1 + m_3 + m_4, C_{12}' = m_0 + m_1 + m_2 \\ &C_{13}' = m_0 + m_1 + m_2 + m_4, C_{14}' = m_0 + m_1 + m_2 + m_3, C_{15}' = m_0 + m_1 + m_2 + m_3 + m_4 \end{aligned}\right\} \quad (4.87)$$

Using the matrix G_1, one can now form eight new check sums for each of the message bits $\{m_1, m_2, m_3, m_4\}$. From these check sums, the values of $\{m_1, m_2, m_3, m_4\}$ can be estimated correctly (in fact it is only required to use 4 check sums for these estimates with single errors in the code vector). Now subtracting the contributions of $\{m_1, m_2, m_3, m_4\}$ from the vectors $\{C_0', C_1', ..., C_{15}'\}$, the resultant vectors $\{C_0'', C_1'', ..., C_{15}''\}$ are equal to m_0 only. A final majority vote gives the correct value of m_0. Thus a succession of majority votes, starting with the information bits in the rth section, completes the decoding process.

For a (16, 5) RM code, one uses $[G] = \begin{bmatrix} G_0 \\ G_1 \end{bmatrix}$, and the decoder has an eight input majority logic gate. Thus correct decisions are obtained with $t \le 3$. Although, the minimum distances of RM codes for a given n, are slightly less than those of comparable BCH codes, RM codes for moderate n are attractive because of the simple decoding process. For large n, however, RM codes are definitely inferior to BCH codes. Useful parameters of the RM codes for moderate n are given in Table 4.13 and these may be compared with the parameters of BCH codes given in Table 4.11.

Table 4.13: Parameters of RM codes
(L = number of orthogonalization steps)

n	15	15	31	31	31	63	63	63	63	127	127	127	127
k	11	5	26	16	6	57	42	22	7	99	64	29	8
t	1	3	1	3	7	1	3	7	15	3	7	15	31
L	1	1	1	3	2	1	2	2	2	5	4	3	2

4.6 CONVOLUTIONAL CODES [11]

So far, the error-correcting codes discussed have been block codes, where the data is grouped into k-bit words, and coded into n-bit words after adding $(n - k)$ parity check bits. Decoding is done word by word and there is no coupling between successive words. In contrast with this technique, a coder for convolutional codes (also called **R**ecurrent codes and is a subclass of Tree codes) transforms k bits of data into n-bit codes, but the coder output depends on the earlier blocks of data that have been processed by the coder; thus the coder memory couples the currently processed data with a few earlier data blocks. The size of the coder memory is indicated by the constraint length K; and the code is designated by (n, k, K) along with the code rate $R = k/n$. Thus, a convolutional code designated as (3, 2), $K = 3$, means that $k = 2$, $n = 3$, $R = 2/3$ and the memory length is 3. Again in contrast with block codes, k, n and K are generally small integers for convolutional codes.*

*Because of the constraint length K, a convolutional code is designated, by some authors, as: (nK, kK) code, i.e., a (3, 2, 3) code is also called as (9, 6) convolutional code [7].

Consider a simple convolutional encoder, shown in Fig. 4.26(a), where the parameters are: $R = 1/2$, $n = 2$, $k = 1$ and constraint length $K = 2$. The parity bit P is generated by the mod-2 sum of the SR-outputs as:

$$P = (X + 1)$$

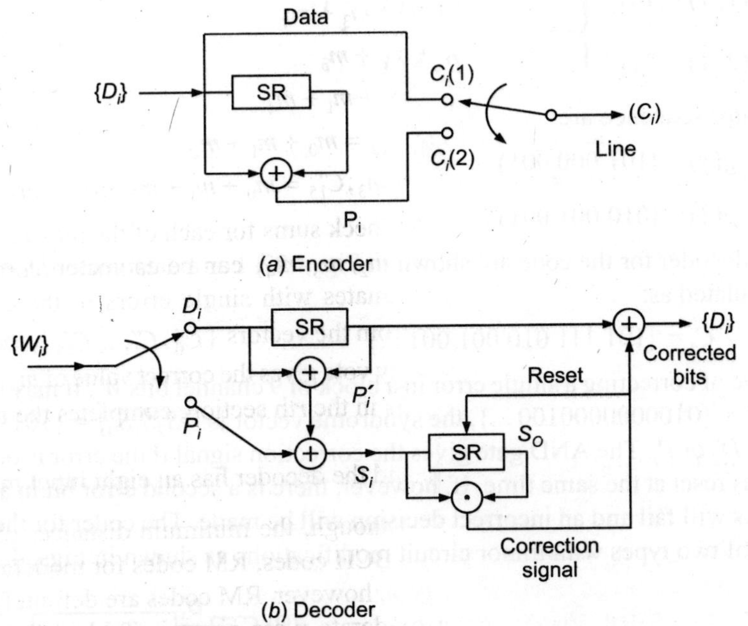

Figure 4.26: Encoder and decoder for $g(1, 1) = (11)$

Conventionally, the subgenerator for P is also written as: $g(1, 1) = (1\ 1)$, and the generator sequence $g(1) = (11\ 01)$, which gives the coder output for a data input: {100 ... 0}. For an arbitary data input D_i: {1011001100 ... 0}, the coder output C_i is: {11 01 11 10 01 00 11 10 01 00 ...}. This is a systematic code as the first bit of each pair in C_i is the data bit. Non-systematic convolutional codes will be discussed later.

The code corrects a single error in a span of 4 bits of the received data plus parity bits, $\{W_i\}$ (i.e., one error in 2 bits of information). Decoding is done through the calculation of syndrome bits by comparing the received parity bits P_i with the locally generated parity bits P_i' in the decoder as shown in Fig. (4.26)(b). It is easily checked that for an error vector e_i = {1000 ...}, with the error in data bits, the syndrome content will be: $\{S_1S_O\}$ = {11} and the AND gate will give a correction signal when the corresponding data bit will be shifted out of the coder, as seen in the circuit of Fig. 4.26(b). Simultaneously, the syndrome register will be reset to zero. If the error is in the parity bit, then the syndrome content will be either {01} or {10} and the AND gate will give zero output. Thus error patterns of the form {0100010} will be corrected by the decoder, but not the patterns of the form {0100100}, {0101000}, {011000}...*

---

*It will be seen that the ordering of the data and error vectors in this section is from leftmost bit onwards, i.e., the leftmost bit is first processed and transmitted. This convention is in contrast with the convention used for cyclic codes. The reason is that in convolutional codes, the starting point in coding and decoding is the first bit transmitted and the coder memory uses this information for subsequent coding/decoding. The polynomial form of the generators, however, is unchanged.

Example 4.23: Consider a (3, 2) rate 2/3 code with $K = 3$ (also called as (9, 6) code), given by the subgenerators:

$$\left.\begin{aligned} g(1, 1) &= 101 \\ g(2, 1) &= 011 \end{aligned}\right\} \rightarrow \quad \begin{aligned} P_1 &= (X^2 + 1) \quad (4.88) \\ P_2 &= (X + 1) \quad (4.89) \end{aligned}$$

and the two generator sequence are:

$$g(1) = (101\ 000\ 001)$$

$$g(2) = (010\ 001\ 001)$$

The coder and the decoder for the code are shown in Fig. 4.27. For a data vector $D_i = \{10\ 11\ 01\ ...\}$, the code vector is calculated as:

$$C_i = \{101\ 111\ 010\ 001\ 001\ ...\}$$

The coder is capable of correcting a single error in a block of 9 channel bits W_i. It may be checked that for an error vector $\{e_i\} = \{0100000000100\ ...\}$, the syndrome vector is: $\{S_2S_1S_0\} = \{101\}$ or $\{110\}$ or (001) for the error on D_1, D_2 or P_i. The AND gate gives the correction signal if the error is on D_1 or D_2, and the syndrome register is reset at the same time. If, however, there is a second error bit in a span of 9 channel bits, then the checks will fail and an incorrect decision will be made. The coder for the generation of the parity bits may be of two types with minor circuit modifications as shown in Figs. 4.27(a) and (b).

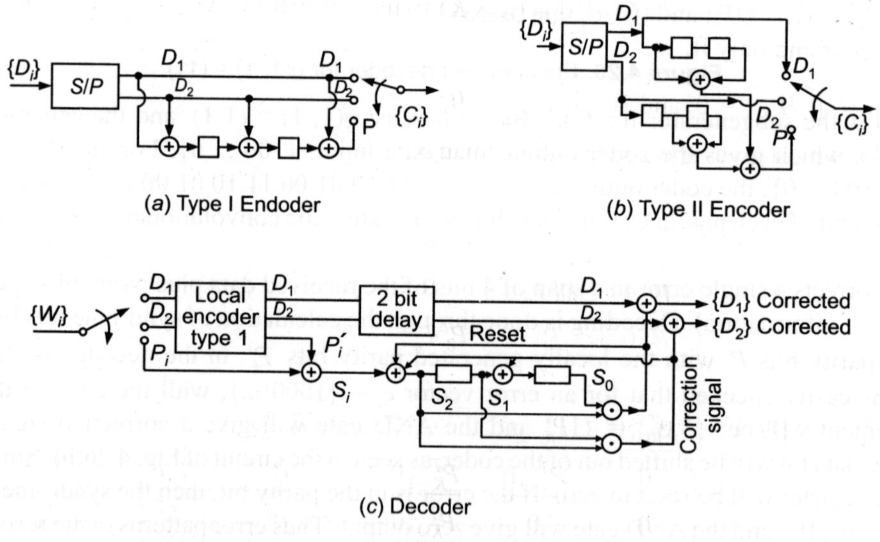

Figure 4.27: Endoder and decoder for $R = 2/3$ code

4.6.1 General Description

From the above examples, it is seen that for systematic convolutional codes, the coded output consists of Kk bits of data and $K(n - k)$ bits of parity check per block of nK bits of $\{C_i\}$. Moreover, the number of subgenerators is equal to k and the number of SR's required in the encoder is $(K - 1)$ in Type I circuits (in Type II circuits, the number is $(K - 1)$ per data bit). In conformity with Block codes, the generator matrix G_∞ for convolutional code may be written as:

$$[G_\infty] = \begin{bmatrix} \underline{I_k P_0} & \underline{OP_1} & \underline{OP_2} & \cdots & \underline{OP_{K-1}} & 0 & 0 & 0 & \cdots & \infty \\ 0 & \underline{I_k P_0} & \underline{OP_1} & \cdots & \underline{OP_{K-2}} & \underline{OP_{K-1}} & 0 & 0 & \cdots & \infty \\ \vdots & & & & & & & & & \\ 0 & 0 & 0 & \cdots & \underline{I_k P_0} & \underline{OP_1} & \underline{OP_2} & \underline{OP_3} & \cdots & \infty \\ \vdots & & & & & & & & & \\ \infty & \infty & \infty & \infty & \infty & \infty & \infty & \infty & \cdots & \infty \end{bmatrix} \quad (4.90)$$

where, $I_k = k \times k$ unity matrix, $0 = k \times k$ zero matrix, $P_i = k \times (n - k)$ parity matrix. For a given message vector $[m]$, the code vectors are:

$$[C] = [m]\cdot[G_\infty] \quad (4.91)$$

For a code block of length nK, $[G_\infty]$ may be written in a truncated form as:

$$[G(nK)] = \begin{bmatrix} G_0 & G_1 & G_2 & \cdots & G_{K-1} \\ 0 & G_0 & G_1 & \cdots & G_{K-2} \\ 0 & 0 & G_0 & \cdots & G_{K-3} \\ \vdots & & & & \vdots \\ 0 & 0 & 0 & & G_0 \end{bmatrix} \quad (4.92)$$

where $G_0 = I_k P_0$; $G_j = OP_j$; and $[G(nK)]$ is $(K \times K)$ triangular matrix. Again, as in Block codes, $[C]$ is a code sequence if and only if,

$$[C]\cdot\left[H_\infty^T\right] = 0; \text{ and } [G_\infty]\cdot\left[H_\infty^T\right] = 0; \quad (4.93)$$

where $[H_\infty]$ is the corresponding parity check matrix, written as:

$$[H_\infty] = \begin{bmatrix} \underline{P_0^T I_{n-k}} & 0 & 0 & & 0 & 0 & 0 & \cdots \\ \underline{P_1^T 0} & \underline{P_0^T I_{n-k}} & 0 & & 0 & 0 & 0 & \cdots \\ \underline{P_2^T 0} & \underline{P_1^T 0} & \underline{P_0^T I_{n-k}} & & 0 & 0 & 0 & \cdots \\ \vdots & \vdots & \vdots & & & 0 & 0 & \cdots \\ \underline{P_{K-1}^T 0} & \underline{P_{K-2}^T 0} & \cdots & \cdots & \underline{P_0^T I_{n-k}} & 0 & 0 & \cdots \\ 0 & \underline{P_{K-1}^T 0} & \underline{P_{K-2}^T 0} & \cdots & \underline{P_1^T 0} & \underline{P_0^T I_{n-k}} & 0 & \cdots \\ 0 & 0 & \underline{P_{K-1}^T 0} & \cdots & \underline{P_2^T 0} & \underline{P_1^T 0} & \underline{P_0^T I_{n-k}} & \cdots \\ \vdots & \vdots & \vdots & & \vdots & \vdots & \vdots & \\ 0 & 0 & 0 & 0 & 0 & 0 & 0 & \end{bmatrix} \quad (4.94)$$

Comparing eqn. (4.90) with eqn. (4.89) of Example 4.23, it is seen that the first row of $[G_\infty]$ is the generator sequences given by $\begin{bmatrix} g(1) \\ g(2) \end{bmatrix}$, which again is derived from the subgenerator polynomials of eqn. (4.88). The code vector for a given data input is obtained by using eqn. (4.91) as in Block codes. Since the received vector with errors is $[W] = [C] + [E]$, the syndrome vectors are calculated as:

$$[S] = [W]\cdot[H_\infty^T] = [C+E]\cdot[H_\infty^T]$$

$$= [E]\cdot[H_\infty^T], \text{ since } [C\cdot H^T] = 0 \tag{4.23a}$$

The syndrome contains the information regarding the error sequence and is obtained at the decoder by reencoding the received vector $\{W_i\}$ and adding the resultant P_i' bits to the received P_i bits, as shown in Figs, 4.26 and 4.27. Based on the above discussions, a generalised encoder structure for systematic codes is shown in Fig. 4.28, where parity bits are generated through SR delay units and mod-2 sums, as per the given sub-generator polynomials. The outputs of the subgenerators and data bits of the same block are multiplexed as $\{C_i\}$ through the parallel to serial converter. Type I encoder will have the modification as in Fig. 4.27(a). The complementary generalised decoder is shown in Fig. 4.29, where the auxiliary parity bits P_i' are generated through a local encoder and the syndrome register stores the syndrome bits $\{S_i\}$ given by $S_i = (P_i + P_i')$. The $[S]$ vectors are checked against the allowable test patterns precalculated for allowable error patterns in $\{D_i\}$, and a correction signal is sent to correct $\{D_1, D_2, ..., D_k\}$ whenever an allowable pattern is detected; at the same time resetting the S-register. The principle of the decoder is somewhat similar to that of the Meggitt decoder, and because of the feedback connection for syndrome resetting, the decoder is called a Feedback decoder.

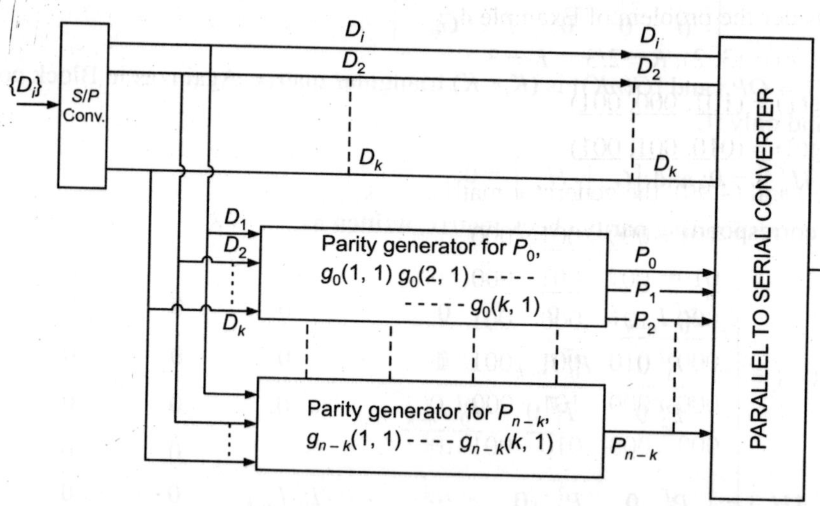

Figure 4.28: Encoder (Type II) for systematic codes, having k-bit data and $(n - k)$-bit parity

A major disadvantage in feedback decoding is the error propagation, whenever a wrong decision is made in the control logic for testing S-patterns. One of the methods to control propagation is to use a synchronization sequence of $k(K - 1)$ bits (usually an all-zero sequence) after a few data-blocks of, say, kL bits. This resynchronisation sequence, whenever detected at the decoder, resets the syndrome to all-zero, and thus avoids the error propagation in future data bits. The transmission rate, of course, decreases due to the addition of the resynchronization sequence and the new rate is:

$$R' = \frac{L}{L+K-1}R, \quad \text{where } R = k/n \tag{4.95}$$

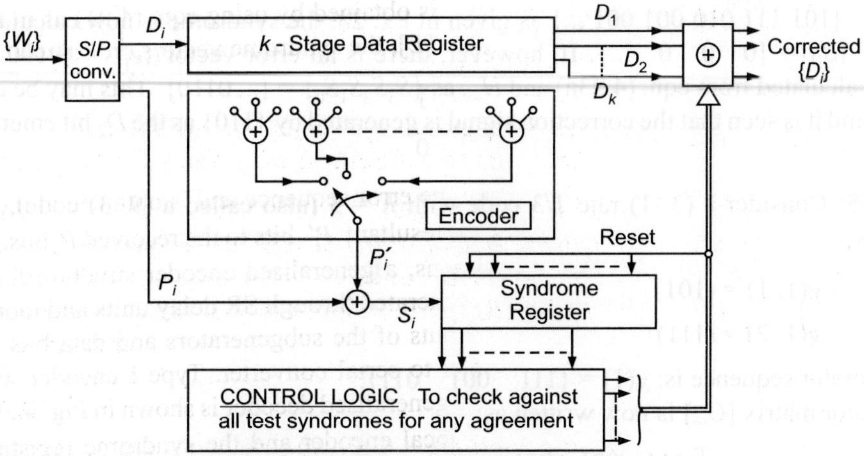

Figure 4.29: A generalised feedback decoder

Thus it is necessary to make $L \gg K$, to maintain $R' \simeq R$. In general, however, other types of decoders, e.g., majority-logic decoder, Viterbi decoder etc., are preferred for large-K codes, but for simple codes, feedback decoders are easily implemented.

Example 4.24: Consider the problem of Example 4.23. The coder specifications are:

$$(n, k) = (3, 2), R = 2/3, \quad K = 3$$

$$g(1) = (\underline{101}, \underline{000}, \underline{001})$$

$$g(2) = (\underline{010}, \underline{001}, \underline{001})$$

Based on eqns. (4.89) and (4.90), the generator matrix of the code may be written as:

$$[G_\infty] = \begin{bmatrix} \underline{101} & \underline{000} & \underline{001} & 000 & 000 & \cdots \\ \underline{010} & \underline{001} & \underline{001} & 000 & 000 & \cdots \\ 000 & 101 & 000 & 001 & 000 & \cdots \\ 000 & 010 & 001 & 001 & 000 & \cdots \\ 000 & 000 & 101 & 000 & 001 & \cdots \\ 000 & 000 & 010 & 001 & 001 & \cdots \\ & & & & \downarrow & \\ & & & & \infty & \end{bmatrix}$$

and the corresponding $[H_\infty]$ is:

$$[H_\infty] = \begin{bmatrix} 101 & 000 & 000 & \cdots & \cdots \\ 010 & 101 & 000 & \cdots & \cdots \\ 110 & 010 & 101 & 000 & \cdots \\ 000 & 110 & 010 & 101 & 000 \\ 000 & 000 & 110 & 010 & 101 \\ \downarrow & & & & \downarrow \\ \infty & & & & \infty \end{bmatrix}$$

Taking $\{C_i\} = \{101\ 111\ 010\ 001\ 001\ ...\}$ as given in Ex. 23, the syndrome is now calculated using the above H_∞ as: $\{S_i\} = \{0\quad 0\quad 0\quad ...\}$. If, however, there is an error vector $\{e_i\} = \{0100\ ...\}$, then the syndrome is calculated from eqn. (4.23a) and H_∞, as $\{S_3S_2S_1S_0\} = \{...\ 0110\}$. This may be checked from Fig. 4.27(b), and it is seen that the correction signal is generated by $\{110\}$ as the D_2 bit emerges out of the buffer.

Example 4.25: Consider a (3, 1) rate 1/3 code with $K = 3$ (also called a (9, 3) code), given by the subgenerators:

$$g(1, 1) = (101)$$
$$g(1, 2) = (111)$$

Then the generator sequence is: $g(1) = \{111\quad 001\quad 011\}$.

The generator matrix $[G_\infty]$ is now written as:

$$[G_\infty] = \begin{bmatrix} \underline{111} & \underline{001} & \underline{011} & \cdots & \cdots & \cdots & \cdots \\ \cdots & 111 & 001 & 011 & \cdots & \cdots & \cdots \\ \cdots & \cdots & 111 & 001 & 011 & \cdots & \cdots \\ \cdots & \cdots & \cdots & 111 & 001 & 011 & \cdots \\ \cdots & \cdots & \cdots & \cdots & \vdots & \vdots & \vdots \end{bmatrix}$$

The corresponding parity check matrix is:

$$[H_\infty] = \begin{bmatrix} 110 & \cdots & \cdots & \cdots \\ 101 & \cdots & \cdots & \cdots \\ 000 & 110 & \cdots & \cdots \\ 100 & 101 & \cdots & \cdots \\ 100 & 000 & 110 & \cdots \\ 100 & 100 & 101 & \cdots \\ \cdots & 100 & 000 & 110 \\ \cdots & 100 & 100 & 101 \end{bmatrix}$$

The syndrome bits can be calculated now using the $[H_\infty^T]$, and it should be noted that there will be two sets of syndrome bits, requiring two syndrome registers for further processing of $\{S_i\}$ bits. The product of $\{W_i\}$ with alternate coloumns of H_∞^T corresponds to the two syndromes $S(1)$ and $S(2)$. The error correction is obtained through a majority-logic decoder and the code corrects 2 errors in a block of 9 channel bits.

4.6.2 Syndrome Decoding: Wyner-Ash Codes [12]

One of the examples of syndrome/feedback decoding is the single-error correcting Wyner-Ash code, which is somewhat similar to Hamming codes. The code is specified by:

For any $m > 0$, $n = 2^m$; $k = 2^m - 1$; $K = (m + 1)$ and

$$R = \frac{2^m - 1}{2^m} \tag{4.96}$$

The $(2^m - 1)$ subgenerators for the code are given by:

$$g(i, 1) = (1, i_0, i_1, i_2, ..., i_{m-1})$$

where $i = 1, 2, 3, ..., 2^m - 1$, and $(i_0, i_1, ..., i_{m-1})$ is the binary representation of i. There is only one parity bit for each code, and it corrects one single error in Kn channel bits.

As for example, Rate (3/4), Wyner-Ash code is specified by:

$$m = 2, n = 4, k = 3 \text{ and } K = 3;$$

and the three subgenerators are:

$$\begin{aligned} g(1, 1) &= (101) \\ g(2, 1) &= (110) \\ g(3, 1) &= (111) \end{aligned} \tag{4.97}$$

and

$$\begin{aligned} g(1) &= (1001 \quad 0000 \quad 0001) \\ g(2) &= (0101 \quad 0001 \quad 0000) \\ g(3) &= (0011 \quad 0001 \quad 0001) \end{aligned}$$

giving

$$[G_\infty] = \begin{bmatrix} 1001 & 0000 & 0001 & \cdots & \cdots \\ 0101 & 0001 & 0000 & \cdots & \cdots \\ 0011 & 0001 & 0001 & \cdots & \cdots \\ \cdots & 1001 & 0000 & 0001 & \cdots \\ \cdots & 0101 & 0001 & 0000 & \cdots \\ \cdots & 0011 & 0001 & 0001 & \cdots \\ \cdots & \cdots & 1001 & 0000 & 0001 \\ \cdots & \cdots & 0101 & 0001 & 0000 \\ \cdots & \cdots & 0011 & 0001 & 0001 \end{bmatrix} \tag{4.98}$$

The corresponding $[H_\infty]$ may be written as

$$[H_\infty] = \begin{bmatrix} 1111 & \cdots & \cdots & \cdots & \cdots \\ 0110 & 1111 & \cdots & \cdots & \cdots \\ 1010 & 0110 & 1111 & \cdots & \cdots \\ \cdots & 1010 & 0110 & 1111 & \cdots \\ \cdots & \cdots & 1010 & 0110 & \cdots \end{bmatrix} \tag{4.99}$$

On calculating syndrome bits $\{... S_2S_1S_0\}$ by using $[H_\infty]$ for error patterns {1000 ...}, {0100 ...}, (00100 ...) and {000100 ...}, the results are

$$\{S_2S_1S_0\} = \{101\}; (011); (111); (001)$$

for the above error patterns respectivley. The syndrome, therefore, recognises the error positions 10, 01, 11 as 1st, 2nd and 3rd data bits of the block, while $S_0 = 1$ for all indicates an error in 0th block of $\{W_i\}$. The syndrome (001) indicates that the check bit is in error. The code corrects a single error in a span of 12 channel bits.

Based on above discussions, the coder and decoder for the (4, 3) Wyner-Ash code is shown in Fig. 4.30, where the syndrome bits are formed by using an auxiliary coder and $P_i \oplus P_1'$ circuit. If any bit in the 0th block is in error, then $S_0 = 1$ gives a start signal to the counter in Fig. 4.30(b), and its output is matched with the next two syndrome bits (S_2S_1) to give the error location. The correction signal is then generated through the NOR and AND gates, and the serial data bits are corrected as required; at the same time the syndrome is reset to all-zero. An alternative Address Decode logic circuit is shown in Fig. 4.30(c), where the syndrome outputs, which are directly related to the error patterns, control three AND gates to generate the individual correction signals. The data bits D_1, D_2, D_3 are corrected in parallel. This concept can be extended to higher order codes with $m = 3, 4, ...,$ etc. An example of (8, 7) code is given below.

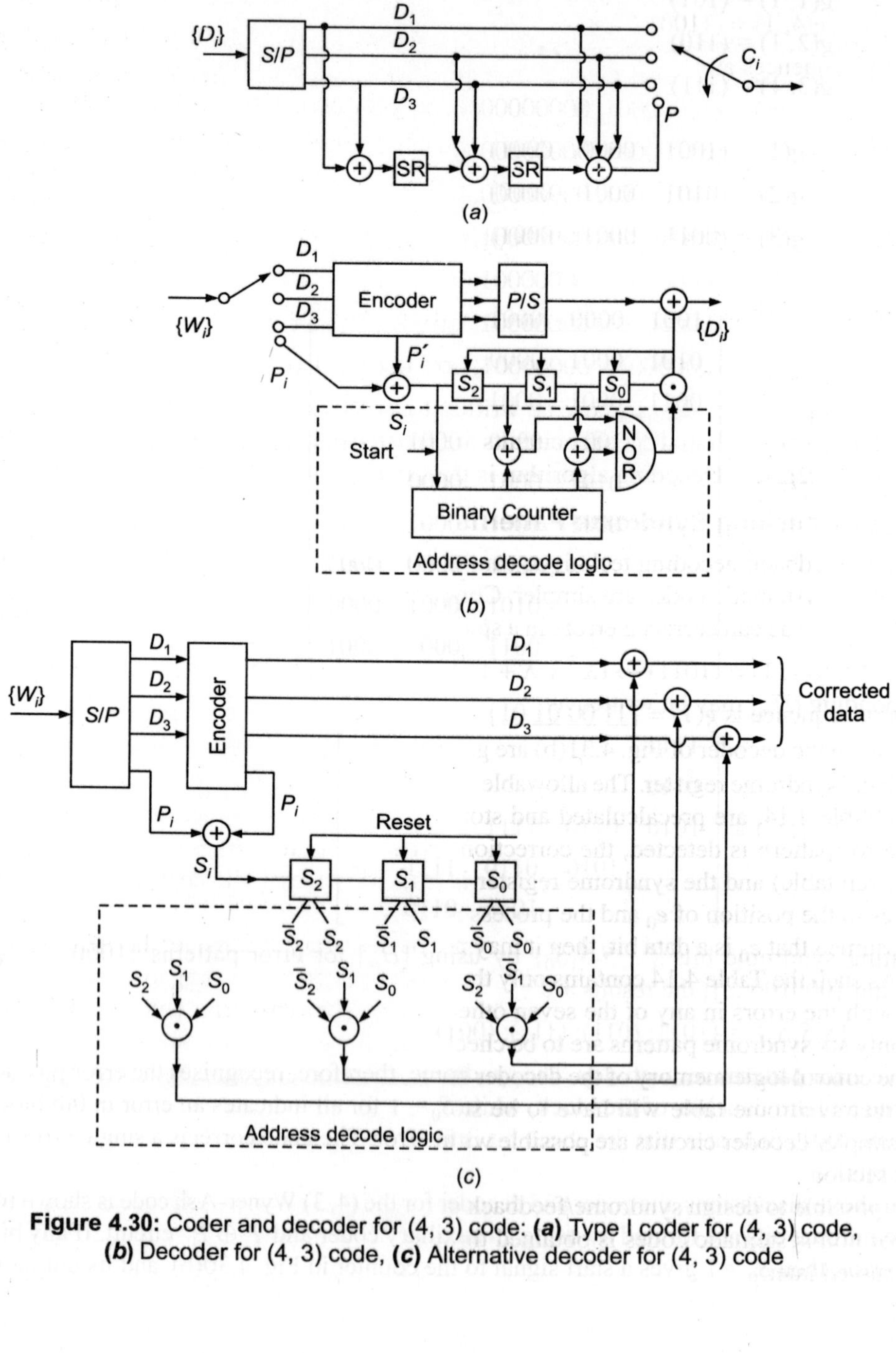

Figure 4.30: Coder and decoder for (4, 3) code: **(*a*)** Type I coder for (4, 3) code, **(*b*)** Decoder for (4, 3) code, **(*c*)** Alternative decoder for (4, 3) code

Example 4.26: Consider the Wyner-Ash code with $m = 3, n = 8, k = 7$ and $K = 4$. This (8, 7) code corrects one error in a span of $nK = 32$ channel bits. The subgenerators for the code are given by

$$\begin{aligned} &g(1, 1) = (1001);\ g(5, 1) = (1101)\\ &g(2, 1) = (1010);\ g(6, 1) = (1110)\\ &g(3, 1) = (1011);\ g(7, 1) = (1111)\\ &g(4, 1) = (1100); \end{aligned} \tag{4.100}$$

The generator sequences are:

$$g(1) = (10000001, 00000000, 00000000, 00000001)$$
$$g(2) = (01000001, 00000000, 00000001, 00000000)$$
$$g(3) = (00100001, 00000000, 00000001, 00000001)$$
$$g(4) = (00010001, 00000001, 00000000, 00000000)$$
$$g(5) = (00001001, 00000001, 00000000, 00000001)$$
$$g(6) = (00000101, 00000001, 00000001, 00000000)$$
$$g(7) = (00000011, 00000001, 00000001, 00000001)$$

The corresponding G_∞ and H_∞ can now be written by comparing the above sequences and the eqns. (4.98) and (4.99). The coder and decoder circuits are now extended with 3-SR's for the coder and 4-SR's for the syndrome register. Decoding algorithm is the same as for (4, 3) code.

Decoding by Checking Syndrome Patterns

The syndrome/feedback decoding technique can be extended to other convolutional codes as well, but the decoders for systematic codes are simpler. Consider a rate 1/2, (2, 1), code with $K = 4$ (also called a (8, 4) code). This code can correct 2 errors in a span of 8 bits in $\{W_i\}$, and has the subgenerator given by:

$$g(1, 1) = (1011) \rightarrow (X^3 + X + 1)$$

The generator sequence is $g(1) = \{\underline{11}\ \underline{00}\ \underline{01}\ \underline{01}\}$. The encoder for the code is given in Fig. 4.31(a). The syndrome bits in the decoder of Fig. 4.31(b) are generated in the usual way by comparing P_i and P_i', and are stored in the syndrome register. The allowable error patterns and the corresponding syndrome patterns, as given in Table 4.14, are precalculated and stored in the memory of the control logic unit. Once an allowable error pattern is detected, the correction signal is sent to correct the corresponding data bits (e_0 in the given table) and the syndrome register is reset. After reset, the next data bit error (i.e., e_2 the table) moves in the position of e_0 and the process is repeated.

If it is assumed that e_0 is a data bit, then it may not be necessary to consider the errors in e_1 which is a parity bit. As such the Table 4.14 contains only the syndromes corresponding to errors in the 1st data bit (e_0) along with the errors in any of the seven other bits, as allowable error patterns. It is seen from the table that only six syndrome patterns are to be checked for generating the correction signal and these are stored in the control logic memory of the decoder for comparison purposes. For correcting the parity bits on e_1, a larger syndrome table will have to be stored, and this will detect many more errors as well. However, simpler decoder circuits are possible with majority-logic decoding and this will be disscussed in the next section.

It is also possible to design syndrome/feedback decoders for non-systematic codes; but more efficient decoding for non-systematic codes is obtained through Viterbi/Sequential decoding techniques, which will be discussed later.

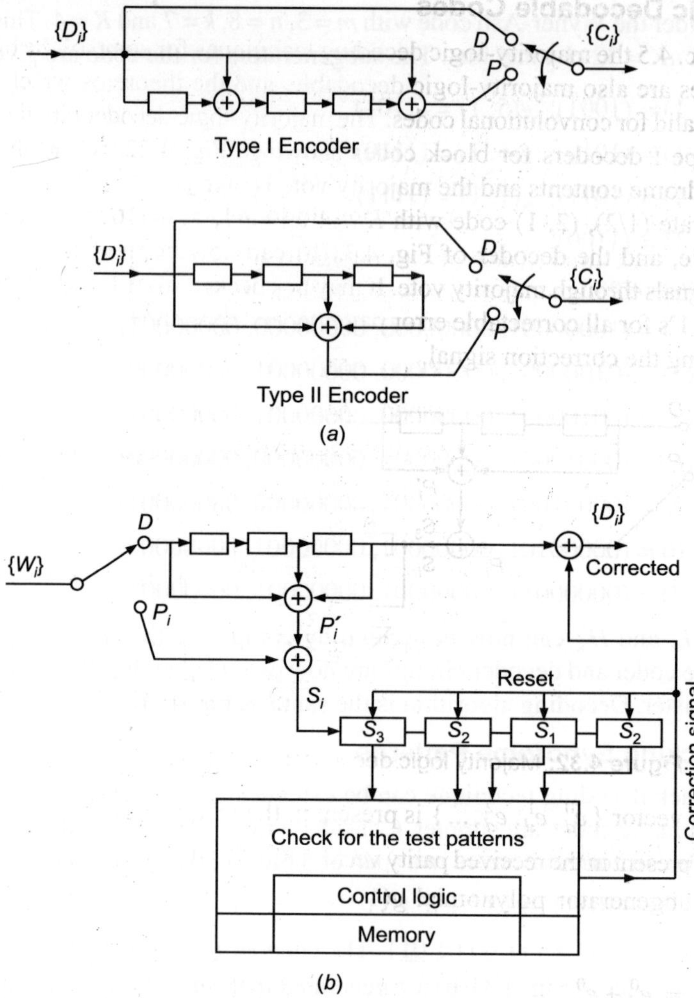

Figure 4.31: Encoder and decoder for (2, 1) code: ***(a)*** Encoder ***(b)*** Syndrome decoder

Table 4.14: (2, 1) Code with $K = 4$

Error pattern ($e_0, e_1, ..., e_7$)	Syndrome pattern (S_0, S_1, S_2, S_3)
10000000	1 0 1 1
11000000	0 0 1 1
10100000	1 1 1 0
10010000	1 1 1 1
10001000	1 0 0 1
10000100	1 0 0 1
10000010	1 0 1 0
10000001	1 0 1 0

4.6.3 Majority-Logic Decodable Codes

We have discussed in Sec. 4.5 the majority-logic decoding techniques for some of the block codes. Some of the convolutioal codes are also majority-logic decodable, and the theorems which were established for block codes are also valid for convolutional codes. The majority-logic decoder circuits for convolutional codes are similar to Type I decoders for block codes shown in Fig. 4.22, where the check sums are calculated from the syndrome contents and the majority vote is taken.

As an example, the rate (1/2), (2, 1) code with $K = 4$ and $g(1, 1) = (1011)$, as discussed above, is majority-logic decodable, and the decoder of Fig. 4.31(b) can now be modified, as in Fig. 4.32, for generating correction signals through majority vote. It may be checked from Table 4.14 that the syndrome $(S_0S_2S_3)$ has at least two 1's for all correctable error patterns, and thus, the majority vote will identify the e_0-error bit, for generating the correction signal.

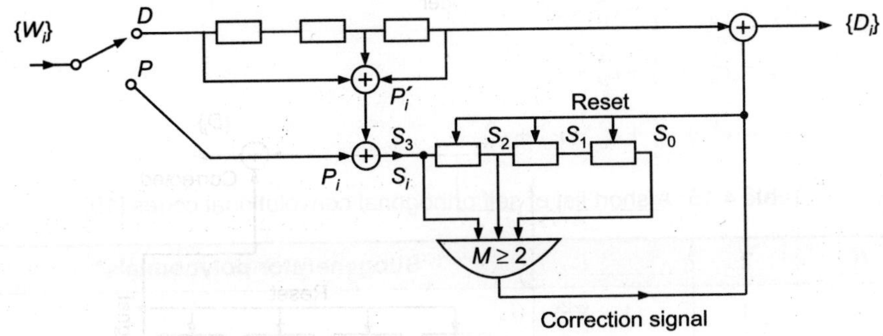

Figure 4.32: Majority logic decoder for (2, 1), $K = 4$, code

Assume that an error vector $\{e_d^0, e_d^1, e_d^2, ...\}$ is present in the received data vector $\{D'_i\}$ and an error vector $\{e_p^0, e_p^1, e_p^2 ...\}$ is present in the received parity vector $\{P''_i\}$ where $\{W_i\} = \{D''_1P_1, D'_2P''_2 ... D'_l P''_l ...\}$. Then from the given subgenerator polynomial $g(1, 1) = (X^3 + X + 1)$, the syndrome equations are written as:

$$\begin{aligned} S_0 &= e_d^0 + e_p^0 \\ S_1 &= e_d^1 + e_p^1 \\ S_2 &= e_d^0 + e_d^2 + e_p^2 \\ S_3 &= e_d^0 + e_d^1 + e_d^3 + e_p^3 \end{aligned} \tag{4.101}$$

since $[CH_\infty^T] = 0$. It is now seen that S_0, S_2, S_3 check on e_d^0 and no other error bit appears in more than one equation. Thus the error bit e_d^0 can be estimated by majority vote, as has been explained earlier from the syndrome pattern of Table 4.14. The code is now known as a Self-orthogonal code, and is decoded by the circuit of Fig. 4.32. Some of the majority-logic decodable convolutional codes are self-orthogonal and others are L-step orthogonalizable, as discussed in Sec. 4.5. As in the case of block codes, the number of errors correctable in a convolutional code is $j/2$, when j parity check sums (= sum of syndrome equations) are orthogonal on an error bit e_d^1. A short list of self-orthogonal codes obtained through computer search is given in Table 4.15 [10]. A few more examples are discussed below.

Example 4.27: Consider rate (1/2), (2, 1) code with $K = 7$ and subgenerator given by:

$$g(1, 1) = (1100101) \rightarrow (X^6 + X^5 + X^2 + 1)$$

Similar to eqn. (4.101), the syndrome equations are formed as:

$$
\begin{aligned}
S_0 &= e_d^0 + e_p^0 \\
S_1 &= e_d^0 + e_d^1 + e_p^1 \\
S_2 &= e_d^1 + e_d^2 + e_p \\
S_3 &= e_d^2 + e_d^3 + e_p^3 \\
S_4 &= e_d^0 + e_d^3 + e_d^4 + e_p^3 \\
S_5 &= e_d^1 + e_d^4 + e_d^5 + e_p^5 \\
S_6 &= e_d^0 + e_d^2 + e_d^5 + e_d^6 + e_p^6
\end{aligned}
\qquad (4.102)
$$

Table 4.15: A short list of self orthogonal convolutional codes [10]

(n, k)	*K*	*t*	Subgenerator polynomials*
(2, 1)	3	1	$(1 + X^2)$
"	7	2	$(1 + X^3 + X^5 + X^6)$
"	18	3	$(1 + X^2 + X^7 + X^{13} + X^{16} + X^{17})$
"	36	4	$(1 + X^7 + X^{10} + X^{16} + X^{18} + X^{30} + X^{31} + X^{35})$
"	56	5	$(1 + X^2 + X^{14} + X^{21} + X^{29} + X^{32} + X^{45} + X^{49} + X^{54} + X^{55})$
(3, 2)	3	1	$(1 + X)$; $(1 + X^2)$
"	14	2	$(1 + X^8 + X^9 + X^{12})$; $(1 + X^6 + X^{11} + X^{13})$
"	41	3	$(1 + X^3 + X^{15} + X^{28} + X^{35} + X^{36})$; $(1 + X^2 + X^6 + X^{24} + X^{29} + X^{40})$
(3, 1)	–	–	Subgenerators $g(1, 1)$ and $g(1, 2)$ are the same as for (3, 2) codes; but *t* is now larger
(4, 3)	4	1	$(1 + X)$; $(1 + X^2)$; $(1 + X^3)$
"	20	2	$(1 + X^3 + X^{15} + X^{19})$; $(1 + X^8 + X^{17} + X^{18})$; $(1 + X^6 + X^{11} + X^{13})$
(4, 1)	–	–	Subgenerators same as for (4, 3) codes
(5, 4)	5	1	$(1 + X)$ $(1 + X^2)$; $(1 + X^3)$; $(1 + X^4)$

It is now seen that S_0, S_1, S_4, S_6 are orthogonal on e_d^0, and no other error bit appears in more than one equation. The code is, therefore, self-orthogonal and corrects 2 errors in a span of 14 W_i-bits. The coder and decoder for the code are similar to those of Figs. 4.31(a) and 4.32, but requires 6 SR's in the coder and another 6 SR's in the syndrome register. It may be further noted that syndrome taps used for majority

*As in Ex. 27, the equivalent subgenerator in binary form is written as:
For $R = 1/2$ code with $K = 18$, $t \leq 3$

$$g(1, 1) = (110010000010000101).$$

The compliments of $f_n(X)$ are also equally valid subgenerator polynomials, where $f_n(X) = X^n.\ f_n(X^{-1})$. As for example, $X^6(1 + X^{-2} + X^{-5} + X^{-6}) = (X^6 + X^4 + X + 1)$ is an equally good subgenerator polynomial.

vote in Fig. 4.32 and for this decoder, appear to form a matched-filter for the coder (since both coder and decoder may be considered to be digital filters) and the syndrome bits are processed through a matched-filter to give the optimum output.

Example 4.28: Consider a rate (2/3), (3, 2) code with $K = 14$. The subgenerator polynomials for the code are given in Table 4.15. It is also possible to use the complement of the subgenerators for generating the (3, 2) code. Assume that:

$$g(1, 1) = (01001100000001) \qquad g(2, 1) = (10100001000001) \tag{4.103}$$

Coder for this is shown in Fig. 4.33(a) and the decoder in Fig. 4.33(b). Using eqn. (4.103), the syndrome equations are formed on the basis that the error vector in $\{D_1\}$ is:

$$\{ e_d^0(1), e_d^1(1), e_d^2(1)... \}$$

and in (D_2):

$$\{ e_d^0(2), e_d^1(2), e_d^2(2)... \}.$$

The local parity bit $P_i' = P'(1) \oplus P'(2)$ and is compared with the received parity P_i'' to form the syndrome bits $\{S_0, S_1, ..., S_{13}\}$. The syndrome equations are:

Figure 4.33: Encoder and decoder for (3, 2), $K = 14$, code **(a)** Encoder for (3, 2), $K = 14$, code, **(b)** Majority Logic Decoder

$$S_0 = e_d^0(2) + e_p^0$$

$$S_1 = e_d^0(1) + e_d^1(2) + e_p^1$$

$$S_2 = e_d^1(1) + e_d^0(2) + e_d^2(2) + e_p^2$$

$$S_3 = e_d^2(1) + e_d^1(2) + e_d^3(2) + e_p^3$$
$$S_4 = e_d^0(1) + e_d^3(1) + e_d^2(2) + e_d^4(2) + e_p^4$$
$$S_5 = e_d^0(1) + e_d^1(1) + e_d^4(1) + e_d^3(2) + e_d^5(2) + e_p^5$$
$$S_6 = e_d^1(1) + e_d^2(1) + e_d^5(1) + e_d^4(2) + e_d^6(2) + e_p^6$$
$$S_7 = e_d^2(1) + e_d^3(1) + e_d^6(1) + e_d^0(2) + e_d^5(2) + e_d^7(2) + e_p^7$$
$$S_8 = e_d^3(1) + e_d^4(1) + e_d^7(1) + e_d^1(2) + e_d^6(2) + e_d^8(2) + e_p^8$$
$$S_9 = e_d^4(1) + e_d^5(1) + e_d^8(1) + e_d^2(2) + e_d^7(2) + e_d^9(2) + e_p^9$$
$$S_{10} = e_d^5(1) + e_d^6(1) + e_d^9(1) + e_d^3(2) + e_d^8(2) + e_d^{10}(2) + e_p^{10}$$
$$S_{11} = e_d^6(1) + e_d^7(1) + e_d^{10}(1) + e_d^4(2) + e_d^9(2) + e_d^{11}(2) + e_p^{11}$$
$$S_{12} = e_d^7(1) + e_d^8(1) + e_d^{11}(1) + e_d^5(2) + e_d^{10}(2) + e_d^{12}(2) + e_p^{12}$$
$$S_{13} = e_d^0(1) + e_d^8(1) + e_d^9(1) + e_d^{12}(1) + e_d^0(2) + e_d^6(2) + e_d^{11}(2) + e_d^{13}(2) + e_p^{13}.$$

It is now seen that S_1, S_4, S_5, S_{13} are orthogonal on $e_d^0(1)$, and S_0, S_2, S_7, S_{13} are orthogonal $e_d^0(2)$. Thus the code is self-orthogonal and corrects 4 errors in a span of 42 W_i-bits. As in earlier decoder circuits for self-orthogonal codes, the present decoder also behaves as a matched-filter as seen from the tap connections of the syndrome register. Thus for self-orthogonal codes, the decoder may be drawn from the inspection of the coder-generator polynomial.

For the rate $(n, 1)$ convolutional codes, the subgenerators are the same as for the dual rate $(n, n-1)$ convolutional codes, and the $(n, 1)$ code is self-orthogonal if the dual code is self-orthogonal if the dual code is self-orthogonal. As such the subgenerators for a (3, 1), $K = 14$ code may be written from eqn. (4.103) as:

$$g(1, 1) = (01001100000001)$$
$$g(1, 2) = (10100001000001) \tag{4.104}$$

The code has one data bit and two parity bits given by eqn. (4.104). It may be easily calculated that the syndrome bits orthogonal on e_d^0 are $S_1(1)$, $S_4(1)$, $S_5(1)$, $S_{13}(1)$ from the 1st-syndrome register and on $S_0(2)$, $S_2(2)$, $S_7(2)$, $S_{13}(2)$ form the 2nd syndrome register. Since there are eight parity check sums orthogonal on e_d^0, the code corrects 4 errors in a span of 42 W_i-bits. The block schematic of the decoder is shown in Fig. 4.34.

Orthogonalizable Codes

Apart from the self-orthogonal convolutional codes, given in Table 4.15 and discussed above, there are many orthogonalizable codes which have better error correcting capabilities. Consider the rate (1/2), $K = 6$, convolutional code with subgenerator give by:

$$g(1, 1) = (100111)$$

and (4.105)

$$g(1) = (\underline{11}\ \ \underline{00}\ \ \underline{00}\ \ \underline{01}\ \ \underline{01}\ \ \underline{01})$$

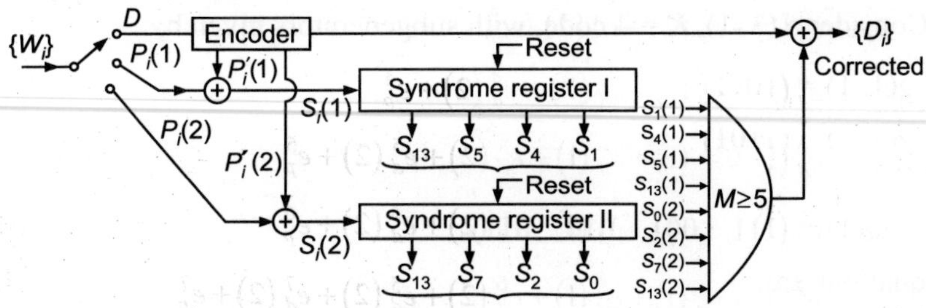

Figure 4.34: Decoder for (3, 1), $K = 14$, code

The coder and decoder circuits for the code are similar to those of Fig. 4.31 and Fig. 4.32, and uses 5 SR's in the syndrome register. Similar to eqn. (4.102), syndrome equations are:

$$\begin{aligned} S_0 &= e_d^0 + e_p^0 \\ S_1 &= e_d^1 + e_p^1 \\ S_2 &= e_d^2 + e_p^2 \\ S_3 &= e_d^0 + e_d^3 + e_p^3 \\ S_4 &= e_d^0 + e_d^1 + e_d^4 + e_p^4 \\ S_5 &= e_d^0 + e_d^1 + e_d^2 + e_d^5 + e_p^5 \end{aligned} \tag{4.106}$$

and

$$A_1 = S_1 + S_5 = e_d^0 + e_d^2 + e_d^5 + e_p^1 + e_p^5$$

It is now seen that the syndrome bits S_0, S_3, S_4 and check sum $A_1 (= S_1 + S_5)$ are all orthogonal on e_d^0 and no other error bit appears in more than one equation. The error bit may now be located by a majority vote in a circuit, shown in Fig. 4.35, and the code will correct $t \le 2$ errors in a span of 12 W_i-bits. Such a code is known as orthogonalizable and Majority-logic decodable as well.

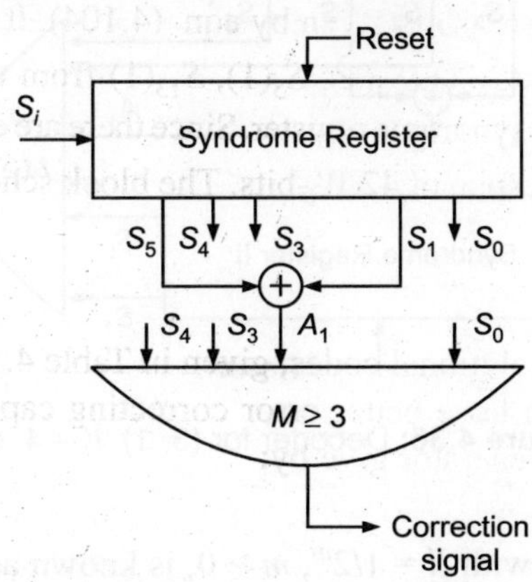

Figure 4.35: Decoder for (2, 1), $K = 6$, code

Example 4.29: Consider a (3, 1), $K = 4$ code, with subgenerators given by:

$$g(1, 1) = (1011)$$
$$g(1, 2) = (1101) \tag{4.107}$$

and

$$g(1) = (111 \quad 001 \quad 010 \quad 011)$$

The syndrome equations are:

$$\left.\begin{aligned} S_0(1) &= e_d^0 + e_p^0(1) \\ S_1(1) &= e_d^1 + e_p^1(1) \\ S_2(2) &= e_d^0 + e_d^2 + e_p^2(1) \\ S_3(1) &= e_d^0 + e_d^1 + e_d^3 + e_p^3(1) \end{aligned}\right\}$$

$$\left.\begin{aligned} S_0(2) &= e_d^0 + e_p^0(2) \\ S_1(2) &= e_d^0 + e_d^1 + e_p^1(2) \\ S_2(2) &= e_d^1 + e_d^2 + e_p^2(2) \\ S_3(2) &= e_d^0 + e_d^2 + e_d^3 + e_p^3(2) \end{aligned}\right\}$$

and

$$A_1 = S_1(1) + S_3(1) = e_d^0 + e_d^3 + e_p^1(1) + e_p^3(1)$$

It is seen that $S_0(1)$, $S_2(1)$, $S_0(2)$, $S_1(2)$ and A_1 are all orthogonal on e_d^0; then the majority vote locates the error in the data bit. Since $j = 5$, the code corrects 2 errors in a span of 12 W_i-bits. The decoder schematic is shown in Fig. 4.36.

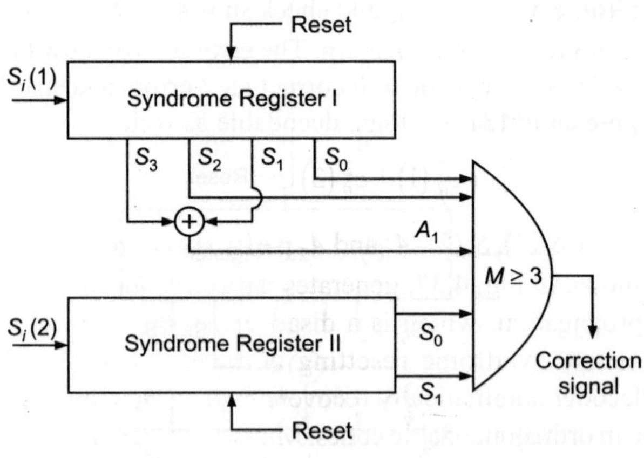

Figure 4.36: Decoder for (3, 1), $K = 4$, code

Uniform Code

A low rate orthogonalizable code with $R = 1/2^m$, $m > 0$, is known as a uniform code and is the dual of Wyner-Ash codes. The parameters of the codes are:

$$n = 2^m,\ k = 1,\ K = m + 1,\ m > 0,$$

and the code is designated as $(2^m, 1)$ code. The $(2^m - 1)$ subgenerators for the codes are the same as those for the corresponding dual Wyner-Ash codes, discussed in Sec. 4.6.2. It has been shown that the codes are orthogonalizable with $j = [(m + 2)2^{m-1} - 1]$ parity check sums. Thus the code corrects $t \leq [(m + 2)2^{m-2} - 1]$ errors in a span of $[(m + 2)\cdot 2^m - 1]$ channel bits.

Example 4.30: Consider a uniform code with $m = 2$, $n = 4$, $R = 1/4$, $K = 3$, this corrects 3 errors or less in a span of 12 W_i-bits. Referring to eqn. (4.97), the subgenerators are:

$$\begin{aligned} g(1,1) &= (110) \\ g(1,2) &= (101) \\ g(1,3) &= (111) \end{aligned} \tag{4.108}$$

and

$$g(1) = (1111 \quad 0101 \quad 0011)$$

The syndrome equations are formed as:

$$\left.\begin{aligned} S_0(1) &= e_d^0 + e_p^0(1) \\ S_0(2) &= e_d^0 + e_p^0(2) \\ S_0(3) &= e_d^0 + e_p^0(3) \end{aligned}\right\}$$

$$\begin{aligned} S_1(1) &= e_d^0 + e_d^1 + e_p^1(1) \\ S_1(2) &= e_d^1 + e_p^1(2) \\ S_1(3) &= e_d^0 + e_d^1 + e_p^1(3) \\ A_1 &= S_1(1) + S_1(2) = e_d^0 + e_p^1(1) + e_p^1(2) \end{aligned}$$

$$\left.\begin{aligned} S_2(1) &= e_d^1 + e_d^2 + e_p^2(1) \\ S_2(2) &= e_d^0 + e_d^2 + e_p^2(2) \\ S_2(3) &= e_d^0 + e_d^1 + e_d^2 + e_p^2(3) \\ A_2 &= S_2(1) + S_2(3) = e_d^0 + e_d^2(1) + e_p^2(3) \end{aligned}\right\}$$

It is seen that $S_0(1)$, $S_0(2)$, $S_0(3)$, $S_1(3)$, $S_2(2)$, A_1 and A_2 parity check sums are all orthogonal on e_d^0. Thus, the majority logic circuit shown in Fig. 4.37, generates the correction signal for the data bits.

The problem of error propagation, which is a disadvantage in feedback decoders, also persists in majority-logic decoders where syndrome resetting is done. It has shown that in self-orthogonal convolutional codes, the decoder automatically recovers from a decoding error over a short span of the received bits; but it is not so in orthogonalizable codes. Moreover, the error propagation may be eliminated if the decoder operates in the Definite decoding mode, i.e., when the feedback connection is removed from the syndrome resetting points. This, however, reduces the error correcting capability of the self-orthogonal codes by a factor of two approximately and the reduction is considerable in orthogonalizable codes. However, the technique of resynchronisation after kL bits, $L \gg K$, can always be used when required. The orthogonalizable codes have been designed mostly by computer search technique, and a short list is given in Table 4.16 [10].

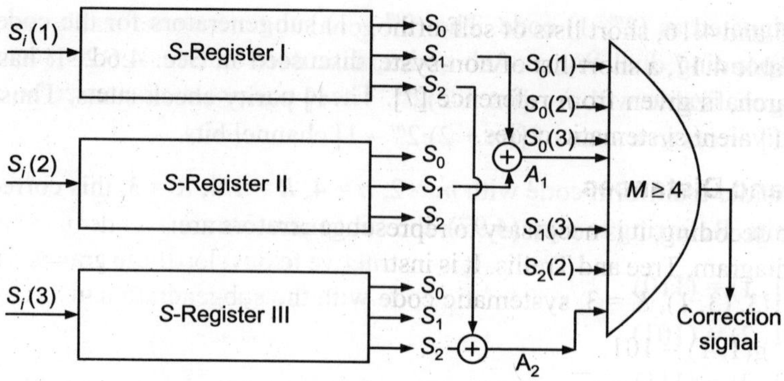

Figure 4.37: Decoder for (4, 1), $K = 3$, code

4.6.4 Probabilistic Decoding: Non-systematic Codes

So far, the decoding of convolutional codes has been discussed on the basis of the algebraic structure of the systematic codes. These decoding techniques, e.g., feedback decoding and majority-logic decoding, are not applicable to non-systematic codes. It is, however, shown that non-systematic codes are surperior to systematic codes. More powerful decoding techniques utilize the statistics of a channel in such a way that the probability of a decoding error can be made arbitrarily small by increasing the constraint length of the code. The decoding methods are now referred as Probabilistic decoding and the two important techniques are known as (a) Viterbi decoding algorithm [13] and (b) Sequential decoding algorithm [14]. Viterbi's maximum-likelihood decoding scheme is very effective for codes with short constraint length, say $K \leq 7$, and has several advantages over the sequential decoding scheme. But with large constraint lengths, where the coding efficiency is much better, the complexity of Viterbi decoding makes it impractical, but the sequential decoding, using computer-like hardware, can be implemented at a modest cost.

Table 4.16: A short list of Orthogonalizable Convolutional Codes [10]

(n, k)	K	t	Subgenerator polynomials*
(2, 1)	6	2	$(1 + X + X^2 + X^5)$
"	12	3	$(1 + X + X^2 + X^4 + X^5 + X^{11})$
"	22	4	$(1 + X + X^2 + X^4 + X^5 + X^8 + X^{10} + X^{21})$
(3, 1)	3	2	$(1 + X^2); (X + X^2)$
"	5	3	$(1 + X + X^2 + X^4); (X^3 + X^4)$
"	8	4	$(1 + X^6 + X^7) (X + X^3 + X^4 + X^5 + X^7)$
"	11	5	$(1 + X^9 + X^{10}); (X + X^2 + X^5 + X^7 + X^8 + X^9 + X^{10})$
"	18	6	$(X+X^2+X^3+X^{16}+X^{17}); (X+X^4+X^5+X^8+X^{10}+X^{11}+X^{12}+X^{13}+X^{17})$
(5, 1)	2	3	$(1 + X); (1 + X); (X); (X)$
"	3	4	$(1 + X + X^2); (X + X^2); (1 + X^2); (X^2)$
"	4	5	$(1 + X + X^2 + X^3); (X^2 + X^3); (X + X^3); (1 + X^3)$
"	6	6	$(X + X^2 + X^3 + X^4 + X^5); (X^4 + X^5); (1 + X^3 + X^5); (1 + X^2 + X^5)$
$(2^m, 1)$	$(m + 1)$	$[(m + 2)2^{m-2} - 1]$	Uniform codes; dual of Wyner-Ash codes. Subgenerators same as for Wyner-Ash codes.

*Contrary to the polynomials of Table 4.15, the compliments of these polynomials are not valid as subgenerator polynomials.

In Tables 4.15 and 4.16, short lists of self-orthogonal and orthogonalizable convolutional codes have been given. In Table 4.17, a short-list of non-systematic convolutional codes, which have been obtained by computer search, is given from reference [7]. These codes have better error-correcting capabilities compared to equivalent systematic codes.

Trees, Trellis and Distances

For probabilistic decoding, it is necessary to represent a convolutional code in the form of special graphs, known as State diagram, Tree and Trellis. It is instructive to develop these graphs with a specific example. Consider a $R = 1/3$, (3, 1), $K = 3$, systematic code with the subgenerators given by:

$$\begin{aligned} g(1,1) &= 101 \\ g(1,2) &= 110 \end{aligned} \tag{4.109}$$

and

$$g(1) = (111 \quad 001 \quad 010)$$

The coder is shown in Fig. 4.38. The output of the coder for an input

$$\{D_i\} = \{10110 \ldots\}$$

is:

$$\{C_i\} = \{111 \quad 001 \quad 101 \quad 110 \quad 011 \quad 010 \quad 000 \quad \ldots\}$$

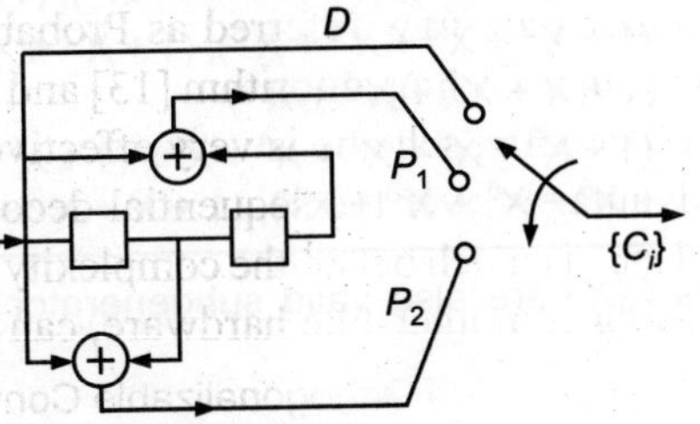

Figure 4.38: Coder for (3, 1) code

This code is majority-logic decodable and its syndrome equations are:

$$\left.\begin{aligned} S_0(1) &= e_d^0 + e_p^0(1) \\ S_1(1) &= e_d^1 + e_p^1(1) \\ S_2(1) &= e_d^0 + e_d^2 + e_p^2(1) \end{aligned}\right\}$$

$$\left.\begin{aligned} S_0(2) &= e_d^0 + e_p^0(2) \\ S_1(2) &= e_d^0 + e_d^1 + e_p^1(2) \\ S_2(2) &= e_d^1 + e_d^2 + e_p^2(2) \end{aligned}\right\}$$

It is seen that $S_0(1)$, $S_2(1)$, $S_0(2)$ and $S_1(2)$ are orthogonal on e_d^0 and this error may be located and corrected. Since the number of orthogonal parity sums $j = 4$, the code corrects all errors $t \leq 2$.

The coder can be represented as a 4-state machine, where the states of the SR's {00, 01, 10, 11} are designated as the machine states, as shown in Fig. 4.39. The state diagram depicts the transition of an earlier state to the present one by the transition branch on which the present bit as well as the corresponding coder output is shown. If, for example, the initial condition of the SR's is (11) and the present data input is '0', then state (11) changes to (01) and the coder output is (011) as indicated in the Fig. 4.39. Thus a four-state directed graph can be used to represent uniquely the input-output relation of a coder of $K = 3$.

Table 4.17: A short list of non-systematic convolutional codes

(n, k)	K	d_{min}	Subgenerator polynomials*
(2, 1)	3	5	$(1+X^2), (1+X+X^2)$
"	4	6	$(1+X+X^3), (1+X+X^2+X^3)$
"	5	7	$(1+X^3+X^4), (1+X+X^2+X^4)$
"	7	10	$(1+X^2+X^3+X^6), (1+X+X^2+X^3+X^6)$
"	10	12	$(1+X^3+X^4+X^5+X^7+X^8+X^9), (1+X+X^3+X^4+X^7+X^9)$
"	12	15	$(1+X^4+X^5+X^7+X^8+X^9+X^{11}), (1+X^2+X^3+X^4+X^5+X^7+X^{10}+X^{11})$
(3, 1)	3	8	$(1+X^2), (1+X+X^2), (1+X+X^2)$
"	4	10	$(1+X^2+X^3), (1+X+X^3), (1+X+X^2+X^3)$
"	5	12	$(1+X^2+X^4), (1+X+X^3+X^4), (1+X+X^2+X^3+X^4)$
"	7	15	$(1+X^2+X^3+X^5+X^6), (1+X+X^4+X^6), (1+X+X^2+X^3+X^4+X^6)$
"	10	20	$(1+X^3+X^6+X^7+X^8+X^9), (1+X^2+X^3+X^4+X^5+X^7+X^9), (1+X+X^2+X^5+X^6+X^8+X^9)$
(4, 1)	3	10	$(1+X^2), (1+X+X^2), (1+X+X^2), (1+X+X^2)$
"	5	16	$(1+X^2+X^4), (1+X+X^3+X^4), (1+X+X^2+X^4), (1+X+X^2+X^3+X^4)$
"	7	20	$(1+X^2+X^3+X^4+X^6), (1+X^2+X^3+X^4+X^6), (1+X+X^4+X^5+X^6), (1+X+X^2+X^5+X^6)$
"	10	27	$(1+X^3+X^6+X^7+X^8+X^9), (1+X^2+X^3+X^4+X^5+X^7+X^9), (1+X+X^2+X^5+X^6+X^8+X^9), (1+X+X^2+X^4+X^6+X^8+X^9)$

*Compliments of the polynomials $f_n(X)$ are also valid subgenerator polynomials, as in the case of the polynomials in Table 4.15.

The state diagram can be redrawn as a tree graph, representing the coder, as shown in Fig. 4.40. Here the convention is that: if the input is '0', then the upper path is followed and if the input is '1', then the lower path is followed. A vertical line is called a Node and a horizontal line is called a Branch, and the output codeword for each input bit is shown on the branches. Following the above rule, the coder output can be easily traced through the tree paths and the particular path traversed for $\{D_i\} = \{10110\ldots\}$ for the above example is shown in the Fig. 4.40 by a heavy line. The coder output is also checked with the above example. It is further seen that the tree structure is repetitive beyond the third node level; this is because the coder output depends only on the two previous inputs and the current input. The repetitive structure of the tree leads to the redrawing of the tree diagram as a Trellis diagram shown in Fig. 4.41. The trellis is called as such because it is a tree-like structure with remerging branches. In the figure, the convention adopted is that: the code branches produced by a '0' input are shown as solid lines and the branches produced by '1' input are shown as dotted lines. The completely repetitive structure of the trellis diagram suggests a further reduction in the representation of the code to the state diagram of Fig. 4.39. The states of the state diagram have been levelled according to the nodes of the trellis, and the states correspond to the last two input bits to the coder. Thus the tree, the trellis and the state diagram all represent the input-output relation of the coder and are used for decoding coder data bits by tracing the most likely path that has been traversed while the code was generated. Specially the trellis-graph is the most useful for probabilistic decoding.

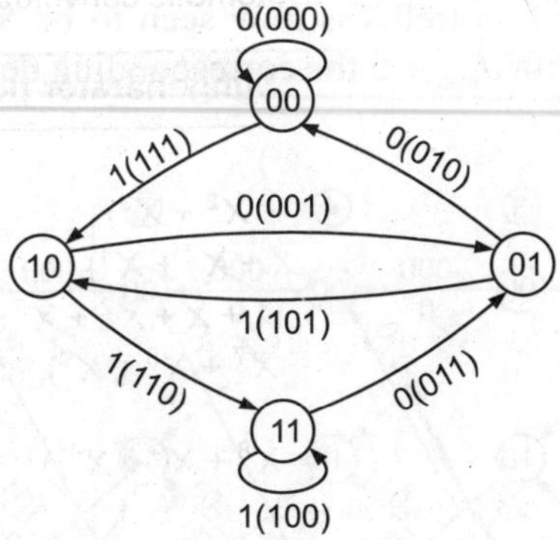

Figure 4.39: State diagrma of the coder of Fig. 4.38

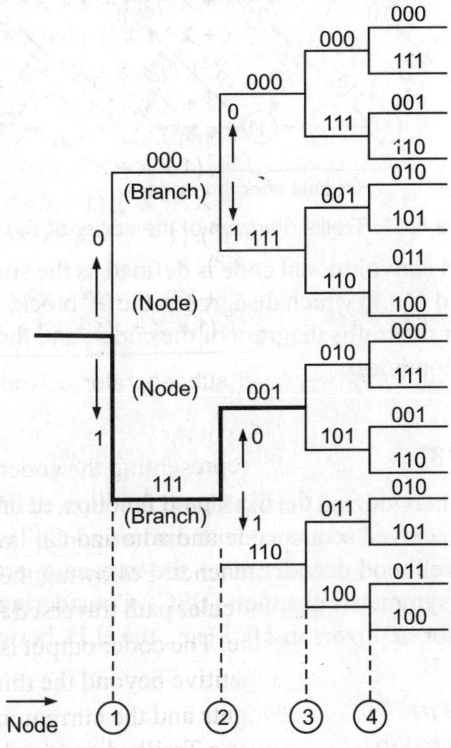

Figure 4.40: Tree graph of the coder

As in block codes, the distance property of convolutional codes is also important in evaluating their error correcting capabilities. One can compute the distances of code words from the all-zero code word, as this set of distances is the same as the set of distances from any specific code word to all others (as in the case of block codes). Referring to the trellis diagram of Fig. 4.41, consider all the paths that start from the first (00) node and merge with the all-zero path at some later node, say j. Then it is seen from the diagram that:

(a) there is one path at distance 5 from the all zero path, which diverged 3 branches earlier, (b) there is one path at distance 8 which diverged 4 branches earlier, (c) there is another path at distance 8 which diverged 5 branches earlier and so on. The code word corresponding to the above distances are {00100 . . .}, {001100 . . .} and {0010100 . . .} respectively. The minimum distance among all paths for the given code and its trellis is now seen to be 5, and this is sometimes called the minimum 'free' distance. With $d_{min} = 5$ the corresponding decoder can correct all $t \leq 2$.

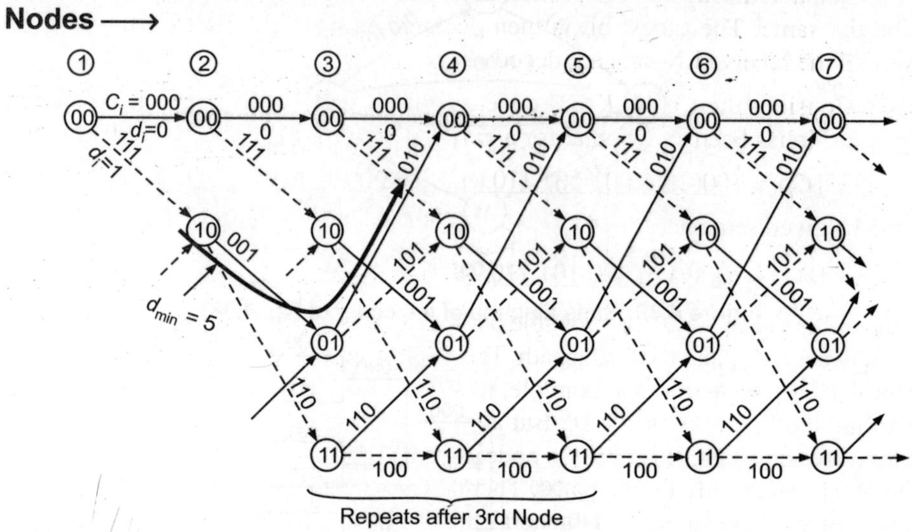

Figure 4.41: Trellis diagram of the coder of Fig. 4.38

In general, the d_{min} of an (n, k) convolutional code is defined as the smallest H.D. = $d(C_1, C_2)$ between two initial code vectors $C_1(X)$ and $C_2(X)$ which disagree in the 0th block. The d_{min} can now be determined from the inspection of the tree or the trellis diagram of the coder; and the number of correctible errors in nK bits of $\{W_i\}$ is given by the condition:

$$(2t + 1) \leq d_{min}.$$

4.6.5 Viterbi Decoding [13]

The Viterbi decoding is based on maximizing the likelihood function, defined as the conditional probability pr. $(W_i/C_i^{(N)})$, where $\{W_i\}$ is the received sequence with errors and $C_i^{(N)}$ is one of the possible transmitted sequences. Such a maximum-likelihood decoder gives the minimum error in decoding for memoryless noisy channels including binary symmetric channels (BSC). Considering a BSC with error probability p, $p < 0.5$, the likelihood function for 'd' errors in $\{W_i\}$ (i.e., the H.D. between $\{W_i\}$ and $\{C_i\} = d$ for N bit long sequence) is given by:

$$pr.(W_i/C_i^{(N)}) = p^d(1-p)^{N-d}$$

and taking logarithm, log $pr.$ $(W_i/C_i^{(N)})$

$$= -d \cdot \log\,[(1-p)/p] + N \log\,(1-p) = -Ad + B \tag{4.110}$$

where A and B are constants. It is now clear that the log-likelihood function (also known as 'path metric') will be maximum when 'd' is minimum for a given p and N. Thus it is sufficient to find a path in the trellis which has the minimum 'd' for a given $\{W_i\}$. Since the trellis has all the paths corresponding to all possible transmitted squences, the minimum-d path gives the most likely path for $\{W_i\}$, and hence the minimum path is taken as the transmitted sequence $\{C_i\}$. Such a decoder using minimum distance 'd_{min}' as the criterion for decoding is also known as a Minimum distance decoder.

To calculate the most-likely path, the H.D. between W_i bits and the branches of the trellis are calculated and added to the H.D.'s of the previous nodes in order to compute the H.D. of the nodes at the next stage. Those paths are said to survive which result in the minimum H.D. at the nodes. These paths (and their history) and the H.D. of the corresponding nodes are stored and similar computations are repeated for other nodes. It can be shown that the paths which have survived at any stage of the computations emerge from one old branch at a distance of $3K$ or $4K$ bits in the past, and thus the oldest bit of all the paths is expected to be the same. The oldest bit is then declared to be the decoded data bit. This process is continued until all the received W_i bits are decoded.

Example 4.31: Consider the (3, 1), $K = 3$ code, given by eqn. (4.109) and the corresponding coder shown in Fig. 4.38. It has been shown that for a given $\{D_i\} = \{010110\ldots\}$, the coder output is:

$$\{C_i\} = \{000\ 111\ 001\ 101\ 110\ 011\ 010\ 000\ \ldots\}.$$

Assume tha the received sequence

$$\{W_i\} = \{000\ 110\ 000\ 101\ 110\ 011\ 010\ 000\ldots\}$$

Referring to the trellis of Fig. 4.41, the probable paths of $\{W_i\}$ may now be traced, computing at the same time the H.D. between $\{W_i\}$ and the trellis path. The corresponding data bit and the H.D. at each node are shown in Table 4.18. In this particular example, it is seen that at nodes 5 and 6, no definite decision can be made regarding the first two bits of D_i. But at node 7, it is observed that $D_i = 101100$ has H.D. = 2 which is the least among all the H.D.'s at the node, and further 1011 occurs in 4 probable paths, while the other data bits 0011, occur only in two paths. Then a majority vote is taken and $D_i = 1011$ is decided as the correct transmitted information bits. In general, however, this ambiguity does not occur with a decoding window of say $4K$ or more bits. The code of this example corrects $t \leq 2$ in 9 W_i-bits, and this may be checked by introducing two more errors, say in the 5th and 6th blocks of digits in $\{W_i\}$, thus producing 4 errors in 18 W_i-bits. The tentative decisions regarding the initial bits of $\{D_i\}$ may be taken at the 4th/5th node as well, and a final check is made regarding the minimum value of the H.D. of the accepted sequence after sufficient data bits are decoded.

Table 4.18: Survivor paths and their H.D.

$$D_i = 101100;\ C_i = 111\ 001\ 101\ 110\ 011\ 010\ 000$$

$$W_i = 110\ 000\ 101\ 110\ 011\ 010\ 000$$

SR state \ nodes →	2	3	4	5	6	7
00	0(2)	00(2)	000(4) ~~100(5)~~	0000(6)		101100(2) 001100(3)
10	1(1)	~~01(5)~~	001(3) 101(2)			101101(5) 001101(6)
01		10(2)	~~110(5)~~	~~0010(6)~~ ~~1010(5)~~ ~~1110(6)~~	10110(2) 00110(3)	101110(6)
11		11(3)	111(4)	1011(2) 0011(3) ~~1111(5)~~	10111(5) ~~00111(6)~~	101111(7)

Consider now the non-systematic convolutional codes, which have better distance properties as compared to a systematic code with same K, as is evident from Tables 4.15, 4.16, and 4.17. Take for example, the (2, 1) code with $K = 4$, $d_{\min} = 6$, and

$$g(1, 1) = (1101)$$
$$g(2, 1) = (1111) \tag{4.111}$$

The coder and the state diagram for the code is shown in Fig. 4.42. It is now possible to write $\{C_i\}$ for a given $\{D_i\}$ from the state diagram. Since $d_{\min} = 6$, the code can correct $t \leq 2$ errors in a span of 8 W_i-bits. However, it should be noted that these codes are designed for random error correction, and as such are not every efficient in correcting pairs or triplets of errors.

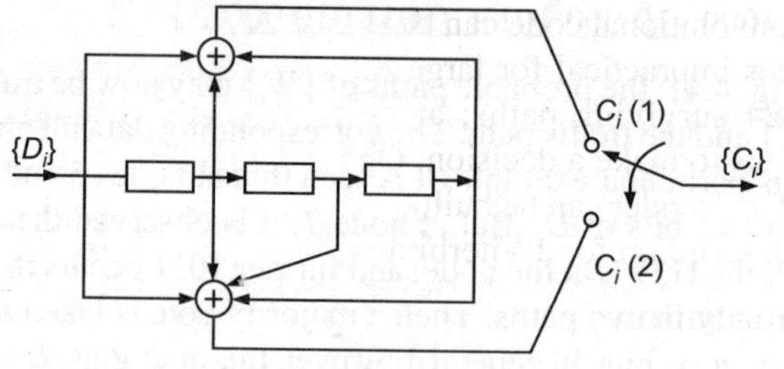

Figure 4.42(*a*): Coder for the (2, 1), $K = 4$ code

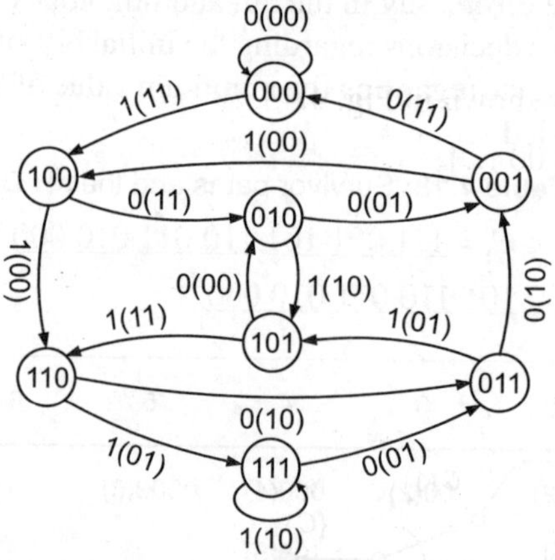

Figure 4.42(*b*): State diagram of coder in (a)

Example 4.32: Consider the above coder for rate 1/2, $K = 4$, convolutional code. Without loss of generality, one can assume $\{C_i\} = \{00 \ldots 00\}$ as the transmitted vector and introduce certain number of errors in $\{W_i\}$ to check the effectiveness of the code. Assume that the received sequence is given by:

$$\{W_i\} = \{\underline{00}\ \underline{01}\ \underline{10}\ \underline{00}\ \underline{00}\ \underline{01}\ \underline{00}\ \underline{01}\ \underline{00}\ \underline{00}\}$$

and the corresponding coder and the state diagram is as shown in Fig. 4.42. By following the different probable paths through the state diagram, the different probable sequences along with their H.D. with refrence to $\{W_i\}$ are given below:

(a) $\{\underline{00}\ \underline{11}\ \underline{11}\ \underline{01}\ \underline{11}\ \underline{00}\ \underline{00}\ \underline{00}\ \underline{00}\ \underline{00}\}$: H.D. = 7

(b) $\{\underline{00}\ \underline{11}\ \underline{00}\ \underline{01}\ \underline{01}\ \underline{10}\ \underline{11}\ \underline{00}\ \underline{00}\ \underline{00}\}$: H.D. = 9

(c) $\{\underline{00}\ \underline{11}\ \underline{11}\ \underline{01}\ \underline{00}\ \underline{00}\ \underline{01}\ \underline{01}\ \underline{10}\ \underline{11}\}$: H.D. = 8

(d) {00 11 00 10 10 11 00 00 00 00} : H.D. = 6

(e) {00 11 11 10 00 01 11 00 00 00} : H.D. = 6

(f) {00 00 00 00 00 00 00 00 00 00} : H.D. = 4

It is observed that the sequence with $d_{\min} = 4$ is the all-zero sequence, and the decoder has corrected 4 errors in 16 W_i-bits.

*It is possible to have ambiguities with a particular error pattern and the solution for such cases is to arbitrarily discard the lower trellis branch. It is, however, convenient to truncate the sequence at certain intervals by introducting 000 in $\{D_i\}$ after blocks of KL bits. This ensures that the final state of the code must be 000, and as such, the ultimate survivor at node (000). The corresponding trellis, therefore, is truncated in the same way as it was begun.

The performance of a convolutional code can be improved by increasing the constraint length K, but the Viterbi decoder becomes impractical for large K, say $K > 10$. With $K = 10$, the decoder has 1024 nodes and must store 1024 surviving paths; at the same time, the trellis must be searched over approximately hundred levels to make a decision. On the other hand, Viterbi decoders can be very fast with K small, and for $K = 7$, a decoder can be built to iterate 10^7 times per second with 128 nodes at each level. A typical hardware description for a Viterbi decoder is given by Sinha and Babu [15].

Catastrophic Error Propagation

Consider a coder with:

$$\begin{aligned} g(1, 1) &= 110 \\ g(2, 1) &= 101 \end{aligned} \tag{4.112}$$

The coder and its state diagram is shown in Fig. 4.43. Suppose that:

$$\{D_i\} = \{11\ 00\ 00\ \ldots\},$$

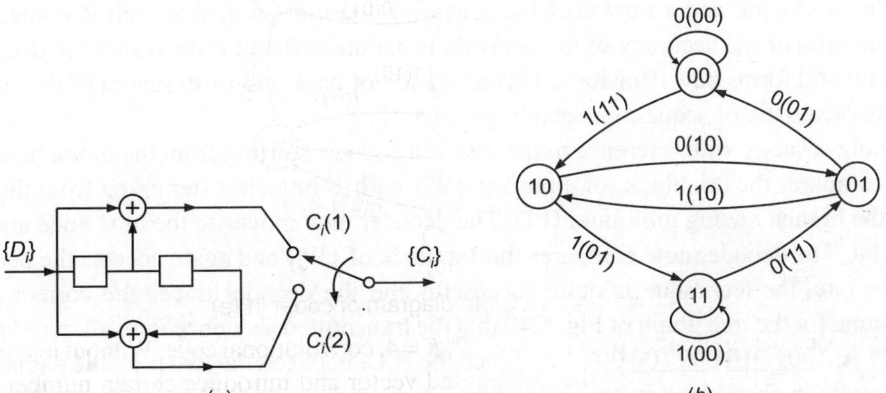

Figure 4.43: Example of catastrophic error propagation: ***(a)*** Coder for eqn. (4.112), ***(b)*** State diagram for coder of eqn. (4.112)

then,

$$\{C_i\} = \{11\ 01\ 11\ 01\ 00\ 00\ \ldots\}$$

---

*In this example, if $\{W_i\} = \{00\ 01\ 10\ 00\ 00\ 01\ 01\ 00\ \ldots\}$, then it may be checked that H.D. for a probable sequence: {00 11 11 10 00 01 11 00 00 00} is 4 also. Thus there is a tie with the all-zero sequence. Since the all zero sequence is at the top of the trellis, this sequence is accepted as the transmitted $\{C_i\}$.

If the received sequence $\{W_i\} = \{11\ 01\ 01\ 11\ 00\ 00 \ldots\}$, then the decoder may follow either of the paths:

(a) {11 01 11 01 00 00} with H.D. = 2
(b) {11 01 00 11 01 00} with H.D. = 2

Since each of the paths has H.D. = 2, there is an ambiguity, and the decoder may follow the incorrect path (b). This may lead to a large number of errors in decoding a long sequence, and the situation is known as the catastropic error propagation. In terms of the generator polynomials, the catastropic error propagation occurs when:

(a) the generator polynomials have a common factor, e.g., in the coder of Fig. 4.43, $(1 + X)$ is the factor.
(b) any closed loop path in the state diagrams has a zero weight, e.g., in the figure the self-loop at the node (11) has the output (00).
(c) for binary tree-code with $R = 1/n$, each of the adder in the coder has an even number of connections, as in the case of Fig. 4.43. This results in an all-zero output for a self-loop corresponding to all ones state.

These conditions should be checked before designing a coder for a non-systematic code. However, this situation does not occur in a systematic code, as may be verified with certain examples.

4.6.6 Sequential Decoding [14]

Since the Viterbi decoder is impractical for convolutional codes with large K, and for better performance one has to use large K, an alternative decoding technique has been developed in the form of sequential decoding. In contrast with the Viterbi decoding, where the trellis paths are maintained through all possible SR-states and the decision is made at a later node; in sequential decoding, the improbable paths are discarded at each node and only one most probable path is decided upon. This path is then carried forward to the next node and a new decision regarding the most probable path is made. However, a continuous measure of the accuracy of the decision is maintained and if necessary the decoder retraces an ignored path and forms new decisions. This provision of back and forth search of the correct path is the special characteristic of sequential decoding.

The decoding strategy with reference to the tree of a code is: starting from the initial node of the tree, the decoder compares the 0th block (of n bits) of $\{W_i\}$ with 2^k branches stemming from the initial node and accepts the branch having minimum H.D. The decoder then moves to the next node and decides on the message bit. The decoder now compares the 1st block of $\{W_i\}$ and again accepts the nearest branch. For low error rate, the technique is quite successful and the decoder traces the correct path. As for example, assume for the tree graph of Fig. 4.40 that the transmitted sequence is an all-zero vector and the received $\{W_i\}$ is: {001 010 000 100 000 ...}. Since the H.D. between the two possible paths at each node is 3, and the error in each block is one only, then a correct decision will be made at each node by the decoder. If, however, $\{W_i\}$ is: {000 000 011 000 ...}, then at the 3rd node, the path decided upon by the decoder will be wrong. This node at which the wrong decision is made, is known as the Node of Separation.

The above branch-by-branch search strategy is not very efficient, because after the node of separation, the decoder will continue to make errors without returning to the stable state. It is thus necessary to have a decoding algorithm which will: (a) recognise with high probability whenever the decoder moves to an incorrect path, and (b) initiate a systematic search for the node of separation when (a) happens. Fano's algorithm [16] and its various refinements provide such sets of rules for efficient sequential decoding.

For the purpose of recognising the node of separation, one defines a 'tilted distance' measure as:

$$\lambda(l) = p'nl - d[C'(l), W(l)] \tag{4.113}$$

where $p < p' < 1/2$, p = BER in channel and $d[\ ]$ = measured H.D. between the received $W(l)$ and the selected path $C'(l)$ through the tree, l being number of blocks decoded. For the correct path, $d[\ ] \simeq pnl$ and $\lambda(l)$ is positive and increasing with l as shown in Fig. 4.44. As long as $\lambda(l)$ is increasing, the decoder continues to move from node to node and follow the most probable path. If $\lambda(l)$ starts decreasing in value, then the decoder concludes that an incorrect path has been chosen, and as such it retraces its path testing alternate paths at earlier nodes. Once it is able to locate the node of separation, then a new most probable paths is followed with $\lambda(l)$ increasing with l. If $\lambda(l)$ starts decreasing again, then the 2nd node of separation is located through back and forth search, and this procedure continues till the $\{W_i\}$ is completely decoded.

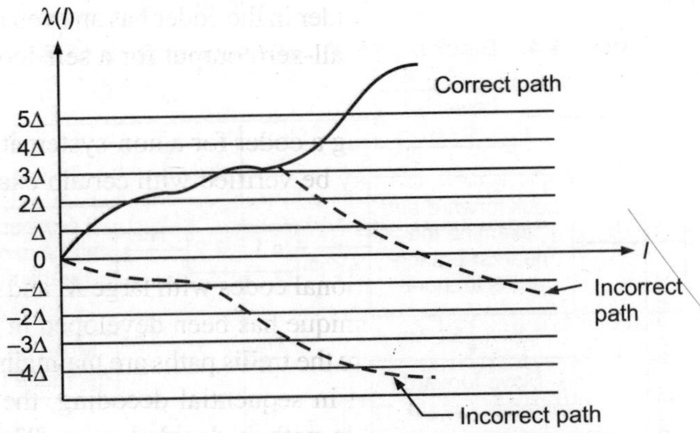

Figure 4.44: Behaviour of $\lambda(l)$ with l

To be able to recognise when $\lambda(l)$ is decreasing, the decoder uses a running threshold $\gamma^{th} = b\Delta$, when b = positive integer, and Δ = thershold increment. At the initial node, $\gamma^{th} = 0$, but γ^{th} increases with l, maintaining that $\gamma^{th} < \lambda(l)$. To facilitate the search procedure, the 2^k branches leaving a node are numbered according to their H.D. with reference to $W(l)$, i.e., the branch nearest to the received branch is ordered first, and so on. The decoder, while searching for the node of separation, examines the branches of each node according to the above ordering. The decoder mainly consists of a replica encoder with auxiliary storage register as shown in Fig. 4.45, and it is controlled through a set of rules given by Fano's algorithm, shown in simplified form in Fig. 4.46. The decoder attempts to simulate the transmitted sequence by feeding the replica encoder with trial data bits generated by the commands of the control section (simulating Fano's algorithm) of the decoder. When the channel is quiet and errors are few, the decoder moves from node to node, at the same time increasing γ^{th} (as $\lambda(l)$ increases proportionately), and its control moves around the leftmost loop of Fig. 4.46. but when the channel is noisy, the threshold condition i.e., $\lambda(l) > \gamma^{th}$, will fail and the decoder will move backward by one node. If the decoder finds another untried branch satisfying the threshold condition, then it moves forward. But if all branches at the node are tried and rejected, then the decoder moves backward by another node, at the same time decreasing γ^{th} if necessary. To increment Δ and p' are design parameters which are optimized such that the average number of branches searched by the decoder is minimum. With the above algorithm, no branch is examined twice with the same γ^{th}, so that the decoder is not trapped in a loop. However, in case of large noise in the channel, the decoder will be required to search over a large number of branches, thus slowing down the speed of decoding considerably. This will also require large buffer memories and a large buffer memories and a larger decoding-window width. Improvements of Fano's algorithm have been given by Jelinek and others [17].

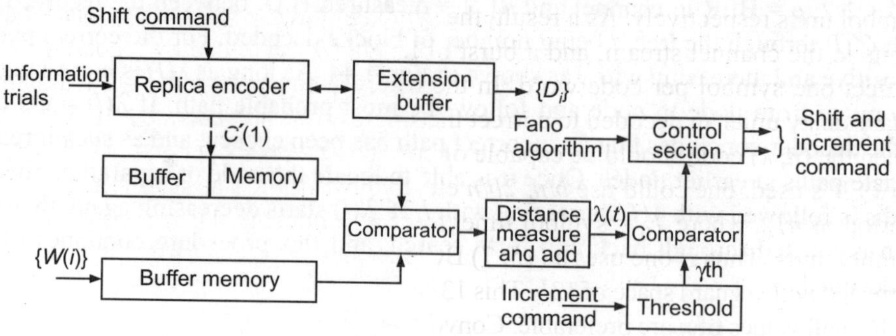

Figure 4.45: Block schematic of a sequential decoder

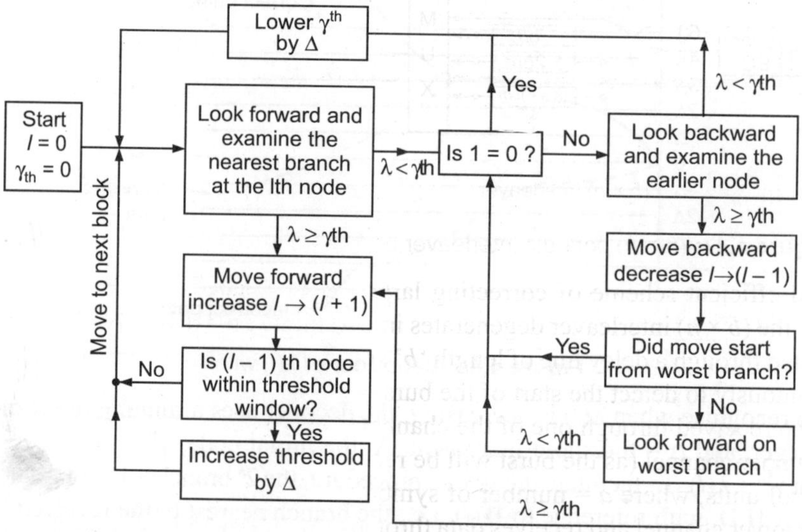

Figure 4.46: Simplified flow chart for Fano's algoritham

4.7 BURST-ERROR CORRECTING CODES [18]

So far, the FEC codes have been discussed on the assumption that the channel is memoryless and the errors occur independently with stationary statistics. With the exception, perhaps, of the space channel, all real channels tend to have error bursts. An error burst of length 'b' is defined as a sequence of b error symbols, the first and last of which are non-zero, e.g., an error vector (0010110011000) is a burst of $b = 8$. For proper decoding, the 'b' symbol bursts have to be separated by a guard space of 'g' symbols. Various specific block and convolutional codes, e.g., Burton, Fire, RS, Berlekamp-Preparata-Massey, Iwadare and Adaptive Gallager codes, have been designed for correcting bursts. However, Interleaved cyclic or convolutional codes offer the advantage of simultaneous correction of random and burst errors, and are almost insensitive to channel statistics.

A simple interleaver is a block interleaver, in which the (n, k) coded words are placed in the rows of a $(b \times n)$ matrix, and read out from the columns. In the receiver, the code words are reassembled in a complimentary way, thereby breaking the burst of errors into t correctable errors per coded word. A somewhat simpler and more effective type of interleaver is a periodic (or convolutional) interleaver, shown in Fig. 4.47, where the successive symbols of an (n, k) code word are delayed by $\{0, b, 2b \ldots (n-1)b\}$ symbol units respectively. As a result, the symbols of one codeword are placed at distances of b-symbol units in the channel stream, and a burst of length 'b' separated by a guard space of $(n-1)b$ units only affect one symbol per code word. In the receiver, the codewords are reassembled through complementary delay units an decoded to correct the single errors so generated. If the burst length $l > b$ but $l \leq 2b$, then the (n, k) code should be capable of correcting 2 errors per codeword. To economise on the number of SR's used, one could use b/n, $2b/n$ etc. number of SR's as shown in Fig. 4.47, and clock them at a period of nT_0, where T_0 = symbol time, thereby producing the required delay of $\{b, 2b, \ldots, (n-1)b\}$ symbol units. Thus if one uses a (15, 7) BCH code with $t \leq 2$ for FEC, then a burst fo length $\leq 2b$ can be corrected with a guard space of $13b$. This 13-to-2 guard-space-to-burst ratio is rather too large, and the codes with small values of n are preferable. Convolutional codes with interleaver may also be used.

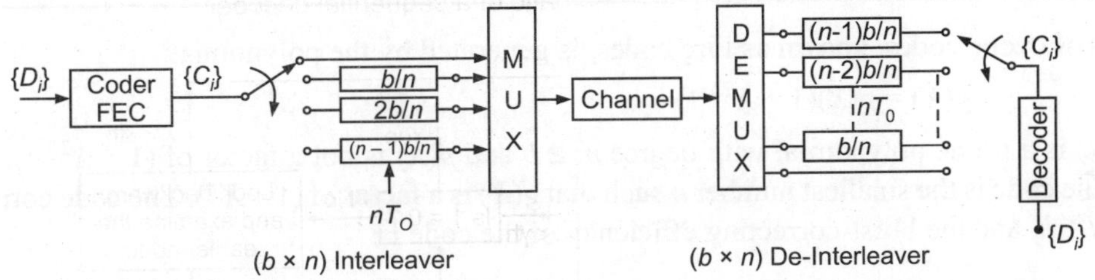

Figure 4.47: $(b \times n)$ periodic interleaver and de-interleaver: T_0 = symbol duration

A simple and efficient scheme of correcting large length bursts with a rate 1/2 code is shown in Fig. 4.48, where the $(b \times n)$ interleaver degenerates into a simple time diversity of order 'b'. In the coder, the data is repeated through a delay line of length 'b' and at the decoder the upper and lower channels are compared continuously to detect the start of the burst whenever it occurs. Initially, when the channel is quiet, data can be received through one of the channels, but as soon as a burst is detected, the receiver switches to the upper channel (as the burst will be received first in the lower channel) and remain there for $(b - a)$ symbol units, where a = number of symbols required to detect the burst. Then the receiver swithches to the lower channel and receives data through the lowre channel for b units, and the burst will pass out of the upper delay line by this time. The receiver now reverts to the quiet state. The scheme corrects bursts of length $(b - 2a - 1)$ channel symbols and requires a guard space of $(b + 2a + 1)$ symbols. The probability of a decoding failure is 2^{-a}, and $a \ll b$, but $a \geq 15$ (say). This scheme has been successfully used for data transmission in telephone channels. The technique may be further sophisticated by using erasure-correcting (n, k) RS code with $(b \times n)$ interleaver, and burst lengths $\leq (n-k)b$ are corrected with a guard space of kb symbols only.

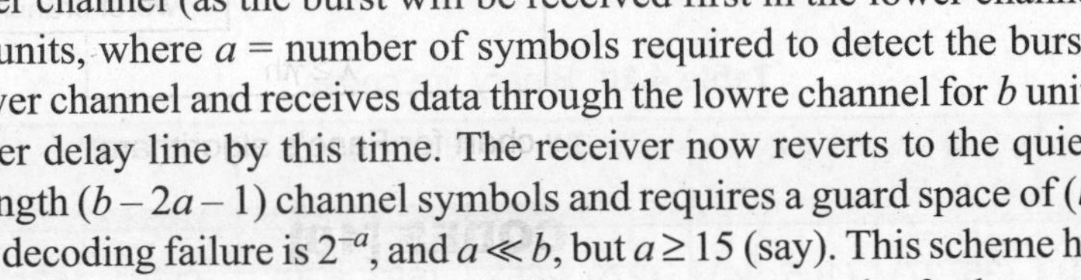

Figure 4.48: A time-diversity scheme

4.7.1 Cyclic Codes for Correcting Bursts

Among all the block codes, the cyclic codes have been found to be easily implemented for burst-correction. A short list of high rate burst-correcting cyclic codes, found by computer search, is given in Table 4.19.

Table 4.19: Burst correcting cyclic codes

(n, k)	'b'	Generator polynomial
(7, 3)	2	$1+X^2+X^3+X^4$
(15, 10)	2	$1+X^2+X^4+X^5$
(15, 9)	3	$1+X^3+X^4+X^5+X^6$
(31, 25)	2	$1+X^4+X^5+X^6$
(63, 56)	2	$1+X^2+X^3+X^5+X^6+X^7$
(63, 55)	3	$1+X^3+X^6+X^7+X^8$
(511, 499)	4	$1+X^3+X^5+X^8+X^{12}$
(1023, 1010)	4	$1+X^2+X^4+X^5+X^6+X^7+X^{10}+X^{13}$

A class of cyclic codes, known as Fire codes, is generated by the polynomial:

$$g(X) = p(X)(1 + X^{2b-1}) \tag{4.114}$$

where $p(X)$ is a prime polynomial with degree $m \geq b$ and $p(X)$ is not a factor of $(1 + X^{2b-1})$. The block length of the code is the smallest number n such that $g(X)$ is a factor of $(1 + X^n)$. The code corrects bursts of length $l \leq b$, and the burst-correcting efficiency of the code is:

$$z = \frac{2b}{m+2b-1} = \frac{2m}{3m-1}, \text{ if } b = m \tag{4.115}$$

where for block codes, z is defined as:

$$(n-k) - 2b = z \geq 0.$$

A short list of useful high rate binary Fire codes is given in Table 4.20. The decoding of the burst -correcting cyclic codes, including Fire codes can be performed by using a modified error-trapping decoder, discussed earlier in Sec. 4.4.3.

Table 4.20: Binary fire codes

Burst length b = m	2	3	4	5	6	7	10
(n, k)	(9, 4)	(35, 27)	(105, 94)	(279, 265)	(693, 676)	(1651, 1631)	(19437, 19408)

Consider a Fire code with $b = m = 4$ and $(n, k) = (105, 94)$. The generator polynomial is:

$$g(X) = (1 + X + X^4)(1 + X^7) = (1 + X + X^4 + X^7 + X^8 + X^{11}).$$

The length of the code $n = 7(2^4 - 1) = 105$ and the code has 11 check symbols making $k = 94$. The encoder for the code is shown in Fig. 4.49(a) and the decoder in Fig. 4.49(b). The decoder has a local encoder and the received $\{W_i\}$ is read into the syndrome generator with G_1 closed and G_2 open. After the syndrome bits have been formed, the $\{W_i\}$ bits are read out from the buffer bit by bit and the syndrome bits are also shifted with zero input. When all-zero bits appear in the first 7 syndrome positions, the test circuit indicates the all-zero pattern; at the same time it is estimated that the 4-bit burst is in the last 4 positions of SR's. The erroneous bits are now ready to be read out of the buffer and they are corrected by the last 4 syndrome bits through the addition of $(W_i \oplus S_i)$ while shifted bit by bit with switches G_2 closed and G_1 open. The basic principle in this decoder has been that the shifted syndrome calculated by the local encoder is identical to the $(n - k)$ high order positions of the error pattern and thus the all-zero pattern at the low-order position of the syndrome indicates the location of the burst in the high order positions. The decoder is appropriate for all burst-correcting cycling codes.

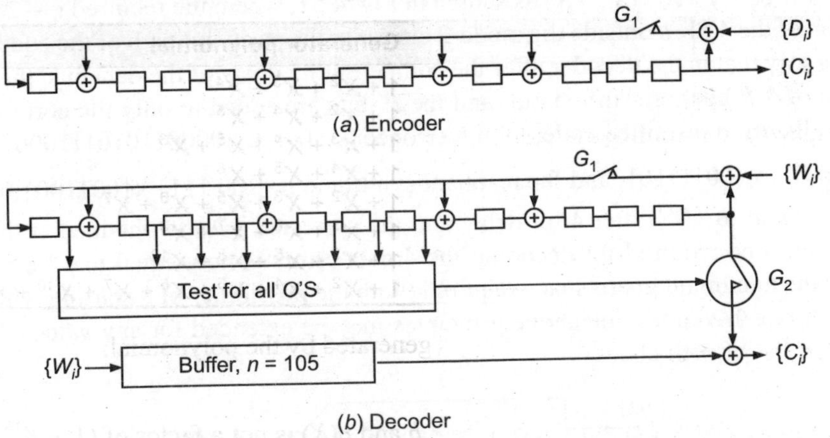

Figure 4.49: Coder and decoder for fire code, (105, 94), $b = 4$

The range of burst correction of above cyclic codes may be extended through interleaving as explained earlier. Any (n, k) burst-correcting cyclic code is extended by interleaving to form a $(\lambda n, \lambda k)$ code and this will correct t bursts of length λ, if the error correcting capability of the original code is t and λ is the degree of interleaving. Since the interleaved code is also cyclic, the $(\lambda n, \lambda k)$ code can be implemented by the generator polynomial $g(X^\lambda)$, if the generator polynomial of the original (n, k) code is $g(X)$. Instead of using interleaving, this method is simpler to implement and requires that each of the SR's in the original coder/decoder circuits is now replaced by λ stages of SR's. This technique is effective not only for deriving long powerful single-burst-correcting codes from short optimal codes, but also is useful for deriving powerful random-cum-burst-correcting codes from short cyclic codes.

4.7.2 Convolutional Codes for Correcting Bursts

As in the case of block codes, the burst-correcting capability of convolutional codes may also be extended by interleaving, and interleaving could be by either bit or block interlacing. The bit interlacing is obtained through the interleaver circuit given in Fig. 4.47, and the resultant code can correct a combination of bursts and random errors.

Example 4.33: Consider from Table 4.16, the (2, 1), $K = 6$, $t \le 2$, orthogonalizable code with $g(X) = (1 + X + X^2 + X^5)$. By using bit interleaving of the degree $\lambda = 6$ with the circuit of Fig. 4.47, the (12, 6) code, so generated, will correct any burst of length $\lambda = 12$ within a span of 72 channel bits. At the same time, it will correct many random error patterns. However, for only burst-correction, other techniques are more efficient, as the guard space required $(g = 60)$ in this case is too large.

The block interleaving can be implemented by noting that "an interleaved (n, k, K) burst-correcting convolutional code generates another convolutional code with the new constraint length $K' = (K - 1)\lambda + 1$. The new code is generated by the generator polynomial $g(X^\lambda)$, if $g(X)$ is the generator polynomial of the original code." As in the case of cyclic block codes, the implementation of the interlacing requires that each of the SR's in the original coder/decoder circuit is now replaced by λ stages of SR's without changing any other connection.

Consider now the single error-correcting (2, 1) convolutional code of Fig. 4.26. Its generator polynomial is $g(X) = (X + 1)$, $K = 2$. With a simple modification, the coder-decoder can be made to correct a burst of $b = 2$ in a span of 8 bits. The modified $g(X) = X^{-1}(X + 1)$, $K = 3$, and the coder and decoder are shown in Fig. 4.50. Now to obtain a coder/decoder with $b = 8$, the above coder may be interleaved to a degree $\lambda = 4$, and the new $g(X^{\lambda}) = X^{-4}(X^4 + 1)$ as shown in Fig. 4.51, where the required $g = 25$. It is observed that the function of the 4 SR's outside the mode-2 circuit in the parity bit path of the coder is to delay the P-bits in such a way that in the decoder after the compensating delay of the data bits, the errors in the P_i bits (maximum of 4 P-bits) are shifted out, and the P'-bits are added to only the correct P_i-bits. As an example, with all-zero transmitted code, let the erroneous $\{W_i\} \rightarrow \{000011010111000\}$. Then $\{D'\} \rightarrow \{0010010\}$ and $P_i \rightarrow \{0011110\}$, and the resultant syndrome is $\{(00\ \underleftarrow{1111)}(10011001)0000\}$. This will give correction signal to $\{D'\}$ after 4 bit delay and the corrected $\{D_i\}$ is obtained at the output of the decoder. The total time required for decoding the 4 data errors is now equal to 12 bits, equivalent to $24W_i$-bits; thus the maximum guard space required (with the possibility of g starting with a parity bit) $g = 25$ bits. For all rate 1/2 codes, the above principles may be extended for any value of 'b', and Table 4.21 gives a short list of such codes.

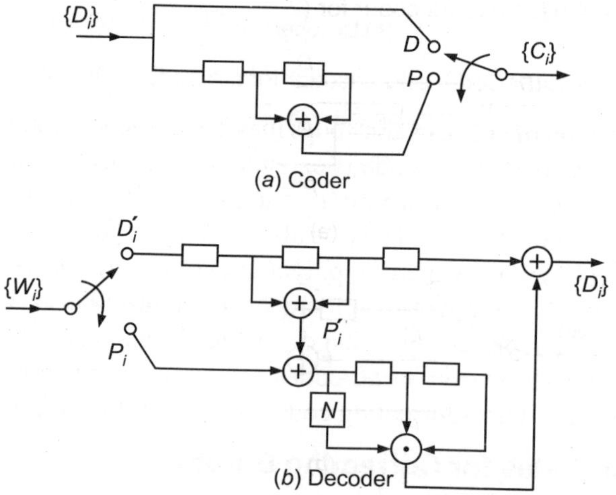

Figure 4.50: Coder/decoder for (2, 1) code, $K = 3$, $b = 2$, $g = 7$, N = NOT circuit

For low rate codes, e.g. rate 1/3, 1/4 etc., a simple and efficient burst correcting technique is possible by using interleaving with delayed repeat codes. Consider a rate 1/3 code with $K = 3$ and generator polynomials given by:

$$\begin{aligned} g(1, 1) &= (010) \\ g(1, 2) &= (001) \end{aligned} \tag{4.116}$$

The coder, shown in Fig. 4.52(a), simply rearranges the data bits and the corresponding parity bits with delay of one or two units, as shown in (4.117) below. It is seen that the matrix on the left with data and corresponding parity bits in the same column is converted to the matrix on the right with shifted parity bits. The coded bits are now transmitted column by column. In the decoder of Fig. 4.52(b), the data and parity bits are regrouped as in the matrix on the left, and as such the errors in parity bits are shifted out before the corresponding data bits are decoded. The decoding is done by simple coincidence of the syndrome bits as shown in the figure. As an example, if the data D_0 is wrong, the P'_0 and P''_0 locally

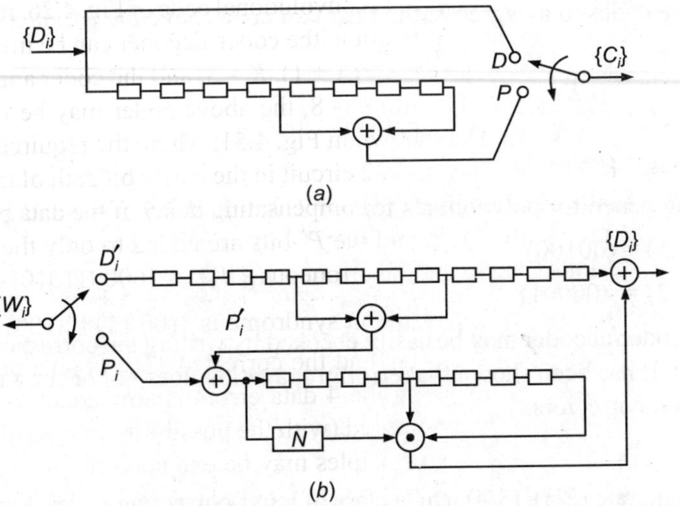

Figure 4.51: Coder/decoder for (2, 1) code, $K = 9$, $b = 8$, $g = 25$

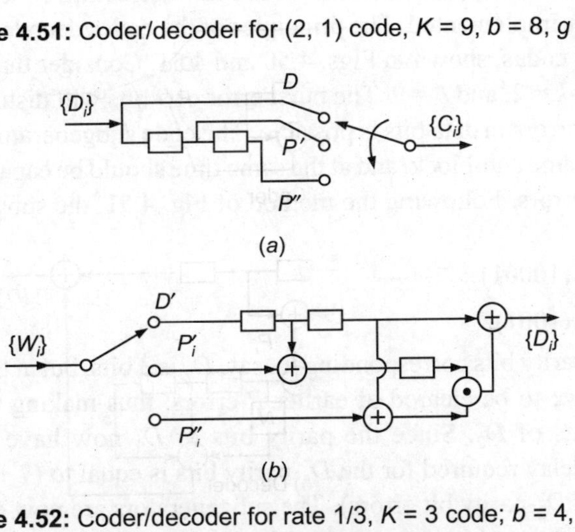

Figure 4.52: Coder/decoder for rate 1/3, $K = 3$ code; $b = 4$, $g = 8$

generated are also wrong and the AND gate output will given a correction signal. If, however, D_0 is correct and one of the received P'_0 or P''_0 is wrong, then the AND gate output is zero. From the matrix above, it is further observed that with a burst length $b = 4$ covering any four contiguous bits in 1st and 2nd columns, the possible errors in D_0 and D_i bits are corrected with a gap length $g = 8$ bits. As a result, the coder-decoder corrects one 4-bit burst in a span of 12 channel bits.

$$\begin{bmatrix} D_0 & D_1 & D_2 & D_3 \\ P'_0 & P'_1 & P'_2 & P'_3 \\ P''_0 & P''_1 & P''_2 & P''_3 \end{bmatrix} \rightarrow \left[\begin{array}{cccc|ccc} D_0 & D_1 & D_2 & D_3 & D_4 & D_5 & \cdots \\ P'_{-1} & P'_0 & P'_1 & P'_2 & P'_3 & P'_4 & \cdots \\ P''_{-2} & P''_{-1} & P''_0 & P''_1 & P''_2 & P''_3 & \cdots \end{array}\right] \tag{4.117}$$

For any value of n, $n \geq 3$ and rate $1/n$ of the $(n, 1)$ code, the parameters are now generalised as:

$$\text{With} \quad K = n = 3,\ b = \frac{nK-1}{2} = \frac{(3n-1)}{2}$$

$$\text{and} \quad g = (nK - 1) \tag{4.118}$$

Interleaving the above codes to a degree λ, the resultant codes have the parameters:

$$K_I = (n-1)\lambda + 1 = 2\lambda + 1$$
$$b = \frac{(nK_I - 1)}{2} \tag{4.119}$$
$$g = (nK_I - 1)$$

and the corresponding generator polynomials for $n = 3$, $\lambda = 2$, $K = 5$, $b = 7$ and $g = 14$ are:

$$g(1, 1) = (00100)$$
$$g(1, 2) = (00001) \tag{4.120}$$

The operation of the coder/decoder may be easily checked by writing the corresponding matrices similar to that in eqn. (4.117). It has been shown that error propagation does not occur with these codes and the codes are optimal burst correctors.

Iwadare Codes

For high rate codes, e.g., rate (2/3), (3/4), etc. a class of burst-correcting codes, known as Iwadare codes, is very efficient and simply implemented. The principles of the coder/decoder for these codes are similar to those of the interleaved codes, shown in Figs. 4.50 and 4.51. Consider that a burst-correcting code is to be designed for $n = 3$, $= k = 2$, and $b = 9$. The burst error of 9 bits will disturb at most 4 sub-blocks, of 3 bits, i.e., maximum of 3 errors in data bits is possible. The code subgenerators should be distinct for the data bits D_1 and D_2 of the same sub-block, and at the same time should be capable of generating syndrome bits which will correct 3 errors. Following the method of Fig. 4.51, the subgenerator polynomials may be:

$$g(1, 1) = (10001) \tag{4.121a}$$
$$g(2, 1) = (1001)$$

The delay required in the parity bits corresponding to, say, D_2 is 3 bits, but at the same time the syndrome register of (3 + 1) bits have to be cleared of earlier P-errors, thus making the total of 7 bits of delay required for the parity bits of D_2. Since the parity bits of D_1 now have to be distinct from those corresponding to D_2, the delay required for the D_1-parity bits is equal to (7 + 6) bits (6 bits for clearing the syndrome generator of D_2-parity bit errors). The subgenerators are now given by:

$$g(1, 1) = (000000000000010001) \tag{4.121b}$$
$$g(2, 1) = (000000010010000000)$$

The last 7-zeros in $g(2, 1)$ are to bring D_1 and D_2 in the same time frame at the decoder output. The decoder, of which the local coder is the same as that in the transmitter, is shown in Fig. 4.53, where D_2-bits are first corrected, and then D_1-bits. (But the order may be reversed). It can now be calculated that for $b = 9$, the guard space required in the decoder is 53 bits.

The parameters of the $(n, n-1)$ Iwadare codes may now be generalised as:

$$\text{For any } \lambda,\ \lambda > 0,\ K = \lambda(2n-1) + (1/2)n(n-1)$$
$$b = \lambda n \tag{4.122}$$
$$g = nK - 1$$

and the subgenerator polynomials are given by:

$$g(i, 1) = (\underbrace{00\ldots00}_{a_0}\underbrace{1000\ldots}_{b_0}\underbrace{1000\ldots00}_{c_0}) \tag{4.123}$$

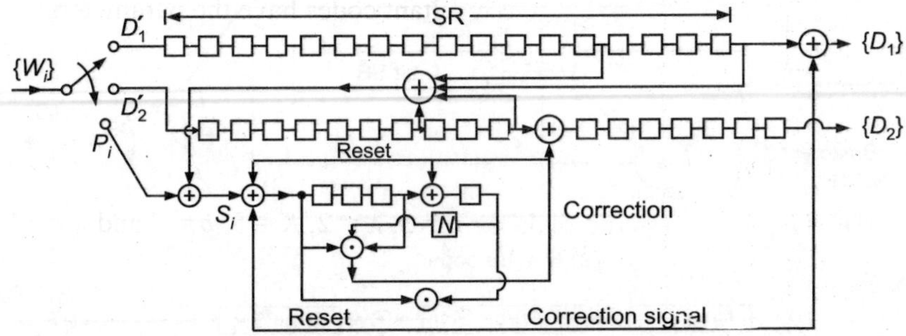

Figure 4.53: Decoder of Iwadare code (3, 2), $b = 9$, $g = 53$

where the number of zeros a_0, b_0 and c_0 are specified by:

$$a_0 = \left(\frac{n-i}{2}\right)(4\lambda + n - i - 3) + (n - 1)$$

$$b_0 = (n + \lambda - i - 2) \tag{4.124}$$

and $$c_0 = (K - a_0 - b_0 - 2)$$

A short list of burst-correcting codes and their parameters is given in Table 4.21.

Table 4.21: Parameters of burst-correcting codes

(n, k)	(2, 1), Ref. Figs. 4.50 and 4.51						(3, 1), Ref Fig. 4.52 and eqn. (4.119)			(3, 2), Ref. Fig. 4.53 and eqn. (4.122)			(4, 3) Ref. eqn. (4.122)		
K	3	5	7	9	11	13	5	7	11	13	18	23	20	27	34
b	2	4	6	8	10	12	7	10	16	6	9	12	8	12	16
g	7	13	19	25	31	37	14	20	32	38	53	68	79	107	135

Example 4.34: Consider that a (4, 3) code with $b = 12$ is required. Then, from eqn. (4.122), $\lambda = 3$, $K = 27$ and $g = 107$. The subgenerators are specified by eqn. (4.123) with the values of a_0, b_0 and c_0, as given by:

$$a_0(1, 1) = 21, \quad b_0(1, 1) = 4, \quad c_0(1, 1) = 0$$
$$a_0(2, 1) = 14, \quad b_0(2, 1) = 3, \quad c_0(2, 1) = 8$$
$$a_0(3, 1) = 8, \quad b_0(3, 1) = 2, \quad c_0(3, 1) = 15$$

Then, the subgenerator polynomials are:

$$g(1, 1) = (000000000000000000000100001)$$
$$g(2, 1) = (000000000000001000100000000)$$
$$g(3, 1) = (000000001001000000000000000)$$

The coder and decoder may now be drawn based on Fig. 4.53.

4.7.3 Adaptive Gallager Code [19]

Many communication channels are composite having both random and burst errors. The adaptive Gallager code is an effective technique for simultaneously correcting both random and burst errors in the channel. The rate 1/2 coder, shown in Fig. 4.54(a), consists of two parts: (*a*) an inner coder which uses a random-error correcting code, say, an orthogonalizable convolutional code, for a design value of t, and (*b*) the outer coder which uses the basic principles of burst correction by the technique of Fig. 4.51. The decoding

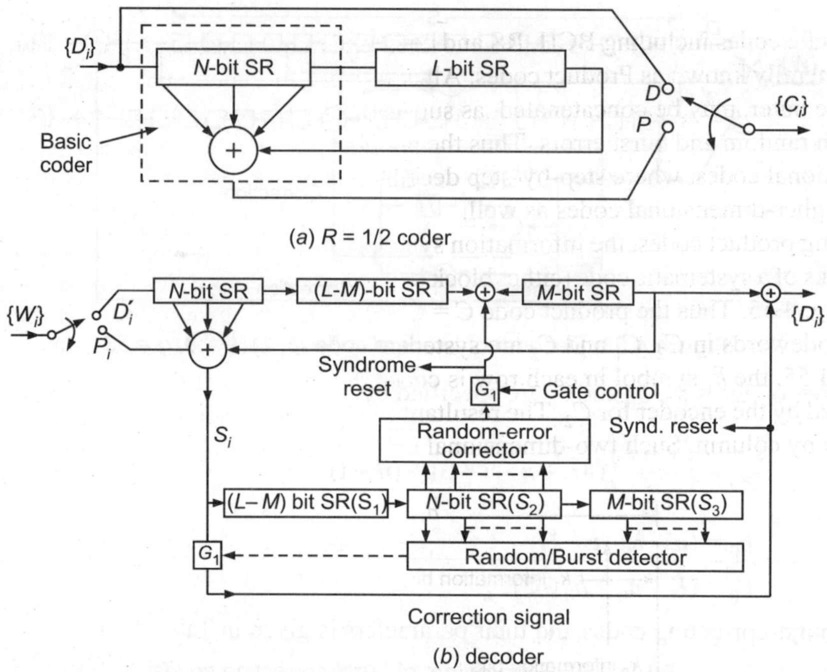

Figure 4.54: Coder/decoder for $R = 1/2$ adaptive code: $b \leq 2L$, $g = 2L + N$

scheme, shown in Fig. 4.54(b) is an adaptive one, where the decoder is usually in the random-correcting mode with gate G_1 open, and t-random errors are corrected through the N-bit-S_2 syndrome register and the associated logic circuit, say, the majority logic. When a large burst occurs in the channel, the S_2 and S_3 registers with the associated burst detector sense the arrival of the burst and switch the decoder to the burst mode, i.e., gate G_2 is opened and G_1 closed. Since the burst is sensed after $(L + N)$ syndrome bits, $N \leq L$, the erroneous data bits of length $(L + N)$ are stored simultaneously in the upper information registers of length $[N + (L - M) + M]$ bits. If the burst length $b \leq 2L$, then the incoming $\{W_i\}$ is now error free, since all erroneous L-bit data and L-bit parity are by this time shifted into different registers. If the guard space is little more than $2L$, then erroneous bits shifting out of the upper M-bit register will now be corrected by the syndrome bits through G_2. At the end of the burst, few all-zero syndrome bits in S_2 and S_3 will restore the decoder to the random mode. The lengths of the registers, i.e., L, M, N and t are chosen to suit the requirements of the channel, with the restriction that $N \ll L$, and $M < L$. The guard space required in the scheme is: $g = 2L + N$, and the ratio $b/g \simeq 1$, showing that the code is optimal. However, all bursts of $b \leq 2L$ are not corrected in the decoder, as some burst-error patterns may not be detected by the burst detector, but this is compensated by the code's capability of correcting bursts of length b with guard space of $g \simeq b$. The above rate 1/2 adaptive coding scheme may be generalised for other code rates as well.

4.8 TWO-DIMENSIONAL CODES [5, 20]

In Sec. 4.2.1, it has been shown that an array of data vectors may be simultaneously checked by simple parity checks along the rows and columns, and the resultant coded array can be processed in the decoder to correct single or double errors. This iteration technique of Elias [5] can be extended to more powerful codes, e.g., cyclic codes including BCH, RS and convolutional codes. Such codes using an array of data vectors are generally known as Product codes. Alternatively, two codes, one called as the inner code and the other as the outer, may be concatenated, as suggested by Forney [20], to form a very powerful code correcting both random and burst errors. Thus the product code and the concatenated code are in a class of two-dimensional codes, where step-by-step decoding is performed. The techniques may be extended to three and higher-dimensional codes as well.

In developing product codes, the information symbols $k = k_1 k_2$ are arranged in rows and columns, and the parity checks of a systematic code (either block or convolutional) are added to these rows and columns, as shown in Fig. 4.55. Thus the product code $C = C_1 \otimes C_2$, where the rows are codewords in C_1 and the columns are codewords in C_2. C_1 and C_2 are systematic, then C is also a systematic code. As is evident from the Fig. 4.55, the k_1 symbol in each row is coded by the encoder for C_1 and the k_2 symbol in each column is coded by the encoder for C_2. The resultant code array is then transmitted serially either row by row of column by column. Such two-dimensional codes have the following properties:

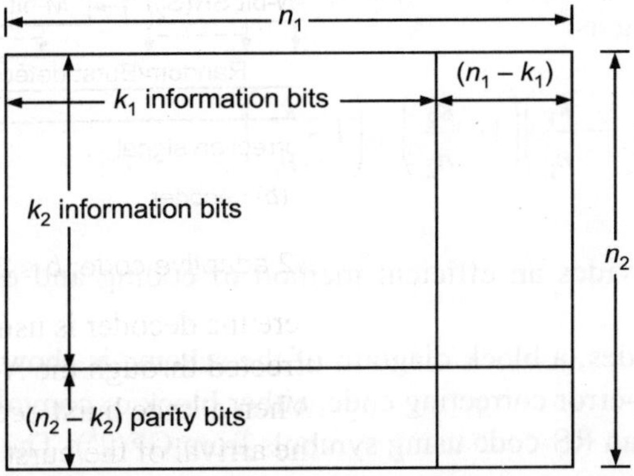

Figure 4.55: Structure of a systematic product code

(a) If the minimum distances of the codes C_1 and C_2 are d_1 and d_2 respectively, then the d_{min} for the Code C is $(d_1 d_2)$.
(b) The code C will correct all error patterns of weight $t \le (d_1 d_2 - 1)/2 = (2t_1 t_2 + t_1 + t_2)$.
(c) The code C also corrects all bursts of length upto $b = \max(n_1 t_2, n_2 t_1)$, where t_1 and t_2 are the error correcting capabilities of C_1 and C_2 respectively.
(d) For a BSC, the error probability in the receiver using the array code C is better or equal to $f_2[f_1(p)]$, if $f_1(p)$ and $f_2(p)$ are the error probabilities obtained through the individual codes C_1 and C_2, and p = transition probability in the BSC.

The properties (a) and (b) are evident from the array organisation. The property (c) is obtained from the consideration that the array is equivalent to the interleaving of the coded words as discussed in Sec. 4.7 and illustrated in Fig. 4.47. With $t = t_1$ for C_1 and $t = t_2$ for C_2, the length of the bursts that will be corrected by the interleaved array is either $n_1 t_2$ or $n_2 t_1$, whichever is the larger. It is then appropriate to use the code C_2 with parameter (n_2, k_2) as a burst correcting code (assuming that the array is transmitted column by column) and the code C_1 for (n_1, k_1) as a random-error correcting code. The residual burst plus the random errors will now be corrected by C_1; however, the decoders now will have to be more intelligent than the single decoders discussed so far. Such composite decoders have been discussed in Ref. 7, Chapter 10.

The overall error probability for a BSC using two-dimensional array codes has been given in eqn. (4.10), which conforms to the property (*d*). More specifically, if SEC Hamming codes with an extra parity for double-error detection is used as C_1, then the residual probability of error p_1 per codeword is

$$p_1 \le n_1 p^2 \tag{4.125}$$

If now Elias's iteration technique is used for $n_2, n_3, \ldots n_r$ blocks, and the blocks are coded as $C_2, C_3, \ldots, C_r$, then,it can be shown that the residual error probability after each iteration is given by:

$$p_2 \le n_2 p_1^2 < n_2 n_1^2 p^4$$

$$p_3 \le n_3 p_2^2 < n_3 n_2^2 p_1^4 < n_3 n_2^2 n_1^4 p^8 \tag{4.126}$$

and

$$p_r \le (n_r n_{r-1}^2 n_{r-2}^4 \ldots n_1^{2(r-1)}) p^{2^r}$$

Then, $p_r \to 0$ as $r \to \infty$, for $n_r > n_{r-1} \ldots > n_1$ and $n_1 p < 1/2$. At the same time, the information rate in the channel decreases according as:

$$R_r = \left(1 - \frac{k_1}{n_1}\right)\left(1 - \frac{k_2}{n_2}\right) \ldots \left(1 - \frac{k_r}{n_r}\right) \tag{4.127}$$

but $R_r > 0$.

Thus the technique provides an efficient method of coding and decoding for binary symmetric channels.

For the concatenated codes, a block diagram of the scheme is shown in Fig. 4.56, where the inner code is generally a random-error correcting code, either block or convolutional, and the outer code is a burst-correcting code, say, an RS-code using symbols from GF(2^k). The coding scheme is as follows:

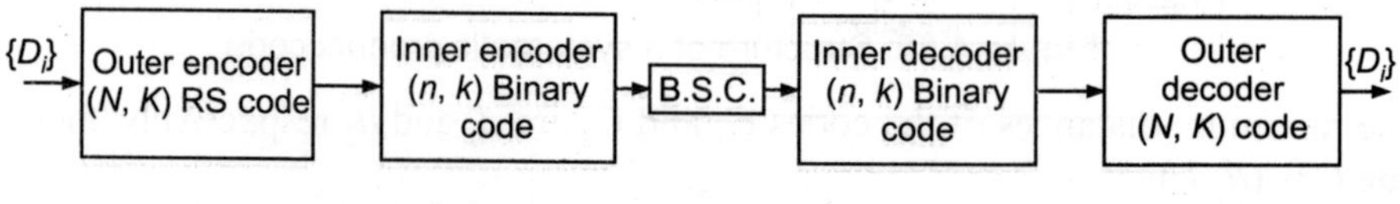

Figure 4.56: A scheme for a concatenated code

The input data $\{D_i\}$ is divided into K bytes of k bits, and each byte of k bits is considered to be a symbol in GF(2^k). The K bytes are then coded into an N-byte codeword in the outer encoder. Each of the N bytes of k bits is now considered as a k-bit word for encoding in the inner encoder, which encodes the k-bit word into n-bit codewords. The set of N codewords of the (n, k) inner code is then a codeword of the concatenated (Nn, Kk) code.

As an example, one may use the (15, 7) RS code with symbols from GF(2^4) as the outer code, which will correct t errors, $t \le 4$ (i.e., 4 bytes, each of 4 bits, out of 15 bytes in the codeword will be corrected). The inner code now has $k = 4$, and the (7, 4) binary Hamming code may be used. Thus 28 bits of $\{D_i\}$ are taken as 7 bytes of 4 bits each, and converted into the (105, 28) code, whose rate ≃ 1/4. For decoding, the inner code is decoded first delivering 4-bit codewords will be corrected in the inner coder, and the residual errors in the output will be in the form of erroneous bytes delivered to the outer coder. If the number of wrong bytes ≤ 4 out of 15 bytes, then all the errors will be corrected by the outer decoder. If there are bursts of length $b \le 28$ bits in 105 channel bits, then this will give a 4-byte error at the input to the outer decoder, and hence will be corrected.

Alternatively, a combination of an inner convolutional code with a simple decoder and an outer RS code is also very effective. The inner code will correct random errors and the burst-errors will be corrected by the outer RS-code. The outer code can also be any other efficient burst-correcting code, say, the Iwadare code. Thus a very powerful code is obtained through the concatenation of simpler codes, specially for Gaussian-noise channels.

4.9 ARQ

For data transmission, the received data has to be 100% correct but this cannot be achieved even with a very good channel. Error rates in ordinary telephone lines is of the order of 10^{-5} and in digital networks, the error rates are in the range of 10^{-6} to 10^{-8}. To eliminate these errors, the usual method is to use an additional feedback channel, through which requests for retransmission of the erroneous data blocks are sent to the transmitter. This technique, known as ARQ (Automatic Request), makes use of forward error detection, as distinct from the forward-error correction (FEC). Of course, FEC may be used to improve the raw forward channel error rate, and FEC is a must in cases where the total path delays are large, as in space channels. The major elements in an ARQ system are shown in Fig. 4.57, where a temporary buffer is required to store the data, till the confirmation is received from the destination. ARQ systems are of two types: (a) stop-and-wait system, where the transmitter sends one block of encoded data at a time and waits for an acknowledgement (ACK) from the receiver before sending the next block; (b) continuous system, where the transmitter continuously sends the coded data and when an error is indicated by the NACK signal via the feedback channel, the transmitter repeats the block in error and subsequent blocks. The number of blocks retransmitted now depends on the round-trip delay of the channel and is usually more than one. Although more efficient, the continuous method is more complex to organize and requires more storage.

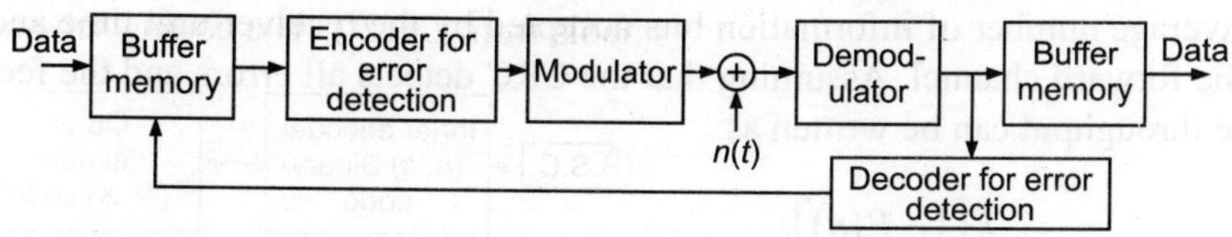

Figure 4.57: Block diagram of an ARQ system

Many types of error-detecting codes have been used in ARQ, and the simplest of them is the odd/even parity check of a data block, applicable for short blocks only. The method is improved by using two-dimensional block coding, where columns and rows are each coded through parity checks, as discussed in Sec. 4.2.1. A further improvement is obtained by adding the binary counts of all marks in each block in addition to the parity check, and the characters thus attached are called the Block Check Sequence (BCS). Various subclasses of the cylic codes are also used, and any cyclic code having H.D. = d is capable of detecting all errors $t \le (d-1)$. A particularly popular block check method, known as Cyclic Redundancy Check (CRC), uses the generator polynomial

$$g(X) = X^{16} + X^{12} + X^5 + 1 \tag{4.128a}$$

or its modified version as a Frame check sequence. As in the case of encoding data blocks $D(X)$ with cyclic codes, discussed in Sec. 4.3, the coding is performed by the equation

$$X^{n-k} \cdot D(X) = Q(X)\, g(X) + R(X)$$

where $R(X)$ is the remainder obtained by dividing $X^{n-k}D(X)$ by $g(X)$. Then the (n, k) code is given by:

$$C(X) = X^{n-k}\,D(X) + R(X) \tag{4.128b}$$

In the receiver, the $C(X)$ is checked with $g(X)$ and the remainder is zero for no error. Since $R(X)$ contains 16 bits of parity check, the code is known as CRC-16. The code detects error bursts of length $\le (n-k)$ with probability = 1, bursts of length = $(n-k+1)$ with probability of $(1-2^{-(n-k-1)})$, and bursts of length $> (n-k+1)$ with probability of $(1-2^{-(n-k)})$.

The average error probability in a block of n chips is:

$$P(n) = 1-(1-p)^n \simeq np,\ p \ll 1$$

where p is the transition probability in the BSC. Then the number of times, N_r, that the block has to be repeated is given by:

$$N_r = \frac{P(n)}{1-P(n)} \tag{4.129}$$

which increases almost linearly with n. The block size n, then, has to be kept within reasonable values so that the repeat requests are not too many.

The effective Rate of transmission R_e in an ARQ system is lower than the Information rate R out of the buffer due to: (a) the round-trip delay time and (b) slight increase of the transmission rate (Chip rate) by factor n/k. The Transmission efficiency or the Throughput T is now given by:

$$T = \frac{R_e}{R \cdot n/k} \tag{4.130}$$

where R_e = Average number of information bits accepted by the receiver/unit time and $R.n/k$ = modem chip rate in the forward channel. Assuming that the CRC detects all errors and the feedback channel is error-free, the throughput can be written as:

$$T(\text{stop and wait}) = \frac{k[1-P(n)]}{n+D}$$

and (4.130a)

$$T(\text{continuous}) = \frac{k[1-P(n)]}{n+P(n)\cdot D}$$

where D = round-trip delay in chips while receving the ACK message. Since for practical values of $P(n)$, the influence of $n > k$ is small, one can write:

$$R_e = TR{\cdot}n/k \simeq TR \tag{4.131}$$

In practical circuits, with 1200 $b/s < R <$ 4800 b/s, the throughput T has been achieved in the range of 0.80 to 0.98.

4.10 PERFORMANCE OF CODES

The methods of generation of codewords and their decoding techniques have been discussed so far. In this section, the error-correcting capabilities and their bounds will be briefly discussed. Consider the simple single-error correcting block codes, say, Hamming codes, where the probability of correct decoding of a codeword is (assuming binary symmetric channels with pr. (1) = pr. (0) = 0.5, pr. (1/1) = pr. (0/0) = q and pr. (1/0) = pr. (0/1) = $p = (1 - q)$)

$$P_c = \text{pr. (correct decoding)} = q^n + \binom{n}{1} p.q^{n-1}$$

Then the word-error probability $P_w = 1 - P_c$

$$= 1 - (1-p)^n - np(1-p)^{n-1}$$

$$\simeq n(n-1)p^2, \; p \ll 1 \tag{4.132}$$

The bit-error probability $P_b = P_w/k$ and with no error correction $P_w \simeq kp$, where k-bit words are coded into n-bit words. A figure of merit F may now be defined as:

$$F = \frac{P_b\,(\text{uncoded})}{P_b\,(\text{coded})} = \frac{kp}{n(n-1)\,p^2} = \frac{k}{n(n-1)\,p} \text{ for SEC.} \tag{4.133}$$

Based on eqn. (4.133), the values of F for (n, k) SEC block codes are given in Table 4.22.

Table 4.22: Values of F and P_b for SEC block codes

(n, k)	(3, 1)	(7, 4)	(15, 11)	(31, 26)	(63, 57)
F	$1/6p$	$2/21p$	$11/210p$	$13/465p$	$19/1302p$
P_b	$6p^2$	$42p^2$	$210p^2$	$930p^2$	$3906p^2$

For multi-error correcting codes with $t > 1$, the probability of correct decoding is given by:

$$P_c = q^n + \binom{n}{1} pq^{n-1} + \binom{n}{2} p^2 q^{n-2} + \ldots + \binom{n}{t} p^t q^{(n-t)} \tag{4.134}$$

and

$$P_w = (1 - P_c) \geq \binom{n}{t+1} p^{(t+1)} q^{(n-t-1)}, \; t \ll n, \; p \ll 1$$

$$\simeq n^{(t+1)} p^{(t+1)} \tag{4.135}$$

For systematic convolutional codes, similar equations are derived based on the equivalent codeword length = nK. As for example, for $R = \frac{1}{2}$, $K = 4$, systematic convolutional code, 2 errors in 8 bits are corrected. The P_w for the code is:

$$P_w = 1 - \left[q^n + \binom{n}{1} pq^{(n-1)} + \binom{n}{2} p^2 q^{(n-2)} \right], \text{ with } n = 8.$$

$$\simeq \frac{n^2 p^2}{2} + \frac{n^3 p^3}{2} \simeq 3 \times 10^{-3} \text{ for } p_e = 10^{-2} \text{ in the channel}$$

But for $p_e = 10^{-4}$, $P_w \simeq 3 \times 10^{-7}$ and $P_b \simeq (3/8) \times 10^{-7}$

For non-systematic codes, however, such simple relations for P_w are not available, and most of the error performances of such codes are obtained through computer simulation. A short table of the figure of merit F for some typical codes is given in Table 4.23, where the complexity of the decoder is also indicated in terms of the shift-registers (SR) required [21]. It is seen that low rate codes given better values of F and F is larger for lesser values of channel errors. For low rate, R = 1/2, 1/4, ..., 1/16, convalutional codes, the improvement at the decoder output is indeed spectacular, as is shown in Fig. 4.58. It is possible to recover data with $P_b < 10^{-5}$, even with a channel error rate $p_e \simeq 0.4$, (but $p_e < 0.5$), if one uses a composite codec with $R = 1/32$. Generally, $R = 1/4$ codecs are built as a single unit and the tandem connections of such codecs produce the low rate codec as desired. An alternative evaluation of error performance of the codes is obtained by noting the coding gain in terms of E_b/n_0 at a given P_b. As for example, P_e (uncoded PSK modulation) = 10^{-4} at E_b/n_0 = 8.5 dB. Referring to Table 4.23, the figure of merit F for $R = 1/2$, convolutional code with Viterbi decoding is 10^6, i.e., P_b at the decoder output is 10^{-10}, but the new E_b/n_0 (coded) = 11.5 dB, and the value of P_b at E_b/n_0 (coded) = 8.5 dB is calculated to be 10^{-7}. The coding gain at $P_b = 10^{-7}$ is seen to be 3 dB. The coding gains of different codes may be obtained from the curves of Fig. 4.59, which shows the variation of P_b (coded) vs. E_b/n_0 (the results obtained through computer simulation). It is seen that in general, the convolutional codes with Viterbi/sequential decoding perform much better than the block codes. However, for non-Gaussian channels, interleaved or adaptive codes have to be used and their performances have to be individually calculated.

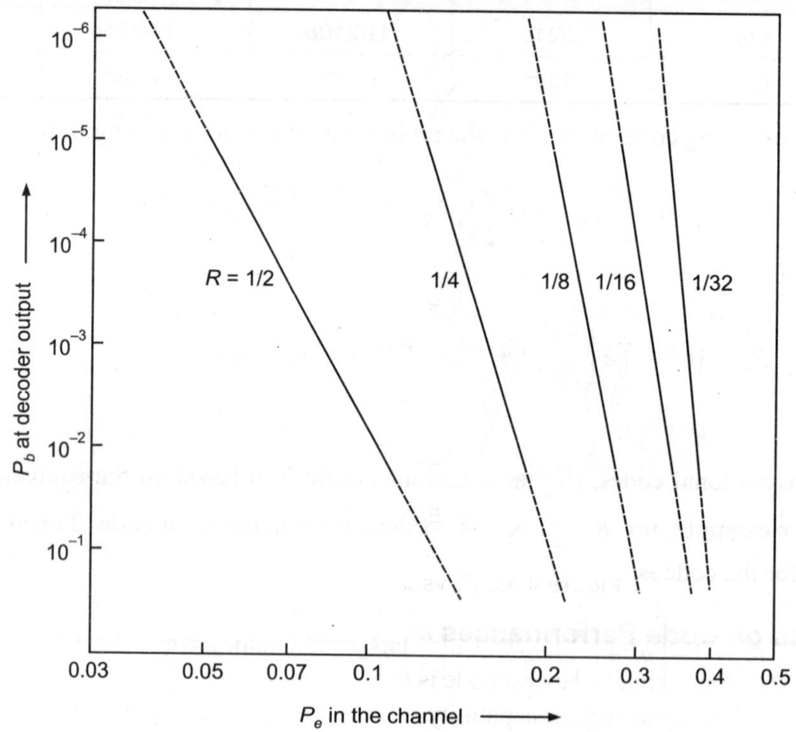

Figure 4.58: P_b vs. P_e(ch) for low-rate convolutional code

Table 4.23: Values of F for some FEC codes

Codes	Rate	No.of SR's required	F for P_b (uncoded) = 10^{-2}	10^{-3}	10^{-4}
Block code $t = 1$	26/31	93	1/10	1	10
BCH, $t = 2$	7/15	87	10	10^3	10^5
Golay code $t = 3$	12/24	53	10^2	10^5	10^7
Convolutional Codes: Wyner-Ash, $t = 1$, Interleaved	1/2	23	10	10^2	10^3
CC-W-A, $t = 1$, Interleaved	3/4	43	5	25	2×10^2
CC-W-A, $t = 1$, Interleaved	7/8	87	1.2	10	10^2
Self-orthogonal CC, Interleaved, $t = 2$	1/2	51	10^2	10^3	$>10^5$
As above	3/4	209	10	5×10^2	$>10^4$
As above	7/8	873	1	50	$>10^3$
Convolutional code: Viterbi Decoding, Hard decision	1/2	–	5×10^2	10^5	10^7
Convolutional code: $K = 7$, Viterbi Decoding, Soft decision	1/2	–	10^6	10^9	–

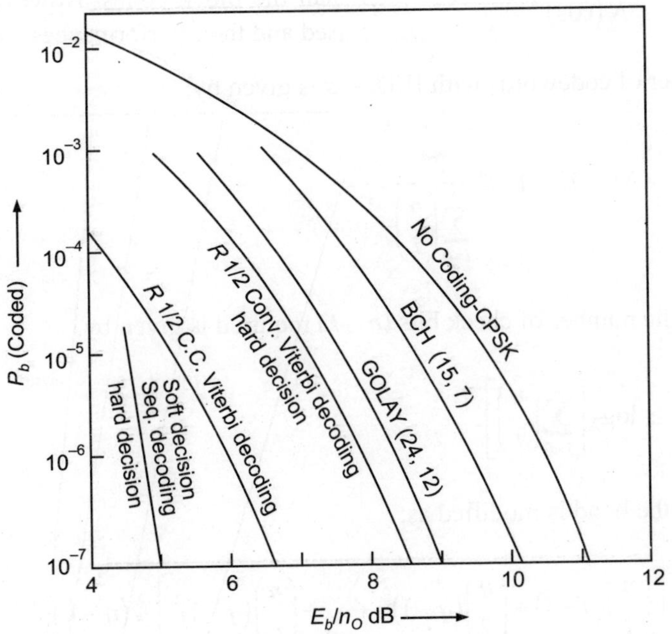

Figure 4.59: P_b vs E_b/n_0 various codes

4.10.1 Bounds on Code Performances

Among (n, k) linear (Group) codes, the best code is the one which gives the maximum H.D. = $d_{\max}$, as d is directly related to the error correcting capability $t \leq (d - 1)/2$. Unfortunately, the value of $d_{\max}(n, k)$ is known only for $k \leq 5$, and for the codes with $d_{\max} \leq 3$, including a few other isolated cases. Thus it is necessary to obtain bounds on d_{max}, and consequently on P_w for (n, k) codes.

Considering first the (n, k) group codes, it is noted that the codes have 2^k message words (= code vectors) and 2^{n-k} cosets. For a code correcting all combinations of t errors or less, all vectors of weight t (= number of 1's) or less must be coset leaders (Refer to Sec. 4.2.4). Thus the number of vectors of weight t or less ≤ number of cosets = 2^{n-k}. But the number of coset leaders $N(\cos)$ is given by:

$$N(\cos) = 1 + \binom{n}{1} + \binom{n}{2} + \dots + \binom{n}{t} \le 2^{n-k}$$

or

$$\sum_{i=0}^{t} \binom{n}{i} \le 2^{n-k} \tag{4.136}$$

Therefore, number of codewords available

$$N_c = \frac{2^n}{N(\cos)} \ge 2^k \tag{4.137}$$

In terms of d, the number of codewords with H.D. = d is given by:

$$N(n, d) = N(n, 2t+1) \le \frac{2^n}{\sum_{i=0}^{t} \binom{n}{i}} \tag{4.138}$$

and from eqn. (4.136), the number of check bits $(n - k)$ required is given by:

$$(n-k) \ge \log_2 \left[\sum_{i=0}^{t} \binom{n}{i} \right] \tag{4.139}$$

For symbols for GF(q), the bond is modified as:

$$\log_q \left[1 + \binom{n}{1}(q-1) + \binom{n}{2}(q-1)^2 + \dots + \binom{n}{t}(q-1)^t \right] \le (n-k) \tag{4.140}$$

The eqn. (4.140) is known as the sphere-packing bound or the Hamming upper bound. It also applies to nonlinear codes. For binary codes, the asymptotic value of the bound is obtained by defining, $t = \lambda n$, and

$$B_H(n) = \frac{1}{n} \log_2 \left[\sum_{i=0}^{n} \binom{n}{i} \right] \le \left(\frac{n-k}{n} \right) = \left(1 - \frac{k}{n} \right) \tag{4.141}$$

In the limit $n \to \infty$,

$$(1 - k/n) \geq \underset{\text{limit } n\to\infty}{B_H(n)} = H(\lambda) = H(t/n), \tag{4.142*}$$

where $H(\lambda)$ = entropy function $= -\lambda \log \lambda - (1 - \lambda) \log (1 - \lambda)$. Other bounds for k/n (for binary codes only) are:

(1) Plotkin's upper bound on d:

$$(n - k) \geq (2d - 1) - 1 - \log_2 d \tag{4.146}$$

and $n \to \infty$, $(1 - k/n) \geq 2d/n$.

This bound is based on the very simple notion that the minimum weight of a codeword in a (n, k) linear code is at the most as large as its average weight.

(2) Elias's upper bound on d:

For any (n, k) block code, the minimum distance is bound by:

$$\lim_{n\to\infty} d \leq 2t_h\left(1 - \frac{t_h}{n}\right)$$

where t_h is given by:

$$\left(1 - \frac{k}{n}\right) = H(t_h/n) \tag{4.147}$$

---

*The limiting value of $B_H(n)$ is derived as follows [2, 6]; Using Sterling approximation,

$$\binom{n}{\lambda n} = \frac{n!}{(\lambda n)!\,(n-\lambda n)!} \simeq \left[\frac{1}{\lambda^{\lambda}(1-\lambda)^{1-\lambda}}\right]^n \cdot \left[\frac{1}{2\pi\lambda(1-\lambda)n}\right]^{1/2} = \left(\frac{A}{\sqrt{n}}\right)2^{nH(\lambda)}, \tag{4.143}$$

where $A/\sqrt{n} < 1$ for $n \gg 1$ and $H(\lambda)$ is as defined above.

Also

$$\binom{n}{\lambda n+1} \simeq \frac{A}{\sqrt{n}}.2^{nH\left(\lambda+\frac{1}{n}\right)} \simeq \left(A/\sqrt{n}\right)^{2nH(\lambda)}, \text{ as } n \to \infty.$$

Now in a Binomial series,

$$\sum_{i=0}^{\lambda n}\binom{n}{i} < \binom{n}{\lambda n+1} \simeq \left(A/\sqrt{n}\right)2^{nH(\lambda)}, \text{ for } \lambda < \tfrac{1}{2}$$

Thus it may be shown that:

$$\sum_{i=0}^{\lambda n}\binom{n}{i} \leq 2^{nH(\lambda)}, \text{ for } \lambda < 1/2, \text{ and } n \geq \frac{1-\lambda}{2\pi\lambda(1-2\lambda)^2} \tag{4.144}$$

Therefore,

$$1/n \log_2\left[\Sigma\binom{n}{i}\right] \leq H(\lambda)$$

or $\quad B_H(n) \leq H(\lambda)$ and in the limit $n \to \infty$, $B_H(n) = H(\lambda)$ $\quad$ (4.145)

(3) The above bounds indicate that the d_{min} of linear (n, k) codes cannot be greater than the upper bound. The lower bound on d is given by Varsharmov-Gilbert bound as:

$$(1 - k/n) \geq 1 + \left(\frac{d-2}{n-1}\right) \xrightarrow[\lim n \to \infty]{} H(d/n) \tag{4.148}$$

The eqns. (4.142), (4.146), (4.147) and (4.148) are plotted in Fig. 4.60, where it is seen that Elias's bound is tighter than others and the rate $k/n \to 0$ for $d/2n \geq 0.25$ (i.e., for $t/n \geq 0.25$). For large rates, $t/n \ll 1$ only will satisfy the bounds.

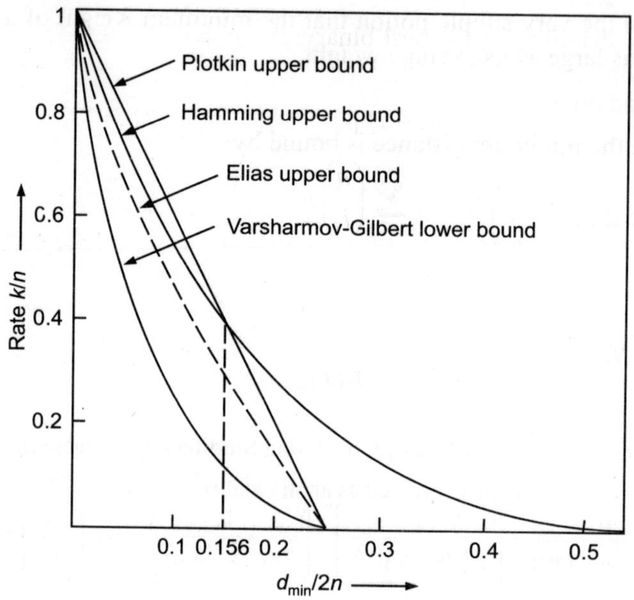

Figure 4.60: Bounds on d_{min} for best binary block codes ($n \to \infty$)

For convolutional codes, the bound for d_{min} is approximately the same for large n. When n is not large, d_{min} for best convolutional codes is slightly larger than the d_{min} for best block codes. Thus for moderate values of n, convolutional codes are slightly superior to block codes.

4.10.2 Bounds on Error Performances

In Sec. 4.2.4, perfect and quasi-perfect codes have been defined. Accordingly, a quasi-perfect (n, k) linear code corrects all t or less errors and also corrects some $(t + 1)$ errors. That is, its coset leaders contain all combinations of t or less 1's (weight t or less), but contains no vector with weight $> t + 1$. The code is then said to be optimum. The probability of correct decoding of a codeword is now given by:

$$P_c = q^n + \binom{n}{1} pq^{(n-1)} + \binom{n}{2} p^2 q^{(n-2)} + \ldots + \binom{n}{t} p^t q^{(n-t)} + a_{t+1} p^{(t+1)} q^{(n-t-1)} \tag{4.149}$$

where a_{t+1} = number of coset leaders with weight $(t + 1)$. Further,

$$1 + \binom{n}{1} + \binom{n}{i} + \ldots + \binom{n}{t} + a_{t+1} = 2^{n-k}.$$

Therefore,

$$a_{t+1} = 2^{n-k} - \left[\sum_{i=0}^{t}\binom{n}{i}\right] \geq 0 \tag{4.150}$$

and t is the greatest integer satisfying the above eqn. (4.150). The word-error probability $P_w = (1 - P_c)$, where P_c is the upper bound for correct decoding, and $P_b = P_w/k$. To show the asymptotic behaviour of P_w, define a quantity E, called reliability, as:

$$E = \lim_{n\to\infty} (-1/n)\,[\log_2 P_w] \tag{4.151}$$

where P_w = word error probability for the best binary group code. Denoting E_s as the sphere-packing bound on reliability, one obtains:

$$\begin{aligned} E \leq E_s &= \lim_{n\to\infty} (-1/n)\log_2[1 - P_c] \\ &= \lim_{n\to\infty} (-1/n)\log_2\left[\sum_{i=t+2}^{n}\binom{n}{i}p^i q^{n-i}\right] \\ &= E(t/n, p) \\ &= E(\lambda, p) > 0,\ \lambda = t/n. \end{aligned} \tag{4.152}$$

where $E(\lambda, p)$ is defined as:

$$E(\lambda, p) = H(p) + (\lambda - p)H'(p) - H(\lambda), \tag{4.154*}$$

---

*The tail of a binomial distribution may be expressed as an inequality as [2] (cf. eqn. (4.143)):

$$\sum_{i=\lambda n}^{n}\binom{n}{i}p^i q^{n-i} \leq \lambda^{-\lambda n}\mu^{-\mu n}p^{\lambda n}q^{\mu n}, \text{ for}$$

$$\lambda > p,\ \mu < q,\ (\lambda + \mu) = 1,\ (p + q) = 1.$$

But

$$1/n \log(\lambda^{-\lambda n}\mu^{-\mu n}) = -\lambda\log\lambda - \mu\log\mu \triangleq H(\lambda),$$

Therefore,

$$-\frac{1}{n}\log(\lambda^{-\lambda n}\mu^{-\mu n}p^{\lambda n}q^{\mu n}) = -H(\lambda) - \lambda\log p - \mu\log q.$$

Since,

$$\begin{aligned} -\lambda\log p - \mu\log q &= -p\log p - q\log q + (p - \lambda)\log p - (\mu - q)\log q \\ &= H(p) + (p - \lambda)\log p - (\mu - q)\log q \\ &= H(p) + (\lambda - p)[\log q - \log p],\ (\lambda - p) = (q - \mu) \\ &= H(p) + (\lambda - p)\log[(1 - p)/p] \\ &= H(p) + (\lambda - p)H'(p), \quad \text{by eqn. (4.154)} \end{aligned}$$

Therefore,

$$(-1/n)\log_2\left[\sum_{i=\lambda n}^{n}\binom{n}{i}p^i q^{n-i}\right] \leq H(p) + (\lambda - p)H'(p) - H(\lambda) = E(\lambda, p) \tag{4.153}$$

and in the limit $n \to \infty$, equality holds.

where

$$H'(p) = \frac{dH(p)}{dp} = \log\left(\frac{1-p}{p}\right)$$

Therefore

$$P_w = 2^{-n(E(\lambda,\, p))} = \delta \to 0 \text{ in the limit } n \to \infty. \tag{4.155}$$

This is the ultimate limit of error probability that may be obtained for best linear codes. From eqn. (4.141),

$$\frac{1}{n}\log_2\left[\sum_{i=0}^{t}\binom{n}{i}\right] \le (1-k/n) \le \frac{1}{n}\log_2\left[\sum_{i=0}^{t+1}\binom{n}{i}\right]$$

and in the limit $n \to \infty$

$$(1 - k/n) = H(\lambda) \doteq H(t/n) \tag{4.156}$$

This value of t has to be used in eqn. (4.152) for evaluating E_s.

The above bounds for d_{min} and the reliability function E are hardly achieved by any practical codes. However, the convolutional codes, specially with soft decision Viterbi decoding and sequential decoding, performe much better than the block codes, as has been shown in Fig. 4.59. The bounds have been evaluated on the assumption of a B.S.C and Gaussian noise statistic. The performance of the codes in non-Gaussian channels, e.g., in channels with composite random and burst noise, can only be evaluated by computer simulation. For burst-error correcting codes, the requirements of partity check symbols have been discussed in Sec. 4.7. In general, the burst-error correcting capability of convolutional codes is superior to that of block codes, for large (n, k).

PROBLEMS

1. Write the Addition and Multiplication tables for the field of four-elements {0, 1, 2, 3}. Using the table, solve the equations

$$2X + Y = 3$$
$$X + 2Y = 3$$

2. Find the roots of (a) $f(X) = X^3 + \alpha^6 X^2 + \alpha^9 X + \alpha^9$ in $GF(2^4)$ (Refer to Table 4.12), (b) $f(X) = X^4 + X^3 + 1$.
3. Form tables, similar to Table 4.1 and Table 4.12, for $GF(2^3)$ and $GF(3^2)$.
4. Draw a shift register circuit for simultaneous multiplication by $B(X)$ and division by $D(X)$, where

$$B(X) = 1 + X^2 + X^5 + X^6$$
$$D(X) = 1 + X^3 + X^4 + X^5 + X^8$$

If the input to the circuit $A(X) = 1 + X^7$, calculate the quotient and the remainder.

5. Form the standard array for a group code using the generator matrix:

$$[G] = \begin{bmatrix} 1 & 1 & 0 & 0 & 1 & 1 & 0 \\ 0 & 1 & 1 & 0 & 0 & 1 & 1 \\ 0 & 0 & 1 & 0 & 1 & 1 & 0 \\ 0 & 0 & 0 & 1 & 1 & 1 & 1 \end{bmatrix}$$

Show that the code generated is a SEC perfect code. Draw a decoder circuit for the code. If two errors occur in the received $W(X)$, what is the result of the parity check equations?

6. The SEC (7, 4) Hamming code can be converted into a double-error detecting and single-error correcting (8, 4) code by using an extra parity check. Construct the generator matrix for the code, Show that the code in quasi-perfect. Design a decoder for the code.
7. Show that in a binary group code, either all the codewords have even weight or half have even weight, half odd.
8. Consider the (15, 11) Hamming code and its generator matrix of eqn. (4.36). Write the corresponding $[H]$ matrix and form a new $[H_1]$ matrix by adding a raw of

$$[H_1] = \left[\begin{array}{ccc|c} & & & 0 \\ & H & & 0 \\ & & & 0 \\ & & & \vdots \\ \hline 1 & 1 & \cdots & 1 \end{array}\right]$$

$\{1\ 1\ 1 \ldots 1\ 1\}$ and a column of $\{0\ 0 \ldots 1\}$ as shown. Specify the new code and its $[G]$. What is the error correcting capability of the new code?
9. Write the $[H]$ matrix for the (15, 11) cyclic code using $g(X) = (1 + X + X^2 + X^3 + X^4)$. Calculate the code polynomial for $m(X) = 1 + X^3 + X^7 + X^{10}$. Construct the decoder for the code.
10. Write the generator polynomial for (31, 26) SEC cyclic code, also the $[G]$ and $[H]$ matrices for the code. Find the code polynomial for the message $m(X) = 1 + X + X^4 + X^{20}$. Draw the Meggitt decoder circuit for the code.
11. Write the $[G]$ matrix for the (15, 11) SEC code. Show that by successively removing some of the rows of $[G]$, one obtains (14, 10), (13,9), (12, 8), (11, 7), (10, 6) etc. SEC codes. Write $[H]$ matrices for the (10, 6) code. Construct the coder and decoder for the codes.
12. The (15, 7) cyclic code is generated by $g(X) = 1 + X^4 + X^6 + X^7 + X^8$. Is $C(X) = 1 + X + X^2 + X^3 + X^4 + X^5 + X^7 + X^9 + X^{14}$ a code polynomials? If not, find the syndrome for $C(X)$.
13. Construct a Meggitt decoder for the (15, 7) BCH code. What are the error patterns which will form the test syndromes for the decoder? Is it possible to reduce the test syndromes to only 8 error patterns as:

$$\begin{bmatrix} 0 & 0 & 0 & 0 & 0 & 0 & 0 & 1 \\ 0 & 0 & 0 & 0 & 0 & 0 & 1 & 1 \\ 0 & 0 & 0 & 0 & 0 & 1 & 0 & 1 \\ 0 & 0 & 0 & 0 & 1 & 0 & 0 & 1 \\ 0 & 0 & 0 & 1 & 0 & 0 & 0 & 1 \\ 0 & 0 & 1 & 0 & 0 & 0 & 0 & 1 \\ 0 & 1 & 0 & 0 & 0 & 0 & 0 & 1 \\ 1 & 0 & 0 & 0 & 0 & 0 & 0 & 1 \end{bmatrix}$$

What is the modification required in the now decoder? What is the total number of shifts required to complete the decoding of a block? [Ref. (7)].

14. The triple-error-correcting Golay (23, 12) code may be constructed with the generating polynomial:

$$g(X) = X^{11} + X^9 + X^7 + X^6 + X^5 + X + 1.$$

Find the parity check polynomial $H(X)$. Construct a decoder for the code. What are the test syndromes to be checked in the decoder?

15. Find the generator polynomial for the (7, 5) RS code over GF(8). Draw the SR-encoder for the code. If the message vector $m(X) = \alpha^4X^3 + \alpha^3X^2 + 1$, the error vector $e(X) = \alpha^4X^4$, determine the code vector $C(X)$ and the received vector $W(X)$. Show the decoding process to correct $W(X)$. [*Hint:* $g(X) = X^2 + \alpha^4X + \alpha^3$].

16. Design the coder for the (31, 16) BCH code, $t \le 3$. Decode the received vector $W(X) = 1 + X + X^{11}$.

17. Find the $g(X)$ for an RS code over GF(2^5), $t \le 3$. If the received $W(X) = \alpha X^3 + \alpha^{11}X^7$, what is the transmitted $C(X)$ where α = primitive element in GF(2^5)?

18. Consider a (10, 5) linear code whose $[H]$ is given as:

$$[H] = \begin{bmatrix} 1 & 1 & 1 & 0 & 0 & 1 & 0 & 0 & 0 & 0 \\ 0 & 1 & 1 & 0 & 0 & 0 & 1 & 1 & 1 & 1 \\ 0 & 0 & 1 & 1 & 0 & 1 & 1 & 0 & 0 & 1 \\ 0 & 0 & 0 & 1 & 0 & 0 & 1 & 1 & 0 & 0 \\ 0 & 0 & 0 & 0 & 1 & 0 & 0 & 0 & 1 & 1 \end{bmatrix}$$

What is the minimum distance of the code? Show that the code is 1-step majority-logic decodable. Design the coder and decoder for the code. [This corrects one error in $W(X)$].

19. Consider (31, 21), $t \le 2$ and (31, 16), $t \le 3$, BCH codes. Write the parity-check matrices for the codes.
    (a) Check if the codes are majority-logic decodable.
    (b) Find the syndrome bits in terms of error bits.
    (c) Find the parity-check sums orthogonal on as many pairs of error bits as possible.
    (d) Construct a decoder for the (31, 16) BCH code.

20. Find the generator polynomial for the (31, 5) simplex code. Write the syndrome equations. Find the parity check sums, which are orthogonal on the error bits e_{29} and e_{30}. Calculate the H.D. of the code and check that the code is 1-step majority-logic decodable.

21. Write the generating matrix $[G]$ for the (31, 6) RM code. Explain the decoding equations for the code and check that the code corrects all $t \le 7$. How many majority gates will be required in the decoder?

22. A rate 1/2, $K = 6$ convolutional code has the generator polynomial given by:

$$g(1, 1) = (1 + X^2 + X^4 + X^5)$$

    (a) Find the generator sequence g(l), the generator matrix $[G_\infty]$, and the parity check matrix $[H_\infty]$ of the code.
    (b) Construct Type I and Type II encoders for the code.
    (c) Determine the H.D. and t for the code.
    (d) Is this code majority-logic decodable?
    (e) Calculate the syndrome patterns for 2 errors in 12 bits of $\{W_i\}$ and check that the code can correct 2 errors in 12 bits (Refer to Table 4.14)
    (f) Draw a possible decoder for the code.

23. A rate 1/3, $K = 6$, convolutional code is given by the generator polynomials:

$$g(1, 1) = 1 + X^2 + X^3 + X^5$$
$$g(1, 2) = 1 + X + X^4 + X^5$$

(a) Write $g(1)$ and the matrices $[G_\infty]$ and $[H_\infty]$.
(b) Determine H.D. and t for the code.
(c) Draw a possible decoder for the code, after checking if the code is majority-logic decodable.

24. For the Wyner-Ash code with $m = 3$, $K = 4$, $R = 7/8$, construct the seven subgenerator polynomials. Draw the coder and decoder. Check the values of minimum H.D. and t.

25. Construct a rate 1/4 convolutional code from Table 4.15. Write its generator polynomials, draw the coder and decoder-circuits. Show that the code corrects $t \leq 3$.

26. For the rate 1/3 systematic convolutional code given by:

$$g(1, 1) = (1\ 1\ 0\ 0\ 0)$$
$$g(1, 2) = (1\ 0\ 1\ 1\ 1)$$

(a) Write its $g(1)$ and $[H_\infty]$.
(b) Show that the code is majority-logic decodable and corrects $t \leq 3$.

27. For the rate 1/3 systematic code given by:

$$g(1, 1) = (1\ 0\ 1)$$
$$g(1, 2) = (1\ 1\ 1)$$

(a) Draw the tree graph, trellis and state diagram.
(b) Find d_{min} and t for the code.
(c) For the message $\{D_i\} = \{0\ 0\ 1\ 1\ 0\ 1\ 0\ 0\ \ldots\}$, find $\{C_i\}$.
(d) If the received vector
$\{W_i\} = \{001\ 110\ 110\ 010\ 100\ 001\ 011\ 000\}$,
find the transmitted $\{D_i\}$ by using the trellis diagram.
(e) Is this code majority-logic decodable?

28. Repeat the above for a systematic code given by:

$$g(1, 1) = (1\ 0\ 1\ 1)$$
$$g(1, 2) = (1\ 1\ 0\ 1)$$

29. For a non-systematic rate 1/2 code given by:

$$g(1, 1) = (1\ 1\ 1)$$
$$g(1, 2) = (1\ 0\ 1)$$

(a) Draw tree graph, trellis and state diagram.
(b) Find d_{min} and t for the code.
(c) Show, by an example, that the code corrects 2 errors in 6 channle bits.

30. A rate 1/3 non-systematic code is given by the subgenerators:

$$g(1, 1) = (1\ 1\ 0\ 1)$$
$$g(2, 1) = (1\ 0\ 0\ 1)$$
$$g(3, 1) = (1\ 1\ 1\ 0)$$

(a) Construct the coder and write $g(1)$.
(b) Draw tree graph, trellis and state diagram for the code.

(c) Find d_{min} and t for the code.
(d) Tabulate the survivor paths and their H.D. for a given error vector, $\{e_r\} = \{0010101010000\}$.

31. Construct a rate 1/5 burst-correcting convolutional code and write its generator polynomials. Then interleave the code to a degree $\lambda = 4$ and find the new generator polynomials. Determine the values of K, b and g.
32. Design a Iwadare code with rate 4/5, and the correctable burst length $b = 20$. Calculate values of K and g.
33. For the adaptive Gallager code, design a practical coder/decoder for correcting 2 random errors and bursts of length $b = 20$.
34. (a) Consider the binary Fire code for $b = 3$, $(n, k) = (35, 27)$. Obtain the generator polynomial $g(X)$ and design the coder/decoder circuits.
    (b) Interleave the code with $\lambda = 4$, and obtain the new $g(X)$. Estimate the values of b and g for the new code. Compare this with a convolutional code of similar capability.
35. Consider a product code using Elias's iteration technique. If r iterations of SEC Hamming code is used with $n_r = 2n_{r-1} = 4n_{r-2} \ldots = 2^{r-1}n_1$, show that $p_r \leq (2n_1 p)^{2^r}$, where p = transition probability in the BSC. Calculate p_r and the rate of transmission R_r for $r = 10$. [Refer to eqns. (4.126) and (4.127)].
36. (a) Consider the concatenated code useing the (15, 7) RS code as the outer code and the (7, 4) Hamming code as the inner code. If the BSC has $p_e = 10^{-2}$, estimate the overall bit error rate p_b at the decoder output.
    (b) Repeat the problem using a rate 1/2 convolutional code as the inner coder and a rate 2/3 Iwadare code as the outer code.
37. Calculate the figure of merit F for
    (a) 2-error correcting BCH codes.
    (b) 3-error correcting convolutional codes.
38. Calculate the E_b/n_0 required for the codes of Table 4.23 for $p_b(\text{coded}) = 10^{-4}$, using coherent BPSK modulation in the channel. Assume Gaussian noise in the channel.

REFERENCES

1. Shannon, C.E. and Weaver. W., *Mathematical Theory of Communication*, Univ. of Illinois Press, Urbana, 1949.
2. Peterson, W.W. and Weldon Jr., E. J., *Error Correcting Codes*, 2nd Ed., MIT press, Cambridge, Mass., 1971.
3. Lin, S. *An Introduction to Error Correcting Codes*, Prentice-Hall, N.J., 1970.
4. Berlekamp E.R., *Algebraic Coding Theory*, McGraw-Hill, N. Y., 1968.
5. Elias, P., 'Error-free coding', *IRE Trans. on Information Theory*, Vol. IT-1, No. 4, pp. 29–37, 1954.
6. Hamming, R.W., 'Error detecting and error correcting codes', *BSTJ*, Vol. 29, pp. 147–160, 1950.
7. Blahut, R.E., *Theory and Practice of Error Control Codes*, Addison-Wesley Publishing Co., Mass., 1983.
8. Bose, R.C. and Raychaudhuri, D. K., 'On a class of error correcting binary group codes', *Information and Control,* Vol. 3, pp. 68–79, 1960.
9. Reed, I.S. and Solomon G., 'Polynomial codes over certain finite fields', *J, Soc. Indust. App. Math.*, Vol. 8, pp. 300–304, 1960.
10. Massey, J.L., *Thereshould Decoding,* The MIT Press, Cambridge, Mass., 1963.
11. Elias, P., 'Coding for noisy channels', *IRE Conv. Record*, part 4, pp. 37–47. 1955.
12. Wyner, A.D. and Ash, R.B., 'Analysis of recurrent codes', *IEEE Trans. on Inf. Theory,* Vol. IT-9 pp. 143–156, 1963.

13. Viterbi, A.J., 'Convolutional codes and their performance in communication systems', *IEEE Trans.,* Vol. COM-19, Pt. II, pp. 751–771, 1971.
14. Wozencraft, J.M. and Reiffen, B., *Sequential Decoding*, MIT Press, Cambridge, Mass., 1961.
15. Sinha, V. and Babu, G.V., 'Simplex organisation for Viterbi decoding of short binary convolutional codes', *J. IETE*, Vol. 25, No. 4, April 1979.
16. Fano, R.M., 'A heuristic discussion of probabilistic decoding', *IEEE Trans.,* Vol. IT-9, pp. 64–74, 1963.
17. Jelinek, F., 'A fast sequential decoding algorithm using a stack', *IBMJ. Res. Develop.,* Vol. 13, pp. 675–685, 1969.
18. Forney Jr., G.D., 'Burst correcting codes for the classic bursty channels', *IEEE Trans. Comm.,* Vol. COM-19, pp. 772–780, 1971.
19. Gallager, R.G., *Information Theory and Reliable Communication*, McGraw-Hill, N. Y., 1968.
20. Forney Jr. G.D., *Concatenated Codes,* MIT Press, Cambridge, Mass., 1967.
21. Matsushita, I., Tsuji, Y. and Sohma, Y., 'Applications of error-correcting codes to TDMA Satellite communications', *Fujitsu Sc. and Tech. Journal,* Vol. 9(3), pp. 87–110, 1973.

CHAPTER

5

Information Theory

5.0 INTRODUCTION [1–8]

With the pioneering work of Wiener, Shannon, Gabor and others starting in mid-40's, the study of information and communication has been formalised in the form of statistical communication theory or the communication-oriented Information theory. This enables theoreticians as well as practical designers to evaluate existing and futuristic communication systems, to evolve new codes for signal design and to optimize information transfer in the transmitter-channel-receiver path. For a generalized communication link, the information is generated in a source, transformed and coded in a suitable signal, transmitted to a channel and then transferred to a sink through a decoder-receiver. For efficient transmission of information, it is necessary to transform the information at many stages as dictated by the characteristics of the channel and the optimum receiver.

The communication channels are broadly classified as: (a) discrete noiseless channel, (b) discrete channel with discrete noise, (c) discrete channel with continuous noise, and (d) continuous channel with continuous noise. For the discrete noiseless channel, e.g., in coding language information, one is led to Shannon-Fano-Huffman codes. For the discrete channel disturbed by discrete noise, the information is protected by error-correcting codes, as discussed in the last chapter. For discrete channel with continuous noise (mostly white and Gaussian noise), e.g., in binary and multilevel PAM transmission, the optimum receiver uses a correlation detector and the coding problem is a search of waveforms, which are mutually as uncorrelated as possible, e.g., orthogonal codes. For the continuous channel with continuous noise, solutions are found in the form of wideband modulation systems, e.g., FM, PPM, PAM-FM, PCM-AM, PCM-FM, and coherent detection techniques using matched filters and correlators.

The major developments in the area of information theory started with Shannon's classical theorems on 'Code capacity' and 'Channel capacity', which specify the maximum information that may be generated by a set of symbols, and the information that may be transmitted by a physical noisy channel. Shannon's theorems did not provide the methods of achieving the ideal channel capacity, given the statistics of the channel, but only indicated a yardstick, with which all practical communication systems can be tested to evaluate their efficiency. Since the time of publication of Shannon's results in 1948, the theoreticians and the designers have been constantly trying to find methods of coding the information in such a way as to approach the so-called Shannon's limit. Large amount of developments in coding techniques and optimum signal design is the result of this effort.

This chapter discusses the information measure of sources, including Markov sources, language capacity, code capacity, and channel capacity. Fundamental theorems of Shannon are discussed and the results used to evaluate some of the existing communication systems. Noiseless coding as well as coding for noisy channels have been illustrated. Exchange of bandwidth for signal-to-noise ratio has also been discussed.

5.1 INFORMATION AND ENTROPY [1, 2]

The quantitative measure of information is based on out intuitive notion of the word 'information', i.e., the more unexpected an event is (with a priori probability p small), the more information is obtained when the event occurs (with a posteriori probability of the event = 1). Conveniently this information is expressed as:

$$I(x) = \log \frac{1}{p(x)} = -\log p(x) \tag{5.1}$$

where $p(x)$ = probability of occurrence of x, and unit of $I(x)$ depends on the base for the log [ ] in eqn. (5.1).

Different units are known as:

$$\begin{aligned} I(x) &= -\log_2 p(x) \text{ bits,} \\ &= -\log_e p(x) = -\ln p(x) \text{ nats,} \\ &= -\log_{10} p(x) \text{ Hartleys} \end{aligned} \tag{5.2}$$

and in general,

$$I(x) = -\log_D p(x) \; D\text{-ary units.}$$

The information measure can be any monotonically decreasing function of p, but the special choice of $-\log p$ is justified because of its following useful properties:

(i) The information content of a message increases as $p(x)$ decreases and *vice versa.*
(ii) $I(x)$ is a continuous function of $p(x)$.
(iii) The total information of two or more independent messages/events is the sum of individual informations.

The property (iii) is evident from the following example: If the two independent events (x_1, x_2) have apriori probabilities of p_1 and p_2, then the joint probability of their occurrence = $p = p_1p_2$ and the information associated with the events

$$I(x_1, x_2) = -\log p = -\log p_1 - \log p_2 = -\sum \log p_i.$$

Consider a zero-memory information source S delivering messages from its alphabet $X = \{x_1, x_2, \ldots, x_i, \ldots x_m\}$ with probabilities $p\{x\} = \{p_1, p_2, \ldots, p_i, \ldots p_m\}$. If N messages are delivered, $(N \to \infty)$, then symbol x_i occurs Np_i times and each occurrence conveys information $-\log p_i$. Hence, the total information due to Np_i messages = $-Np_i \log p_i$ and the total information due to all N messages = $-N\sum_1^m p_i \log p_i$.

The Entropy $H(X)$ of the source S is now defined as the averge amount of information per source symbol and,

$$H(X) \triangleq -\sum_{1}^{m} p_i \log p_i \text{ with } \sum p_i = 1 \tag{5.3}$$

Thus the entropy is associated with a source, whereas information is associated with a message. $H(X)$ is also referred to as Marginal entropy, and $p(x_i)$ as Marginal probability.

The entropy, as defined in eqn. (5.3), has the following important properties:

(i) $H(X) = 0$, if all p_i's except one, are zero.

(ii) $H(X) = -\sum_{1}^{m} p_i \log p_i = -\sum_{1}^{m} \frac{1}{m} \log \frac{1}{m} = \log m;$

$$\text{if } p_1 = p_2, \ldots = p_m = \frac{1}{m}. \tag{5.4}$$

and $\log m$ is maximum for any possible set of p_i's.

This can be shown by writing the entropies in nats,

$$\log_e m - H(X) = \sum_{i=1}^{m} p_i \log_e m + \sum_{i=1}^{m} p_i \log_e p_i, \;\; p_i \neq p_j$$

$$= \sum_{1}^{m} p_i \log_e mp_i.$$

By using the inequality,

$$\ln (1/x) \geq (1 - x), \tag{5.5}^*$$

$$\ln m - H(X) \geq \sum_{1}^{m} p_i \left(1 - \frac{1}{mp_i}\right)$$

$$\geq \sum_{1}^{m} p_i - \frac{1}{m}\sum_{1}^{m}(p_i/p_i)$$

$$\geq 0$$

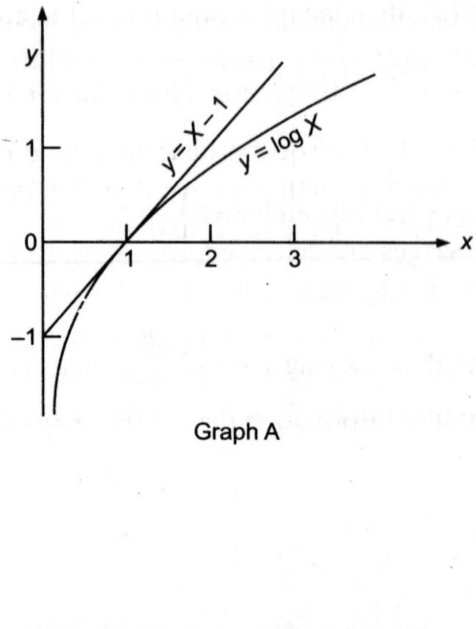

Graph A

*The inequality of eqn. (5.5) is obtained from the well known inequality: $\ln x \leq x - 1$, as can be seen from graph A. For $x = 1$, $\ln x = (x - 1)$. By multiplying both sides by –1, the inequality obtained is:

$$\ln(1/x) \geq (1 - x) \tag{5.5a}$$

and

$$\ln m = H(X), \text{ if and only if } p_i = p_j.$$

For a binary source with alphabet {0, 1}, and probabilities $(p, 1 - p)$,

$$H(X) = p \log\frac{1}{p} + (1-p)\log\frac{1}{1-p}$$

$$\triangleq H(p) \tag{5.6}$$

A plot of $H(p)$ is shown in Fig. 5.1 where it is seen that $H(p)$ is maximum with $H(p) = 1$ bit, only when $p = (1 - p) = 0.5$. For m-symbol source, the maximum Entropy $H(X)_{max}$ increases only as log m, and for any other $p(i) \neq p(j)$, the entropy of the source is less, as shown for the following examples.

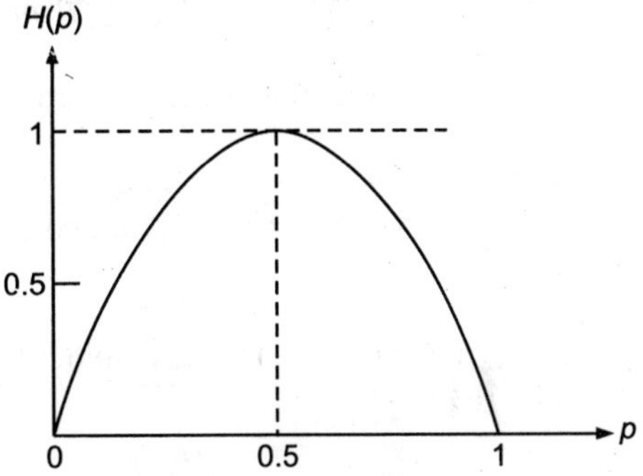

Figure 5.1

Example 5.1: A source delivers 4 symbols x_1, x_2, x_3, x_4 with

$$p\{X\} = \left\{\frac{1}{2}, \frac{1}{4}, \frac{1}{8}, \frac{1}{8}\right\} \text{ then, } H(X) = -\sum p_i \log p_i$$

$$= -\left[\frac{1}{2}\log\frac{1}{2} + \frac{1}{4}\log\frac{1}{4} + \frac{2}{8}\log\frac{1}{8}\right]$$

$$= 1.75 \text{ bits.}$$

If however, all $p(i)$'s were equal having $p(i) = \frac{1}{4}$, then max $H(X) = \log_2 4 = 2$ bits.

Example 5.2: In English language, the probabilities of occurrence of the 27 symbols (26 letters + space) are not equal and they are tabulated in various texts. The symbol $p(i)$'s in the descending order are as follows:

Symbols →	Space	*E*	*T*	*A*		*D*	*L*		*B*	*V*		*Q*	*Z*
$p(i)$	0.187	0.107	0.086	0.067	…	0.031	0.028	…	0.012	0.008	…	0.001	0.0006

and $H(x) = -\sum_{1}^{27} p_i \log p_i = 4.065$ (as calculated from all $p(i)$'s).

If, however, we choose the letters at random and with equal probability, then the information value of all letters are the same and $H(X)_{max}$ of the source $= -\sum_{1}^{27} \frac{1}{27} \log(1/27) = 4.76$ bits/symbol. This means that the length of messages could be shortened by the ratio of $4.065/4.76 = 0.854$, if equal probability random coding is used in forming English words. In other words, approximately 17 letters ($\approx$ Antilog$_2$ 4.065) would be sufficient to express English messages, if the choice of the letters are ideally random.

5.1.1 Entropy of Several Alphabets

Consider two sources S_1 and S_2 delivering symbols x_i and y_j and there may be some dependence (correlation) between x's and y's. In conformity with the definition of joint probability $p(x_i, y_j)$, the entropy of the joint event (x_i, y_j), also known as the Joint Entropy of the sources S_1 and S_2, is defined as:

$$H(X, Y) = -\sum_{i=1}^{m}\sum_{j=1}^{n} p(x_i, y_j) \log p(x_i, y_j) \tag{5.7}$$

where $\quad p(x_i, y_j) = p(x_i)\, p(y_j/x_i) \quad$ (5.7a)

If $\{X\}$ and $\{Y\}$ are independent of each other, then $p(x_i, y_j) = p(x_i)p(y_j)$,
and $H(X, Y) = H(X) + H(Y) \quad$ (5.8)

Similarly, the Conditional Entropy $H(Y/X)$, the entropy of Y given that X has occurred, is defined as:

$$H(Y/X) = -\sum_{i=1}^{m}\sum_{j=1}^{n} p(x_i, y_j) \log p(y_j / x_i) \qquad (5.9)*$$

If $\{X\}$ and $\{Y\}$ are independent, then,

$$\begin{aligned} H(Y/X) &= -\sum_{1}^{n}\sum_{1}^{m} p(x_i, y_j) \log p(y_j) \\ &= -\sum_{1}^{n} p(y_j) \log p(y_j) \\ &= H(Y) \end{aligned}$$

The interrelations between the above entropies, defined by eqns. (5.3), (5.7) and (5.9), are given as,

(i) $H(X, Y) = H(X) + H(Y/X) \quad$ (5.10a)
$\qquad\qquad = H(Y) + H(X/Y) \quad$ (5.10b)
(ii) $H(X) \geq H(X/Y) \quad$ (5.10c)
$\qquad H(Y) \geq H(Y/X) \quad$ (5.10d)
(iii) $H(X, Y) \leq H(X) + H(Y) \quad$ (5.10e)

The relation (*i*) is easily shown by expanding $H(X, Y)$ as:

$$\begin{aligned} H(X, Y) &= -\sum_{i}\sum_{j} p(x_i)p(y_j/x_i) \log [p(x_i)p(y_j/x_i)] \\ &= -\sum_{i}\sum_{j} p(x_i)p(y_j/x_i) \log p(x_i) \\ &\quad -\sum_{i}\sum_{j} p(x_i)p(y_j/x_i) \log p(y_j/x_i) \\ &= H(X) + H(Y/X) \end{aligned} \tag{5.10a}$$

---

*Equation (5.9) is derived by first defining $H(Y/x_i) = -\sum_{j=1}^{n} p(y_j / x_i).\ \log p(y_j/x_i)$ and then averaging this over all values of x_i, giving,

$$\begin{aligned} H(Y/X) = E\{H(Y/x_i)\} &= -\sum_{i=1}^{m} p(x_i)H(Y / x_i) \\ &= -\sum_{m}\sum_{n} p(x_i)p(y_j/x_i) \log p(y_j/x_i) \\ &= -\sum_{m}\sum_{n} p(x_i, y_j) \log p(y_j/x_i) \end{aligned} \tag{5.9}$$

Since, $-\sum_i[\sum_j p(x_i)p(y_j/x_i)\log p(x_i)]$

$$= -\sum_i p(x_i)\log p(x_i) = H(X)$$

Similarly, eqn. (5.10b) is proved, by using the relation:

$$p(x_i, y_j) = p(x_i)p(y_j/x_i) = p(y_j)p(x_i/y_j)$$

The relations (*ii*) is proved by using the inequality of eqn. (5.5). By writing the entropies in nats,

$$H(X) - H(X/Y) = -\sum_i p(x_i)\log p(x_i) + \sum_i\sum_j p(x_i, y_j)\log p(x_i/y_j)$$

$$= \sum_i\sum_j p(x_i, y_j)\log\left[\frac{p(x_i/y_j)}{p(x_i)}\right]$$

By eqn. (5.5),

$$\text{L.H.S.} \geq \sum_i\sum_j p(x_i, y_j)\left[1 - \frac{p(x_i)}{p(x_i/y_j)}\right]$$

$$\geq \sum_i\sum_j [p(x_i, y_j) - p(y_j)p(x_i)]$$

$$\geq \sum_i [p(x_i) - p(x_i)]$$

$$\geq 0 \quad (5.10c)$$

Equality occurs when $\{X\}$, $\{Y\}$ are independents, giving $H(X/Y) = H(X)$.
Similarly eqn. (5.10d) is proved. The relation (5.10e) follows from the eqns. (5.10a) and (5.10d).

The above results for two variables may be extended to three variables and more, giving the relations:

$$H(X, Y, Z) = H(X, Y) + H(Z/X, Y)$$

$$H(X, Y) \geq H(X, Y/Z) \quad (5.11)$$

and for q-variables,

$$H(S_1, S_2, \ldots, S_q) = H(S_1, S_2, \ldots, S_{q-1}) + H(S_q/S_1, S_2, \ldots, S_{q-1})$$

where $H(S_1, \ldots, S_q)$ is the joint entropy of all sources $S_1 \ldots$ to S_q.

Example 5.3: Consider that two sources S_1 and S_2 emit messages x_1, x_2, x_3 and y_1, y_2, y_3 with the joint probability $p(X, Y)$ as shown in the matrix form. Calculate $H(X)$, $H(Y)$, $H(X/Y)$ and $H(Y/X)$. Given:

$p(X, Y) \rightarrow$	y_1	y_2	y_3
x_1	3/40	1/40	1/40
x_2	1/20	3/20	1/20
x_3	1/8	1/8	3/8

By adding the entries in rows and in columns, we get,

$$p(x_1) = 1/8, p(x_2) = 1/4 \text{ and } p(x_3) = 5/8$$
$$p(y_1) = 1/4, p(y_2) = 3/10 \text{ and } p(y_3) = 9/20$$

The matrices $p(Y/X)$ and $p(X/Y)$ are calculated by using the eq. (5.7a), and they are:

$p(Y/X) \rightarrow$	y_1	y_2	y_3
x_1	3/5	1/5	1/5
x_2	1/5	3/5	1/5
x_3	1/5	1/5	3/5

and

$p(X/Y) \rightarrow$	y_1	y_2	y_3
x_1	3/10	1/12	1/18
x_2	1/5	1/2	1/9
x_3	1/2	5/12	5/6

Using eqns. (5.3), (5.7) and (5.9),

$$H(X) = 1/8 \log 8 + 1/4 \log 4 + 5/8 \log 8/5 = 1.3 \text{ bits/symbol}$$
$$H(Y) = 1/4 \log 4 + 3/10 \log 10/3 + 9/20 \log 20/9$$
$$= 1.54 \text{ bits/symbol}$$
$$H(Y/X) = [3/40 \log 5/3 + 1/40 \log 5 + 1/40 \log 5 + 1/20 \log 5 + 3/20 \log 5/3 + 1/20 \log 5 + 1/8 \log 5 + 1/8 \log 5 + 3/8 \log 5/3]$$
$$= 1.37 \text{ bits/symbol.}$$
$$H(X/Y) = 1.13 \text{ bits/symbol}$$

and

$$H(X, Y) = 2.67 \text{ bits/symbol} < H(X) + H(Y)$$

The example illustrates the inter-relations between different entropies. The matrix $p(Y/X)$ is also known as the channel/transition/noise matrix and is graphically shown in Fig. 5.2.

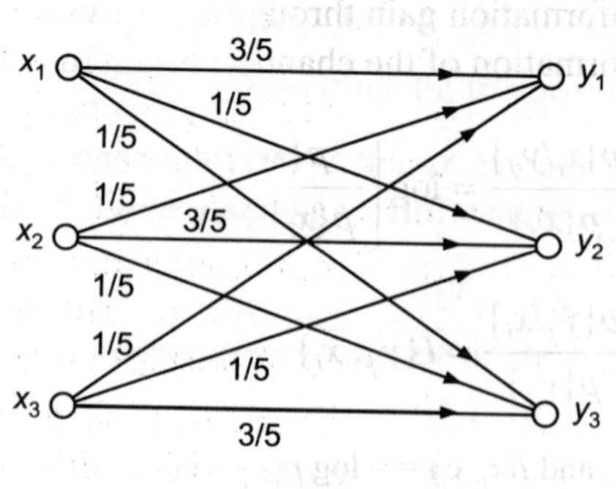

Figure 5.2

5.1.2 A Communication Network

An useful interpretation of the different entropies, defined so far, may be given in terms of the input/output of a coder/communication channel, as shown in Fig. 5.3. A coder normally changes the statistics of the source symbols $\{x_i\}$ and is characterized by the transition matrix $p(Y/X)$. Similarly in an actual transmission channel, the statistics of the transmitted symbols is changed due to the noise present and the received symbols $\{y_i\}$ have to be optimally interpreted from the inverse probability matrix $p(X/Y)$. We can now interpret the different entropies as:

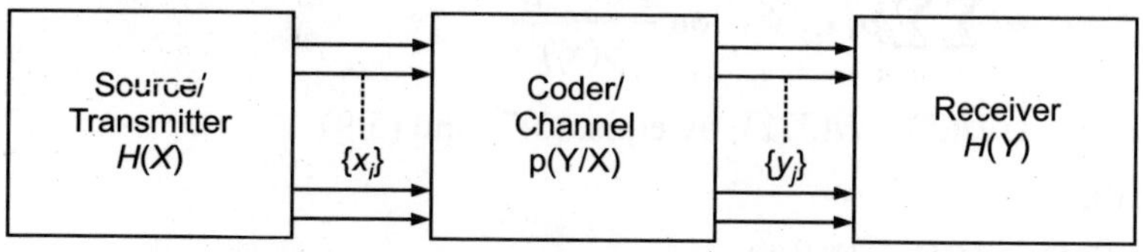

Figure 5.3: A communication system model

$H(X)$ = average information/symbol of the source, i.e., the entropy of the source (also known as Marginal entropy).

$H(Y)$ = entropy of the receiver

$H(X, Y)$ = average information per pairs of transmitted and received symbols (x_i, y_j), equivalently the average uncertainty of the communication link as a whole (i.e., the Joint entropy of (x_i, y_j))

$H(Y/X)$ = the entropy of the received symbols when the transmitter state is known. This conditional entropy is the measure of information given by $\overline{H(Y/x_i)}$ averaged over all $\{x_i\}$.

$H(X/Y)$ = the entropy of the source when the state of the receiver is known. This measure of information given by $\overline{H(X/y_j)}$ averaged over $\{y_j\}$ is also known as 'Equivocation'.

In spite of the change of the transmitter statistics $p\{x_i\}$ due to noise in the channel, certain probabilistic information regarding the transmitter states is available, once $p\{x_i/y_j\}$ is calculated at the receiver. The initial uncertainty regarding $\{x_i\}$ is $-\log p\{x_i\}$ and the final uncertainty about $\{x_i\}$ after the reception of $\{y_j\}$ is $-\log p\{x_i/y_j\}$; thus the Information gain through the channel (also known as Mutual Information contained in $\{x_i/y_j\}$ or Transinformation of the channel) is given by:

$$I(n_i, y_j) = \log\frac{p\{x_i/y_j\}}{p\{x_j\}} = \log\left[\frac{p\{x_i, y_j\}}{p\{x_i\}\cdot p\{y_j\}}\right]$$

$$= \log\frac{p\{y_j/x_i\}}{p\{y_j\}} = I\{y_j, x_i\} \tag{5.12*}$$

*For noiseless channel, $p(x_i/y_i) = 1$, and $I(x_i, y_i) = -\log p(x_i)$; and this information is usually known as:

Self-Information $= I(x_i, x_i) = -\log p(x)$ of eqn. (5.1). It then follows that $H(X) = \sum_1^m p_i I(x_i, x_i)$, which is the same as eqn. (5.3).

Averaging eqn. (5.12) over all possible values of x_i and y_j, one obtains the Average Information gained by the receiver, and is given as

$$I(X, Y) = E[I(x_i, y_j)]$$

$$= \sum_{j=1}^{n}\sum_{i=1}^{m} p(x_i, y_j) I(x_i, y_j) \text{ bits/symbol}$$

$$= \sum_{j}\sum_{i} p(x_i, y_j) \log \frac{p(x_i/y_j)}{p(x_i)} \tag{5.13}$$

$$= H(X) - H(X/Y), \text{ by eqns. (5.7) and (5.9).}$$

By using eqn. (5.10),

$$I(X, Y) = H(Y) - H(Y/X)$$

$$= H(X) + H(Y) - H(X, Y) \text{ bits/symbol} \tag{5.14}$$

where $H(X/Y)$ is specially called 'Equivocation', meaning thereby the loss of information due to channel noise.

Further for a noise-free channel, $H(Y/X) = H(X/Y) = 0$, and

$$I(X, Y) = H(X) = H(Y) = H(X, Y) \tag{5.15}$$

For a channel where $\{x_i\}$ and $\{y_j\}$ are independent and have no correlation between them, $H(X/Y) = H(X)$ and $H(Y/X) = H(Y)$ and

$$I(X, Y) = H(X) - H(X/Y) = 0 \tag{5.16}$$

thereby no information is transmitted through the channel.

Example 5.4: Using the channel matrix, given in Ex. 3, one now obtains:

$$I(X, Y) = H(Y) - H(Y/X) = 1.54 - 1.37 = 0.17 \text{ bits/symbol}$$

$$= H(X) - H(X/Y) = 1.3 - 1.13 = 0.17 \text{ bits/symbol.}$$

Further discussions of the channel, e.g., the Rate of Transmission, Channel capacity etc. will be taken up in a later section.

5.1.3 Extension of Zero-memory Sources [5, 6]

If multialphabet source-outputs are to be coded into words of a lesser alphabet, then it is necessary to have an extension of the coder alphabets. Consider that 8 messages are to be coded with a binary alphabet (0/1), then it is necessary to have the binary words as: {000, 001, 010, ..., 110, 111}, leading to the third extension of the source coder. In general, for m messages, with $m = 2^n$, an nth extension of the binary alphabet is necessary. Thus if zero-memory source S has the source alphabets $\{x_1, x_2, \ldots, x_m\}$, then the nth extension of S, called S^n, is a zero-memory source with m^n symbols $\{y(1), y(2), \ldots, y(m^n)\}$ as its higher-order alphabet. The corresponding statistics of the extended source S^n is given by:

$$p\{y(j)\} = \{p(x_{j1}) \cdot p(x_{j2}) \ldots p(x_{jn})\},$$

since each of $y(j)$ corresponds to a word made of n alphabets of $\{x_j\}$, i.e.,

$$y(j) = (x_{j1}, x_{j2}, \ldots, x_{jn})$$

The condition that $\sum_{j=1}^{m^n} p\{y(j)\} = 1$, is satisfied, since,

$$\sum_{S^n} p\{y(j)\} = \sum_{j=1}^{m^n}\{p(x_{j1})\cdot p(x_{j2})\dots p(x_{jn})\}$$

$$= \sum_{j1=1}^{m}\sum_{j2=1}^{m}\dots\sum_{jn=1}^{m}\{p(x_{j1})\cdot p(x_{j2})\dots p(x_{jn})\}$$

$$= \sum_{j1=1}^{m} p(x_{j1})\sum_{j2=1}^{m} p(x_{j2})\dots\sum_{jn=1}^{m} p(x_{jn})$$

$$= 1 \text{ (since each of } \sum^{m} = 1).$$

The entropy of the extended source S^n is now given by:

$$H(S^n) = \sum_{j=1}^{m^n} p\{y(j)\}\log\frac{1}{p\{y(j)\}}$$

$$= nH(S) \tag{5.17}$$

Thus the entropy of source S^n is n times that of the primary source $H(S)$.

Example 5.5: Consider the third extension S^3 of a binary source with $p(0) = 0.25$ and $p(1) = 0.75$, having the primary source entropy $H(S) = 1/4 \log 4 + 3/4 \log 4/3 = 0.815$ bit.

The extended source S^3 has the alphabets and the corresponding $p\{y(j)\}$ given by

Symbols of S^3	y_1	y_2	y_3	y_4	y_5	y_6	y_7	y_8
Alphabets	000	001	010	011	100	101	110	111
$p\{y(j)\}$	1/64	3/64	3/64	9/64	3/64	9/64	9/64	27/64

It is observed that:

$$\sum_{j=1}^{8} p\{y(j)\} = 1$$

and

$$H(S^3) = 1/64 \log 64 + 3 \times 3/64 \log 64/3 + 3 \times 9/64 \log 64/9 + 27/64 \log 64/27$$

$$= 6/64 + 9/64 \times (6 - 1.58) + 27/64 \times (6 - 3.17) + 27/64\,(6 - 4.75)$$

$$= 2.44 = 3H(S).$$

Example 5.6: Consider a source with alphabets x_1, x_2, x_3, x_4 and a priori probability $p\{x_i\} = \{1/2, 1/4, 1/8, 1/8\}$. Thus a typical message will be:

$$\{\dots x_1\ x_1\ x_2\ x_1\ x_3\ x_1\ x_2\ x_4\ x_2\ x_1\ x_3\ x_4\ x_1\ x_2\ x_1\ x_1 \dots\}$$

satisfying $p\{x_i\}$. The entropy of the source $H(X)$ is:

$$H(S) = H(X) = -\sum_{i=1}^{4} p(x_i)\log p(x_i) = 1.75 \text{ bits/symbol.}$$

Suppose the source is extended to deliver messages with two symbols, then the new alphabets and the corresponding $p\{y(j)\}$ are given as:

Symbols of S^2	y_1	y_2	y_3	y_4	y_5	y_6	y_7	y_8	y_9	y_{10}	y_{11}	y_{12}	y_{13}	y_{14}	y_{15}	y_{16}
Alphabets	x_1x_1	x_1x_2	x_1x_3	x_1x_4	x_2x_1	x_2x_2	x_2x_3	x_2x_4	x_3x_1	x_3x_2	x_3x_3	x_3x_4	x_4x_1	x_4x_2	x_4x_3	x_4x_4
$p\{y(j)\}$	$\frac{1}{4}$	$\frac{1}{8}$	$\frac{1}{16}$	$\frac{1}{16}$	$\frac{1}{8}$	$\frac{1}{16}$	$\frac{1}{32}$	$\frac{1}{32}$	$\frac{1}{16}$	$\frac{1}{32}$	$\frac{1}{64}$	$\frac{1}{64}$	$\frac{1}{16}$	$\frac{1}{32}$	$\frac{1}{64}$	$\frac{1}{64}$

and

$$H(S^2) = 1/4 \log 4 + 2/8 \log 8 + 5/16 \log 16 + 4/32 \log 32 + 4/64 \log 64$$
$$= 3.5 \text{ bits/symbol pair} = 2H(s).$$

5.2 SOURCE ENCODING: NOISELESS CODING [4, 5]

In practical communication systems, it is necessary to transform the m-ary source alphabets $\{x_i \dots x_m\}$ to a convenient form, say, binary or D-ary digits to match the channel. Since the rate of transmission of information in a noisy channel is maximum if the source probabilities $p(x_i)$ are all equal i.e., $H(x) = \log m$ (this will be shown in a later section), it is desirable that the transformed code alphabets $\{C_1, C_2, \dots, C_D\}$ have equal probability of occurrence, hence the technique is also known as Entropy coding.

To avoid ambiguity in deciphering the codes, as well as, to minimise the time of transmission of the message, the codes should have the following properties:

1. Codes should be 'separable' or uniquely decodable. Also they should be 'comma-free', i.e., no synchronising signal should be required to recognise the words. This restricts the selection of codes in such a way that no shorter code can be a prefix of a longer code. Such codes are also called Instantaneous codes.
2. The average length of the code words

$$\bar{L} = \Sigma p_i \cdot L_i \tag{5.18}$$

L_i = length of ith code word, should be minimum, and $\bar{L}$ should approach the value $[H(X)/\log D]$, with the condition $\bar{L} \geq H(X)/\log D$. (This will be proved later). Since $\Sigma p_i L_i$ is to be minimum, the code words with larger L_i should have smaller p_i, i.e.,

$$L_i \propto \log (1/p_i) \tag{5.19}$$

3. The code efficiency η is defined as

$$\eta = \frac{H(X)}{\bar{L} \log D} \text{ and } \eta \to 1 \text{ for the optimum (also called compact) codes.}$$

Further, Redundancy of codes is defined as $R = 1 - \eta$.

Example 5.7: Consider a source alphabet of $A, B, C, \dots, G, H$ having $p(x_i)$ given as:

$$p(x_i) = 1/2, 1/4, 1/16, 1/16, 1/32, 1/32, 1/32, 1/32.$$

If we have to code the alphabet with binary (1/0) digits, then the code generating process is shown in Table 5.1.

The symbols are arranged according to their decreasing probability. Then the 1st selection is made such that the symbols are divided into groups such that $\Sigma_p(x_i)$ in each group = 1/2, thus symbol A is designated by 1 and the rest by 0.

Table 5.1

Symbols	$p(x_i)$	No. of selections					Code	Complement of the code
		1st	2nd	3rd	4th	5th		
A	1/2	1					1	0
B	1/4	0	1				01	10
C	1/16	0	0	1	1		0011	1100
D	1/16	0	0	1	0		0010	1101
E	1/32	0	0	0	1	1	00011	11100
F	1/32	0	0	0	1	0	00010	11101
G	1/32	0	0	0	0	1	00001	11110
H	1/32	0	0	0	0	0	00000	11111

Then the symbols $\{B \ldots H\}$ are divided into two groups B, and $\{C, D, \ldots, H\}$, so that $\Sigma p(x_i) = 1/4$ in each group, and B is now designated by 1 and the rest by 0. This process continues till we complete designating the last symbol. The resulting code is shown in the Table. The following properties of the code is evident.

(a) Number of bits required for any code = log $1/p(x_i)$, and, say, for E, a 5 bit code is used and log (1/32) = 5.

(b) $\bar{L} = \sum p_i L_i = 1/2 + 1/2 + 1/4 + 1/4 + \dfrac{5 \times 4}{32} = 2.125$

$H(X) = \sum p(x_i) \log 1/p(x_i) = 2.125$ bits/symbol $= \bar{L}$.

(c) In a long message derived out of binary codes, number of occurrences of '1' = 1/2 + 1/4 + 1/8 + 1/16 + 1/16 + 1/32 + 1/32 = $1\frac{1}{16}$ and number of occurrence of '0' = $1\frac{1}{16}$.

Therefore, by normalising, $p(0) = (17/16)/(34/16) = 0.5 = p(1)$, and the coder output has maximum $H(X)$ in terms of binary digits.

(d) If the 8 alphabets were directly coded into binary digits, then codes would be {000, 001, 010, ..., 110, 111}, each symbol requiring 3 digits, and given $\sum p_i L_i = 3$ bit/symbol $> H(X)$.

The above technique of grouping the symbol probabilities in two halves and then designating each half with 1/0 is also known as Shannon-Fano technique. We have however considered a very convenient set of $p(x_i)$, where log $1/p(x_i)$ was an integer. In a general case, where log $1/p(x_i)$ is not an integer, Shannon [4] has suggested that L_i should have the bounds (for binary coding),

$$\log \frac{1}{p(x_i)} \leq L_i \leq \log 1/p(x_i) + 1 \tag{5.20}$$

If the alphabet size m is large, then the resultant code alphabet is quite efficient, but for small m, the code is not very efficient, as will be seen from the following example. Black [15] has suggested that $L_i \simeq \log 1/p(x_i)$, may be taken as the code-length.

Example 5.8: Consider again 8-alphabet source with $p(x_i)$ as shown in Table 5.2. According to Shannon-Fano technique, as shown in Example 5.7, the binary codes are generated as shown. In the table, the values of log $1/p(x_i)$ are also shown, and it is seen that $L_i \simeq \log 1/p(x_i)$, as suggested by Black. However, L_i given by eqn. (5.20) is larger than the actural values in the Table.

Table 5.2

Symbols	$p(x_i)$		Code selection					Codes	$\log 1/p(x_i)$
A	0.30	$\sum = .5$	0	0				00	1.7
B	0.20		0	1				01	2.3
C	0.15		1	0	0			100	2.7
D	0.12	$\sum = .5$	1	0	1			101	3.05
E	0.10		1	1	0			110	3.3
F	0.07		1	1	1	0		1110	3.8
G	0.04		1	1	1	1	0	11110	4.6
H	0.02		1	1	1	1	1	11111	5.6

$$H(X) = \Sigma p(x_i) \log 1/p(x_i) = 2.663 \text{ bits/symbol}$$

$$\overline{L} = \Sigma p_i L_i = 2.69$$

$$\eta(\text{with } D = 2) = \frac{2.663}{2.69} = 99\%$$

At the coder output, $p(0) = 0.53$ and $p(1) = 0.47$.

Huffman [11] has suggested a simple method of constructing separable codes with minimum redundancy (i.e., minimum $\overline{L}$). The rules of construction are:

1. The symbols are arranged according to their decreasing probability.
2. For minimum $\overline{L}$, $L(i)$ should be greater or equal to $L(i-1)$. Only in case of last two symbols, $L(m) = L(m-1)$, and they should differ only in the last digit.
3. Select the last two symbols x_{m-1} and x_m, and assign 0/1 to them. Form a composite symbol with probability $= p(x_{m-1}) + p(x_m)$, and construct a reduced source with $(m-1)$ alphabets and order them according to their decreasing probability.
4. Assign 0/1 to the last symbols of the new column and again construct a 2nd reduced source with the ordering as above.
5. Continue this till all symbols are assigned 0/1 codes.

Example 5.9: Huffman coding with binary alphabet.

Consider a source with 8 alphabets as given in Table 5.3. The methods of constructing reduced sources are shown below. By tracing the code path backwards, the assigned code is defined as shown.

$H(X) = 2.75$ bits and $\overline{L} = \Sigma p_i L_i = 2.8$, giving $\eta = 2.75/2.8 = 98.2\%$.

Shannon-Fano method also gives a similar code with $\overline{L} = 2.8$.

Table 5.3

Assigned Code	Symbol	$p(x_i)$	Reduced sources					
10	A	0.22	0.22	0.22	→0.25	→0.33	→0.42	→0.58(0)
11	B	0.20	0.20	0.20	0.22	0.25	0.33(0)	0.42(1)
000	C	0.18	0.18	0.18	0.20	0.22(0)	0.25(1)	
001	D	0.15	0.15	0.15	0.18(0)	0.20(1)		
011	E	0.10	0.10	→0.15(0)	0.15(1)			
0100	F	0.08	0.08(0)	0.10(1)				
01010	G	0.05(0)	→0.07(1)					
01011	H	0.02(1)						

Example 5.10: Huffman coding with $D = 4$.

For a code alphabet $D > 2$, the rules are exactly similar, except that each combination now contains D symbols. However, the first composite symbol should be formed out of the last r source symbols such that:

$r = [m - k(D - 1)]$, k is the highest integer, and the last reduced source has only D entries in the last column. In the above example, for $D = 4$,

$$r = (8 - 3k) = 2, k = 2$$

The codewords with symbols {0, 1, 2, 3}, can be constructed as given in Table 5.4.

Table 5.4

			Symbols	Codes
0.22	0.22	→0.40(0)	A	1
0.20	0.20	0.22(1)	B	2
0.18	0.18	0.20(2)	C	3
0.15	0.15(0)	0.18(3)	D	00
0.10	0.10(1)		E	01
0.08	0.08(2)		F	02
0.05(0)	→0.07(3)		G	030
0.02(1)			H	031

From the table,

$$\bar{L} \cdot \log D = 2 \times \Sigma p_i L_i = 2.94;$$

and η of this code is worse than that of the binary code. However, for large values of m, both binary and quatenary code are equally efficient.

Example 5.11: Coding of an extended source S^n.

Consider a two-alphabet source with $p(x_1) = 0.25$ and $p(x_2) = 0.75$ as discussed in Example 5.5. The best binary code for this source is:

$$x_1 = 0, x_2 = 1, \text{ giving } H(X) = 0.815 \text{ bits and } \eta = 0.815.$$

However, if we consider the extended source S^3 with alphabets: {000, 001, 010, …, 110, 111}, then the codes for the new set of symbols (by using Huffman technique) are shown in Table 5.5.

Table 5.5

Codes	Symbols	$p(y_j)$ (from Ex. 5)	Reduced sources					
1	y_8	27/64	27/64	27/64	27/64	27/64	27/64	→ 37/64(0)
001	y_7	9/64	9/64	9/64	→ 10/64	→18/64	→19/64(0)	27/64(1)
010	y_6	9/64	9/64	9/64	9/64	10/64(0)	18/64(1)	
011	y_4	9/64	9/64	9/64	9/64(0)	9/64(1)		
00000	y_5	3/64	→ 4/64	→ 6/64(0)	9/64(1)			
00001	y_3	3/64	3/64(0)	4/64(1)				
00010	y_2	3/64(0)	3/64(1)					
00011	y_1	1/64(1)						

We have now $H(S^3) = 2.44$ bits, and $\overline{L} = \sum p_i L_i = 2.47$ bits, giving $\eta = 2.44/2.47 = 98.8\%$, showing remarkable improvement over the coding without extension. Further extension of the source gives better η (e.g., η for $S^4 = 99.1\%$), but the rate of improvement is slower.

5.3 SHANNON'S FIRST FUNDAMENTAL THEOREM [2, 3, 4]

It has been shown, by earlier examples and coding techniques, that by efficient (noiseless) encoding, the average length $\overline{L}$ of the codewords at the coder output approaches $H(X)$/symbol of the source, particularly with S^n, the nth extension of the source. Since average information per source symbol is $H(X)$ bits, a message of length n source symbols gives $nH(X)$ bits of information and equivalently the source generates $M_s = 2^{nH(X)}$ messages. If the duration of each symbol is t_0 sec, then in time T sec, the number of source-messages generated $M_s = 2^{H(X)T/t_0}$, $(n = T/t_0)$. At the coder output, the averge information per codeword is $\overline{L}$ log D bits and in T sec, the number of code-messages generated $M_c = 2^{n\overline{L}\log D}$, where, $n = T/t_0$, and t_0 = average duration of codewords. It is intuitively evident that M_c has to be greater or equal to M_s, if the overflow of the buffer memory at the coder input is to be avoided. Thus we have the bound:

$$M_c = 2^{n\overline{L}\,\log D} \geq M_s = 2^{nH(X)}$$

Therefore,

$$\overline{L} \geq \frac{H(X)}{\log D} \tag{5.21}$$

What is now necessary to show is that with $n \to \infty$, $\overline{L} \to H(X)/\log D$. Shannon has formalised these results in the following theorem:

'Suppose that there are a zero-memory source S with the entropy $H(X)$/symbol and a noiseless channel with a capacity C bits/message. Then it is possible to encode the output of S (with the encoder delivering messages to the channel) so that all the information generated in S is transmitted through the channel without loss if, and only if $C \geq H(X)$.

If $C < H(X)$, then all messages generated by S in T sec, cannot be transmitted through the channel in T sec, without buffer memory overflow'. [Here the channel capacity of C bits/message is equivalent to the code capacity of the coder = $\overline{L}$ log D bits per codeword, since the coder has to be matched to the channel].

It has been shown in Ex. 7, that $\overline{L} = H(X)$, i.e., optimum binary coding is obtained if we have $L_i = \log 1/p(X_i)$ where $\log 1/p(x_i)$ is an integer. If $\log 1/p(x_i)$ is not an integer, one uses eqn. (5.20) for binary encoding. Thus with D-ary encoding, $D > 2$, we have the bound:

$$\log 1/p(x_i) \leq L_i \log D \leq \log 1/p(x_i) + 1 \tag{5.22}$$

This also satisfies the Kraft inequality which states that*:

*The proof that eqn. (5.22) also satisfies eqn. (5.23) is as follows: The left inequality of eqn. (5.22) can be written as: $\log_D 1/p(x_i) \leq L_i$, or $1/p(x_i) \leq D^{L_i}$, or $p(x_i) \geq D^{-L_i}$. Summing over all i, we get

$$1 = \sum_{i=1}^{m} p(x_i) \geq \sum_{1}^{m} D^{-L_i} \tag{5.23a}$$

A necessary and sufficient condition for the existence of an instantaneous code with word length L_i is that:

$$\sum_{i}^{m} D^{-L_i} \leq 1 \tag{5.23}$$

As such L_i, as given in eqn. (5.22), is acceptable as the word lengths of an instantaneous code. Now multiplying eqn. (5.22) by $p(x_i)$ and summing over all i's we get

$$\sum_{i=1}^{m} p(x_i) \log 1/p(x_i) \leq \sum_{i=1}^{m} p_i L_i \log D \leq \sum_{i=1}^{m} p(x_i) \log 1/p(x_i) + 1$$

or

$$\frac{H(X)}{\log D} \leq \bar{L} \leq \frac{H(X)}{\log D} + 1 \tag{5.24}$$

To obtain better efficiency, one may use the nth extension of S, giving $\bar{L}_n$ as the new codelengths. Since eqn. (5.24) is valid for any zero-memory source, then this is also valid for (S^n) having $H(S^n) = nH(x)$, leading to:

$$\frac{nH(X)}{\log D} \leq \bar{L}_n \leq \frac{nH(X)}{\log D} + 1$$

or

$$\frac{H(X)}{\log D} \leq \bar{L}_i/n \leq \frac{H(X)}{\log D} + 1/n$$

and with $n \rightarrow \infty$,

$$\lim_{n\rightarrow\infty} \bar{L}_n/n = \frac{H(X)}{\log D} \tag{5.25}$$

Here, $\bar{L}_n/n$ is the average number of code symbols used per single symbol of S, when the input to the coder is n-symbol messages for the extended source S^n. $\bar{L}_n/n \neq \bar{L}$, where $\bar{L}$ is the average code length for the source S and (see Ex. 11) in general, $\bar{L}_n/n \leq \bar{L}$. The code capacity is now $(\bar{L}_n/n) \log D = C$ bits/message of the channel, and for successful transmission of messages through the channel, we have:

$$H(X) \leq (\bar{L}_n/n) \log D = C \quad \text{bits/message} \tag{5.26}$$

The theorem is also valid for sources with finite memory, i.e., for Markov sources and this will be discussed later. Moreover, all code symbols may not be of equal duration t_0 as assumed above and this leads to a further modification of the 1st Fundamental Theorem, as shown in Sec. 5.4.3.

5.4 SOURCES WITH FINITE MEMORY: MARKOV SOURCES [5, 6]

So far we have considered only zero memory sources, i.e., sources where there is no intersymbol influence and in a long message, the symbols occur independently according to their $p(x_i)$. In real-life sources, however, there are intersymbol influence (i.e., probabilistic constraints) present such that the occurrence of x_i in the zero-th position s_0 of the message depends on the previous q symbols $\{s_1, s_2, \ldots, s_q\}$. Such a source is known as a qth order Markov source S_q, (English or any other language is an excellent example of such sources) and these sources are generally specified by the set of conditional probabilities:

$$p(x_i/s_1, s_2, \ldots, s_q)$$

where x_i is the symbol in the s_0 position and each s has an m-symbol alphabet $\{x_1, x_2, \ldots, x_m\}$. Since $p(x_i)$ now depends on the earlier q symbols, the transitional probabilities given by eqn. (5.26) may be shown in a state diagram, where there will be m^q possible states previous to s_0, and the behaviour of the Markov source may be predicted from the state-diagram. Such a state diagram with four states and the transitional probabilities are shown in Fig. 5.4.

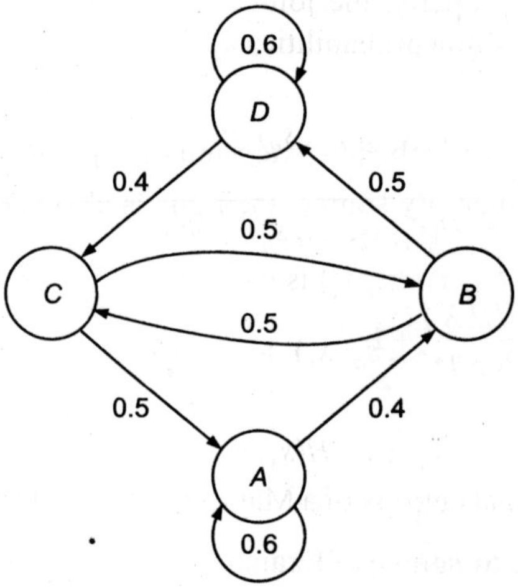

Figure 5.4: State diagram of a second order Markov source

Example 5.12: The four states may represent the binary digits: $A = 00, B = 01, C = 10, D = 11$, and the system would then become a 2nd-order Markov source with the transitional probabilities $p(x_i/s_1, s_2)$,

$$p(0/00) = p(1/11) = 0.6$$
$$p(1/00) = p(0/11) = 0.4$$
$$p(0/01) = p(0/10) = p(1/01) = p(1/10) = 0.5.$$

In this simple case, the state probabilities are calculated from the equations:

$$p(A) = 0.6p(A) + 0.5p(C)$$
$$p(B) = 0.4p(A) + 0.5p(C)$$
$$p(C) = 0.5p(B) + 0.4p(D)$$
$$p(D) = 0.6p(D) + 0.5p(B)$$

By using the relation $p(A) + p(B) + p(C) + p(D) = 1$, and the above equations, the state probabilities are calculated as:

$$p(A) = p(D) = 5/18$$
$$p(C) = p(B) = 2/9 \qquad (5.27)$$

Because of the symmetry of the state diagram of Fig. 5.4, the first-order symbol probabilities are: $p(0) = p(1) = 0.5$. Based on the above probabilities, a typical message is:

$$\ldots CBC \,\vdots\, AAAAABCBCBDDDDDCBC \,\vdots\, AAA\ldots$$

which satisfies the given conditions.

While calculating the state probabilities and finding a typical sequence as above, it has been assumed that the Markov source is ergodic, where the intersymbol influence lasts only for a finite number of earlier q symbols ($q = 2$ in the above example), i.e., the memory of the source is finite. Moreover, if the source is allowed to deliver symbols for a long time, the sequence of symbols will become 'typical' and the state probabilities given as calculated above. The distribution of 'states' in a typical output sequence does not depend upon their initial distribution and finally, a stationary distribution is obtained. The state probabilities given in eqn. (5.27) are calculated on the assumption of the 'stationary distribution'.

Based on the above stationary property, the joint probability and different entropies may now be calculated from the given state transition probabilities by extending the results of eqns. (5.7) and (5.11). Thus we have:

$$p(s_1, s_2, \ldots, s_q, x_i) = p(s_1, s_2, \ldots, s_q)\cdot p(x_i/s_1, s_2, \ldots, s_q)$$

and

$$H(s_1, s_2, \ldots, s_q, x_i) = H(s_1, s_2, \ldots, s_q) + H(x_i/s_1, s_2, \ldots, s_q) \tag{5.28}$$

where the conditional entropy $H(x_i/s_1, s_2, \ldots, s_q)$ is given by

$$H(s_q) = -\sum_{s_q+1} p(s_1, s_2, \ldots, s_q, x_i)\, \log p(x_i/s_1, s_2, \ldots, s_q) \tag{5.29a}$$

$$= H(s_1, s_2, \ldots, s_q, x_i) - H(s_1, s_2, \ldots, s_q) \tag{5.29b}$$

We may now calculate the conditional entropy of a Markov source by using either eqn. (5.29a) or (5.29b).

Example 5.13: Consider the Markov source of Example 5.12. The different probabilities are calculated as shown in Table 5.6.

Table 5.6

s_1, s_2, x_i	$p(x_i/s_1, s_2)$	$p(s_1, s_2)$	$p(s_1, s_2, x_i)$
000	0.60	5/18	1/6
001	0.4	5/18	1/9
010	0.5	2/9	1/9
011	0.5	2/9	1/9
100	0.5	2/9	1/9
101	0.5	2/9	1/9
110	0.4	5/18	1/9
111	0.6	5/18	1/6

$$H(s_1, s_2) = \frac{2\times 5}{18}\log 18/5 + \frac{2\times 2}{9}\log 9/2 = 2.0$$

$$H(s_1, s_2, x_i) = \frac{2\times 1}{6}\log 6 + \frac{6\times 1}{9}\log 9$$

$$= 2.97 \text{ bits}$$

Therefore, $H(x_i/s_1, s_2) = 2.97 - 2.0 = 0.97$ bits/symbol.

5.4.1 Extension of a Markov Source

As in the case of an extended zero-memory source S^n, a qth-order Markov source may also be extended to S_q^n-source, with m^n symbols and reduced intersymbol influence. Thus the nth extension of S_q is a μth-order Markov source, where $\mu = q/n$, and a qth-order Markov source can be converted into a first-order Markov source by making $n = q$. This means that the intersymbol influence is between the two adjacent n-symbol words only, and as such the coding problem is much simplified. In similarity with eqn. (5.17) for zero-memory S^n, the entropy of the extended first-order Markov source S_1^n is:

$$H(S_1^n) = nH(S_1) \tag{5.30}$$

For a first-order Markov source, the set of states are the same as the set of source symbols $\{x_1, x_2, \ldots, x_m\}$ and the stationary distribution of the states gives the first order (unconditional) probability distribution of the source symbols directly. Given a qth-order Markov source S_q with the source alphabet $\{x_1, x_2, \ldots, x_m\}$, an Adjoint source, $\overline{S}_q$, is defined as a zero-memory source with the same source alphabet and with the first order symbol probabilities:

$\{p_1, p_2, \ldots, p_m\} = \{p(x_1), p(x_2), \ldots, p(x_m)\}$. Then the two source S_q and $\overline{S}_q$, have identical first order probabilities, but because of the intersymbol constraint is S_q,

$$H(S_q) \le H(\overline{S}_q) \tag{5.31}$$

The intersymbol constraints, therefore, tend to decrease the average amount of information flowing out of the source, as is seen from eqn. (5.10c) and Ex. 13.

For an extended first order Markov source S_1^n, it may be shown that [5]

$$\begin{aligned} H(\overline{S_1^n}) &= (n-1)H(S_1) + H(\overline{S}_1) \\ &= nH(S_1) + [H(\overline{S}_1) - H(S_1)] \end{aligned} \tag{5.32}$$

where $H(\overline{S}_1) > H(s_1)$.

For an extended qth-order Markov source, S_q^n, the equation is modified as:

$$H(\overline{S_q^n}) = nH(S_q) + \epsilon, \; \epsilon > 0$$

or

$$\frac{H(\overline{S_q^n})}{n} = H(S_q) + \epsilon/n$$

and for $n \to \infty$, $\lim\limits_{n\to\infty} \dfrac{H(\overline{S_q^n})}{n} = H(S_q)$. (5.33)

In other words, for large n, the intersymbol constraints in the Markov source become less and less significant and the average entropy per n-symbol words tends to be equal to n-times the entropy of the Markov source.

Shannon's first Fundamental theorem, given by eqn. (5.25), may now be shown to hold for Markov sources as well. Since the First-roder Markov source S_1 and its adjoint $\overline{S}_1$ have identical first-order symbol probabilities, $\{p(x_1), p(x_2), \ldots, p(x_m)\}$ and $\overline{S}_1$ is a zero-memory source, with $H(S_1) < H(\overline{S}_1)$, the average codelength for both the sources would be indentical. Thus, eqn. (5.21) also holds for $\overline{S}_1$, giving

$$\bar{L} \log D \geq H(\bar{S}_1) > H(S_1) \tag{5.34}$$

Using the nth extension of S_1, the eqn. (5.25) is now modified as:

$$H(\overline{S_1^n}) \leq \bar{L}_n \log D \leq H(\overline{S_1^n}) + 1$$

Further from eqn. (5.32), and dividing by n,

$$H(S_1) + [H(\bar{S}_1) - H(S_1)]/n \leq \frac{\bar{L}_n \log D}{n} \leq H(S_1) + [H(\bar{S}_1) - H(S_1) + 1]/n$$

This gives the bound for the average codelength of a 1st order Markov source and for large n,

$$\lim_{n\to\infty} \frac{\bar{L}_n}{n} = H(S_1)/\log D$$

The result may be extended to the qth order Markov sources, giving

$$\lim_{n\to\infty} \frac{\bar{L}_n}{n} = \frac{H(S_q)}{\log D} \tag{5.35}$$

5.4.2 Entropy of English Language [12]

In Ex. 2, it was shown that the probability of English letters while forming the words are not equal and therefore, H/letter = 4.065 bits, and not 4.76 bits with all letters equiprobable. There are further constraints in forming words such that some combinations, e.g., th, qu, etc., are more frequent, and in many cases the next letter after a few earlier letters in a word may be easily predicted. Thus the English language source, and all other language sources, are Markov sources with finite memory, since $p(x_i/s_1, s_2, \ldots, s_q)$ for various letters are well established. It is then possible to calculate the digram, trigram and q-gram entropies for the language by using eqn. (5.29).

Consider initially that the language source is only a first order Markov source i.e., the intersymbol influence extends only to the earlier letter in words. Then, the digram entropy, $H(S_1) = H(x_i/s_1) = H(s_1, x_i) - H(s_1)$. Statistics of digrams, trigrams etc. for English are available, and from tabulated values, $H(s_1, x_i) = 7.70$ bits, $H(s_1, s_2, x_i) = 11.0$ bits. Thus:

$$H(S_1) = 7.7 - 4.065 = 3.635 \text{ bits/letter}$$

and the digram structure has reduced the information value of the language to 3.64/4.76 = 76.5% of the maximum possible. Considering the intersymbol influence to extend to earlier two letters, the source is now a 2nd order Markov and the trigram entropy is:

$$\begin{aligned} H(S_2) &= H(x_i/s_1, s_2) = H(s_1, s_2, x_i) - H(s_1, s_2) \\ &= 11.0 - 7.7 = 3.3 \text{ bits/letter} \end{aligned}$$

The q-gram entropy is now calculated by using eqn. (5.29), but all such statistics are not easily available. It has been shown that 8-gram entropy, i.e., $H(S_7)$, is approximately 2.3 bits/letter. Shannon [12] has suggested an alternative method of determining the entropy of English words. From the tabulated word frequencies, as given by Zipf, the most common word 'The' is assigned the word-order number 1, the second most common word 'Of' is assigned word number 2 and so on. The probability of words p_n

against the word numbers n are plotted in a log-scale, as shown in Fig.5.5. It is seen that the locus of the points is approximately a straight line, whose slope is (–1). Thus p_n of the nth most common word is given by:

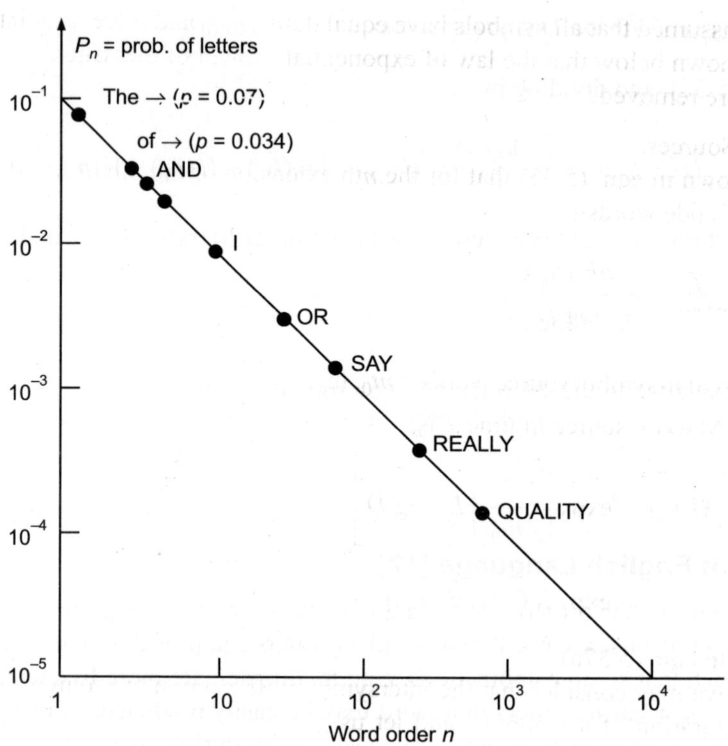

Figure 5.5: Word probability in English (after Zipf)

$$p_n = 0.1/n \tag{5.36}$$

and $\sum p_n = 1$ for maximum $n = 8727$.

We may neglect $n > 8727$ and the law holds good for many languages. The entropy per word is now:

$$H/\text{Word} = -\sum_{1}^{8727} p_n \log p_n = 11.82 \text{ bits/word}$$

Taking each word ≃ 4.5 letters, H/letter ≃ 11.82/4.5 = 2.62 bits. Including space symbol, each word has approximately 5.5 letters, and H/letter ≃ 2.14 bits.

Finally, if one considers all possible intersymbol influences in English then, H/letter ≃ 1.5 bits only. Thus the English language has an efficiency = 1.5/4.76 = 30% and a redundancy of 70%.

5.4.3 Source Entropy and Code-channel Capacity [1, 5]

It has been shown in Section 5.3 that the number of source messages M_s and that of code messages M_c grow exponentially with time T and are given as:

$$M_s = 2^{H(X)T/t_0} \tag{5.37a}$$

and

$$M_c = 2^{(T/t_0)\bar{L}\log D} \tag{5.37b}$$

However, it was assumed that all symbols have equal duration t_0 and there is no intersymbol probability constraint. It is shown below that the law of exponential growth of messages still hold when the above two restrictions are removed.

Case I: Markov Sources.

It has been shown in eqn. (5.35) that for the nth extension of the qth order Markov sources, S_q, the average length of code words is:

$$\bar{L}_{\lim n\to\infty} = \frac{nH(S_q)}{\log D}$$

Since the duration of n-symbol source words $= nt_0$, the duration of $\bar{L}_n = nt_0$, and the number of messages generated by the Markov source in time T is:

$$M(S_q) = \exp_2\left[\left(\frac{T}{nt_0}\right)\bar{L}_n \log D\right]$$

$$= \exp_2\left[(T/t_0)H(S_q)\right] \tag{5.38}$$

which is similar to eqn. (5.37a).

Alternatively we may consider that the intersymbol influence in the source words is of finite range, say less than a duration of q symbols, and let us group the T-duration messages into γ groups of q symbols each. Then the number of different sequences of q-symbol duration is: $m^q = \alpha$, m = size of the source alphabet. Further we may assume that the occurrence of these sequences are somewhat independent, their internal intersymbol influences being absorbed in their overall probability distribution*. Now, consider a very long sequence, β symbols long, made up of $\beta/q = \gamma$ groups of q-symbol words. Then if the probabilities of these α-sequences are $\{p_1, p_2, \ldots, p_\alpha\}$, then the probability of occurrence of β-long sequences is, by law of large numbers,

$$P = (p_1)^{p1\gamma}\,(p_2)^{p2\gamma}\ldots(p_\alpha)^{p\alpha\gamma} \tag{5.39}$$

and the probability is the same for all ergodic sequences of length β. If the sequence is not ergodic, then its probability is vanishingly small and may be considered zero. The number of possible ergodic sequences is then:

$$M(S_q) = 1/P$$

and

$$\log M(S_q) = -\log P = -\gamma \sum_{i=1}^{\alpha} p_i \log p_i + B \tag{5.40}$$

*This also gives an excellent method of coding q-symbol words by using Huffman method, as discussed in Ex. 11.

The constant B is due to the intersymbol constraints within q-symbol groups and varies slowly with γ. For pure random sequences, $B = 0$. Thus $M(S_q)$ may be written as

$$M(S_q) = A \exp_2\left[-\gamma\sum_{i=1}^{\alpha} p_i \log p_i\right]$$

$$= A_m 2^{\gamma H(q)} \qquad (5.41)$$

where A = constant, and $H(q)$ = entropy/q-symbol word. It is seen that the number of possible sequences $M(S_q)$ grows exponentially with the length of the sequence β, even when the intersymbol constraints are present, provided that the sequences are ergodic. In terms of T-sec duration of β-long sequences,

$$M_T(S_q) = A_m \exp_2 [(T/qt_0)H(q)];\ \beta t_0 = T,$$

$$= A_m R_m^T \qquad (5.42)$$

which is similar to eqn. (5.38).

Asymptotic source (language) capacity may now be defined as:

$$\lim_{T\to\infty}\left[\frac{\log M_T(S_q)}{T}\right] = \frac{\log A_m + T\log R_m}{T}$$

$$= \log R_m \qquad (5.43)$$

Moreover, the total language information of β long sequences is $-\log P$, and therefore, language information/symbol

$$\approx -\frac{\gamma}{\beta}\sum_i p_i \log p_i = -\frac{1}{q}\sum_i p_i \log p_i$$

$$= H(q)/q \ \ldots \text{(from eqn. (5.41))} \qquad (5.44)$$

Case II: Unequal Duration of Code Symbols [1, Appx. 1–4].

It has been assumed in Sec. 5.3, that the code symbols are all of equal duration = $t_0/\overline{L}$. However, if the symbols are not of equal duration, e.g., in Pulsewidth modulated signals, then the number of possible code messages generated in time T, is given as:

$$N(T) = N(T - t_1) + N(T - t_2)\ldots + N(T - t_D) \qquad (5.45)$$

where $t_1 \neq t_2 \ldots \neq t_D$, and t_i = duration of the symbol d_i, for a code alphabet of D symbols. The solution of the above difference equation for large T is of the form:

$$\lim_{T\to\infty} [N(T)] = A_R R_0^T \qquad (5.46)*$$

where A_R is a constant and R_0 is the largest real root of the characteristic equation:

$$X^{-t_1} + X^{-t_2} + \ldots + X^{-t_D} = 1$$

*The equation is also true for source codes, e.g., Morse code in telegraphy, where there are certain specific fixed constraints for transition from one state to the other [see Goldman, [1] Appx. 1].

Thus it is seen that $N(T)$ grows exponentially with T for $T \to \infty$, and is similar to eqn. (5.37b), giving:

$$M_c(T) \rightleftarrows N(T)$$

The asymptotic code capacity can now be defined as:

$$\lim_{T\to\infty}\left[\frac{\log N(T)}{T}\right] = \frac{\log A_R + T \log R_0}{T}$$

$$= \log R_0 \tag{5.47}$$

Source Entropy vs. Code Capacity

In view of eqns. (5.43) and (5.47), it is now evident that for successful transmission of all the information generated by the source in time T, the condition to be satisfied is:

$$N(T) \geq M_T(S_q)$$

and for $T \to \infty$,

$$R_0 \geq R_m \tag{5.48}$$

Shannon's first fundamental theorem may now be restated for general sources and code-channels as:

'If the code capacity of a code-channel and encoder exceeds the average information per unit time generated in the information source, i.e., iff $R_0 \geq R_m$, then there is a method of encoding all the information into the channel (even if one has to wait for a long time T). There is no method of encoding all the information if it is generated at a rate exceeding the code capacity, i.e., if $R_0 < R_m$'.

The proper encoding, of course, will require a large buffer memory at the input to the coder and a second memory at the output if the codes of unequal length and duration are to be transmitted through the channel in a synchronous mode.

5.5 DISCRETE CHANNEL WITH DISCRETE NOISE [4, 6, 8]

In digital communication, binary or multilevel (D-ary) signals are normally transmitted, and the channel noise produces errors at the output. The channel characteristic is normally defined by the error rate in the channel or equivalently by the transition/noise matrix of the channel, as shown in Fig. 5.2. Mutual information $I(X, Y)$ has been defined by eqn. (5.3) in Sec. 5.1.2, and is given by:

$$I(X, Y) = H(X) - H(X/Y)$$
$$= H(Y) - H(Y/X)$$
$$= H(X) + H(Y) - H(X, Y)$$

$I(X, Y)$ is the rate of transmission in bits/symbol, and the rate of transmission in bits per second is defined as

$$R(X, Y) = I(X, Y)/t_0 \tag{5.49}$$

where t_0 = duration of each symbol.

In Ex. 3, the different entropies of a multiinput/multioutput channel have been calculated and in Ex. 4, the value of mutual information has been calculated, based on any of the formulae of eqns. (5.13), (5.14). In general, the channel may have m inputs and n outputs, and all the entropies may be calculated using eqns. (5.13), (5.14).

Example 5.14: Consider a channel with two inputs x_1, x_2 and three outputs y_1, y_2, y_3, and the noise matrix of the channel is as given below. Calculate $I(X, Y)$, with $p(x_1) = p(x_2) = 0.5$.

Given:

$p(Y/X) \rightarrow$	y_1	y_2	y_3
x_1	3/4	1/4	0
x_2	0	1/2	1/2

Then

$$p(X, Y) = \begin{bmatrix} 3/8 & 1/8 & 0 \\ 0 & 1/4 & 1/4 \end{bmatrix}$$

and

$$p(X/Y) = \begin{bmatrix} 1 & 1/3 & 0 \\ 0 & 2/3 & 1 \end{bmatrix}$$

where, $\quad p(y_1) = 3/8,\ p(y_2) = 3/8,\ p(y_3) = 1/4$

Then,

$$H(y) = -\,3/8 \log 3/8 - 3/8 \log 3/8 - 1/4 \log 1/4$$
$$= 1.56 \text{ bits/symbol}$$
$$H(X) = 1 \text{ bit/symbol}$$
$$H(Y/X) = 0.90 \text{ bits/symbol, and } H(X/Y) = 0.34 \text{ bits/symbol}$$

Therefore,

$$I(X, Y) = H(Y) - H(Y/X) = 1.56 - 0.90 = 0.66 \text{ bits/symbol}$$
$$= H(X) - H(X/Y) = 1.0 - 0.34 = 0.66 \text{ bits/symbol.}$$

Example 5.15: Consider another example with $m \neq n$, whose noise matrix is given below. Given that:

$$p\{X\} = \{1/4, 2/5, 3/20, 1/20\},$$

calculate $I\{X, Y\}$.

$X \downarrow$ \ $Y \rightarrow$	Y_1	Y_2	Y_3	Y_4
x_1	1	0	0	0
$p(Y/X) \rightarrow x_2$	1/4	3/4	0	0
x_3	0	1/3	2/3	0
x_4	0	0	1/3	2/3
x_5	0	0	1	0

Using $p\{X\}$ as given, the $p(X, Y)$ matrix is:

$$p(X, Y) = \begin{bmatrix} 1/4 & 0 & 0 & 0 \\ 1/10 & 3/10 & 0 & 0 \\ 0 & 1/20 & 1/10 & 0 \\ 0 & 0 & 1/20 & 1/10 \\ 0 & 0 & 1/20 & 0 \end{bmatrix}$$

and

$$p(X/Y) = \begin{bmatrix} 5/7 & 0 & 0 & 0 \\ 2/7 & 6/7 & 0 & 0 \\ 0 & 1/7 & 1/2 & 0 \\ 0 & 0 & 1/4 & 1 \\ 0 & 0 & 1/4 & 0 \end{bmatrix}$$

This gives:

$$p\{Y\} = \{0.35, 0.35, 0.2, 0.1\}$$

Then

$$H(X) = 2.06, \qquad H(Y) = 1.86,$$
$$H(X, Y) = 2.665, \qquad H(Y/X) = 0.61, H(X/Y) = 0.81$$

Therefore,

$$I(X, Y) = H(Y) - H(Y/X) = 1.86 - 0.61 = 1.25 \text{ bits/symbol}$$
$$= H(X) - H(X/Y) = 2.06 - 0.81 = 1.25 \text{ bits/symbol.}$$

Channel Capacity and Efficiency

The channel capacity C of a channel is defined as the maximum of the mutual information that may be transmitted through the channel. As such:

$$C = I(X, Y)_{max} \text{ bits/symbol}$$
$$= R(X, Y)_{max} \text{ bits/sec.} \tag{5.50}$$

From eqns. (5.13), (5.14), it is evident that: given the noise matrix $p(Y/X)$, C may be obtained by maximizing $H(Y)$, and the resultant $H(X)$ gives the values of $p\{X\}$.

The transmission efficiency or the channel efficiency η is then defined as:

$$\eta = I(X, Y)/C \tag{5.51}$$

and Redundancy of the channel

$$= 1 - \eta = \frac{C - I(X,Y)}{C} \tag{5.52}$$

Since the maximization problem as indicated by eqn. (5.50) is rather difficult, we consider first some of the simpler channels, e.g., the Binary channel and its extension.

5.5.1 Binary Symmetric Channel (BSC)

One of the most common and widely used channels is a BSC, whose transition matrix, and the channel diagram are shown in Fig. 5.6. Mutual Information of the channel,

$$I(X, Y) = H(Y) - H(Y/X)$$

where

$$H(Y) = H(X) = 1 \text{ bit},$$

and

$$H(Y/X) = p \log 1/p + q \log 1/q$$
$$= H(p) \text{ of eqn. (5.6).}$$

Therefore,

$$I(X, Y) = 1 + p \log p + q \log q$$
$$= 1 - H(p)$$

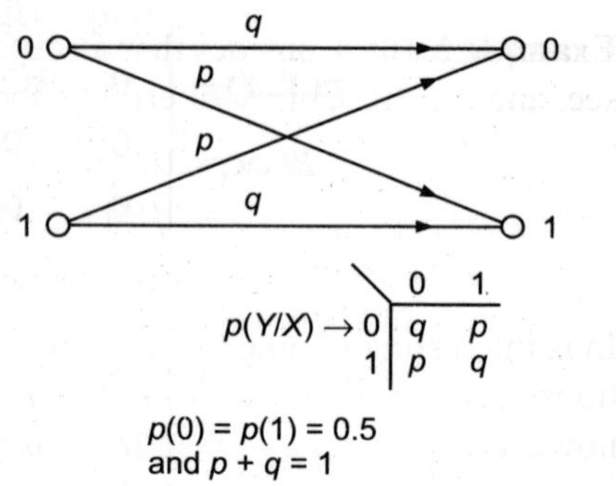

Figure 5.6: A BSC and its channel matrix

Channel capacity is defined as the maximum of $I(X, Y)$ and for the BSC with

$$p(0) = p(1) = 0.5, C = 1 + p \log p + q \log q \tag{5.53}$$

In this particular symmetric case, the 'equivocation' $H(X/Y) = H(Y/X)$. An interesting interpretation of the equivocation may be given, if we consider an idealized communication system with the above symmetric noise matrix and an observer/auxiliary noiseless channel as shown in Fig. 5.7. The observer compares the transmitted and the received symbols, and whenever there is an error, a correction signal in the form of '1' is sent to the receiver. Otherwise the observer continues to send '0' and no change is made in the received symbols. The correction signal transmitted by the observer is the additional information supplied to the receiver to compensate for the loss due to the noise in the channel and is equal to the equivocation $H(X/Y)$. This may be calculated as follows:

Pr. of sending 1 by the observer (with Pr. $(x = 1) =$ Pr. $(x = 0) = 0.5$),

$=$ Pr. of error in the channel $= (1/2)p + (1/2)p = p$

and

Pr. of sending 0 by the observer $= 1 - p = q$.

Therefore, the additional information supplied

$$= p \log 1/p + q \log 1/q = H(X/Y)$$

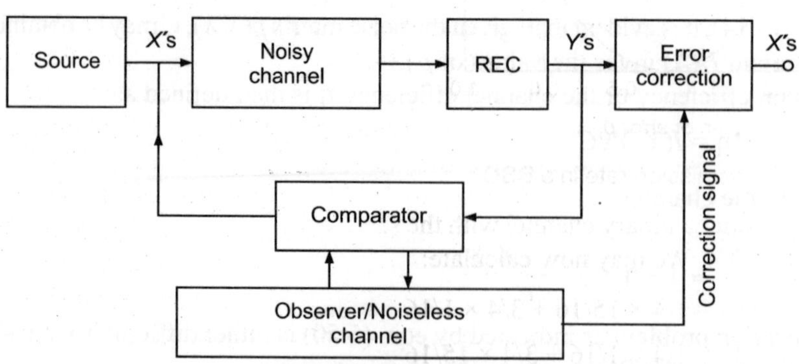

Figure 5.7: An idealised binary communication system

and the Transinformation of the channel

$$= H(X) - H(X/Y) = H(Y) - H(Y/X)$$

$$= 1 + p \log p + q \log q \qquad (5.54)$$

Example 5.16: Consider that a source is transmitting equiprobable 1/0 symbols at the rate of 10^3 bits/sec, and the probability of error in the channel, $P_e = 1/6$. Then the Rate of transmission R/sec is:

$$R/\text{sec} = 10^3[1 + 15/16 \log 15/16 + 1/16 \log 1/16]$$
$$= 10^3 [1 - 0.087 - 0.25]$$
$$= 663 \text{ bits/sec.}$$

In is interesting to note that if p and q are interchanged in the channel matrix, then also the same rate of transmission is obtained. But with $p \rightarrow 0.5$, R decreases and $R = 0$ for $p = 0.5$, as is shown in Fig. 5.8. If, however, $p(x_1) \neq p(x_2)$, then one has to use the eqn.

$$R = H(Y) - H(Y/X),$$

and for a given value of error probability p,

$$p(y = 0) = p_1 q + p_2 p$$

$$p(y = 1) = p_1 p + p_2 q$$

$$H(Y/X) = p \log 1/p + q \log 1/q = H(p)$$

Therefore,

$$R = H(Y) - H(P)$$

and is maximum for $p_1 = p_2 = 0.5$. The variation of R with $p(x_i)$ is shown in Fig. 5.9.

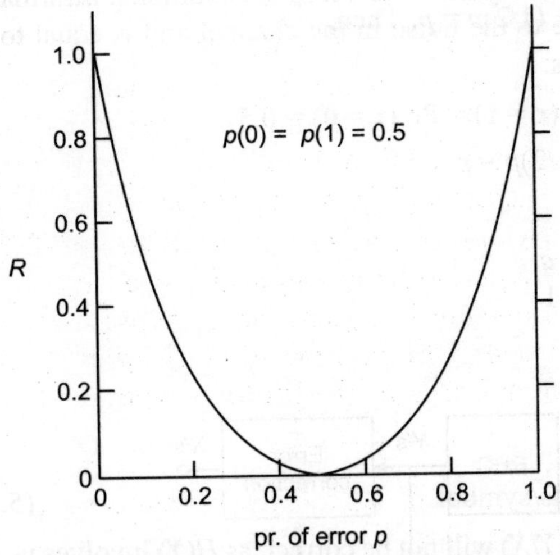

Figure 5.8: Transmission rate in a BSC

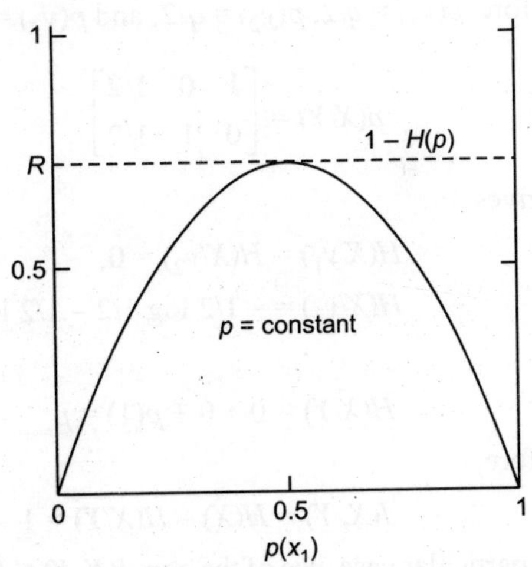

Figure 5.9: Mutual information for a BSC

Example 5.17: Consider a binary channel with the same noise matrix, as in Example 5.16, but $p_1 = p(0) = 1/4$ and $p_2 = p(1) = 3/4$. We may now calculate:

$$p(y = 0) = 1/4 \times 15/16 + 3/4 \times 1/16 = 9/32$$

$$p(y = 1) = 1/4 \times 1/16 + 3/4 \times 15/16 = 23/32$$

Therefore,

$$H(Y) = 9/32 \log 32/9 + 23/32 \log 32/23 = 0.86$$

and

$$I(X, Y) = H(Y) - H(p) = 0.86 - 0.337, \text{ (}H(p)\text{ from Ex. 16)}$$
$$= 0.523$$

Therefore, for transmission rate of 10^3 bits/sec the information rate R/sec = 523 bits/sec.

Binary Erase Channel (BEC)

In data communication, it is a common practice to use ARQ (automatic request for retransmission) techniques for 100% correct data recovery. In the receiver, an error detecting circuit is used along with some parity checks transmitted with the data and the erroneous data is rejected asking for retransmission. The transition matrix of BEC is shown in Fig. 5.10.

Assume that $p(x_1) = p(x_2) = 0.5$, and

$p(Y/X) \rightarrow$

	y_1	y_2	y_3
x_1	q	0	p
x_2	0	q	p

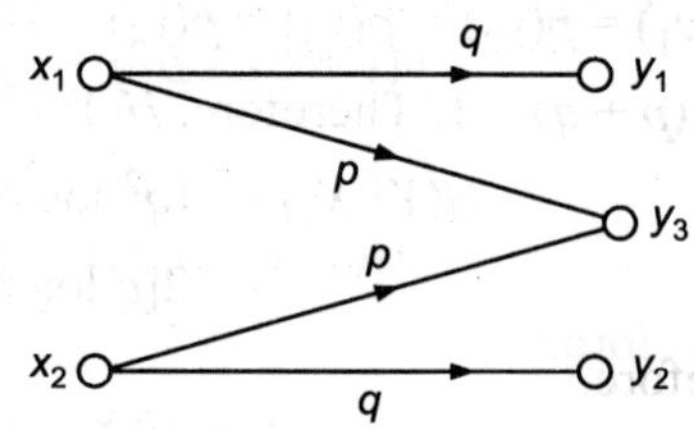

Figure 5.10: A BEC

Then,

$$p(X, Y) = \begin{bmatrix} q/2 & 0 & p/2 \\ 0 & q/2 & p/2 \end{bmatrix}$$

Therefore, $p(y_1) = q/2$, $p(y_2) = q/2$, and $p(y_3) = (1/2)p + (1/2)p = p$. Then,

$$p(X/Y) = \begin{bmatrix} 1 & 0 & 1/2 \\ 0 & 1 & 1/2 \end{bmatrix}$$

This gives

$$H(X/y_1) = H(X/y_2) = 0,$$
$$H(X/y_3) = -1/2 \log 1/2 - 1/2 \log 1/2 = 1$$

and

$$H(X/Y) = 0 + 0 + p(1) = p$$

Therefore,

$$I(X, Y) = H(X) - H(X/Y) = 1 - p = q \text{ bits/symbol} \tag{5.55}$$

In this particular case, use of the eqn. $I(X, Y) = H(Y) - H(Y/X)$ will not be correct, as $H(Y)$ involves y_3, but the information given by y_3 is rejected at the receiver.

5.5.2 Extension of Binary Channels [5]

It has been shown in Sec. 5.1.3 that for the nth extension of the source $H(S^n) = nH(S)$. Similarly, the mutual information of the nth extension of the channel may be shown to be:

$$I(X^n, Y^n) = nI(X, Y) \tag{5.56}$$

Consider again the channel matrix given in Fig. 5.6 for a BSC. For the second extension of the channel, the channel matrix is now expanded as

$X \downarrow$ \ $Y \rightarrow$	00	01	10	11
00	q^2	pq	pq	p^2
$p(Y/X) \rightarrow$ 01	pq	q^2	p^2	pq
10	pq	p^2	q^2	pq
11	p^2	pq	pq	q^2

For $p(0) = p(1) = 0.5, p(x_1) = p(x_2) = p(x_3) = p(x_4) = 0.25$.

Similarly,

$$p(y_1) = p(y_2) = p(y_3) = p(y_4) = 1/4(p^2 + q^2 + 2pq) = 0.25,$$

since $(p + q) = 1$. Therefore, $H(Y^2) = 2$ bits, and

$$H(Y^2/X^2) = -(q^2 \log q^2 + p^2 \log p^2 + 2\,pq \log pq)$$
$$= -2[q \log q + p \log p] \text{ bits}$$

Therefore,

$$I(X^2, Y^2) = 2[1 + q \log q + p \log p] = 2I(X, Y) \tag{5.57}$$

For the nth extension, $H(Y^n) = nH(Y)$, and because of the symmetry of the channel, $H(Y^n/X^n) = nH(Y/X)$, giving the eqn. (5.56).

Cascaded Channels

Sometimes the channels are cascaded for operational reasons, and if both the channels are noisy, there is an information loss in the final outcome.

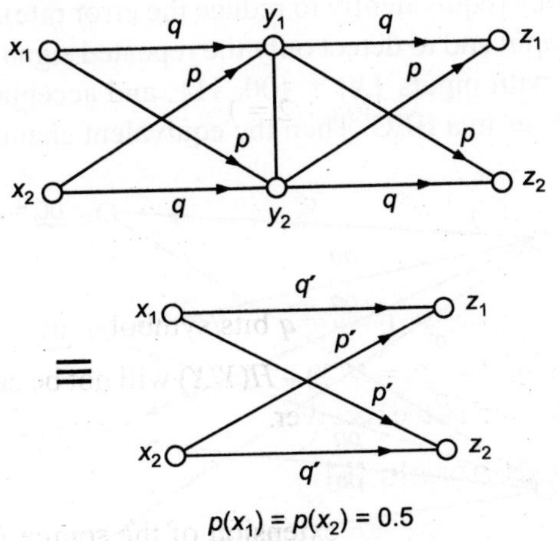

Figure 5.11: Two BSC's in cascade

Consider that two BSC's are cascaded, giving the overall channel matrix, as shown in Fig. 5.11.

$$p(Z/X) = \begin{matrix} & Z_1 & Z_2 \\ X_1 & q' & p' \\ X_2 & p' & q' \end{matrix}$$

where,

$$q' = p^2 + q^2 = 1 - 2pq$$
$$p' = 2pq$$

Thus the cascaded channel is equivalent to a single BSC with error probability = $2pq$.

We have,

$$I(X, Y) = 1 - H(p)$$

and therefore,

$$I(X, Z) = 1 - H(2pq) \tag{5.58}$$

Similarly for a cascade of three BSC's, giving the output $\{U_1, U_2\}$, the mutual information is:

$$I(X, U) = 1 - H(3pq^2 + p^3) \tag{5.59}$$

where the equivalent $q' = q^3 + 3p^2q$

$$p' = p^3 + 3pq^2$$

and

$$(p' + q') = (p + q)^3 = 1.$$

It is seen that the equivalent error probability increases with cascading of more channels and therefore the overall mutual information decreases. For a number of non-identical channels cascaded, the general relations are:

$$I(X, Y) \geq I(X, Z) \geq I(X, U) \tag{5.60}$$

and equality holds when the later channels are noiseless (except in some special cases).

Repetition of Signals: *Additivity of Mutual Information*

To improve the channel efficiency (equivalently to reduce the error rate), an useful technique is to repeat the signals $\{X\}$ at the channel input and to detect only the repeated signals as $\{Y\}$, as shown in Fig. 5.12. Consider the BSC of Fig. 5.6, with inputs $\{X\}$ = {00, 11}, and acceptable as $\{Y\}$ = {00, 11}, thereby discarding the outputs {01, 10}, as in a BEC. Then the equivalent channel matrix is:

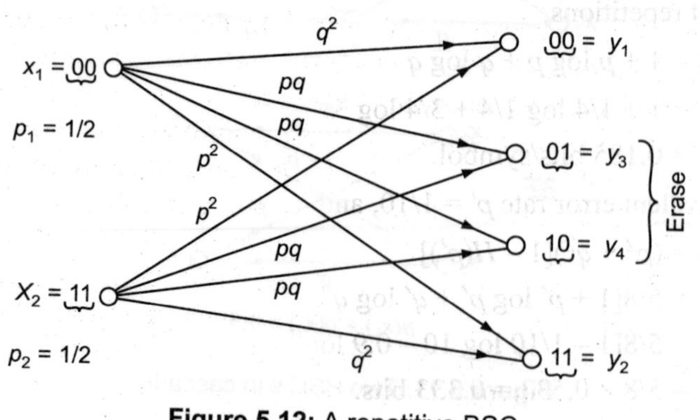

Figure 5.12: A repetitive BSC

$$p(Y/X) = \begin{array}{c} \\ x_1 \\ x_2 \end{array}\begin{array}{c} \begin{array}{cccc} y_1 & y_2 & y_3 & y_4 \end{array} \\ \begin{bmatrix} q^2 & p^2 & pq & pq \\ p^2 & q^2 & pq & pq \end{bmatrix} \end{array}$$

$$p(y_1) = p(y_2) = \frac{p^2+q^2}{2};\ p(y_3) = p(y_4) = pq$$

Therefore,

$$H(Y)_2 = (p^2+q^2)\log\frac{2}{p^2+q^2} + 2pq\log\frac{1}{pq}$$

$$H(Y/X)_2 = q^2 \log 1/q^2 + p^2 \log 1/p^2 + 2pq \log 1/pq$$

$$I(X, Y)_2 = H(Y)_2 - H(Y/X)_2 = (p^2+q^2)\left(\log\frac{1}{p^2+q^2}+1\right) + q^2\log q^2 + p^2\log p^2$$

$$= (p^2+q^2)\left[1+\frac{q^2}{p^2+q^2}\log\frac{q^2}{p^2+q^2}+\frac{p^2}{p^2+q^2}\log\frac{p^2}{p^2+q^2}\right]$$

$$= (p^2+q^2)[1 - H(p^2/(p^2+q^2))] \qquad (5.61)$$

It is seen that the channel is now equivalent to a BSC with error probability $p' = \dfrac{p^2}{p^2+q^2}$ and a normalising factor of $(p^2 + q^2)$ = probability of observing either 00 or 11 at the output. Since $p' < p$, $I(X, Y)_2$ is now greater than the original value of $[1 - H(p)]$ of a BSC. It may be shown that with three repetitions of the input, the mutual information is given as:

$$I(X, Y)_3 = (p^3+q^3)\left[1-H\left(\frac{p^3}{p^3+q^3}\right)\right] + 3pq[1-H(p)] \qquad (5.62)$$

Example 5.18: Assume a BSC with $p = 1/4$ and $q = 3/4$ and $p(0) = p(1) = 0.5$. Calculate the improvements in rate of transmission by 2 and 3 repetitions of the input.

For the BSC, without repetitions,

$$I(X, Y) = 1 + p\log p + q\log q$$
$$= 1 + 1/4 \log 1/4 + 3/4 \log 3/4$$
$$= 0.185 \text{ bits/symbol.}$$

With 2 repetitions, equivalent error rate $p' = 1/10$, and

$$I(X, Y)_2 = (p^2 + q^2)[1 - H(p')]$$
$$= 5/8[1 + p' \log p' + q' \log q']$$
$$= 5/8[1 - 1/10 \log 10 - 0.9 \log 10/9]$$
$$= 5/8 \times 0.533 = 0.333 \text{ bits.}$$

With 3 repetitions, the error rate $p'' = 1/28$ and $q'' = 27/28$

$$I(X, Y)_3 = (p^3 + q^3)[1 - H(p'')] + 3pq[1 - H(p)]$$
$$= (7/16)[1 - 1/28 \log 28 - 27/28 \log 28/27] + (9/16) \times 0.185$$
$$= 7/16 \times 0.78 + 0.104 = 0.444 \text{ bits/symbol.}$$

It is thus seen that the mutual information improves considerably with repetitions. A plot of eqns. (5.61) and (5.62) is shown in Fig. 5.13, for various values of p. This technique of improving the error rate is somewhat similar to error-correcting codes and is widely used in radar-signal reception.

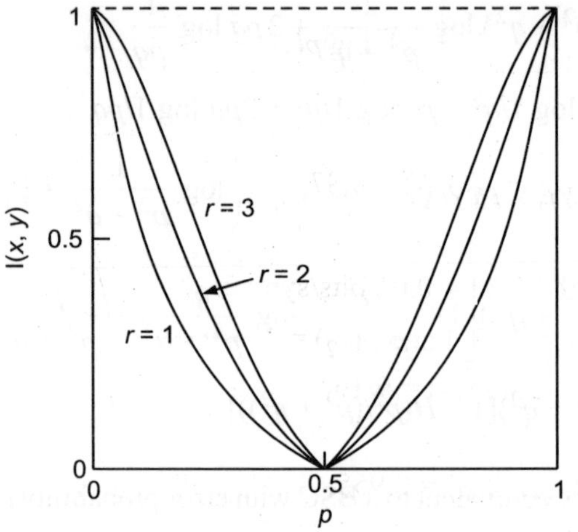

Figure 5.13: I(x, y) for BSC with r-repetition

5.5.3 Channel Capacity of Channels: General Case [4, 6]

Consider a simple binary channel, given by a noise matrix which is not symmetric. In this case, $I(X, Y)$ is to be calculated from eqn. (5.13). If now one attempts to maximize $I(X, Y)$, then it is necessary to use some variational technique so that $H(Y)$ and $H(Y/X)$ are simultaneously optimised, and at the same time a realizable $p\{X\}$ is obtained. $p\{X\}$ are restricted by the constraints: $p(x_i) \geq 0$ and $\sum_i p(x_i) = 1$. Only in certain restricted channels, e.g., in the symmetric/uniform channels, the maximization can easily be done. We shall discuss the techniques in the following. First of all, let us consider a general binary channel and see how $I(X, Y)$ varies with $p(x_i)$ for a given $p(Y/X)$.

Example 5.19: Given a channel matrix:

$$p(Y/X) \rightarrow \begin{bmatrix} 1/2 & 1/2 \\ 1/4 & 3/4 \end{bmatrix}$$

and $p(x_i) = 1/2 = p(x_2)$, the matrix $p(X/Y)$ is obtained as:

$$p(X, Y) = \begin{bmatrix} 1/4 & 1/4 \\ 1/8 & 3/8 \end{bmatrix} \text{ and } p(X/Y) = \begin{bmatrix} 2/3 & 2/5 \\ 1/3 & 3/5 \end{bmatrix}$$

giving $p(y_1) = 3/8$, $p(y_2) = 5/8$.

Therefore, $H(X) = 1$, $H(Y) = 3/8 \log 8/3 + 5/8 \log 8/5 = 0.96$

$$H(Y/X) = 1/4 \log 2 + 1/4 \log 2 + 1/8 \log 4 + 3/8 \log 4/3 = 0.91$$

$$H(X/Y) = 1/4 \log 3/2 + 1/4 \log 5/2 + 1/8 \log 3 + 3/8 \log 5/3 = 0.95.$$

Therefore,

$$I(X, Y) = H(Y) - H(Y/X) = 0.96 - 0.91 = 0.05 \text{ bits/symbol}$$
$$= H(X) - H(X/Y) = 1 - 0.95 = 0.05. \text{ bits/symbol.}$$

If now, $p(x_i)$ is changed to: $p(x_1)$ 1/4 and $p(x_2) = 3/4$, then,

$$p(X, Y) = \begin{bmatrix} 1/8 & 1/8 \\ 3/16 & 9/16 \end{bmatrix}, \text{ giving } p(y_1) = 5/16 \text{ and } p(y_2) = 11/16.$$

Then,

$$H(Y) = 0.89 \text{ and } H(Y/X) = 0.87$$

giving,

$$I(X, Y) = 0.89 - 0.97 = 0.02 \text{ bits/symbol.}$$

If again, $p(x_i)$ is changed to: $p(x_1) = 3/4$ and $p(x_2) = 1/4$, then,

$$p(y_1) = 7/16 \text{ and } p(y_2) = 9/16,$$

and

$$H(Y) = 0.99, H(Y/X) = 0.955,$$

giving,

$$I(X, Y) = 0.035 \text{ bit/symbol.}$$

Thus it is seen that $p(x_1) = p(x_2) = 0.5$ is near about optimum, giving $I(X, Y) = 0.05$ bits/symbol. It will be shown more rigorously later that it is so.

Symmetric/Uniform Channels

Consider a generalised $[m \times n]$ channel matrix as below:

$$p(Y/X) = [p_{ij}] \rightarrow \begin{array}{c|cccccc} & y_1 & y_2 & \cdots & y_j & \cdots & y_n \\ \hline x_1 & p_{11} & p_{12} & \cdots & p_{1j} & \cdots & p_{1n} \\ x_2 & p_{21} & p_{22} & \cdots & p_{2j} & \cdots & p_{2n} \\ \vdots & \vdots & \vdots & & \vdots & & \vdots \\ x_i & p_{i1} & p_{i2} & \cdots & p_{ij} & \cdots & p_{in} \\ \vdots & \vdots & \vdots & & \vdots & & \vdots \\ x_m & p_{m1} & p_{m2} & \cdots & p_{mj} & \cdots & p_{mn} \end{array}$$

and

$$p(X, Y) = [p_i][p_{ij}] \rightarrow \begin{array}{c|ccccccc} & y_1 & y_2 & \cdots & y_j & \cdots & y_n \\ \hline x_1 & p_1p_{11} & p_1p_{12} & \cdots & p_1p_{1j} & \cdots & p_1p_{1n} \\ x_2 & p_2p_{21} & p_2p_{22} & \cdots & p_2p_{2j} & \cdots & p_2p_{2n} \\ \vdots & \vdots & \vdots & & \vdots & & \vdots \\ x_i & p_ip_{i1} & p_ip_{i2} & \cdots & p_ip_{ij} & \cdots & p_ip_{in} \\ \vdots & \vdots & \vdots & & \vdots & & \\ x_m & p_mp_{m1} & p_mp_{m2} & \cdots & p_mp_{mj} & \cdots & p_mp_{mn} \end{array} \quad (5.63)$$

where

$$p\{X\} = \{p_1, p_2, \ldots, p_i, \ldots, p_m\},$$

and

$$p\{Y\} = \{p'_1, p'_2, p'_j, \ldots, p'_n\},$$

Now in matrix form, the values of $\{p_i\}$ and $\{p'_j\}$ are related as:

$$\underline{p_1, p_2, \ldots, p_i, \ldots, p_m} \times [p_{ij}] = \underline{p'_1, p'_2 \ldots, p'_j, \ldots, p'_n}$$

and

$$[p_i] = [p'_j][p_{ij}]^{-1} \quad (5.64)$$

But this solution of $[p_i]$ is possible only when $m = n$. Further we set two restrictions as:

(i) If $\sum_{i=1}^{m} p(y_j / x_i) = \sum_{i=1}^{m} p_{ij}$ is independent of j, i.e., the sums of all columns of $p(Y/X)$ are the same.

Then all receiver probabilities $\{p'_1, p'_2, \ldots, p'_j, \ldots, p'_n\}$ will be equal if all transmitter probabilities, $\{p_1 \ldots p_m\}$ are equal.

(ii) If we further restrict the values of $[p_{ij}]$ in such a way that:

$$H(Y/x_i) = -\sum_{j=1}^{n} p(y_j/x_i) \log p(y_j/x_i)$$

$$= k \text{ (constant) for all } i\text{'s},$$

then

$$H(Y/X) = \sum_{i=1}^{m} p(x_i)H(Y / x_i) = k \sum_{i} p(x_i) = k \quad (5.65)$$

Thus,

$$I(X, Y) = H(Y) - H(Y/X) = H(Y) - k,$$

can be maximised for $H(Y)$ only. Maximum $H(Y)$ is obtained for all p_j's equal and $H(Y) = \log n$, $m = n$.

Such a restricted channel is called a symmetric/uniform channel and the restrictions are satisfied, if the entries in every row and every column of the matrix $[p_{ij}]$ consist of an arbitrary permutation of the entries in the first row. Now the channel capacity of such a uniform channel is given by:

$$C = I(X, Y)_{max} = \log n - k \tag{5.66}$$

The equation may be used for cases where $m \neq n$, but other restrictions hold. In such cases, there is no guarantee of the equiprobable p_i's at the source end. With $m = n$, i.e., for square symmetric matrix of the channel, it can be shown that:

$$p_i = p'_j \text{ for all } i\text{'s and } j\text{'s}$$

and

$$H(X) = H(Y)$$
$$H(Y/X) = H(X/Y).$$

The binary symmetric channels (BSC) and their properties have been already discussed in Sec. 5.5.1. We shall now discuss some examples with higher order channel matrix.

Example 5.20: Consider the channel matrix:

$$p(Y/X) = \begin{bmatrix} p/2 & p/2 & (1-p)/2 & (1-p)/2 \\ (1-p)/2 & (1-p)/2 & p/2 & p/2 \end{bmatrix}$$

which is a symmetric matrix fulfilling the condition (ii) of eqn. (5.65). Therefore, from eqn. (5.66),

$$C = \log n - k,$$

where

$$k = -p \log p/2 - (1-p) \log (1-p)/2 = p + (1-p) + H(p)$$
$$= 1 + H(p)$$

Therefore,

$$C = \log 4 - k = 2 - 1 - H(p) = 1 - H(p)$$
$$= C \text{ for a binary channel with error probability } p.$$

$$\text{Receiver } p'_j = 0.25 \text{ and the transmitter probability} = p_i = 1/2.$$

The binary symmetric channel may be generalised with m inputs and m outputs giving m-ary symmetric channel (mSC), whose channel matrix is:

$$p(Y/X) \to \begin{bmatrix} q & \frac{p}{m-1} & \frac{p}{m-1} & \cdots & \frac{p}{m-1} \\ \frac{p}{m-1} & q & \frac{p}{m-1} & \cdots & \frac{p}{m-1} \\ \vdots & \vdots & \vdots & \vdots & \vdots \\ \frac{p}{m-1} & \cdots & \cdots & \cdots & q \end{bmatrix}$$

Also, $p + q = 1$.
Since the matrix is symmetric,

$$C = \log m - k,$$

$$\text{where } k = -q \log q - p \log \left(\frac{p}{m-1}\right)$$

$$= p \log (m-1) + H(p)$$

Therefore,

$$C = \log m - p \log (m-1) - H(p).$$

Example 5.21: Consider the channel matrix:

$$p(Y/X) \rightarrow \begin{bmatrix} 3/5 & 1/5 & 1/5 \\ 1/5 & 3/5 & 1/5 \\ 1/5 & 1/5 & 3/5 \end{bmatrix}$$

Since the matrix is symmetric, using eqn. (5.66),

$$C = \log 3 - k,$$

where $k = H(Y/x_1) = -[3/5 \log 3/5 + 2/5 \log 1/5]$

$$= 1.37 \text{ bits.}$$

Therefore, $C = 1.585 - 1.37 = 0.215$ bit/symbol, and $p'_j = 1/3 = p_i$ for all i's and j's.
If, however,

$$p\{X\} = \{0.125, 0.25, 0.625\}$$

then $p\{Y\}$ is obtained as: $\{1/4, 3/10, 9/20\}$, giving

$$H(Y) = 1.54 \text{ bits and } H(Y/X) = H(Y/x_1) = 1.37 \text{ bits.}$$

Therefore $I(X, Y) = 1.54 - 1.37 = 0.17$ bit, which is less than $C = 0.215$ bits obtained for all $p_i = p'_j$.

$$\text{The channel efficiency } \eta = \frac{0.17}{0.215} = 0.79$$

and

$$\text{Redundancy} = 1 - 0.79 = 21\%.$$

Unsymmetric Channels [4, 13]

If the symmetry of the channel matrix is not present, then the eqn. (5.66) for channel capacity cannot be used. In such situations, provided that $m = n$, one may use the method suggested by Muroga [13]. Consider first an unsymmetric binary channel with:

$$p(Y/X) = \begin{bmatrix} p_{11} & p_{12} \\ p_{21} & p_{22} \end{bmatrix}$$

and

$$p(x_1) = p_1, p(x_2) = p_2, p(y_1) = p'_1 \text{ and } p(y_2) = p'_2.$$

Introducing new variables Q_1 and Q_2, we can write:

$$\begin{bmatrix} p_{11} & p_{12} \\ p_{21} & p_{22} \end{bmatrix} \begin{bmatrix} Q_1 \\ Q_2 \end{bmatrix} = \begin{bmatrix} p_{11} \log p_{11} + p_{12} \log p_{12} \\ p_{21} \log p_{21} + p_{22} \log p_{22} \end{bmatrix} \tag{5.67}$$

Then,

$$\begin{aligned} R &= I(X, Y) = H(Y) - H(Y/X) \\ &= -(p'_1 \log p'_1 + p'_2 \log p'_2) + p_1(p_{11}Q_1 + p_{12}Q_2) + p_2(p_{21}Q_1 + p_{22}Q_2) \\ &= -(p'_1 \log p'_1 + p'_2 \log p'_2) + p'_1Q_1 + p'_2Q_2, \end{aligned} \tag{5.68}$$

where,

$$p'_1 = p_1p_{11} + p_2p_{21}$$
$$p'_2 = p_1p_{12} + p_2p_{22}$$

To maximize $I(X, Y)$, one uses the calculus of variations (method of Langrangian multipliers) and maximize the function:

$$F = I(X, Y) + \lambda(p'_1 + p'_2) \text{ through a proper selection of } \lambda.$$

Differentiating with respect to p'_1 and p'_2, the condition to be satisfied is:

$$\frac{\partial F}{\partial p'_1} = -(\log_2 e + \log p'_1) + Q_1 + \lambda = 0$$

$$\frac{\partial F}{\partial p'_2} = -(\log_2 e + \log p'_2) + Q_2 + \lambda = 0$$

Solving the equations:

$$\lambda = -Q_1 + (\log_2 e + \log p'_1) = -Q_2 + (\log_2 e + \log p'_2)$$

and

$$Q_1 = Q_2 + \log p'_1 - \log p'_2.$$

Using these values in eqn. (5.68),

$$\begin{aligned} C = R_{\max} &= -p'_1 \log p'_1 - p'_2 \log p'_2 + p'_1Q_2 + p'_1 \log p'_1 - p'_1 \log p'_2 + p'_2Q_2 \\ &= Q_2 - \log p'_2, \text{ since } (p'_1 + p'_2) = 1 \\ &= Q_1 - \log p'_1, \end{aligned}$$

where Q_1 and Q_2 are obtained by solving eqn. (5.67). Further,

$$p'_1 = \exp(Q_1 - C) = e^{Q_1}/e^C$$
$$p'_2 = \exp(Q_2 - C) = e^{Q_2}/e^C$$

and summing p'_1 and p'_2,

$$\frac{e^{Q_1} + e^{Q_2}}{e^C} = 1.$$

Therefore,

$$\begin{aligned} C &= \log\,[e^{Q_1} + e^{Q_2}],\ \text{nats/symbol} \\ &= \log\,[2^{Q_1} + 2^{Q_2}],\ \text{bits/symbol} \end{aligned} \tag{5.69}$$

Example 5.22: Consider the channel matrix of Example 5.19. The solution for Q_1 and Q_2 gives:

$$\left.\begin{aligned} Q_1 &= -1.38 \\ Q_2 &= -0.62 \end{aligned}\right\},\ \text{where } p(Y/X) = \begin{bmatrix} 1/2 & 1/2 \\ 1/4 & 3/4 \end{bmatrix}$$

Therefore,

$$C = \log\,(2^{-1.38} + 2^{-0.62}) = 0.05\ \text{bit/symbol}$$

and

$$p_1' = \exp\,(-\,1.38 - 0.05) = 0.375$$

$$p_2' = 1 - 0.375 = 0.625.$$

These results agree with the results of Ex. 19.

The method is generalised for a $(m \times m)$ channel matrix by using the equation:

$$\begin{bmatrix} p_{11} & p_{12} & \cdots & p_{1m} \\ p_{21} & p_{22} & \cdots & p_{2m} \\ \vdots & & & \\ p_{m1} & p_{m2} & \cdots & p_{mm} \end{bmatrix} \begin{bmatrix} Q_1 \\ Q_2 \\ \vdots \\ Q_m \end{bmatrix} = \begin{bmatrix} \sum_{j=1}^{m} p_{1j} \log p_{1j} \\ \sum_{j=1}^{m} p_{2j} \log p_{2j} \\ \vdots \\ \sum_{j=1}^{m} p_{mj} \log p_{mj} \end{bmatrix} \tag{5.70}$$

Solving the eqn. (5.70), Q_i is obtained. Then the rate of transmission becomes:

$$R = I(X, Y) = -\sum_{j=1}^{m} p_j' \log p_j' + \sum_{i,\ j=1}^{m}\sum p_j' Q_i$$

Maximizing the value of R leads to:

$$\begin{aligned} C &= Q_1 - \log p_j', \quad i = j \\ &= \log \sum_{i=1}^{m} 2^{Q_i} \end{aligned} \tag{5.71}$$

and

$$p' = \exp\,(Q_i - C), \quad i = j.$$

The values of C may not necessarily lead to realizable values of $\{p_i\}$, and also the solution of eqn. (5.70) may not exist. However, within certain limits of the noise matrix values, the method will provide a satisfactory solution.

Example 5.23: Consider a (3 × 3) channel matrix as given:

$$p(Y/X) \rightarrow \begin{bmatrix} 0.4 & 0.6 & 0 \\ 0.5 & 0 & 0.5 \\ 0 & 0.6 & 0.4 \end{bmatrix}$$

Using eqn. (5.70),

$$Q_1 = -1,\ Q_2 = -0.95 \text{ and } Q_3 = -1,$$

and

$$C = \log\,[2^{-1} + 2^{-0.95} + 2^{-1}] = \log\,(1.5) = 0.58 \text{ bits}$$
$$p_1' = 0.35,\ p_2' = 0.3,\ p_3' = 0.35,$$
$$p_1 = 0.25,\ p_2 = 0.5, \text{ and } p_3 = 0.25 \text{ (by using eqn. (5.64))}.$$

If now the channel matrix is changed to:

$$p(Y/X) \rightarrow \begin{bmatrix} 0.8 & 0.2 & 0 \\ 1/2 & 0 & 1/2 \\ 0 & 0.2 & 0.8 \end{bmatrix}$$

then,

$$Q_1 = -1,\ Q_2 = 0.4,\ Q_3 = -1, \text{ and } C = 1.2 \text{ bits;}$$
$$p_1' = 0.22,\ p_2' = 0.56, \text{ and } p_3' = 0.22.$$

Evaluation of p_i requires to satisfy the equations:

$$0.8p_1 + 0.5p_2 = 0.22$$
$$0.2p_1 + 0.2p_3 = 0.56$$
$$0.5p_2 + 0.8p_3 = 0.22.$$

Using 1st and 3rd eqns. it is found that $p_1 = p_3$, and then from 2nd equation $p_1 = p_3 = 1.4$, which violates the condition that $p_i < 1$.

Thus the channel capacity equations for the modified $p(Y/X)$ do not result into realizable values of p_i. It may be shown that for realizable p_i, $p_{11} \le 0.64$, so that the above technique may be used for evaluation of R_{max}.

5.6 SHANNON'S SECOND FUNDAMENTAL THEOREM ON CODING FOR MEMORYLESS NOISY CHANNELS [2, 3, 5]

The theorem states that:

Given that a source has a transmission rate of R bits/symbol and the capacity of the discrete noisy channel without memory is C bits/symbol, then it is possible to transmit the messages through the channel with as small a probability or error (or equivocation) as desired, only if $R < C$ and the coded message length $n \to \infty$. Thus for $n \to \infty$, $R = (C - \in)$, $\in > 0$, we have $P_e \to 0$. Conversely if $R > C$, then $P_e \to 1$, and reliable transmission is not possible.

Consider a binary symmetric channel, with $p(1/1) = q = p(0/0)$ and $p(0/1) = p(1/0) = p$, and $p + q = 1$. By using the nth extension of the binary channel, one may use a n-bit word to designate a k-bit message, such that the total number of messages generated $M = 2^k = 2^{nR}$, and $R = k/n$. This leads to the error correcting techniques discussed in Chapter IV. Consider that a set of messages (codes) $\{x_1, x_2, \ldots, x_m\}$, with the Hamming distance D, are transmitted. In the receiver, the messages will be received as y_i's with their weights altered due to noise, and the optimum decoding procedure is to decide y_j as the transmitted x_j, where $D(y_j, x_j)$ is minimum, as shown in the Fig. 5.14. The conditional probability of receiving y_j, given that x_i is transmitted, is:

$$p(y_j/x_i) = p^D \cdot q^{n-D}, \tag{5.72}$$

where $D = D(y_j, x_i)$ (since D errors occurs in transmitting x_i).

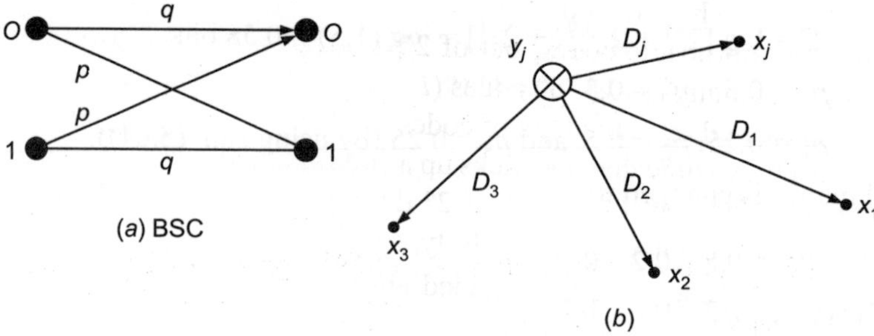

Figure 5.14: (*a*) BSC **(*b*)** Maximum-likelihood decision rule for (BSC)^n

For $p < 1/2$ (for any real channel), $p(y_j/x_i)$ decreases with increasing D, meaning thereby that further y_j is from the transmitted sequence, the less likely it is to be received. The procedure is also equivalent to the maximum likelihood decision rule where the codeword, which maximizes $p(y_j/x_i)$, is chosen as the correct message, as discussed in Chapter III.

Consider now that M codewords, $\{X_1, X_2, \ldots, X_M\}$ are transmitted with n-bits/word. In the receiver, we can visualise an N-dimensional space as in Fig. 5.15, where each code will have a sphere $S_j(np)$ of radius np, within which the code must lie, if the error in the channel is less or equal to np. (Since in a block of n bits, the average number of error $\simeq np$). There is, of course, a finite probability that $X_i(i \neq j)$ may fall within the sphere around Y_j, if the original distance between the codes (X_j, X_i) is reduced due to noise in the channel. Thus there are two types of errors possible: (a) where $D(Y_j, X_i)$ increases due to

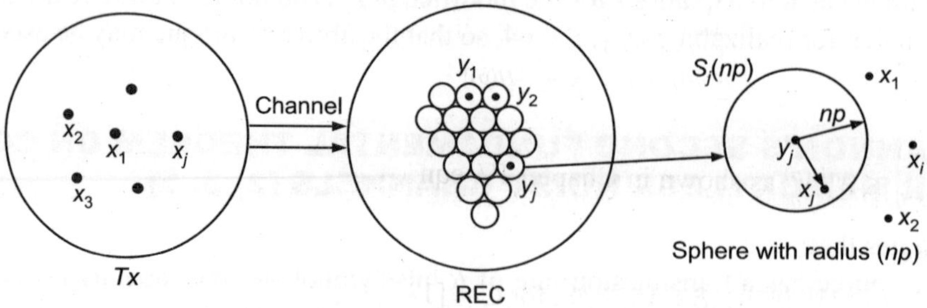

Figure 5.15: A *N*-dimensional space for receiver operation

noise and the received vector falls outside the np-sphere around Y_j, and (b) where $D(Y_j, X_i)$, $i \neq j$, is reduced due to noise such that X_i falls within the sphere around Y_j. These are similar to the errors that occur in radar signal reception and are known as probability of miss, P_m and probability of false alarm, P_f respectively. The overall probability of error may now be written as:

$$P_e = P_r \cdot \{X_j \notin S_j(np)\} + P_r \cdot \{X_i \in S_j(nq)\},\ i \neq j$$
$$= P_m + P_f \tag{5.73}$$

where, $\notin$ and $\in$ are read as 'is not contained in' and 'is contained in' respectively. The value of P_m is dependent on the efficiency of the code used and the block length n. It has been shown in the Chapter IV on FEC that for deterministic group codes.

$$P_m = 2^{-nE(t/n,\, p)} = \delta, \text{ where } E(t/n, p) > 0. \tag{5.74}$$

and $\delta \to 0$, for $n \to \infty$.

Thus with large n, $P_m < \delta$, where $\delta \to 0$.

Since there are M equiprobable codes, out of 2^n possible codes, that are transmitted by the source encoder, any one of/or as many as $(M-1)$ codes $(i \neq j)$ may fall within $S_j(np)$-sphere around Y_j. Thus P_f is function of $(M-1)$ as well as of the type of codes used. To obtain a bound on P_f, Shannon's Random coding argument is used. Assume that one picks up a code at random from the 2^n possible codes and thus makes M separate selections out of a total of 2^n different possibilities. P_f is now averaged over the ensemble of codes, thus generated and, hopefully, some code in the ensemble must be as good as the average. This average $\overline{P}_f$ may now be bounded and then, there will be at least one member of the ensemble of codes which has P_f less than $\overline{P}_f$.

The overall average $\overline{P}_e$ is now given by

$$\overline{P}_e \leq \delta + (M-1)\ \overline{P}_f$$
$$\leq \delta + M\overline{P}_f.$$

Since all sequences with $D \leq np$ from Y_j will fall within $S_j(np)$ and this number $= \sum_{r=0}^{np} \binom{n}{r}$ and the total number of possible codes $= 2^n$, the average $\overline{P}_f$ is calculated as:

$$\overline{P}_f = \overline{p_r\{X_i \in S_j(np)\}},\quad i \neq j$$
$$= \text{number of sequences in } S_j(np)/(2^n)$$
$$= \sum_{r=0}^{np} \binom{n}{r} \Big/ 2^n,$$
$$\leq 2^{nH(p)}/2^n = 2^{-n[1-H(p)]} \tag{5.75}$$

since $\sum_{r=0}^{np} \binom{n}{r} \leq 2^{nH(p)}$ as shown in Chapter IV, eqn. (4.145).

Therefore,

$\overline{P}_e \leq \delta + M \cdot 2^{-n[1-H(p)]} = \delta + M \cdot 2^{-nC}$, for $C = [1 - H(p)]$ in BSC. Assume that the rate of transmission $R = (C - \epsilon)$, then

$$M = 2^{nR} = 2^{n(C-\epsilon)}, \tag{5.76}$$

and the average overall $\bar{P}_e \le [\delta + 2^{-n\epsilon}] \to 0$ with $\lim n \to \infty$ and $\epsilon > 0$ but $\to 0$. Thus if $M = 2^{n(C-\epsilon)}$, i.e., $R = (C - \epsilon)$, $\epsilon > 0$, then only, with random coding, $P_e \to 0$, for $n \to \infty$. The ultimate channel capacity is given by:

$$\lim_{n\to\infty} \frac{\log M}{n} \to C \text{ with } \bar{P}_e \to 0 \tag{5.77}$$

If, however, $R > C$, say $R = C + \epsilon$, then $\bar{P}_e \le \delta + 2^{n\epsilon}$, which tends to ∞ with $n \to \infty$, and error free transmission is not possible. This is also known as the converse to Shannon's 2nd Fundamental theorem. The above theorem can be characterised as an existence proof only as the theorem does not show us how to construct a good code to obtain the rate of transmission $R \simeq C$. The random coding technique only gives a bound on the error probability of the optimum code among the ensemble of possible codes. P. Elias and others [14], on the other hand, have obtained exponential bounds for the minimum probability of error for BSC and BEC, using block codes (n, k), and keeping $R = (C - \epsilon)$. The bounds are given as:

$$K_2 2^{-nE_2} \le \bar{P}_e \le K_1 2^{-nE_1} \tag{5.78}$$

where E_1 and E_2 are functions of C and R, but independent of n. K_1 and K_2 vary very slowly with n. Further, E_1 and E_2 are positive for $R < C$, and vanish for $R = C$. Thus it is possible to make $\bar{P}_e \to 0$ as $n \to \infty$, while $R = (C - \epsilon)$, $\epsilon \to 0$, thereby proving the 2nd fundamental theorem. The equation is further illustrated in Fig. 5.16 for a BSC with $p = 0.05$, $C = 0.7136$, $R = 1/2$. For small p, $E_1 \simeq E_2$, and the curves become parallel for large n. It is further seen that for $R \to C$, the code length n required for $P_e \to 0$, will be very large indeed. For the proof of the theorem for a general discrete channel, one may refer to Abramson [5].

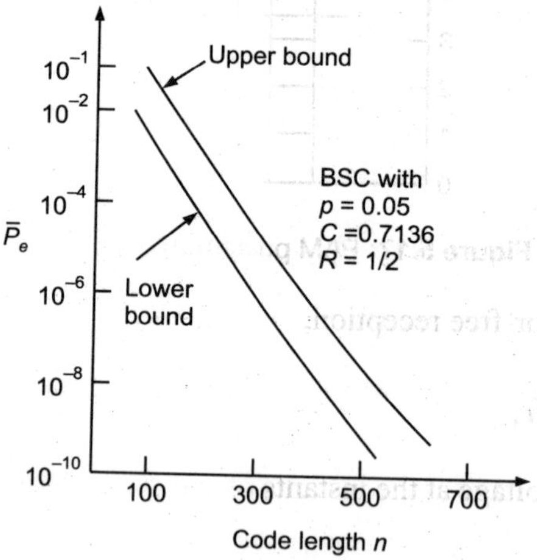

Figure 5.16: $\bar{P}_e$ vs n for a BSC, $R = 1/2$

5.7 DISCRETE CHANNELS WITH CONTINUOUS NOISE

In the earlier sections, we have discussed the transmission rate and channel capacity for discrete channels with discrete noise, and have shown how the error probability may be reduced through coding (FEC), given that $R < C$. In practical channels, the disturbing noise is generally continuous (mostly white and Gaussian), and bandwidth expansion is sought for to obtain better noise-margin. In digital communication, binary and multilevel PAM, coded waveforms, e.g., *M*-ary signals and other orthogonal signal sets, are used as the signalling waveforms. It is therefore necessary to understand the exchange of bandwidth with interference threshold, excess power required to achieve transmission rates equal to the channel capacity and the use of coded waveforms for efficient signalling through a noisy channel. These will be discussed below.

5.7.1 Multilevel Discrete PAM: Interference Threshold vs. Bandwidth

According to sampling theorem, $2W$ pulses per second may be transmitted in a lowpass bandwidth of W Hz. If a pulse of height A is divided into L levels, $(0, 1, 2, .., L-1)$, then the amount of information that may be transmitted in a noiseless channel is:

$$R = 2W \log_2 L \text{ bits/sec.} \tag{5.79}$$

If the channel has peak-limited noise and the noise peaks are limited to ± half a quantum step of the PAM pulse, then the pulses will be received without error as in Fig. 5.17. Thus half a quantum step is the noise threshold for error free reception (e.g., in case of crosstalk and impulsive noises).

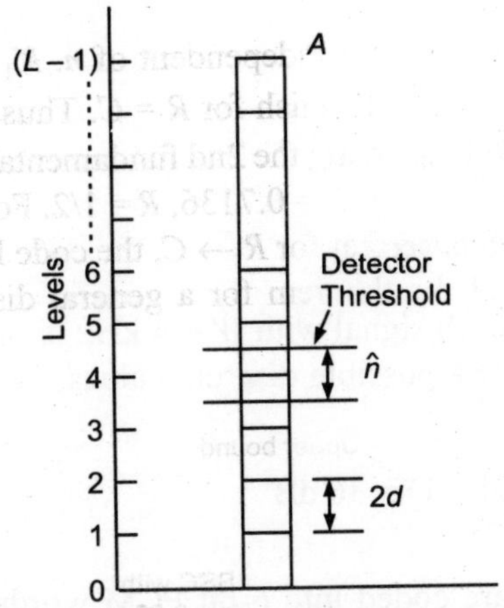

Figure 5.17: PAM pulse parameters

Referring to Fig. 5.17, for error-free reception,

$$\frac{A}{L-1} = 2d \geq \hat{n},$$

where $\hat{n}$ = peak-to-peak noise voltage at the instants of sampling. Then the interference threshold in dB is denined as:

$$I_{\text{th}} = 20 \log \frac{A}{\hat{n}} = 20 \log (L-1) \text{ dB} \tag{5.80}$$

and from eqn. (5.79), the rate of transmission is:

$$R = 2W \log_2 L = 2W \log_2 \left(1 + \frac{A}{\hat{n}}\right) \text{ bits/sec.} \tag{5.81}$$

For small error rates, say $P_e \leq 10^{-4}$, the crest factor (= peak to r.m.s. value) for Gaussian noise $\simeq 4$. Then using $\hat{n} \geq 8\sigma$, one gets the approximate R in the channel.

To reduce the peak power required in the channel, or equivalently to reduce I_{th}, one may convert the PAM pulses to binary/q-ary words and thereby increase the bandwidth and decrease the peak power required for a given noise power σ^2 in the channel. Consider that the L-level pulses are converted to m-bit PCM words, then:

$$L = 2^m, I_{th} = 0 \text{ dB and channel BW required} = mW \text{ Hz}.$$

In general, if PAM pulses are coded into q-ary words, giving,

$$L = q^x,$$

then,

$$I_{th} = 20 \log (q - 1) = 20 \log (L^{1/x} - 1) \text{ dB}$$

and

$$\text{channel BW} = xW \text{ Hz} \qquad (5.82)$$

The rate of transmission now is:

$$\begin{aligned} R &= 2xW \log_2 q \\ &= 2xW \log_2 L^{1/x} \\ &= 2W \log_2 L \text{ bits/sec.} \end{aligned} \qquad (5.81)$$

Example 5.24: Consider that the speech signal with $W = 4$ kHz is sampled at a rate of 8 kHz/sec. and each sample is quantized into one of 64 possible discrete values, i.e., $L = 64$. Then for the equal-step 64'ary codes,

$$I_{th} = 20 \log_{10} (64 - 1) = 36 \text{ dB}$$

and the required BW = 4 kHz.

If, however, 64 quantized levels are coded into 6-bit PCM words, then, the number of pulses/sec = 6 × 8 = 48 Kb/s, and

$$\text{B.W. required} = 24 \text{ kHz}$$

and

$$I_{th} = 0 \text{ dB}.$$

In case the bandwidth reduction is called for, then, two 64-level PAM pulses may be combined into one pulse having

$$L = (64)^2 = 4096 \text{ levels},$$

and

$$\text{BW required} = 2\text{kHz},$$

but $I_{th} = 20 \log_{10} (4096 - 1) = 72.2$ dB.

The same 4 kHz speech signal may be transmitted by using $q = 4, 8$, levels also and the corresponding $I_{th} = 9.5$ and 16.9 dB, for BW = 12 and 8 kHz respectively. These relations showing the exchange of bandwidth with I_{th} are shown in Fig. 5.18.

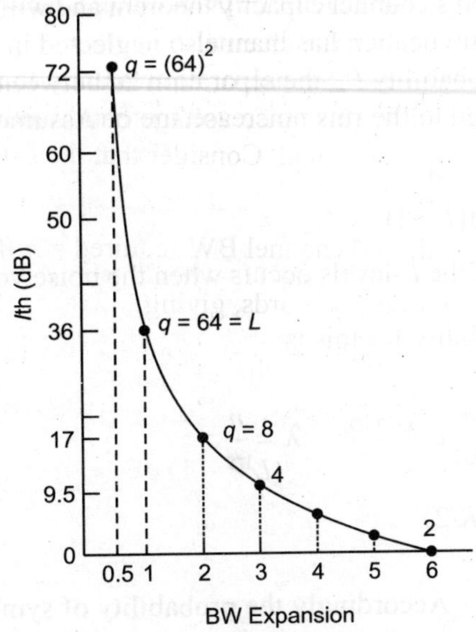

Figure 5.18: Exchange of BW with interference threshold for 64-level speech sample

5.7.2 Channel Capacity and Excess Power in PAM

In channels disturbed with random noise, the condition that $2d \geq \hat{n}$ cannot be maintained even with large signal power A^2 and it is necessary to consider the finite error probability in the channel to calculate the channel capacity of the communication link. It has been assumed that the number of distinguishable levels per sample in PAM $= L = (1 + A/\hat{n})$. Then the total number of possible messages per two samples (per Hz per second) is:

$$L^2 = (1 + A/n)^2 = \left(1 + \frac{\hat{s}}{\hat{n}}\right)^2, A = \hat{s} = \text{peak value of the samples.}$$

For a given error rate $P_e(P_e \to 0)$, $|\hat{s}| = K_1\sigma_s$ and $|\hat{n}| = K_2\sigma_n$, where K_1 and K_2 are constants and σ is the r.m.s. value of signal/noise samples. Assuming the same pdf for both s and n, (say, Gaussian), $K_1 = K_2$, giving

$$|\hat{s}|/|\hat{n}| = \sigma_s/\sigma_n, \text{ and } L^2 = \left[1 + |\hat{s}|^2/|\hat{n}|^2 + 2\hat{s}/\hat{n}\right].$$

Averaging over a large number of samples,

$$\overline{L^2} = \langle 1 + (\sigma_s/\sigma_n)^2 + 2\hat{s}/\hat{n}\rangle \simeq (1 + S/N), S = \sigma_s^2, N = \sigma_n^2,$$

since $\langle \hat{s}/\hat{n}\rangle = 0$, with s and n uncorrelated and of zero mean. For a bandwidth W, total number of possible messages is L^{2W}, and

The rate of transmission $= \log (L^{2W}) = W \log (\overline{L^2})$

then,

$$C = W \log (1 + S/N) \text{ bits/sec.} \tag{5.82}$$

This is a heuristic proof of Shannon's channel capacity theorem and will be discussed further in Sec. 5.8. The loss of information due to equivocation has been also neglected in the above discussion.

For a given acceptable error probability P_e, the separation $2d$ between the levels of the PAM pulse has to be sufficiently large as compared to the rms noise voltage σ. Assume $2d = K\sigma$, $K > 1$; then

$$K = \frac{2d}{\sigma} = \frac{A}{\sigma(L-1)}.$$

The error in the decision of any of the L-levels occurs when the noise voltage exceeds the noise margin, i.e., $n(t) > d = \dfrac{K\sigma}{2}$ and the probability for this is:

$$P\left(n > \frac{K\sigma}{2}\right) = \frac{2}{\sqrt{2\pi}}\int_{K/2}^{\infty} e^{-\lambda^2/2}\, d\lambda, \qquad \lambda = \frac{n}{\sigma},$$

$$= 2\ \text{erfc}\ (K/2) \tag{5.83}$$

where, $\text{erfc}\ (x) = \dfrac{1}{\sqrt{2\pi}}\displaystyle\int_x^{\infty} e^{-\lambda^2/2\, d\lambda}$. Accordingly the probability of symbol error in a PAM channel has been shown to be (refer to eqn. 3.142)

$$P_e = 2\left(1 - \frac{1}{L}\right)\text{erfc}\ (d/\sigma)$$

$$= 2\left(1 - \frac{1}{L}\right)\text{erfc}\left[\left(\frac{3}{L^2-1}\frac{S}{N}\right)^{1/2}\right] \tag{5.84}$$

where $N = \sigma^2$, $S = \dfrac{A^2}{12}\left(\dfrac{L+1}{L-1}\right) \simeq \dfrac{A^2}{12}$ for $L \gg 1$ and the signal amplitude is in the range $-A/2$ to $+A/2$*.

The bit error rate is:

$$P_b \simeq P_e/(\log_2 L)$$

For small P_e, $\sigma^2 = N = \dfrac{A^2}{K^2(L-1)^2}$ and $S = \dfrac{A^2}{12}\left(\dfrac{L+1}{L-1}\right)$. Therefore,

$$S/N = \frac{K^2}{12}(L^2-1) \simeq \frac{K^2}{12}L^2,\ L \gg 1.$$

---

*Assuming uniform distribution of all levels and bipolar pulses of maximum height $\pm A/2$, the average signal power

$$S = \frac{A^2}{4L}\sum_{j=0}^{L-1} s_j^2, \text{ and } s_j = \frac{A}{2}\left(1 - \frac{2j}{L-1}\right)$$

$$= \frac{A^2}{12}\left(\frac{L+1}{L-1}\right) \simeq A^2/12 \text{ for } L \gg 1$$

If, however, unipolar pulses in the range of $0 - A$ volts are used, then

$$S = \frac{A^2}{3}\left(\frac{L+1}{L-1}\right) \simeq A^2/3,\ \ L \gg 1$$

The rate of transmission is now given by:

$$R \leq W \log L^2 = W \log \left[1 + \frac{12}{K^2}(S/N)\right] \tag{5.85}$$

It is seen that the actual rate of transmission (neglecting equivocation) is much less than the channel capacity C of eqn. (5.82), if no coding is used. To reach the capacity C, the signal power has to be increased by a factor of $K^2/12$ (K defined by the acceptable P_e of eqn. (5.83)), in case of binary and multilevel PCM systems. ($K^2/12$) is known as the excess power required to reach the ideal capacity.

Example 5.25: If $P_e = 10^{-5}$ in a binary system, then from Tables, $K = 8.4$, and excess power required = $10 \log_{10} (K^2/12) = 7.8$ dB. For $P_e = 10^{-6}$, $K = 10$, excess power required 9 dB. Using eqns. (5.83) and (5.85), the capacity per Hz = C/W for $P_b = 10^{-5}$ has been plotted in Fig. 5.19, where it is seen that the L-level PCM systems require about 8 dB higher S/N to obtain the ideal C/W. The values of discrete PPM systems, also shown in the figure, are within 3 dB of the ideal values for small values of S/N.

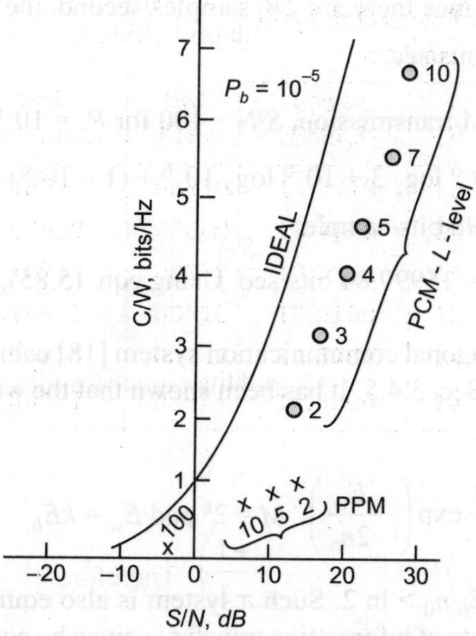

Figure 5.19: Comparison of practical PCM and PPM systems with ideal channel capacity

5.7.3 Information Rate in L-ary Symmetric Channels

In deriving eqn. (5.85), the information loss (equivocation) due to symbol (sample) error rate P_e was neglected. In PAM, assume that all L-levels are equiprobable and the error is equally distributed among all $(L-1)$ levels not transmitted*, then the system is equivalent to an L-ary symmetric channel as discussed under Ex. 20. Its transition diagram is shown in Fig. 5.20. Since the matrix is symmetric, the channel capacity is:

* This is not strictly true for multilevel signalling with random noise in the channel, since the adjacent levels will be more affected than the distant levels. However, when data is transmitted through word codes (e.g., PCM/orthogonal codes), then the channel becomes an L-ary symmetric channel.

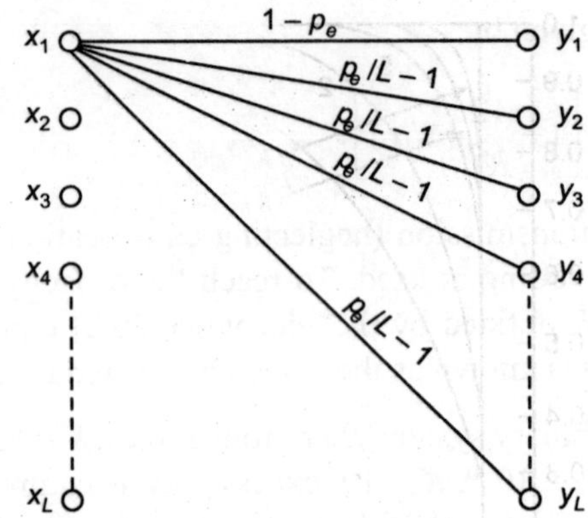

Figure 5.20: *L*-ary symmetric channel

$$C(\text{in bits/sample}) = \log L - P_e \log (L-1) + P_e \log P_e + (1 - P_e) \log (1 - P_e) \quad (5.86)$$

with P_e defined by eqn. (5.84). Since there are $2W$ samples/second, the information rate/sec is:

$$R = 2WC \text{ bits/sec.} \quad (5.87)$$

Example 5.26: In a 4-level PAM transmission, $S/N = 100$ for $P_e = 10^{-6}$, then the capacity/sample

$$= 2 - 10^{-6} \log_2 3 + 10^{-6} \log_2 10^{-6} + (1 - 10^{-6}) \log_2 (1-10^{-6})$$
$$\simeq 1.99998 \text{ bits/sample.}$$

For a bandwidth of 4 kHz, $R = 15999.84$ bits/sec. Using eqn. (5.85), $R \simeq 4.10^3 \log_2 (13) = 15801.76$ bits/sec.

Consider now the *M*-ary orthogonal communication system [18] using *M*-orthogonal alphabets as the transmitted words, discussed in Sec. 3.4.5. It has been shown that the word-error probability is bounded by:

$$P_e(w) \leq \frac{M-1}{2} \exp\left(-\frac{E_w}{2n_0}\right), \; M = 2^k \text{ and } E_w = kE_b \quad (5.88)$$

and $P_e(w) \to 0$, for $M \to \infty$, if $E_b/n_0 > \ln 2$. Such a system is also equivalent to an *M*-ary symmetrical channel, as in Fig. 5.20, and the rate of information transfer is given by eqn. (5.86). If the data transmission rate is $1/T_b$ bits/sec, then the word transmission rate is $1/kT_b$ symbols/sec, and R is given by

$$R = \frac{1}{kT_b} C \text{ bits/sec, where } L = M, P_e = P_e(w) \quad (5.89)$$

Since $P_e(w) \to 0$, as $k \to \infty$ $R \to \dfrac{1}{kT_b} \log M = \dfrac{1}{T_b}$. Using eqns. (5.86), (5.88) and (5.89), the values of R for different E_b/n_0 and k is plotted in Fig. 5.21, where it is seen that for large k, R approaches $1/T_b$ with small values of E_b/n_0, given that $E_b/n_0 > \ln 2$.

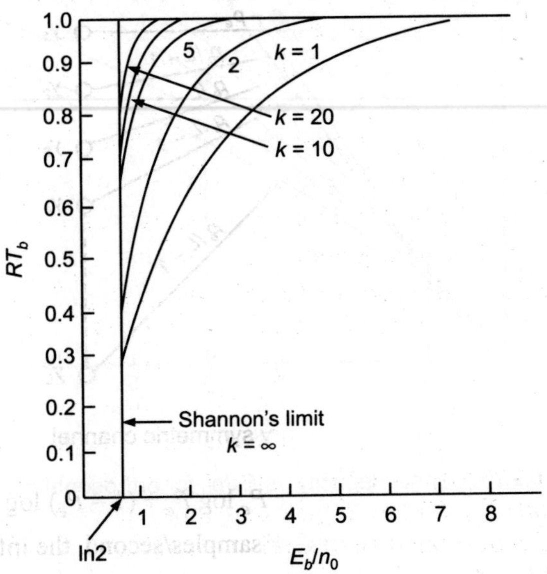

Figure 5.21: Variation of R with E_b/n_0 for orthogonal codes [18]

Considering the channel capacity theorem of eqn. (5.82), C may be expressed as a function of E_b/n_0 as:

$$C = W \log (1 + S/N)$$

$$= \frac{2^k}{kT_b} \log_2 \left[1 + \frac{k}{2^k}(E_b/n_0)\right] \text{ bits/sec, since } R(\text{bits}) = \frac{1}{T_b};$$

and in the limit $k \to \infty$,

$$\log_e \left[1 + \frac{k}{2^k}\frac{E_b}{n_0}\right] \simeq \frac{k}{2^k}\frac{E_b}{n_0},$$

giving,

$$C \simeq \frac{1}{T_b \ln 2}(E_b/n_0) \text{bits/sec,} \tag{5.90}$$

since, $\log_e(1 + x) \simeq x$, $x \ll 1$ and $\frac{k}{2^k}(E_b/n_0) \to 0$ for large k. Channel capacity for various values of E_b/n_0 is now plotted in Fig. 5.22. Observe that with $k \geq 10$, the limiting value of C is reached. It is also observed from eqns. (5.89) and (5.90) that for large k and $E_b/n_0 = \ln 2$, $P_e(w) \to 0$ and the bit transmission rate of $1/T_b$ is achieved. But for higher values of E_b/n_0, R now remains fixed, but C increases linearly. Thus the channel efficiency defined as:

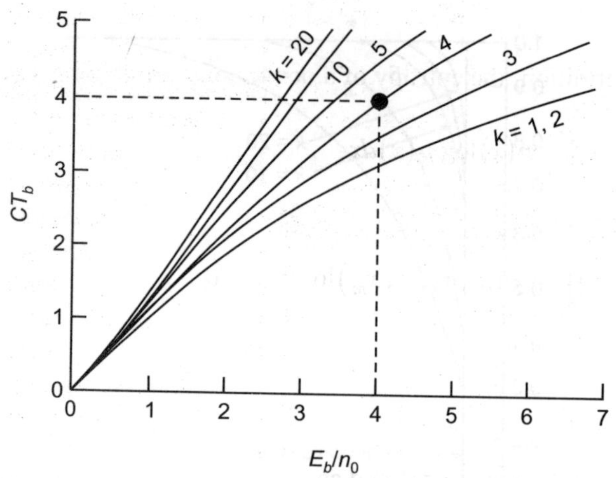

Figure 5.22: Channel capacity vs E_b/n_0 for orthogonal codes [18]

$$\eta_{(ch)} = R/C \tag{5.91}$$

decreases with larger values of E_b/n_0. Table 5.7 gives some values of η vs. E_b/n_0 and $\eta = 1$ only for $k = \infty$ and $E_b/n_0 = \ln 2$.

Table 5.7

E_b/n_0	$\eta_{(ch)}$				
	$k = 1$	2	3	5	10
1	0.32	0.45	0.50	0.53	0.58
2	0.30	0.40	0.42	0.40	0.38
3	0.28	0.35	0.34	0.28	0.26
5	0.25	0.27	0.25	0.18	0.16
7	0.23	0.23	0.20	–	–

It is finally observed that of all possible signalling systems, the *M*-ary orthogonal systems are the only ones which lead to a rate of transmission *R* equal to the channel capacity *C*, for a large alphabet size.

5.8 CONTINUOUS CHANNEL WITH CONTINUOUS NOISE

We have so far discussed communication systems using digital signalling. But many of the naturally occurring signals, e.g., speech, television and telemetry signals, are analog and most of the classical modulation techniques, e.g., AM, FM, PM, are also analog in nature. As such it is necessary to analyse the rate of transmission of information when such signals are corrupted by continuous noise. It is assumed in the following that the channel noise is mostly additive white Gaussian and the transmitted signals are either peak-power limited or average power limited, as is the case in practical systems. Shannon's channel capacity theorem will be discussed and other systems evaluated with reference to the ideal.

5.8.1 Entropy of Continuous Signals

In case of discrete signals, the Entropy was defined as:

$$H(X) = -\sum_{m} p_i \log p_i,\ \{x_i\} = \{x_1, x_2, ..., x_m\} \tag{5.3}$$

In conformity with this definition, the entropy of a continuous variable may be defined as:

$$H(X) = -\int_{-\infty}^{\infty} p(x)\log p(x)dx \tag{5.92)*}$$

and for m variables,

$$H(m) = -\int \dots \int p(x_1, x_2, \dots, x_m)\log p(x_1, x_2, \dots, x_m).dx_1\, dx_2 \dots dx_m \tag{5.93}$$

where

$$\int_{-\infty}^{\infty} p(x_1, x_2, \dots, x_m)dx_1\, dx_2 \dots dx_m = 1.$$

The other related entropies may now be defined as:

$$H(Y) = -\int_{-\infty}^{\infty} p(y)\log p(y)dy$$

$$H(X, Y) = -\int_{-\infty}^{\infty}\int_{-\infty}^{\infty} p(x, y)\log p(x, y)dx\, dy \tag{5.94}$$

$$H(X/Y) = -\int_{-\infty}^{\infty}\int_{-\infty}^{\infty} p(x, y)\log p(x/y)dx\, dy$$

$$H(Y/X) = -\int_{-\infty}^{\infty}\int_{-\infty}^{\infty} p(x, y)\log p(y/x)dx\, dy$$

In the discrete case, all entropies were positive, since the probabilities were bounded as: $0 < p_i < 1$. In the continuous case, the only condition to be satisfied is that

$$\int_{-\infty}^{\infty} p(x)dx = 1$$

and

$$\int_{-\infty}^{\infty}\int_{-\infty}^{\infty} p(x, y)dx\, dy = 1$$

---

*Since probability that x lies between x_i and $(x_i + \Delta x_i) = p(x_i)\, \Delta x_i$, the entropy of continuous variables is written as:

$$H = \lim_{\Delta x \to 0} -\sum_i p(x_i)\Delta x_i \log\left[p(x_i)\Delta x_i\right]$$

$$= -\int p(x)\log p(x)dx - \log \Delta x_i . \int p(x)dx$$

$$= -\int p(x)\log p(x)dx - \log \Delta x_i, \text{ since} \int p(x)dx = 1 \quad (\Delta x_i \to 0)$$

The last term tends to infinity as $\Delta x_i \to 0$ and the integral does not converge. However, assume $\Delta x = 1$ arbitrarily, thereby specifying a particular coordinate for measuring H. Thus the concept of self-information associated with $H(X)$ in the discrete case is no longer valid for the continuous case. $H(X)$ is now simply the entropy function. However, when the difference of two entropies is taken, as in case of mutual information, the ambiguities are automatically resolved.

then for certain distributions, the entropy may turn out to be negative. However, the mutual information, as discussed latter, is never negative.

Example 5.27: Consider a rectangular distribution

$$p(x) = \begin{cases} 5, & -0.1 < x < 0.1 \\ 0, & \text{otherwise} \end{cases}$$

Then,

$$H(X) = -\int_{-1/10}^{1/10} 5 \log 5 \, dx = -\log 5 \text{ bits/sample}$$

showing negative entropy $H(X)$.

Example 5.28: Consider a rectangular distribution,

$$p(x) = \begin{cases} 1/4 & 0 < x < 4 \\ 0, & \text{otherwise} \end{cases}$$

then

$$H(X) = -\int_0^4 1/4 \log 1/4 \, dx$$

$$= -\log 1/4 = 2 \text{ bits/sample.}$$

If now the variable x is amplified by 4, giving the new distribution,

$$p(x) = \begin{cases} 1/16, & 0 < x < 16 \\ 0, & \text{otherwise} \end{cases}$$

then,

$$H(X) = -\log 1/16 = 4 \text{ bits/sample.}$$

It now appears that the amplifier has generated new information, which is untenable. This has happened because the coordinate system has changed i.e., Δx was different in two cases. As such, self-information of $H(X)$ is not valid here. Similarly, other entropy functions, of eqn. (5.94) will also change with the amplification of the variables. Hence, it is only the 'Mutual information' which is meaningful in the continuous case. ❖

Entropy Maximization

In the case of discrete signals, it was shown in Sec. 5.1, that the entropy $H(X) = \sum_m p_i \log p_i$ was maximum when all p_i's were equal. In case of continuous signals, this is equivalent to a rectangular distribution leading to the entropy,

$$H(X) = -\frac{1}{A} \log \frac{1}{A} \int_0^4 dx$$

$$= \log A,$$

where

$$p(x) = 1/A, \quad 0 \le x \le A$$

In practical systems, the sources, e.g., radio transmitters, are either average-power or peak-power limited, and the entropy has to be maximized under such restrictions. In general, the constraints are:

$$\int_{-\infty}^{\infty} p(x)dx = 1, \qquad \text{Normalising condition}$$

$$\int_{-\infty}^{\infty} x^2 p(x)dx = \text{constant}, \qquad \text{Average power limitation}$$

$$\int_{-A}^{A} p(x)dx = 1, \qquad \text{Peak signal limitation}$$

$$\int_{0}^{\infty} xp(x)dx = \text{constant}, \qquad \text{Average signal limitation}$$

$$\int_{0}^{\infty} x^2 p(x)dx = \text{constant}, \qquad \text{Average power limitation with unidirectional distribution.}$$

The problem of maximizing $H = -\int p(x)\log p(x)dx$ is a particular case of the so-called isoperimetric problem of the Calculus of Variations. The problem here is to determine optimum $p(x)$, so that the integral $I = \int_a^b F(x, p)dx$ yields a maximum (or minimum), subject to the conditions (i.e., the given constraints) that:

$$\begin{aligned} \int_a^b \phi_1(x, p)dx &= K_1 \\ \int_a^b \phi_2(x, p)dx &= K_2 \\ &\vdots \\ \int_a^b \phi_n(x, p)dx &= K_n, \end{aligned} \tag{5.95}$$

where $K_1, K_2, ..., K_n$ are preassigned constants. The form of $p(x)$ which satisfies the above constraints and makes I maximum (or minimum) is obtained by solving the equation:

$$\frac{\partial F}{\partial p} + \lambda_1 \frac{\partial \phi_1}{\partial p} + \lambda_2 \frac{\partial \phi_2}{\partial p} + ... + \lambda_n \frac{\partial \phi_n}{\partial p} = 0 \tag{5.96}$$

The quantities, $\lambda_1, \lambda_2, ..., \lambda_n$ are the Lagrangian (undetermined) multipliers, and they are determined by substituting the value of $p(x)$ in eqn. (5.95).

Case I: Symmetric distribution with peak power (and peak signal) limitation which occurs in practical systems, e.g., in AM, FM and Pulse Modulation transmitters.

Suppose that the signal is peak limited to $\pm A$, equivalent to the peak power limitation of S. Then we have

$$H = -\int_{-A}^{A} p(x)\log p(x)dx$$

subject to the normalising constraint:

$$\int_{-A}^{A} p(x)\,dx = 1.$$

From (5.95), we have; $p = p(x)$

$$F(x, p) = -p \log p;\ p = p(x)$$

and

$$\phi_1(x, p) = p$$

Then,

$$\frac{\partial F}{\partial p} = -(1+\log p);\quad \frac{\partial \phi_1}{\partial p} = 1$$

Substituting these in eqn. (5.96), the value of p is obtained by solving:

$$-(1+\log p) + \lambda_1 = 0$$

or

$$p(x) = \exp(\lambda_1 - 1) \text{ for maximum } H.$$

The value of λ_1 is obtained from:

$$\int_{-A}^{A} p\,dx = \int_{-A}^{A} \exp(\lambda_1 - 1)\,dx = \exp(\lambda_1 - 1)\int_{-A}^{A} dx$$

$$= \exp(\lambda_1 - 1)\, 2A = 1$$

Therefore, $p(x) = \exp(\lambda_1 - 1) = \frac{1}{2A}$, giving a rectangular distribution.

Now,

$$H_{\max} = -\int_{-A}^{A} \frac{1}{2A} \log \frac{1}{2A}\,dx = \log(2A) \text{ per sample.}$$

$$= (1/2)\log\left(4\hat{S}\right), \text{ where } \hat{S} = A^2.$$

This entropy is per sample and hence H/sec in a bandwidth of W is:

$$H/\text{sec} = W \log\left(4\hat{S}\right) = W \log\left(12\overline{S}\right) \tag{5.97}$$

where Average power $\overline{S} = \hat{S}/3$.

Case II: **Unidirectional distribution with average signal limitation, as applicable in PAM, PLM or in AM with average carrier amplitude imitation.**

The problem now is to maximize $H = -\int_0^{\infty} p(x)\log p(x)\,dx$ subject to the constraints:

$$\int_0^{\infty} p(x)\,dx = 1$$

and

$$\int_0^{\infty} xp(x)\,dx = \beta.$$

Using eqns. (5.95) and (5.96),

$$F(x, p) = -p \log p; \qquad \frac{\partial F}{\partial P} = -(1+\log p)$$

$$\phi_1(x, p) = p; \qquad \frac{\partial \phi_1}{\partial p} = 1$$

$$\phi_2(x, p) = xp; \qquad \frac{\partial \phi_2}{\partial p} = x$$

Then,

$$-(1 + \log p) + \lambda_1 + \lambda_2 x = 0,$$

giving,

$$p(x) = \exp(\lambda_2 x) \cdot \exp(\lambda_1 - 1).$$

The values of λ_1 and λ_2 are determined from:

$$\int_0^{\infty} \exp(\lambda_2 x) \cdot \exp(\lambda_1 - 1)\,dx = 1,$$

giving,

$$-\lambda_2 = \exp(\lambda_1 - 1),\ \lambda_2 \text{ negative}$$

and

$$\int_0^{\infty} x \exp(\lambda_2 x) \cdot \exp(\lambda_1 - 1)\,dx = -\beta,$$

giving,

$$\exp(\lambda_1 - 1) = -\lambda_2 = \frac{1}{\beta}.$$

Thus,

$$p(x) = \frac{1}{\beta} e^{-x/\beta} \ldots x \geq 0$$

and

$$H_{\max} = -\int_0^{\infty} p\left[-\log\beta - \frac{x}{\beta}\right]dx$$

$$= \log \beta + 1$$

$$= \log (e\beta) \text{ per sample.}$$

Therefore, the entropy per second in a bandwidth W

$$= 2W \log(e\beta). \tag{5.98}$$

Case III: Symmetrical distribution with average power limitation, as applicable for random noise with given variance σ^2, audiofrequency telephony and similar systems.

The problem is to determine $p(x)$ such that:

$$H = -\int_{-\infty}^{\infty} p(x)\log p(x)\,dx$$

is maximum, subject to the constraints:

$$\int_{-\infty}^{\infty} p(x)\,dx = 1$$

and

$$\int_{-\infty}^{\infty} x^2 p(x)\,dx = \sigma^2$$

Using eqns. (5.95) and (5.96),

$$\frac{\partial F}{\partial p} = -(1+\log p), \frac{\partial \phi_1}{\partial p} = 1, \frac{\partial \phi_2}{\partial p} = x^2.$$

Then, eqn. (5.96) becomes

$$-(1 + \log p) + \lambda_1 + \lambda_2 x^2 = 0.$$

Therefore,

$$p(x) = \exp(\lambda_2 x^2 + \lambda_1 - 1) = \exp(\lambda_1 - 1).\ \exp(\lambda_2 x^2)$$

From the 1st condition:

$$\int_{-\infty}^{\infty} \exp(\lambda_1 - 1)\cdot\exp(\lambda_2 x^2)\,dx = 1$$

or

$$2\exp(\lambda_1 - 1)\int_{0}^{\infty} \exp(\lambda_2 x^2)\,dx = 2\exp(\lambda_1 - 1).\frac{1}{2}\sqrt{-\frac{\pi}{\lambda_2}} = 1$$

Therefore,

$$\exp(\lambda_1 - 1) = \sqrt{-\lambda_2/\pi},\ \lambda_2 \text{ negative.}$$

Substituting in the 2nd condition,

$$\int_{-\infty}^{\infty} x^2\sqrt{\frac{-\lambda_2}{\pi}}\exp(\lambda_2 x^2)\,dx = 2\int_{0}^{\infty}\sqrt{\frac{-\lambda_2}{\pi}}.x^2.\exp(\lambda_2 x^2)\,dx$$

$$= 1/2\sqrt{-\frac{\lambda_2}{\pi}}\sqrt{\frac{\pi}{(-\lambda_2)^3}} = -\frac{1}{2\lambda_2} = \sigma^2$$

then,

$$\lambda_2 = -\frac{1}{2\sigma^2} \text{ and } \exp(\lambda_1 - 1) = \sqrt{\frac{1}{2\pi\sigma^2}},$$

and

$$p(x) = \frac{1}{\sqrt{2\pi\sigma^2}} \exp\left(-x^2/2\sigma^2\right).$$

This gives the maximum entropy,

$$H_{max} = -\int_{-\infty}^{\infty} p(x) \log \frac{1}{\sqrt{2\pi\sigma^2}} dx + \int_{-\infty}^{\infty} \frac{p(x) x^2 \, dx}{2\sigma^2}$$

$$= \log\sqrt{2\pi\sigma^2} + \frac{1}{2}$$

$$\left[\text{Since} \int p(x)\,dx = 1 \text{ and } \int x^2 p(x)\,dx = \sigma^2\right]$$

$$= \log\sqrt{2\pi\sigma^2} + \log\sqrt{e}$$

$$= \log\sqrt{2\pi e\sigma^2} \text{ per sample.}$$

Therefore,

$$\text{entropy per sec} = W \log(2\pi e\sigma^2) \tag{5.99}$$

It is seen that the give $p(x)$ is the Gaussian pdf of WGN and this has the maximum entropy for a given average power $\sigma^2 = \bar{N}$, among all realistic pdf's for continuous signals.

Case IV: Undirectional distribution with average power limitation, as applicable in AM with average carrier power limitation. The constraints are:

$$\int_0^{\infty} p(x)\,dx = 1, \int_0^{\infty} x^2 p(x)\,dx = P_0$$

Then,

$$p(x) = \sqrt{\frac{2}{\pi P_0}} \exp\left(-x^2/2P_0\right), \quad 0 \le x \le \infty$$

and

$$H_{max}/\text{sample} = \frac{1}{2} \log\left(\frac{\pi e P_0}{2}\right)$$

$$H/\text{sec} = W \log\left(\frac{\pi e P_0}{2}\right) \tag{5.100}$$

This entropy is seen to be less than that of eqn. (5.99) with Gaussian distribution by 1 bit/sample.

5.8.2 Entropy Power

It has been shown in Case III above that the entropy of Gaussian noise = (1/2) log ($2\pi e\sigma^2$)/sample, and is maximum among all pdf's of continuous signals. For calculation of Rate of transmission and Channel capacity, it is convenient to determine the equivalent entropy of signals and interference which are not random or Gaussian. The entropy power N_e of an ensemble of samples is defined as the average power of a Gaussian random signal, limited to the same W and T as the original ensemble, and having the same entropy H/sample. Then,

$$H/\text{sample} = \log \sqrt{2\pi e N_e}$$

Therefore,

$$N_e = e^{2H}/2\pi e \tag{5.101}$$

Since H for random noise is maximum for a given $\bar{N}$, then N_e of any arbitrary signal $\leq$ the actual average power of the signal.

Example 5.29: Impulse noise—Impulse noise or burst noise, e.g., the noise generated by thunderstorms or automobile ignition, has uniform frequency spectrum and the components are all in the phase. As a result, the peak amplitude of impulses $\propto B$ (IF BW) and width of noise pulses $\propto 1/B$. Thus the amplitude of the impulse noise $\propto B$, whereas the amplitude of random noise $\infty \sqrt{B}$. The crest factor for impulse noise $\simeq \sqrt{2}$ (when the impulses are separately distinguishable from each other), as compared to 4 for the Gaussian noise.

Consider the impulses of the form shown in Fig. 5.23, where the rise time $t_r = 1/B$, and $p(x)dx = dx/X$. Then

$$H/\text{sample} = -\int_0^x \frac{1}{X}\log\frac{1}{X}\,dx = \log X.$$

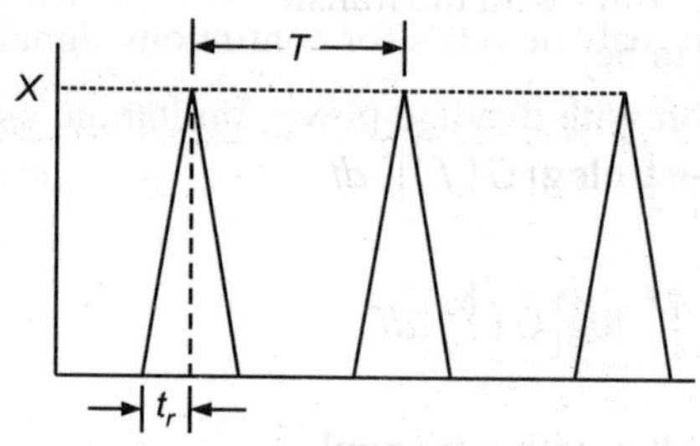

Figure 5.23: Repetitive impulses

Entropy power $N_e = X^2/2\pi e$

$$\text{Average power } \bar{N} = \frac{2}{T}\int_0^{t_r} \frac{X^2 t^2}{t_r^2}\,dt = \frac{2X^2 t_r}{3T}$$

$$= \frac{2X^2}{3TB}$$

If the pulses are close enough such that $2t_r = T$, then $\bar{N} = X^2/3$ is the average power of a triangular waveform. Actually, due to the limited bandwidth, the waveform will not be triangular but somewhat rounded, thus reducing the crest factor from $\sqrt{3}$ as above to a pracitcal value of $\sqrt{2}$. Therefore,

$$N_e : \bar{N} = 2 : 2\pi e \simeq 2 : (4)^2,$$

equivalent to the ratio of the squares of crest factors of impulses and random noise.

Coloured Gaussian Noise

Suppose a WGN is passed through a filter with non-uniform transfer function $G(f)$. Then the output noise power in the band $(f_2 - f_1)$ is:

$$N_0 = \int_{f_1}^{f_2} n_0 |G(f)|^2 \, df, \quad n_0 = \text{noise spectral density of WGN.}$$

$$= \int_{f_1}^{f_2} n(f) df, \quad n(f) = \text{spectral distribution of output noise} = n_0 |G(f)|^2$$

If $n(f)$ is approximated by a step function, i.e., $n(f_i)$ is constant for the ith interval of Δf, then the entropy for the noise power in the band Δf around f_i is:

$$H_i(\Delta_f) = (1/2) \log (2\pi e n(f_i) \cdot \Delta_f)$$

and

$$\text{the overall } H/\text{sec} = \sum_{i=1}^{m} (1/2) \log\left(2\pi e n(f_i) \cdot \Delta f\right) \tag{5.102}$$

where $(f_2 - f_1) = m \cdot \Delta f$.

Since in any passive filter, the output noise power will always decrease due to filtering, H/sec also decreases due to such filtering.

Entropy Loss in Linear Filters [1, Appx. IX]

If a signal is passed through a linear filter with the transfer function, $G(f)$, then the entropy H' per sample at the output of the filter is shown to be:

$$H' = H + \frac{1}{2W} \int_0^W \log |G(f)|^2 \, df \tag{5.103}$$

$$= H + \frac{1}{W} \int_0^W \log |G(f)| \, df$$

Example 5.30: Consider a linear filter with a triangular transfer function, as shown in Fig. 5.24,

$$|G(f)| = \begin{cases} (1 - f/W), & 0 < f < W \\ 0, & \text{otherwise} \end{cases}$$

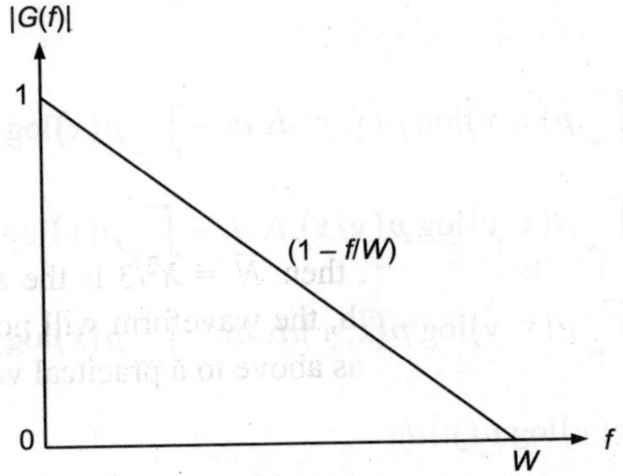

Figure 5.24

Then,

$$\int_0^W \log(1 - f/W) = -W.$$

Therefore, $H = (H - 1)$ bits/sample at the output of the filter. The difference in entropy given by eqn. (5.103) is a property of the filter only and does not depend on the statistics of the signal transmitted through the filter.

Entropy of Sum of Signals

If two independent random (not necessarily Gaussian) signals are linearly added, then the resultant signal distribution is given by the convolution theorem:

$$p(Z) = \int_{-\infty}^{\infty} p_2(Z - X)p_1(X)dX,$$

where $p_2(Z - X) = p_2(Y)$ and $p_1(Z - Y) = p_1(X)$, $Z = X + Y$. When incoherent ergodic ensembles are superimposed, they tend to become more incoherent and more like a random-noise ensemble. The intersymbol constraints, if any, become weaker and the distribution tends to be Gaussian, as given by the central limit theorem. Thus the entropy power of $\{Z\}$ will increase, at least not decrease. We then have:

$$P_e(X) + P_e(Y) \le P_e(Z) \le P(X) + P(Y) \tag{5.104}$$

where $P_e(X)$, $P_e(Y)$ are entropy powers of $\{X\}$ and $\{Y\}$

$P_e(Z)$ = entropy power of the sum signal $\{Z\}$

P = average power of $\{X\}$, $\{Y\}$ or $\{Z\}$

and $\quad P(X) + P(Y) = P(Z).$

5.8.3 Mutual Information of a Continuous Noisy Channel

As in the discrete case, the mutual information of a continuous noisy channel is also defined as: (see eqns. (5.13) and (5.14)):

$$\begin{aligned} I(X, Y) &= H(X) - H(X/Y) \\ &= H(Y) - H(Y/X) \\ &= H(X) + H(Y) - H(X, Y) \end{aligned} \tag{5.105}$$

Using eqn. (5.94),

$$\begin{aligned} I(X, Y) &= \int_{-\infty}^{\infty}\int_{-\infty}^{\infty} p(x, y)\log p(x/y)dx\,dy - \int_{-\infty}^{\infty} p(x)\log p(x)dx \\ &= \int_{-\infty}^{\infty}\int_{-\infty}^{\infty} p(x, y)\log p(y/x)dx\,dy - \int_{-\infty}^{\infty} p(y)\log p(y)dy \\ &= \int_{-\infty}^{\infty}\int_{-\infty}^{\infty} p(x, y)\log p(x, y)dx\,dy - \int_{-\infty}^{\infty} p(x)\log p(x)dx \\ &\quad - \int_{-\infty}^{\infty} p(y)\log p(y)dy \end{aligned} \tag{5.106}$$

It was shown in examples 5.27 and 5.28 that the entropies associated with a continuous variable may be negative and may change with the linear transformation of variable. Incase of mutual information, however, $I(X, Y)$ remains invariant under all linear transformations at the input and output of the channel. Consider that the variables X and Y are transformed as:

$$X_1 = aX + c$$

$$Y_1 = bY + d.$$

Then,

$$p(X_1) = p(X)\left|\frac{dX_1}{dX}\right| = \frac{p(X)}{|-a|}$$

$$p(Y_1) = \frac{p(y)}{|b|}$$

$$p(X_1, Y_1) = \frac{p(X,Y)}{\begin{vmatrix} a & 0 \\ 0 & b \end{vmatrix}} = \frac{p(X,Y)}{|ab|}$$

Therefore,

$$H(X_1) = H(X) + \log|a|$$

$$H(Y_1) = H(Y) + \log|b|$$

$$H(X_1, Y_1) = H(X, Y) + \log|ab|$$

and

$$I(X_1, Y_1) = H(X) + H(Y) - H(X, Y) + \log|a| + \log|b| - \log|ab|$$

$$= I(X, Y)$$

Non-negativeness of $I(X, Y)$ may be proved by using the inequality of eqn. (5.5), as was used for proving eqn. (5.4). The other elementary properties of $I(X, Y)$ are the same as given for discrete signals through eqn. (5.10).

Rate of Transmission

If the noise in the channel is additive and statistically independent of the transmitted signal, then $p(Y/X)$ depends on $(Y - X)$ only, and not on $\{X\}$ or $\{Y\}$. If n is the channel noise (random) variable, then $y = x + n$. If x is now held constant at x_0, then,

$$p[(y_1 < y \le y_2)/x = x_0] = p[(n_1 < n \le n_2)/x = x_0]$$

where $y_1 = x_0 + n_1$ and $y_2 = x_0 + n_2$.

Alternatively, we may write

$$\int_{y_1}^{y_2} p(y/x)\,dy = \int_{n_1}^{n_2} p(n/x)\,dn$$

$$= \int_{n_1}^{n_2} p(n)\,dn = \int_{y_1}^{y_2} p(n)\,dn$$

since n and x are independent variables, and $(y_1 - y_2) = (n_1 - n_2)$.

Therefore, $p(y/x) = p(y - x_0) = p(n)$, subject to the normalisation of the received signal $\{Y\}$ with respect to the transmitted signal $\{X\}$. The conditional entropy $H(Y/X)$ may now be written as:

$$H(Y/X) = -\int_{-\infty}^{\infty}\int_{-\infty}^{\infty} p(X)\,p(Y/X)\log p(Y/X)\,dx\,dy$$

$$= -\int_{-\infty}^{\infty}\int_{-\infty}^{\infty} p(x)\,dx\,p(n)\log p(n)\,dr$$

$$= H(n), \text{ since } \int_{-\infty}^{\infty} p(x)\,dx = 1 \tag{5.107}$$

Therefore, the Rate of transmission/sample = Mutual Information

$= H(Y) - H(n)$

and

equivocation $= H(X/Y) = H(X) - H(Y) + H(Y/X)$

$= H(X) - H(Y) + H(n).$

Since Channel capacity is defined as R_{max},

$$C = R_{max} = [H(Y) - H(n)]_{max}. \tag{5.108}$$

One has to maximize $H(Y)$ under the given constraints and $H(n)$ will be specified by its entropy power.

Mutual Information with Input-Output Correlation

Consider that input and output signal pdf's of a channel are given by:

$$p(x) = \frac{1}{\sqrt{2\pi\sigma_x^2}}\exp\left(-x^2/2\sigma_x^2\right)$$

$$p(y) = \frac{1}{\sqrt{2\pi\sigma_y^2}}\exp\left(-y^2/2\sigma_y^2\right)$$

and the correlation between input and output signals is:

$$\rho = \frac{E(XY)}{\sigma_x\sigma_y}.$$

Then the joint pdf is:

$$p(x, y) = \frac{1}{2\pi\sigma_x\sigma_y\sqrt{1-\rho^2}}\exp\left[-\frac{1}{2(1-\rho^2)}\left(\frac{x^2}{\sigma_x^2} - \frac{2\rho xy}{\sigma_x\sigma_y} + \frac{y^2}{\sigma_y^2}\right)\right],$$

$$|\rho| \neq 1$$

From eqn. (5.99), then different entropies are:

$$H(X) = \log\sqrt{2\pi e\sigma_x^2},\quad H(Y) = \log\sqrt{2\pi e\sigma_y^2}$$

and from eqn. (5.94)

$$H(X, Y) = \int_{-\infty}^{\infty}\int_{-\infty}^{\infty} p(x, y)\log 2\pi\sigma_x\sigma_y\sqrt{1-\rho^2}\,dxdy$$

$$+\int_{-\infty}^{\infty}\int_{-\infty}^{\infty} p(x, y)\frac{1}{2(1-\rho^2)}\left(\frac{x^2}{\sigma_x^2} - \frac{2xy}{\sigma_x\sigma_y} + \frac{y^2}{\sigma_y^2}\right)dxdy$$

Since

$$\iint p(x, y)\,dxdy = 1, \iint p(x, y)x^2 dx\ dy = E(X^2) = \sigma_x^2,$$

$$\iint p(x, y)y^2 dxdy = E(Y^2) = \sigma_y^2, \iint p(x, y)xy\ dxdy = E(XY),$$

$$H(X, Y) = \log 2\pi\sigma_x\sigma_y\sqrt{1-\rho^2} + \frac{1}{2(1-\rho^2)}(2-2\rho^2)$$

$$= \log 2\pi\sigma_x\sigma_y e\sqrt{1-\rho^2}$$

Therefore,

$$I(X, Y) = H(X) + H(Y) - H(X, Y)$$

$$= \log 2\pi e\sigma_x\sigma_y - \log 2\pi e\sigma_x\sigma_y\sqrt{1-\rho^2}$$

$$= \log\frac{1}{\sqrt{1-\rho^2}} = -(1/2)\log(1-\rho^2) \quad (5.109)$$

The equation gives the measure of mutual information in a Gaussian channel and it is seen that the rate of transmission of information depends only on the correlation coefficient between X and Y. If $\rho = 0$, then $I(X, Y) = 0$, and in the limit $\rho \to 1$, $I(X, Y) \to \infty$. In practical channels, due to the interfering noise, the correlation between X and Y decreases and the decrease in ρ depends on S/N in the channel.

5.8.4 Channel Capacity with AWGN and Average Power Limitation of Signals

In eqn. (5.108), it was shown that the channel capacity is given by:

$$C = R_{\max} = [H(Y) - H(n)]_{\max} \quad (5.108)$$

In eqn. (5.99), it has been shown that for maximum entropy the signal pdf has to be Gaussian, if its average power is limited to say S. With the Gaussian noise added to the signal, the statistics of the received y's is also Gaussian with the average power $S + N$, where $N = \sigma^2$ = noise power. The entropy $H(Y)$ is now:

$$H(Y) = (1/2)\log 2\pi e\,(S + N)/\text{sample}$$

and (5.109)

$$H(n) = (1/2)\log 2\pi e N/\text{sample}.$$

Therefore,

$$C/\text{sample} = (1/2)\log(1 + S/N)\text{ bits}$$

and C per second is in bandwidth W

$$= W \log_2(1 + S/N)\text{bits} \tag{5.110}$$

The overall capacity with a signal duration of T sec. and $BW = W$,

$$C = WT \log_2(1 + S/N) \text{ bits.}$$

This is the well known Shannon's Channel Capacity theorem, which is an ideal limit for a given SNR in the channel. Comparing this with eqn. (5.79), it is seen that the number of distinguishable levels in the received signal is ideally $L = \sqrt{1 + S/N}$, which is ot realizable in practice as is evident from eqn. (5.85). However, by using orthogonal codes, the channel capacity may be reached with $W \to \infty$, as has been discussed in Sec. 5.7.3. Further properties of the theorem will be discussed in Sec. 5.9.

Channel with Arbitrary Additive Noise

For noise with arbitrary pdf, H/sample = log $\sqrt{2\pi e N_e}$ where N_e is the entropy power defined by eqn. (5.101). With the addition of this noise to the signal, the power of the received signal = $S + N$, where N = average power of noise. If $S/N > 1$, the statistics of Y is marginally affected by the noise, and hence $H(Y) \le (1/2) \log 2\pi e(S + N)$, where the transmitted signal with average power limitation has the Gaussian distribution for maximum entropy.

Therefore,

$$R_{\max} \le (1/2)\log\left(\frac{S+N}{N_e}\right).$$

On the other hand, from eqn. (5.104), the total entropy power of Y is:

$$P_e(Y) \ge P_e(X) + N_e$$

where $P_e(X)$, $P_e(Y)$ are entropy powers of X and Y respectively.

Therefore,

$$R_{\max} \ge (1/2)\log\frac{P_e(X)+N_e}{N_e} = (1/2)\log\left(\frac{S+N_e}{N_e}\right)$$

since X is Gaussian and $P_e(X) = S$. Thus the channel capacity = $R_{\max}$ is bounded as:

$$W\log\left(\frac{S+N_e}{N}\right) \le C \le W\log\left(\frac{S+N}{N_e}\right) \tag{5.111}$$

For

$$S/N \gg 1, C \simeq W\log(S/N_e)$$

and C is minimum if the noise is Gaussian and $N_e = \bar{N}$.

Example 5.31: Consider the baseband transmission of speech signals. The S/N required in a toll-quality transmission is better than 32 dB with a dynamic range ≥ 40 dB. The nominal bandwidth of speech is 4 kHz. Thus the channel capacity C_M of a baseband transmission system is (assuming Gaussian density of the variables):

$$C_M = W \log_2(1 + S/N) \text{ bits/sec.}$$
$$= 4 \times 10^3 \log_2(1.6 \times 10^3)$$
$$= 4 \times 10^3 \times 10.64$$
$$= 4.25 \times 10^4 \text{ bits/sec.}$$

But using eqn. (5.85),

$$\text{the Rate of transmission} \le W \log_2 \left[1 + \frac{12}{K^2} S/N\right]$$

Assuming $(12/K^2) \simeq 1/8.5$ for $P_e = 10^{-4}$,

$$R \simeq 4.10^3 \log_2 \left(\frac{1.6}{8.5} 10^3\right)$$
$$= 30.16 \times 10^3 \text{ bits/sec.}$$

This result is more realistic since multichannel ADM systems operate at 32 kb/s only, as discussed in Chapter II. The more realistic pdf for speech is the Laplacian density function given by:

$$p(x) = \frac{1}{\sqrt{2}\sigma_x} \exp\left(-\frac{\sqrt{2}|x|}{\sigma_x}\right), -\infty \le x \le \infty,$$

and for only positive values of x,

$$p(x) = 1/\beta \exp\left(-\frac{2x}{\beta}\right),$$

where $\quad \beta = \sqrt{2}\sigma_x$ and $x \ge 0$, $\int_0^\infty p(x)dx = 1/2$.

Using eqn. (5.98), the entropy/sec in a $BW = W$ and for both $\pm x$ values,

$$H/\text{sec} = 2W \left\{-\int_{-\infty}^{\infty} p(x)\left[-\log\beta - \frac{2x}{\beta}\right] dx\right\}$$
$$= 2W \log \beta, \text{ since } \int_{-\infty}^{\infty} xp(x)dx = 0$$
$$= W \log\left(2\sigma_x^2\right) = W \log(2S)$$

and C_M now is:

$$C_M = W \log_2 \left(1 + \frac{2S}{2\pi eN}\right)$$
$$= W \log_2 \left(1 + \frac{S}{\pi eN}\right)$$

and the rate of transmission is now further reduced by 3 bits/Hz/sec.

The effective information content in speech signals is then approximately 16 kb/s and the intersymbol constraints further reduce this rate to approximately 10 kb/s or so.

Channel with Coloured Gaussian Noise

Consider that the interfering noise in the channel has Gaussian pdf, but its spectrum is not flat. Under this condition, eqn. (5.102) gives the entropy of this coloured noise.

Since the signal also has Gaussian pdf for maximum entropy with average power limitation, the entropy of Y in the band Δf around f_i is

$$H_i(Y)_{\Delta f} = \Delta f \log \{2\pi e[s(f_i) + n(f_i)]\Delta f\}$$

$$H_i(n)_{\Delta f} = \Delta f \log [2\pi e n(f_i)\Delta f]$$

Therefore,

$$R_i(\Delta f) = \Delta f \log\left[\frac{s(f_i)+n(f_i)}{n(f_i)}\right]$$

If Δf is sufficiently small, then R in the frequency band $(f_2 - f_1)$ is given by:

$$R = \int_{f_1}^{f_2} \log\left[\frac{s(f)+n(f)}{n(f)}\right] df \tag{5.112}$$

It may be noted that R in eqn. (5.112) is not $R_{max} = C$ and $s(f)$ need not be constant. To obtain the channel capacity, the method of calculus of variations has to be used to match $s(f)$ with $n(f)$ such that R in eqn. (5.112) is maximum. This will be further discussed in Sec. 5.8.6. The eqn. (5.112) is useful in cases where $s(f)$ is not constant, e.g., in speech and T.V. signals. Assuming channel noise to be white, R may be calculated for the given spectrum (assuming approximate Gaussian statistics of the signal).

5.8.5 Channel Capacity with AWGN and Peak Power Limitation of Signals [1, Appendix IX]

It has been shown in Sec. 5.8.1, Case I, that the maximum entropy per sample for an ensemble with peak power limitation at the sample points is given by:

$$H_{max}/\text{sample} = (1/2) \log 4\,\hat{S} \tag{5.97}$$

While reconstructing the continuous signal by an ideal lowpass filter, it is possible that the signal peaks occasionally exceed the peaks of the sample points and the entropy due to larger peaks will have to be subtracted from that given by eqn. (5.97). Shannon has shown that any signal which is peak limited at the sample points may also be made peak limited throughout all time T by passing the signal through a triangular filter with the transfer function (see Ex. 30):

$$|G(f)| = \begin{cases} |1 - f/W|, & 0 < f < W \\ 0, & \text{otherwise} \end{cases}$$

The output of the triangular filter with signal samples $\{a_n\}$ as the input is:

$$S_0(t) = \sum_{n=0}^{2TW} \frac{a_n}{2} \frac{\sin^2(\pi W t - n\pi/2)}{(\pi W t - n\pi/2)^2} \tag{5.113}$$

Since eqn. (5.113) is positive for all t and n, the peak value of the signal $S_0(t)$ will occur only when all a_n's are added in phase, which is possible only at $f = 0$, and since $|G(0)| = 1$, the output is the same as the input, giving the peak of the output = the peak of the input = $A = \sqrt{\hat{S}}$. However, the filter causes an entropy loss to the signal, as shown in Ex. 30. The entropy of the signal thus filtered is now giving by eqn. (5.103), and the entropy/sec in the bandwidth W is:

$$\begin{aligned} H/\text{sec} &= W \log (4\hat{S}) - 2W \\ &= W \log (4\hat{S}/e^2) \end{aligned} \tag{5.114}$$

The actual entropy for the peak-limited signal is then bounded by:

$$W \log (4\hat{S}/e^2) \le H/\text{sec (peak-limited)} \le W \log(4\hat{S}).$$

and the channel capacity is also bounded by:

$$W \log\left(1 + \frac{4\hat{S}}{2\pi e^3 N}\right) \le C \le W \log\left(1 + \frac{4\hat{S}}{2\pi e N}\right) \tag{5.115}$$

An Alternative Technique for Peak-limitation of Signals

Instead of using a triangular filter after the reconstruction filter, one may use a limiting amplifier, as shown in Fig. 5.25. Assume the amplifier gain function is specified as:

$$y = K\frac{x}{|x|}\left(1 - e^{-\alpha x^2/|x|}\right), \quad -\infty < x < \infty$$

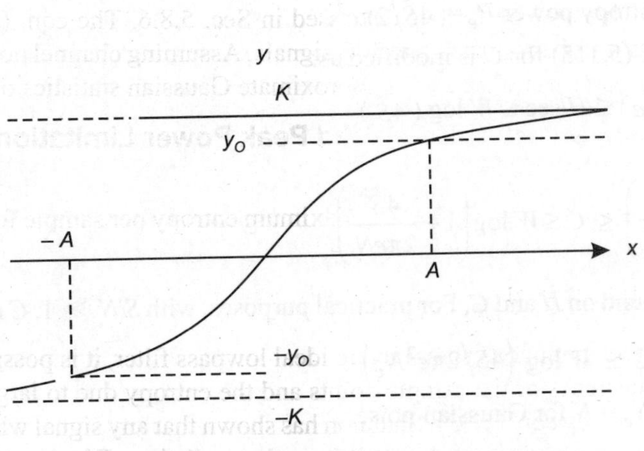

Figure 5.25: Characteristics of a limiting amplifier

Since the characteristic is symmetric, one may use only:

$$\left.\begin{aligned} y &= K\left(1 - e^{-\alpha x}\right) \\ y_0 &= K\left(1 - e^{-\alpha A}\right) \end{aligned}\right\}, \quad 0 < x < \infty$$

and the new pdf is:

$$p(y) = p(x)\Big/\frac{dy}{dx} = \frac{1}{2A(K-y)\alpha},$$

where $p(x) = 1/2A$, $dy/dx = K\alpha e^{-\alpha x} = \alpha(K-y)$.

Now the entropy per sample is:

$$H/\text{smaple} = -2\int_0^{y_0} \frac{1}{2A\alpha(K-y)} \log \frac{1}{2A\alpha(K-y)}\,dy$$

$$+\left[-2\int_{y_0}^{\infty} p(y)\log p(y)\,dy\right]$$

Since $p(x)$ is negligible outside $\pm A$, the second integral does not contribute any significant information, and thus (using a change of the variable and integrating):

$$H = \frac{1}{2A\alpha}\left[(\log 2AK\alpha)^2 - (\log 2AK\alpha\, e^{-\alpha A})^2\right]$$

$$= \frac{1}{2A\alpha}\left[(\log 4A^2K^2\alpha^2\, e^{-\alpha A})\cdot \log e^{\alpha A}\right],$$

$$= \frac{1}{2}\log\left(\frac{4A^2}{e}\right) = \frac{1}{2}\log\left(4\hat{S}/e\right), \text{ assuming that } A = K \text{ and } \alpha A = 1.$$

The corresponding entropy power $P_e = 4\hat{S}/2\pi e^2$.

Therefore, the eqn. (5.115) for C is modified as:

$$W\log\,(4\hat{S}/e) \le H/\text{sec} \le W\log\,(4\hat{S}\,)$$

and

$$W\log\left(1+\frac{4\hat{S}}{2\pi e^2 N}\right) \le C \le W\log\left(1+\frac{4\hat{S}}{2\pi e N}\right) \tag{5.116}$$

This gives a tighter bound on H and C. For practical purposes, with $S/N \gg 1$, C may be approximated as:

$$C \simeq W\log\left(4\hat{S}/2\pi e^2 N_e\right) \tag{5.117}$$

where $N_e = N$ for Gaussian noise.

Example 5.32: (Analog modulation systems)

The practical transmitters using AM and FM are generally peak-power limited, as such their entropies will be governed by eqns. (5.115), (5.116) and (5.117). In SSB, the channel signal spectrum and the message spectrum are the same, and the SNR in the received message is the same as that in the channel. Then the channel capacity is bounded by:

$$W\log\left(1+\frac{4\hat{S}}{2\pi e^2 N}\right) \le C_{(\text{SSB})} \le W\log\left(1+\frac{4\hat{S}}{2\pi e N}\right) \tag{5.118}$$

In case of DSB-AM, the information is in the sidebands only and the effective sideband power = 1/3 of the total transmitted power, but each sideband peak power $\hat{S}_1 = \hat{S}/4$. If $S/N \gg 1$, then the information of the two sidebands add coherently, add thus,

$$2W \log\left(1+\frac{\hat{S}}{2\pi e^2 N}\right) \ll C_{(\text{DSB-AM})} \le 2W \log\left(1+\frac{\hat{S}}{2\pi e N}\right) \tag{5.119}$$

In DSB-SC, there is no carrier and the sideband power = 1/2 of the total transmitter power. The eqn. (5.119) is then modified with $\hat{S}$ replaced by ($2\hat{S}$), and the approximate channel capacity is now given by:

$$C(\text{DSB-SC}) \simeq 2W \log\left(2\hat{S}/2\pi e^2 N\right) \tag{5.120}$$

In FM the carrier power is constant and the information is carried through the instantaneous changes in the transmitted frequency. Since the channel $BW = 2(D+1)W$, W = message BW, D = diviation ratio, the ideal capacity of the RF channel is:

$$C(FM) = 2(D+1)\, W \log_2(1 + S/N) \tag{5.121}$$

where S/N = average SNR in the RF channel. However, the message capacity in the receiver output will be different than that of eqn. (5.121). This will be discussed further in Sec. 5.10.

5.8.6 Spectrum Matching and Optimum Filtering

In eqn. (5.112), the rate of transmission of information in a channel with non-uniform signal and noise spectra has been shown to be:

$$R = \int_{f_1}^{f_2} \log\left[\frac{s(f)+n(f)}{n(t)}\right] df \tag{5.112}$$

and

$$S = \int_{f_1}^{f_2} s(f)\, df.$$

If the signal power S and the coloured noise spectrum $n(f)$ are given, then the method of calculus of variations may be used to optimise R with given constraints of S and $n(f)$. Following the technique used for entropy maximization problems of Sec. 5.8.1, we have:

$$F = \log\left[1+\frac{s(f)}{n(f)}\right], \quad \phi_1 = s(f)$$

and

$$\frac{\partial F}{\partial s(f)} + \lambda_1 \frac{\partial \phi_1}{\partial s(f)} = 0$$

Substituting the values of F and ϕ_1,

$$\frac{\partial}{\partial s} \log\left(1 \div \frac{s}{n}\right) + \lambda_1 \frac{\partial s}{\partial s} = 0$$

or

$$[s(f) + n(f)] = -\frac{1}{\lambda_1}, \tag{5.122}$$

where λ_1 is a constant, independent of f. The value of λ_1 is obtained from the condition: $\int_{f_1}^{f_2} s(f)df = S$. From eqn. (5.122), it is observed that for maximum R_{max}, the signal power spectrum $s(f)$ should be adjusted so that $[s(f) + n(f)]$ is a constant. Graphically this means that $s(f)$ should be complementary to $n(f)$, as shown in Fig. 5.26.

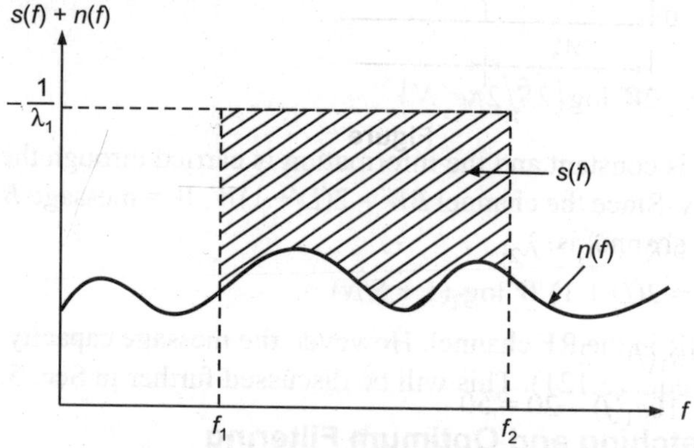

Figure 5.26: Complementary nature of $s(f)$

Further it may be shown that R is minimum for AWGN. From eqns. (5.112) and (5.122),

$$R = (f_2 - f_1)\log\left(-\frac{1}{\lambda_1}\right) - \int_{f_1}^{f_2} \log n(f)\, df$$

The first term is positive and constant and is independent of $n(f)$. R is minimum if the second term is maximum and this happens with the condition:

$$\frac{\partial}{\partial n}\{\log[n(f)]\} + \lambda_2\frac{\partial n}{\partial n} = 0$$

or

$$n(f) = -\frac{1}{\lambda_2} = \text{constant} \tag{5.123}$$

Thus for AWGN, R is minimum and $s(f)$ is also constant, since $n(f)$ is constant.

Example 5.33: Consider that $n_1(f) = 1$, $n_2(f) = 3$, $W_1 = 5$ and $W_2 = 10$, with a total $BW = 15$ and $S = 50$, as shown in Fig. 5.27.

Then using eqn. (5.122),

for W_1, $s_1(f) + 1 = 1/\lambda_1$

and

for W_2, $s_2(f) + 3 = -1/\lambda_2$

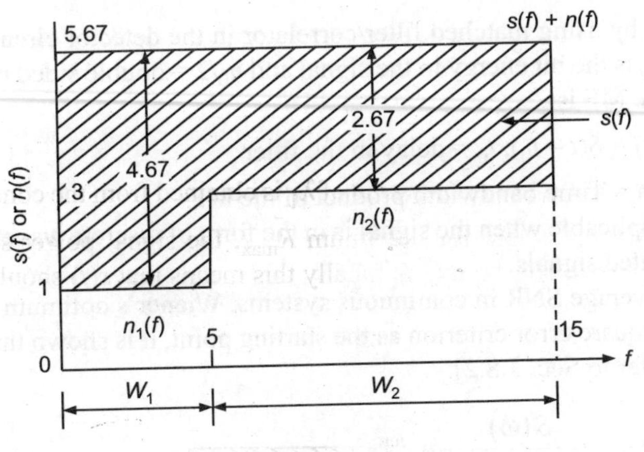

Figure 5.27

Also,

$$5s_1(f) + 10s_2(f) = 50; \quad \lambda_1 = \lambda_2.$$

Solving these eqns.

$$s_2(f) = s_1(f) - 2$$

$$5s_1(f) = 10s_1(f) - 20 = 50$$

Therefore,

$$s_1(f) = 70/15 = 4.67$$

and

$$s_2(f) = 2.67,$$

and

$$s_1(f) + n_1(f) = 5.67 = s_2(f) + n_2(f)$$

Signal power S_1 in the band $W_1 = 5 \times 4.67 = 23.30$ and S_2 in band $W_2 = 26.7$.

The channel capacity

$$C = C_1 + C_2 = 5 \log_2\left(1 + \frac{23.3}{5}\right) + 10 \log_2\left(1 + \frac{26.7}{30}\right) \simeq 21.6 \text{ bits/sec.}$$

If all signal power is transmitted in band W_1 and no power in W_2, then

$$C_1 = 5 \log_2(1 + 50/5) = 17.3 \text{ bits},$$

and if all power is in W_2 and none in W_1, then

$$C_2 = 10 \log_2(1 + 50/30) = 14.2 \text{ bits}.$$

It is then that if the interference or noise spectrum confines itself to a part of the channel bandwidth, then the obvious rule is: 'to put the signal where the noise is not present', and in general more signal power is to be concentrated where the noise power is less.

Optimum Filters

A complementary problem is to maximize the SNR in the received signal, so that R in the received message is also maximized. In Chapter III, it has been shown that the peak SNR, with AWGN in the channel, is maximized by using matched filter/correlator in the detector circuit. Using an MF, the peak $S/N = 2E_b/n_0$, where E_b is the bit energy in the signal and $n_0/2$ = double sided noise spectral density. The impulse response of the MF is:

$$h(t) = S(t - t_0),\ t_0 = \text{delay in the filter}$$

and the processing gain = Time bandwidth product of the MF.
The MF, however, is applicable when the signal is in the form of discrete waveforms, but cannot be used for continuous transmitted signals.

For improving the average SNR in continuous systems, Wiener's optimum filter [1, Chapter VIII] is used. Using the mean-square error criterion as the starting point, it is shown that the transfer function of te optimum filter is [refer to Sec. 3.8.2]:

$$H(\omega) = \frac{S(\omega)}{S(\omega)+N(\omega)} e^{-j\omega\tau_0}. \tag{5.124}$$

and the mean square error is:

$$\overline{\epsilon^2} = \frac{1}{2\pi j}\int_{-\infty}^{\infty} \frac{S(\omega)N(\omega)}{S(\omega)+N(\omega)} d\omega$$

where $S(\omega)$, $N(\omega)$ are signal and noise spectra, and τ_0 is the delay in the filter. However, the filter in general is not physically realizable and a modified realizable filter is designed using Bode-Shannon technique. Since $S(\omega)$ and $N(\omega)$ are given for the channel, the filtering changes the spectra to the values:

$$S'(\omega) = \frac{S^2(\omega)}{S(\omega)+N(\omega)} \text{ and } N'(\omega) = \frac{S(\omega)N(\omega)}{S(\omega)+N(\omega)}$$

giving the sum of the spectra as: $[\,S'(\omega)+N'(\omega)\,] = S(\omega)$, which is somewhat different than the condition of eqn. (5.122).

Example 5.34: Given that

$$S(\omega) = \frac{1}{1+\omega^2},\quad N(\omega) = \frac{1}{4+\omega^2},$$

the optimum filter (non-physical) has $H(\omega)$ as:

$$H(\omega) = K\frac{4+\omega^2}{5/2+\omega^2}, \quad \text{and} \quad \epsilon^2 = 0.157 \text{ with } K = 1/2.$$

The signal power

$$S = \frac{1}{\pi}\int_0^{\infty} \frac{d\omega}{1+\omega^2} = 1/2,$$

and the noise power $N = \dfrac{1}{\pi}\displaystyle\int_0^{\infty} \frac{d\omega}{4+\omega^2} = 1/4$ without filter,

and with filter $N' = \dfrac{1}{\pi}\displaystyle\int_0^{\infty} \frac{1/2\, d\omega}{5/2+\omega^2} = 0.157$

Therefore, S/N without filter = 2, and S/N' with filter ≃ 0.5/0.157 = 3.2, showing 50% improvement. Using realizable filter, the SNR improvement is approximately the same.

5.9 CHANNEL CAPACITY THEOREM: SOME PROPERTIES [15, 16]

The channel capacity with Gaussian signals and AWGN in the channel, has been shown in Sec. 5.8.4 to be:

$$C = R_{\max} = W \log_2(1 + S/N) \text{ bits/sec in a bandwith of } W \quad (5.110)$$

This gives the upper limit for channel capacity. By using a geometric interpretation of eqn. (5.110), it will now be shown that 'by proper encoding, it is possible to transmit information at this rate with an error $\in \to 0$'. In time T seconds, the number of messages $M(T)$ that can be generated (as discussed in Sec. 4.5.3) is:

$$\begin{aligned} M(T) &= 2^{TR_{\max}} \\ &= \exp_2[WT \log_2(1 + S/N)] \\ &= \left(\sqrt{(S+N)/N}\right)^{2TW} \end{aligned} \quad (5.125)$$

This shows that $M(T)$ has $2TW$ dimensions and it is known from the Sampling theorem that any signal of duration T and bandwidth W is completely represented by $2TW$ sample values.

It is now possible to consider a signal space (a hypersphere) of $2TW$ dimensions, in which every signal is represented by a different point, whose $2TW$ coordinates are the sample values of the signal. Considering that the coordinate axes of this space are straight lines at right angles to each other, the distance r from the origin to the point representing the signal is given by:

$$r^2 = 2W \cdot \text{Energy of the signal} = 2TWS$$

and

$$r = \sqrt{2TWS} \quad (5.126)$$

Then each transmitted signal may be represented as a point on the surface or very near the surface of the hypersphere of radius $\sqrt{2TWS}$. Similarly the received signal vectors with power $(S + N)$ will lie on the surface of a hypersphere with radius $\sqrt{2TW(S+N)}$*. The volume of this hypersphere is given by:

$$V(S + N) = k[2TW(S + N)]^{TW}, \quad k = \text{constant} \quad (5.127)$$

*In similarity with the relation that the volume of a sphere = $4/3\pi r^3$ in three dimensions, the hypervolume of an n-dimensional sphere $V(r) = [\pi^{n/2}\, r^n/\Gamma(n/2 + 1)]$ where r = radius and $\Gamma(\ )$ = Gamma function. Having two hyperspheres with radii r_1, r_2, $r_1 > r_2$, the hypervolume between two such concentric hyperspheres (i.e., the annular space) is give by:

$$V(r_1 - r_2) = [1 - (r_2/r_1)^n] \cdot V(r_1).$$

As $n \to \infty$, $(r_2/r_1)^n \to 0$, and hence $V(r_1 - r_2) \simeq V(r_1)$ = volume of the larger hypersphere. This means that most of the volume of the hypersphere is concentrated close to its n-dimensional surface. Therefore, the tips of the signal vectors lie on or very near the surface of the hypersphere with $r = \sqrt{2TWS}$. Similarly, for the noise vectors and the (signal + noise) vectors. As an example, if $n = 500$, then only 0.6% of the hypervolume with radius = r_1 is within a smaller hypersphere of $r_2 = 0.99r_1$. Hence, 99.4% of the volume of the larger hypersphere is contained in the annular space between the two hyperspheres, even though their radii by 1% only.

Similarly the volume of noise-hyperspheres = $k(2TWN)^{TW}$. Now the noise voltage produces an uncertainty at the tip of the signal vector (lying on the surface of the hypersphere of $r = \sqrt{2TWS}$), and this produces small noise-hyperspheres around this tip. The larger hypersphere of $r = \sqrt{2TW(S+N)}$, then, contains a number of noise-hyperspheres, and these smaller noise-hyperspheres should not overlap so that there is no ambiguity between signal samples. The maximum number $M(T)$ of noise-hyperspheres that may be contained in the larger hypersphere without overlap must be less than the ratio of the two volumes, thus,

$$M(T) \leq \frac{\text{Vol. with } r = \sqrt{2TW(S+N)}}{\text{Vol. with } r = \sqrt{2TWN}} = \frac{k[2TW(S+N)]^{TW}}{k[(2TWN)]^{TW}}$$

$$= (1 + S/N)^{TW}$$

$M(T)$ is now the number of distinguishable signal vectors contained in the larger hypersphere, and then,

$$R_{\max} = \frac{\log_2 M(T)}{T} \leq W \log_2(1 + S/N) \tag{5.128}$$

The eqn. (5.128) shows the upper bound of $R_{\max}$, but unlike the proof for eqn. (5.110), the arguments leading to eqn. (5.128) do not have the constraint of Gaussian signals and AGWN.

Consider now a cross-section of different hyperspheres, shown in Fig. 5.28, where the signal vector OB is received as OA due to noise vector AB. Assume that the signal vectors $M(T)$ are chosen at random (refer to the Random coding argument in Sec. 5.6) and the noise-hyperspheres may overlap occasionally, producing errors in the received signal. The noise-hypersphere overlaps the signal hypersphere forming a lens-shaped region having $2TW$ dimensions, and this is the region of uncertainty regarding the actual transmitted vectors. This overlap region contains the tips of all possible transmitted signal vectors (e.g., OC and OC') which might have produced the $(S + N)$ vector OA, which is actually received.

Since the transmitted signals are chosen at random i.e., randomly distributed over the surface of the hypersphere, the probability of the tip of a signal vector to be within the lens-shaped region

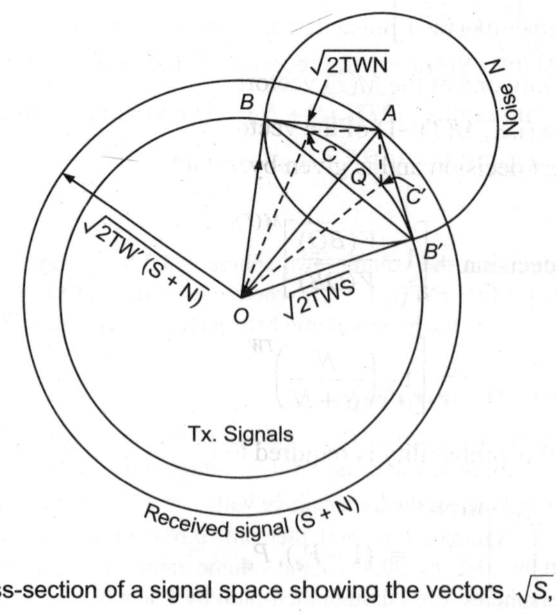

Figure 5.28: Cross-section of a signal space showing the vectors $\sqrt{S}$, $\sqrt{N}$ and $\sqrt{S+N}$

(= the probability of any transmited signal other than OB giving rise to OA) is given by the ratio of the hypervolume $V(\overline{BQ})$ of the lens-shaped region to the hypervolume $V(\overline{OB})$ of the signal hypersphere. It is seen that the hypervolume of the lens-shaped region is less than the hypervolume of a hypersphere having the same maximum radius of $\overline{BQ}$.

Since,

$$2TW(S+N) = 2TWS + 2TWN,$$

therefore,

$$\overline{OA^2} = \overline{OB^2} + \overline{BA^2}$$

and

$$\angle OBA = \pi/2$$

Then the area of $\Delta OBA = (1/2)\overline{AB}\cdot\overline{OB} = (1/2)\overline{OA}\cdot\overline{BQ}$

Therefore, $\overline{BQ} = \dfrac{(\overline{AB}\cdot\overline{OB})}{\overline{OA}} = \sqrt{2TW\dfrac{SN}{S+N}}$

and the ratio of the volumes:

$$\frac{V(BQ)}{V(OB)} \leq \left[\frac{2TW\dfrac{SN}{S+N}}{2TWS}\right]^{TW} = \left[\left(\frac{N}{S+N}\right)\right]^{TW} \tag{5.129}$$

Then, the probability that the tip of any one of the $M(T)$ signal vectors is outside (i.e., not inside) the lens hypersphere is given by:

$$\left[1 - \frac{V(BQ)}{V(OB)}\right]$$

Now the probability that only one of the $M(T)$ vectors shall have its tip inside $V(\overline{BQ})$ is the same as the probability that all but one (i.e., $M(T)-1$) of the vector tips shall lie outside $V(\overline{BQ})$; and this probability is the probability of correct decision and is given by:

$$\text{Probability of correct decision} = \left[1 - \frac{V(\overline{BQ})}{V(\overline{OB})}\right]^{M(T)-1}$$

$$= \left[1 - \left(\frac{N}{S+N}\right)^{TW}\right]^{M(T)-1}$$

For error-free reception, this probability is required to approach unity, i.e.,

$$\left[1 - \left(\frac{N}{S+N}\right)^{TW}\right]^{M(T)-1} = (1 - P_e),\ P_e \to 0$$

where P_e is the error in the decisions.

Since $\left(\frac{N}{S+N}\right)^{TW} \ll 1$, using Binomial expansion,

$$(1-x)^n > (1-nx), \quad \text{for } x \ll 1, n > 0,$$

the following inequality is obtained:

$$(1-P_e) = \left[1-\left(\frac{N}{S+N}\right)^{TW}\right]^{M(T)-1} > \left\{1-[M(T)-1]\left[\frac{N}{S+N}\right]^{TW}\right\}$$

Assume that P_e is bounded as:

$$P_e < M(T)\left(\frac{N}{S+N}\right)^{TW} \leq \in$$

then,

$$M(T) \leq \in \left(\frac{S+N}{N}\right)^{TW}$$

Therefore, the rate of transmission

$$R = \frac{\log_2 M(T)}{T} \leq W \log_2 (1+S/N) + \left(\frac{\log_2 \in}{T}\right)$$

For T large,

$$\frac{\log \in}{T} \to 0,$$

and

$$R_{\max} = C \leq W \log_2 (1 + S/N) \text{ bits/sec.} \tag{5.130}$$

The eqn. (5.130) is independent of the error bound $P_e \leq \in$, and gives the upper bound for C with random coding and large T, leading to a very large set of alphabets $M(T)$.

5.9.1 Ideal Communication System

Based on the assumption of random coding and a large set of alphabets $M(T)$ for large T, Shannon has suggested an idealised communication system where M different samples of white noise waveform of duration T and average power S are used as the M alphabets for transmission as shown in Fig. 5.29. Each of these waveforms are also known at the receiver, and the receiver subtracts these M waveforms from the received (signal + noise) waveforms either in parallel or in a serial form. Then squaring and integrating the remainder, the net power at the receiver output would be only N if the correct waveform has been subtracted, otherwise the output power is $(2S + N)$. Thus, the decision circuit will be able to decide which alphabet was transmitted by comparing all the outputs. If all M messages are equiprobable, then

$$H(X) = \frac{\log M}{T} \text{ bits/sec.}$$

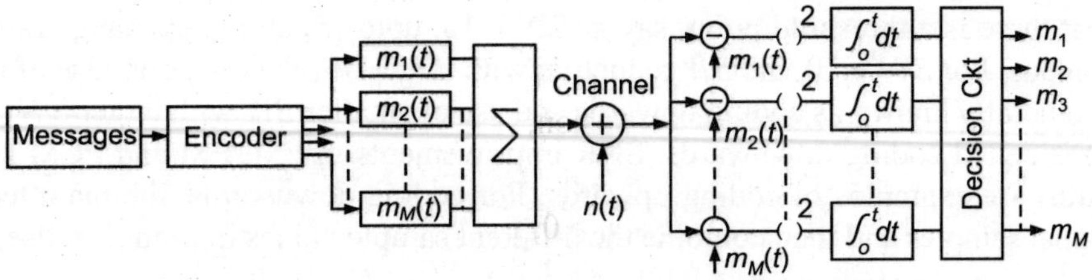

Figure 5.29: An idealised communication system

To maintain $H(X)$ constant, (i.e., the rate of transmission constant), $M(T)$ has to grow exponentially with T. It has been shown that for $P_b = 10^{-5}$, with $S/N = 10$, the number of symbols required to achieve a rate $R = 0.96\ C$, is equal to $(2^{10})^4 = 10^{12}$, indeed a very large number! [17].

The system, however, is similar to the M-ary orthogonal system discussed in Chapter 3. Since different samples of a noise waveform have been taken as the alphabets, the alphabets are orthogonal, and correlation/MF detection with maximum-likelihood decision circuits are optimum for minimum P_b. Then a sim ple modification of Fig. 5.29 results into the M-ary orthogonal system. However, because of the instrumentation problem, it will be difficult to store and operate with multilevel noise waveforms for correlation and detection. The digital orthogonal signals generated through Hadamard matrices or Walsh functions are easier to implement and hence, practically useful. As discussed in Sec. 5.7.3, these digital orthogonal systems also give ideal transmission rate with large alphabet size, given that $E_b/n_0 > \ln 2$.

5.9.2 Important Properties

The following important properties of the Channel capacity theorem are useful in understanding the various implications of the results.

I. The channel capacity eqn. (5.130) shows that the same capacity may be obtained by increasing W if S/N is small. The eqn. (5.130) may now be written in the form

$$\frac{W}{C} \cdot \log_2 (1 + S/N) = 1$$

and a plot of W/C vs. S/N is shown in Fig. 5.30.

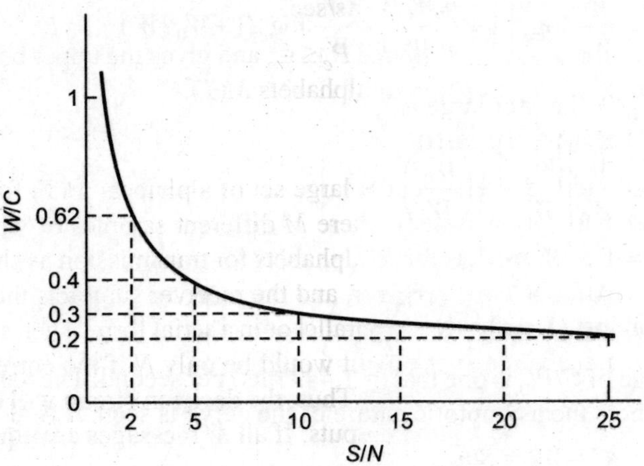

Figure 5.30

It is seen that there is a threshold point, say at $S/N \simeq 10$, upto which the exchange rate of BW with S/N is advantageous. For $S/N > 10$, the BW reduction with larger S/N is very poor. Use of larger BW for smaller S/N is generally known as coding upwards and use of smaller BW with larger S/N (equivalent to BW-reduction) is called coding downwards. SNR improvements in FM, PM and PCM systems using larger bandwidths are examples of coding upwards. For coding downwards, the main technique is to quantize the signal samples and then combine the different sample values in a single pulse, as discussed in Sec. 5.7.1.

As an example, if the original signal samples are quantized into 8 levels, then by using 64-level pulses, two signal samples may be combined into a single pulse and the corresponding bandwidth is halved, of course at the expense of higher S/N required in the channel. The process may be pushed further if sufficient power is available.

Example 5.35: Suppose S/N in the channel = 10, then $C/W = \log_2(1 + S/N) = \log_2 11$. If BW is halved and the same C is maintained, then

$$C/W = 2\log_2 11 = \log_2 (11)^2 = \log_2 (1 + S'/N')$$

Therefore, the new value of $S'/N' = 120 = 12$ times S/N.

Thus the coding downwards is not very advantageous.

II. In practical channels, the noise spectral density n_0 is generally constant and $N = n_0 W$. Then the capacity per Hz, C/W, is an indication of the efficiency of the system, but the rate of change of C/W with W is found to be negative, as is shown by:

$$\eta = C/W = \log_e\left(1 + \frac{S}{n_0 W}\right)$$

and

$$\frac{\partial \eta}{\partial W} = -\frac{1}{W}\frac{S/n_0 W}{(1 + S/n_0 W)} \tag{5.131}$$

This indicates that η decreases with the increase of W, as is seen in FM and PM systems.

Further, assume that $S = n_0 W_0$, i.e., W_0 is the bandwidth for which $N = S$, then

$$\frac{C}{W_0} = \frac{W}{W_0}\log_e\left(1 + \frac{n_0 W_0}{n_0 W}\right) = \frac{W}{W_0}\log_e(1 + W_0/W)$$

For large W and small W_0/W (i.e., for large n_0),

$$C/W_0 \simeq \frac{W}{W_0}\left[\frac{W_0}{W} - \frac{1}{2}\left(\frac{W_0}{W}\right)^2 + \ldots\right] \tag{5.132}$$

$$\simeq 1 \ldots W \gg W_0.$$

$$\left[\text{using the approximation } \log_e(1 + x) = x - \frac{x^2}{2} + \frac{x^3}{3} \ldots x < 1\right].$$

Thus, the limiting value of C/W_0 is one that (= 1.443 bits) per second. The variation of C/W_0 vs. W/W_0 is shown in Fig. 5.31, where the asymptotic nature of the curve is seen. It is also seen that:

$$\underset{\lim W \to \infty}{C = W_0} = S/n_0,$$

which may be stated as:

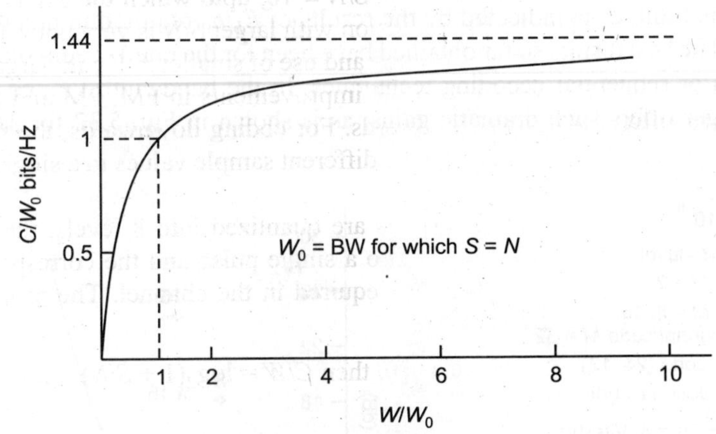

Figure 5.31

'The maximum signalling rate for a given S is 1.443 bits/sec/Hz in the bandwidth over which the signal power can be spread without its falling below the noise level.'

Since $\log_2 3 = 1.58$, the graph also indicates that a ternary system approaches the ideal C/W_0 and a ternary communication system should be more efficient than the binary system. However, due to the difficulties in fabricating ternary logic devices, binary systems have been more popular.

It may appear somewhat anomalous that communication is possible, even when $S/N = W_0/W \ll 1$, as in the case of radar signals. This is possible by using correlation detection and integration of signals over a large T, thereby improving the SNR at the receiver output considerably.

III. The channel capacity equation may also be written as:

$$C/W = \log_2 [1 + (S/n_0C)(C/W)]$$

$$= \log_2 [1 + (E_b/n_0)(C/W)], \tag{5.133}$$

where $S/(n_0C) = E_b/n_0$, since C is the transmission rate in bits/sec, and E_b = energy per bit. This equation is plotted in Fig. 5.32, where it is seen that for a fixed S/n_0, more efficient communication is possible as $W \to \infty$. In the limit $W \to \infty$,

$$C/W = (\log_2 e) \log_e [1 + (E_b/n_0)(C/W)]$$

$$\to (\log_2 e)\left(\frac{E_b}{n_0}\frac{C}{W}\right), \qquad \text{since } \left(\frac{E_b}{n_0}\frac{C}{W}\right) \ll 1$$

Therefore, efficient communication is possible for

$$E_b/n_0 \underset{\text{limit } W\to\infty}{} \to \ln 2 = 0.69 = -1.6 \text{ dB} \tag{5.134}$$

The value of $E_b/n_0 = -1.6$ dB is the so-called Shannon's limit for transmission at capacity and the communication fails if $E_b/n_0 < 0.69$ (also discussed in Sec. 5.7.3). Although the E_b/n_0 limit is reached only at $W \to \infty$, even at 1/2 bit/Hz (or for a code rate of 1/4 with binary PSK), we are practically there. Since the ideal PSK signalling with no coding requires $E_b/n_0 = 9.6$ dB for $P_b = 10^{-5}$, it is seen that the capacity theorem promises a potential coding gain of 11 dB, as it should be possible to obtain $P_b \to 0$ with $E_b/n_0 = -1.6$ dB only with ideal coding and $W \to \infty$. However, no practical system has been so far designed to reach this limit, as is indicated by the results of E_b/n_0 required in practical coded systems, plotted in Fig. 5.32. The best results so far obtained have been for the rate 1/2 convolutional coding using Viterbi soft decision or sequential decoding techniques. In the bandwidth limited case, i.e., for large C/W, coding no longer offers such dramatic gains, as is shown in Fig. 5.32 for M-level AM or PSK signalling.

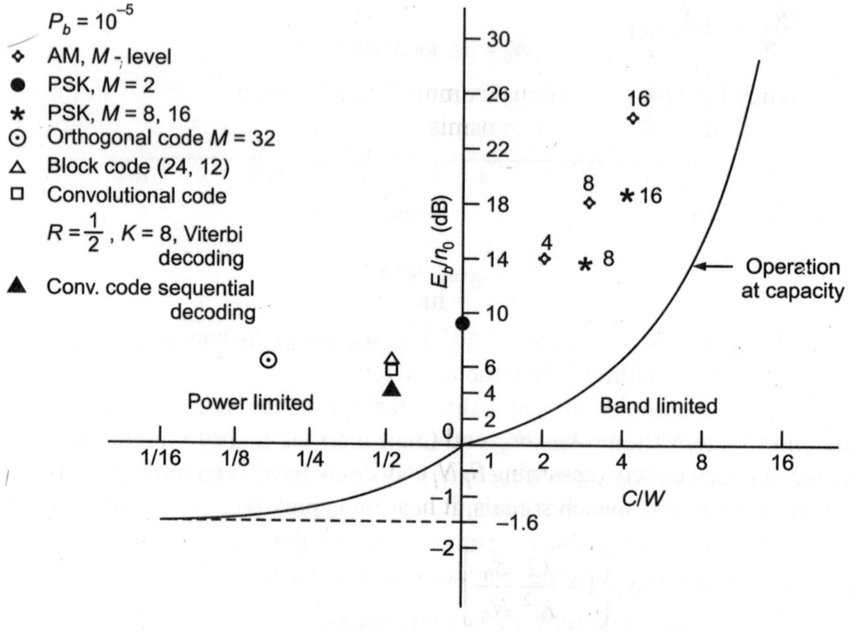

Figure 5.32: C/W vs E_b/n_0 [22]

5.10 EFFICIENCIES OF DIFFERENT COMMUNICATION SYSTEMS

It has been shown in Sec. 5.5 and Sec. 5.7.3, that in practical communication systems, the rate of transmission of information, R, in a channel is always less than the channel capacity C. The source/message transmission rate is also less than capacity, as discussed in Sec. 5.2 and Sec. 5.4.2. Based on these observations, it is convenient to define Message Efficiency $\eta_{(M)}$ and Channel (Transmission) Effficiency $\eta_{(ch)}$ [already defined in eqn. (5.91)], as:

$$\eta_{(M)} = \frac{R_{(M)}}{C_M}$$

and

$$\eta_{(ch)} = \frac{R_{(ch)}}{C_{ch}} \tag{5.135}$$

In Ex. 31, the baseband transmission of speech signals has been discussed and it has been shown that the actual $R_{(M)}$ for speech is approximately 24 kb/s, whereas for a channel with $W = 4$ kHz and $S/N \geq 32$ dB, $C_M = 4.10^4$ bit/s. Thus, the message efficiency of the channel is

$$\eta_{(M)} = \frac{24}{40} = 60\% \text{ only.}$$

Similarly, for Television signals, $W = 5$ MHz and $S/N \geq 40$ dB, giving $C_M \simeq 66.5$ Mb/s, whereas it has been shown in Chapter 2, that excellent TV pictures can be obtained at a rate of 24–32 Mb/s.

$$\text{Thus,} \quad \eta_{(M)} \leq \frac{32}{66.5} = 48\% \text{ only}$$

The channel efficiency for M-ary orthogonal communication systems has been discussed in Sec. 5.7.3 and $\eta_{(ch)}$ vs. E_b/n_0 is given in Table 5.7. The transmission rates and $\eta_{(ch)}$ for FM, PPM and other systems are discussed below using the above relations.

5.10.1 FM and PPM Signals

In an FM system, the system parameters are:

$$W_{(ch)} = 2(D + 1)W_M, \qquad W_M = \text{message bandwidth}$$

and

$$(S_0/N_0)_M = 3D^3(S_i/N_i)_{ch},$$

where D = deviation ratio, $N_i = 2(D + 1)n_0W_M$ and $(S_0/N_0)_M$, $(S_i/N_i)_{ch}$ refer to the average SNR's in the message and the channel respectively (assuming S_i/N_i to be above threshold). Considering that the message signal has similar statistics as the speech signals, it has been shown in Ex. 31, that $R_{(M)}$ is given by:

$$R_{(M)} \simeq W_M \log_2\left[1 + \frac{12}{K^2}\frac{S_0}{N_0}\right]$$

$$\simeq 30 \text{ kb/s for speech signals.}$$

$R_{(M)}$ may also be expressed as:

$$R_{(M)} \leq W_M \log_2\left[1 + \frac{3D^3}{8.5}(S_i/N_i)_{ch}\right]$$

$$\simeq W_M \log_2\left[\frac{3D^2}{8.5}\frac{S_i}{2n_0W_M}\right] \tag{5.136}$$

The ideal capacity of the RF channel is:

$$C_{(ch)} = 2(D + 1)W_M \log_2 (1 + S_i/N_i)$$

and when the signal in the channel is considered to be peak limited, then,

$$C_{(ch)} \simeq W_{(ch)} \log_2\left[1 + \frac{4S_i}{2\pi e^2 N_i}\right]$$

Thus, the channel efficiency may be written as:

$$\eta_{(ch)} = \frac{R_{(ch)}}{C_{(ch)}} = \frac{R_{(M)}}{C_{(ch)}}$$

$$\simeq \frac{N}{2(D+1)} \cdot \frac{\log_2\left[\frac{3D^2}{8.5} S_i / 2n_0 W_M\right]}{\log_2\left[\frac{S_i}{2(D+1)n_0 W_M}\right]}, D > 1 \tag{5.137}$$

A plot of $\eta_{(ch)}$ vs. $(S_i/n_0 W_M)$ is shown in Fig. 5.33, for various values of D. It is seen that η decreases with the increase of D i.e., with the increase of $W_{(ch)}$, as was shown in eqn. (5.131). It is also seen from eqn. (5.136), that $R_{(M)}$ increases with W only as 2 log D, whereas $C_{(ch)}$ increases directly as $W_{(ch)}$. In addition, the information rate suffers a penalty of 9 dB due to the factor $(12/K^2)$, but this is partly offset by 4.75 dB (= 10 log 3) due to the gain through the triangular noise spectrum in FM detector output. With $S_i/N_i \gg 1$,

$$\eta_{(ch)} \leftarrow \frac{1}{2D}.$$

In PPM systems, the improvement in SNR is given by [20]:

$$(S_0/N_0)_M = \frac{2\Delta d_m^2 W_{(ch)}^3}{W_M} (S_i/N_i)_{ch} \tag{5.138}$$

where, Δd_m = maximum time shift of the transmitted pulses due to modulation. The channel efficiency is now given as:

$$\eta_{(ch)} = \frac{R_{(M)}}{C_{(ch)}} = \frac{W_M \log_2\left[1 + \frac{1}{8.5}(S_0/N_0)_M\right]}{W_{(ch)} \log_2[1 + (S_i/N_i)]} \tag{5.139}$$

where $(S_0/N_0)_M$ is given by eqn. (5.138). It will be again seen that with increasing $W_{(ch)}$, $\eta_{(ch)}$ will decrease as in Fig. 5.33.

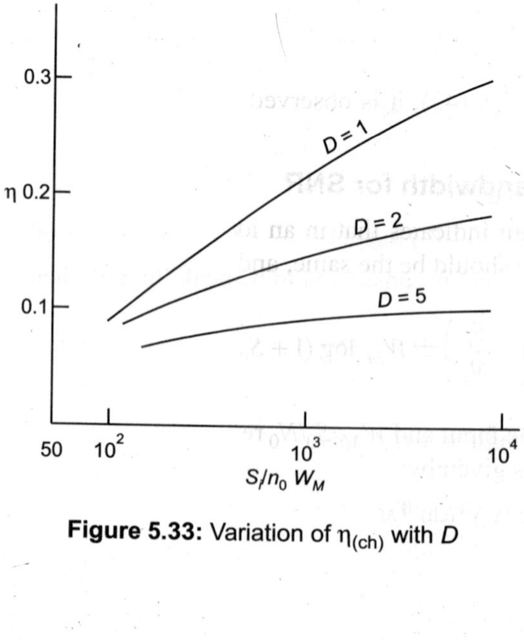

Figure 5.33: Variation of $\eta_{(ch)}$ with D

Multilevel PCM

Channel capacity and transmission rate of multilevel PAM/PCM have been discussed in Secs. 5.7.2 and 5.7.3. If number of levels L is large, then the signal may be considered to be almost continuous with peaks of $\pm A/2$ volts. Then using eqn. (5.97),

$$H(X) = W \log_2 (4\hat{S})$$

$$= W \log (12\overline{S}),$$

where $\hat{S} = A^2/4$, and $\overline{S} = \frac{A^2}{12} = \hat{S}/3$.

For Gaussian noise,

$$H(n) = W \log (2\pi eN).$$

Then

$$\text{the rate of transmission } R \le W \log_2 \left(1 + \frac{12\overline{S}}{2\pi eN}\right)$$

$$\le W \log_2 \left[\frac{6}{\pi e}(\overline{S}/N)\right], \overline{S}/N \gg 1 \tag{5.140}$$

Sanders [19] has given a similar equation, for R in continuous PAM channels, and the eqn. is:

$$R \simeq W \log_2 (S/N) + 2W(0.75)\sqrt{N/S} - 0.51\,W \tag{5.141}$$

Equation (5.141) gives approximately the same result as the eqn. (5.140). The channel efficiency is now given as (assuming that $W_{(ch)}$ is the same as W_M),

$$\eta_{(ch)} = \frac{\log_2 \left[\frac{6}{\pi e}(\overline{S}/N)\right]}{\log_2 \left[\overline{S}/N\right]} \tag{5.142}$$

Comparing eqns. (5.137) and (5.142), it is observed that the channel efficiency of multilevel PCM is much better than that of FM.

5.10.2 Exchange of Bandwidth for SNR

The channel capacity theorem indicates that in an ideal receiver (assuming no information loss), the input-output information rate should be the same, and thus,

$$W_{(ch)} \log \left(1 + \frac{S_i}{N_i}\right) \rightleftarrows W_M \log (1 + S_0/N_0) \tag{5.143}$$

where, $W_{(ch)}$, S_i/N_i refer to the input and W_M, S_0/N_0 refer to the output of the receiver. For $S_i/N_i \gg 1$ and $S_0/N_0 \gg 1$, the output SNR is given by:

$$S_0/N_0 \simeq (S_i/N_i)^{W_{(ch)}/W_M} \tag{5.144}$$

This indicates the maximum possible exchange between input and output SNR's for an ideal receiver. In practical modulation schemes, this ideal is never achieved and is approached only by coded communication systems [more realistic eqn. (5.85) for R may also be considered]. From known results, the ideal and actual SNR values for some systems are given in Table 5.8. For PCM, it is normally assumed that the effective noise at the output due to quantization and that due to the channel errors are equal, i.e., $\sigma_q = \sigma_N$, thus, the output SNR is given as:

$$\frac{S_0}{N_0} = 2^{2m} = 6m \text{ dB},$$

and the corresponding BW-expansion for PCM-PSK is $2m$ (m = number of bits in the PCM-word). It is seen from the table that PCM gives the best result among the usual modulation schemes. However, the coded systems given better results as discussed in Sec. 5.7.3.

Table 5.8

System	BW expansion	Ideal S_0/N_0	Actual S_0/N_0
SSB	1	S_i/N_i	S_i/N_i
AM-DSB-SC	2	$(S_i/N_i)^2$	$2S_i/N_i$
WBFM	$2(D+1)$	$(S_i/N_i)^{2(D+1)}$	$3D^3(S_i/N_i)$
PPM	$W_{(ch)}/W_M$	$(S_i/N_i)^{W_{(ch)}/W_M}$	$KW^3_{(ch)}(S_i/N_i)$
PCM-PSK	m	$(S_i/N_i)^m$	2^{2m}

Example 5.36: Assume that 64-level PAM pulses are converted into 6 bit PCM words. The $S_i/N_i \simeq 10$ for binary transmission (for $P_e \simeq 10^{-4}$). For the bandwidth expansion of 6, the ideal $S_0/N_0 = (10)^6$, whereas the actual $S_0/N_0 = 2^{12} = 36$ dB. If the PAM pulses were transmitted as a multilevel signal, then $S_i/N_i \geq 4 \times 10^4$ ($\equiv 46$ dB); thus m-bit PCM has the advantage of requiring lower S_i/N_i, as discussed in Sec. 5.7.1; although the ideal improvement is not obtained. The channel η for the binary channel is:

$$\eta_{ch}(\text{binary}) = \frac{R(M)}{C_{(ch)}} = \frac{12}{6 \log_2 (1+10)} \simeq 0.6.$$

and for PAM,

$$\eta_{ch}(\text{PAM}) = \frac{12}{\log_2 (4.10^4)} \simeq 0.8$$

5.10.3 Communication Efficiency [19, 20]

We have discussed above the message efficiency and channel efficiency of some of the commonly used communication systems. The overall system efficiency is related to the values of S_i/N_i and the channel BW, $W_{(ch)}$, for a given bit error rate P_b or the output SNR S_0/N_0. Accordingly, Sanders [19] has defined a composite efficiency parameter β, called here as the Communication efficiency, and β is given as:

For a given P_b or S_0/N_0,

$$\beta = \frac{E_{b(\min)}}{n_0} = \frac{S_{(\min)}}{n_0 R} = \frac{S_i}{N_i} \cdot \frac{W_{(ch)}}{R}$$

$$= \frac{S_i}{N_i}(\alpha) \qquad (5.145)$$

where $E_{b(\min)}$ is the minimum signal energy required per message bit, n_0 = noise spectral density in the channel, R = actual information rate of the received digital/analog message and $\alpha = W/R$ = bandwidth expansion due to the particular modulation scheme. A lower bound on β may be obtained from the channel capacity theorem as:

$$R = W \log_2\left(1+\frac{S_{(\min)}}{n_0 W}\right)$$

or,

$$R/W = \log_2 e \cdot \log_e\left[1+\frac{\beta_{\min}}{\alpha}\right] \tag{5.146}$$

and in the limit $\alpha \to \infty$, $\frac{1}{\alpha} = \log_2 e \cdot \frac{\beta_{\min}}{\alpha}$, which leads to the condition that: $\beta_{\min} \geq \log_e 2 = 0.69$, as has been shown in eqn. (5.134).

For digital modulation schemes, e.g., in PCM-PSK, PCM-FSK and PCM-ASK, β for a given P_e is calculated from the results of Chapter 3. As an example, for Binary-coherent PSK, $\alpha = W/R = 1$ and S_i/N_i for $P_e = 10^{-5}$ is 10 dB, then $\beta = 10$. In Binary incoherent ASK, S_i/N_i for $P_e = 10^{-5}$ is 17 dB, $\alpha = 1$, then $\beta = 50$. Similarly, in narrow-band FSK with $\alpha = 1$, S_i/N_i for $P_e = 10^{-5}$ is 13 dB; and $\beta = 20$. For orthogonal M-ary communication systems, E_b/n_0 and α for various P_e's are given in Chapter 3 and the β values are directly obtained from these results. Using the results of Sanders, β-efficiency for ASK, FSK*, PSK and orthogonal M-ary systems with $P_b = 10^{-6}$ are shown in Fig. 5.34.

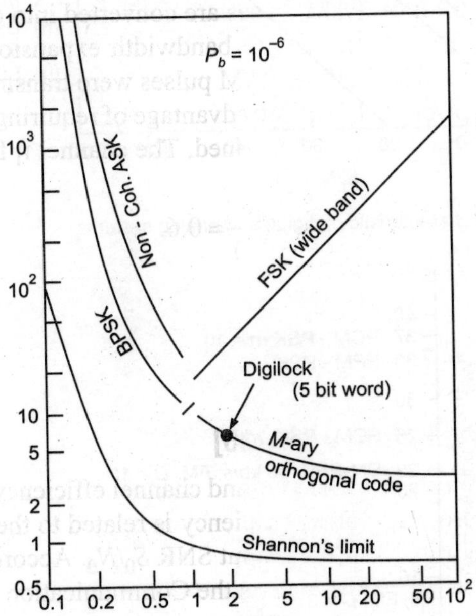

Figure 5.34: β-effciencies vs. α [19]

*For FSK with $\alpha > 1$, the threshold of S_i/N_i increases proportionately with the deviation ratio D and the receiver has to operate above threshold only. Thus, β increases proportionately with α. At threshold, β is given by: $\beta = 16\alpha$, and this value is plotted in Fig. 5.34.

Along with the results of the practical systems, Shannon's limit of β values, given by eqn. (5.146), are also shown in Fig. 5.34. It is observed that for BW-expansion $\alpha > 1$, *M*-ary orthogonal systems are the only modulation schemes which give decreasing values of β with increase of α, as predicted by Shannon's Channel capacity theorem. FM/FSK with $\alpha > 1$, on the other hand, has increasing β, thus showing lesser channel efficiency, as indicated in Fig. 5.33.

For analog modulation systems, a similar comparison can be made by using the relation,

$$\beta = \frac{S_i}{N_i}(\alpha) \quad \text{for a given } S_0/N_0 \tag{5.147}$$

and the corresponding ideal R may be calculated from

$$R = W_M \log_2\left[1 + \frac{S_0}{N_0}\right] = 40.7 \text{ kb/s}$$

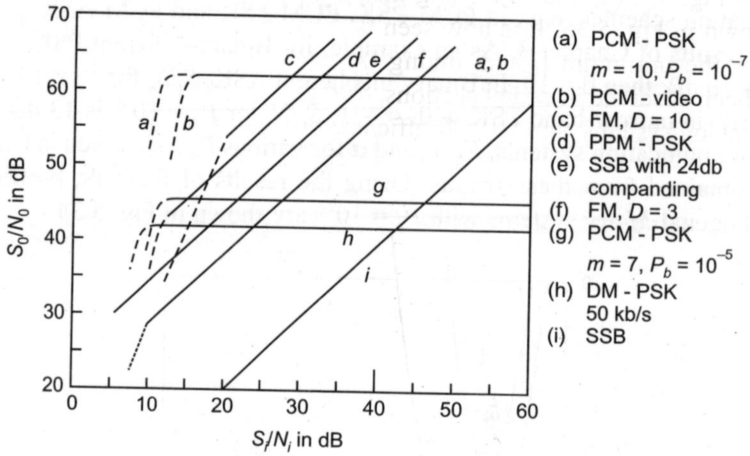

(*a*) SNR characteristic of digital and analog systems

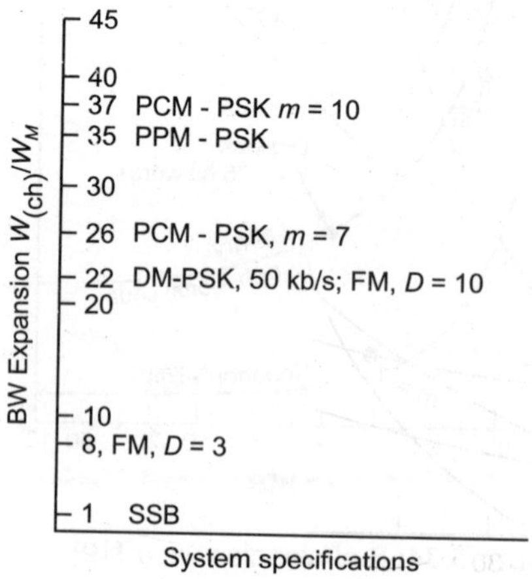

(*b*) Relative bandwidths [20]

Figure 5.35

for a complex baseband signal with $W_m = 3.5$ kHz and $S_0/N_0 = 35$ dB. The digital systems e.g., PCM and DM, may also be compared for a given S_0/N_0, in the receiver output, using the equivalence between P_b and S_0/N_0, as discussed in Chapter 2. The relations are:

$$(S_0/N_0)_{\text{PCM}} = \frac{1}{4P_b}$$

and

$$(S_0/N_0)_{\text{DM}} = (f_s/f_m)\,\frac{10^{-2}}{P_b} \tag{5.148}$$

where f_s = bit rate and f_m = message frequency.

Using now the known results of S_0/N_0 vs. S_2/N_2, and the corresponding α, for different systems, β-efficiency may be calculated. For ready reference, SNR characteristics and BW-expansion in typical systems are shown in Fig. 5.35(a) and (b). Using these results, β-efficiency of these systems [20] for different S_0/N_0 is shown in Fig. 5.36. It is now seen that among the analog systems, SSB with 24 dB companding requires the minimum β and among the digital systems, n-stage-DDM has the best β-efficiency. As has been observed in earlier sections, that the channel efficiencies of digital systems are better than those of analog systems; similarly, β-efficiencies are also seen to be better in case of digital systems.

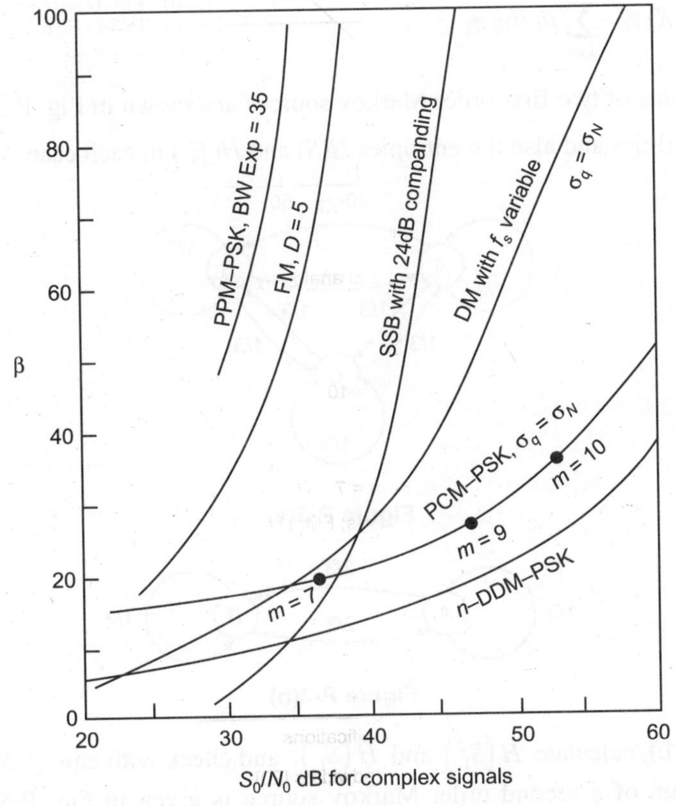

Figure 5.36: β-efficiencies for different modulation schemes [20]

PROBLEMS

1. The joint probabilities of the transmitted and received messages of a communication system is given as:

$$p(X,Y) \rightarrow \begin{array}{c|cccc} & y_1 & y_2 & y_3 & y_4 \\ \hline x_1 & 1/4 & 0 & 1/10 & 0 \\ x_2 & 0 & 1/4 & 0 & 1/20 \\ x_3 & 0 & 0 & 1/10 & 1/20 \\ x_4 & 0 & 1/20 & 0 & 1/10 \\ x_5 & 0 & 0 & 0 & 1/20 \end{array}$$

Calculate $p\{x_1, x_2, ..., x_5\}$, $p\{y_1, y_2, y_3, y_4\}$, $H(X)$, $H(Y)$, $H(Y/X)$, $H(X/Y)$, $H(X, Y)$ and the mutual information $I(X, Y)$.

2. Two sources emit messages $\{x_1, x_2, ..., x_n\}$ with probabilities $\{p_1, p_2, ..., p_n\}$ and $\{y_1, y_2, ..., y_n\}$ with probabilities $\{q_1, q_2, ..., q_n\}$.
Prove that:

$$H(X) \le -\sum_{k=1}^{n} p_k \log q_k$$

3. The state diagrams of two first-order Markov sources are shown in Fig. P-3(a) and (b). Calculate the state probabilities and also the entropies $H(S)$ and $H(\bar{S})$ in each case.

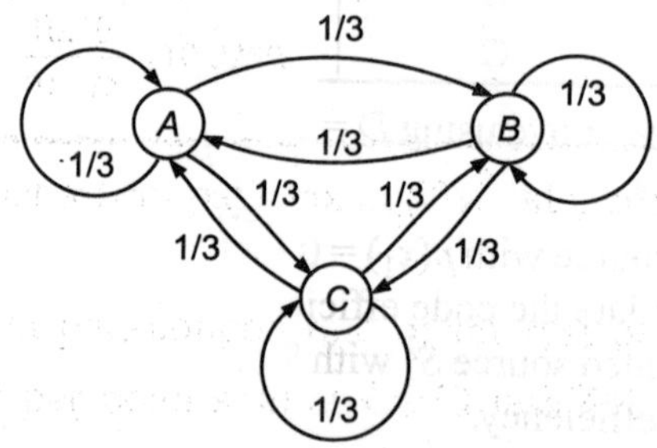

Figure P-3(a)

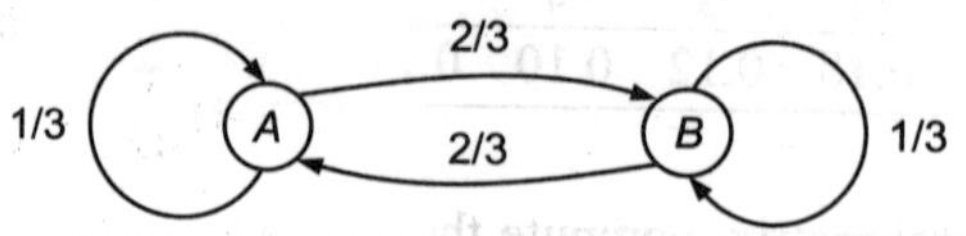

Figure P-3(b)

4. For the Fig. P-3(b), calculate $H\left(S_1^2\right)$ and $H\left(S_1^3\right)$; and check with eqn. (5.30).
5. The state diagram of a second order Markov source is given in Fig. P-5. Calculate the state probabilities, tabulate $p(s_1, s_2)$, $p(s_1, s_2, x_i)$ and then calculate $H(S)$ $H(s_1, s_2, x_i)$.

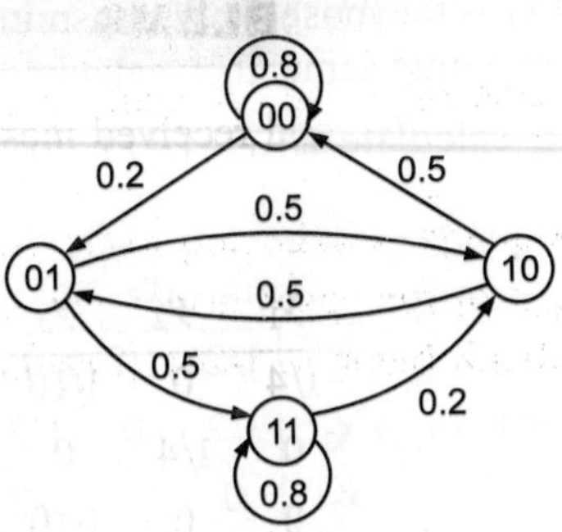

Figure P-5

6. A source emits 3 messages x_1, x_2, x_3 which are transmitted in the channel through PLM and the pulse widths are: 1, 2, 3 msecs respectively. Calculate the code capacity of the transmitter for very long messages [use eqns. (5.45), (5.46) and (5.47)].
7. Calculate the code capacity of a telegraph channel using Morse codes (Refer [1], Appx. I).
8. The Probabilities of English letters are given as [4]:

Letters	p_i	Letters	p_i	Letters	p_i
Space	0.1859	I	0.0575	R	0.0484
A	0.0642	J	0.0008	S	0.0514
	0.0127	K	0.0049	T	0.0796
C	0.0218	L	0.0321	U	0.0228
D	0.0317	M	0.0198	V	0.0083
E	0.1031	N	0.0574	W	0.0175
F	0.0208	O	0.0632	X	0.0013
G	0.0152	P	0.0152	Y	0.0164
H	0.0467	Q	0.0008	Z	0.0005

Design efficient codes for the above using $D = 2, 3, 4$ and 5. Compute the code efficiencies in each case and compare the results.

9. (a) Consider a 3-alphabet source with $p(x_1) = 0.4$, $p(x_2) = 0.3$ and $p(x_3) = 0.3$. Design the best code for the source and calculate the code efficiency.
   (b) Consider now the extended source S^2 with 9 alphabets. Design a new code for these alphabets and calculate the code efficiency.

10. For the given source probabilities, design all possible compact codes and compare their efficiencies.

$x_i \rightarrow$	x_1	x_2	x_3	x_4	x_5	x_6	x_7	x_8	x_9	x_{10}
$p(x_i) \rightarrow$	0.20	0.18	0.12	0.10	0.10	0.08	0.06	0.06	0.06	0.04

11. (a) For the given channel matrix, compute the mutual information $I(X, Y)$ with $p(x_1) = 1/2$ and $p(x_2) = 1/2$

$p(Y/X) \rightarrow$

	y_1	y_2	y_3
x_1	2/3	1/3	0
x_2	0	1/6	5/6

(b) If the receiver decides that x_1 is the most likely transmitted symbol when y_2 is received, other conditional probabilities remaining same, then calculate the new value of $I(X, Y)$.

12. For the given channel matrix, calculate $H(X)$, $H(Y)$, $I(X, Y)$ and C, given that $p(x_1) = 0.6$, $p(x_2) = 0.3$, $p(x_3) = 0.1$.

$$p(Y/X) \rightarrow \begin{array}{c|ccc} & y_1 & y_2 & y_3 \\ \hline x_1 & 1/2 & 1/2 & 0 \\ x_2 & 1/2 & 0 & 1/2 \\ x_3 & 0 & 1/2 & 1/2 \end{array}$$

13. Compute the channel capacities for the following channel matrices. In each case, compute $p(x_i)$ and check that they are realizable.

(a) $\begin{bmatrix} 0.8 & 0.1 & 0.1 \\ 0.1 & 0.8 & 0.1 \\ 0.1 & 0.1 & 0.8 \end{bmatrix}$.

(b) $\begin{bmatrix} 3/4 & 1/8 & 1/8 & 0 \\ 1/8 & 3/4 & 0 & 1/8 \\ 1/8 & 1/8 & 3/4 & 0 \\ 0 & 1/8 & 1/8 & 3/4 \end{bmatrix}$.

(c) $\begin{bmatrix} 0.5 & 0.25 & 0.25 & 0 \\ 0.25 & 0.25 & 0.25 & 0.25 \\ 0 & 0 & 1 & 0 \\ 0.5 & 0 & 0 & 0.5 \end{bmatrix}$.

14. For the following channel matrices, calculate C, and then obtain the limits of p_{11} and p_{12} such that $p(x_i)$ are realizable.

(a) $\begin{bmatrix} p_{11} & p_{12} & 0 \\ 0.5 & 0 & 0.5 \\ 0 & p_{12} & p_{11} \end{bmatrix}$.

(b) $\begin{bmatrix} p_{11} & 0 & p_{12} \\ 0 & 1.0 & 0 \\ p_{12} & 0 & p_{11} \end{bmatrix}$.

15. Calculate the capacity of the following channels and show how C varies with p.

(a) $\begin{bmatrix} 1 & 0 & 0 \\ 0 & p & (1-p) \\ 0 & (1-p) & p \end{bmatrix}$.

(b) $\begin{bmatrix} p & (1-p) & 0 & 0 \\ 0 & p & (1-p) & 0 \\ 0 & (1-p) & p & 0 \\ 0 & 0 & (1-p) & p \end{bmatrix}$.

16. (a) A source produces 8-level PAM pulses and the coder converts them into 3-bit binary pulses for transmission in the channel. If the error rate in the channel is 10^{-1}, calculate the rate of transmission in the channel, if all p_i's = 1/8.
    (b) If the p_i's are unequal and

    $$p(X) = \{1/4, 1/4, 1/8, 1/8, 1/8, 1/16, 1/32, 1/32\},$$

    then calculate $H(X)$, $H(X/Y)$ and $I(X, Y)$.
    (c) Encode the messages into an optimum binary code and then use the binary channel for transmission. Calculate again $H(X)$, $H(X/Y)$ and $I(X, Y)$. Compare the three results.

17. Three BSC's, each with error probability = 0.1, are cascaded as shown in Fig. P-17, and $p(0) = 1/4$, $p(1) = 3/4$. Calculate $H(Y)$, $H(Z)$, $H(U)$ and $I(X, Z)$, $I(X, U)$.

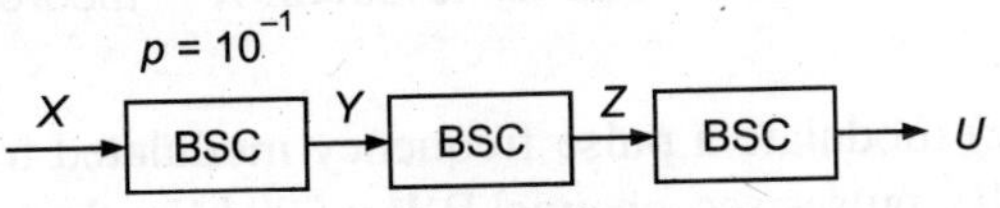

Figure P-17

18. A user has m channels available for transmission of alphabets. The probability of using ith channels is p_i. Show that the capacity of the combination of channels is given by:

    $$2^C = \sum_{i=1}^{m} 2^{C_i}$$

    and that to achieve capacity, $p_i = 2^{C_i - C}$ (see Ref. [21], p 100).

19. A BSC has the error probability $p = 0.2$, and the input to the channel consists of 4 equiprobable messages:

    $$x_1 = 000,\ x_2 = 001,\ x_3 = 011,\ x_4 = 111$$

    Calculate (a) $p(0)$ and $p(1)$ at the input
    (b) Efficiency of the code
    (c) Channel capacity

20. The input and output of a noisy continuous channel is related by the joint PDF, given by:

    $$p(x, y) = \begin{cases} \dfrac{1}{\pi X_0 Y_0} & \text{for } \left(\dfrac{x^2}{X_0^2} + \dfrac{y^2}{Y_0^2}\right) < 1, \quad X_0 > Y_0 > 0 \\ 0, & \text{otherwise.} \end{cases}$$

    Calculate $H(X)$, $H(Y)$, $H(X/Y)$ and $I(X, Y)$.

21. A continuous noisy channel has the joint PDF

$$p(x, y) = \frac{1}{2\pi\sigma^2\sqrt{3}}\exp\left[-\left(\frac{X^2+Y^2-XY}{3\sigma^2}\right)\right]$$

$$-\infty < x < \infty, -\infty < y < \infty.$$

Calculate $H(X)$, $H(Y/X)$ and $I(X, Y)$.

22. For the given channel characteristics, calculate $I(X, Y)$:
    (a) $p(x) = e^{-x}, \quad 0 \le x \le \infty$
    (b) $p(x, y) = \frac{1}{2x^2 y}$, $p(y/x) = x \cdot e^{-xy}, 0 \le y \le \infty, \quad 1 \le x \le \infty,$
    $1/x \le y \le x.$

23. In a L-ary symmetric pulse transmission system, with $L = 32$ and $W = 10^3$ Hz, the average $\bar{S}/\bar{N} = 33$ dB in the channel.
    (a) Calculate the error rate P_e.
    (b) Rate of transmission R in bits/sec.
    (c) Excess power required in the transmitter to obtain R = theoretical channel capacity for the above calculated P_e.

24. Audio speech (0–4 kHz) modulate a pulse frequency modulated transmitter with $D = 5$ and the mean PRF of the Tx. = 10^5 pulses/sec, channel BW = 500 kHz, the received power = 1 mW, and n_0 in the channel = 2.10^{-9} W/kHz. Calculate the bounds of channel capacity for the system. Compare this capacity with that of an SSB system having the same S/n_0 ratio at the receiver input.

25. (a) A speech signal of BW = 4 kHz is converted to 6 bit PCM and the resulting Bipolar signal is transmitted in a continuous channel disturbed by AWGN. Obtain the S/N required in the channel for $P_b \le 10^{-4}$ in the receiver output and also the channel BW required.
    (b) To reduce the channel BW, the signalling pulses are converted to L-ary pulses, $L = 8$, and the same bit error rate is maintained. Calculate the S/N required in the channel. Also calculate the equivocation and the information rate in the channel.
    (c) For the reduced channel BW and the required S/N, calculate the ideal channel capacity and the information rate taking into consideration the excess power required to achieve channel capacity. Compare this rate with the rate obtained in (b).
    (d) Considering that the signals in (b) are approximately peak-limited continuous signals, calculate the rate of transmission obtainable in the channel and compare this with that of (b) and (c).

26. A picture-phone signal contains 6 Mb/s of information. A channel with $W = 750$ kHz and $S/N = 28$ dB, is available for transmission of this information. Give a scheme for the transmission system and (a) Calculate the error rate in the receiver, (b) Calculate the actual rate of transmission, (c) Compare the results with theoretical channel capacity.

REFERENCES

1. Goldman, S., *Information Theory*, Prentice Hall, 1953.
2. Shannon, C.E., 'A Mathematical Theory of Communication', *BSTJ*, pp. 379–423, July 1948 and pp. 623–654, Oct. 1948.
3. Shannon, C.E. and Weaver, W., '*The Mathematical Theory of Communication*', Univ. of Illinois Press, 1949.
4. Reza, F.M., *An Introduction to Information Theory,* McGraw-Hill, 1961.
5. Abramson, N., *Information Theory and Coding*, McGraw-Hill, 1963.
6. Ash, R., *Information Theory,* Interscience Pub., 1965.
7. Bell, D.A., *Information Theory and its Engineering Applications*, Isaac Pitman, 1953.
8. Sharma, P.D., *Introduction to Modern Communication Theory*, Prem Chand and Bros., 1971.
9. Fano, R.M., *Transmission of Information*, John Wiley, 1961.
10. Feinstein, A., *Foundations of Information Theory*, McGraw-Hill, 1958.
11. Huffman, D.A., 'A method of the construction of minimum redundancy codes', *Proc. IRE*, Vol. 40, pp. 1098–1101, Sept. 1952.
12. Shannon, C.E., 'Prediction and entropy of printed English', *BSTJ*, Vol. 30, pp. 50–64, Jan. 1951.
13. Muroga, S., 'On the capacity of a discrete channel', *J. Phys. Soc. Japan*, Vol. 8, pp. 484–494, 1953.
14. Elias, P., 'Coding for noisy channels', *IRE Convention Record,* Pt. 4, pp. 37–46, 1955.
15. Black, H.S., *Modulation Theory,* Van Nostrand, N. J., 1953.
16. Young, J.K., *Information Theory,* Butterworth, London, 1971.
17. Rice, S.O., 'Communication in presence of noise', *BSTJ*, Vol. 29, pp- 60–93, 1950.
18. Viterbi, A.J., *Principles of Coherent Communication*, McGraw-Hill, NY, 1966.
19. Sanders, R.W., 'Communication efficiency comparison of several communication systems'. *Proc. IRE*, Vol. 48, pp. 575–588, 1960.
20. Das, J., 'Optimum detection and communication in presence of noise Pt. V', *IETE Students' Journal,* Vol. 14, pp. 222–237, 1973.
21. Jones, D.S., *Elementary Information Theory,* Clarendon Press, Oxford, 1979.
22. Ristenbatt, M.P., 'Alternatives in digital communications', *Proc, IEEE*, Vol. 61, pp. 703–721, 1973.

APPENDIX

A

Statistical Characteristics of Speech and Image Signals

A.1 SIGNAL THEORY OF SPEECH

The mechanism of speech production may be described adequately in terms of the physiological structure of the vocal tract. The mouth and throat form an acoustic cavity which is bounded by the tongue and the roof of the mouth on bottom and top, by the lips and teeth on the front, and the vocal cords and glottis on the back. This cavity is a multiple resonant system whose poles or resonances are functions of its dimension. In the production of speech, the shape of this cavity is varied relatively slowly by gestures of the tongue and jaw, thereby changing its resonances, and their various positions correspond to the succession of sounds which make up speech. During the production of vowel sounds, the cavity is excited by a series of nearly periodic pulses generated by the vocal cords. These pulses evoke a repeated acoustic transient which, after radiation from the mouth, comprises the speech wave. Sounds excited by this mode are designated as 'Voiced'. In the case of consonants, excitation is provided by air passing turbulently through constrictions in the tract, and such sounds are called 'Unvoiced'. Most of the sounds can be characterized as either voiced or unvoiced, except in a few cases, e.g., the 'z' in the 'azure', which are composite sounds. Thus, speech production may be thought of as a low-frequency modulation of either a pulse or noise carrier wave [1].

It would appear from above that there are two fundamental principles involved in speech-processing. The first is that speech is a message, which, in its initial expression from the body, is represented by a group of muscular vibrations equivalent to a multiplicity of telegraph signals. The second principle is that these 'multiple telegraph signals' are made audible to the ear by modulating a set of audible carrier frequency components, which range from an inflected, buzz-like tone for voiced sounds to a hiss-like noise for unvoiced sounds, including whispers. Based on these considerations, Dudley, in his classical work, has shown that the voiced sound may be represented as [2]:

$$F_v(\omega, t) = s(t)\sum_{k=1}^{n} r(\omega, t)A_k \cdot \cos\left\{kP\int_0^t p(t)\,dt + \theta_k\right\} \tag{A.1}$$

where the carrier is composed of n audible harmonics of relatively high frequencies having amplitude A_k, frequency kP and phase θ_k; $s(t)$ is the switching function, $p(t)$ the inflecting factor and $r(\omega, t)$ the effect of selective transmission. The three message functions $s(t)$, $p(t)$ and $r(\omega, t)$ produce the necessary modulation processes on the carrier at the relatively low syllabic rate (at about 50 Hz maximum). The unvoiced sound is only a degenerate case of eqn. (A.1), given by:

$$F_u(\omega, t) = s(t)\int_{\omega_1}^{\omega_2} r(\omega, t)\,A(\omega)\cdot\cos\left[\,\omega t + \theta(\omega)\right]d\omega \qquad \text{(A.2)}$$

where P of eqn. (A.1) approaches zero and n tends to infinity, giving an integral instead of a finite sum over the audible frequency range $(\omega_2 - \omega_1)$. The unvoiced sound has no inflecting factor but does have switching and transmitting factors to make up the message impressed on the carrier.

The representation of speech signals by equations (A. 1) and (A. 2) is not easy to be broken up into the basic components for signal processing and various other simpler but approximate representations have been sought to make the instrumentation and synthesis practicable. For any arbitrary random signal, where the limitations on time and frequency are not very critical, an approximate representation in the time-frequency plane is of the form [3]

$$f(\omega, t) = \sum_{m,n} A_{mn}\cdot S_{mn}(t) \qquad \text{(A.3)}$$

where A_{mn} are the co-efficients of the sampling functions $S_{mn}(t)$ corresponding to the cross-points of any grid laid on the t-ω plane for which the separation of the cross-lines is θ in time and $1/\theta$ in frequency as shown in Fig. A.1. The signal may then be represented by $(m \times n)$ numbers, once the choice of θ has been made. If the smapling is done either in the time domain or in the frequency domain, eqn. (A.3) will ideally represent the two sampling theorems of Shannon, and a signal of bandwidth W and duration T will require $2TW$ number to specify it completely. Further approximations in signal representation are possible if one considers that any information-bearing process will have to be somewhat random in nature, but it will also have finite auto-correlation in time and frequency domains. Because of this finite correlation, some portions of the signal either in the t-plane or in the ω-plane may be discarded, thus leading to simple sampling of signals in time or in frequency, as represented by:

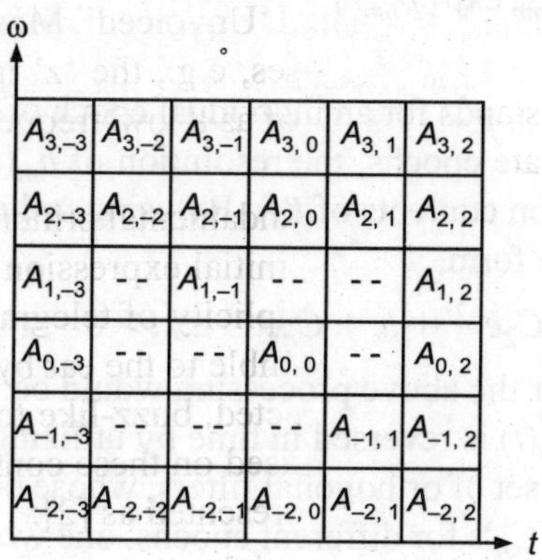

Figure A.1: Sampling in the t-ω plane

$$f(\tilde{t}) = \sum_{n/2} f(t)\{U(t-t_n) - U(t-t_{n+1})\} \tag{A.4}$$

and

$$f(\tilde{\omega}) = \sum_{n/2} f(\omega)\{U(\omega-\omega_n) - U(\omega-\omega_{n+1})\} \tag{A.5}$$

To obtain a reasonable approximation in $f(t)$ and $f(\omega)$, the intersample gaps should be less than the signal correlation in t- or in ω-plane.

If the message functions $s(t)$, $p(t)$ and $r(\omega, t)$ in eqn. (A. 1) are simple enough, then the signal may be considered to be composed of the two important parameters, viz. (a) the amplitude variation, (b) the frequency variation. Thus, one may write,

$$f(\tilde{t}) = A(\omega, t) \cos [p(\omega, t) + \theta] \tag{A.6}$$

where A and p are the variable amplitude and frequency respectively of the signal. In simple cases, it is possible to extract the two signals proportional to the variations in A and p with time, and reconstruct an $f(t)$ at the receiver, as is done in narrow-band speech vocoders.

From a different point of view, the speech signal may be considered as composed of an impulsive forcing function $e(t)$ and a system memory function $h(t)$ so that $f(t)$ is the convolution given by

$$f(t) = e(t) * h(t) = \int_{-\infty}^{t} h(\tau), e(t-\tau)\, d\tau \tag{A.7}$$

Here $e(t)$ may be regarded as a series of successive impulses whose epochs and intensities determine the temporal attributes of $f(t)$. A typical example of $e(t)$, $h(t)$ and $f(t)$ alongwith their spectral characteristics is shown in Fig. A.2. The system function $h(t)$, responsible largely for the structural attributes of $f(t)$, may be represented by an exponential series. The composite signal may now be written as [4, 5]

$$f(t) = \sum_{n} C_{1n} \cdot \exp(p_{1n} t) + \sum_{n} C_{2n} \cdot \exp(p_{2n} t) + \ldots + \sum_{n} C_{mn} \exp(p_{mn} t)$$

$$= \sum_{m}\sum_{n} C_{mn} \exp(p_{mn} t) \tag{A.8}$$

where each of the summation term stands for an individual epoch of $e(t)$ ranging from t_1 to t_m. If now the signal is segmented into approximate epochs, the resolution of $h_m(t)$ into elementary signals would be sufficient to indicate the information contents of $f(t)$. By using orthogonal filters, it is possible to find a good approximation of $h_m(t)$ in the form,

$$h_m(t) \simeq C_1 e^{-t} + C_2 e^{-2t} + \ldots + C_n e^{-nt} \ldots,\ t > 0 \tag{A. 9}$$

Thus, the sequence of operation for the above processing would be: $f(t)$ is first broken into the elements $h_m(t)$ by suitable gates and each $h_m(t)$ is reversed in time by utilizing a recording and reversed-playback technique. This $h_m(-t)$ is fed to the set of orthogonal filters, whose outputs are suitably combined to give the values of C_n. Recombining the C_n's for different epochs, one would get the approximated signal $f(t)$.

It is seen from Fig. A.2, that the voiced sound has discrete frequency components occurring at the harmonics of the pitch frequency and the unvoiced sound has a continuous spectrum. The spectral characteristics of speech sounds have been extensively studied [1, 2]. The short-time spectra of speech, obtained through multiple narrowband bandpass filters, is known as Spectrograms, and one typical

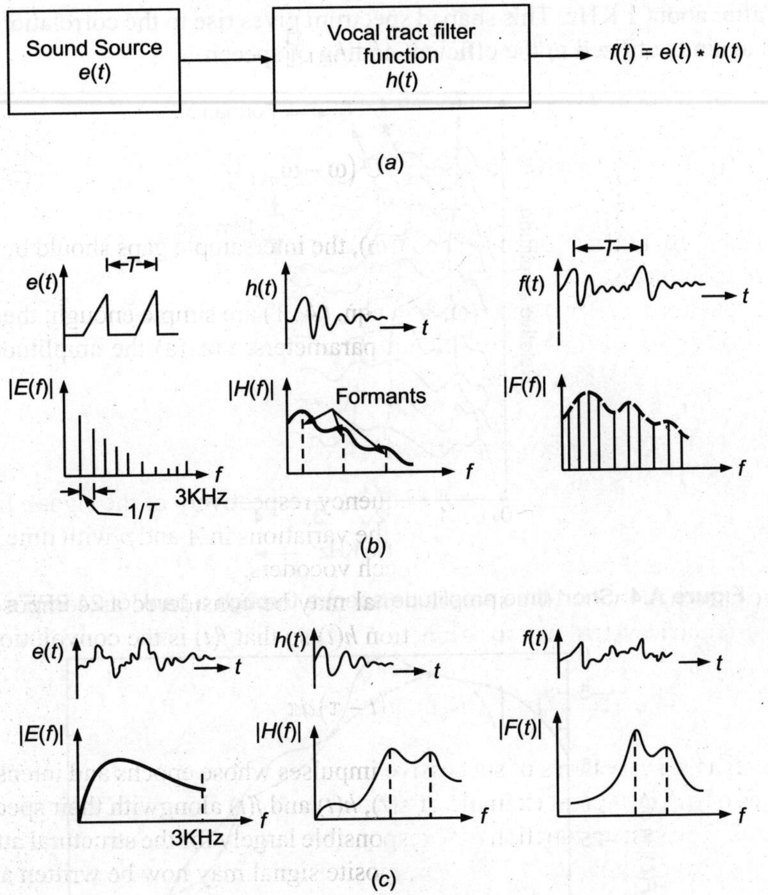

Figure A.2: Source-system model of speech production: **(a)** Block diagram, **(b)** Voiced sound, **(c)** Unvoiced sound

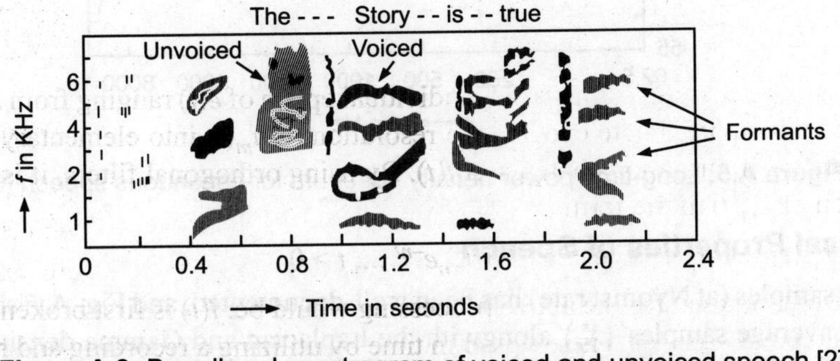

Figure A.3: Broadband spectrogram of voiced and unvoiced speech [2]

spectrogram is shown in Fig. A.3. An alternative representation of the spectrum, obtained by scanning of the outputs of the BPF's (of width = 150 Hz) at 10 msec. intervals, is shown in Fig. A.4, where the formant structure of the voiced sound is explicitly shown. The long-time power density spectrum of speech is shown in Fig. A.5, where it is seen that the spectrum falls at a rate of 10 dB per octave approximately after about 1 KHz. This shaped spectrum gives rise to the correlation between the speech samples, which again is utilized in the efficient coding of speech.

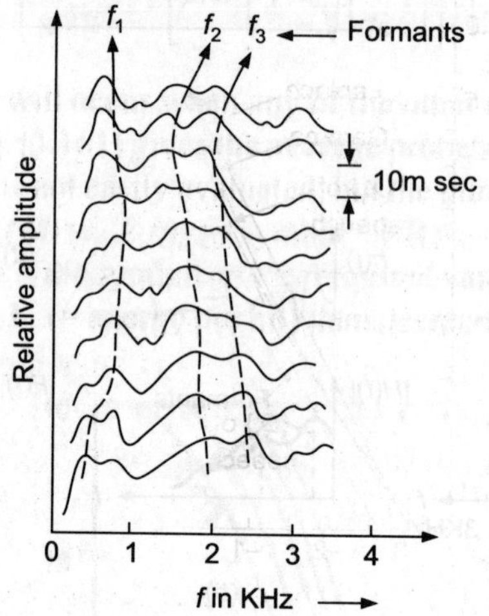

Figure A.4: Short-time amplitude spectra through a bank of 24 BPF's [1]

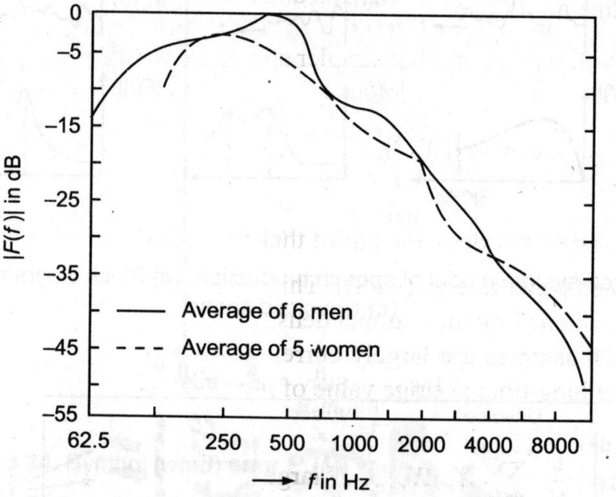

Figure A.5: Long-time power density spectrum for continuous speech [6]

A.1.1 Statistical Properties of Speech

The pdf of speech samples (at Nyquist rate) has been well-documented, and Fig. A.6 shows the pdf of the amplitudes of the average samples $\{X_r\}$ alongwith the Laplacian and Gamma density functions. It has been shown that the pdf may be approximated by a special form of the Gamma density as [7]:

$$p(X_r) = \frac{\sqrt{k}}{2\sqrt{\pi}} \cdot \frac{\exp\left(-k|X_r|\right)}{\sqrt{|X_r|}}, \tag{A.10}$$

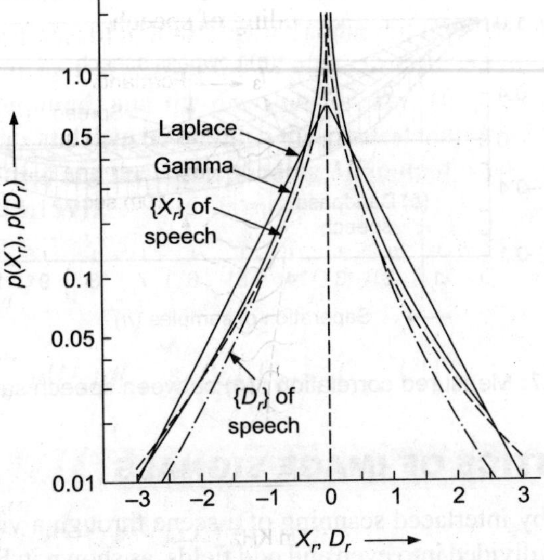

Figure A.6: pdf of speech samples $\{X_r\}$ and the differential $\{D_r\}$ of samples, along with Gamma and Laplace pdf's [7]

where, $k = \sqrt{0.75}/\sigma_x$. A slightly poorer but simpler approximation is given by the Laplacian density as:

$$p(X_r) = \frac{\alpha}{2} \cdot \exp(-\alpha |X_r|), \tag{A.11}$$

where, $\alpha = \sqrt{2/\sigma_x}$. For DPCM codecs, the pdf of the difference signals $D_r = (X_r - X_{r-1})$ has also been experimentally determined [refer to eqn. (2.48)]. This $p(D_r)$ is shown in Fig. A.6 and it is seen that this pdf is also closely approximated by the Gamma density function.

The successive speech samples are largely correlated and this correlation $p(n)$ decreases with the sample separation n. The long-time average value of $p(n)$, with $\rho(0) = 1$, is given by:

$$\rho(n) = \frac{1}{N} \sum_{r=0}^{N-n-1} X_r \cdot X_{r+n}, \; N \text{ large} \tag{A.12}$$

These correlation values for $0 \le n \le 10$ have been experimentally determined and are shown in Fig. A.7, where the curve (a) is for lowpass and the curve (b) is for bandpass speech signals. The spread around the curves is due to different speakers. It is observed that $\rho(n)$ is large for $n \le 3$, and this property has been used in estimating the speech samples for the DPCM codecs, discussed in Chapter 2. The long-time power-spectral density $F(f)$ of speech signals may now be calculated from the above $\rho(n)$ by using discrete Fourier transform and the calculated values of $|F(f)|$ agree well with the experimental values as shown in Fig. A.5. The overall conclusion from the above results is that the speech signals are highly redundant and contain approximately 2–4 K bits of information only, as against 64 K bits of channel capacity used in a log-PCM codec.

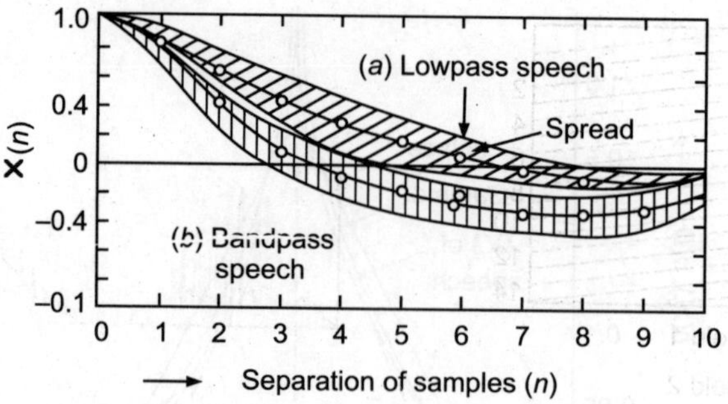

Figure A.7: Measured correlation ρ(*n*) between speech samples [8]

A.2 CHARACTERISTICS OF IMAGE SIGNALS

Video signals are generated by interlaced scanning of a scene through a video camera, and a complete image scan, called a frame, is divided into even and odd fields, as shown in Fig. A.8(a). A frame contains 525/625 scan lines, the frame time is 1/30 (or 1/25) sec. and each field scan period is 1/60 (or 1/50) sec. The camera output signal $e_q(t)$ per frame is approximately of the form shown in Fig. A.8(b), and may be written as:

$$e_q(t) = \sum_{k=0}^{\infty} E_k \cos\left(k\omega_h t + \theta_k\right) \tag{A.13}$$

where $\omega_h = 2\pi/t_h$, and

t_h = horizontal scan time.

The continuous output for many frames repeated at the frame frequency $f_v = \omega_v/2\pi$, is then written as:

$$e_0(t) = \sum_{q=0}^{\infty}\left\{\sum_{k=0}^{\infty} \frac{E_k E_q}{2}\left[\cos\left(k\omega_h t + q\omega_v t + \theta_k - \theta_q\right)\right.\right.$$

$$\left.\left. + \cos\left(k\omega_h t + q\omega_v t + \theta_k + \theta_q\right)\right]\right\} \tag{A.14}$$

The spectrum $F(f)$ of $e_0(t)$ is now shown in Fig. A.8(c), and it is seen that the spectrum is descrete, rather than continuous, because of the line and frame scanning processes. The spectrum consists of a series of spectral clusters separated by the line scanning frequency and each cluster is composed of spectral components separated by the field scanning frequency. When the scenes contain moving objects, the individual spectral lines are broadened, as shown in Fig. A.8(d) [9]. It may be shown that for ideal resolution, 1300 × 1300 pixels per frame are required and the effective spectrum spread of $e_0(t)$ is approximately 22 MHz. However for commercial systems, 512 × 212 pixels per frame are sufficient, and the approximate signal bandwidth is now 5 MHz. However this bandwidth is poorly utilized as the spectral amplitudes fall rapidly after 1 MHz, as shown in Fig. A.9. This property leads to large correlation between successive pixels and the consequent redundancy reduction techniques, discussed in Chapter 2, have been immensely successful. Specifically refer to Fig. 2.57 and eqn. (2.75) which describe the operation of 2*D*-ADM codec based on the prediction of S_0 as a combination of $\{S_1, S_2, S_3\}$ pixels.

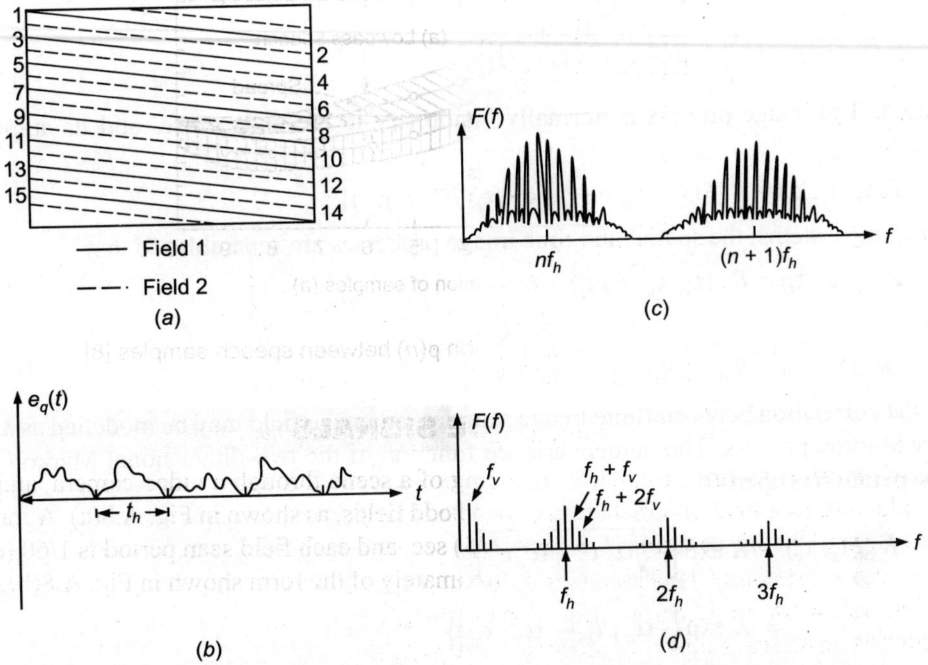

Figure A.8: Image signal and its spectrum: **(*a*)** Interlaced scanning, **(*b*)** $e_q(t)$ per frame, **(*c*)** Spectrum of $e_0(t)$ (with no scene movement), **(*d*)** Amplified detail of (c) with scene movement

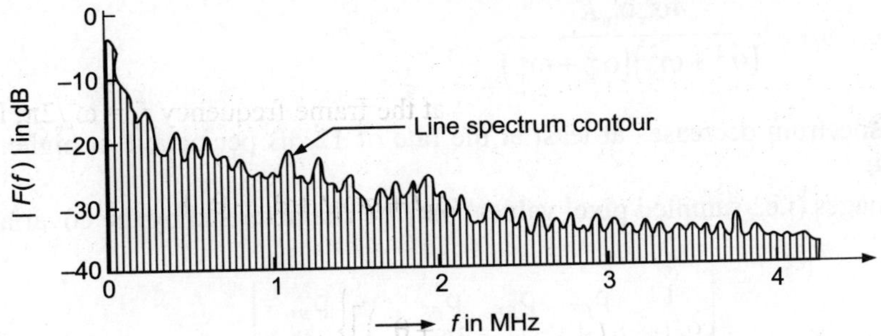

Figure A.9: Approximate contour of the Luminance signal spectrum of a typical image

A.2.1 Statistical Properties [10]

A moving image may be characterised with a three-dimensional function $F(x, y, t)$ where (x, y) are the spatial variables and t, the time variable. If the luminance parameter is sufficiently random, then the stochastic process $F(x, y, t)$ can be specified by its joint probability density for all sample points. In general, higher order joint probability densities of images are not known, but the first order density $p\{F(x, y, t)\}$ has been widely discussed in literature. Since the luminance measure is always positive, its pdf has to be a one-sided density. The Rayleigh, log-normal and exponential density functions have been used for modelling $p\{F(x, y, t)\}$; but of these the most useful has been the exponential pdf given by

$$p\{F(x, y, t)\} = K \exp\{-\alpha F(x, y, t)\}, \tag{A.15}$$

where $F \geq 0$, and K, α are constants. For the pdf of the differences between successive image samples, the Laplacian density function, given by

$$p\{F(x, y, t)\} = \frac{\alpha}{2}\exp\{-\alpha|F(x, y, t)|\} \tag{A.16}$$

has been used. The image process is normally 'stationary in the wide sense' and its autocorrelation function is:

$$R(\tau_x, \tau_y, \tau_t) = E\{F(x+\tau_x, y+\tau_y, t+\tau_t)\cdot F^*(x, y, t)\} \tag{A.17}$$

In many imaging systems, the spatial and time image processes are separable, so that

$$R(\tau_x, \tau_y, \tau_t) = R_{xy}(\tau_x, \tau_y)\, R_t(\tau_t)$$

and further,

$$R_{xy}(\tau_x, \tau_y) \simeq R_x(\tau_x)\cdot R_y(\tau_y) \tag{A.18}$$

Because of the correlation between finite image samples, an image field may be modelled as a sample of a first-order Markov process. The autocovariance function of the two-dimensional Markov process is (using the separation property):

$$K_{xy}(\tau_x, \tau_y) = K\exp\left\{-\sqrt{\alpha_x^2\tau_x^2 + \alpha_y^2\tau_y^2}\right\}$$

$$\simeq K\exp\{-\alpha_x|\tau_x| - \alpha_y|\tau_y|\} \tag{A.19}$$

where, K, α_x, α_y are constants.

The power spectrum of the process is then,

$$S(\omega_x, \omega_y) = \frac{4\alpha_x\alpha_y K}{\left(\alpha_x^2+\omega_x^2\right)\left(\alpha_y^2+\omega_y^2\right)} \tag{A.20}$$

Thus, the power spectrum decreases at least at the rate of 12 dB per octave at higher frequencies, as shown in Fig. A.9.

The discrete images (i.e., sampled pixel values) may be characterised by the covariance matrix,

$$[K_{ik}] = \sigma_{px}^2 \begin{bmatrix} 1 & \rho_{px} & \rho_{px}^2 & \rho_{px}^3 & \cdots & \rho_{px}^{N-1} \\ \rho_{px} & 1 & \rho_{px} & \rho_{px}^2 & \cdots & \rho_{px}^{N-2} \\ \vdots & & & & & \vdots \\ \rho_{px}^{N-1} & . & . & . & \cdots & 1 \end{bmatrix} \tag{A.21}$$

where the adjacent pixels along an image row have a correlation of ($0 \leq \rho_{px} \leq 1$), a self-correlation of unity and a variance of σ_{px}^2. The above relation has been verified by measuring the correlation of pixels along a line, between the lines and along a diagonal of a typical image, and the measured data is shown in Fig. A.10[10]. It is seen that the data points are related with the Markov covariance function having

(a) $\rho_{px} = 0.953$ for pixels along a line
(b) $\rho_{px} = 0.965$ for pixels between lines
(c) $\rho_{px} \simeq 0.92$ for pixels along the diagonal

For the diagonal case, $\rho_{px} = 0.92 = (0.953 \times 0.965)$ for upto five pixel separations, i.e., the horizontal and vertical correlations are separable for $n \leq 5$. For $n > 5$, $\rho_{px} \simeq 0.938$. This correlation property has been used in designing the predictors for D-PCM codecs, as is discussed in Chapter 2.

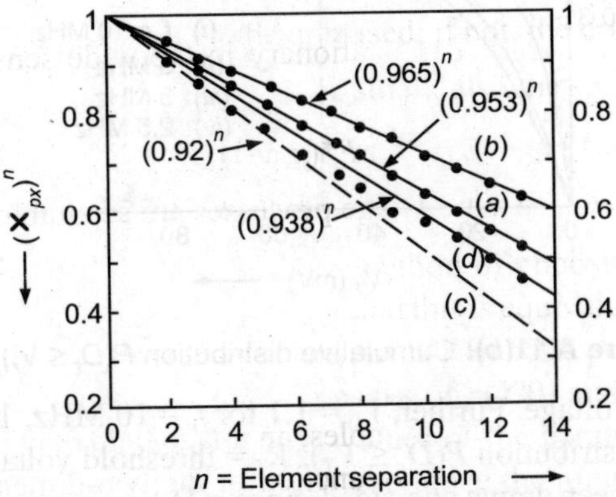

Figure A.10: Measured correlation $(\rho_{px})^n$ of pixels of a typical scene. Values indicated are the equations for the $(n - \rho_{px})$ lines. (*a*) $(\rho_{px})^n$ along a line, (*b*) $(\rho_{px})^n$ between lines, and (*c*) and (*d*) for $(\rho_{px})^n$ along the diagonal [10]

Frei et al. [11] have measured the statistics of the amplitude differences $D_r(1) = (X_r - X_{r-1})$ between consecutive samples in a busy picture. Their results for $p(D_r)$ are shown in Fig. A.11(a), where different sampling rates f_r of 10, 5 and 3 MHz for a 5 MHz TV system (625 lines/frame, 25 frames/sec) and 2.5 MHz for a 1.2 MHz video-phone system (315 lines/frame, 25 frames/sec) have been used. It is easily shown that $p(D_r)$ may be approximated by the Laplacian density function:

$$p(D_r) = \frac{1}{2V_0} e^{-|D_r|/V_0} \tag{A.22}$$

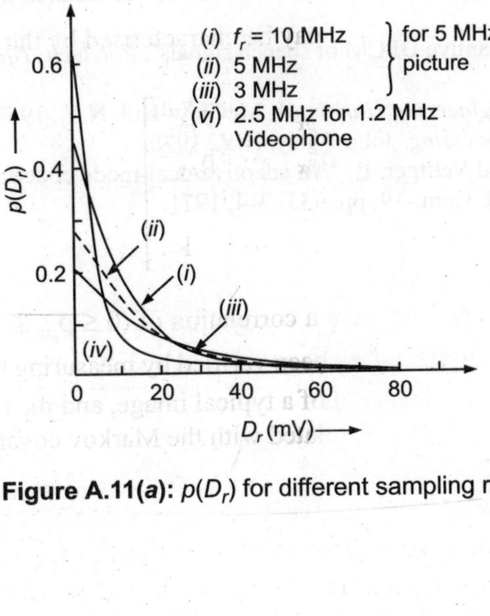

Figure A.11(*a*): $p(D_r)$ for different sampling rates f_r

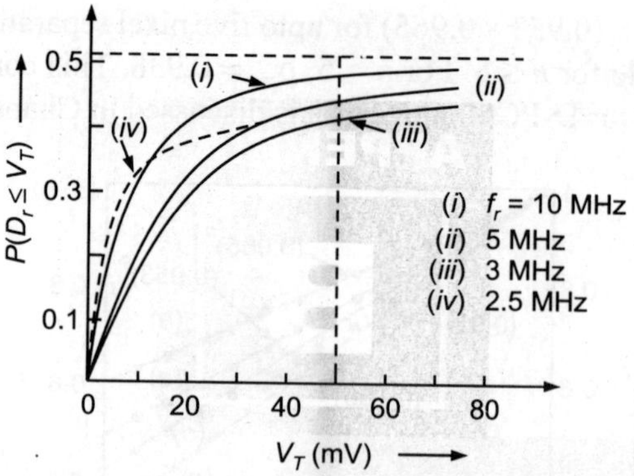

Figure A.11(*b*): Cumulative distribution $P(D_r \le V_r)$ [11]

where V_0 is the characteristic voltage. Further, $V_0 = 1.1$ for $f_r = 10$ MHz, 1.85 for $f_r = 5$ MHz and 2.5 for $f_r = 3$ MHz. The cumulative distribution $P(D_r \le V_T)$, V_T = threshold voltage level, is also shown in Fig. A.11(b). (Both the curves have been drawn one-sided, since $p(D_r)$ is assumed symmetrical). The cumulative distribution indicates that, even in a DPCM codec, the generated codewords are not equiprobable and there is a scope for further entropy coding, as discussed in Sec. 2.14.

REFERENCES

1. Flanagan, J.L., *Speech Analysis, Synthesis and Perception,* Academic Press, N. Y., 1965.
2. Dudley, H.W., 'Fundamentals of speech synthesis', *J. Audio Engg. Soc.,* Vol. 3, pp. 170–185, 1955.
3. Lerner, R.M., 'The representation of signals', *IRE Trans.,* Vol. CT-6, pp. 197–205, 1959.
4. Das, J., 'Some approximations of signals and informations', *Ind. Jour. Physics,* Vol. 38, pt. 6, pp. 311–325, 1964.
5. Das, J., 'A review of speech processing', *IETE St. Journal,* Vol. 19, pp. 77–85, 1978.
6. Dunn, H.K. and White, S.D., 'Statistical measurements on conversational speech', *J. Acoust. Soc. Am.,* Vol. 11. pp. 278–288, 1940.
7. Paez, M.D. and Glisson, T.H., 'Minimum mean-squared-error quantization in speech', *IEEE Trans. Comm.,* Vol. 20, pp. 225–230, 1972.
8. Noll, P., 'Non-adaptive and adaptive DPCM of speech signals', *Polytech. Tijdschr. Ed. Electrotech/Electron (Netherlands),* No. 19, 1972.
9. Fink, D.G., Ed., *Television Engineering Handbook,* McGraw-Hill, N.Y., 1957.
10. Pratt, W.K., *Digital Image Processing,* John Wiley, N.Y., 1978.
11. Frei, A.H., Schindler, H.R. and Vettiger, P., 'An adaptive dual-mode coder/decoder for television signals', *IEEE Trans. Comm. Tech.,* Vol. Com.-19, pp. 933–944, 1971.

APPENDIX

B

A Short List of Minimum Polynomials Over GF(2^m)

As discussed in Sec. 4.4.1, the generator polynomials for BCH codes are obtained from the minimum polynomials, $M_i(X)$, where $M_1(X)$ is the primitive polynomial $p(X)$ over GF(2^m). The successive polynomials $M_3(X)$, $M_5(X)$ etc. are the minimum polynomials of α^3, α^5 etc. for the given m and $p(X)$. The generator polynomial is then given by eqn. (4.41). The table below gives the minimum polynomials for $m = 3, 4, 5, 6, 7$ and 8, and is based on the table of irreducible polynomials, given in Peterson and Weldon [1]. Clark and Cain [2] have also given a list of code generators for BCH codes. It is to be noted that the reciprocal polynomial of an irreducible polynomial is also irreducible, and the reciprocal of the primitive polynomial is also primitive.

Table of Minimum Polynomials over GF(2^m)

m	$M_i(X)$	
3	$p(X) = M_1(X) = X^3 + X + 1$;	$M_3(X) = X^3 + X^2 + 1$;
4	$p(X) = M_1(X) = X^4 + X + 1$; $M_5(X) = X^2 + X + 1$;	$M_3(X) = X^4 + X^3 + X^2 + X + 1$; $M_7(X) = X^4 + X^3 + 1$;
5	$p(X) = M_1(X) = X^5 + X^2 + 1$; $M_5(X) = X^5 + X^4 + X^2 + X + 1$; $M_{11}(X) = X^5 + X^4 + X^3 + X + 1$;	$M_3(X) = X^5 + X^4 + X^3 + X^2 + 1$; $M_7(X) = X^5 + X^3 + X^2 + X + 1$; $M_{15}(X) = X^5 + X^3 + 1$;
6	$p(X) = M_1(X) = X^6 + X + 1$; $M_5(X) = X^6 + X^5 + X^2 + X + 1$; $M_9(X) = X^3 + X^2 + 1$; $M_{13}(X) = X^6 + X^4 + X^3 + X + 1$; $M_{21}(X) = X^2 + X + 1$; $M_{27}(X) = X^3 + X + 1$;	$M_3(X) = X^6 + X^4 + X^2 + X + 1$; $M_7(X) = X^6 + X^3 + 1$; $M_{11}(X) = X^6 + X^5 + X^3 + X^2 + 1$; $M_{15}(X) = X^6 + X^5 + X^4 + X^2 +1$; $M_{23}(X) = X^6 + X^5 + X^4 + X + 1$; $M_{21}(X) = X^6 + X^5 + 1$;
7	$p(X) = M_1(X) = X^7 + X^3 + 1$; $M_5(X) = X^7 + X^4 + X^3 + X^2 + 1$; $M_9(X) = X^7 + X^5 + X^4 + X^3 + X^2 + X + 1$;	$M_3(X) = X^7 + X^3 + X^2 + X + 1$; $M_7(X) = X^7 + X^6 + X^5 + X^4 + X^2 + X + 1$; $M_{11}(X) = X^7 + X^6 + X^4 + X^2 + 1$;

(Contd.)

	$M_{13}(X) = X^7 + X + 1;$	$M_{15}(X) = X^7 + X^6 + X^5 + X^3 + X^2 + X + 1;$
	$M_{19}(X) = X^7 + X^6 + X^3 + X + 1;$	$M_{21}(X) = X^7 + X^6 + X^5 + X^2 + 1;$
	$M_{23}(X) = X^7 + X^6 + 1;$	$M_{27}(X) = X^7 + X^6 + X^4 + X + 1;$
	$M_{29}(X) = X^7 + X^5 + X^3 + X + 1;$	$M_{31}(X) = X^7 + X^6 + X^5 + X^4 + 1;$
	$M_{43}(X) = X^7 + X^5 + X^2 + X + 1;$	$M_{47}(X) = X^7 + X^5 + X^4 + X^3 + 1;$
	$M_{55}(X) = X^7 + X^6 + X^5 + X^4 + X^3 + X^2 + 1;$	$M_{63}(X) = X^7 + X^4 + 1;$
8.	$p(X) = M_1(X) = X^8 + X^4 + X^3 + X^2 + 1;$	$M_3(X) = X^8 + X^6 + X^5 + X^4 + X^2 + X + 1;$
	$M_5(X) = X^8 + X^7 + X^6 + X^5 + X^4 + X + 1;$	$M_7(X) = X^8 + X^6 + X^5 + X^3 + 1;$
	$M_9(X) = X^8 + X^7 + X^5 + X^4 + X^3 + X^2 + 1;$	$M_{11}(X) = X^8 + X^7 + X^6 + X^5 + X^2 + X + 1;$
	$M_{13}(X) = X^8 + X^5 + X^3 + X + 1;$	$M_{15}(X) = X^8 + X^7 + X^6 + X^4 + X^2 + X + 1;$
	$M_{17}(X) = X^4 + X + 1;$	$M_{19}(X) = X^8 + X^6 + X^5 + X^2 + 1;$
	$M_{21}(X) = X^8 + X^7 + X^3 + X + 1;$	$M_{23}(X) = X^8 + X^6 + X^5 + X + 1;$
	$M_{25}(X) = X^8 + X^4 + X^3 + X + 1;$	$M_{27}(X) = X^8 + X^5 + X^4 + X^3 + X^2 + X + 1;$
	$M_{29}(X) = X^8 + X^7 + X^3 + X^2 + 1;$	$M_{31}(X) = X^8 + X^5 + X^3 + X^2 + 1;$

REFERENCES

1. Peterson, W.W. and Weldon, E.J., Error-correcting Codes, MIT Press, Cambridge, Mass, 1972.
2. Clark, G.C. and Cain, J.B., 'Error-correction Coding for Digital Communications', Plenum Press, N.Y. 1981.

APPENDIX C

Table of Entropy Function $H(p)$ (Base 2)

$$H(p) = -p \log_2 p - (1 - p) \log_2 (1 - p)$$

p	$H(p)$	p	$H(p)$
1.10^{-5}	16.6×10^{-5}	10^{-2}	.08512
2.10^{-5}	33.2×10^{-5}	2.10^{-2}	.14051
3.10^{-5}	49.8×10^{-5}	3.10^{-2}	.19312
4.10^{-5}	63.1×10^{-5}	4.10^{-2}	.24070
5.10^{-5}	79.7×10^{-5}	5.10^{-2}	.28449
6.10^{-5}	93.0×10^{-5}	6.10^{-2}	.32528
7.10^{-5}	106.3×10^{-5}	7.10^{-2}	.36350
8.10^{-5}	119.6×10^{-5}	8.10^{-2}	.39953
9.10^{-5}	132.8×10^{-5}	9.10^{-2}	.43358
10^{-4}	146.1×10^{-5}	0.1	.46589
2.10^{-4}	288.9×10^{-5}	0.2	.71716
3.10^{-4}	408.5×10^{-5}	0.3	.87549
4.10^{-4}	512.5×10^{-5}	0.4	.96456
5.10^{-4}	634.5×10^{-5}	0.5	1.0000
6.10^{-4}	740.8×10^{-5}		
7.10^{-4}	847.1×10^{-5}		
8.10^{-4}	950.1×10^{-5}		
9.10^{-4}	1053.1×10^{5}		
10^{-3}	.01152		
2.10^{-3}	.02109		
3.10^{-3}	.02986		
4.10^{-3}	.03751		
5.10^{-3}	.04538		
6.10^{-3}	.05288		
7.10^{-3}	.06012		
8.10^{-3}	.06719		
9.10^{-3}	.07403		

APPENDIX D

Table of Complementary Error Function

Define:

$$\text{erf}\,(x) \triangleq \int_{-\infty}^{x} \frac{1}{\sqrt{2\pi}} e^{-z^2/2} dz$$

and

$$\text{erfc}(x) = [1 - \text{erf}\,(x)] = \int_{0}^{\infty} \frac{1}{\sqrt{2\pi}} e^{-z^2/2} dz \quad (1)$$

An alternative definition of error function, used by some authors, is:

$$\text{erf}_*(x) \triangleq \frac{2}{\sqrt{\pi}} \int_{0}^{x} e^{-z^2} dz.$$

and

$$\text{erfc}_*\,(x) = [1 - \text{erf}_*(x)] = 2\,\text{erfc}\,(\sqrt{2x}\,) \quad (2)$$

Eqns. (1) and (2) relate the two definitions. Further, erfc(x) may be approximated as:

$$\text{erfc}(x) \simeq \frac{1}{\sqrt{2\pi x}} \exp\left(-\frac{x^2}{2}\right) \quad \text{for } x \geq 4$$

and

$$\text{erfc}\,(x) \simeq \frac{1}{2}\left[1 - \sqrt{\frac{2}{\pi}} x\right] \quad \text{for } x \text{ small } (x \leq 0.1).$$

Table below gives the values of erfc (x) for $0.01 \leq x \leq 5.0$.

x	erfc (x)	x	erfc (x)	x	erfc (x)
0.00	0.5000	0.15	0.4404	0.30	0.3821
0.01	0.4960	0.16	0.4364	0.31	0.3783
0.02	0.4920	0.17	0.4325	0.32	0.3745
0.03	0.4880	0.18	0.4286	0.33	0.3707
0.04	0.4840	0.19	0.4247	0.34	0.3669
0.05	0.4801	0.20	0.4207	0.35	0.3632
0.06	0.4761	0.21	0.4168	0.36	0.3594
0.07	0.4721	0.22	0.4129	0.37	0.3557
0.08	0.4681	0.23	0.4090	0.38	0.3520
0.09	0.4641	0.24	0.4052	0.39	0.3483
0.10	0.4602	0.25	0.4013	0.40	0.3446
0.11	0.4562	0.26	0.3974	0.41	0.3409
0.12	0.4522	0.27	0.3936	0.42	0.3372
0.13	0.4483	0.28	0.3897	0.43	0.3336
0.14	0.4443	0.29	0.3859	0.44	0.3300
0.45	0.3264	0.87	0.1922	1.29	0.0985
0.46	0.3228	0.88	0.1894	1.30	0.0968
0.47	0.3192	0.89	0.1867	1.31	0.0951
0.48	0.3156	0.90	0.1841	1.32	0.0934
0.49	0.3121	0.91	0.1814	1.33	0.0918
0.50	0.3085	0.92	0.1788	1.34	0.0901
0.51	0.3050	0.93	0.1762	1.35	0.0885
0.52	0.3015	0.94	0.1736	1.36	0.0869
0.53	0.2981	0.95	0.1711	1.37	0.0853
0.54	0.2946	0.96	0.1685	1.38	0.0838
0.55	0.2912	0.97	0.1660	1.39	0.0823
0.56	0.2877	0.98	0.1635	1.40	0.0808
0.57	0.2843	0.99	0.1611	1.41	0.0793
0.58	0.2810	1.00	0.1587	1.42	0.0778
0.59	0.2776	1.01	0.1563	1.43	0.0764
0.60	0.2743	1.02	0.1539	1.44	0.0749
0.61	0.2709	1.03	0.1515	1.45	0.0735
0.62	0.2676	1.04	0.1492	1.46	0.0721
0.63	0.2643	1.05	0.1469	1.47	0.0708
0.64	0.2611	1.06	0.1446	1.48	0.0694
0.65	0.2578	1.07	0.1423	1.49	0.0681
0.66	0.2546	1.08	0.1401	1.50	0.0668
0.67	0.2514	1.09	0.1379	1.51	0.0655
0.68	0.2483	1.10	0.1357	1.52	0.0643
0.69	0.2451	1.11	0.1335	1.53	0.0630
0.70	0.2420	1.12	0.1314	1.54	0.0618
0.71	0.2389	1.13	0.1292	1.55	0.0606
0.72	0.2358	1.14	0.1271	1.56	0.0594
0.73	0.2327	1.15	0.1251	1.57	0.0582
0.74	0.2297	1.16	0.1230	1.58	0.0571
0.75	0.2266	1.17	0.1210	1.59	0.0559
0.76	0.2236	1.18	0.1190	1.60	0.0548
0.77	0.2206	1.19	0.1170	1.61	0.0537
0.78	0.2177	1.20	0.1151	1.62	0.0526

(Contd.)

0.79	0.2146	1.21	0.1131	1.63	0.0516
0.80	0.2119	1.22	0.1112	1.64	0.0505
0.81	0.2090	1.23	0.1093	1.65	0.0495
0.82	0.2061	1.24	0.1075	1.66	0.0485
0.83	0.2033	1.25	0.1056	1.67	0.0475
0.84	0.2005	1.26	0.1038	1.68	0.0465
0.85	0.1977	1.27	0.1020	1.69	0.0455
0.86	0.1949	1.28	0.1003	1.70	0.0446
1.71	0.0436	2.04	0.0207	2.66	0.0039
1.72	0.0427	2.06	0.0197	2.68	0.0037
1.73	0.0418	2.08	0.0188	2.70	0.0035
1.74	0.0409	2.10	0.0179	2.72	0.0033
1.75	0.0401	2.12	0.0170	2.74	0.0031
1.76	0.0392	2.14	0.0162	2.76	0.0029
1.77	0.0384	2.16	0.0154	2.78	0.0027
1.78	0.0375	2.18	0.0146	2.80	0.0026
1.79	0.0367	2.20	0.0139	2.82	0.0024
1.80	0.0359	2.22	0.0132	2.84	0.0023
1.81	0.0351	2.24	0.0125	2.86	0.0021
1.82	0.0344	2.26	0.0119	2.88	0.0020
1.83	0.0336	2.28	0.0113	2.90	0.0019
1.84	0.0329	2.30	0.0107	2.92	0.0018
1.85	0.0322	2.32	0.0102	2.94	0.0016
1.86	0.0314	2.34	0.0096	2.96	0.0015
1.87	0.0307	2.36	0.0091	2.98	0.0014
1.88	0.0301	2.38	0.0087	3.00	0.00135
1.89	0.0294	2.40	0.0082	3.10	0.00095
1.90	0.0287	2.42	0.0078	3.20	0.00069
1.91	0.0281	2.44	0.0073	3.30	0.00050
1.92	0.0274	2.46	0.0069	3.40	0.00035
1.93	0.0268	2.48	0.0066	3.50	0.00025
1.94	0.0262	2.50	0.0062	3.60	0.000159
1.95	0.0256	2.52	0.0059	3.70	0.00010
1.96	0.0250	2.54	0.0055	3.80	0.000072
1.97	0.0244	2.56	0.0052	3.90	0.000050
1.98	0.0239	2.58	0.0049	4.00	0.000032
1.99	0.0233	2.60	0.0047	4.50	0.000003
2.00	0.0228	2.62	0.0044	5.00	0.000003
2.02	0.0217	2.64	0.0041		

APPENDIX

E

Table of $I_0(x)$

$$I_0(x) = \frac{1}{2\pi}\int_0^{2\pi} \exp(x \cos \alpha)\, d\alpha$$

x	$I_0(x)$	x	$I_0(x)$
0.0	1	2.75	3.9
0.5	1.1	3.0	4.6
1.0	1.2	3.25	5.7
1.25	1.35	3.5	7.0
1.5	1.5	3.75	8.7
1.75	1.8	4.0	10.4
2.0	2.2	4.25	13.2
2.25	2.7	4.5	17.0
2.5	3.2	4.65	20

APPENDIX

F

Logarithms to the Base 2

$\log_2 e = 1.4419$

x	*log x*	*x*	*log x*	*x*	*log x*
1	0.0000	35	5.1293		
2	1.0000	36	5.1699	69	6.1085
3	1.5850	37	5.2095	70	6.1293
4	2.0000	38	5.2479	71	6.1497
5	2.3219	39	5.2854	72	6.1699
6	2.5849	40	5.3219	73	6.1898
7	2.8074	41	5.3576	74	6.2095
8	3.0000	42	5.3923	75	6.2288
9	3.1699	43	5.4263	76	6.2479
10	3.3219	44	5.4594	77	6.2668
11	3.4594	45	5.4919	78	6.2854
12	3.5850	46	5.5236	79	6.3038
13	3.7004	47	5.5546	80	6.3219
14	3.8074	48	5.5850	81	6.3399
15	3.9069	49	5.6147	82	6.3576
16	4.0000	50	5.6439	83	6.3750
17	4.0875	51	5.6724	84	6.3923
18	4.1699	52	5.7004	85	6.4094
19	4.2479	53	5.7279	86	6.4263
20	4.3219	54	5.7549	87	6.4429
21	4.3923	55	5.7814	88	6.4594
22	4.4594	56	5.8074	89	6.4757
23	4.5236	57	5.8329	90	6.4919
24	4.5850	58	5.8580	91	6.5078
25	4.6439	59	5.8826	92	6.5236
26	4.7004	60	5.9069	93	6.5392

(*Contd...*)

27	4.7549	61	5.9307	94	6.5546
28	4.8074	62	5.9542	95	6.5699
29	4.8579	63	5.9773	96	6.5850
30	4.9069	64	6.0000	97	6.5999
31	4.9542	65	6.0224	98	6.6147
32	5.0000	66	6.0444	99	6.6294
33	5.0444	67	6.0661	100	6.6439
34	5.0875	68	6.0875		

Index